COMPARATIVE CHEMICAL MUTAGENESIS

ENVIRONMENTAL SCIENCE RESEARCH

Recent Volumes in this Series

A Continuation Order Plan is available for this series. A continuation order will bring delivery of each new volume immediately upon publication. Volumes are billed only upon actual shipment. For further information please contact the publisher.

COMPARATIVE CHEMICAL MUTAGENESIS

EDITED BY

FREDERICK J. DE SERRES

AND

MICHAEL D. SHELBY

National Institute of Environmental Health Sciences
Research Triangle Park, North Carolina

PLENUM PRESS • NEW YORK AND LONDON

Library of Congress Cataloging in Publication Data

Main entry under title:

Comparative chemical mutagenesis.

(Environmental science research ; v. 24)

"Proceedings of the Workshop on Comparative Chemical Mutagenesis, sponsored by NIEHS and held October 30-November 4, 1977, at the Crabtree Valley Mall, Raleigh, North Carolina"—T.p. verso.

Bibliography: p.

Includes index.

1. Chemical mutagenesis—Congresses. I. De Serres, Frederick J. II. Shelby, Michael D. III. Workshop on Comparative Chemical Mutagenesis (1977 : Raleigh, N. C.) IV. National Institute of Environmental Health Sciences. V. Series. [DNLM: 1. Mutagenicity tests—Congresses. 2. Mutagens—Analysis—Congresses. 3. Mutation—Drug effects—Congresses. W1 EN986F v.24 / QH 465.C5 W926c 1977]

QH465.C5C65 575.2 81-17887

ISBN-13:978-1-4613-3411-8 e-ISBN-13: 978-1-4613-3409-5 AACR2

DOI: 10.1007/978-1-4613-3409-5

Proceedings of the Workshop on Comparative Chemical Mutagenesis, sponsored by NIEHS and held October 30-November 4, 1977, at the Crabtree Valley Mall, Raleigh, North Carolina

softcover reprint of the hardcover 1st edition 1981
A Division of Plenum Publishing Corporation
233 Spring Street, New York, N.Y. 10013

CONTENTS

RISK ESTIMATION

CHAPTER 1

INTRODUCTION

Frederick J. de Serres, Ph.D.

Office of the Associate Director for Genetics

National Institute of Environmental Health Sciences

Research Triangle Park, North Carolina (U.S.A.) 27709

The Workshop on Comparative Chemical Mutagenesis was organized to begin the process of problem identification and resolution concerning our needs to evaluate the data on test chemicals arising from assays for mutagenic activity on laboratory organisms.

In the past, data on chemical mutagens has been generated and published in the scientific literature on a more or less random basis. Individual chemicals enjoy a brief period of "popularity" that leads to a burst of publications in the same or sometimes related assay systems. The incompleteness of the data base, in many of these cases, makes comparative mutagenesis difficult or impossible. In our attempts to compare the genetic effects of a given chemical over a wide range of assay systems, we are often interested in making quantitative as well as qualitative comparisons. To restate the first comparison: is the chemical under question a weak, moderate or potent mutagen over a wide range of assay systems--or alternatively, does the level of response vary markedly? To make the second comparison, what is needed is information on the spectrum of genetic alterations produced as well as whether this spectrum is consistent over a wide range of organisms. In other words, if a chemical shows marked specificity with regard to the production of a particular class or type of genetic damage, is this specificity consistent over a wide range of organisms.

The questions are important not only in the evaluation of the efficacy of individual assay systems under development (do they give the right answer?) but also to evaluate our current capability with respect to extrapolation of data from laboratory organisms to man to estimate the genetic effects of exposure to a given chemical on the human population.

A conference of this type has never been organized nor attempted even on the modest scale in the original program plan. In the selection of candidate chemicals, it was obvious that since very few chemicals have been tested in whole animal systems that selection of chemicals from such lists would be more appropriate than chemicals tested in E. coli, Salmonella, Neurospora, Drosophila or other organisms. Preliminary printouts of references in the Environmental Mutagen Information Center (EMIC) data bank were extremely useful in this selection of test chemicals. A list of the chemicals selected for evaluation is given in Table 1.

The Workshop was organized into two main sessions. In the first session the task of the working groups was to review and evaluate the data in the published literature on individual assay systems. This was to be a comprehensive evaluation of the overall utility of the data base published in terms of protocol, chemical purity of the test chemical (if possible), presence of simultaneous positive or negative controls, data for reproducibility of the effects reported, etc. Since many of the papers were published with different objectives in mind, this was often a frustrating process.

Having achieved an evaluation of the data on individual chemicals as a function of assay systems, the next portion of the workshop was devoted to a general evaluation of the genetic effects of each chemical over a wide range of assay systems. Again, there has been little experience in this exercise (see the report of Committee 17, Drake et al., 1975, Science 187: 503). In these papers an attempt has been made to compare the quantitative and qualitative effects of exposure to a given chemical mutagen over a wide range of assay systems. This process has been hampered in many cases because of the lack of a critical data base. If this Workshop serves no other purpose, it is to demonstrate, objectively, the paucity of data in the literature in chemical mutagenesis, even when a biased sample of chemicals is selected. One effect of this realization will be to stimulate research on the chemicals selected for study so that a more comprehensive evaluation could be made at some point in the future.

This whole exercise would not have been possible without an outstanding and essentially super-human level of effort on the part of EMIC staff and particularly the Director, Mr. John Wassom and his associates, Dr. Michael D. Shelby and Mrs. Elizabeth Von Halle. EMIC had the primary responsibility not only for the initial literature reviews and providing a bibliography of suitable papers (as well as reprints) to the investigators but they were also involved in the preparation and distribution of the individual manuscripts. In addition, the editors would like to acknowledge the contribution of Ruth Krigman as copy editor for this volume.

In the period during which the more formal publication of the papers in this book has taken place, EMIC has used these papers as a mechanism for responding to numerous requests for information on the comparative mutagenesis of the chemicals selected for evaluation and study.

This book represents another milestone in the rapidly developing field of environmental mutagenesis. It is only through such exercises in comparative mutagenesis that we can hope to tackle the more difficult task of risk estimation for the human population. However, incomplete and inadequate in terms of what needs to be done, it is another step towards developing a capacity for predicting the genetic consequences of exposure of the human population to the numerous chemical mutagens in our environment.

Table 1. Chemicals selected for evaluation at the Comparative Chemical Mutagenesis Workshop

Compound	CAS registry number
Aflatoxin B1	1162-65-8
Cadmium compounds	
Cyclophosphamide	50-18-0
DEN	55-18-5
DMN	62-75-9
EMS	62-50-0
Epichlorohydrin	106-89-8
Ethylenimine	151-56-4
ICR-170	146-59-8
MeTEPA	57-39-6
Mitomycin C	50-07-7
MMS	66-27-3
MNNG	70-25-7
Myleran	55-98-1
Natulan	366-70-1
TEM	51-18-3
TEPA	545-55-1
Thio-TEPA	52-24-4
Trenimon	68-76-8
Vinyl chloride	75-01-4

CHAPTER 2

MUTAGENICITY OF SELECTED CHEMICALS IN ESCHERICHIA COLI DNA REPAIR DEFICIENT ASSAYS

Herbert S. Rosenkranz

Department of Microbiology, New York Medical College

Valhalla, New York 10595

"Normal" cells exposed to noxious agents which alter the cellular DNA will attempt to overcome this effect by excising portions of modified DNA and resynthesizing the correct sequence. The enzyme DNA polymerase has been implicated in this repair process (both in the repair replication step and the excision step in excision repair). It is to be expected, therefore, that cells lacking this repair enzyme will be more sensitive to the action of agents which react with cellular DNA. The availability of *E. coli* mutants (pol A_1-) lacking this enzyme (1) has permitted verification of this prediction.

RESULTS AND DISCUSSION

Using normal (pol A+) and DNA polymerase-deficient (pol A_1-) strains of *E. coli*, it was shown that pol A_1- was much more sensitive than pol A+ to a large number of agents, including known mutagens and carcinogens, known to alter cellular DNA (Table 1) (1-10).

Other bacterial strains with mutations in uvrA, uvrB, uvrC, recA, recB, and recC genes have also been shown to exhibit an enhanced sensitivity to a number of chemical and physical agents known to alter the cellular DNA.

Based upon these observations a simple assay procedure was developed (10). Bacteria (pol A+ or pol A_1-) are spread onto the surface of agar plates, and discs containing the substance to be tested are placed on the plates. After incubation (12 hr) the

Table 1. Effect of a DNA-modifying agent on the survival of normal and DNA polymerase-deficient *E. coli*[a]

Time (hr)	Additions	Survival (%) Pol A+	Pol A_1-
0	None	100	100
4	None	1375	4416
1	Hydroxyurethane, 0.05M	118	2.7
2	Hydroxyurethane, 0.05M	8.1	0.3
4	Hydroxyurethane, 0.05M	2.5	0.007

[a]Bacteria were brought to the exponential growth phase in medium HA+T (10) (1.6×10^8 and 1.2×10^8 bacteria/ml for strains pol A+ and pol A_1-, respectively) whereupon portions of the culture were supplemented with N-hydroxyurethane (final concentration 0.05M). The number of viable cells was determined at intervals.

diameters (or areas) of the zones of inhibition are measured. Some agents known to alter cellular DNA were found (Table 2, Group II) to produce larger zones of inhibition of the pol A_1- plates than on the corresponding pol A+ ones. Agents known not to alter the cellular DNA (e.g., cycloserine, chloramphenicol, methicillin, etc. Table 2, Group I) gave equal zones of inhibition on both sets of plates.

Some known mutagens and carcinogens do not, however, inhibit either strains when tested by this procedure (i.e., "no test"; Table 2, Group III). This lack of activity could be due to the fact that these agents require metabolic activation by mammalian enzymes or that because of solubility, size, shape or charge they do not diffuse readily in the agar.

Metabolic activation can be achieved by using an agar overlay (0.8% HA+T agar) (10) seeded with the bacteria and a microsomal activation mixture (S-9 + cofactors) (11). A disc containing the test agent is again placed on the surface of the agar plate.

For substances that do not diffuse readily, a simplification of the standard liquid suspension procedure was devised (12): growing bacteria are diluted to a density of approximately 10^4/ml, and 1 ml aliquots are exposed to the test agent for predetermined

Table 2. Effects of various agents on the growth of a DNA polymerase-deficient strain

Agent	Amount	Size of Zone of Inhibition (mm) Parent (pol A+)	Size of Zone of Inhibition (mm) DNA Polymerase-deficient (pol A_1-)
Group I			
Penicillin	3 units	9	8
Erythromycin	15 μg	9	9
Cycloserine	50 μg	62	62
Chloramphenicol	30 μg	30	30
Streptomycin	10 μg	26	26
Kanamycin	30 μg	18	18
Group II			
Methyl methanesulfonate	20 μl	44	60
Ethyl methanesulfonate	10 μl	0	20
N-Methyl-N-nitrosourea	0.5 μmole	45	85
N-Ethyl-N-nitrosourea	0.5 μmole	0	13
N-Methyl-N-nitrosourethane	0.1 μmole	2	46
N-Ethyl-N-nitrosourethane	0.5 μmole	0	16
N-Hydroxyurethane	20 μmole	12	21
Nitrosofluorene	0.5 μmole	0	15
N-Hydroxylaminofluorene	0.5 μmole	0	12
1,2-Dimethylhydrazine	250 μg	0	12
NFTF[a]	60 μg	25	38
1-phenyl-3,3-dimethyl-triazine	250 μg	24	47
Natulan	250 μg	16	22
Chloroquine	0.2 μmole	9	15
Acridine Orange	0.2 μmole	0	9
Miracil D	1 μmole	7	19
Auramine O	250 μg	9	14
p-Rosaniline	250 μg	6	10
Group III			
N-Acetoxy-N,2-fluorenyl-acetamide	250 μg	0	0
2-Nitronaphthalene	250 μg	0	0
Urethane	20 μmole	0	0
2-Aminofluorene	250 μg	0	0
Dimethylnitrosamine	10 μl	0	0

[a]N-[4-(5-Nitro-2-furyl)thiazolyl] formamide.

periods of time whereupon 0.1 ml portions are plated onto the surface of agar plates for the determination of the number of viable cells. In this procedure results are shown as survival indices (SI) (% survivors pol A_1-/% survivors pol A+). SI values below 1.0 indicate a preferential killing of the pol A_1- strain (Table 3). On the other hand, streptomycin, although it induces lethality in both indicator strains, does not preferentially kill the pol A_1- strain (SI = 1.12). It should be noted that this procedure is compatible with metabolic activation. Thus, the procarcinogen 2-aminofluorene, which requires metabolic activation by hepatic enzyme, does not preferentially inhibit the pol A_1- strain in the absence of rat liver microsomes, but does so in the presence of this preparation (Table 3).

By using the disc assay procedure described herein or simple modifications thereof, the results summarized in Table 4 were recorded. In most of the publications listed, no dose responses were given. Accordingly, the lowest or the only concentration tested is reported. It should be noted that for chemicals requiring no metabolic activation (i.e., direct-acting agents), the results appear reliable. There is no disagreement among the authors. On the other hand, chemicals requiring metabolic activation by microsomal enzymes yield differing results in some laboratories. This appears to be mainly a reflection of the techniques used in preparing the microsomes and of the procedure used for incorporating the microsomes plus cofactors into the assay (e.g., into agar overlay, surface application, well).

Some of the test chemicals were also tested in liquid suspension, and the preferential killing of the DNA repair deficient strains was determined (Table 5). For chemicals tested at several levels, the lowest effective concentration (lec) reported or determined from the published data are included.

MATERIALS AND METHODS

Media

HA+T Agar (1.5%). Two flasks are prepared. Flask 1 contains KH_2PO_4, 3.0 g; K_2HPO_4, 7.0 g; $(NH_4)_2SO_4$, 1.0 g; sodium citrate, 0.5 g; $MgSO_4$, 0.4 ml of 1.0 M solution; water, 500 ml. Flask 2 contains Bacto-agar, 15 g; dextrose, 5.0 g; water, 500 ml.

The contents of flasks 1 and 2 are autoclaved separately and mixed. (This medium is also available commercially as Davis-Mingioli Medium.) The medium is supplemented with casein hydrolyzate (1% solution, Nutritional Biochemicals Corp.), 10 ml/liter; thiamine hydrochloride, 5 mg/liter; and thymine: 5 mg/liter. The agar is cooled to 45°C (water-bath), and 25 ml-portions of this agar are dispensed into plastic petri plates. When the agar has

Table 3. Preferential inhibition of pol A_1- cells by DNA-modifying agents (liquid suspension)

Additions	Amount	Enzyme[a]	Pol A+		Pol A_1-		SI[b]
			Survivors Per Plate	Percent Survivors	Survivors Per Plate	Percent Survivors	
None	0	-	1873	100.0	1632	100.0	1.00
Methyl methanesulfonate	0.025M	-	292	15.6	44	2.7	0.17
N-Acetoxy-N-2-fluorenylacetamide	10 μg/ml	-	543	29.0	208	12.8	0.44
2-Aminofluorene	10 μg/ml	-	1703	90.0	1514	92.8	1.02
2-Aminofluorene	10 μg/ml	+	910	48.6	579	35.5	0.73
None	0	+	1927	102.9	1670	102.3	1.0
Streptomycin	10 μg/ml	-	270	14.4	263	16.1	1.12

[a]Enzyme: microsomal (S-9) preparation supplemented with cofactors; exposure period: 90 min.

[b]% Survivors pol A_1-/% survivors pol A+.

Table 4. DNA repair-deficient *E. coli* (disc assays)

Agent	Amount per disc	S-9	Strain	Result	Reference
Aflatoxin B_1	0.02 μg	+	recB recC	+	13
Cyclophosphamide	250 μg	+	pol A_1-	+	14
DEN	1 μg	+	recB recC	+	13
	5-50 μl	+	pol A_1-	-	15,16
	50 μl	+	pol A_1-	+	10
DMN	0.05 μg	+	recB recC	+	13
	50 μl	+	pol A_1-	-	15
	50 μl	+	pol A_1-	+	10
EMS	5-25 μl, 0.1 μmole	-	pol A_1-	+	10,15,17
Mitomycin C	1 μg	-	pol A_1-	+	8,18,19
MMS	5 μg, 10-25 μl, 0.1 μmole	-	pol A_1-	+	10,12,15, 16,17,20
MNNG	150-250 μg	-	pol A_1-	+	18-21
	50 μg	-	recA	+	20
	100 μg	-	recA, recF	+	13
	100 μg	-	recL	+	13
	100 μg	-	sbcB, recB, recC	+	13
	100 μg	-	recA, recB, recC	+	13
Myleran	10 μg	-	pol A_1-	+	14
Natulan	250 μg	-	pol A_1-	+	21
Vinyl chloride	20 μl, 20% in DMSO	+	pol A_1-	-	14
Epichlorohydrin	1 μl	-	pol A_1-	+	14,18

Table 5. DNA repair-deficient *E. coli* differential killing, liquid suspension

Agent	Concen-tration	lec[a]	S-9	Strain	Result	Refer-ence
Cyclophosphamide	10 μg/ml	No	+	pol A_1-	+	14
DEN	0.1 μg/ml	No	+	pol A_1-	+	14
DMN	0.1 μg/ml	No	+	pol A_1-	+	14
EMS	0 02M	No	-	pol A_1-	+	14
Mitomycin C	0.05 μg/ml	Yes	-	uvrA,	+	22
		No	-	pol A_1-	+	14
MMS	0.01%(v/v)	Yes	-	uvrA,exrA	+	23
			-	recA	+	23
			-	exrA,uvrA,	+	23
			-	pol A,uvrA	+	23
			-	pol A, uvrA-,exrA	+	23
	0.01-0.02M	No	-	pol A_1-	+	12,21,24
MNNG	2.5 μg/ml	Yes	-	pol A_1-	+	22
	5 μg/ml	Yes	-	recA	+	22
Myleran	2 μg/ml	No	-	pol A_1-	+	14
Natulan	2 μg/ml	No	-	pol A_1-	+	14
Vinyl chloride	0.1%(v/v gas)	No	+	pol A_1-	+	14
Epichlorohydrin	0.01 μg/ml	No	-	pol A_1-	+	14

[a]lec indicates whether the concentration indicated is approximately the lowest effective concentration based upon dose-response curves.

solidified the plates are incubated at 37°C for 16 hr, whereupon they are used for the assay. The purpose of this preincubation is (a) to allow for adequate drying and (b) to detect bacterial contaminants.

Studies in our laboratory have indicated that the composition of the liquid and semisolid medium is of great importance in these assays. Moreover, the results reflect differential inhibitions dependent upon the diffusion of the test agent in the agar. It is, therefore, essential that the volume, moisture content, age of the agar plates be controlled carefully. This is best done by pouring exact amounts of agar (25 ml). Upon solidification of the agar, the plates are incubated overnight at 37°C (to remove water of condensation and to eliminate contaminated plates). The plates are then stored at 4°C in plastic bags. Only plates from a single prepared batch can be used at one time.

HA+T Soft Agar. The medium is the same as above but contains half the quantity of agar (7.5 g per liter).

HA+T Liquid Medium. This medium is the same as HA+T Agar but lacks the agar.

Preservation of Tester Strains

The tester strains can be shipped either in Amies Transport Medium (a sterile swab impregnated with a liquid culture of the tester strain is placed into the semisolid transport medium) or a slant composed of HA+T agar inoculated with the tester strains. The slants are incubated overnight at 37°C and then examined for the presence of visible growth.

Upon receipt, the slants or transport media are processed for single-colony isolation on HA+T agar plates. Single colonies are picked, inoculated into 10 ml of HA+T liquid medium, grown overnight at 37°C and then portions of the overnight cultures used as indicated below. (The plates containing the single colonies of pol A+ and pol A_1- are saved until the results of procedure C below are known.)

Procedure A. Slants containing HA+T are inoculated, incubated overnight at 37°C, and transferred to a refrigerator.

Procedure B. Replicate 1-ml portions are supplemented with glycerol (final concentration: 10%) and stored at -20°C.

Procedure C. The cultures are used to determine the response in the standard assay to chloramphenicol and methyl methanesulfonate.

Procedure D. The identity of the cultures as strains of *E. coli* must be confirmed by gram-staining and appropriate biochemical tests, performed, preferably, in a clinical microbiology laboratory.

Care should be taken to label the slants and 1-ml ampules with strain designation and the date of subculture.

If the strains show the expected response to chloramphenicol and to methyl methanesulfonate and are identified as strains of *E. coli*, the original plate containing the single colonies, as well as the overnight cultures can be discarded. In the event that the correct response is not achieved, the technician should be advised to repeat the procedure using another set of single colonies.

Slants containing pol A+ or pol A_1- can be used to initiate overnight liquid cultures for 1-1.5 months, provided that the slant is in a screw-cap tube. Every 1-1.5 months, portions of the cultures kept at -20°C are retrieved (with sterile toothpick or spatula; thawing of tube is not necessary), tested for correct response to chloramphenicol and methyl methanesulfonate, and dispensed into slants.

Overnight cultures of pol A+ or pol A_1- are used only once.

Preparation of Cultures for Testing

An overnight culture of either pol A+ or its pol A_1- derivative in medium HA+I is diluted 10,000-fold in medium HA+T, and 0.1-ml portions of the cultures are spread (with a glass spreader) onto the surface of the agar plates.

It is essential that the bacterial strains be grown in medium HA+T and that the test plates also be of the same composition. Failure to do so may give unreliable results, as in rich media the pol A+ strain appears to have a selective advantage and hence the occasional pol A+ revertants present in pol A_1- cultures will overgrow.

An alternate to the procedure described above, is to place 0.1 ml of the culture into 2 ml of HA+T soft agar (at 45°C), to mix this suspension (Vortex mixer) and to pour into the surface of the agar plates. This procedure allows for more homogeneous distribution of the bacteria and results in a more uniform zone of growth inhibition. This procedure is slightly more time-consuming than the one described above.

Important Precautions

Great caution must be taken in laboratories handling E. coli pol A+/pol A_1- as well as strains of Bacillus subtilis or other spore-formers. Glass spreaders are sterilized by dipping into alcohol and flaming. This is, however, not sufficient to destroy such spore-forming bacteria. Therefore, care must be taken not to use glass spreaders which have been used for spreading B. subtilis or other similar species.

Procedure for the Pol A Assay

When the surface of the agar plates has dried, filter discs (antibiotic type, 6.35 mm in diameter) impregnated with known amounts of test agents are deposited onto the surface of the agar plates and, following incubation at 37°C in the dark for 7-12 hr, the diameters of the zones of growth inhibition are determined. On occasions the edges of the zones of growth inhibitions are not well-delineated; for this reason, it is convenient to allow the plates to remain at room temperature (approx. 23°C) for an additional 24 hr and to measure zones of growth inhibition once more.

Every set of assays includes "positive" as well as "negative" controls, i.e., an agent known to interfere with cellular DNA and one that affects a structure other than the cellular DNA. Convenient standards are methyl methanesulfonate (10 µl) and chloramphenicol (30 µg), respectively. Unless these agents give the expected results, the data are discounted. Every agent is tested in duplicate on at least three differential occasions.

Preparation of Test Agents. Water or dimethyl sulfoxide are the solvents of choice to carry out this bioassay procedure. The test agent (or dilutions thereof) are placed onto filter discs (antibiotic sensitively type); 6.35 mm in diameter) by means of Eppendorf pipets with disposable tips.

Standards. Because the assay depends upon the differential inhibition of tester strains, the use of positive as well as negative controls is mandatory. For practical purposes it is best to use commercial discs containing 30 µg chloramphenicol and freshly prepared discs containing 10 µl of methyl methanesulfonate. (Typical diameters of zones of growth inhibition are given in Table 2.)

Modification of the Pol A_1- Assay (Liquid Suspension)

The standard E. coli pol A_1- assay depends upon the rate of diffusion of the test agent from a disc placed at the center of the plate. Results and their interpretation are then dependent upon the differential inhibition of the pol A_1- strain by DNA-modifying agents. This is expressed as differences in the size of the zones of growth inhibition. Substances which because of size or solubility properties do not diffuse rapidly will not be recognized in this assay even though they may possess DNA-modifying activity (Table 2, group III). To overcome this effect, it is possible to run the assay in suspension. In this procedure portions of the test agent are added to liquid cultures and survivors are enumerated.

Procedure. Overnight cultures of pol A+ or pol A_1- are prepared in medium HA+T. Fresh medium HA+T in Erlenmeyer flasks (250 ml) is inoculated with portions of these overnight cultures (0.05 ml overnight per 10 ml of fresh medium). The cultures are placed on a shaking water bath (37°C) and allowed to reach early exponential growth phase (approximately 2×10^8 cells/ml, which corresponds to an optical density at 450 nm of 0.2, 1 cm light path). The cultures are then diluted to a density of approximately 20,000 cells/ml. Then 1 ml portions of these cultures are distributed into tubes which are supplemented with dilutions (0.1 ml) of the test agent or solvent. The suspensions are incubated at 37°C for 90 min, whereupon replicate 0.1-ml portions of the treated cultures are deposited and spread onto the surface of HA+T agar plates which are incubated at 37°C for 16-18 hr. Colonies are then enumerated. Results are expressed as the survival index which is the ratio, % survivors pol A_1-/% survivors pol A+. A value below 1.0 indicates preferential killing of the pol A_1- strain, which indicates DNA-modifying activity (e.g., methyl methanesulfonate, N-acetoxy-N-2-fluorenyl-acetamide, Table 3). On the other hand, an agent which is bactericidal but devoid of DNA-modifying potentiality will devitalize the two strains to the same extent (e.g., streptomycin, Table 3).

E. coli Pol A+ Assay Coupled to Microsomal Activation

Some carcinogens are inactive in microbial assay systems because they require metabolic activation by mammalian enzymes to the ultimate active form. This activation can be duplicated, to some extent, by incorporating into the microbial system cell-free extracts derived from mammalian liver.

Preparation of Rat Liver S-9 Fraction. The preparation of the S-9 fraction from Araclor-induced rat liver has been described by Ames et al. (11).

Plate Assay. The protein concentration of the S-9 mix (S-9 fraction + cofactors) is adjusted to contain 400 μg protein/0.1 ml of S-9 mix, and 0.1-ml portions of this preparation are incorporated into the 2 ml of soft agar overlay (HA+T Soft Agar). The temperature of the soft agar is maintained at 43-45°C. The indicator bacteria are added next, the mixture dispersed (Vortex mixer), and the soft agar deposited on top of HA+T agar plates. Upon solidification of the agar overlay, filter discs impregnated with test agents are deposited on the surface of the agar. The plates are incubated and processed as described earlier.

Liquid Assay Procedure. The procedure is carried out as described earlier, except that following distribution of the 1-ml portions into test tubes, some of the tubes are supplemented with 0.1 ml S-9 mix containing 100 μg protein. The remainder of the procedure is carried out as described above.

It should be noted that when performing the suspension assay coupled to the S-9 mix, the test agent must be tested both in the presence and in the absence of S-9. Thus, in the data presented above, 2-aminofluorene in the absence of S-9 is toxic to the cell as evidenced by a slight drop in the viable count (Table 3); however, this effect is not indicative of the preferential inhibition of the pol A_1- strain, as the survival index is 1.02. On the other hand, in the presence of the S-9 activation mixture, there is evidence of DNA modification, as shown by a survival index below 1.0 (i.e., 0.73).

REFERENCES

1. De Lucia, P., and Cairns, J. Nature 224: 1164-1166 (1969).
2. Monk, M., Peacey, M., and Gross, J.D. J. Mol. Biol. 58: 623-630 (1971).
3. Gross, J., and Gross, M. Nature 224: 1166-1168 (1969).
4. Kanner, L., and Hanawalt, P. Biochem. Biophys. Res. Commun. 39: 149-155, (1970).
5. Paterson, M. C., Boyle, J. M., and Setlow, R. B. J. Bacteriol. 107: 61-67 (1971).
6. Town, C. D., Smith, K. C., and Kaplan, H. S. Science 172: 851-854 (1971).
7. Mullinix, K P., and Rosenkranz, H. S. J. Bacteriol. 105: 565-572 (1971).
8. D'Alisa, R. M., Carden, G. A., III, Carr, H. S., and Rosenkranz, H. S. Mol. Gen. Genet. 110: 23-26 (1971).

9. Winshell, E. B., and Rosenkranz, H. S. J. Bacteriol. 104: 1168-1175 (1970).

10. Slater, E. E., Anderson, M. D., and Rosenkranz, H. S. Cancer Res. 31: 970-973 (1971).

11. Ames, B. N., McCann, J., and Yamasaki, E. Mutat. Res. 31: 347-364 (1975).

12. Rosenkranz, H. S., Gutter, B., and Speck, W. T. Mutat, Res. 41: 61-70 (1976).

13. Ichinotsubo, D., Mower, H. F., Setliff, J., and Mandel, M. Mutat. Res. 46: 53-62 (1977).

14. Rosenkranz, H. S., and McCoy, E. C., in preparation.

15. Fluck, E. R., Poirier, A., and Ruelius, H. W. Chem. Biol. Interactions 15: 219-231 (1976).

16. Longnecker, D. S., Curphey, T. J., James, S. T., Daniel, D. S. and Jacobs, N. J. Cancer Res. 34: 1658-1663 (1974).

17. Rosenkranz, H. S. Cancer Res. 33: 458-459 (1973).

18. Rosenkranz, H. S., McCoy, E. C., Anders, M., Speck, W. T., and Bickers, D. In: Application of Short-Term Bioassays in the Fractionation and Analysis of Complex Environmental Mixtures, Plenum Press, New York, 1979, pp. 3-42.

19. Bamford, D., Sorsa, M., Gripenberg, U., Laamanen, I., and Meretoja, T. Mutat. Res. 40: 197-202 (1976).

20. Nagao, M., and Sugimura, T. Cancer Res. 32: 2369-2375 (1972).

21. Rosenkranz, H. S., and Poirier, L. A. J. Natl. Cancer Inst. 62: 873-892 (1979).

22. Ishii, Y., and Kondo, S. Mutat. Res. 27: 27-44 (1975).

23. Bridges, B. A., Mottershead, R. P., Green, M. H. L., and Gray, W. J. H. Mutat. Res. 19: 295-303 (1973).

24. Rosenkranz, H. S., and Carr, H. S. Antimicrob. Agents Chemotherapy 2: 376-372 (1972).

CHAPTER 3

MUTAGENICITY OF SELECTED CHEMICALS IN THE REC-ASSAY IN *BACILLUS SUBTILIS*

Tsuneo Kada

Department of Induced Mutation, National Institute of Genetics, Misima, Sizuoka-ken 411, Japan

INTRODUCTION

Since DNA damages of different types are subjected to cellular recombination repair, recombinationless bacteria are usually more sensitive than wild ones. Agents showing increased lethal activity on Rec^- over Rec^+ cells may have damaged DNA. The rec-assay (1) is extremely simple and inexpensive, and positive chemicals can easily be pooled. These samples should be examined for their capacity to induce mutations in bacteria. *Bacillus subtilis* is profitable in the assay of chemicals because its cellular membrane is more permeable to chemicals than enteric bacteria. A number of new environmental mutagens among food additives (1-3) (such as phloxine and AF2), pesticides (4-6) (such as Dexon and Vamidothion), or heavy metals (7,8) (chromium and vanadium) have been detected as a result of the isolation and use of radiation supersensitive strains (9).

Chemical mutagens detected by the rec-assay involves those of both base-change and frameshift types. Those inducing deletions are also detectable by the present assay. The rec-assay positive chemicals are also often inducers of chromosome aberrations in mammalian cells cultured in vitro.

In vitro metabolic activation was carried out by using S9 extracted from PCB-pretreated S.D. rats. Procedures of extraction and preparation of the S9 mixture were those originally described (10).

OUTLINE OF METHODS

Rapid Plate Screenings

In rapid plate screenings (1,11), strains H17 Rec^+ and M45 Rec^- are grown overnight in broth (wet meat extract 10 g, polypeptone dry powder 10 g, NaCl 10 g, water 1000 ml, pH adjusted to 7.0). To 3 ml of full-grown bacterial broth culture is added 1 ml of 50% glycerol (v/v) and this stocked at -80°C. On the day of experiments each culture is streaked radically on the "dry" surface of broth agar (15 g/ℓ. of agar added in the above liquid broth) and a paper disk (diameter 16 mm) containing a solution of the test chemical is so placed as to cover starting points of the streaks (Fig. 1). All the plates are kept at 4-5°C for 24 hr, then incubated at 37°C for about 20 hr. The length of inhibition zones is then measured.

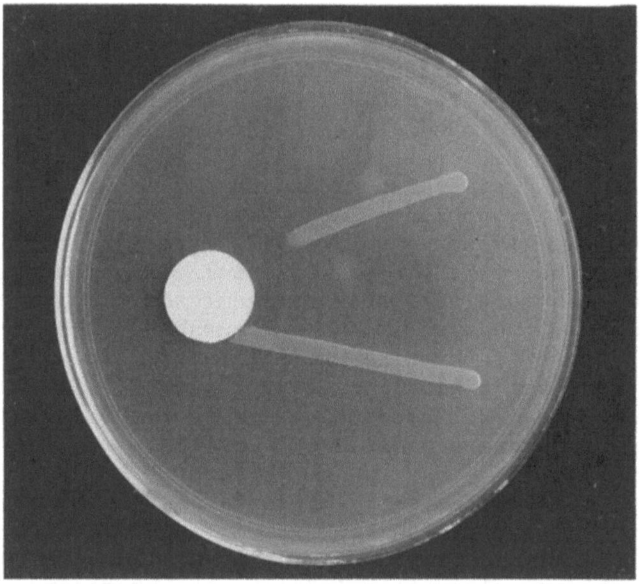

Figure 1. A typical rec-assay positive plate.

Quantitative Liquid Method with Metabolic Activation

A 0.20-ml portion of broth containing drug at the maximum concentration is placed into the first hole of a plastic microtiter plate (Fig. 2); 0.10-ml portions of broth are added into all other holes. First, 0.10 ml of the drug broth is taken from the first hole and poured into the second one; thereafter, 0.10 ml is removed from the second hole and poured into the third one. Thus continuing, we have a gradient of stepwise decreasing concentrations of drug.

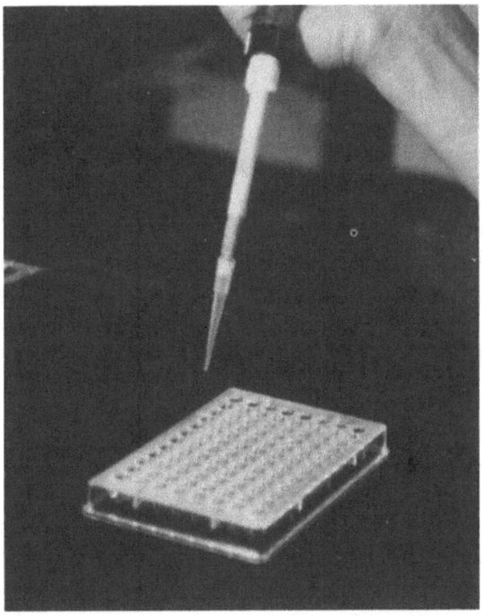

Figure 2. Quantitative rec-assay in liquid by using a microtiter plate.

The overnight broth culture of the strains H17 Rec^+ and M45 Rec^- is stocked at -80°C by supplementing with 12.5% glycerol. On the day of experiments, the stock is lysed and diluted 10 times with fresh broth; 0.01-ml portions are inoculated into each hole, covered with aluminum foil, and incubated overnight at 37°C. The presence or absence of bacterial growth is observed, and the minimum inhibition concentrations (MIC) are compared among Rec^+ and Rec^- strains.

For metabolic activation, 0.02 ml of the S9 mixture is added to each hole in which drug and bacteria have already been added.

CHARACTERISTICS OF THE STRAINS

The strains used in the rec-assay were selected in the following way. We collected 50 repair-deficient mutant strains of *B. subtilis*, *E. coli*, and *Salmonella* possessing *hcr*, *rec*, *pol* or *dna* character and studied their sensitivities to thirty chemical mutagens such as 4NQO, AF2 MMS, etc. The results showed that *B. subtilis* strains possessing rec-45 deficiency has the widest spectrum of sensitivity to the mutagens tested and also the highest sensitivity to them.

The strain M45 (rec-45) had been isolated originally from H17 (*try*, *arg*, *rec*$^+$) by treating with *N*-methyl-*N'*-nitro-*N*-nitrosoguanidine and by examining γ-ray sensitivities of many clones (9). The rec-45 strains of *B. subtilis* have phenotypes similar to those of *E. coli* rec A strains as to UV-induced DNA degradation, prophage noninducibility, etc. (12). Deficiencies in genetic recombination of strains possessing rec-45 were shown by the experiments on their capacity in transformation, SPO2 transfection, and PBS1 phage transduction as well as on their radiation and drug sensitivities and their Hcr$^+$ capacity for UV-exposed phage M2. The rec-45 was mapped near rec-Al in *B. subtilis* (9).

CHARACTERISTICS OF THE METHOD

The rate of MIC(Rec$^+$)/MIC(Rec$^-$) indicates the magnitude of total lethal DNA damage produced by a given chemical. Since any mutation has a chance, when accumulated, to produce lethality, the MIC rate should be proportional to the mutagenic capacity if the "minimum chemical damage" is purely mutagenic. On the other hand, there may exist chemicals producing predominantly lethal damage that has nothing to do with mutations. In any case, evaluation of results of the quantitative rec-assay may be easily done with chemicals of a defined group, such as nitrofurans, nitrosamines, etc.

The rec-assay on plates is recommended for rapid prescreening of large numbers of chemicals for mutagenicity. It is possible for one person to assay several hundred chemicals in a week. The samples positive by rec-assay are then subjected to the Ames reversion assay by use of strains of different mutation specificities (13). The rec-assay possitive (even weakly) samples give positive results with high probability in the Ames assay. Use of the rec-assay speeds up very much the Ames assay. In screening of pesticides, of 194 samples tested, 23 gave the positive results in the rec-assay; of these, 16 were found to be positive in the Ames assay. On the other hand, no positive sample was obtained in the Ames assay among 171 samples negative to the rec-assay, even with S9 activation (5).

Table 1. Quantitative rec-assays of the samples

Sample	Minimum growth inhibition concentration (MIC) without S9		$\frac{MIC(Rec^+)}{MIC(Rec^-)}$	Minimum growth inhibition concentration (MIC) with S9		$\frac{MIC(Rec^+)}{MIC(Rec^-)}$	Conclusion
	17A Rec^+	45T Rec^-		17A Rec^+	45T Rec^-		
Aflatoxin B1	50	12.5	4	50	6	8	+
Cadmium chloride	4	1	4	0.5	0.5		+
Epichlorhydrin	3	0.1	32	1.6	1.6	1	+
Cyclophosphamide	1000	250	4	1000	1000		+
DEN	50000	25000	2				
DMN	25000	12500	2	25000	6250	4	+
Ethylenimine	800	400	2	800	200	4	+
EMS	1600	800	2	1600	200	8	+
ICR-170	6.4	3.2	2	25	6.4	4	+
MMS	1760	440	4	3520	880	4	+
Mitomycin C	0.1	0.006	16	0.1	0.012	4	+
MNNG	100	50	2	25	6.3	4	+
TEM	500	250	2	500	125	4	+

RESULTS

Literature on testing the 20 chemicals under consideration in the B. subtilis rec-assay is very limited. Available reports are based exclusively on quantitative results obtained by rapid screenings on plates. We therefore carried out new experiments in order to evaluate quantitatively the DNA-damaging capacity of the samples. The results (Table 1) indicate the minimum concentrations to inhibit growth at 37°C (MIC) of either wild strain or recombination-deficient (rec-45) strain in liquid broth. The rate, MIC(Rec^+)/MIC(Rec^-), indicates the degree of DNA damage that was nonreparable in the repair-deficient strain in the presence or absence of S9 metabolic activation. The experiments were repeated three or four times for each case, and the mean values were calculated.

CONCLUSIONS

Aflatoxin B1 has been tested by different microbial systems including the rec-assay as a sample for a U.S.-Japan cooperative program on the mutagenicity of chemical carcinogens and the results have been discussed (14,15). Cadmium chloride gave clearly positive results in the rec-assay by plating with cold incubation and was found to be positive with other metal compounds such as beryllium, cadmium, cobalt, chromium, platinum, rhodium, selenium, tellurium, vanadium, and mercury (7). In the case of metal compounds, the correlation between positive results in the rec-assay and carcinogenicity is about 80%. Epichlorhydrin and DMN have been included in test samples for a project of short-term carcinogenicity testing of the Japanese Ministry of Welfare, and the results on different experimental systems have been reported elsewhere (16,17). Mitomycin C, MNNG, and MMS have been shown to be typical positive samples in the rec-assay system using B. subtilis strains H17 Rec^+ and M45 Rec^- (1,9).

The involvement of recombination repair in induced mutagenesis was discussed (18,19) in a symposium at the International Congress of Genetics (Tokyo, 1968). The method of checking mutagenicity by recombination repair was later developed. Similar studies using different repair-deficiencies other than rec have been fruitful (20,21).

ACKNOWLEDGEMENTS

The author is indebted to Dr. Y. Sadaie, Dr. K. Hirano, and Miss M. Hara (National Institute of Genetics, Misima) for their collaboration in the preparation of bacterial strains as well as in carrying out the present experiments. The studies have been aided in part by grants from the Ministry of Welfare and the Ministry of Education, Japan.

REFERENCES

1. Kada, T., Tutikawa, K., and Sadaie, Y. Mutat. Res. 16: 165-174 (1972).

2. Kada, T. Japan. J. Genet. 48: 301-305 (1973).

3. Kada, T. In: Screening Tests in Chemical Carcinogens, R. Montesano, H. Bartsch, L. Tomatis, Eds., Sci. Publ. No. 12, Lyon, 1975, pp. 105-115.

4. Kada, T., Moriya, M., and Shirasu, Y. Mutat. Res. 26: 243-248 (1974).

5. Shirasu, Y., Moriya, M., Kato, K., and Kada T. Mutat. Res. 40: 19-30 (1976).

6. Shirasu, Y., Moriya, M., Kato, K., Lienard, F., Tezuka, H., Teramoto, S., and Kada, T. In: Origins of Human Cancer, Cold Spring Harbor, in press.

7. Kanematsu, N., Hara, M., and Kada, T. Mutat. Res. 77: 109-116 (1980).

8. Nishioka, H. Mutat. Res. 31: 185-189 (1975).

9. Sadaie, Y., and Kada, T. J. Bacteriol. 125: 489-500 (1976).

10. Ames, B. N., Lee, F. D., and Durston, W. E. Proc. Natl. Acad. Sci. 70: 782-786 (1973).

11. Kada, T. Mutat. Res. 38: 340 (1976).

12. Kada, T., Noguti, T., and Sadaie, Y. Anais Acad. Brasil Ciencias 45: 179-184 (1973).

13. Ames, B. N. In: Chemical Mutagens, A. Hollaender, Ed., Plenum Press, New York, 1971, pp. 267-282.

14. De Serres, F. J. Report on U.S.-Japan Workshop on the Mutagenicity of Chemical Carcinogens, Honolulu, 1975.

15. Ueno, Y., and Kubota, K. Cancer Res. 36: 445-451 (1976).

16. Kawachi, T. Report on Studies of Mutagenicity Screening of Chemical Carcinogens, 1976.

17. Odashima, S. Report on Studies of Mutagenicity Screening of Chemical Carcinogens, 1975.

18. Kondo, S. Proc. 12th Internatl. Congr. Genet. 2: 126-127 (1968).

19. Witkin, E. M. Proc. 12th Inter. Congr. Genet. 3: 225-245 (1969).

20. Nagao, M., and Sugimura, T. Cancer Res. 32: 2369-2374 (1972).

21. Slater, E. E., Anderson, M. D., Rosenkranz, H. S. Cancer Res. 31: 970-973 (1971).

CHAPTER 4

MUTAGENICITY OF SELECTED CHEMICALS IN THE *SALMONELLA*/MICROSOME MUTAGENICITY TEST

Lynne Haroun and Bruce N. Ames

Department of Biochemistry, University of California at Berkeley, Berkeley, California 94720

The *Salmonella*/microsome mutagenicity test provides a simple, economical method for detecting chemical mutagens. Compounds are tested on Petri plates with several specially constructed mutants of *Salmonella typhimurium* which have been selected for their sensitivity and specificity in being reverted from a histidine requirement back to prototropy. Liver homogenates can be added directly to the Petri plates, thus incorporating an important aspect of mammalian metabolism into the *in vitro* test. The system has been recently reviewed (1) and the test method described in detail (2).

The test system has recently been validated for the detection of carcinogens as mutagens (3-6). There is a high correlation between carcinogenicity and mutagenicity: in one study (3,4), 90% of the carcinogens were mutagenic in the test, while few of the noncarcinogens showed any degree of mutagenicity. The simplicity, accuracy, and sensitivity of this test make it a useful method for screening the large number of environmental sources of potential carcinogens and mutagens.

In the present review, summary data is provided for each of the 20 compounds under consideration. Unless otherwise noted, the data for the dose-response curves are from this laboratory. Previous reports by other laboratories and independent observations are cited in the tables; however, we have not attempted to provide a comprehensive review of the literature.

MATERIALS AND METHODS

The following describes procedures used in this laboratory. Protocols followed in other laboratories will vary.

Tester Strains

The strains used for mutagenesis testing are shown in Table 1. The standard bacterial tester strains are mutants of Salmonella typhimurium LT-2 and contain histidine missense (TA1535 and TA100) or frameshift (TA1537, TA1538, and TA98) mutations. (These strains have previously been described in detail (7,8)). TA1538 has a repetitive -C-G-C-G-C-G-C-G- sequence near the site of the histidine mutation (9) and TA1537 appears to have a run of C's at the site of the mutation (7). The strains also contain uvrB and rfa (deep rough) mutations which greatly increase their sensitivity to reversion by a wide variety of mutagens: uvrB causes loss of the excision repair system and rfa causes loss of the lipopolysaccharide barrier. In addition, strains TA100 and TA98 contain the resistance transfer factor (R factor), pKM101.

Two additional strains, TA92 (hisG46, pKM101) and TA94 (hisD3052, pKM101), are being used for the detection of chemicals which require an intact uvrB system to be detected as mutagens,

Table 1. Genotype of the TA strains used for mutagenesis testing[a]

Histidine mutation			Additional mutations		
hisG46	hisC3076	hisD3052	LPS	Repair	R factor
TA1535	TA1537	TA1538	rfa	ΔuvrB	-
TA100		TA98	rfa	ΔuvrB	+R
hisG46			+	+	-
TA92		TA94	+	+	+R
TA1530			Δgal	ΔuvrB	-

[a]All strains were originally derived from S. typhimurium LT2. Wild-type genes are indicated by +. The deletion (Δ) through uvrB also includes the nitrate reductase (chl) and biotin (bio) genes. The Δgal strains (and rfa uvrB strains) have a single deletion through gal chl bio uvrB.

[e.g., mitomycin C (8) and malondialdehyde (1,10)]. In addition, TA92, hisG46, and TA1530 are more sensitive to the short-chain nitrosamines, dimethylnitrosamine and diethylnitrosamine. We are currently considering introducing one or more of these strains for general mutagenesis screening.

Liver Homogenate

The S-9 fraction was prepared as previously described (2,11). Male rats (Sprague-Dawley/Bio-1 strain, Horton Animal Laboratories) were induced with sodium phenobarbital (0.1% sodium phenobarbital in the drinking water one week before sacrifice) or with a polychlorinated biphenyl (PCB) mixture (Aroclor 1254) (single IP injection, 500 mg/kg, 5 days before sacrifice). The livers were homogenized in 0.15 M KCl (3 ml/g wet tissue) and centrifuged at 9000 g for 10 mins.

The S-9 mix contains per milliliter: S-9 (0.04-0.60 ml), $MgCl_2$ (8 μmole), KCl (33 μmole), glucose 6-phosphate (5 μmole), NADP (4 μmole), and sodium phosphate, pH 7.4 (100 μmole).

Mutagenicity Assays

Standard Plate Incorporation Assay. To 2 ml molten top agar (45°C) are added 0.1 ml of the bacterial tester strain ($2-3 \times 10^9$ cells/ml), 10-100 μl of the compound to be tested (diluted in dimethyl sulfoxide or water), and 0.5 ml of the S-9 mix (if required). The contents are mixed and poured on minimal glucose agar plates. The number of his+ revertant colonies are determined after a 48-hr incubation at 37°C.

Liquid Assay. The above protocol is followed with the slight modification that the top agar is added last, after preincubating the other components for 20 min at room temperature.

Spot Test. The protocol described for the standard plate incorporation assay is followed, with the slight modification that the test compound is not added until after the plate has been poured. A few crystals (or microliters of a liquid) are then placed directly on the agar surface.

RESULTS AND DISCUSSION

Results for various compounds are summarized in Table 2. The results for individual compounds are detailed in Tables 3-10. In

Table 2. Summary: mutation induction in <u>Salmonella</u>

Compound	CAS Registry Number	Mutagenicity[a]	Rev/nmole[b]	S-9[c]	Tester Strain[d]
Aflatoxin B1	1162-65-8	+	6953	+	TA100
Cadmium sulfate	10124-36-4	-	<0.007	±	TA100, TA98, TA1537
Cyclophosphamide	50-18-0	+	1.4	+	TA1535
Diethylnitrosamine	55-18-5	+[e]	0.01[g]	+	TA100
Dimethylnitrosamine	62-75-9	+[e]	0.02[g]	+	TA100
Eipchlorohydrin	106-89-8	+[f]	h	-	TA1535
Ethylenimine	151-56-4	+	2.23	-	TA1535
Ethyl methanesulfonate	62-50-0	+	0.17[g]	-	TA100
ICR-170	146-59-8	+	193	-	TA1537
MeTEPA	57-39-6	+	0.22	-	TA100
Methyl methanesulfonate	66-27-3	+	0.30	-	TA100
MNNG	70-25-7	+	1375[g]	-	TA100
Mitomycin C	50-07-7	+	121	-	TA94
Myleran	55-98-1	+	0.22	-	TA1535
Natulan	366-70-1	-	<0.002	±	TA100, TA1535, TA1537, TA98
TEPA	545-55-1	+	1.52	-	TA100
Thio-TEPA	52-24-4	+	0.16	-	TA1535
Trenimon	68-76-8	+	128	-	TA100
Triethylenemelamine	51-18-3	+	4.75	-	TA1535
Vinyl chloride	75-01-4	+[f]	h	+	TA100

[a]Mutagenicity: the standard plate incorporation assay was used except where indicated; (+) mutagenic; (-) nonmutagenic.

[b]Revertants per nmole is calculated from the revertant colonies per plate (determined from dose response curves) and the molecular weight of the test compound. The value represents the mutagenic potency of the particular chemical in reverting the indicated strain. However, the relationship between the mutagenic potency of a compound to a particular strain and the overall mutagenic potential of the compound for DNA remains to be determined. Comparisons between potencies of different chemicals must be undertaken with caution (1,3).

[c]S-9: (+) S-9 required for activity; (-) S-9 not required for activity; (±) tested in the presence and absence of S-9 (for nonmutagenic compounds), or that compound is slightly more active in the presence of S-9 (vinyl chloride).

[d]The Salmonella tester strain for which dose response data are reported. For nonmutagenic compounds, indicates strains on which the compound was tested.

[e]Liquid assay.

[f]Desiccator experiment.

[g]Nonlinear dose response curve.

[h]Test compound a gas or volatile liquid; the test was done in a desiccator and quantitation is difficult.

---

these tables, references were taken primarily from the EMIC literature searches and have been grouped within the tables by author(s). No attempt has been made to determine priority, and references in which data are reported on the nonstandard tester strains have, in general, not been cited.

The dose is given in micrograms per plate (standard plate incorporation assay and spot test) or as the concentration (millimolar) in the incubation medium (liquid assay). Where it has not been possible to convert to these units, the dose is given as it appears in the cited reference.

Procedures for the standard plate incorporation assay, spot test, and liquid assay (as followed in this laboratory) are described in the Methods section. Protocols followed in other laboratories will vary. Significant changes have been noted in the tables.

The source of the S-9 (species and tissue) and pretreatment of animals with enzyme inducers is indicated. Purified microsomes were used in place of S-9 in some studies.

In general, results are reported only on the standard tester strains TA1535, TA100, TA1538, TA98, and TA1537. These strains detect almost all compounds which were previously found to be positive on the earlier Salmonella strains. Exceptions are discussed in the text. Mutagenicity (in the presence or absence of S-9) is reported as positive or negative, as determined by the authors.

Aflatoxin B1

Aflatoxin B1 is a potent mutagen in Salmonella, reverting the frameshift tester strain TA1538, and the two R factor strains, TA100 and TA98 (Table 3, Fig. 1). Mutagenic activity requires the presence of a mammalian microsomal fraction (11,12) and is significantly higher in those strains with the R factor plasmid (8). Recent studies (13) suggest that aflatoxin B1-2,3-oxide is the active carcinogenic and mutagenic metabolite. Swenson et al. (13) have shown that aflatoxin B1-2,3-dichloride, a model for the putative 2,3-oxide metabolite, is a direct acting mutagen in the Salmonella tester strains TA98 and TA100. Further studies utilizing Salmonella to determine the mutagenic potential of other possible metabolites and to investigate the ability of S-9 fractions prepared from various mammalian species, strains, and tissues to activate aflatoxin to its mutagenic form would be of interest.

Cadmium Sulfate

Cadmium sulfate, tested under standard conditions, was not mutagenic on strains TA100, TA98, and TA1537 in the presence and absence of S-9 (Table 3). However, the test system is not suitable for the detection of inorganic salts due to the presence of magnesium salts, citrate, and phosphate in the minimal media. Cadmium sulfate was further tested in several assays in which the concentration of the magnesium, citrate, and phosphate in the medium was varied, but was not shown to be mutagenic under these conditions. However, recent studies (14) have shown that cadmium acetylacetonate is weakly mutagenic (0.015 rev/nmole) on strain TA1538 when tested in the standard plate incorporation assay.

Cyclophosphamide

Cyclophosphamide is a mutagen in Salmonella, reverting the

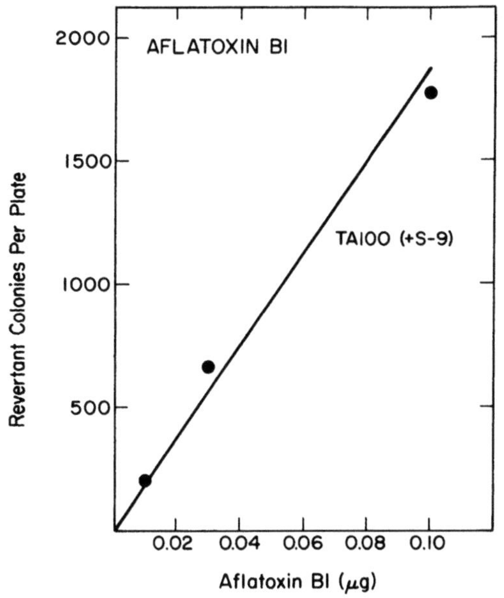

Figure 1. Mutagenicity of aflatoxin B1 in the standard plate incorporation assay on S. typhimurium TA100. Spontaneous revertant colonies have been subtracted. Liver homogenate (50 μl/plate) from phenobarbital induced rats was added as described in the text. The mutagenic potency in revertants/nmole (3) is 6953, and in revertants/μg is 22267.

missense tester strains TA1535 and TA100 (Fig. 2). The negative results of Ellenberger and Mohn (16) (Table 3) may be due to the assay method employed (liquid assay), the use of uninduced liver as a source of S-9, or the use of an insufficient quantity of S-9. In work from this laboratory (18), activation of cyclophosphamide required approximately 10-fold more S-9 than normally used. However, in the same study, cyclophosphamide induced point mutations in other bacterial test systems following metabolic activation.

Results from several laboratories (Table 3) indicate that cyclophosphamide, which is biologically inactive per se, is mutagenic only in the presence of S-9. However, Hanna and Dyer (17) reported that cyclophosphamide was directly active when assayed using a spot test. As the dose used was quite high (crystals were placed on the plate), it is possible that the mutagenic activity observed was due to the presence of impurities in the sample tested. These results confirm an independent study (20) in which cyclophosphamide was found to be weakly active on the Salmonella strain TA1950, without activation, when assayed

Table 3. Mutagenicity of aflatoxin B1, cadmium sulfate, and cyclophosphamide

		S-9			Mutagenicity in Salmonella										
					TA1535		TA100		TA1538		TA98		TA1537		
Dose	Method	Species	Tissue	Induction	S-9 +	-	+	-	+	-	+	-	+	-	Reference
AFLATOXIN B1 (CAS registry number: 1162-65-8)															
1.0 μg per plate	Standard plate incorporation assay	Rat	Liver	Pb	-	-			+	-			+	-	11
0.1 μg per plate	Standard plate incorporation assay	Rat	Liver	PCB	-	-	+	-	+	-	+	-			3,8
0.01-0.1 μg per plate	Standard plate incorporation assay	Rat	Liver	Pb	-		+		+						1
CADMIUM SULFATE (CAS registry number: 10124-36-4)															
104-2085 μg per plate	Standard plate incorporation assay	Rat	Liver	PCB			-	-			-	-	-	-	15

CYCLOPHOSPHAMIDE (CAS registry number: 50-18-0)

0.5-20 mM	Liquid assay	Rat, mouse	Liver	Unin-duced	-	-			16
Crystal	Spot test	Tested without S-9				+			17
100-500 μg per plate	Standard plate incorporation assay	Rat	Liver	PCB	+	-	+	-	18
50-1000 μg per plate	Standard plate incorporation assay	Rat	Liver	Pb	+	-			19

using a spot test. Further studies using purified samples would be useful to help clarify the observed discrepancies.

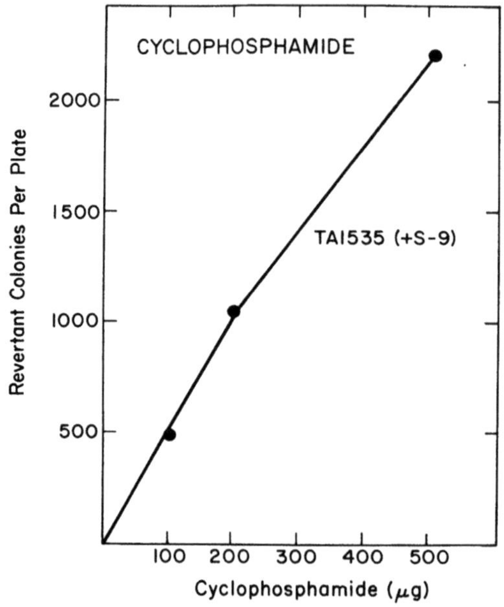

Figure 2. Mutagenicity of cyclophosphamide in the standard plate incorporation assay on S. typhimurium TA1535. Spontaneous revertant colonies have been subtracted. Liver homogenate (330 μl/plate) from Aroclor 1254-induced rats was added as described in the text. The mutagenic potency in revertants/nmole (3) is 1.4, and in revertants/μg is 5.4.

The metabolic pathway of cyclophosphamide has recently been elucidated (21) and several of the metabolites have been shown to be mutagenic (22). 4-Hydroperoxycylophosphamide, the synthetic precursor of 4-hydroxycyclophosphamide (the initial product of cyclophosphamide metabolism), nornitrogen mustard, and the two urinary metabolites, carboxyphosphamide and 4-ketocyclophosphamide, show direct mutagenic activity in Escherichia coli. It would be of interest to confirm the mutagenicity of these metabolites in Salmonella on strains TA1535 and TA100.

Diethylnitrosamine

Diethylnitrosamine (DEN) was first shown to be a mutagen in

Salmonella on strain hisG46 by Malling (23), and subsequently Bartsch, Malaveille, and Montesano (24) showed that it was mutagenic on strain TA1530 (Table 4). Activity required preincubating the bacteria and DEN in the presence of a liver microsomal fraction.

The dose-response curve given here (Fig. 3) is from work by Yahagi et al. (25) on the standard tester strain TA100 using a liquid assay; see also Sugimura et al. (26). However, preliminary results from this laboratory indicate that DEN is significantly more active on strains TA1530 and hisG46 than on the standard tester strains TA1535 and TA100, when tested in the liquid assay. While previously we have been unable to demonstrate mutagenic activity in the plate incorporation assay when using the standard tester strains, a positive response, although considerably weaker than that obtained in the liquid assay, is observed with these more sensitive strains. We are considering introducing one of these tester strains for general mutagenesis screening.

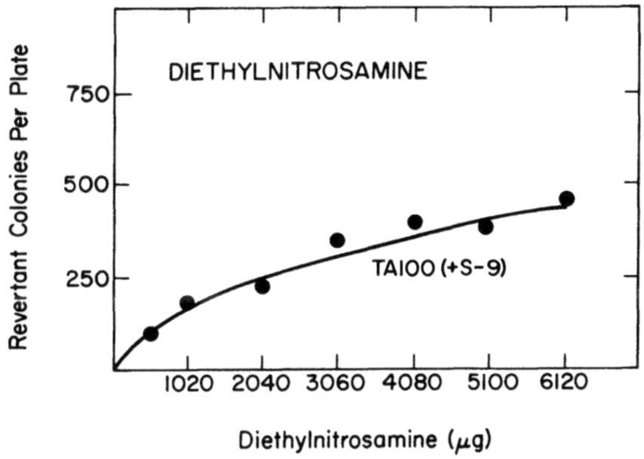

Figure 3. Mutagenicity of diethylnitrosamine in the liquid assay on S. typhimurium TA100. Spontaneous revertant colonies have been subtracted. Liver homogenate from phenobarbital induced rats was added as described in the text. The mutagenic potency in revertants/nmole (3) is 0.01, and in revertants/μg is 0.90. Data are from Yahagi et al. (25).

Dimethylnitrosamine

Dimethylnitrosamine (DMN) was first shown to be a mutagen in Salmonella on strains hisG46 and TA1530 by Malling (33), and

Table 4. Mutagenicity of diethylnitrosamine (CAS registry number: 55-18-5)

		S-9			Mutagenicity in Salmonella										
					TA1535		TA100		hisG46		TA1530		TA98		
Dose	Method	Species	Tissue	Induction	S-9 +	-	+	-	+	-	+	-	+	-	Reference
25-200 mM	Liquid assay	Mouse	Liver	Unind.					+						27,23
200 mM	Liquid assay	Mouse	Liver	Unind., Pb, 3-MC[a]					+						27,25
45 mM	Liquid assay	Rat	Liver	Unind.					±						28,29
2.5-250 mM	Liquid assay	Rat	Liver	Pb							+				24
50 mM	Liquid assay	Rat, mouse, hamster	Liver	Pb							+				24,30
50 mM	Liquid assay	Rat, mouse, hamster	Liver	Unind.							-[b]				24,30
50 mM	Liquid assay	Rat, mouse, hamster	Lung	Unind.,Pb							-				24
10-50 μmole/ml of agar overlay	Standard plate incorporation assay	Rat	Liver	Pb							-				31

150 mM	Liquid assay	Mouse	Liver	Unind.			+			32
150 mM	Liquid assay	Mouse	Kidney, lung	Unind.			-[b]			32
7-86 mM	Liquid assay	Rat	Liver	Pb	+	-			+	25,26
15-150 mM	Liquid assay	Rat	Liver	PCB			+	+		15
940-9400 μg per plate	Standard plate incoporation assay	Rat	Liver	PCB			+	+		15

[a]Mutagenic activity is increased in the presence of S-9 from induced versus control animals.

[b]Very weak or no activity is observed.

subsequently, by Bartsch, Malaveille, and Montesano (24). Activity required preincubating the bacteria and DMN in the presence of a liver microsomal fraction. Several studies (23,29,34) have since examined various factors affecting the metabolism and mutagenicity of DMN (see Table 5). A correlation between DMN demethylase activity and mutagenicity has been clearly demonstrated, and in addition, the effects of incubation time, enzyme cofactors, microsomal protein concentration, and microsomal enzyme inducers have been examined. The capacity of microsomal fractions from different species and tissues to activate DMN to a mutagen is shown in the table, although data concerning strain differences has not been included.

The dose-response curve given here (Fig. 4) is from work by Yahagi et al. (25) on the standard tester strain TA100 using a liquid assay; see also Sugimura et al. (26). However, preliminary results from this laboratory indicate that DMN is significantly more active on strains hisG46, TA92, and TA1530 than on the standard tester strains TA1535 and TA100 when tested in the liquid assay. While previously we have been unable to demonstrate mutagenic activity in the plate incorporation assay when using the standard tester strains, a positive response, although considerably weaker than that obtained in the liquid assay, is observed with strains TA92 and hisG46. We are considering introducing one of these strains for general mutagenesis screening.

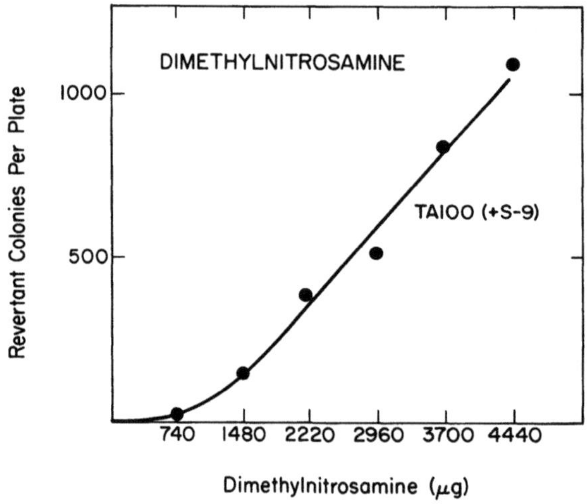

Figure 4. Mutagenicity of dimethylnitrosamine in the liquid assay on S. typhimurium TA100. Spontaneous revertant colonies have been subtracted. Liver homogenate from phenobarbital induced rats was added as described in the text. The mutagenic potency in revertants/nmole (3) is 0.02, and in revertants/μg is 0.25. Data are from Yahagi et al. (25).

Table 5. Mutagenicity of dimethylnitrosamine (CAS registry number: 62-75-9)

Dose	Method	S-9 Species	S-9 Tissue	S-9 Induction	Mutagenicity in Salmonella TA1535 +	TA1535 −	TA100 +	TA100 −	hisG46 +	hisG46 −	TA1530 +	TA1530 −	TA92 +	TA92 −	Reference
45 mM	Liquid assay	Mouse	Liver	Unind.					+	-	+	-			33
5-200 mM	Liquid assay	Mouse	Liver	Unind.					+						27,28
100 mM	Liquid assay	Mouse, rat	Liver	Unind., Pb, 3-MC[a]					+						27
5-500 mM	Liquid assay	Rat	Liver	Pb					+		+				24
50 mM	Liquid assay	Rat, mouse, hamster	Liver	Unind., Pb[a]							+				24,30
50 mM	Liquid assay	Rat, mouse, hamster	Lung	Unind., Pb							-				24
10-50 μmole/ml of agar overlay	Standard plate incorporation assay	Rat	Liver	Pb							-				31
0-150 mM	Liquid assay	Mouse	Liver	Unind.					+						35

(continued)

Table 5. (continued)

Dose	Method	S-9 Species	S-9 Tissue	S-9 Induction	Mutagenicity in Salmonella: TA1535 S-9 +	TA1535 S-9 −	TA100 S-9 +	TA100 S-9 −	hisG46 S-9 +	hisG46 S-9 −	TA1530 S-9 +	TA1530 S-9 −	TA92 S-9 +	TA92 S-9 −	Reference
100 μmole/ml in top agar overlay	Standard plate incorporation assay[b]	Human, monkey, rabbit, mouse, guinea pig, rat	Liver	Unind.					+						35
100 mM	Liquid assay	Mouse	Liver, lung, kidney	Unind.					+						32
100 mM	Liquid assay	Mouse	Spleen, testes, stomach	Unind.					-						32
100 mM	Liquid assay	Human	Liver		+										12
6.25–200 mM	Liquid assay	Mouse	Liver	Unind., PCB[a]	+	-									36
100 mM	Liquid assay	Mouse	Liver	Unind., Pb, 3-MC, PCB[a]							+				37
14-86 mM	Liquid assay	Rat	Liver	Pb			+	-							25,26

740-4440 µg per plate	Standard plate incorporation assay	Rat	Liver	Pb		-				25
11-45 mM	Liquid assay	Rat	Liver	PCB	+	+	+	+	+	15
10^3-10^4 µg per plate	Standard plate incorporation assay	Rat	Liver	PCB	-	-	+		+	15

[a]Mutagenic activity is increased in the presence of S-9 from induced versus control animals.

[b]Bacteria and DMN were plated in the usual manner; 0.1 ml of a microsomal system consisting of an S-9 fraction and appropriate cofactors were then added to the center of the plate.

Epichlorohydrin

Epichlorohydrin is a direct-acting mutagen in Salmonella reverting the missense tester strains TA1535 and TA100 (Table 6). While dose-response data are available, the compound is quite volatile and quantitative results are difficult to obtain. In experiments by Simmon (38) (Fig. 5), the bacteria were placed in desiccators containing varying amounts of epichlorohydrin. Srám et al. (39) tested epichlorohydrin in a liquid assay, although sufficient details concerning the experimental procedure are not provided. Additional experiments providing more quantitative data would be useful in assessing the mutagenic potency of this compound.

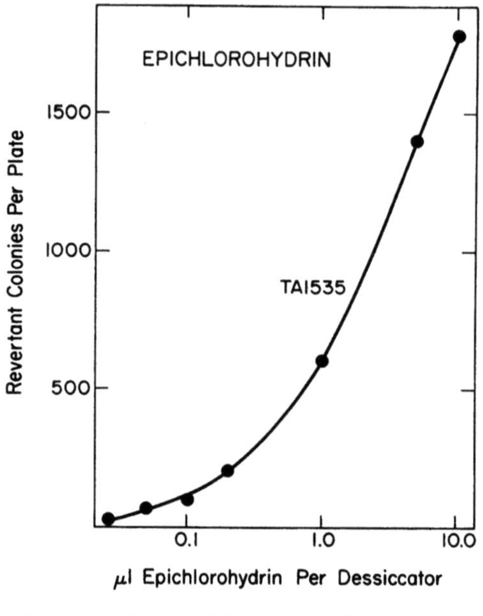

Figure 5. Mutagenicity of epichlorohydrin, assayed in a desiccator experiment on S. typhimurium TA1535. Spontaneous revertant colonies have been subtracted

Ethylenimine

Ethylenimine is a direct-acting mutagen in Salmonella, reverting the missense tester strains TA1535 and TA100 (Table 6, Fig. 6). This compound has not been tested in the presence of S-9 or on the frameshift tester strains TA1537 and TA1538, although it is unlikely that these strains would be reverted by this chemical.

Table 6. Mutagenicity of direct-acting mutagens

Dose	Method	S-9	Mutagenicity in Salmonella										Reference
			TA1535		TA100		TA1538		TA98		TA1537		
		S-9	+	−	+	−	+	−	+	−	+	−	
EPICHLOROHYDRIN (CAS registry number: 106-89-8)													
0.5-50 μl per plate	Spot test	Not required				+		−				−	39
1.08-108 mM	Liquid assay	Not required				+							39
0.025-10 μl per desiccator	Desiccator experiment[a]	Not required		+		+							38
ETHYLENIMINE (CAS registry number: 151-56-4)													
5-20 μg per plate	Standard plate incorporation assay	Not required		+		+							3,15
ETHYL METHANESULFONATE (CAS registry number: 62-50-0)													
5000 μg per plate	Standard plate incorporation assay	Not required		+		+		−		−			8

(continued)

Table 6. (continued)

Dose	Method	S-9	S-9	Mutagenicity in Salmonella TA1535		TA100		TA1538		TA98		TA1537		Refer- ence
				+	-	+	-	+	-	+	-	+	-	
2000-10000 μg per plate	Standard plate incorporation assay	Not required			+		+							3,15
ICR-170 (CAS registry number: 146-59-8)														
2-10 μg per plate	Standard plate incorporation assay	Not required											+	3,15

[a]Bacteria were plated in the usual manner. Plates were placed in desiccators containing 0.025-10 μl epichlorohydrin and incubated at 37°C for 7 hr, removed, and incubated an additional 41 hr.

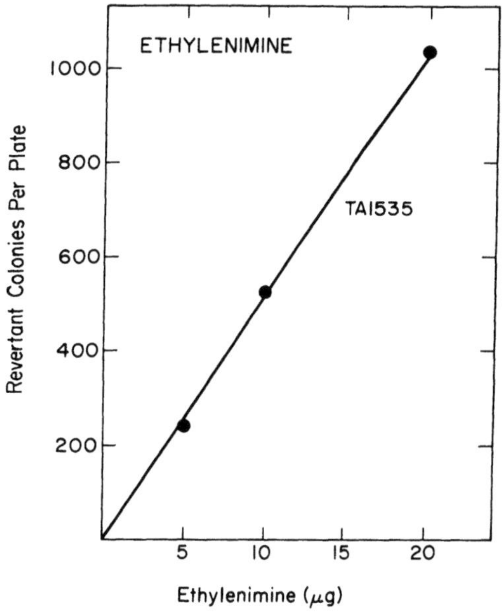

Figure 6. Mutagenicity of ethylenimine in the standard plate incorporation assay on S. typhimurium TA1535. Spontaneous revertant colonies have been subtracted. The mutagenic potency in revertants/nmole (3) is 2.23, and in revertants/μg is 51.7.

Ethyl Methanesulfonate

Ethyl methanesulfonate is a direct-acting mutagen in Salmonella, reverting the missense tester strains TA1535 and TA100 (Table 6, Fig. 7). This compound has not been tested in the presence of S-9 or on the frameshift tester strain TA1537, although it is unlikely that this strain would be reverted by this chemical.

ICR-170

ICR-170 is a direct-acting mutagen in Salmonella, reverting the frameshift tester strain TA1537 (Table 6, Fig. 8). It has not been tested in the presence of S-9 or on any of the other standard tester strains, although it is unlikely that these strains would be reverted by this agent.

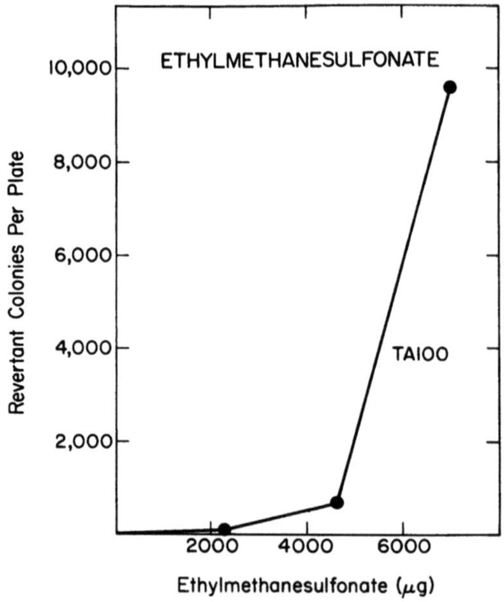

Figure 7. Mutagenicity of ethylmethanesulfonate in the standard plate incorporation assay on S. typhimurium TA100. Spontaneous revertant colonies have been subtracted. The mutagenic potency in revertants/nmole (3) is 0.17, and in revertants/μg is 1.4.

MeTEPA

Studies of the mutagenic activity of MeTEPA in Salmonella are complete (Table 7, Fig. 9). The compound is a direct-acting mutagen, reverting the missense tester strain TA1535, and the R factor strain TA100. The observed activity is neither increased nor decreased in the presence of S-9.

Methyl Methanesulfonate

Methyl methanesulfonate (MMS) is a direct-acting mutagen in Salmonella, reverting the R factor strain TA100 (Table 7, Fig. 10). It has not been tested in the presence of S-9 or on the frameshift tester strain TA1537, although it is unlikely that this strain would be reverted by this compound.

Table 7. Mutagenicity of METEPA, MMS, and MNNG

Dose	Method	S-9 Species	S-9 Tissue	S-9 Induction	Mutagenicity in Salmonella: TA1535 (S-9 +)	TA1535 (S-9 −)	TA100 (S-9 +)	TA100 (S-9 −)	TA1538 (S-9 +)	TA1538 (S-9 −)	TA98 (S-9 +)	TA98 (S-9 −)	TA1537 (S-9 +)	TA1537 (S-9 −)	Reference
METEPA (CAS registry number: 57-39-6)															
5-10 µl per plate	Spot test	Not required				+									17
50-800 µg per plate	Standard plate incorporation assay	Rat	Liver	PCB	+	+	+	+			-	-	-	-	15
METHYL METHANESULFONATE (CAS registry number: 66-27-3)															
570 µg per plate	Standard plate incorporation assay	Not required				-		+		-		-			8
130-650 µg per plate	Standard plate incorporation assay	Not required						+							8

(continued)

Table 7. (continued)

Dose	Method	S-9 Species	S-9 Tissue	S-9 Induction	TA1535 +	TA1535 -	TA100 +	TA100 -	TA1538 +	TA1538 -	TA98 +	TA98 -	TA1537 +	TA1537 -	Reference
N-METHYL-N'-NITRO-N-NITROSOGUANIDINE (CAS registry number: 70-25-7)															
0.5-2.0 μg per plate	Standard plate incorporation assay	Not required				+		+		-		-			8,3,15
0.02 mM	Liquid assay	Human[a]	Liver		+	+									12
Not specified		Not required				+		+		-				+	6

[a]Mutagenic activity is decreased in the presence of S-9.

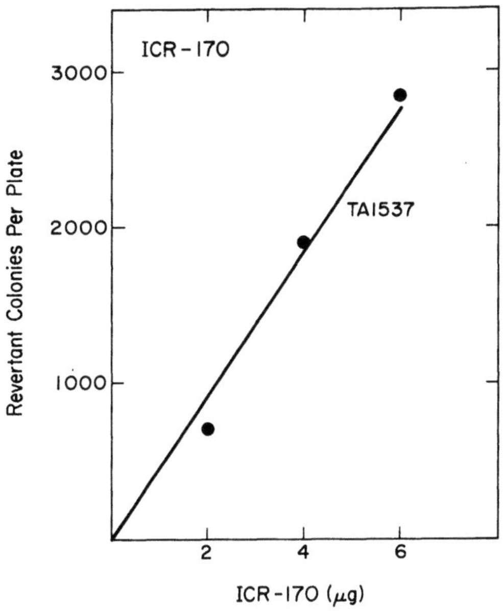

Figure 8. Mutagenicity of ICR-170 in the standard plate incorporation assay on S. typhimurium TA1537. Spontaneous revertant colonies have been subtracted. The mutagenic potency in revertants/nmole (3) is 193, and in revertants/μg is 475.

N-Methyl-N'-nitro-N-nitrosoguanidine (MNNG)

MNNG is a potent mutagen in Salmonella, reverting the missense tester strains TA1535 and TA100 (Table 7, Fig. 11). Weak activity is observed with the frameshift tester strain TA1537. Czygan et al. (40) tested MNNG in the presence of human liver microsomes and have shown that the mutagenic activity is decreased. This decrease in activity is dependent on the presence of an NADPH generating system. While MNNG gives rise to alkylating intermediates nonenzymatically, little information is available concerning the metabolism of this compound by the microsomal enzymes. It would be of interest to further examine the relationship between enzymatic deactivation and mutagenicity in Salmonella.

Mitomycin C

Mitomycin C, a carcinogen and a DNA crosslinking agent, does not revert the standard tester strains, but does revert strains TA92 and TA94 which have excision repair (Table 8, Fig. 12).

Table 8. Mutagenicity of Mitomycin C (CAS registry number: 50-07-7)

Dose	Method	S-9 Species	S-9 Tissue	S-9 Induction	S-9	TA1535 +	TA1535 -	TA100 +	TA100 -	TA1538 +	TA1538 -	TA98 +	TA98 -	TA1537 +	TA1537 -	TA92 +	TA92 -	TA94 +	TA94 -	Reference
1.0 μg per plate	Standard plate incorporation assay	Not required					-		-		-		-		-		+			3,8
0.02-1.0 μg per plate	Standard plate incorporation assay	Not required															+		+	1,15
0.05-1.0 μg per plate	Standard plate incorporation assay	Rat	Liver	Pb		-	-			-	-			-	-					19

Excision repair is required for mutagenesis with mitomycin C in Escherichia coli (41), and it has been suggested that mutation can be detected only if potentially lethal crosslinks are broken by the excision repair system (41,42). We are considering introducing one of these strains for the detection of crosslinking agents.

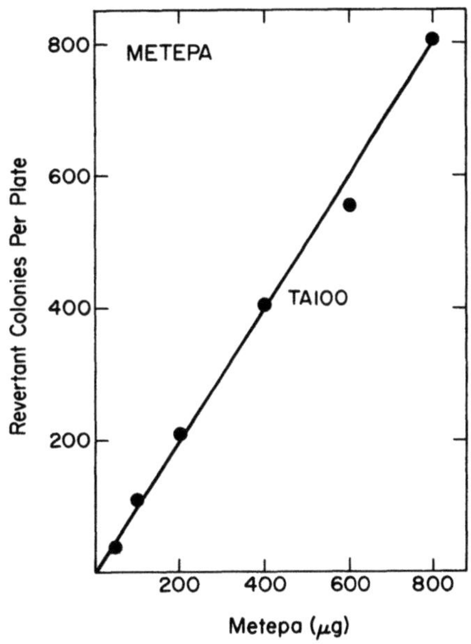

Figure 9. Mutagenicity of METEPA in the standard plate incorporation assay on S. typhimurium TA100. Spontaneous revertant colonies have been subtracted. The mutagenic potency in revertants/nmole (3) is 0.22, and in revertants/μg is 1.01.

Myleran

Myleran is a direct-acting mutagen in Salmonella, reverting the missense tester strains TA1535 and TA100 (Table 9, Fig. 13). It has not been tested in the presence of S-9 or on the frameshift tester strains TA1537 or TA1538, although it is unlikely that these strains would be reverted by this compound.

Natulan

Natulan, which has been thoroughly tested in this laboratory, has not been shown to be a mutagen in Salmonella (<0.0015 revertants/nmole) (Table 9). As discussed previously (4), the enzymes

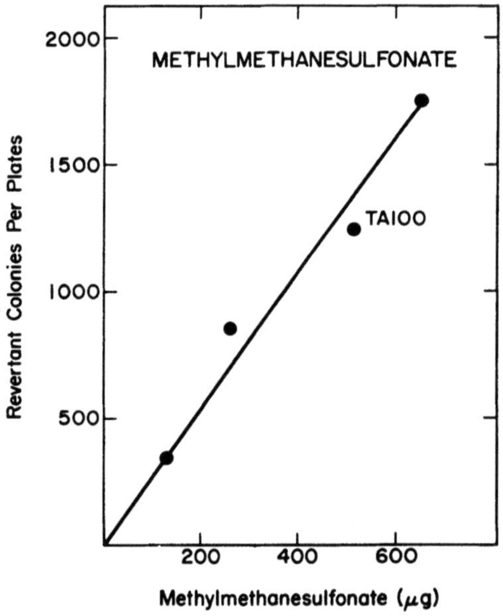

Figure 10. Mutagenicity of methylmethanesulfonate in the standard plate incorporation assay on S. typhimurium TA100. Spontaneous revertant colonies have been subtracted. The mutagenic potency in revertants/nmole (3) is 0.30, and in revertants/μg is 2.75.

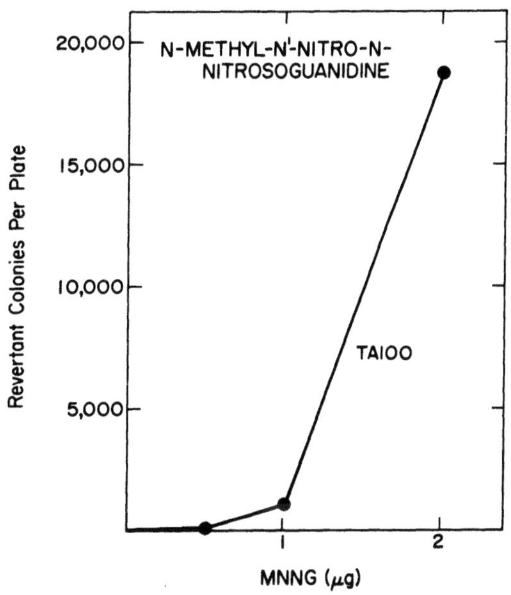

Figure 11. Mutagenicity of MNNG in the standard plate incorporation assay on S. typhimurium TA100. Spontaneous revertant colonies have been subtracted. The mutagenic potency in revertants/nmole (3) is 1375, and in revertants/μg is 9351.

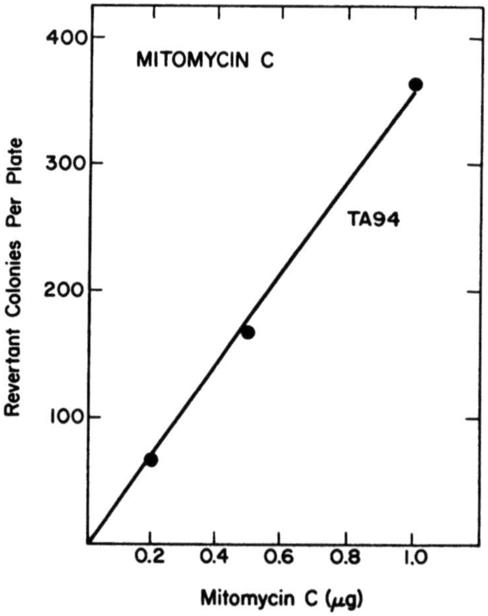

Figure 12. Mutagenicity of mitomycin C in the standard plate incorporation assay on S. typhimurium TA94. Spontaneous revertant colonies have been subtracted. The mutagenic potency in revertants/nmole (3) is 121, and in revertants/μg is 363.

Table 9. Mutagenicity in *Salmonella*

Dose	Method	S-9 Species	S-9 Tissue	S-9 Induction	Mutagenicity in *Salmonella*: TA1535 S-9 +	TA1535 S-9 −	TA100 +	TA100 −	TA1538 +	TA1538 −	TA98 +	TA98 −	TA1537 +	TA1537 −	Reference
				MYLERAN (CAS registry number: 55-98-1)											
50-400 µg per plate	Standard plate incorporation assay	Not required				+		+							15
				NATULAN (CAS registry number: 366-70-1)											
0.075-75 mM	Liquid assay	Rat	Liver	PCB	-	-	-	-							15
10-500 µg per plate	Standard plate incorporation assay	Rat	Liver	PCB	-	-	-	-			-	-	-	-	3,15
				TEPA (CAS registry number: 545-55-1)											
10-200 µg per plate	Standard plate incorporation essay	Rat	Liver	PCB	+	+	+	+			-	-	-	-	15
Crystal	Spot test	Not required				+									17

(continued)

Table 9. (continued)

Dose	Method	S-9 Species	S-9 Tissue	S-9 Induction	TA1535 S-9 +	TA1535 S-9 −	TA100 S-9 +	TA100 S-9 −	TA1538 S-9 +	TA1538 S-9 −	TA98 S-9 +	TA98 S-9 −	TA1537 S-9 +	TA1537 S-9 −	Reference
THIO-TEPA (CAS registry number: 52-24-4)															
Crystal	Spot test	Not required				+									17
100-1000 μg per plate	Standard plate incorporation assay	Not required				+		+							3,43
TRENIMON (CAS registry number: 68-76-8)															
0.2-1.0 μg per plate	Standard plate incorporation assay	Rat	Liver	PCB	+	+	+	+			+	+	-	-	15
TRIETHYLENEMELAMINE (CAS registry number: 51-18-3)															
5-25 μg per plate	Standard plate incorporation assay	Not required				+		+							15

required for the activation of natulan may have only very low activity in our S-9 fractions. It is possible that by employing more sensitive strains or changing the assay conditions and by examining the various factors affecting the metabolic activation of natulan, it will be detected as a mutagen in Salmonella.

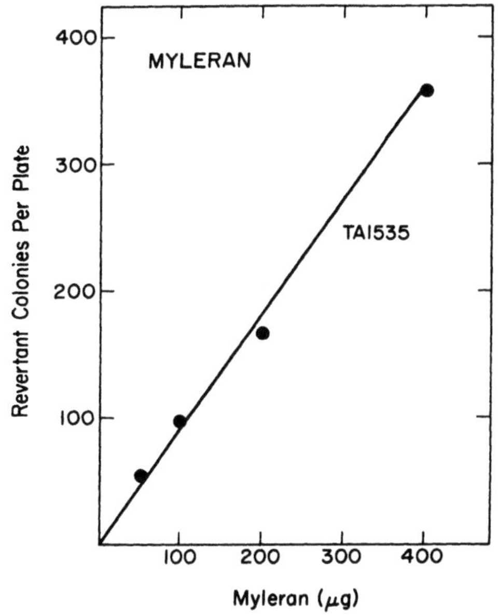

Figure 13. Mutagenicity of myleran in the standard plate incorporation assay on S. typhimurium TA1535. Spontaneous revertant colonies have been subtracted. The mutagenic potency in revertants/nmole (3) is 0.22, and in revertants/μg is 0.89.

TEPA

Studies of the mutagenic activity of TEPA in Salmonella are complete (Table 9, Fig. 14). The compound is a direct-acting mutagen, reverting the missense tester strain TA1535, and the R factor strain TA100. The observed activity is neither increased nor decreased in the presence of S-9.

Thio-TEPA

Thio-TEPA is a direct-acting mutagen in Salmonella, reverting the missense tester strains TA1535 and TA100 (Table 9, Fig. 15).

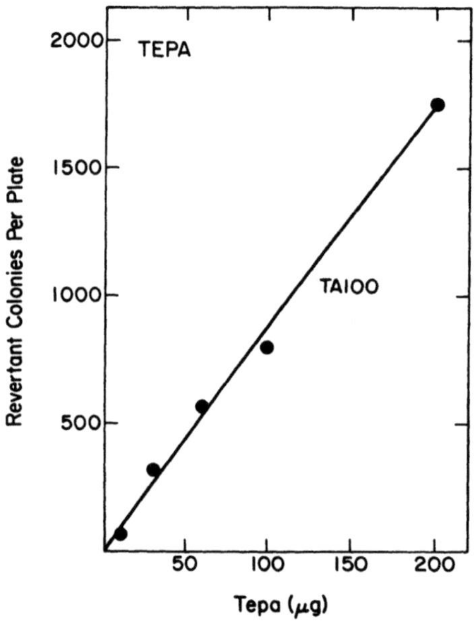

Figure 14. Mutagenicity of TEPA in the standard plate incorporation assay on S. typhimurium TA100. Spontaneous revertant colonies have been subtracted. The mutagenic potency in revertants/nmole (3) is 1.52, and in revertants/μg is 8.8.

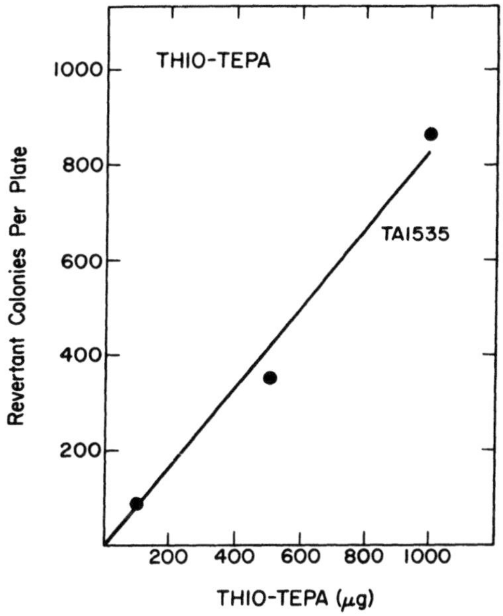

Figure 15. Mutagenicity of thio-TEPA in the standard plate incorporation assay on S. typhimurium TA1535. Spontaneous revertant colonies have been subtracted. The mutagenic potency in revertants/nmole (3) is 0.16, and in revertants/μg is 0.86.

It has not been tested in the presence of S-9 or on the frameshift tester strains TA1537 and TA1538, although it is unlikely that these strains would be reverted by this compound.

Trenimon

Studies of the mutagenic activity of Trenimon in Salmonella are complete (Table 9, Fig. 16). The compound is a direct-acting mutagen, reverting the missense tester strain TA1535, and the two R factor strains, TA100 and TA98. The observed activity is neither increased nor decreased in the presence of S-9.

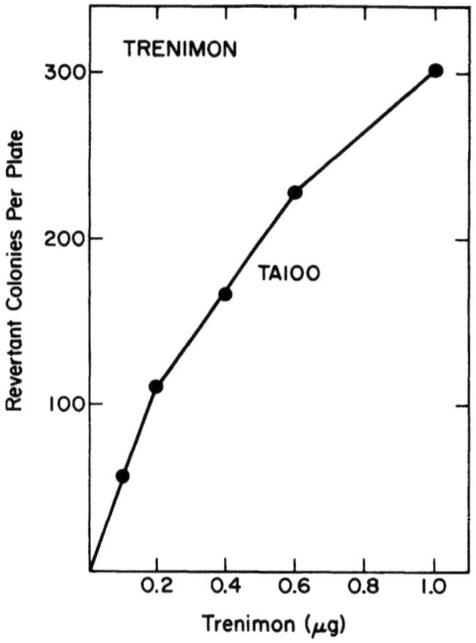

Figure 16. Mutagenicity of Trenimon in the standard plate incorporation assay on S. typhimurium TA100. Spontaneous revertant colonies have been subtracted. The mutagenic potency in revertants/nmole (3) is 128.3, and in revertants/μg is 555.

Triethylenemelamine

Triethylenemelamine is a direct-acting mutagen in Salmonella, reverting the missense tester strains TA1535 and TA100 (Table 9, Fig. 17). It has not been tested in the presence of S-9 or on the frameshift tester strains TA1537 and TA1538, although it is likely that these strains would be reverted by this compound.

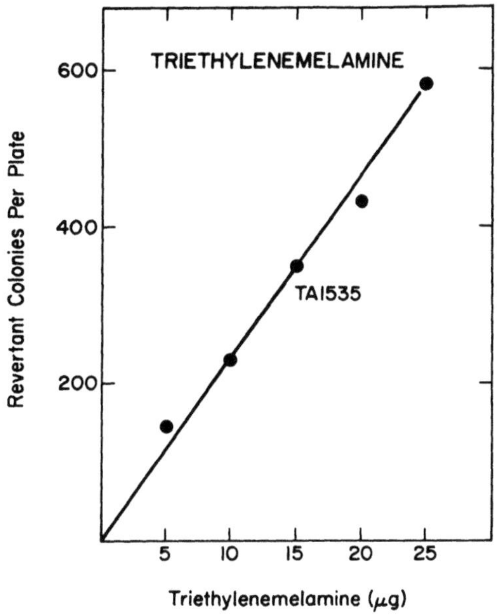

Figure 17. Mutagenicity of triethylenemelamine in the standard plate incorporation assay on S. typhimurium TA1535. Spontaneous revertant colonies have been subtracted. The mutagenic potency in revertants/nmole (3) is 4.75, and in revertants/μg is 23.2.

Vinyl Chloride

Vinyl chloride is a direct-acting mutagen in Salmonella, reverting the missense tester strains TA100 and TA1535 (Table 10, Fig. 18). Bartsch et al. (44,45) have shown that liver S-9 fractions from phenobarbital-pretreated and untreated rats and mice are capable of increasing the mutagenic activity of vinyl chloride, although the activity was not significantly increased by S-9 fractions prepared from the lung or kidney. Human liver S-9 fractions were also shown to increase the mutagenic response, although there was a wide range of variability among different samples, ranging from a 2- to 7-fold increase over controls. The enzyme activity was shown to be localized in the microsomal fraction and to be dependent on an NADPH generating system.

In contrast to the work by Bartsch, Garrol et al. (46) have shown that while the mutagenic activity of vinyl chloride is enhanced by mouse or liver S-9 fractions, this increase does not require an NADPH generating system and is still present, although at a reduced level, in liver extracts in which the microsomal mixed function oxidase system has been heat inactivated. Further

Table 10. Mutagenicity of vinyl chloride (CAS registry number: 75-01-4)[a]

	S-9			Mutagenicity in Salmonella										
				TA1535		TA100		TA1538		TA98		TA1530		
Species	Tissue	Induction	S-9	+	-	+	-	+	-	+	-	+	-	References
Rat	Liver	Unind.		+	-			-						47
Mouse	Liver	Pb[b]		+					-			+	+	44,45,48
Mouse, rat	Liver	Pb, unind.										+		44,45,48
Human	Liver											+		44,45,48
Rat	Liver	Pb[b]		+	+	+	+							18

[a]Bacteria and S-9 were plated in the usual manner. Plates were placed in desiccators containing an atmosphere of vinyl chloride in air (0.2% to 20%, v/v) for 48 hr (44,45) or containing a constant concentration of vinyl chloride (20% v/v) for varying exposure times (18,47).

[b]Mutagenic activity is higher in the presence of S-9.

evidence suggests that the mutagenic effect of vinyl chloride may involve a free radical mechanism.

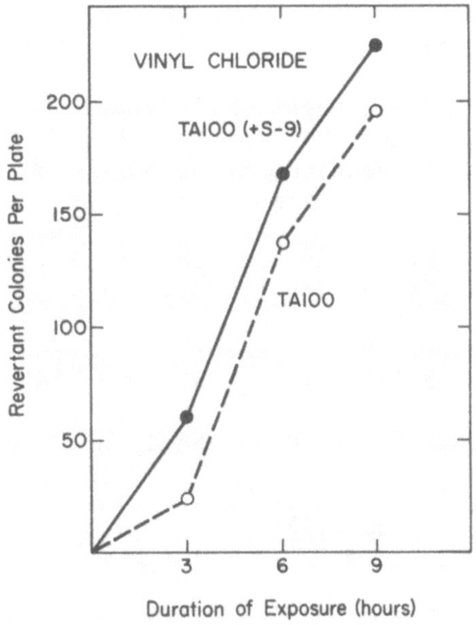

Figure 18. Mutagenicity of vinyl chloride in S. typhimurium TA100. Spontaneous revertant colonies have been subtracted. Liver homogenate (150 μl/plate) from phenobarbital-induced rats was added as described in the text.

Further work investigating the role of the S-9 fraction and the details of vinyl chloride metabolism are needed to help resolve the discrepancies between the two studies cited.

REFERENCES

1. McCann, J., and Ames, B. N. In: Origins of Human Cancer, H. Hiatt, J. D. Watson, and J. A. Winsten, Eds., Cold Spring Harbor Laboratory, N.Y., 1977, pp. 1431-1450.

2. Ames, B. N., McCann, J., and Yamasaki, E. Mutat. Res. 31: 347-364 (1975).

3. McCann, J., Choi, E., Yamasaki, E., and Ames, B. N. Proc. Natl. Acad. Sci. (U.S.) 72: 5135-5139 (1975).

4. McCann, J., and Ames, B. N. Proc. Natl. Acad. Sci. (U.S.) 73: 950-954 (1975).

5. Purchase, I. F. H., Longstaff, E., Ashby, J., Styles, J. A., Anderson, D., Lefevre, P. A., and Westwood, F. R. Nature 264: 624-627 (1976).

6. Sugimura, T., Sato, S., Nagao, M., Yahagi, T., Matsushima, T., Seino, Y., Takeuchi, M., and Kawachi, T. In: Fundamentals in Cancer Prevention, P. N. Magee, et al., Eds., University Park Press, Baltimore, 1976, pp. 191-215.

7. Ames, B. N., Lee, F. D., and Durston, W. E. Proc. Natl. Acad. Sci (U.S.) 70: 782-786 (1973).

8. McCann, J., Spingarn, N. E., Kobori, J., and Ames, B. N. Proc. Natl. Acad. Sci (U.S.) 72: 979-983 (1975).

9. Isono, K., and Yourno, J. Proc. Natl. Acad. Sci. (U.S.) 71: 1612-1617 (1974).

10. Mukai, F. H., and Goldstein, B. D. Science 191: 868-869 (1976).

11. Ames, B. N., Durston, W. E., Yamasaki, E., and Lee, F. D. Proc. Natl. Acad. Sci. (U.S.) 70: 2281-2285 (1973).

12. Garner, R. C., and Wright, C. M. Brit. J. Cancer 28: 544-551 (1973).

13. Swenson, D. H., Miller, J. A., and Miller, E. C. Cancer Res. 35: 3811-3823 (1975).

14. G. Lofroth, personal communication.

15. Unpublished results, this laboratory.

16. Ellenberger, J. and Mohn, G. Arch. Toxicol. 33: 225-240 (1975).

17. Hanna, P. J., and Dyer, K. F. Mutat. Res. 28: 405-420 (1975).

18. McCann, J., Simmon, V., Streitwieser, D., and Ames, B. N. Proc. Natl. Acad. Sci. (U.S.) 72: 3190-3193 (1975).

19. Minnich, V., Smith, M.E., Thompson, D., and Kornfeld, S. Cancer 38: 1253-1258 (1976).

20. Braun, R., and Schoeneich, J. Mutat. Res. 31: 191-194 (1975).

21. Colvin, M., Brundrett, R. B., Kan, M. N., Jardine, I., and Fenselau, C. Cancer Res. 36: 1121-1126 (1976).

22. Ellenberger, J., and Mohn, G. R. Mutat. Res. 38: 120-121 (1976).

23. Malling, H. V., and Frantz, C. N. Environ. Health Perspect. 6: 71-82 (1973).

24. Bartsch, H., Malaveille, C., and Montesano, R. Cancer Res. 35: 644-651 (1975).

25. Yahagi, T., Nagao, M., Seino, Y., Matsushima, T., Sugimura, T., and Okada, M. Mutat. Res. 48: 121-130 (1977).

26. Sugimura, T., Yahagi, T., Nagao, M., Takeuchi, M., Kawachi, T., Hara, K., Yamasaki, E., Matsushima, T., Hashimoto, Y., and Okada, M. In: Screening Tests in Chemical Carcinogenesis, R. Montesano, H. Bartsch, and L. Tomatis, Eds., IARC, Lyon, 1976.

27. Frantz, C. N., and Malling, H. V. Cancer Res. 35: 2307-2314 (1975).

28. Malling, H. V. Mutat. Res. 26: 465-472 (1974).

29. Frantz, C. N., and Malling, H. V. Mutat. Res. 31: 365-380 (1975).

30. Bartsch, H., Malaveille, C., and Montesano, R. Chem. Biol. Interact. 10: 377-382 (1975).

31. Bartsch, H., Camus, A., and Malaveille, C. Mutat. Res. 37: 149-162 (1976).

32. Weekes, U., and Brusick, D. Mutat. Res. 31: 175-183 (1975).

33. Malling, H. V. Mutat. Res. 13: 425-429 (1971).

34. Weekes, U. Y. J. Natl. Cancer Inst. 55: 1199-1201 (1975).

35. Gletten, F., Weekes, U., and Brusick, D. Mutat. Res. 28: 113-122 (1975).

36. Czygan, P., Greim, H., Garro, A. J., Hutterer, F., Schaffner, F., Popper, H., Rosenthal, O., and Cooper, D. Y. Cancer Res. 33: 2983-2986 (1973).

37. Guttenplan, J. B., Hutterer, F., and Garro, A. J. Mutat. Res. 35: 415-422 (1976).

38. V. Simmon, personal communication.

39. Sram, R. J., Cerna, M., and Kucerova, M. Biol. Zentralbl. 95: 451-462 (1976).

40. Czygan, P., Greim, H., Garro, A. J., Hutterer, F., Rudick, J., Schaffner, F., and Popper, H. J. Natl. Cancer Inst. 51: 1761-1764 (1973).

41. Kondo, S., Ichikawa, H., Iwo, K., and Kato, T. Genetics 66: 187-217 (1970).

42. Murayama, I., and Otsuji, N. Mutat. Res. 18: 117-119 (1973).

43. Benedict, W. F., Baker, M. S., Haroun, L., Choi, E., and Ames, B. N. Cancer Res. 37: 2209-2213 (1977).

44. Bartsch, H., Malaveille, C., and Montesano, R. Int. J. Cancer 15: 429-437 (1975).

45. Malaveille, C., Bartsch, H., Barbin, A., Camus, A. M., Montesano, R., Croisy, A., and Jacquignon, P. Biochem. Biophys. Res. Commun. 63: 363-370 (1975).

46. Garro, A. J., Guttenplan, J. B., and Milvy, P. Mutat. Res. 38: 81-88 (1976).

47. Rannug, U., Johansson, A., Ramel, C., and Wachtmeister, C. A. Ambio 3: 194-197 (1974).

48. Bartsch, H., and Montesano, R. Mutat. Res. 32: 93-114 (1975).

CHAPTER 5

MUTAGENICITY OF SELECTED CHEMICALS IN *ESCHERICHIA COLI* TEST SYSTEMS

Georges R. Mohn

Department of Radiation Genetics and Chemical Mutagenesis, State University of Leiden, Wassenaarsweg 72, Leiden, The Netherlands

INTRODUCTION

General Remarks

Mutagenicity test systems using *Escherichia coli* are characterized by the great diversity of tester strains and substrains (e.g., several different repair-proficient and repair-deficient strains), of genetic endpoints (several different forward- and back-mutation types, lysogenic induction), and of testing methods (spot test, liquid suspension test, fluctuation test). There is no single set of experimental conditions where all the substances of this study would have been assayed. It is, therefore, difficult quantitatively to compare the results that have been gained under the influence of different experimental factors. As an example, the lowest detectable mutagenic activity of MMS is about 1×10^{-3} mole/l. in usual liquid suspension tests, whereas newly developed fluctuation tests (1) allow the detection of MMS concentrations as low as 45×10^{-6} mole/l. Nevertheless, attempts have been made to estimate roughly the relative mutagenic potencies of the test substances as indicated further in the text. Induction of vegetative growth of prophage in lysogenic cells has been added as a further genetic endpoint, since it was repeatedly demonstrated to be correlated to mutagenic activity.

The following results were selected as the most significant or relevant data giving information on the lowest reported effective doses, the type of induced genetic alterations, and the experimental conditions. Attempts have been made to standardize the dosage data (which are given in tenths of moles/liter and in milligrams or micrograms/milliliter) but it must be noted that the concentrations given do not represent the actual effective concentrations, especially when spot tests or plate tests were performed. It must also be noted that treatment time varied greatly among different experiments.

Types of Genetic Alterations Which Can be Detected

The availability of a wide variety of mutation systems allows the detection of practically all known types of genetic alterations leading to gene or point-mutation, such as single base-pair changes (substitutions, reading frame shifts), gene deletions, duplications, and insertions. Mutations in structural genes as well as regulatory genes can be tested. The specific effects of chemicals on certain DNA bases or base configurations are best characterized in back-mutation assays (e.g., reversion from auxotrophy to prototrophy), whereas the overall effects on several loci within genes are assayed in forward-mutation tests (e.g., resistance to antimetabolites). In general, it can be stated that the back-mutation assays (that is, those with no interference of extragenic suppressors) are very sensitive because of the usually low back-mutation frequencies but also very specific in the detectable mutation spectrum which makes the use of several different back-mutation systems in routine testing necessary, and it is probable that deletions will not be detected there. Certain forward-mutation systems have the advantage of detecting simultaneously different types of mutagenic agents (e.g., frameshift-inducing as well as base-pair substitution-inducing chemicals) but their sensitivity is usually reduced because of the higher spontaneous frequency as compared with back-mutation tests. Taken together, it seems that a combination of both back- and forward-mutation systems will be most appropriate in determining the effects of mutagenic chemicals.

Mutagenicity Testing Procedures

Forward and back mutations can be detected with the same techniques, since both make use of selective semisolid agar media where only the rare mutants among large populations of nonmutated cells can grow and form colonies. In general, three techniques are used in only slight modification of those already described in the 1950s by Iyer and Szybalski (2) and Demerec (3,4).

Spot Test. The bacterial population is spread homogeneously over the surface of the selective agar medium, and, after drying, a small amount of the chemical being tested is added in the middle of the plate. After a 2- to 4-day incubation period at 37°C, the chemical has diffused through the agar and, in the case of mutagenicity, induces a ring of mutant colonies around the drop. Besides their great simplicity, spot tests offer the possibility of reducing, under appropriate experimental conditions, the selection of pre-existing spontaneous mutants in the culture and allow the detection of chemicals active preferentially on replicating DNA. Furthermore, a wide range of concentrations can be tested on a single Petri plate. One disadvantage of testing unknown compounds with spot test techniques is the possibility of obtaining false negative results; for example, when the chemical is water-insoluble or active at a pH different from that of the selective medium. The test sensitivity can furthermore be reduced when the mutagenic concentration of the test compound is close to the inactivating one. Since false positive results may also be observed under certain circumstances, it seems to be necessary to confirm spot test results by performing liquid suspension tests or fluctuation tests as described below.

Liquid Suspension Test. Several variations of the test exist, but, in principle, the treatment of the stationary or growing bacterial population is done for a limited period of time of ca. 0.5-4 hr in a suspension containing a known amount of the test chemical. Treatment is terminated by washing the cells and/or diluting in buffer. The survival of colony-forming ability and the mutation frequency is determined by plating aliquots of the appropriately diluted bacterial suspension on complete medium and selective mutation medium respectively. Most compounds of this study have been assayed in liquid test procedures and these have served as the basis for comparing the relative mutagenic efficiencies, although it should again be noted that the different *E. coli* strains used, certainly have different intrinsic sensitivities to chemicals.

Fluctuation Test. This technique was used very early to compare the spontaneous mutation rates of certain genes in strains of *E. coli*. Recent developments by Green et al. (1) allow one to apply the fluctuation test in routine chemical mutagenicity testing. In principle, several aliquots of a bacterial population are allowed to grow for a known, restricted number of generations in a liquid, selective growth medium. Only in those cultures where one or more mutation occurred (e.g., from auxotrophy to prototrophy) will full growth be observed by turbidity or by acid production measurements. An increasing proportion of full grown cultures with increasing concentrations of the test chemical will thus indicate mutagenicity. Although this method may not be as flexible as liquid tests to study different treatment conditions such as time of exposure, pH of treatment, or temperature, it seems to be extremely sensitive to the mutagenic action of certain chemicals and is certainly an appropriate method for detecting lowest effective concentrations.

Use of Metabolic Activation Systems

At present, metabolic activation systems are an integrated part of routine mutagenicity testing in E. coli. Greatest experience has been gained in liquid test procedures combined with mammalian organ homogenates. The tests are performed in essentially the same way as liquid tests, except that microsomes or S-9 fractions are added to the bacterial incubation mixtures which also contain an NADPH-generating system and cofactors. In the present comparative study all chemicals which require mammalian metabolic activation have been successfully detected by using E. coli as an indicator; it was also shown that the effects of mammalian microsomal enzymes on directly acting mutagens such as MNNG and EMS can be accurately determined.

Detailed Description of Assay Systems

Spot tests and plate tests have been detailed by Iyer and Szybalski (5), by Mohn (6), and by Green and Muriel (7). Liquid tests with and without metabolic activation were also described by Green and Muriel (7) and by Mohn and Ellenberger (8). Fluctuation tests are reported in detail by Voogd et al. (9) and by Green and Muriel (7). Heinemann (10) and Moreau et al. (11) have reported on prophage induction in lysogenic bacteria as a method of detecting potential mutagenic, carcinogenic, carcinostatic, and teratogenic agents.

RESULTS AND DISCUSSION

The results with Escherichia coli are summarized in Table 1. Data on individual chemicals are detailed in Table 2. Results with x-radiation are shown in Table 3.

Aflatoxin B1

Aflatoxin B1 does not induce genetic effects in E. coli unless metabolic activation through mammalian enzymes is provided (12,13). After suitable activation - which probably leads to an epoxy derivation of Aflatoxin B1 - a linear, concentration-dependent increase of mutation frequency is observed. The experiments performed in a repair-proficient strain indicate that base-pair substitution mutations are induced, as well as induction of vegetative growth of prophage lambda (12). The lowest reported effective concentration was 1.6×10^{-5} mole/l for 90 min at 37°C on growing cells.

Cadmium

The only published report of experiments with cadmium (14) did not indicate a genetic effect, although other metals such as Cr^{2+} and As (III) were active in inducing trp^+ reversions in E. coli B/r. Since modification of treatment procedures may profoundly alter the response of E. coli to metal ions, the negative result with cadmium is not a definitive indication of nonmutagenicity of this chemical toward E. coli.

Cyclophosphamide

Cyclophosphamide, a bifunctional derivative of nitrogen mustard, is not mutagenic to E. coli in liquid suspension tests unless metabolic activation through mammalian microsomal enzymes is provided. Known mammalian metabolites of cyclophosphamide such as carboxyphosphamide, 4-ketocyclophosphamide, phosphoramide mustard, and nornitrogen mustard are directly mutagenic to E. coli. Cyclophosphamide itself is mutagenic to E. coli in a concentration-dependent manner after activation through rat or mouse liver microsomal fractions. When tested in a repair-proficient strain, the induced mutations are of the base-pair substitution type. The lowest reported effective concentration is 5×10^{-4} mole/l for 180 min at 37°C in stationary cells (15,16).

Diethylnitrosamine

Diethylnitrosamine is not mutagenic to E. coli unless metabolic activation through mammalian microsomal enzymes is provided (17,18). After suitable activation a concentration-dependent, linear increase of mutation frequency is observed. In repair-proficient strains of E. coli the induced mutations are of the base-pair substitution type. The lowest reported concentration is 1×10^{-4} mole/l. for 180 min at 37°C on stationary cells (18).

Dimethylnitrosamine

Dimethylnitrosamine is not mutagenic to E. coli unless metabolic activation through mammalian microsomal enzymes is provided, but inactivation of colony-forming ability has been reported after treatment with high doses (2.5×10^{-1} mole/l.). After suitable activation, a concentration-dependent, linear increase of mutation frequency is observed. In repair-proficient strains of E. coli the induced mutations are of the base-pair substitution types. Lowest reported effective concentration is 1×10^{-4} mole/l. for 180 min at 37°C in stationary cells (18,19).

Table 1. Summary of results with *Escherichia coli*

Chemical	Lowest required Effective concn. (mole/l.)	Genetic effect observed[a]			Activation Required	Comparison with	
		BPS	FS	λ-IND		X-rays	EMS
Aflatoxin B1	1.6×10^{-5}	+		+	yes	>	<
Cadmium salt	Not effective[b]					<	<
Cyclophosphamide	5.0×10^{-4}	+	-		yes	>	>
Diethylnitrosamine	1.0×10^{-4}	+	-		yes	>	>
Dimethylnitrosamine	1.0×10^{-4}	+	-		yes	>	>
Epichlorohydrin	5.0×10^{-2}	+	-		no	<	<
Ethyleneimine	1×10^{-4} mole/plate	+			no	>	<
EMS	5.0×10^{-3}	+		+	no[c]	>	
ICR-170	1.1×10^{-5}	-	+		no	>	>
MeTEPA	5×10^{-7} mole/plate	+	-		no	>	>
Mitomycin C	8.0×10^{-8}	+[d]	-		yes[e]	>	>
MMS	4.5×10^{-6}	+	-	+	no	>	>
MNNG	1.1×10^{-6}	+	(+)[f]	+	no[g]	>	>

Myleran	2.0×10^{-3}	+		+	no	>	>
Natulan	No data[h] on *E. coli* so far available						
TEPA	2.3×10^{-4}	+	-	+	no	>	>
Thio-TEPA	2.6×10^{-3}	+	-	+	no	>	>
TEM	1.0×10^{-6}	+	-	+	no	>	>
Trenimon	2.2×10^{-3}	+	-	+	no	>	>
Vinyl chloride	1.1×10^{-2}	+	-		yes	<	<
X-rays	8.5 rad	+	-	+	no		<

[a] BPS = base pair substitution; FS = reading frameshift; λ-IND = induction of vegetative growth of prophage in lysogenic cells.

[b] Not effective in *E. coli* strains tested to date.

[c] Shows increased mutagenic activity after incubation through mammalian liver S-9.

[d] Also induces DNA crosslinking.

[e] Metabolic activation of Mitomycin C performed also by *E. coli* cells.

[f] Also induces duplications and single-strand DNA breaks.

[g] Shows decreased mutagenicity after incubation with mammalian liver S-9.

[h] Mutagenicity without mammalian activation in *E. coli* 343/113 has been recently reported (76); lowest reported effective concentration: ≧31 mM, treatment time 120 min; at 73°C.

Table 2.
Comparative chemical mutagenesis of *Escherichia coli* test systems

Mutagen	Genetic effect observed	Treatment conditions: pH	T(°C)	Lowest reported effective concn (mole/l.)	Treatment time (min)
Aflatoxin B1 (CAS 1162-65-8)	Induction of vegetative growth of prophage in *E. coli* K-12 (λ)	7	37	1.6×10^{-5} (5.0 μg/ml)	90
	Back mutations from arg^- to prototrophy in *E. coli* K-12/343/113	7.0	37	6.4×10^{-5} (20 μg/ml)	150
Cadmium (salt) (CAS 10108-64-2)	Back mutations from trp^- to prototrophy in *E. coli* B/r WP2	7	37	Crystal (ca. 10 mg) per plate	Ca. 2 days
Cyclophosphamide (CAS 50-18-0)	Back mutations from arg^- to protorophy and forward mutations to MTR resistance in *E. coli* K-12/343/113	7.4	37	5×10^{-4} (140 μg/ml)	180
	Forward and back mutations to arg^+, gal^+, and MTR in *E. coli* 343/113	7.4	37	5×10^{-5} (13.95 μg/ml)	180

Testing method	Remarks	References
Liquid test with rat liver S-9 homogenate	Phenobarbital-pretreated male Wistar rats used No detectable effect of Aflatoxin B1 without mammalian activation	12
Spot test without mammalian activation	No mutagenic effect detected while Cr^{2+} and AS(III) ions are active in this system	14
Liquid test with activation through mammalian liver S-9 homogenates	S-9 supernatants of livers of not pretreated NMRI mice and SIV 40 rats used Dose-dependent increase in both mutation systems observed Mouse liver S-9 was about twice as active as rat liver S-9 No mutagenic effects detected without S-9 addition	15
Liquid test without mammalian activation	All known major mammalian metabolites of cyclophosphamide tested All exert certain mutagenic effects without further mammalian activation	16
Liquid test with purified rat liver microsomes	Dose-dependent effect with microsomes Reduced mutagenic effect when supernatant of 105,000 g centrifugation is used No detectable mutagenicity without mammalian activation	17

(continued)

Table 2. (continued)

Mutagen	Genetic effect observed	Treatment conditions: pH	T(°C)	Lowest reported effective concn (mole/l.)	Treatment time (min)
Diethylnitrosamine (CAS 55-18-0)	Back mutations from str dependence to independence in *E. coli* Sd-B(TC)	7	37	3×10^{-4} (30.6 μg/ml)	30
	Forward and back mutations to gal^+, arg^+, and MTR in *E. coli* K-12/343/113	7.4	37	1×10^{-4} (10.2 μg/ml)	180
Dimethylnitrosamine (CAS 62-75-9)	Forward and back mutations to gal^+, arg^+, and MTR in *E. coli* K-12/343/113	7.4	37	1×10^{-4} (7.4 μg/ml)	180
	Back mutations to str independence in dependent *E. coli* Sd-B(TC)	7	37	6×10^{-4} (44.4 μg/ml)	30
	Back mutations to trp^+ in trp^- strains of *E. coli* B/r WP2 with different repair capacities	7.4	37	1.35×10^{-1} (10 mg/ml)	30

Testing method	Remarks	References
Liquid test with activation through mouse liver S-9 homogenate	Dose-dependent increase of mutation frequency in all systems investigated No detectable effects without mammalian activation	18
Liquid test with activation through mouse liver S-9 homogenate	Dose-dependent increase of mutation frequency in all systems studied No effect without mammalian activation	18,19
Liquid test with activation through rat liver microsomes	Dose-dependent effect with microsomes Reduced mutagenic effect when supernatant of 105,000 g centrifugation is used No detectable mutagenicity without mammalian activation	17.20
Liquid test with activation through rat liver S-9	Dose-dependent increase in mutation frequency No effect when S-9 homogenat is omitted	21
Liquid test without mammalian activation	Concentration-dependent decrease of colony-forming ability *E. coli* B/s-1 more sensitive than B/r	22

(continued)

Table 2. (continued)

Mutagen	Genetic effect observed	Treatment conditions: pH	T(°C)	Lowest reported effective concn (mole/l.)	Treatment time (min)
	Lethal effect in *E. coli* B/r and repair-deficient derivatives	7.3	37	2.7×10^{-1} (18.5 mg/ml)	60
Epichlorohydrin (CAS 106-89-8)	Back mutations from trp^- to prototrophy in *E. coli* B/r	7.2	37	5×10^{-2} (4.63 mg/ml) (not completely dissolved)	20
Ethylenimine (CAS 151-56-4)	Back mutations from str dependence to independence in *E. coli* B/Sd-4	7	37	1×10^{-4} mole/plate (8.32 mg/plate)	ca. 2 days
EMS (CAS 62-50-0)	Back mutations to str independence in *E. coli* B/Sd-4	7.0	37	2.7×10^{-1} (33.48 mg/ml)	60
	Forward mutations to Col B resistance in *E. coli* B/NG_{30}, recA	6.8	37	8×10^{-2} (10 mg/ml)	60
	gal^- and mtl^- mutations in *E. coli* B/r, WP2/RH-53 and mut H1 derivative	7	37	2×10^{-1} (24.8 mg/ml)	60
	Back mutations from lys^- to lys in *E. coli* K-12	7.0	37	1×10^{-2} (1.24 mg/ml)	60

Testing method	Remarks	References
Liquid test without mammalian activation.	Concentration-dependent decrease of colony forming ability *E. coli* B/s-1 more sensitive than B/r.	22
Liquid test without mammalian activation	Dose-dependent increase of trp^+ revertants	23
Spot test without mammalian activation	Mutagenic effect; dose-dependent increase of revertants	24
Liquid test without mammalian activation	Relative mutagenic effectiveness at LD_{50}: diethyl sulfate > ethyl methmethanesulfonate > ethyl methanesulfonate > methyl methanesulfonate	25
Liquid test without mammalian activation	Mutagenic effectiveness: ethyl methanesulfonate ≃ MNNG > UV irradiation ≃ 4-nitroquinoline 1-oxide > hydroxylamine > mitomycin C ≃ x-rays	26,27
Liquid test without mammalian activation	Dose-dependent induction of both types of mutations Mutagenic effectiveness: MMS > EMS Enhancement of mutagenic effects through mutator gene	28
Liquid test without mammalian activation	Dose-dependent increase of lys revertants	18

(continued)

Table 2. (continued)

Mutagen	Genetic effect observed	Treatment conditions pH	T(°C)	Lowest reported effective concn (mole/l.)	Treatment time (min)
	Back mutations from arg^- to prototrophy and induction of vegetative growth of prophage Φ 80 in *E. coli* H/r-30	6.8	37	2.5×10^{-2} (3.1 mg/ml)	30
	Various back mutations in auxotrophic strains of *E. coli* K-12	7.0	37	5.6×10^{-6} mole/plate (0.7 mg/plate)	ca. 2 days
	Back mutations to lac^+ in *E. coli* K12/U120	7	37	2.9×10^{-4} mole/plate (36 mg/plate)	ca. 2 days
	Back mutations of amber mutants of *E. coli* K-12 S-26	7.4	37	4.96×10^{-1} (61.5 mg/ml)	2.5
	Forward mutations to lac^- in *E. coli* K-12/M1 308	7.0	37	1×10^{-1} (12.4 mg/ml)	30
	Forward mutations to azide resistance in *E. coli* K-12	7.0	37	2.5×10^{-1} (31 mg/ml)	15
	Duplication of gly T / pur D region in *E. coli* K-12/CH-408	7	37	4.8×10^{-5} 50 mg on paper disk on plate	ca. 2 days

Testing method	Remarks	References
Liquid test without mammalian activation	Dose-dependent increase of arg⁻ revertants and of lytic centers	29
Spot test without mammalian activation	Mutagenic effectiveness: MNNG > β-propiolactone > EMS > TEM	30
Spot test without mammalian activation	Mutagenic effect significant	31
Liquid test without mammalian activation	Mostly GC→AT transitions induced	32
Liquid test without mammalian activation	Preferentially GC→AT transitions induced	33
Liquid test without mammalian activation	Dose-dependent increase in mutation frequency Mutability enhanced in recA strains	34
Spot test without mammalian activation	Dose-dependent induction of duplications	35

(continued)

Table 2. (continued)

Mutagen	Genetic effect observed	Treatment conditions: pH	T(°C)	Lowest reported effective concn (mole/l.)	Treatment time (min)
	Forward mutations to phage T5 resistance in *E. coli* B:r,WP2/RH-53	7	37	5×10^{-3} 50 mg on paper disk on plate	60
	Back mutations to leu^+ in *E. coli* SE-35	7.0	37	1×10^{-1} (12.4 mg/ml)	10
	Forward and back mutations to gal^+, arg^+ and MTR in *E. coli* K-12/343/113	7.0	37	6×10^{-3} (0.74 mg/ml)	120
ICR-170 (CAS 146-59-8)	Back mutations to ara^+ in proflavine-induced ara^- mutants of *E. coli* K-12/AB 2271	7	37	1.1×10^{-5} (5 μg/ml)	30
	Back mutations to ara^+ in *E. coli* K-12/AB 2271	7	37	2.3×10^{-7} mole/plate (100 μg)	ca. 2 days
MeTEPA (CAS 57-39-6)	Back mutations to trp^+ in exrA derivative of *E. coli* B:WP2 trp^-	7.0	37	4.7×10^{-7} mole/plate (100 μg/plate)	ca. 3 days

Testing method	Remarks	References
	Time-dependent increase of phage T5 resistant mutants	36
Continuous liquid culture without mammalian activation		
Liquid test without mammalian activation	Time-dependent increase of leu^+ revertants	37
Liquid test with and without mammalian activation using mouse liver S-9	Dose-dependent increase of mutation frequency in all mutation systems tested	24
	Enhanced mutagenic potential of EMS after incubation with mouse liver homogenate	
Liquid test without mammalian activation	Dose-dependent increase of ara^+ mutation frequency, presumably due to frameshift	38
Spot test without mammalian activation	Induced back mutations to ara^+ presumably due to frameshift	39
Spot test without mammalian activation	Dose-dependent increase of trp^+ revertants	40

(continued)

Table 2. (continued)

		Treatment conditions			
Mutagen	Genetic effect observed	pH	T(°C)	Lowest reported effective concn (mole/l.)	Treatment time (min)
Mitomycin C (CAS 50-07-7)	Induction of vegetative growth of prophage in *E. coli* K-12 (λ)	7.0	37	2×10^{-6} (0.60 μg/ml)	60
	Host-range mutations in phage T2 and prophage induction in *E. coli* K-12 (λ)	7.0	37	4.42×10^{-10} mole/plate (0.05 ml of 3 μg/ml solution	ca. 24 hr
	Back mutations to $argF^{+}$ and induction of prophage Φ 80 in *E. coli* H/30	7.0	37	8×10^{-8} (0.025 μg/ml)	60
	Forward mutations to lac^{-} in *E. coli* S,lac^{+}	7.0	37	1×10^{-6} (0.3 μg/ml)	60
	DNA crosslinking in several bacterial species	7	37	6.6×10^{-6} (2 μg/ml)	1
	Back mutations to prototropy in *E. coli* K-12/AB1157, arg^{-}, his^{-}	7.0	37	3.3×10^{-7} (0.6 μg/ml)	48
	Back mutations to prototrophy in *E. coli*	7.0	37	1.6×10^{-7} (0.05 μg/ml)	ca. 3 days

Testing method	Remarks	References
Liquid test without mammalian activation	Dose-dependent increase in lytic centers	41,42, 43
Spot test without mammalian activation	Dose-dependent increase in phage mutations and lysogenic induction of prophage	44
Liquid test without mammalian activation	Dose-dependent increase of arg^+ revertants and of lytic centers	29,45
Liquid test without mammalian activation	Dose-dependent increase of lac^+ mutants	46
Liquid test without mammalian activation	Crosslinking of DNA in vitro occurs only after activation of Mitomycin C, e.g., through cell lysates of *Sarcina lutea*, *Bacillus subtillis*, or *E. coli*	47
Liquid test without mammalian activation	Dose-dependent increase in mutation frequency	48
Fluctuation test withou mammalian activation	Induction of trp^+ revertants significant	1

(continued)

Table 2. (continued)

Mutagen	Genetic effect observed	pH	T(°C)	Treatment conditions: Lowest reported effective concn (mole/l.)	Treatment time (min)
	Forward mutations to Col B-resistance in strain *E. coli* B/016, polA$^-$	6.8	37	6×10^{-7} (0.18 μg/ml)	60
	Back mutations to lac$^-$ in *E. coli* C 600	7	37	6.6×10^{-7} (0.2 μg/ml)	ca. 2 days
MMS (CAS 66-27-3)	Back mutations to trp$^+$ in *E. coli* WP2, trp$^-$	7.0	37	4.54×10^{-6} (0.5 μg/ml)	ca. 3 days
	Back mutations to str independence in *E. coli* B/Sd-4	7.0	37	2×10^{-3} (220 μg/ml)	60
	Forward mutations to str-resistance in strain *E. coli* P3478 pol A$^+$ and pol A$^-$	7.0	37	1.2×10^{-2} (1.3 mg/ml)	15
	Forward mutations to gal$^-$ and mtl$^-$ in *E. coli* WP2/RH53 and mut H1 derivative	7.0	37	5×10^{-2} (5.5 mg/ml)	60
	Back mutations to trp$^+$ in *E. coli* WP2 and repair-deficient derivatives	7.0	37	4.7×10^{-2} (0.52 mg/ml)	60

Testing method	Remarks	References
Liquid test without mammalian activation	Mutagenic effectiveness: EMS ≃ MNNG > UV-irradiation ≃ 4-nitroquinoline 1-oxide > hydroxylamine > Mitomycin C ≃ x-rays	27
Plate test without mammalian activation	Dose-dependent increase of lac^+ mutants	49
Fluctuation test without mammalian activation	Dose dependent-increase in trp^+ mutation frequency	1
Liquid test without mammalian activation	Relative mutagenic effective LD_{50}: diethyl sulfate > isopropyl methanesulfonate > EMS > MMS	25
Liquid test without mammalian activation	Mutagenic activity significant Evidence that single strand breaks in DNA occurred	50,51
Liquid test without mammalian activation	Dose-dependent mutagenic effect Effectiveness: MMS ≥ EMS Enhancement of mutagenicity through mutator gene	28
Liquid test without mammalian activation	Dose-dependent increase Enhanced mutability pol A^- derivatives	52

(continued)

Table 2. (continued)

Mutagen	Genetic effect observed	Treatment conditions: pH	T(°C)	Lowest reported effective concn (mole/l.)	Treatment time (min)
	Forward mutations to MTR resistance in E. coli K-12/343/113	7.0	37	1×10^{-3} (110 μg/ml)	120
	Forward mutations to valine-resistance E. coli K-12/AB 1621	7.0	37	2.5×10^{-3} (275 μg/ml)	120
MNNG (CAS 70-25-7)	Back mutations to arg^+ and prophage induction in E. coli H/r-30	6.8	37	1.4×10^{-5} (2 μg/ml)	60
	Deletions and base-pair changes in argF and Col B genes of E. coli K-12/3-016	6.8	37	2.7×10^{-5} (4 μg/ml)	60
	Forward and back mutations to gal^+, arg^+, and MTR in E. coli K-12/343/113	7	37	1.1×10^{-6} (0.16 μg/ml)	120
	Various back mutations in auxotrophic strains of E. coli K-12	7.0	37	6.4×10^{-8} mole/plate (20 μg/plate)	120
	Reversions of amber mutants in E. coli K-12:S-26	7.0	37	6.8×10^{-4} (100 μg/ml)	2

Testing method	Remarks	References
Liquid test without mammalian activation	Time- and concentration-dependent increase of MTR mutation frequency Effectiveness: MNNG > MMS	53,54
Liquid test without mammalian activation	Concentration-dependent increase of val^r mutants	55
Liquid test without mammalian activation	Dose-dependent increase	29
Liquid test without mammalian activation	Dose-dependent increase	26,27
Liquid test with and without mouse liver S-9 homogenate	Dose-dependent increase of mutation frequency in systems tested MNNG mutagenicity is eliminated after addition of mouse liver S-9 homogenate	24,53
Spot test without mammalian activation	Mutagenic effectiveness: MNNG > β-propiolactone > EMS > TEM	30
Liquid test without mammalian activation	Mostly AT → GC and AT → TA changes induced	32

(continued)

Table 2. (continued)

Mutagen	Genetic effect observed	Treatment conditions: pH	T(°C)	Lowest reported effective concn (mole/l.)	Treatment time (min)
	Forward mutations to phage T5 resistance in *E. coli* RH 53	7.0	37	6.8×10^{-6} (1 μg/ml)	ca. 600
	Back mutations to arg^+ and pro^+ in *E. coli* TAU bar	4.0-7.5	37	6.8×10^{-4} (100 μg/ml)	5
	Forward (auxotrophic) mutations in *E. coli* K-12:DB-1032	5.5	37	6.8×10^{-4} (100 μg/ml)	30
	Back mutations from auxotrophy in *E. coli* TAU bar	5.5	37	6.8×10^{-4} (100 μg/ml)	5
	Auxotrophic and valine-resistant mutants in *E. coli* K-12/AB 1621	6.0	37	2.0×10^{-4} (30 μg/ml)	5
	Various forward and back mutations in *E. coli* ara^-; *E. coli* S; *E. coli* A58	5.3-7.9	37	2.0×10^{-6} (0.3 μg/ml)	25
	Streptomycin-resistant mutations in *E. coli* K-12 pol A^+ and pol A^-	7.0	37	3.4×10^{-4} (50 μg/ml)	30

Testing method	Remarks	References
Continuous liquid culture without mammalian activation	Time-dependent increase of $T5^r$ mutants	36
Liquid test without mammalian activation	Evidence that diazomethane is responsible for MNNG mutagenesis at pH $\geq$ 5	56
Liquid test without mammalian activation	Evidence for nonrandom mutation induction through MNNG in stationary phase cell	57
Liquid test without mammalian activation	Majority of MNNG-induced mutations are located in the replication point region of the *E. coli* chromosome	58
Liquid test without mammalian activation	Time- and concentration-dependent increase of mutation frequency	
Liquid test without mammalian activation	Time- and concentration-dependent increase in mutation frequency Mutagenicity optimum at pH $\geq$ 7.0 Both transitions and transversions induced	46,60-
Liquid test without mammalian activation	Evidence that DNA single-strand breaks are produced	50

(continued)

Table 2. (continued)

Mutagen	Genetic effect observed	pH	T(°C)	Lowest reported effective concn (mole/l.)	Treatment time (min)
		Treatment conditions			
	MNNG-resistant mutants in *E. coli* S	5.5	37	1.4×10^{-5} (2 μg/ml)	ca. 2 days
	Back mutations, prophage induction, and duplications in various *E. coli* strains	7.0	37	6.8×10^{-6} (1 μg/ml)	60
Myleran (CAS 55-98-1)	Back mutations to str^- independence in *E. coli* B/Sd-4	7.0	37	4×10^{-7} mole/plate (100 μg/plate)	ca. 2 days
	Induction of vegetative growth of prophage in *E. coli* K-12 (λ)	7.0	37	2.03×10^{-3} (500 μg/ml)	120
Natulan (CAS 366-70-1)	No data available				
TEPA (CAS 545-55-1)	Back mutations to str^- independence in *E. coli* B/Sd-4	7.0	37	5.8×10^{-7} mole/plate (100 μg/plate)	ca. 2 days
	Induction of vegetative growth of prophage in *E. coli* K-39 (λ)	7.4	37	2.3×10^{-4} (39.8 μg/ml)	60

Testing method	Remarks	References
Fluctuation test without mammalian activation	Significant increase in mutation frequency	64
Liquid test without mammalian activation	Time- and concentration-dependent increase Mutations concentrated in small segment of the *E. coli* chromosome	35,65, 66
Spot test without mammalian activation	Significant increase over control	67,68
Liquid test without mammalian activation	Time-dependent increase in frequency of lytic centers	69
Spot test without mammalian activation	Significant increase over control	68
Liquid test without mammalian activation	Dose-dependent increase of lytic centers	70

(continued)

Table 2. (continued)

Mutagen	Genetic effect observed	Treatment conditions: pH	T(°C)	Lowest reported effective concn (mole/l.)	Treatment time (min)
	Back mutations to trp^+ in exrA derivative of E. coli B/r,WP2 trp^-	7.0	37	5.8×10^{-7} mole/plate (100 g/plate)	ca. 3 days
Thio-TEPA (CAS 52-24-4)	Back mutations to str^- independence in E. coli B/Sd-4	7.0	37	5.3×10^{-7} mole/plate (100 μg/plate)	ca. 2 days
	Induction of vegetative growth of prophage	7.0	37	2.6×10^{-3} (500 μg/ml)	120
	Back mutations to trp^+ in E. coli B/r,WP2 trp^-	7.0	37	5.3×10^{-7} mole/plate (100 μg/plate)	ca. 2 days
TEM (CAS 51-18-3)	Back mutations to str^- independence in E. coli B/Sd-4	7.0	37	4.9×10^{-7} mole/plate (100 μg/plate)	ca. 2 days
	Induction of vegetative growth of prophage in E. coli K-12 (λ)	7.0	37	2.5×10^{-3} (500 μg/ml)	120
	Forward and back mutations to gal^+, arg^+ and MTR in E. coli K-12/343	7.0	37	1×10^{-6} (0.2 μg/ml)	120

Testing method	Remarks	References
Spot test without mammalian activation	Significant increase over control	40
Spot test without mammalian activation	Significant increase over control	68
Liquid test without mammalian activation	Dose-dependent increase of frequency of lytic centers	69,71
Spot test without mammalian activation	No mutagenic effect observed	40
Spot test without mammalian activation	Mutagenic effectiveness: TEM ≡ β-propiolactone ≡ azaserine > nitrogen mustard	2,67, 68
Liquid test without mammalian activation	Dose-dependent increase of frequency of lytic centers	69
Liquid test without mammalian activation	Dose-dependent increase of mutation frequency in all systems tested	41

(continued)

Table 2. (continued)

		Treatment			
Mutagen	Genetic effect observed	pH	T(°C)	Lowest reported effective concn (mole/1.)	Treatment time (min)
	Various back mutations in auxotrophic strains of *E. coli* K-12	7.0	37	3.4×10^{-6} mole/plate (700 μg/plate)	ca. 2 days
Trenimon (CAS 68-78-8)	Mutations to gal^+ in $galR^s$ *E. coli*/	7.0	37	ca. 10^{-4} mole/plate	ca. 2 days
	Induction of vegetative growth of prophage in *E. coli* K-12 (λ)	7.0	37	2.2×10^{-3}	120
Vinyl chloride (CAS 75-01-4)	Forward and back mutations to gal^+, arg^+ and MTR in *E. coli* 343/113	7.0	37	1.1 10^- (663 g/ml)	120

Epichlorohydrin

Epichlorohydrin, an alkylating epoxide, is directly mutagenic to *E. coli* B/r, indicating that probably base-pair substitutions are induced. The lowest reported effective concentration is 5×10^{-2} mole/1 (23).

Ethylenimine

Ethylenimine, a monofunctional alkylating agent, is directly mutagenic to *E. coli*. In a repair-proficient strain of *E. coli* B the induced mutations are probably of the base pair substitution type. The lowest effective concentration is 1×10^{-4} mole/1 mutation plate for 2 days at 37°C (24).

EMS

EMS, a monofunctional alkylating agent, is directly mutagenic to *E. coli*. Experiments with several different *E. coli* all indicate that mutation induction kinetics have an exponential

Testing methods	Remarks	References
Spot test without mammalian activation	Significant increase over control Mutagenic effectiveness: MNNG > β-propiolacton > EMS > TEM	30
Spot test without mammalian activation	Significant increase of gal^+ mutations over control	72
Liquid test without mammalian activation	Dose-dependent increase of frequency of lytic centers	69
Liquid test with mammalian activation through purified mouse liver microsomes	Dose-dependent increase of mutation frequencies No effect without microsomes	73

part at low doses (without inactivation of colony forming ability) followed by a linear portion and, at higher doses, by a pronounced decline in mutation frequency. In repair-proficient strains the induced mutations are due to base repair substitutions whereby mostly GC→AT transitions are induced. After incubation with active mammalian microsomal fractions, an enhanced mutagenic activity of EMS is observed. The lowest effective concentration is 5.6×10^{-6} mole per mutation plate for 2 days at 37°C on stationary cells (30).

ICR-70

ICR-170, a nitrogen mustard derivation of acridine, is directly mutagenic to E. coli. Experiments with a repair-proficient strain of E. coli K-12 indicate that ICR-170 induced mutations are of the frameshift type. The lowest reported effective concentration is 2.3×10^{-7} mole per mutation plate for 2 days at 37°C in stationary cells (39).

Table 3. Comparative chemical mutagenesis of Escherichia coli test systems: reference mutagenic agent = x-rays

Genetic effect observed	Treatment conditions[a]			Method	Remarks	References
	pH	Dose	Dose rate			
Back mutations to Streptomycin independence in E. coli B/r Sd-4	7	3000 rad	30 rad/sec	Liquid	Dose-dependent increase	3
Back mutations to prototrophy in E. coli B/arg-3	7	8.5 rad	130 rad/min	Liquid	Dose-dependent increase; over range linear 17.0-432.0 r	4
Back mutations to prototrophy in several auxotrophs of E. coli B and B/r	7	20,000 rad	600 r/min	Liquid	Mutagenic effectiveness: UV = $MnCl_2$ > DEB = triazine > x-rays > nitrogen mustard > B-PL	4
Back mutations to prototrophy in E. coli B/met-2	7	17.0 rad	180 r/min	Liquid	Dose-dependent increase, linear over range 17.0-4320 r	4
Deletions in lac region of E. coli K12/X-7167 lac_2^-	7	30,000 r		Liquid	Mutagenic effectiveness: HNO_2 > UV > x-rays > 2-AP	75

Point mutations in lac region of *E. coli* K12/ X-7167 lac_{2}^{-}	7	30,000 r		Liquid	Mutagenic effectiveness: UV > 2-AP > HNO_2 > x-rays	75
Deletions and base pair changes in genes arg-F and Col B of *E. coli* H/r30	6.8	700 rad	1.56 rad/min	Liquid	Dose-dependent linear increase Effectiveness: EMS > NTG > UV > NQO > HA > x-rays ≃ MTC	26,27
Prophage induction and base pair changes in *E. coli* H/r 30 and repair-deficient derivatives	6.8	1000 rad	450 rad/min	Liquid	Dose-dependent increase	29

[a]All treatments at room temperature.

MeTEPA

In E. coli, MeTEPA produces mutations without prior metabolic activation in a concentration-dependent manner. Experiments with the exr A strain of E. coli/B indicate that probably base-pair substitutions are induced. The lowest reported effective dose is 4.7×10^{-7} mole per mutation plate for 3 days at 37°C in stationary cells (40).

Mitomycin C

The crosslinking agent Mitomycin C is not active on isolated DNA but extracts of bacterial cells such as E. coli have demonstrated to activate the compound. Thus, Mitomycin C is mutagenic to E. coli without mammalian metabolic activation. Experiments with several E. coli strains indicate that the results of Mitomycin C-induced DNA alterations are probably base-pair substitutions. Lytic growth of prophage lambda is also induced. The lowest reported effective concentration is 4.4×10^{-10} mole/mutation plate for 24 hr at 37°C (44).

MMS

The monofunctional alkylating agent MMS is directly active in E. coli. Experiments in repair-proficient strains indicate that base-pair substitutions mutations are induced and that the effect is enhanced in pol A^- and mut† H1 derivatives. The lowest reported effective concentration is 4.5×10^{-6} mole/l. for 3 days in slowly growing cells (1).

MNNG

The monofunctional alkylating agent MNNG is directly mutagenic in E. coli. Most experiments indicate that MNNG induces predominantly base-pair substitution mutations, but the induction of frameshifts has been reported. The direct mutagenic activity of MNNG is strongly reduced after the addition of active mammalian microsomal fractions. The lowest reported effective dose is 1×10^{-6} mole/l. for 120 min at 37°C in stationary cells (24,53).

Myleran

The alkylating agent Myleran is directly active on E. coli. Both induction of vegetative growth of prophage lamda in lysogenic cells as well as induction of base-pair substitution type mutations has been reported. The lowest effective concentration is 4×10^{-7} mole/mutation plate for 2 days at 37°C in stationary cells (67,68).

Natulan

A recent report (76) indicates that the compound is mutagenic without metabolic activation in E. coli 343/113 (see footnote of Table 1).

TEPA

The alkylating agent TEPA is directly active for E. coli. Both the induction of vegetative growth of prophage lamda in lysogenic cells as well as induction of base-pair substitution type mutations has been reported. The lowest effective concentration is 5.8×10^{-7} mole/mutation plate for 2 days at 37°C in stationary cells (68).

Thio-TEPA

The alkylating agent Thio-TEPA is directly mutagenic in E. coli. Both the induction of negative growth of prophage lambda in lysogenic cells as well as induction of base-pair substitution mutations has been reported. The lowest effective concentration is 5.3×10^{-7} mole/mutation plate for 2 days at 37°C in stationary cells (68).

TEM

The trifunctional alkylating agent TEM is directly mutagenic to E. coli. Both the induction of vegetative growth of prophage lambda in lysogenic cells as well as the induction of base-pair substitution mutations has been reported. Lowest effective concentration is 5×10^{-7} mole/mutation plate for 2 days at 37°C in stationary cells (2,67,68).

Trenimon

The trifunctional alkylating agent Trenimon is directly mutagenic in E. coli. Both the induction of vegetative growth of prophage lambda in lysogenic cells as well as the induction of probably base-pair substitution mutations has been reported. The lowest effective dose is 2.2×10^{-3} mole/l. for 120 min at 37°C in growing cells (69).

Vinyl Chloride

Vinyl chloride monomer is not markedly mutagenic to E. coli unless activation through mammalian microsomal enzymes is provided. The experiments with a repair-proficient strain of E. coli K-12 indicate that, after appropriate activation through microsomes, mostly base-pair substitution type mutations are induced. The lowest reported effective concentration is 1.1×10^{-2} mole/l. for 120 min at 37°C in stationary cells (73).

REFERENCES

1. Green, M. H. L., Muriel, W. J., and Bridges, B. A. Mutat. Res. 38: 33-42 (1976).

2. Iyer, V. N., and Szybalski, W. Proc. Natl. Acad. Sci. (U.S.) 44: 446-456 (1958).

3. Demerec, M. Genetics 36: 585-597 (1951).

4. Demerec, M., and Sams, J. Intern. J. Rad. Biol. Suppl. (Proc. Conf. Immediate and Low.Level Effects of Ionizing Radiation) 00: 283-291 (1959).

5. Iyer, V. N., and Szbalski, W. Appl. Microbiol. 6: 23-29 (1958).

6. Mohn, G. Arch. Toxikol. 28: 93-104 (1971).

7. Green, M. H. L., and Muriel, W. J. Mutat. Res. 38: 3-32 (1976).

8. Mohn, G. R., and Ellenberger, J. In: Handbook of Mutagenicity Test Procedures, B. Kilbey et al., Eds., Elsevier, Amsterdam, 1977. pp. 95-118.

9. Voogd, C. E., vanden Stel, J. J., and Jacobs, J. J. J. A. A. Mutat. Res. 26 483-490 (1974).

10. Heinemann, B. In: Chemical Mutagens: Principles and Methods for Their Detection, Vol. 1, A. Hollaender, Ed., Plenum Press, New York, 1971. pp. 235-266.

11. Moreau, P., Bailone, A., and Devoret, R. Proc. Natl. Acad. Sci. (U.S.) 74: 1378-1382 (1977).

12. Goze, A., Sarasin, A., Moulé, Y., and Devoret, R. Mutat. Res. 28: 1-7 (1975).

13. Callen, D. F., Mohn, G. R., and Ong, T. M. Mutat. Res. 45: 7-11 (1977).

14. Venitt, S., and Levy, L. S. Nature 250: 493-495 (1974).

15. Ellenberger, J., and Mohn, G. Arch. Toxikol. 33: 225-240 (1975).

16. Ellenberger, J., and Mohn, G. Mutat. Res. 38: 120-121 (1976).

17. Nakajima, T., Tanaka, A., and Tojyo, K. I. Mutat. Res. 26: 361-366 (1974).

18. Mohn, G., Ellenberger, J., McGregor, D., and Merker, H. J. Mutat. Res. 29: 221-233 (1975).

19. Mohn, G., Ellenberger, J., and McGregor, D. Mutat. Res. 2: 187-196 (1974).

20. Nakajima, T., and Iwahara, S. Mutat. Res. 18: 121-127 (1973).

21. Green, M. H. L., and Muriel, W. Y. Chem. Biol. Interact. 11: 63-65 (1975).

22. Read, J. Mutat. Res. 33: 107-112 (1975).

23. Strauss, B., and Okubo, S. J. Bacteriol. 79: 464-473 (1975).

24. Mohn, G., Ellenberger, J., and Osterhoff, J. Mutat. Res. 31: 311-312 (1975).

25. Turtoezlay, I., and Ehrenberg, L. Mutat. Res. 8: 229-238 (1969).

26. Ishii, Y., and Kondo, S. Mutat. Res. 16: 13-25 (1972).

27. Ishii, Y., and Kondo, S. Mutat. Res. 27: 27-44 (1975).

28. Hill, R. F. Mutat. Res. 14: 23-31 (1972).

29. Kondo, S., Ichikawa, H., Iwo, K., and Kato, T. Genetics 66: 187-217 (1970).

30. Mohn, G., and Kaplan, R. Mol. Gen. Genet. 99: 191-202 (1967).

31. Brock, R. D. Mutat. Res. 11: 181-186 (1971).

32. Osborn, M., Person, S., Philips, S., and Funk, F. J. Mol. Biol. 26: 437-447 (1967).

33. Schwarz, N. M. Genetics 48: 1357-1375 (1963).

34. Howell-Saxton, E., Zamenhof, S., and Zamenhof, P. J. Mutat. Res. 20: 327-337 (1973).

35. Hill, C. W., and Combriato, G. Mol. Gen. Genet. 127: 197-214 (1973).

36. Nestmann, E. R. Mutat. Res. 28: 323-330 (1975).

37. Guerola, N., and Cerda-Olmedo, E. Mutat. Res. 29: 145-147 (1975).

38. Sesnowitz-Horn, S., and Adelberg, E. A. J. Mol. Biol. 46: 1-15 (1969).

39. Katz, L., and Englesberg, E. J. Bacteriol. 107: 34-52 (1969).

40. Hanna, P. J., and Dyer, K. F. Mutat. Res. 28: 405-420 (1975).

41. Lein, J., Heinemann, B., Gourevitch, A. Nature 196: 783-784 (1962).

42. Endo, H., Ishizawa, M., Kamiya, T., and Sonoda, S. Nature 198: 258-260 (1963).

43. Heinemann, B., and Howard, A. J. Appl. Microbiol. 12: 234-239 (1964).

44. Haeda, Y., and Iijima, T. J. Gen. Appl. Microbiol. 11: 129-135 (1965).

45. Kondo, S. Genetics 73(Suppl.): 109-122 (1973).

46. Zampieri, A., and Greenberg, J. Genetics 57: 41-51 (1967).

47. Iyer, V. N., and Szybalski, W. Science 145: 55-58 (1964).

48. Murayama, I., and Otsiji, N. Mutat. Res. 18: 117-119 (1973).

49. Mazue, G., Vindel, J. A., Landsmann, F., Brunaud, M. Arzneim. Forsch. 24: 1422-1425 (1974).

50. Smirnov, G. B., Favorskaya, Y., and Skavronskaya, A. G. Mol. Gen. Genet. 111: 357-367 (1971).

51. Wild, D. Mutat. Res. 19: 33-41 (1973).

52. Bridges, B., Mottershead, R. P., Green, M. H. L., and Gray, W. J. H. Mutat. Res. 19: 295-303 (1973).

53. Mohn, G. Mutat. Res. 20: 7-15 (1973).

54. Fahrig, R., Grafe, A., and Buselmaier, W. Arch. Toxikol. 34: 9-15 (1975).

55. Suessmuth, R., and Lingens, F. Mutat. Res. 36: 273-282 (1976).

56. Cerda-Olmedo, E., and Hanawalt, P. C. Mol. Gen. Genet. 101: 191-202 (1968).

57. Botstein, D., and Jones, E. W. J. Bacteriol. 98: 847-848 (1969).

58. Cerda-Olmedo, E., Hanawalt, P. C., and Guerola, N. J. Mol. Biol. 33: 705-519 (1968).

59. Adelberg, E. A., Mandel, M., and Chen, G. C. C. Biochem. Biophys. Res. Commun. 18: 788-795 (1965).

60. Silengo, L., Schlessinger, D., Mangiarotti, G., Apirion, D. Mutat. Res. 4: 701-703 (1967).

61. Hince, T. A., and Neale, S. Mutat. Res. 24: 383-387 (1974).

62. Neale, S. Mutat. Res. 14: 155-164 (1972).

63. McCalla, D. R., Voutsinos, D. L., and Olive, P. L. Mutat. Res. 31: 31-37 (1975).

64. Mandell, J. D., and Greenberg, J. Biochem. Biophys. Res. Commun. 3: 575-577 (1960).

65. Zampieri, A., and Greenberg, J. G. Microbiol. 14: 199-203 (1966).

66. Jimenez-Sanchez, A., and Cerda-Olmedo, E. Mutat. Res. 28: 337-345 (1975).

67. Iyer, V. N., and Szybalski, W. Appl. Microbiol. 6: 23-29 (1958).

68. Szybalski, W. Ann. N. Y. Acad. Sci. 76: 475-489 (1958).

69. Specht, I. Arch. Mikrobiol. 51: 9-17 (1965).

70. Zetterberg, G. EMS Newsletter 3: 14-15 (1970).

71. Price, K. E., Buck, R. E., Lein, J. Appl. Microbiology 12: 428-435 (1964).

72. Propping, P., Roehrbein, G., Buselmaier, W. Mol. Gen. Genet. 117: 197-209 (1972).

73. Greim, H., Bonse, G., Radwan, Z., Reichert, D., and Henschler, D. Biochem. Pharmacol. 24: 2013-2017 (1975).

74. Glover, S. W. Carnegie Inst. Wash. Publ. 612: 121-136 (1956).

75. Schwartz, D., and Beckwith, O. Genetics 61: 371-376 (1969).

76. Mohn, G. R., and Ellenberger, J. Arch. Toxicol., in press.

CHAPTER 6

MUTAGENICITY OF SELECTED CHEMICALS IN *NEUROSPORA CRASSA*

H. E. Brockman and C. Y. Hung

Department of Biological Sciences, Illinois State University, Normal, Illinois 61761

F. J. de Serres

National Institute of Environmental Health Sciences, P.O. Box 12233, Research Triangle Park, North Carolina 27709

and

T. M. Ong

National Institute for Occupational Safety and Health, 944 Chestnut Ridge Road, Morgantown, West Virginia 26505

INTRODUCTION

A fairly recent review provides an excellent discussion of the research techniques used for *Neurospora crassa* (1). A section on mutagenesis and mutant selection is included, and many references, including earlier reviews of *N. crassa*, are cited. Of the numerous publications in which mutagens have been used in *N. crassa*, relatively few can be referred to as "mutagenicity" papers. Rather, mutagens were often used simply to induce mutants, which were then studied for other purposes. Among the publications in which study of mutagens was the major objective, a number of mutation-assay systems have been utilized in *N. crassa*, although the majority of the mutagenicity literature in *N. crassa* is from use of the *ad-3* system developed by de Serres. The methodology of this mutation-assay system has been described in detail (2), and many references to the mutagenicity data obtained with this system are cited in this review. The main features of the ad-3 system are discussed briefly below.

N. crassa is a haploid eukaryotic microbe, and therefore the ad-3 system generally has the advantages of microbial mutation assays and also has the advantage that the structures being mutagenized are eukaryotic chromosomes. Forward mutation from ad-3$^+$ to ad-3 can be studied in homokaryotic strains, in two-component heterokaryons, and in these strains carrying repair-deficient genes. Because a two-component heterokaryon has definite advantages over a homokaryon, the remaining remarks on forward mutation will concern primarily heterokaryon-12, whose genetic composition is shown in Table 1.

Table 1. Genetic composition of the two components of heterokaryon-12

Component	Strain Number	Linkage group I	III	IV	V	VI
A	74-OR60-29A	A,hist-2,ad-3A,ad-3B,nic-2,+,	ad-2;	+ ;	inos;	+
B	74-OR31-16A	A, + , + , + , +,al-2,	+;	cot;	+ ;	pan-2

Heterokaryon-12 is used to study many of the same mutational lesions that occur in diploid eukaryotes. Three types of conidia are produced: homokaryotic for component A, homokaryotic for component B, and heterokaryotic. By determining the survival of mutagenized conidia on plates of minimal medium and on plates of minimal medium supplemented with the requirements of the two components, information is obtained as to whether the mechanism of inactivation is nuclear or cytoplasmic. Furthermore, the frequency of recessive-lethal damage occurring over the entire genome of component B can be determined by culturing colonies from the minimal medium plates, plating conidia from these heterokaryotic cultures into minimal medium supplemented with pantothenate, incubating these plates at 35°C, and screening for the presence of cot colonies.

The induction of ad-3 mutations in component B is measured

by scoring for the presence of purpose colonies in the forward-mutation assay. This is a direct method, and an important advantage of the ad-3 system is that leaky mutants are not eliminated as they are in most forward-mutation assays. Various stages of the asexual life cycle--resting conidia, slightly germinated conidia, and growing cultures--can be treated with mutagens. Some agents that are not active in resting conidia are active in growing cultures. Resting conidia can be treated with mutagens that require metabolic activation by using an in vitro activation system or a host-mediated assay. The spontaneous ad-3 forward-mutation frequency is low ($\sim 0.4 \times 10^{-6}$), and dose-response curves for survival and for ad-3 mutation are obtained rather easily.

A major advantage of the ad-3 system is that two major classes of ad-3 mutations that occur in component B of heterokaryon-12 are recovered: intracistronic alterations of all types in the ad-3A or ad-3B cistron, and multicistronic (multilocus) deletions that include one or both of the closely-linked ad-3 cistrons and one or more adjacent cistrons. The frequencies of the two classes can be determined by appropriate tests. Therefore, dose-response curves for intracistronic mutations and for multicistronic deletions at the ad-3 region can be constructed. The deletions can be characterized for size and distribution within the ad-3 region by additional tests. Multicistronic deletions may be a very important class of mutations, but they usually act as recessive lethals and are therefore not isolated in haploid microbial mutation assays. The intracistronic ad-3 mutants can be characterized for the frequency of leakiness, and the intracistronic ad-3B mutants can be characterized for the frequency of intracistronic complementation and for the types of complementation patterns, including the relative frequencies of polarized and nonpolarized complementation patterns. These characteristics of the intracistronic ad-3 mutants indicate the relative frequencies of mutations that result in a nonfunctional gene product and those that result in a partially functional gene product. The intracistronic ad-3 mutants can be further characterized by reversion tests, including spot, overlay, and suspension tests. The mutants can be placed in silica gel stock for future reference.

The study of a particular agent in the ad-3 system is designed according to the information desired. The design may be simply to determine whether the agent is mutagenic, to answer a particular question such as whether the agent induces multicistronic deletions, or to characterize the genetic effects of the agent as completely as possible. The results with the chemicals that have been selected for the Comparative Chemical Mutagenesis Workshop are reviewed below.

RESULTS AND DISCUSSION

Summary

Certain features of the data for the chemicals that have been selected for the Comparative Chemical Mutagenesis Workshop and that have been tested in the ad-3 forward-mutation system of N. crassa are summarized in Table 2. Three other chemicals--epichlorohydrin (ECH), tris(1-aziridinyl)phosphine oxide (TEPA) and tris(2-methyl-1-aziridinyl)phosphine oxide (MeTEPA)--have been shown to be mutagenic in other mutation assays in N. crassa. Vinyl chloride (VCl) has been reported to be nonmutagenic. To our knowledge, there are no publications on tests for mutagenicity in N. crassa of the following chemicals: cadmium, cyclophosphamide, mitomycin-C, myleran, natulan, tris(1-aziridinyl)phosphine sulfide, (thioTEPA), and trenimon.

Abbreviations used for locus designations used in the present review are: ad, adenine; aga, arginase; al, albino; arg, arginine; ars, aryl sulfatase; aza, 8-azaquanine; bd, band; cot, colonial temperature sensitive; ed^r, edeine resistant; frg, frequency; hist, histidine; inos, inositol; iv, isoleucine-valine; me, methionine; mtr, 4-methyltryptophan; nic, nicotinic acid; nuc, nuclease; os, osmotic; pan, pantothenic acid; pho, phosphatase; pyr, pyrimidine; uvs, UV-sensitive.

N-(2-Chloroethyl-N'-(6-chloro-2-methoxy-9-acridinyl)-N-ethyl-1,3-propanediamine Dihydrochloride (ICR-170) (CAS Registry #146-59-8)

The mutagenicity of ICR-170 in N. crassa was first reported in 1965 (3). Since then, ICR-170 has been used rather extensively in N. crassa; 33 references are indicated in the EMIC search, the highest number for this organism among the agents selected for the Comparative Chemical Mutagenesis Workshop. Because of the presumed mutational specificity of ICR-170 for base-repair additions and deletions, it has been used mainly in reverse-mutation analyses in order to identify the types of mutational lesions present in the mutants. It is a potent mutagen under appropriate treatment conditions. All of the studies have been done with ICR-170 synthesized by H. J. Creech and co-workers of the Institute for Cancer Research, Philadelphia. All treatments should be done under red or yellow light to eliminate photoinactivation of the agent and any photodynamic effects.

ad-3. The potent mutagenic activity of ICR-170 at rather high survival was noted in the first report of its mutagenicity in N. crassa (3). The ad-3 forward-mutation frequencies in heterokaryon-12 were 187 and 2287 per 10^6 survivors among conidia treated for 8 hr with 1 μg/ml (2.1 μM) and 5 μg/ml (10.5 μM), respectively. As the average spontaneous forward-mutation freqeuncy at the ad-3 region is 0.4 per 10^6 survivors (4), these frequencies are 468 and 5718-fold increases over the spontaneous frequency. Because no multilocus deletions were found, ICR-170 induces only intracistronic lesions at the ad-3 region (3,5). Furthermore, because of the low frequencies of leakiness, allelic complementation, and nonpolarized complementation patterns found among the ICR-170-induced mutants, one mode of action suggested for this agent was that it causes mainly base-pair additions or deletions (3).

Twenty-four mutants isolated from the treatment of 1 μg of ICR-170/ml were studied for reversion with ICR-170 and nitrous acid (NA) (6). None was reverted by NA, but 16/24 were reverted by ICR-170. It was concluded that at the ad-3 region, ICR-170 induced predominantly single-site, nontransitional alterations, probably of the base-pair addition or deletion type. However, no evidence of the base-pair addition or deletion type. However, no evidence that the ICR-170-induced revertants resulted from intracistronic suppressors was obtained. The original ad-3 mutation was not recovered from revertant X wild-type crosses, and the growth rates of revertants did not differ significantly from that of wild type.

Twelve of the 16 ICR-170-induced mutants that were revertible by ICR-170 (but not by NA) were selected as a presumptive base-pair addition-deletion tester set. A majority of these 12 mutants were reverted by x-ray, ultraviolet radiation (UV), EMS, β-propiolactone (PL), and MNNG (7). The mutant that was most revertible by ICR-170, 12-9-17 (6), was not reverted by these five agents (7).

The mutagenic potency of ICR-170 as measured by ad-3 forward mutation in heterokaryon-12 and by reversion of ad-3 mutant 12-9-17 is greatly increased by low ionic strength of the ICR-170 solution used to treat the conidia, distilled H_2O solutions of ICR-170 giving the highest mutagenicity (8), by anoxic condition of the conidia during treatment with ICR-170 (9,10), and by high pH of the treatment solution (11).

By treating conidia in a solution of ICR-170 in distilled H_2O bubbled with N_2, substantial mutagenic activities can be obtained at low concentrations of ICR-170 that cause no or little killing (12). In heterokaryon-12, the LEDT, 0.1 μg/ml (0.21 μM) for 4 hr, caused 21 ad-3 mutants per 10^6 survivors at 99% survival, a 52-fold increase over the average spontaneous frequency of 0.4 per 10^6 survivors (4). Increasing the dose to 0.3 μg/ml (0.63 μM) for 4 hr resulted in 154 ad-3 mutants per 10^6 survivors at 99% survival, a 385-fold increase over the spontaneous frequency. In reversion studies of the ad-3 mutant 12-9-17, the

Table 2. Summary of mutation data in the ad-3 forward-mutation system.

Mutagen[a]	Muta-genicity[b]	Dose response[c]	Lowest effective dose tested (LEDT) Time, hr	Concn.	Fold increase over spontaneous at LEDT
ICR-170	+	+	4	0.2 μM	52
MNNG	+	+	0.25	25 μM	13-44
EMS	+	+	0.5	80 mM	7
MMS	+	+	1	2 mM	4
AfB_1	+	+	168	30 mM	110
EI	+	+	2	10 mM	10-31
TEM	+	-			
DEN	+	-	2	100 mM	33
DMN	+	+	2	90 mM	5

[a]Abbreviations: ICR-170, N-(2-chloroethyl)-N'-(6-chloro-2-methoxy-9-acridinyl)-N-ethyl-1,3-propanediamine dihydrochloride; MNNG, N-methyl-N'-nitro-N-nitrosoguanidine; EMS, ethyl methanesulfonate; AfB_1, aflatoxin B_1; EI, ethylenimine; TEM, triethylene melamine; DEN, diethylnitrosamine; DMN, dimethylnitrosamine.

[b]+, the chemical has been reported as mutagenic in the ad-3 system.

[c]+, dose-response data have been reported; -, not reported.

Metabolic activation[d]	Host Mediated assay[e]	Leaky, %[f]	Comple-mentation, %[g]	Multilocus deletions, %[h]
N	-	5	24	0
N	-		82	< 2
N	+	7	53	
N	+		52	up to 16
R		22	62	6
N		13	41	3
N				
R	+(?)		70	1
R	+			

[d]R, metabolic activation required for mutagenicity; N, not

[e]+, chemical mutagenic in host-mediated assay; -, not mutagenic.

[f]Percent of ad-3 mutants that were leaky.

[g]Percent of ad-3B mutants that showed intracistronic complementation.

[h]Percent of ad-3 mutants that were multilocus deletions.

LEDT, 0.03 μg/ml (0.06 μM) for 3 hr, caused 8 revertants per 10^7 survivors at 100% survival, and a dose of 0.3 μg/ml (0.63 μM) for 3 hr caused 109 revertants per 10^7 survivors at 94% survival. This mutant is stable spontaneously.

Because each of the components of the ICR-170 molecule (acridine nucleus and nitrogen mustard side chain) is mutagenic by itself in at least some test systems, it was of interest to determine whether the potent mutagenic activity and the apparent mutational specificity for base-pair additions or deletions of ICR-170 are due only to the intact molecule. Forward-mutation (13) and reverse-mutation (6,14) studies show that acridine nuclei are not mutagenic (in the absence of visible light), that the nitrogen mustard side chain (ICR-177) is very weakly mutagenic compared to ICR-170, and that ICR-177 induced mainly base-pair substitutions. Therefore, the potent mutagenic activity and the mutational specificity of ICR-170 are due to the intact molecule and not to either component.

Because of its presumed mutational specificity for base-pair additions or deletions, ICR-170 has been used rather extensively in reversion studies of ad-3 mutants induced by various agents in order to identify the presumed molecular defects of the mutants. It was used for this purpose in a tester set of eight ad-3 mutants (15) and in ad-3 mutants induced by NA (16,17), EMS (18), ICR-177 (13), hydroxylamine (HA) (19), AfB_1 (20) and x-ray (21). More specifically, ICR-170 was one of the agents used in some of these same studies to show that there is a relationship between the type of genetic alteration (base-pair substitution or base-pair addition or deletion) as deduced from reversion studies and the type of complementation pattern (noncomplementing, polarized, or nonpolarized) among ad-3B mutants induced by NA (16) and EMS (18). The methods used for ICR-170-induced reversion in many of these studies have been described (22).

ICR-170 was one of four agents tested for mutagenic activity in heterokaryon-12 in the host (mouse)-mediated assay (23). An in vitro mutation frequency of 1036 ad-3 mutants per 10^6 survivors at an ICR-170 concentration of 4.8 mg/kg was lowered to zero in the host-mediated assay with an ICR-170 concentration of 8 mg/kg. The results indicate that ICR-170 is strongly inactivated by some mechanism in the host animal. The mutagenicities of ICR-170 in these two kinds of assays in N. crassa and in other mutagenicity assays have been compared (24).

hist-3. ICR-170 was used in a reversion test with 68 hist-3 mutants (25). The conclusion is the same as that reached with ad-3 mutants (16,18), i.e., that there is a relationship between the type of genetic alteration as deduced from reversion studies and the type of complementation pattern.

mtr26. ICR-170 was used to induce reversion of the mtr26 allele (26). ICR-170-induced reversion occurred at the mtr locus and at an unlinked suppressor locus. Some of the ICR-170-induced reversions were probably due to intragenic suppressors. The authors concluded from these and other data that mtr26 appears to be a frameshift mutant.

pyr-3. Dose (concentration and time)-response curves for ICR-170-induced pyr-3 mutants have been published (27,28). The mutagenicity was rather weak, perhaps because treatments were in a phosphate buffer. Results from studies on pyr-3 mutants isolated after ICR-170 treatment, as well as on pyr-3 revertants isolated after exposure to ICR-170, have been used to elucidate various aspects of the structure of, and the biochemistry controlled by, the pyr-3 locus

me-1. ICR-170 was one of five agents used in a reversion study of seven NA-induced me-1 mutants in order to identify the genetic alterations present (33).

Sterol Mutants. The author concluded that of four mutagens tested (ICR-170, EMS, NA, and UV), ICR-170 was the most effective mutagen for inducing sterol mutants, but the spontaneous frequency was not given (34).

aza-3. ICR-170 was a very effective mutagen in the induction of aza-3 mutants (35,36).

N-Methyl-N'-nitro-N-nitrosoguanidine (MNNG) (CAS Registry #70-25-7)

Since the first report of the mutagenicity of MNNG in N. crassa in 1967 (37), this agent has been used rather extensively in this organism. The 30 references listed in the EMIC search for MNNG in N. crassa is second in number only to that for ICR-170. MNNG is probably used so extensively because of its potent mutagenicity in some studies in N. crassa and other organisms and because of its presumed mutational specificity for base-pair substitutions. Probably because of its reported potent mutagenicity, authors frequently failed to present control and experimental data that would permit the reader to conclude whether MNNG had actually induced mutation.

ad-3. In the first report (abstract) of the mutagenicity of MNNG in N. crassa (37), its potent mutagenic activity was evident. In heterokaryon-12, the ad-3 forward-mutation frequency was 0.14% (1400 per 10^6 survivors), and survival was 68% among conidia treated for 4 hr with 0.025 mM MNNG at pH 7. Compared to the average spontaneous forward-mutation frequency at the ad-3 region of 0.4 per 10^6 survivors (4), this frequency after exposure to MNNG is a 3500-fold increase over the spontaneous frequency. In an abstract in the following year (38), about the same mutation frequency (1.45×10^{-3}) was reported for a treatment of 2 μM MNNG for 4 hr. Because a latter extensive paper (39) reported 1450 ad-3 mutants per 10^6 survivors after treatment with 25 μM MNNG for 4 hr, the concentration of MNNG in the 1968 abstract (38) may be an error. In the 1968 abstract (38) and in a much more comprehensive paper in the following year (40), the mutagenicities of two methylating agents (MNNG and MMS) and the characteristics of ad-3 mutants induced by these two agents were compared. From these comparisons, the authors proposed the following interesting working hypothesis (40): "Compounds that induce a high percentage of mutants in which the specific gene product has an altered function are strong carcinogens, whereas compounds that induce a high percentage of mutants in which the specific gene product is nonfunctional have only a limited carcinogenic activity."

The most extensive study on the mutagenicity of MNNG in *N. crassa* is on ad-3 mutants induced in conidia of heterokaryon-12 (39). The major results as given in the summary are: "(1) forward-mutation frequency increases as the square of the time of treatment, (2) MNNG is an extremely efficient mutagen, e.g., the frequency of mutation in the ad-3 region (2 loci) was 0.14% after 240 min treatment with 25 μM MNNG at pH 7.0 with 73.4% survival, (3) at least 98.1% of the MNNG-induced ad-3 mutants are point mutations, (4) tests for genotype and allelic complementation showed that (a) the frequency of genotypes was ad-3A = 19.7%, ad-3B = 80.3% and ad-3A ad-3B = 0.0%, and (b) 81.8% of the ad-3B mutants have allelic complementation with 79.9% nonpolarized and 1.9% polarized complementation patterns and 18.2% noncomplementing mutants, and (5) the ratio between mutations in the ad-3A and ad-3B loci and spectrum of complementation patterns among the ad-3B mutants was independent of dose. Comparison of the spectrum of the complementation patterns among ad-3B mutants induced by MNNG with the spectrum among ad-3B mutants induced by 2-aminopurine, nitrous acid, hydroxylamine, and the acridine mustard derivative ICR-170 suggests that the majority of the MNNG-induced mutants have guanine-cytosine at the mutant site." The LEDT (25 μM MNNG for 15 min) caused 17.5 ad-3 mutants per 10^6 survivors at 86% survival. This frequency is a 44-fold increase over the average spontaneous ad-3 forward-mutation frequency of 0.4 per 10^6 survivors (4), and a 13-fold increase over the spontaneous ad-3 frequency of 1.3 per 10^6 survivors found in the experiment reported in this paper (39).

Because of its potent mutagenicity in some studies and its presumed mutational specificity for base-pair substitutions, MNNG has been used in a number of reversion studies of ad-3 mutants induced by various agents in order to identify the presumed molecular defects of the mutants. These reversion studies were on ad-3 mutants induced by HA (19), AfB_1 (20), x-rays (21), and ICR-170 (7). Although the major mutational alteration induced by MNNG at the ad-3 region was deduced to be AT → GC transitions (39), a majority of twelve ICR-170-induced ad-3 mutants that were revertible by ICR-170 but not by NA (5) were reverted by MNNG (7). The induction of reversion by MNNG in these presumed base-pair addition or deletion mutants (3,6) was interpreted as indicating that MNNG was reverting these mutants by the deletion of a single base pair (7).

MNNG was tested for its ability to induce ad-3 mutants in heterokaryon-12 in the host (mouse)-mediated assay (23). An in vitro mutation frequency of 1040 ad-3 mutants per 10^6 survivors with 3.5 mg of MNNG/kg was lowered to zero in the host-mediated assay at a MNNG concentration of 50 mg/kg. The author concluded that MNNG was strongly inactivated by some mechanism in the host animal. The mutagenicities of MNNG in these two kinds of assay in *N. crassa* have been compared to the mutagenicities of MNNG in other assay systems (24).

The mutagenic activity for forward-mutation at the ad-3 region and the lethal activity of MNNG were determined in a repair-sufficient strain and five UV-sensitive strains (upr-1, uvs-2, uvs-3, uvs-5, and uvs-6) that are essentially isogenic (41). The authors concluded that the differences in sensitivities found among the strains with MNNG were similar, but not identical, to those found for UV, and completely unlike those found for gamma irradiation.

Extranuclear. Four extranuclear mutants were isolated after treatment of conidia or mycelium with MNNG, but there was no clear evidence that the mutants were not of spontaneous origin (42,43).

Actidione Resistance and inos. Although mutagenic treatments with MNNG are usually done at pH 7 (other references cited), it has been reported that one of the optimal conditions for mutagenesis of MNNG for actidione resistance and for reversion of inos strain 498a was pH 4-5.2 (44). The authors did not indicate if other pH's were tested.

mtr26. MNNG did not revert the mtr26 allele, even though it did revert a hist-2 mutant (26). From this and other evidence, the authors suggested that mtr26 is a frameshift mutant.

nuc-1, -2. MNNG was one of three agents (UV and x-ray also tested) used in a study of mutants at two loci (nuc-1, -2) that have reduced activity of nuclease N_3 (45). A nuc-1 mutant was no more sensitive than wild type to the killing activity of the three agents, but a nuc-2 mutant was more sensitive than wild type to MNNG and UV. The MNNG survival curves (dose as concentration) of both nuc mutants, wild type, and a revertant of a nuc-1 mutant were exponential.

os. Osmotic mutants were induced with MNNG (46).

uvs. The sensitivity of three uvs mutants to MNNG was tested in experiments that scored for growth rather than percent survival. The author concluded that the uvs-3 mutant was very sensitive to MNNG, whereas the uvs-4 mutant was only slightly sensitive to MNNG (47) and that the uvs-2 mutant was sensitive to MNNG (48). Because the methods of the growth experiments were not identical in the two publications, the degree of sensitivity to MNNG of uvs-2 is difficult to compare to that of the uvs-1 and uvs-3 mutants.

In a paper referred to above, the survivals of conidia after treatment with 20 μM MNNG were reported as: wild type, 80%; upr-1, 75%; uvs-2, 44%; uvs-3, 20%; uvs-6, 7%; and uvs-5, 1% (41). The authors also reported from the results of ad-3 forward-mutation experiments with MNNG that uvs-3 and uvs-5 were slightly less mutable than wild type, that uvs-6 had a mutability similar to wild type, and that upr-1 and uvs-2 were two and five times more mutable than wild type, respectively.

aga. Conidia from a specially-constructed strain were treated with MNNG at pH 5.8 and then subjected to the filtration-enrichment procedure (49). Two mutants lacking arginase (aga) were isolated. Because no control data were presented, it cannot be concluded that the mutants were induced by MNNG.

ars. MNNG was used to treat conidia from two mutants that appeared to be deficient in aryl sulfatase (ars) with the objective of obtaining revertants having an aryl sulfatase with properties distinguishable from those of the wild-type enzyme (50). Revertants were obtained, but no qualitatively altered enzyme was apparent in any of the revertants. One cannot conclude that MNNG induced the revertants, because data on the spontaneous reversion frequency were not included.

pho-1. Wild-type conidia treated with MNNG were subjected to a selection procedure, and one mutant lacking repressible alkaline phosphatase activity (pho-1) was obtained (51). The data presented do not permit one to conclude that the mutant was induced by MNNG.

Sorbose Resistance. Sorbose-resistant strains (209) were isolated after exposure of wild-type conidia to MNNG (52). Because of the absence of mutation frequencies for control and treated populations, no conclusion about the induction of these mutants by MNNG can be made.

pyr-3. Six UV-induced pyr-3 mutants with nonpolar complementation patterns were tested for reversion with four mutagens, including MNNG, in order to determine whether either of two enzymatic functions deficient in these mutants could be restored (31). The author concluded that two of the mutants were revertible by MNNG.

frq. Three mutants were isolated that had altered period lengths of their circadian rhythm of conidiation (frq) after exposure of conidia of a bd strain to MNNG (53). Because control data were not reported, it is not known whether the three mutants were induced by MNNG.

me-1. MNNG was one of five agents used in a reversion study of seven NA-induced me-1 mutants in order to determine the genetic alterations present in the mutants (33).

Citrulline Resistance. Citrulline-resistant mutations were induced by MNNG and selected for in three auxotrophic, citrulline-arginine-sensitive strains (54). The authors stated that the MNNG treatment "... gave a mutational increase estimated to be more than 100 times the spontaneous rate ...", but data for this conclusion were not reported.

ed^r. Edeine-resistant (ed^r) mutants were isolated after exposure of wild-type conidia to MNNG (55). A stock solution of MNNG of 3 mg/ml was filter-sterilized and stored at 4°C, but the authors did not indicate the length of storage. Because neither induced nor spontaneous mutation frequencies were given, it cannot be concluded that the ed^r mutants were induced by MNNG.

Heat Sensitives. Heat-sensitive mutants were isolated from wild-type conidia that were treated with MNNG and then subjected to tritium suicide (56). The data clearly indicate that tritium suicide enriched for heat-sensitive mutants, but because mutation results with unmutagenized conidia were not reported, it cannot be concluded that the mutants were induced by MNNG.

Ethyl Methanesulfonate (EMS) (CAS Registry #62-50-0)

The first report of the mutagenicity of EMS in N. crassa was in 1957 (57). This agent has been used rather extensively in N. crassa; the 28 references listed in the EMIC search for EMS is the third highest number among the agents selected for the Comparative Chemical Mutagenesis Workshop, exceeded only slightly by ICR-170 and MNNG. This extensive use is probably a result of its early use in mutagenesis studies with bacteriophage and its presumed mutational specificity for GC to AT transitions. Compared to some other chemical mutagens, rather high concentrations are required for mutagenicity; and although the majority of EMS-induced mutants are probably base-pair substitutions, the agent is not a highly specific mutagen. As is true with some other mutagens, the absence of adequate control and/or experimental data make conclusions about fold increases over the spontaneous mutation frequency difficult or impossible in some publications.

ad-3. In 1957 (57), EMS was one of 24 chemicals listed in a table of a review paper showing the mutagenic activity of these chemicals as determined by measuring the back-mutation frequency in a purple adenineless strain (#38701). The data in this table are interesting, in that the single back-mutation frequency listed for each chemical is the number of back mutations per 10^6 treated conidia under optimal conditions (defined as the treatment that induced the highest number of back mutations per 10^6 treated conidia). The dose of EMS used (0.1 M for 12.5 min) caused 17 revertants per 10^6 treated conidia at 14% survival. The author also reported results with EMS in an adenineless (#38701), inositolless (#27401) double mutant (57). The proportion $ad^+/inos^+$ varied from 1.5 to 445 for six mutagens tested in this strain, and in 1957 the author concluded that: "The back-mutation pattern of a given mutant gene, as defined by its response to various physical and chemical mutagens, depends upon how the gene was originally damaged."

ad-3 mutants were induced in wild-type strain 74-OR23-1A by exposing conidia to 1% EMS in 0.5 M, pH 7 $KH_2PO_4 \cdot Na_2HPO_4$ buffer for 0.5 to 2 hr at 25°C (58). The LEDT, 1% (0.08 M) for 0.5 hr, caused 2.9 ad-3 mutants per 10^6 survivors at 92% survival. No spontaneous ad-3 mutants were recovered in this experiment, but this frequency is a 7-fold increase over the average spontaneous ad-3 forward-mutation frequency of 0.4 per 10^6 survivors (4). Leaky ad-3 mutants comprised 7.4%, and among the ad-3B mutants 53.2% showed allelic complementation--with 41.5% being of the nonpolarized type and 11.7% of the polarized type. Many of the ad-3B mutants with nonpolarized complementation patterns had the same complementation pattern, indicating that there were hot spots with EMS.

A random sample of 76 of the EMS-induced ad-3 mutants referred to above (58) was studied in reversion tests with O-methylhydroxylamine (OMHA), NA, EMS, and ICR-170 (17). The major results as given in the summary are: "The mutants characterized as to reversion mechanism and type of unidentified genetic alteration consisted of (1) base-pair transitions; AT to GC, 41%; and GC to AT, 17%; (2) base-pair insertions or deletions, 9%; (3) nonrevertible, 7%; (4) mutants which only revert spontaneously and not after treatment with any of the four mutagens, 18%. It is likely that a part of the mutants in the latter class result from a base-pair substitution. Of the mutants which only reverted spontaneously, 13% out of the total of 18% had a nonpolarized complementation pattern and are therefore likely to result from a base-pair substitution. As a result of these considerations, a range from 58% to 79% of the EMS-induced mutants may result from single base-pair substitutions. The correlation between complementation pattern and genetic alteration at the molecular level found previously by us among NA-induced ad-3B mutants was also found within EMS-induced mutants."

Because an anoxia effect had been observed with the alkylating agent ICR-170 (9), two other alkylating agents, including EMS, were tested for the influence of bubbling with nitrogen or oxygen during exposure of conidia to the mutagen (59,60). The killing activity and the mutagenic activity for ad-3 forward mutation of EMS were not different in the presence of the two gases.

Because of its presumed mutational specificity for base-pair substitutions, EMS was used in reversion studies with a tester set of eight ad-3 mutants (15) and with ad-3 mutants induced by NA (16,17,61,62) with the objective of identifying the presumed mutational alterations in the mutants. The methods used for EMS-induced reversion of ad-3 mutants have been described (22).

Although it was concluded that EMS induced mainly base-pair substitutions (18), a majority of twelve ICR-170-induced ad-3 mutants that were revertible by ICR-170 but not by NA (6) were reverted by EMS (7). These twelve mutants are presumed base-pair additions or deletions (3,6), and the induction of reversion by EMS was interpreted as indicating that EMS reverted these mutants by the deletion of one base pair (7).

EMS was tested for mutagenic activity at the ad-3 region in heterokaryon-12 in the host (mouse)-mediated assay (23). A concentration of 3100 mg of EMS/kg in vitro caused 5 ad-3 mutations per 10^6 survivors, whereas a dose of 300 mg/kg in the host-mediated assay caused 10 ad-3 mutations per 10^6 survivors. Therefore, the author concluded that "... EMS is actually ten times more effective in the animal than *in vitro*." The mutagenicities of EMS in these two kinds of assays in *N. crassa* and in other mutagenicity assay systems have been compared (24).

ad-4, ad-8, hist-1, and pan-2. Mutants at the ad-4, ad-8, hist-1, and pan-2 loci were isolated by the filtration-concentration technique after conidia were exposed to EMS (63). No data on increase over the spontaneous forward-mutation frequencies were given. The mutants were placed into four classes after reversion studies: revertible with neither NA nor EMS although they were generally revertible spontaneously, revertible with EMS and NA, revertible with EMS only, and revertible with NA only. The adenylsuccinase activity of most of the revertants of the ad-4 mutants was less than that of wild type. The authors concluded that many mutational changes induced by EMS are more complex than single base-pair changes.

pyr-3. Dose (concentration)-response curves for survival and for the production of arginine-independent colonies from an arg strain (one class of such colonies are pyr-3 mutants that suppress the arg mutation) after exposure to 20-100 mM EMS for 11 hr at pH 9 have been published (64). The logarithm of the surviving fraction was approximately a linear function of the concentration of EMS over three log cycles. A plot of mutants per survivors vs. -logarithm of the surviving fraction was nonlinear and essentially identical to that for UV. The LEDT (20 mM for 11 hr) caused approximately 5 mutants per 10^5 survivors at about 45% survival. Based on the author's stated "... usual value of spontaneous mutants (of the order of 10^{-7}/live microconidium) ...", the LEDT caused approximately a 500-fold increase over the spontaneous frequency. However, the spontaneous frequency for the EMS experiment was not given.

ad-8. EMS was one of a number of chemical mutagens used to induce ad-8 mutants isolated by the filtration-concentration technique. Genetic fine structure and interallelic complementation maps of these mutants have been published (65-67). The author stated that: "Mutants obtained after the treatment with ... EMS involved very few stable mutants and no multisite mutation and are distributed randomly on the ad-8 map" (65), but the reviewers did not find data in these papers indicating what fraction of the ad-8 mutants isolated after EMS treatment were induced by the agent.

hist-3. EMS was one of four agents used in reversion experiments on 68 hist-3 mutants (25). The main conclusion was the same as that reached with ad-3 mutants induced by NA (16) and by EMS (18), i.e., that there is a relationship between the type of genetic alteration as deduced from reversion studies and from the type of complementation pattern.

mtr26. Analyses of EMS-induced revertants of the mtr26 allele showed that reversion occurred at the mtr locus and at an unlinked suppressor locus (26). Because of the reversion pattern of mtr26 with the mutagens tested, the authors suggested that the allele is a frameshift mutant and that EMS is "... able to delete (or add) a base or bases with low frequency."

ad-3A and inos. EMS was one of three agents used in reversion experiments of diauxotrophs of four ad-3A alleles in combination with three inos alleles (68). The author concluded that: "The pattern of allele specificity is too complex for an interpretation solely on the basis of the initial DNA damage. It appears that specificities arising at some later stage in mutagenesis contribute to the observed response pattern." In another study in which six of the above diauxotrophs were treated with EMS (68), the author concluded similarly that: "... the dose-response curves cannot be entirely controlled by the reactions between mutagen and DNA but must include components that represent the effect of the genetic background on secondary steps in mutagenesis."

me-1. EMS was one of five chemical mutagens used in a reversion study of seven NA-induced me-1 mutants with the objective of identifying the genetic alterations present (33).

Sterol Mutants. EMS was one of four mutagens tested for their ability to induce sterol mutants, but the presence of sodium thiosulfate (used to quench the EMS reaction) affected the potency of the antibiotic in the selection medium (34). A few sterol mutants were recovered when sodium thiosulfate was eliminated, but spontaneous and induced frequencies were not included.

aza-3. EMS in pH 7 phosphate buffer at 25°C was one of four mutagens used to induce forward mutation at the aza-3 locus (35, 36). The LEDT (0.1 M EMS for 5 hr) resulted in 289 aza-3 mutants per 10^5 survivors at 74% survival (36). This frequency was a 13-fold increase over the spontaneous frequency of 22 per 10^5 survivors reported for this experiment.

Methyl Methanesulfonate (MMS) (CAS Registry #66-27-3)

Although the first report of the mutagenicity of MMS in N. crassa was in 1968 (38), this agent has not been used very extensively in this organism. Nevertheless, it is a fairly potent mutagen in N. crassa, and ad-3 mutants induced by MMS have interesting characteristics (70).

ad-3. In the first report (abstract) of the mutagenicity of MMS in N. crassa (38), the characteristics of ad-3 mutants induced by two methylating agents (MNNG and MMS) were compared. A much more comprehensive comparison of these two alkylating carcinogens was published the following year (40). The authors concluded that MMS is less mutagenic but more toxic than MNNG. Furthermore, based on the characteristics of the ad-3 mutants induced by these two agents, the authors proposed that compounds which induce a high percentage of mutants whose gene product has an altered function are strong carcinogens, whereas compounds that induce a high percentage of mutants with a nonfunctional gene product have only limited carcinogenic activity (40).

The most extensive study of the mutagenicity of MMS in N. crassa is for ad-3 mutants induced in heterokaryon-12 (70). A 0.1 ml portion of a solution of MMS in 100% methanol was added to 9.9 ml of a conidial suspension in 0.067 M, pH 7 phosphate buffer. The treatments were at 25°C on a rotary shaker for 1 or 5 hr at concentrations of MMS up to 20 mM. Dose (concentration)-survival curves for the heterokaryotic fraction of the conidia and for the total conidial population had a shoulder; the first curve declined at a faster rate than the second, indicating that nuclear killing played an important role in conidial inactivation. Dose (concentration)-response curves for ad-3 mutation induction showed that the overall ad-3 forward-mutation frequency increased in proportion to the 1.9 power of the concentration of MMS, that the frequency of intracistronic ad-3 mutations increased in proportion to the 1.7 power of the concentration, and that the frequency of multilocus deletions at the ad-3 region increased in proportion to the 2.8 power of the concentration. At the highest dose tested (20 mM MMS for 5 hr), 15.5% of the ad-3 mutants were multilocus deletions. Among the ad-3B mutants, 52.1% showed allelic complementation, of which 39.2% and 12.9% had nonpolarized and polarized complementation patterns, respectively. The authors concluded from these complementation data that the majority of the MMS-induced point mutations resulted from mutational events other than missense mutations, i.e., nonsense mutations and/or mutations other than base-pair substitutions. The ratio between point mutations at the ad-3A and ad-3B loci, as well as the spectrum of complementation patterns among the ad-3B mutants, was independent of MMS concentration.

The LEDT (2 mM for 1 hr) caused 1.5 ad-3 mutants per 10^6 survivors at 100% survival (68). There were no spontaneous ad-3 mutants in this experiment, but this frequency is a 4-fold increase over the average spontaneous ad-3 forward-mutation frequency of 0.4 per 10^6 survivors (4). At 5 hr of treatment, the LEDT (3 mM) resulted in 13.3 ad-3 mutants per 10^6 survivors at 85% survival. The concentration of 20 mM for 1 hr caused 40.1 ad-3 mutants per 10^6 survivors at 77% survival, whereas the same concentration for 5 hr caused 367 and 731 ad-3 mutants per 10^6 survivors at 27 and 15% survival, respectively. These results show that the time of treatment with MMS, as well as the concentration, has a large effect on mutation frequencies.

MMS was used in one part of a study on ad-3 mutation induced in conidia during their stay in the peritoneal cavity of mice and rats (71). In order to determine whether dialysis bags were permeable to MMS, the ad-3 mutation frequencies among conidia in dialysis bags surgically placed into the peritoneal cavity, and among conidia injected into the peritoneal cavity, were determined. MMS (100 mg/kg of animal) was injected into the tail vein 8 hr after the conidia were placed into the animals, and the conidia were recovered from the animals 10 hr after injection of MMS. The ad-3 mutation frequency was the same in both groups of conidia; therefore, the dialysis bags were permeable to MMS.

The ability of MMS to induce ad-3 mutants in conidia of heterokaryon-12 in vitro and in the host (mouse)-mediated assay was compared (23). The ad-3 mutation frequency in vitro was 23 ad-3 mutants per 10^6 survivors at a MMS concentration of 330 mg/kg. In the host-mediated assay, 250 mg of MMS per kg caused 23 ad-3 mutants per 10^6 survivors. The author concluded that MMS is equally effective in the host and in vitro. The mutagenicities of MMS in these two kinds of assays in N. crassa and in other mutagenicity assay systems have been compared (24).

Aflatoxin B_1(AfB_1) (CAS Registry #1162-65-8)

All of the results on AfB_1 in N. crassa have been obtained with the ad-3 system. AfB_1 is nonmutagenic in resting conidia without liver microsomes but is mutagenic in resting conidia with liver microsomes or in growing vegetative cultures without liver microsomes. AfB_1 appears to induce a wide spectrum of genetic alterations at the ad-3 region, including multilocus deletions and intracistronic mutations that include frameshifts, base-pair substitutions, and possibly other intragenic changes (72).

ad-3. In the first report on the mutagenicity of AfB_1 in N. crassa, it was reported that the agent was not mutagenic when nongrowing conidia were treated but was mutagenic in growing vegetative cultures (73). Slants of medium containing AfB_1 were inoculated with conidia from heterokaryon-12, and conidia from these 7-day cultures were assayed for ad-3 mutants. In duplicate experiments, 40 μg of AfB_1/ml of culture medium (0.13 mM) caused 69 and 87 ad-3 mutants per 10^6 survivors--an about 200-fold increase over the spontaneous frequency. Selection for spontaneous ad-3 mutants as an explanation of the increased frequency of ad-3 mutants was ruled out by reinoculation experiments.

The mutagenicities of AfB_1 and AfB_2 have been compared, and the characteristics of ad-3 mutants induced by these two agents have been determined (20,74). Again, the aflatoxins were mutagenic in growing vegetative cultures but not in resting conidia. The LEDT of AfB_1, 10 μg/ml (0.03 mM) during the 7 days of growth, resulted in 72% survival and 44 ad-3 mutants per 10^6 survivors. Concentrations of 20 and 40 μg/ml were also tested. The mutation frequency induced by the LEDT for AfB_1 was a 110-fold increase over the average spontaneous frequency observed in these experiments. A random sample of 194 of the AfB_1-induced ad-3 mutants was characterized, and 19 randomly selected ad-3 mutants were studied in reversion tests with NA, MNNG, and ICR-170. The results were interpreted as indicating that in addition to a low frequency of multilocus deletions AfB_1 induced frameshifts, base-pair substitutions, and possible other types of intracistronic alterations.

AfB_1 was one of 16 chemicals, including AfB_2, G_1, and G_2, tested for mutagenic activity at the ad-3 region in heterokaryon-12 (72,75). The authors concluded for these 16 chemicals that there is a positive correlation between carcinogenic activity in animals and mutagenic activity as measured by forward mutation at the ad-3 region of heterokaryon-12.

Because AfB_1 was not mutagenic in resting conidia, conidia from a two-component heterokaryon were treated with 0.67 mM AfB_1 in the presence of mouse or hamster liver microsomes and a NADPH-generating system (76). With hamster liver microsomes, the ad-3 mutation frequency increased to between 29 and 87 per 10^6 survivors, a 76 to 230-fold increase over the spontaneous frequency. No increase over the spontaneous frequency was observed without microsomes, and only a 7- to 9-fold increase over the spontaneous frequency was observed when mouse microsomes were used.

Autoclaving for 15 or 30 min did not decrease the mutagenic activity of AfB_1 as assayed by ad-3 forward mutation (77). The mutagenicity of aflatoxins, including the results in *N. crassa*, has been reviewed (78).

Ethylenimine (EI) (CAS Registry #151-56-4)

Only six papers in which EI was used in *N. crassa* have been published since the initial reports of its mutagenicity in *N. crassa* in 1952 and 1953 (79,80). It is a fairly potent mutagen and appears to induce a variety of genetic alterations in *N. crassa* (81).

ad-3. EI (0.002 M at pH 5.4) decreased survival but did not induce reversion in a purple adenine mutant (79). However, in the same publication, a footnote in a table says for EI that: "Positive effect obtained in recent experiments (note added in proof)." These positive results were published later (80). Dose (time)-response results for 0.05 M EI showed that the longest exposure (50 min) reduced survival to 1% and increased the reverse-mutation frequency from 0.13 to 134 per 10^6 survivors. The LEDT (8 min of 0.05 M) resulted in 4.6 revertants per 10^6 survivors, a 35-fold increase over the spontaneous frequency. EI was listed in a table of a review paper (57) showing the mutagenic activity of 24 chemicals as determined by measuring the back-mutation frequency in the same adenineless strain (#38701) used in the previous references cited. In this table, the single back-mutation frequency given for each chemical is the number of back mutations per 10^6 treated conidia under optimal conditions (defined as the treatment that induced the highest number of back mutations per 10^6 treated conidia). A 30-min treatment with 0.05 M EI resulted in 16 revertants per 10^6 treated conidia at 75% survival.

EI was one of 16 chemicals reported in an abstract as being tested for their ability to induce ad-3 mutants in a two-component heterokaryon (75). In a more complete report on 12 of these agents, the results from two concentrations of EI at 25°C for 2 hr in a pH 7, 0.067 M $Na_2HPO_2 \cdot KH_2PO_4$ buffer were reported (72). The LEDT (10 mM for 2 Hr) caused no decrease in survival and 12.5 ad-3 mutants per 10^6 survivors. The authors indicated that the spontaneous frequency did not exceed 1.3 per 10^6 survivors; this frequency and the average spontaneous ad-3 forward-mutation frequency of 0.4 per 10^6 survivors (4) indicate a 10-fold and a 31-fold increase, respectively, over the spontaneous frequency. Doubling the dose to 20 mM decreased the survival to 68% and increased the ad-3 mutation frequency to 106 ad-3 mutants per 10^6 survivors.

More complete data for EI under the same treatment conditions have been published (81). The LEDT was the same as in the previous publication. The does (concentration)-response curve for survival was one of continually increasing slope over the concentration range of 10 to 40 mM, resulting in 8% survival at 40 mM EI. The frequency of ad-3 mutants increased as the 2.8 power of the EI concentration over the same dose range, with 40 mM EI causing 542 ad-3 mutants per 10^6 survivors. The characteristics of 181 ad-3 mutants selected randomly from ad-3 mutants isolated from a dose of 15 mM EI (causing 85% survival and 29.4 ad-3 mutants per 10^6 survivors) have been determined (81,82). Five (2.7%) of the mutants were multilocus deletions. The frequencies of leakiness, allelic complementation, and nonpolarized complementation patterns among EI-induced ad-3 mutants were similar to those among ad-3 mutants induced by other alkylating agents (ICR-177 and EMS). The authors concluded "... that in addition to multilocus deletions (which occur at a low frequency), EI-induced mutations probably include base-pair substitutions, frameshift mutations, and other types of intragenic alterations." It has been concluded (abstract) from a study of the characteristics of ad-3 mutants induced by nine chemical carcinogens, including EI, that "... potent chemical carcinogens cause predominantly base-pair substitution mutations in eukaryotic organisms (83).

iv-1 and iv-3. EI was one of nine agents used to mutagenize conidia from which isoleucine-valine mutants to be used for complementation studies were obtained by the inositol-less death technique (84). Twelve iv-1 and two iv-3 mutants were isolated after conidia were treated with 0.05 M EI for 40 min. Because control data were not included, no conlusion about the induction of these mutants can be made.

Triethylenemelamine (TEM) (CAS Registry #51-18-3)

The first report of the mutagenicity of TEM in *N. crassa* was in 1957 (57). The only other reference on the mutagenicity of TEM in *N. crassa* reported that it induced point mutations and multilocus deletions at the ad-3 region (85).

ad-3. TEM was one of 24 chemicals listed in the table of a review paper in 1957 (57) that showed the mutagenic activity of these chemicals in a purple adenineless mutant (#38701). The single back-mutation frequency given for each chemical is the number of back mutations per 10^6 treated conidia under optimal conditions (defined as the treatment that induced the highest number of back mutations per 10^6 treated conidia). A dose of TEM of 0.02 M for 50 min caused 0.6 back mutations per 10^6 treated conidia at 55% survival.

Conidia from heterokaryon-12 were treated for 20 to 170 min with 0.01 M TEM in a pH 7 phosphate buffer (85). The inactivation of the conidia by TEM was due in part to nuclear inactivation. On a log-log plot, the frequency of point mutations at the ad-3 region increased with a slope of 1.35, whereas the frequency of multilocus deletions increased with a slope greater than two.

Diethylnitrosamine (DEN) (CAS Registry #55-18-5)

DEN has not been studied very extensively in *N. crassa*, and all of the data have been obtained with the ad-3 system. As in other test organisms, DEN requires activation for mutagenic activity. The most complete study is on the induction and characterization of a sample of ad-3 mutants (86).

ad-3. DEN was first reported to be nonmutagenic in *N. crassa* (85). Later, it was pointed out that this lack of mutagenicity might be due to the inability of *N. crassa* to carry out an enzymatic oxidative dealkylation believed to be necessary to convert DEN to an active form (87). This idea was tested by using an enzyme-free hydroxylation mixture composed of 20 mM ascorbic acid, 2 mM EDTA (Na salt), and 1 mM $FeSO_4$ in a 0.06 M, pH 7.4

phosphate buffer (86). Conidia from an ad-3B mutant (2-17-68) that is a presumptive base-pair substitution were treated with 0.1 mM DEN/ml for 1 to 4 hr in the hydroxylation mixture bubbled with O_2 or N_2. Survival decreased approximately linearly to about 30% at 4 hr, and the survival curves were almost the same in the presence of the two gases. The reversion frequency increased almost linearly up to 3 hr of treatment in the presence of O_2, reaching approximately a 14-fold increase over the spontaneous frequency at 3 hr, but DEN was not mutagenic in the presence of N_2.

Forward mutation at the ad-3 region and survival have been studied in heterokaryon-12 with 100 mM DEN (86) in the same hydroxylation mixture referred to in the previous paper (88). By studying the survival of heterokaryotic and homokaryotic conidia, the authors concluded that the inactivations observed were cytoplasmic. Furthermore, unlike in the previous paper, DEN had much greater killing activity in the presence of N_2 than in the presence of O_2. In a forward-mutation experiment, treatment with 100 mM DEN for 2 hr (the LEDT) in the presence of O_2 resulted in 15.3 ad-3 mutants per 10^6 survivors, a 33-fold increase over the average spontaneous frequency (4). The same dose in the presence of N_2 was not mutagenic. Only one out of 93 mutants analyzed was a multilocus deletion, and the frequencies of complementation and of nonpolarized complementation patterns were fairly high. The authors concluded that their data on DEN-induced ad-3 mutants support their previously-stated working hypothesis (40) that "potent carcinogenic activity is associated with the production of gene products with altered function rather than with gene products with no function."

DEN has been studied in the host (mouse)-mediated assay using conidia from heterokaryon-12 as the indicator organism (23,89,90). There was no increase in the frequency of ad-3 mutants over the spontaneous frequency when DEN (200 mg/kg) was administered IM and conidia were injected IP, but there was a "slight increase" over the control frequency when DEN (150 mg/kg) was administered IP and conidia were injected IV (tail vein) and recovered from the liver. This "slight increase" was from a control value of 0.65 to an experimental value of 1.74 ad-3 mutants per 10^6 survivors. Because percent survivals were not given, we cannot rule out the possiblity of the increase being due to selective killing.

Dimethylnitrosamine (DMN) (CAS Registry #62-75-9)

DMN has not been studied very extensively in N. crassa. All of the data have been obtained with the ad-3 system, and as in other test organisms DMN requires activation for mutagenic activity.

ad 3. The first report of the mutagenicity of DMN in *N. crassa* was the enzyme-free hydroxylation mixture referred to in the section on DEN (88). Conidia from a presumptive base-pair substitution ad-3B mutant (2-17-68) were treated with "0.1 mM DMN per ml" for 1 to 4 hr in the hydroxylation mixture bubbled with O_2 or N_2. The killing activity of DMN was much greater in the presence of O_2 than N_2, the survivals at 4 hr being approximately 3% and 60% in O_2 and N_2, respectively. The survival curve was approximately linear with DMN in the presence of N_2, but the survival curve with DMN in the presence of O_2 had a shoulder. DMN was not mutagenic in the presence of N_2, but in the presence of O_2 and DMN an increase in revertants per 10^6 survivors was observed. The author indicated that: "Since the number of reversions per initial number of conidia exceeded the number of spontaneous reversion by only 26%, an increase in reversion frequency due to selective killing cannot be completely discounted." The author did not indicate the exposure time at which the 26% increase was noted, and our calculations indicate that no increase in revertants per initial number of conidia was observed.

DMN has been studied in the host (mouse)-mediated assay using conidia from heterokaryon-12 as the indicator organism (23,89,90). There was a large increase (approximately 500 times the spontaneous frequency) in the ad-3 mutation frequency when DMN (75 mg/kg) was administered IM and conidia were injected IV (tail vein) and recovered from the liver (23). In two references (89,90), the authors indicated that DMN (100 mg/kg) administered IM did not increase the ad-3 mutation frequency above the control among conidia injected IP. However, in the third reference (23), the author discussed the same data as those in a previous reference (89) and stated for DMN that: "In experiments where the conidia were injected IP and the compound either intravenously or intramuscularly, only a slight increase in the mutation frequency was found."

ad-3 mutation in conidia of heterokaryon-12 was studied with DMN and crude microsomes from livers of mouse strains B6D2F1 and C_3H (90). One concentration of microsomes was used; and 0.09, 0.18, and 0.36 M concentrations of DMN were studied. Five to 15-fold increases over the spontaneous mutation frequencies were observed. DMN or microsomes alone gave no increase. Liver microsomes from phenobarbitol-induced mice were no more active than those from noninduced mice.

Tris(1-aziridinyl)phosphine Oxide (TEPA) (CAS Registry #545-55-1) and Tris(2-methyl-1-azizidinyl) phosphine Oxide (meTEPA) (CAS Registry #57-39-6)

In the only reference listed for TEPA and MeTEPA in *N. crassa* in the EMIC search (91), the authors concluded that the agents cause nuclear inactivation and recessive lethal mutation.

Recessive Lethals. Conidia from a heterokaryon between arg-6 and meth-7, amyc were treated with 0.5% TEPA or 0.8% (92). in a 0.067 M, pH 7 phosphate buffer at 37°C for up to 50 min One other polyimine was also tested. The authors concluded from the survival curves that the lethal activity of the agents was due to nuclear rather than cytoplasmic inactivation. In another experiment, conidia were treated with 0.5% TEPA or MeTEPA for 15 min in water at 25°C. The surviving fraction was in "the neighborhood of 0.1 on minimal medium", and there was 4.0% recessive lethals (24 lethals and 2 semilethals/603) in the case of TEPA and 3.7% recessive lethals (30 lethals and 3 semilethals/804) in the case of METEPA, compared to a control frequency of 0/887.

Epichlorohydrin (ECH) (CAS Registry #106-89-8)

Summary. Only one reference is known to the authors for ECH in *N. crassa* (93). The agent was a potent mutagen in a reversion test of the purple adenineless mutant strain 38701.

ad-3. ECH was one of four monoexpoxides tested for their ability to induce reversion in the strain W.40 "distinctus" A, which is an extreme colonial strain of the purple adenineless mutant 38701 (93). Conidia of this strain were exposed to 0.15 M ECH at 25°C for 15 to 60 min. The longest exposure reduced viability to 0.7% and resulted in a reversion frequency of 411 per 10^6 survivors (corrected for the spontaneous frequency). The LEDT (15 min of 0.15 M) resulted in 95% survival and 8.5 revertants per 10^6 survivors (corrected for the spontaneous frequency).

Vinyl Chloride (VCl) (CAS Registry #75-01-4)

In the only reference known to the authors for VCl in *N. crassa*, mutagenicity was not detected (94).

Acriflavin Resistance and nic-1. The authors conclude in their summary that: "No detectable induction of mutations could be found in two strains of *N. crassa* after their conidia were treated with vinyl chloride, in ethanol solution and in its gaseous form." Reversion in a *nic-1* strain and mutation to acriflavin resistance in a wild-type strain were studied. No data from experiments with VCl in ethanol solution, which were done without mammalian activation, were presented. The experiments with gaseous VCl were done in the absence and presence of either a S-9 microsomal fraction from phenobarbitol-induced rats or "purified" microsomes from uninduced rats. There was no detectable induction of mutation to acriflavin resistance.

Chemicals Not Tested in *N. crassa*

To our knowledge, there are no publications on tests for mutagenicity in *N. crassa* of the following chemicals selected for the Comparative Chemical Mutagenesis Workshop: cadmium (#7440-43-9), cyclophosphamide (#50-18-0), mitomycin C (#50-07-7), myleran (#55-98-1), natulan (#336-70-1), thioTEPA (#54-24-4), and trenimon (#68-76-8).

ACKNOWLEDGMENT

This research was sponsored by ERDA and NIH.

REFERENCES

1. Davis, R. H., and de Serres, F. J. Methods in Enzymology, Vol. 17A, 1970, pp. 79-143.

2. de Serres, F. J., and Malling, H. V. In: Chemical Mutagens: Principles and Methods of Their Detection, A. Hollaender, Ed., Plenum Press, New York, 1971, p. 311-342.

3. Brockman, H. E., and Goben, W. Science 147: 750-751 (1965).

4. Brockman, H.E., and de Serres, F. J. Genetics 48: 597-604 (1963).

5. de Serres, F. J., and Brockman, H. E. Genetics 58: 79-83 (1968).

6. Brusick, D. J. Mutat. Res. 8: 247-254 (1969).

7. Brockman, H. E. Abstr. Ann. Meet. Am. Soc. Microbiol. 74: 58 (1974).

8. Whong, W., and Brockman, H. E. Unpublished data.

9. Brockman, H. E., and Whong, W. Genetics 77: S6-S7 (1974).

10. Brockman, H.E., and Whong, W. Mutat. Res. 31: 305-306 (1975).

11. Whong, W., Ong, T., and Brockman, H. E. Abstracts of 8th Annual Meeting of the Environmental Mutagen Society, 1977, p. 45.

12. Whong, W., and Brockman, H. E. Mutat. Res. 26: 450-451 (1974).

13. Ong, T. Mutat. Res. 9: 183-191 (1970).

14. Malling, H. V. Mutat. Res. 4: 265-274 (1967).

15. Malling, H.V. Mutat. Res. 3: 470-476 (1966).

16. Malling, H.V., and de Serres, F. J. Mutat. Res. 4: 425-440 (1967).

17. de Serres, F. J., and Malling, H. V. Genetics 56: 555 (1967).

18. Malling, H. V., and de Serres, F. J. Mutat. Res. 6: 181-193 (1968).

19. Malling, H. V. Hereditas 68: 219-234 (1971).

20. Ong, T. Mol. Gen. Genet. 111: 159-170 (1971).

21. Malling, H. V., and de Serres, F. J. Radiat. Res. 53: 77-87 (1973).

22. Malling, H. V. Neurospora Newsl. 9: 13-14 (1966).

23. Malling, H. V. In: Molecular and Environmental Aspects of Mutagenesis. Proc. Prakash et al., Eds., Charles C. Thomas Springfield, Ill., 1974, pp. 149-177.

24. Hollaender, A. In: Chemical Mutagens: Principles and Methods for Their Detection, Vol. 2, A. Hollaender, Ed., Plenum Press, New York, 1971, pp. 607-610.

25. Webber, B. B., and Malling, H. V. Genetics 56: 595 (1967).

26. Brink, N. G., Kariya, B., and Stadler, D. R. Genetics 63: 281-290 (1969).

27. Radford, A. Neurospora Newsletter 14: 6-7 (1969).

28. Radford, A. Mutat. Res. 8: 537-544 (1969).

29. Radford, A. Mol. Gen. Genet. 104: 288-294 (1969).

30. Radford, A. Mutat. Res. 12: 57-64 (1971).

31. Radford, A. Mol. Gen. Genet. 107: 97-106 (1970).

32. Radford, A. Mutat. Res. 15: 23-29 (1972).

33. Brown, L. P., and Aiuto, R. Mutat. Res. 21: 251-255 (1973).

34. Grindle, M. Mol. Gen. Genet. 130: 8a-90 (1974).

35. Hoffmann, G. R., and Malling, H. V. Mutat. Res. 26: 450 (1974).

36. Hoffmann, G. R., and Malling, H. V. Mutat. Res. 27: 307-318 (1975).

37. Malling, H. V., and de Serres, F. J. Genetics 56: 575 (1967).

38. Malling, H. V., and de Serres, F. J. Genetics 60: 201 (1968).

39. Malling, H.V., and de Serres, F. J. Mol. Gen. Genet. 106: 195-207 (1970).

40. Malling, H. V., and de Serres, F. J. Ann. N. Y. Acad. Sci. 163: 788-800 (1969).

41. Inoue, H., Ong, T., and de Serres, F. J. Mutat. Res. 31: 307-308 (1975).

42. Bertrand, H., and Pittenger, T. H. Genetics 60: 161-162 (1968).

43. Bertrand, H., and Pittenger, T. H. Genetics 71: 521-533 (1972).

44. Gordon, P. H., and DeBusk, A. G. Genetics 61: S22 (1969).

45. Uno, I., and Ishikawa, T. Mutat. Res. 8: 239-246 (1969).

46. Mays, L. L. Genetics 63: 781-794 (1969).

47. Schroeder, A. L. Mol. Gen. Genet. 107: 305-320 (1970).

48. Schroeder, A. L. Mutat. Res. 24: 9-16 (1974).

49. Morgan, D. H. Mol. Gen. Genet. 108: 291-302 (1970).

50. Metzenberg, R. L., Chen, G. S., and Ahlgren, S. K. Genetics 68: 359-368 (1971).

51. Toh-E, A., and Ishikawa, T. Genetics 69: 339-351 (1971).

52. Russell, P. J., and Srb, A. M. Genetics 71: 233-245 (1972).

53. Feldman, J. F., and Hoyle, M. N. Genetics 75: 605-613 (1973).

54. Thwaites, W. M., Knauert, F. K., Jr., and Carney, S. S. Genetics 74: 581-593 (1973).

55. Teles Grilo, M. L., and Klingmueller, W. Mol. Gen. Genet. 133: 123-133 (1974).

56. Russell, P. J., and Cohen, M. P. Mutat. Res. 34: 359-366 (1976).

57. Westergaard, M. Experientia 13: 224-234 (1957).

58. de Serres, F. J., Brockman, H. E., Barnett, W. E., and Koelmark, H. G. Mutat. Res. 12: 129-142 (1971).

59. Seyfried, T. N. Genetics 77: S59 (1974).

60. Seyfried, T. N. Mutat. Res. 28: 155-162 (1975).

61. Malling, H. V. Mutat. Res. 2: 320-327 (1965).

62. Barnett, W. E., and de Serres, F. J. Genetics 47: 941 (1962).

63. Case, M. E. Genetics 47: 946 (1962).

64. Reissig, J. L. J. Gen. Microbiol. 30: 317-325 (1963).

65. Ishikawa, T. Idengaku Zasshi 42: 43-50 (1967).

66. Ishikawa, T. Genetics 47: 1147-1161 (1962).

67. Ishikawa, T. Genetics 47: 1755-1770 (1962).

68. Allison, M. Mutat. Res. 7: 141-154 (1969).

69. Allison, M. J. Mutat. Res. 7: 297-306 (1969).

70. Malling, H. V., and de Serres, F. J. Mutat. Res. 18: 1-14 (1973).

71. Malling, H. V. Genetics 64: S40-41 (1970).

72. Ong, T., and de Serres, F. J. Cancer Res. 32: 1890-1893 (1972).

73. Ong, T. Mutat. Res. 9: 615-618 (1970).

74. Ong, T. Genetics 64: S48-49 (1970).

75. Ong, T., Matter, B. E., and de Serres, F. J. Genetics 68: S48 (1971).

76. Matzinger, P. K., and Ong, T. Mutat. Res. 38: 121 (1976).

77. Ong, T. EMS Newsletter 6: 10-12 (1972).

78. Ong, T. Mutat. Res. 32: 35-53 (1975).

79. Jensen, K. A., Kirk, I., Koelmark, G., and Westergaard, M. Cold Spring Harbor Symp. Quant. Biol. 16: 245-261 (1951).

80. Koelmark, G., and Westergaard, M. Hereditas 39: 209-224 (1953).

81. Ong, T., and de Serres, F. J. Mutat. Res. 18: 251-258 (1973).

82. Ong, T., and de Serres, F. J. Genetics 71: S45 (1972).

83. Ong. T., and de Serres, F. J. Genetics 74: S204 (1973).

84. Bernstein, H., and Miller, (1961).

84. Bernstein, H., and Miller, A. Genetics 46: 1039-1052 (1961).

85. Malling, H. V., and de Serres, F. J. Abstr. XI Int. Bot. Congr., 1969, p. 139.

86. Malling, H. V., and de Serres, F. J. Cancer Res. 32: 1273-1277 (1972).

87. Marquardt, H., Schwaier, R., and Zimmermann, F. Naturwiss. 50: 135-136 (1963).

88. Malling, H. V. Mutat. Res. 3: 537-540 (1966).

89. Malling, H. V., and Frantz, C. N. Mutat. Res. 25: 179-186 (1974).

90. Malling, H. V., and Frantz, C. N. Environ. Health Perspec. 6: 71-82 (1973).

91. Ong, T., and Malling, H. V. Mutat. Res. 31: 195-196 (1975).

92. Kaney, A. R., and Atwood, K. C. Nature 201: 1006-1008 (1964).

93. Koelmark, G., and Giles, N. H. Genetics 40: 890-902 (1955).

94. Drozdowicz, B. Z., and Huang, P. C. Mutat. Res. 48: 43-50 (1977).

CHAPTER 7

MUTAGENICITY OF SELECTED CHEMICALS IN YEAST: MUTATION-INDUCTION AT SPECIFIC LOCI

N. Loprieno

Laboratorio di Genetica of the University

Via S. Maria 53, 56100 Pisa, Italy

INTRODUCTION

Forward mutations and reversions of known mutations are the basis for mutation testing of environmental mutagens in yeast. Species of two different genera of yeast are commonly used, namely *Saccharomyces cerevisiae* and *Schizosaccharomyces pombe*. Stationary phase, haploid cells of both species usually are exposed to the mutagens, although some diploid strains of *S. cerevisiae* have been developed for testing. In the stationary phase, *S. cerevisiae* is in the G1 stage of the cell cycle and *S. pombe* is in the G2 stage. Mutagen assays have been made with growing cells of yeast, and striking differences between stationary phase and growing cells in their response to mutagens have been observed occasionally.

FORWARD MUTATION SYSTEMS

Forward mutations in genes controlling the synthesis of adenine are useful for assays of mutagens in both *Schizosaccharomyces pombe* and *Saccharomyces cerevisiae*. This test is based on the recognition of white adenine-requiring mutants from strains containing either *ade 1* or *ade 2* in *S. cerevisiae* or *ade 6* or *ade 7* in *S. pombe*. These loci confer a red color to the strain. White adenine-requiring mutants are mutations in one of five genes controlling steps earlier in the adenine biosynthesis pathsay. In *S. pombe* these would be base substitution or frameshift mutations in *ade 1*, *ade 3*, *ade 4*, *ade 5*, or *ade 9*. In *S. cerevisiae* these would be mutations in *ade 3*, *ade 4*, *ade 5*, *ade 6*, or *ade 7* loci. Since it is relatively simple to plate large numbers of cells and search for white colonies or sectored colonies among red colonies, this system is of considerable usefulness. In order to infer the type of molecular change that led to a mutation, it is possible to analyze each white, adenine-requiring mutant in detail.

Other forward mutation systems have been devised for yeast (1), and the one most commonly used for testing is resistance to canavanine, which appears to take place in a gene that encodes an arginine-specific arginase. The sensitivity for such resistant mutants is not much greater than second site reversions of certain auxotrophic strains.

REVERSION SYSTEMS

Reverse mutations in yeast afford a ready opportunity to gather information on the type of molecular event that transpired. Strains have been developed which can be used to simultaneously analyze for frameshift or missense reversions and reversions at a suppressible locus or by mutation of a suppressor. These strains can be used in either the haploid or diploid condition. Strains using genetic defects in one or more mutagen-sensitive pathways have been used to sensitize the strain, or to gather information on mutagens whose lesions are repaired preferentially by one or more repair pathways.

RESULTS AND DISCUSSION

Results of yeast mutation studies are summarized in Table 1. No reports of studies on yeast mutagenesis were found for the following chemicals: mitomycin C, natulan, ethylenimine, triethylenemelamine, aflatoxin B_1, cadmium compounds, myleran, METEPA, Thio-TEPA, and trenimon.

Dimethylnitrosamine (DMN)

Mutation induction data on yeast for DMN are rare. In 1964 Marquardt et al. (13) analyzed the mutagenic effect of 1 M concentration at pH 7.5 in treatments, lasting for 30 min, of strain ade 6-45 of *Saccharomyces cerevisiae*, by analyzing the phenotypic reversions at the ade 6 locus: the results were negative.

In 1971 Mayer (3) analyzed the induction of canavanine-resistant mutants by DMN on the strain D273-10B (CAN^s) of *S. cerevisiae*. The treatment was performed in the presence of a hydroxylation medium (Udenfriend's reaction mixture), DMN was tested at 100 μmole/ml for 6 hr. In the presence of O_2, DMN induced 24.5 CanR/10^6 survivors, whereas in the untreated sample

Table 1. Results of yeast mutation studies

Chemical	Results of yeast mutation studies	Effective exposure	References
DMN	Positive in forward mutation assay	12.5 mg/kg, 15 hr	2
DEN	Positive in forward mutation assay	---	3
MMS	Positive in forward mutation and reversion assays	0.6 mM/hr	4
MNNG	Positive in forward mutation assay	2 mg/ml/min	5
EMS	Positive in forward mutation and reversion assays	10 mM/min	6
Vinyl chloride	Positive in forward mutation assay	2 mM/hr	7
Ethylenimine	Positive in forward mutation assay	50 mM/hr	8
TEPA	Positive in reversion assay in *S. pombe*; negative in *S. cerevisiae*	2.3×10^{-3} M	9
ICR-170	Positive in forward mutation and reversion assays	5 μg/ml, 30 min	10
Epichlorohydrin	Positive in reversion assay		11
Cyclophosphamide	Positive in forward mutation and reversion assays	800 mg/kg	12

a mutation frequency of 10.6 CanR/survivors was obtained.

Fumero et al. (2) have tested DMN in the host-mediated assay, on analyzing forward mutations induced at five adenine loci of Schizosaccharomyces pombe (strain P1). Mice and rats were used as host: the compound was administered orally by gavage and the yeast cells inoculated into the peritoneum.

A dose-dependent relationship was established in the experiments done on mice, with the minimal dose tested of 12.5 mg/kg which has given a significant increase over the spontaneous level (control $1.43 \pm 0.45 \times 10^{-4}$ mutations per survivor; DMN = $5.02 \pm 1.33 \times 10^{-4}$ mutations per survivor. In this experiment the cells were incubated for 15 hr in the animals. In this experiment a mutation specific activity of 0.10×10^{-4} per locus-mg/kg has been obtained.

A time-dependent relationship was established for mice and rats in experiments lasting 0, 3, 6, or 15 hr. In both cases, 3 hr of incubation produced significant increase over the spontaneous level (mouse: control, $2.55 \pm 21.15 \times 10^{-4}$; 3 hr, $25.91 \pm 7.81 \times 10^{-4}$; rat: control, $2.02 \pm 0.82 \times 10^{-4}$; 3 hr, $5.18 \pm 1.51 \times 10^{-4}$). The specific mutation activities in the two species were the following: mouse, 1.19×10^{-6} per locus-hr-mg/kg; rat, 0.46×10^{-6} per locus-hr-mg/kg.

Diethylnitrosamine (DEN)

Forward Mutations. A sixfold increase in the production of can^R mutants in the strain D273-10B of S. cerevisiae over the spontaneous level has been observed after treatment with DEN in a hydroxylation medium (3).

Methyl Methanesulfonate (MMS)

Forward Mutations. In Schizosaccharomyces pombe MMS has been used for the induction of ade 6 and ade 7 purple mutants a 0.1M solution produced, after 1 hour of treatment, 1.36×10^{-4} and 1.24×10^{-4} mutation frequencies respectively at these two loci (14,15). The spontaneous mutation rates at these two loci are respectively 4.10×10^{-7} and 0.09×10^{-7} (4).

Forward mutations at five adenine loci have been induced by MMS in S. pombe SP-198ade6-60,rad10-198,h⁻ strain: a linear dose-response curve has been obtained in the range of 0.6 to 5.0 mM solutions. The specific mutation induction was 5.02×10^{-4} per locus-mM-hr (4 times the spontaneous mutation frequency observed). The minimal dose tested (0.6 mM) produced a 10-fold increase over the spontaneous level.

A value of 7.94×10^{-4} per locus-mM-hr was observed in treatments which were made in the presence of mouse liver purified microsomal preparation; in the host-mediated assay (mouse injected with yeast cells into the peritoneum) a value of 4.00×10^{-4} per locus-mM-hr has been reported (16). A comparison with x-rays is reported in the EMS section.

Reverse Mutations. MMS 0.1%, 40 min treatment, induces 5×10^{-7} reverse mutations (cy1-131→CYC1) in *S. cerevisiae* (17).

In *S. pombe* a linear dose-effect relationship has been observed when reverse mutations at the *arg 1* locus were studied: MMS concentrations were between 3.4 and 27 mM. A specific mutation induction of 770 $arg^+ \times 10^{-7}$/mM was obtained (γ-rays; 170 $arg^+ \times 10^{-7}$/rad (11).

N-Methyl-N'-nitro-N-nitrosoguanidine (MNNG)

Forward Mutations. MNNG in treatments with doses between 0.13 mM and 3.4 mM at pH 5.5 induces auxotrophic mutants in *Saccharomyces cerevisiae* strain S288C. A dose-dependent linear relationship has been obtained; the induction of auxotrophic mutants ranges from 4.2% (100-30% survival) to 30% (0.01-0.003% survival). No spontaneous value of auxotrophic mutants is given (18). In this strain the possible spontaneous mutation frequency of auxotrophic mutants is less than 1.0% (19). A time-dependent induction of auxotrophic mutants by MNNG (1 mg/ml) in *S. rouxii* has been observed: treated free spores produced a higher mutation frequency (20).

Dose-dependent (0, 30, 100, 300, 1000 μg/ml) and a pH-dependent (3.5, 5, 6, 7) relationships of induction of canavanine resistant mutants in *S. cerevisiae* 59R strain, have been obtained (20). It is not possible to draw conclusions about specific mutation inductions: in all cases the minimal dose tested (30 μg/ml) produces a 100-fold increase over the spontaneous value (21).

In strain S288C of *S. cerevisiae* MNNG (2 mg/ml) produces 0.7×10^{-7}/min cychloheximide-resistant mutants: the spontaneous value for two types of mutations are, respectively, 2×10^{-7} and 7×10^{-7} (5). No great differences in the production of auxotrophic mutants has been observed if log. or stat. cells of *S. cerevisiae* S288C are treated (10). In *Schizosaccharomyces pombe*, MNNG (1 mM) produces in treatments lasting less than

30 min a specific mutation induction of 5×10^{-5}/min per locus (spontaneous values for ade 6 and ade 7 are of the order of 10^{-7} 10^{-9}). At higher treatment an exponential relationship is observed (22,23).

A time-dependent relationship in the induction of auxotropic mutants by MNNG (3 mg/ml) in *S. pombe* after 2-deoxyglucose treatment has been reported (24). 100% of induced mutations in *S. pombe* have been shown to be base-pair substitutions (25).

Reverse Mutations. A linear dose-response curve has been obtained in *S. cerevisiae* in the induction of reverse mutations cyc → CYC by MNNG: a dose of 10 μg/ml produces significant increases over the spontaneous reversion frequency in the wild type. rad 6 and rad 9 were no more sensitive to the mutation induction by MNNG (17).

A pH dependence in the induction of reverse mutations $ade\ 6^- \rightarrow ade\ 6^+$ by MNNG (5 mM) has been observed: the pH 6 produces a higher mutation frequency (26). A temperature dependence in the mutation induction in the same system has been reported: the mutagenic effect of MNNG 5 mM increases by a factor of about 3 for every 10°C difference in the range from 10°C to 30°C (27,28).

In *S. pombe*, $met\ 4^+$ reversions have been studied after treatments with MNNG. A linear dose-effect relationship has been observed: the doubling of spontaneous reversion frequencies is obtained with a treatment of 10^{-6} M of MNNG. At the dose of 10^{-3} M a specific mutation induction of 23×10^{-7} survivors/mM-min per locus: the spontaneous reverse mutation frequency is of the order of 0.5×10^{-7} survivors (29).

Ethyl Methanesulfonate (EMS)

Forward Mutations. EMS has been applied to S288C strain of *S. cerevisiae* at a dose in the range between 0.05 M and 0.12 M for 20 hr to induce auxotrophic mutants (0.05% to 4.0%), or at a dose of 0.4 M for 1, 2, or 3 hr. In this case the percentage of mutants was between 0.05% and 4.0% (30). No value of spontaneous background is given.

The haploid strain No. 60615 of *S. cerevisiae* has been used for the induction of auxotrophic mutants by a treatment of EMS at 3% (v/v) concentration for different incubation times. The minimal value of auxotrophic frequency (0.5%) was obtained with 30 min treatment. No value of spontaneous background was given (31). A solution of 3% (v/v) of EMS has been employed for determining the induction of auxotrophic mutants after 30 min treatment at 30°C on S288C strain of *S. cerevisiae* (REV) and three mutant strains (rev 1-1, rev 2-1, rev 3-1): the percentages of auxotrophic mutants were, respectively 5.6, 4.4, 6.1, and 5.2%. The spontaneous values were <1.0%, <1.8%, 12.9%, and <1.0%. For the induction of auxotrophic mutants in strain rev 2-1, the value (6.1%) has been corrected for spontaneous background (19). A treatment of 3% (v/v) of EMS for 30 min on the same strains produced respectively 28.9 × 10^{-4}, 23.7 × 10^{-4}, 16.2 × 10^{-4}, and 14.7 × 10^{-4} of ade -1 and ade 2 mutation frequencies (spontaneous values not reported) (19). In S. pombe forward mutation frequencies induced by EMS have been reported in two genetic systems.

Mutations from wild type to ade 6, and ade 7 mutants (red mutants): EMS 1 M for 2 hr and 25°C produced 5.27 × 10^{-4} mutants in the ade 6 locus and 2.60 × 10^{-4} mutants in the ade 7 locus (15). The spontaneous mutation frequency for these two loci has been evaluated and corresponds to 4.10 × $10^{-7}$and 0.09 × 10^{-7}, respectively (4).

Mutation from red mutants (ade 6) to white mutants (five different loci) has been induced in the wild type and rad 10 mutant of S. pombe by different molar concentrations of EMS in 1 hr treatments. The mutation induction specific activity was: wild type, 1.32 × 10^{-7}/min-10 mM; rad 10-198, 0.39 × 10^{-5}/min-10 mM. The spontaneous mutation rate was: wild type, 0.76 × 10^{-6}/cell division rad 10-198; 1.30 × 10^{-5}/cell division. The minimal dose tested of EMS was 25 mM, and this treatment has given 10^{15} times more mutants than in the control for the rad 10-198; for the wild type 50 mM treatment has resulted in a significant increase over the spontaneous level (6).

To compare with x-ray, the same two strains of S. pombe (wild type and rad 10-198) have observed an induction of 1.42 × 10^{6} per locus - 1000 R and 1.30 × 10^{-5} per locus - 1000 R: the minimal dose which produced significant increase over the spontaneous level in the wild type was 32,000 R and in the rad 10-198 16,000 R (32).

EMS has resulted to produce a forward-mutation induction of 1.88 × 10^{-4} locus/mM-hr in buffer solution, whereas in the presence of purified microsomal preparations from mouse liver a value of 1.58 × 10^{-4} locus/mM-hr in the rad 10-198 strain of S. pombe. In the same system, host-mediated assays with mice (injection of yeast into the peritoneum) have produced a mutation induction of 2.43 × 10^{-4} locus/mM (treatment of 2 hr). Minimal significant dose tested: in buffer, 25 mM (× 47 SP value); in micros., 25 mM (× 40 SP value); in mice, 2.4 mM (× 2.4 SP value). In all cases, dose-effects relationship have been evaluated, and the data submitted to a regression analysis (14,16,33,34).

Reverse Mutations. Reverse mutations have been recently

evaluated in the D7 strain of S. cerevisiae at the ilv 1-92 locus: a log increase of reversion frequency was observed with increasing doses from 40 to 160 mM. The minimal dose tested was 40 mM; the spontaneous level was undetectable ($\times 10^{-6}$) (35). A 40 mM concentration of EMS in treatments which lasted between 1 and 5 hr has been employed for the evaluation of reversion induction at cyc 1-131 locus of S. cerevisiae in the wild type and rad 6 or rad 9 strains: a dose-effect relation was found for the wild type; the remaining strains were almost mutagen-negative (17).

EMS was shown to revert S211 strain of S. cerevisiae by plate method, by inducing base-pair substitution mutations (BPS) (36), to produce 84% of BPS in the ad1-A and 68% of BPS in the ad1-B of S. pombe (25), or 95.8% of transition mutations in the ade 6 locus of S. pombe (14). Reverse mutations have been evaluated in S. pombe on treating arg 1 mutant of S. pombe with different molar concentrations (the minimal value tested was 0.09 M) (11).

Vinyl Chloride

Forward Mutations. Vinyl chloride has been shown to produce forward mutations in S. pombe at five ade 6 loci in the strain SP-198 ade6-60/rad10-198, h^-, in the presence of purified microsomal preparation from mouse liver cells (37).

A linear dose-effect relationship was observed in treatments between 16 and 60 mM (30 min), where a specific mutation induction of 0.2×10^{-4}/mM was observed; spontaneous mutations in these experiments were of the order of $0.2\text{-}0.4 \times 10^{-4}$ (37). In Table 2 comparative mutagenic effects induced by x-rays and VCM are reported.

Vinyl chloride has been shown to induce forward mutations in S. pombe inoculated into the peritoneum of mice. Animals were treated orally with 700 mg/kg of compound dissolved in olive oil; treatment for 6 hr was effective in the production of mutation (37).

It has been possible to determine that during 1 hr of incubation with microsomes a solution containing 50 mM of vinyl chloride is converted to the active metabolite 2-chloroethylene oxide for a fraction of 0.1% only: the specific mutagenic activity of the metabolite is 1000-fold higher than that produced by vinyl chloride in the presence of microsomes (38).

Reverse Mutations. Reverse mutations were not induced in S. cerevisiae in absence of microsomes (39).

Table 2. Comparative mutagenic effects in *S. pombe* (1 hr treatment)[a]

Agent	Dose of mutagen for doubling specific mutation induction
X-rays	9900 rads
EMS	28.0 mM
Vinyl chloride	2.0 mM
MMS	0.25 mM

[a]Data of Loprieno et al. (7).

Ethylenimine

Forward Mutations. In *S. cerevisiae* (Peterhop strains), EI at 0.05 M after a treatment of 10 min induced *ade 1* and *ade 2* mutants. Data on mutation frequencies have been reported (8).

TEPA

Reverse Mutations. In *S. pombe*, TEPA-induced reversions in *met4-D19* strain at 7×10^{-7} survivors. The minimal dose significantly different from the control value was 2.3×10^{-3} (9).

In *S. cerevisiae* a dose of 3.5×10^{-2} M was ineffective in the induction of reverse mutations in the *ade 1* locus (9).

ICR-170

Forward Mutations. ICR-170 (5 μg/ml) for 30 min induced auxotrophic mutants in four haploid strains of *S. cerevisiae* (S288C, S101, 77341, S100), on after treating log. or stat. cells. The percentages of auxotrophs ranged between 0.4 to 27.9 (10). The spontaneous value of auxotrophic mutation frequency is known only for S288C strain, and it is <1.0% (19).

Reverse Mutations. ICR-170 induced reverse-mutations is auxotrophic mutants of *S. cerevisiae* induced originally with the same mutagen (40). In *S. pombe*, *ade* mutants have been induced by ICR-170, and they could be reverted by the same mutagen (5 μg/ml) (25).

Epichlorohydrin

Reverse Mutations. An exponential dose-effect relationship has been obtained on treating *S. pombe* strain *arg 1* for the induction of arg^+ reversion: an induction of 40×10^{-7} mutations/mM-min has been observed (11).

Cyclophosphamide

Forward Mutations. In *S. cerevisiae*, strain S214 α, cyclophosphamide has been tested in the host-mediated assay for the induction of canavanine resistant mutants. A dose-effect increase has been observed (two doses tested) with a mutation induction of $9 \times 10^{-7} \times 100$ mg/kg; the control value was 50×10^{-7}: the minimum dose tested was 800 mg/kg (12).

Reverse Mutations. The S214 α (met^-) has been employed as previously for reverse mutation induction in the host-mediated assay. A dose-effect dependence was observed with a mutation induction of $2 \times 10^{-7} \times 100$ mg/kg; the control value was 13×10^{-7} (12).

REFERENCES

1. Mortimer, R. K., and Manney, T. R. In: Chemical Mutagens, Principles and Methods for Their Detection, Vol. 1, A. Hollaender, Ed., Plenum Press, New York, 1971, pp. 289-310.

2. Fumero, S., Monding, A., Meriggi, G., Silvestri, S., Barale, R., and Loprieno, N. Mutat. Res. 53: 189 (1978).

3. Mayer, V. W. Mol. Gen. Genetics 112: 289-294 (1971).

4. Loprieno, N. Genetics (Suppl.) 73: 161-164 (1973).

5. Brusick, D. J. Bacteriol. 109: 1134-1138 (1972).

6. Loprieno, N., Barale, R., Bauer, C., Baroncelli, S., Bronzetti, G., Cammellini, A., Cinci, A., Corsi, G., Leporini, C., Nieri, R., Nozzolini, M., and Serra, C. Mutat. Res. 25: 197-217 (1974).

7. Loprieno, N., Barale, R., Baroncelli, C., Bronzetti, G., Cammellini, A., Corsi, G., Leporini, C., Nieri, R., and Rossi, A. M. In: Screening Tests in Chemical Carcinogenesis, R. Montesano, Ed., IARC Sci. Publ. 12, IARC Lyon, 1976, pp. 505-519.

8. Aleksandrova, N. N. Genetika 9: 164-167 (1973).

9. Zetterberg, G. Hereditas 68: 245-254 (1971).

10. Brusick, D., and M. S. Legator. EMS Newsletter 4: 31-32 (1971).

11. Heslot, H. Abhandl. Deut. Alsad. Wissen. (Berlin) Klasse Med. 1: 98-105 (1960).

12. Brusick, D., and M. S. Legator. EMS Newsletter 4: 31-32 (1971).

13. Marquardt, H., Zimmermann, F. K., and Schwaier, R. Z. Vererbungsl. 95: 82-96 (1964).

14. Loprieno, N., Guglielminetti, R., Bonatti, S., and A. Abbondandolo. Mutat. Res. 8: 65-71 (1969).

15. Loprieno, N. Mutat. Res. 3: 486-493 (1966).

16. Loprieno, N., Barale, R., Baroncelli, S., Bronzetti, G., Cammellini, A., Corsi, G., Leporini, C., Nieri, R., Nozzolini, M., and Rossi, A. M. Consig. Nazl. Recerchi Lab. Mutaten. Differenz. (No. 109): 1-49 (1974).

17. Prakash, L. Genetics 78: 1101-1118 (1974).

18. Lingens, F., and Oltmanns, O. Naturforsch. 218:660-663 (1966).

19. Lemontt, J. Mol. Gen. Genetics 119: 27-42 (1972).

20. Mori, H. Hakko Kogaku Zasshi 50: 219-220 (1972).

21. Divies, C., and Morsauz, J.-N. Ann. Technol. Agric. 19: 261-266 (1970).

22. Loprieno, N., and Schupbach, M. EMS Newsletter 3: 17 (1970).

23. Loprieno, N., and Schupbach, M. Mol. Gen. Genetics 110: 348-354 (1971).

24. Megnet, R. Mutat. Res. 2: 328-331 (1965).

25. Segal, E., Munz, P., and Leupold, U. Mutat. Res. 18: 15-24 (1973).

26. Zimmermann, F. K., Schwaier, R., and v. Laer, U. Z. Vererbungsl. 97: 68-71 (1965).

27. Schwaier, R., Zimmermann, F. K., and v. Laer, U. Z. Vererbungsl. 97: 72-74 (1965).

28. Zimmermann, F. K., Schwaier, R., and v. Laer, U. Z. Vererbungsl. 98: 152-166 (1966).

29. Loprieno, N., and Clarke, C. H. Mutat. Res. 2: 312-319 (1965).

30. Nesvera, J. Folia Microbiol. 18: 353-360 (1973).

31. Lindegren, G., Hwang, Y. L., Oshima, Y., and Lindegren, C. Can. J. Gen. Cytol. 7: 491-499 (1965).

32. Loprieno, N., Barale, R., Baroncelli, S., Cammellini, A., Melani, M., Nieri, R., Nozzolini, M., and Rossi, A. M. Mutat. Res. 28: 163-173 (1975).

33. Loprieno, N., Barale, R., Baroncelli, S., Bauer, C., Bronzetti, G., Cammellini, A., Ducci, M., Guglielminetti, R., Leporini, C., Nieri, R., and Nozzolini, M. Mol. Environment Aspects Mutagenesis Proc. Publ. Rochester Int. Conf. Environ. Toxic 6th (1973), 143-156 (1974).

34. Loprieno, N., Barale, R., Baroncelli, S., Bauer, C., Bronzetti, G., Cammellini, A., Cinci, A., Corsi, G., Leporini, C., Nieri, R., Nozzolini, M., and Serra, C. IARC Sci. Publ. 10: 183-199 (1974).

35. Zimmermann, F. K., Kern, R., and Rasenberger, H. Mutat. Res. 28: 381-388 (1975).

36. Brusick, D., and E. Zeiger. Mutat. Res. 14: 271-275 (1972).

37. Loprieno, N., Barale, R., Baroncelli, S., Bauer, C., Bronzetti, G., Cammellini, A., Cercignani, G., Corsi, C., Gervasi, G., Leporini, C., Nieri, R., Rossi, A. M., Stretti, G., and Turchi, G. Mutat. Res. 40: 85-96 (1976).

38. Loprieno, N., Barale, R., Baroncelli, S., Bartsch, H., Bronzetti, G., Cammellini, A., Corsi, C., Frezza, D., Nieri, R., Leporini, C., Rosellini, D., and Rossi, A. M. Cancer Res. 36: 253-257 (1977).

39. Shahin, M. M. Mutat. Res. 40: 269-272 (1976).

40. Pittman, D., and Brusick, D. Mol. Gen. Genetics 111: 352-356 (1971).

CHAPTER 8

MUTAGENICITY OF SELECTED CHEMICALS IN YEAST: MITOTIC RECOMBINATION, GENE CONVERSION, AND NONDISJUNCTION

F. K. Zimmermann*

Genetisk Institut, Københavns Universitet,

Øster Farimagsgade 2A, DK-1353 København K, Danmark

YEAST MITOTIC RECOMBINATION AND MITOTIC NONDISJUNCTION

There are two types of mitotic recombination: mitotic crossing-over, the reciprocal type, and mitotic gene conversion, the nonreciprocal type. At present, the term, "mitotic recombination," is used to cover both types of recombination. In early publications, mitotic recombination has been used less discriminantly.

Mitotic Crossing-Over

Mitotic crossing-over in diploid yeast cells can be detected by the appearance of phenotypes caused by recessive markers originally present in a heterozygous condition. The mere expression of recessive markers alone does not provide definite proof for mitotic crossing-over, since such phenotypic changes can be caused by a variety of additional mechanisms such as point mutation of the dominant allele, deletion of the chromosomal segment carrying the dominant allele, loss of the entire chromosome carrying the dominant allele, and mitotic gene conversion of the dominant allele. Mitotic crossing-over is only indicated when the appearance of the recessive phenotype in a sector of a colony is associated with the appearance of another sector in the same colony which is homozygous for the other allele.

---

*Permanent address: Institut für Mikrobiologie, Schnittspahnstr. 10, D-6100 Darmstadt, Federal Republic of Germany.

Two strains are presently in use which allow for this type of analysis. One is D3 constructed by Zimmermann et al. (1). The relevant markers are

centromere	CYH4	ade2-1	his8
	+	+	+

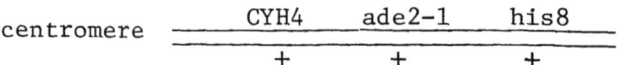

This diploid is resistant to 0.5 ppm cycloheximide because of dominant CYH4, white, and does not require adenine or histidine. Mitotic crossing-over generates red, adenine-requiring colonies or sectors in otherwise white colonies. If the cross-over event had occurred in the interval between the centromere and CYH4, the red sectors would be adenine-requiring (this is the selected phenotype) and would also require histidine, as well being as resistant to cycloheximide. The accompanying white sector will be sensitive to cycloheximide and does not require adenine or histidine. In many publications, such a further analysis has not been performed with more than a few colonies or not at all.

In order to facilitate the unambiguous demonstration of mitotic crossing-over, another strain was constructed by Zimmerman (2). This strain, called D5, carries two alleles of the ade2 gene locus. One allele, ade2-40, is a typical, completely blocked allele which causes a deeply red colony pigmentation on media low in adenine. The other allele, ade2-119, is leaky and causes the formation of pink colonies on low adenine media. Alleles ade2-40 and ade2-119 complement each other, so that D5 is white and does not show an adenine requirement. Mitotic crossing-over anywhere between the centromere and the ade2 gene locus will cause, at a 50% chance, the formation of a twin spotted colony which is partly red and partly pink. Thus, the generation of the two reciprocal products by mitotic crossing-over can be detected visually. Another strain, D7, established by Zimmermann et al. (3), carries the same ade2 alleles in addition to a noncomplementing pair of alleles of the trp5 gene locus which provides a test for mitotic gene conversion.

In order to properly evaluate dose-response relations for mitotic crossing-over, one has to bear in mind that twin spotted colonies can only be detected when both segregants of the treated cell survive. If there is a lethal mutation segregating, one will observe only a red or a pink colony. Moreover, loss of one of the homologs or part of the chromosome carring the ade2 gene or mitotic gene conversion will also produce colonies which are red, pink, red-white, or pink-white. Point mutation can also alter one of the two ade2 alleles in a way as to upset the complementation behavior so that pigment accumulates. Published data clearly show that, in addition to twin spotted colonies, there always is high incidence of other types of aberrant colonies. A true reciprocal recombinational event accompanied by lethal segregation will generate a red or pink colony which cannot be distinguished from the same type of colonies generated by different mechanisms. Therefore, interpretation of dose-response curves is difficult. Dose-response curves for induced mitotic crossing-over always are of a plateau type. However, at lower doses there is a sharp increase in recombinants. Due to the nonselective procedures used, little information is available on the shape of dose response curves at low doses.

Mitotic Gene Conversion

Mitotic gene conversion can be studied by using selective procedures. The basic genetic set-up is a diploid carrying two noncomplementing alleles combined in a diploid so that there is a nutritional requirement. Intragenic recombination between the two mutant sites generates either a double mutant or a wild type allele. The wild type allele is dominant and allows for prototrophic growth. Consequently, recombinants can be selected for on a minimal medium. These recombinants are mostly due to nonreciprocal recombination also called gene conversion.

Basically, any mutants causing a nutritional requirement can be used for a convenient study of mitotic gene conversion. The most commonly used strain introduced by Zimmermann and Schwaier (4) is D4, which is heteroallelic at two loci: trp5-12 and trp5-27 on one hand, and ade2-1 and ade2-2 on the other. Therefore, mitotic gene conversion can be studied at two loci using the same strain. Selection of prototrophs in heteroallelic auxotrophic strains uncovers conversion and reverse mutation. Usually, the vast majority of prototrophs is due to mitotic gene conversion. The contribution of reverse mutation to the total yield of prototrophs is much lower than that of mitotic gene conversion. It can be assessed using strains homoallelic for either of the mutant alleles.

Mitotic gene conversion in yeast could be induced by all known mutagens as far as tested, and no case of mutagen specificity has been observed. Mutagen specificity has been a considerable drawback with reverse mutation. Up to date, several hundred compounds have been shown to be genetically active in inducing mitotic gene conversion in yeast.

Mitotic gene conversion can be induced in stationary phase cells and scored upon immediate plating on selective media without any residual growth as shown by Zimmermann and Schwaier (4). Moreover, D4 has be used by Zimmermann (5) in spot tests. Cells were spread on a medium without tryptophan and the chemical to be tested applied to the center of the plate. In case of genetic activity, there was a ring of convertants surrounding the site of application. For obscure reasons, the spot test procedure has been negative with gene conversion with the ade2-conversion system of D4.

Mitotic Loss of Chromosomes (Nondisjunction)

A strain has been constructed by Parry and Zimmermann (6) to study the mitotic loss of chromosome VII. This strain, D6, has the following relevant genotype:

$$\frac{met13 \quad cyh2 \quad trp5 \quad leu1}{+ \quad + \quad + \quad +} \text{ centromere } \frac{ade3}{+} \quad \frac{ade2\text{-}40}{ade2\text{-}40}$$

D6 forms red colonies because of the homozygous condition of ade2; it does not require methionine, tryptophan, histidine (ade3 does not only cause a requirement for adenine but also for histidine which allows the detection of ade3 in the presence of defective alleles of the other adenine loci); it is also not resistant to cycloheximide (cyh2 is a recessive allele causing resistance to 2ppm cycloheximide).

Loss of the chromosome VII homolog carrying the dominant wild type alleles will result in a monosomic diploid which expresses all the recessive markers on both sides of the centromere. Such monosomics are selected for on a medium with cycloheximide and grow up to white instead of the normal red colonies. These white colonies are then tested for the expression of the other alleles notably of leu1 which is only 5 cM from the centromere. Basically, the same phenotype can be caused by two concomitant mitotic crossover events, one between the centromere and leu1, the other between the centromere and ade3. The much higher incidence of leucine-requiring colonies among the white than the red types is indicative of mitotic chromosome loss. Genetic analysis of this class is usually impossible due to a lack of sporulation. The few sporulating isolates were diploid for chromosome VII. However, DNA determination showed that the nonsporulating, cycloheximide-resistant, leucine-requiring isolates all had less DNA than the original strain.

The treatment procedure involves exposure to the mutagen followed by a 48 hr growth period in nonselective medium before plating on a cycloheximide medium. Immediate plating on cycloheximide does not result in an increase in cycloheximide-resistant cells probably because presence of this inhibitor of protein synthesis does not allow the formation of the repair enzymes necessary for mitotic recombination nor the segregational loss of chromosomes due to an arrest of growth.

A summary of tests performed with selected chemicals is given in Table 1.

Table 1. Summary of Tested Substances: Yeast Mitotic Recombination and Mitotic Chromosome Loss[a]

Agent	Gene conversion	Cross-over	Chromosome loss
X-rays	20-25 rad, LED	1000 rad, LTED	1000 rad, LTED
Cyclophosphamide			
Without activation	Negative (D4!)	10 mg/ml, LTED	No test
Human urinary assay	3.1 mg/kg, LTED	No test	No test
Diethylnitrosamine			
Without activation	Negative	Negative	No test
Udenfriend system	No test	0.1 M, LTED	No test
Dimethylnitrosamine			
Without activation	Negative	Negative	No test
Udenfriend system	No test	0.1 M, LTED	No test
Liver microsomes	0.77 mM, LTED	0.77 mM, LTED	No test
In vivo activation liver, rats	10 mmole/kg, LTED	No test	No test
In vivo activation lung, rats	10 mmole/kg, TRED	No test	No test
In vivo activation testes, rats	10 mmole/kg, negative	No test	No test
Peritoneum, rats	10 mmole/kg, LTED	No test	No test
Peritoneum, mice	10 mmole/kg, LTED	No test	No test
Ethylenimine, no activation	20 mM, LTED	10 mM, LTED	No test
Ethyl methanesulfonate, no activation	25 mM, LTED	48.6 mM, LTED	No test
ICR-170, no activation	1.0 g/ml, LED	10 g/ml, LED	No test

(continued)

Table 1. (continued)

Agent	Gene conversion	Cross-over	Chromosome loss
Methyl methanesulfonate, no activation	1 mM, LED	No test	No test
N-Methyl-N'-nitro-N-nitrosoguanidine, no activation	0.1 mM, LTED	0.1 mM, LTED	No test
Mitomycin C, no activation	1.75 mM, LTED	1.17, LTED	No test
Triethylenemelamine, no activation	10 μM, LED	No test	No test
Trenimon, no activation	4.32×10^{-9} M, LED	No test	No test
Vinyl chloride			
No activation	Negative	No test	No test
Liver microsomes	48 mM, LTED	No test	No test

[a]LED = lowest effective dose. This was rigorously determined only in the case of x-rays. In all other cases the LED has been derived from dose-response curves where data for effective and ineffective doses were reported. Therefore, these values may be much higher than the actually, lowest detectable active dose. It has to be borne in mind that only recently has it become popular to ask for the LED, and most of the work had been done much earlier. Consequently, in most cases, only the lowest tested effective dose, LTED is given. This may be several orders of magnitude higher than the real LED. Cyclophosphamide is active without metabolic activation in a number of yeast strains but not in D4 which has been used for detecting induction of gene conversion. Thus, the discrepancy between the gene conversion and the mitotic crossing-over test is caused by the strain difference not by the different endpoints scored.

INDUCTION OF MITOTIC RECOMBINATION WITH IONIZING RADIATION

Induction of mitotic gene conversion with 50 kVp x-rays was first studied by Manney and Mortimer (7) at a dose rate of 3120 rad/min. The dose response was strictly linear in the range between 1000 and 4000 rad.

The fact that mitotic gene conversion leads to prototrophic cells in an originally auxotrophic culture allows for the use of selective media. Moreover, strong induction of mitotic gene conversion can be observed at doses which do not lead to a reduction of colony forming cells. Consequently, at such low doses, usually below 5 krad, cell killing can be ignored. Experiments can be carried out by subdividing a given cell suspension into various aliquots, one of which serves as a control, others being exposed to different low doses of ionizing radiation. They can be plated on a medium selective for convertants and the actual counts of convertant colonies used to assess the effects of various low doses. Conversion frequencies are usually expressed as revertants per survivors which requires dilution series for each mutagen dose and this increases variation due to technical inaccuracies. At low doses of ionizing radiation, variations in survival platings between different samples can then be considered as of technical origin and pooled.

In order to test the sensitivity of mitotic gene conversion in response to ionizing radiation, Zimmermann and Unrau (unpublished), Chalk River Nuclear Laboratories of Atomic Energy of Canada Limited, have taken advantage of this situation. The strain used was D7 of Zimmermann et al. (3). Stationary phase cells of cultures selected for a low spontaneous background of convertants at the trp5 gene locus (alleles trp5-12/trp5-27) were incubated in fresh medium for 4-6 hr to get back into the logarithmic phase of growth. After washing and suspending in distilled water, aliquots of the resulting suspension were distributed into a number of plastic vials one of which was used as the control, the others being exposed to various low doses of 185 kVp x-rays delivered at a rate of 20 rad/min. Between 2×10^6 and 2×10^7 cells were plated onto 90 mm diameter Petri dishes with tryptophan-free medium selective for convertants. In a given experiment, the number of plates (between 10 and 20) was the same for the control and all doses. This allowed a direct comparison of the actually counted number of convertant colonies. Five dose-response curves were established in independent experiments. Table 2 shows the results obtained with the doses at the lower end up to the dose which gave an increase in mitotic gene conversion at a significance level of 1%. Such a response could be observed starting with 20 rad and an increase of about 15% over the spontaneous background. These results

Table 2. Induction of mitotic gene conversion in *Saccharomyces cerevisiae*, strain D7, by low doses of x-rays

Expt. no.	X-ray dose (rad)	TRP colonies[a]	Expt.-control[b]	TRP per 10^6 [c]
1	Control	117	---	3.307
	25	156	39 ± 16.5	4.410
	50	180	54 ± 17.0	5.088 (+53.7%)[d]
2	Control	381	---	4.573
	25	492	111 ± 29.7	5.626 (+23.0%)[d]
3	Control	923	---	3.429
	25	1010	87 ± 44.0	3.752
	50	1154	231 ± 45.6	4.286 (+25.0%)[d]
4	Control	1444	---	3.908
	14	1582	138 ± 55.0	4.281
	28	1743	299 ± 56.5	4.717 (+20.7%)[d]
5	Control	760	---	3.349
	10	803	43 ± 39.5	3.491
	20	936	176 ± 41.2	3.842 (+14.7%)[d]

[a] TRP colonies = convertants counted.

[b] Expt.-control = increase in convertant colony numbers over control plus standard error of difference

[c] TRP per 10^6 = convertants per 10^6 colony-forming cells.

[d] Induced conversion frequency in percent of control.

document both the sensitivity and also the accuracy of the yeast mitotic gene conversion assay.

A comparative study on the induction of mitotic crossing-over and mitotic gene conversion has been carried out by Schwaier (8). The results were obtained in stationary phase cells of a diploid strain D48 which was heteroallelic at the gene locus ilv1, alleles 1 and 2, and heterozygous for the linked markers CYH4, ade2, and his8 to detect mitotic crossing-over (the same marker arrangement as in D3). Recombinants were scored as red, histidine-requiring colonies or sectors. The data are shown in Figure 1.

Induction of mitotic chromosome loss was investigated by Parry and Zimmermann (6) by using stationary phase cells of strain D6. The radiation used were γ-rays from a 4000 Ci Hotspot Mark IV at a rate of about 10 krad/min. The spontaneous frequencies in three independent experiments were 1.5 ± 0.55, 1.0 ± 0.45, and 1.7 ± 0.60; with 2 krad the results were 9.7 ± 1.45, 8.3 ± 1.41, and 6.8 ± 1.39; with 5 krad 11.8 ± 1.5, 9.4 ± 1.4, and 25.7 ± 2.3 per 10^5, respectively. Doses of 10, 15, and 20 krad did not lead to higher frequencies of monosomics. The lowest detectable effect could have been induced by about 500 rad.

STUDIES ON INDIVIDUAL AGENTS

Cyclophosphamide

Studies Without Activation. Mayer et al. (9) used cyclophosphamide (Mead Johnson and Co.) in 0.067 M phosphate buffer, pH 7.2, at a constant concentration of 10 mg/ml to treat stationary phase yeast cells.

Mitotic gene conversion was investigated in strain D4. Treatments were carried out at 23 and 30°C for periods of 2, 4, and 6 hr. Survival was 77% after 6 hr at 30°C. However, no induction of mitotic gene conversion was observed.

Mitotic crossing-over was studied in strain D5 and detected on the basis of red/pink twin spotted colonies. The temperature was 30°C and survival after 6 hr was 70%. Results are shown in Table 3.

The induction of red sectors and red whole colonies was followed in strain D3. After 6 hr treatment at 23°C there was a 10-fold increase in red sectors and no killing, at 30°C there was still 69% survival and a 16-fold increase in red sectors. A respiratory-deficient derivative of D3 was also used. Survival was 67% at 23°C and 53% at 30°C. The increase of red sectors and colonies was 15-fold at the lower and 77-fold at the higher temperature.

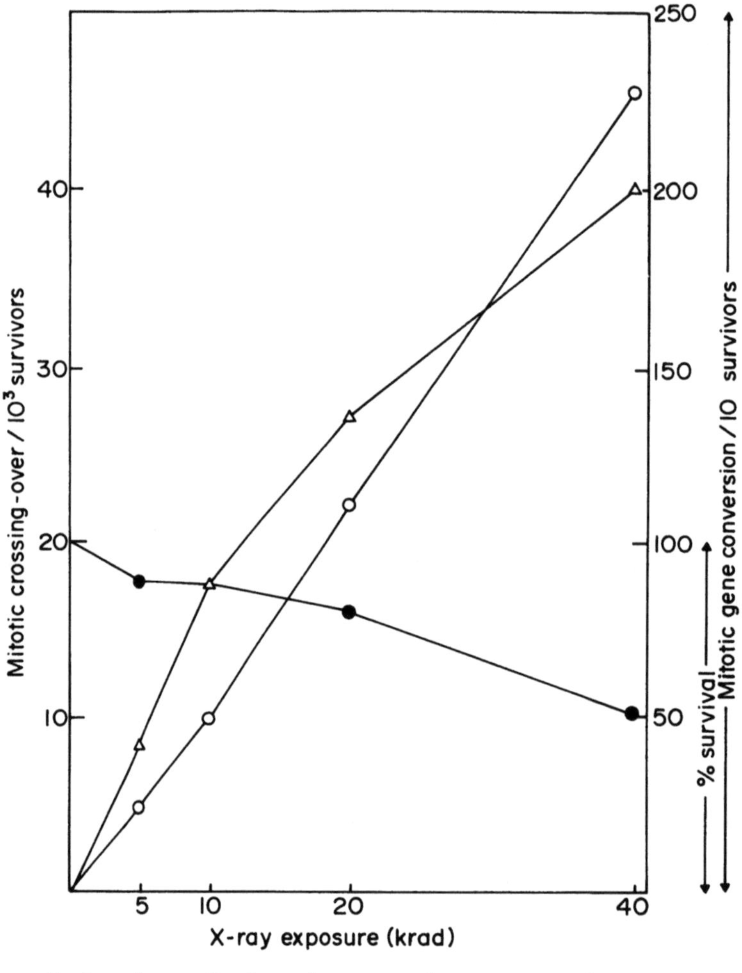

Figure 1. Induction of mitotic crossing-over and gene conversion in strain D48 after x-ray treatment: (●) percent survivors; (o) mitotic gene conversion; (Δ) mitotic crossing-over. Data of Schwaier (8).

It could be demonstrated that incubation of cyclophosphamide in buffer at 30°C resulted in the formation of an active product. Cells of D3 were exposed for a constant period of 30 min to cyclophosphamide solutions of various age. This incubation resulted in an increase of aberrant, red colonies with the age of the drug solution up to 2.5 hr. Pre-incubation beyond that period of time led to no further increase in aberrant colonies.

This showed the formation of an active derivative during about 2.5 hr; thereafter, no further increase seemed to occur. The chemistry of this activation was not explored.

Table 3. Cyclophosphamide and strain D5.

Treatment time (hr)	Twin spots/ survivors (%)	Colonies counted
0	0.005	3
2	0.13	73
4	0.22	111
6	0.30	133

Studies with in Vivo Activation. Siebert (10) used BD IX rats (Druckrey, Freiburg, Germany) as test animals and yeast strain D4-RD II (a respiratory-deficient derivative of D4) as a microbial indicator to study induction of mitotic gene conversion. Cyclophosphamide was given orally at a dose of 500 mg/kg body weight and also a diuretic Lasix to stimulate urine production. Urine was collected during the first 4 hr, mixed with 0.1 M potassium phosphate buffer, pH 7.0, in various ratios to treat yeast cells for 8 hr. After washing, these cells were kept in distilled water for 14 hr, a liquid holding procedure that enhances genetic effects induced in respiratory deficient yeast cells. All incubations were carried out at 25°C. A 10% urine solution showed a significant effect, and a maximum conversion frequency was obtained with 75% urine, the increase was between 100- and 200-fold. A typical host-mediated assay was carried out by injecting yeast cells into the peritoneal cavity for 6 hr, and urine was collected from the same animals. No conversion induction was observed in cells recovered from the peritoneal cavity; however, the conversion-inducing activity of urine was again very high. No significant killing effects were reported under any of these conditions.

Siebert (11) observed a positive response of D4-RD II in the intraperitoneal test only when renal excretion of DB IX rats was blocked by probenicid. Probenicid had no genetic effect. Cyclophosphamide after oral gavage of 170 mg/kg body weight led to a very weak effect which could be enhanced by probenicid. At higher doses, 340 and 500 mg/kg, there was induction of mitotic gene conversion, and again, this effect was increased by the renal blocker. Survival at the highest dose was 57% without probenicid and 56% with the blocker. In spite of a clear demonstration of an active metabolite in the peritoneal cavity, the effects were much weaker than in the urinary assay.

Siebert and Simon (12) performed a urinary assay with human patients treated with therapeutic doses of cyclophosphamide administered intravenously at 3.1 mg/kg body weight. Urine was collected and applied as a 75% solution in 0.1 M potassium buffer, pH 7.0, to cells of strain D4-RDII. Treatment in the urine mixture was performed for 8 hr at 25°C followed by a 12 hr liquid holding regimen in distilled water. The maximum yield of induced mitotic gene conversion was observed in urine collected 12 h after injection of the drug: The induced frequencies were increased 11-fold for the ade2 locus and 18-fold for the trp5 locus. No killing was observed.

Diethylnitrosamine

Diethylnitrosamine (Eastman Organic Chemicals, Rochester, NY) was tested for the induction of mitotic crossing-over by Mayer (13) in strain D3. This chemical is not active per se and induced neither genetic nor lethal effects all by itself. A 0.1 M solution was prepared in the Udenfriend hydroxylation mixture (20 mM ascorbic acid, 2 mM EDTA sodium salt, 1 mM $FeSO_4$ in a 0.06 M phosphate buffer of pH 7.4). Cells from a stationary phase culture were suspended at a density of 5×10^7 cells/ml in this mixture. A control sample was withdrawn at the start of the experiment. Then the suspension was aerated with either oxygen or nitrogen. Samples were withdrawn after 6 hr, diluted, and plated to score for red sectors or whole red colonies. The data are shown in Table 4.

Table 4. Diethylnitrosamine and D3

	Survival (%)	Red sectors per 10^5 survivors
Control	100	12.3
DEN + N_2	109	16.9
DEN + O_2	103	1020.2

Representative samples of red and white sectored colonies were analyzed for the formation of red, histidine-requiring and white, cycloheximide-sensitive sectors. The results clearly indicated the formation of the expected reciprocal products. In addition to that, genetic analysis clearly established the homozygous state of these segregants.

Dimethylnitrosamine

Dimethylnitrosamine (Eastman Organic Chemicals, Rochester, NY) was tested for the induction of mitotic crossing-over by Mayer (13) in strain D3. This chemical is not active per se and induced neither genetic nor lethal effects by itself. A 0.1 M solution was prepared in the Udenfriend hydroxylation mixture (20 mM ascorbic acid, 2 mM EDTA sodium salt, 1 mM $FeSO_4$ in a 0.06 M phosphate buffer of pH 7.4). Cells from a stationary phase culture were suspended at a density of 5×10^7 cells/ml in this mixture. A control sample was withdrawn at the start of the experiment. Then the suspension was aerated with either oxygen or nitrogen. Samples were withdrawn after 6 hr, diluted, and plated to score for red sectors or whole red colonies. The data are shown in Table 5.

Table 5. Dimethylnitrosamine and D3

	Survival (%)	Red sectors per 10^5 survivors
Control	100	12.3
DMN + N_2	105	18.4
DMN + O_2	94	84.4

Representative samples of red and white sectored colonies were analyzed for the formation of red, histidine-requiring and white cycloheximide-sensitive sectors. The results clearly indicated the formation of the expected reciprocal products. In addition to that, genetic analysis clearly established the homozygous state of these segregants.

Brusick and Andrews (14) used dimethylnitrosamine (Eastman Kodak Co., Rochester, NY) to treat strains D3 and D5 (to detect induction of mitotic crossing-over) and D4 (to detect induction of mitotic gene conversion) in combination with a postmitochondrial supernatant activation fraction prepared from the livers of

Table 6. Dimethylnitrosamine and D4, D3, and D5

	D4			D3		D5	
DMN (mM)	Survival (%)	TRP convertants[a] $/10^5$	ADE convertants $/10^5$	Survival (%)	Events	Survival (%)	Events
Control (no chemical)	100	3.9	5.3	100	20.5	100	43.0
0.77	100	9.3	10.5	100	45.2	49	78.0
7.70	98	12.5	15.6	59	45.2	40	66.0
77.0	70	80.6	85.8	56	86.7	49	170.0
770.0	70	9.8	14.0	63	69.9	40	189.0

[a]Mitotic crossing-over (aberrant colonies scored, per 10^5).

random-bred, ICR, Swiss albino male mice (30-35 g body weight). To this preparation was added isocitrate, TPN, $MgCl_2$, Tris buffer, and isocitric dehydrogenase. The cell density was between 1.5 and 1.7×10^7 cells/ml; incubation temperature was 37°C and treatment time 3 hr. Stationary phase cells were used. The suspensions were oxygenated before use, no agitation was applied during treatment. Cells were not washed after treatment. The data are shown in Table 6.

In the experiments of Brusick and Andrews (14), only aberrant colony types were scored. No twin spotted colonies, red and pink doubly sectored colonies, were observed in D5.

Fahrig (15) studied the induction of mitotic gene conversion in strain D4 with dimethylnitrosamine (Schuchardt, Munich) using male Wistar rats (Jautz/Kissleg) (180-200 g body weight) as host organisms. Yeast cells were injected either directly into the testes or alternatively, into the tail vein. A total amount of 10 mmole/kg body weight of dimethylnitrosamine in dimethyl sulfoxide was injected intramuscularly into the back of the animals. Yeast cells were recovered 4 hr after the administration of DMN, washed, and plated. The results are shown in Table 7.

Table 7. Dimethylnitrosamine and D4[a]

	ADE-convertants/10^6 cells	TRP-convertants/10^6 recovered cells
Control testes	10.2 ± 1.9 n=10	8.9 ± 2.7 n=12
Experimental testes	13.1 ± 2.7 n=4	13.1 ± 2.0 n=7
Liver	76.2 ± 24.0 n=12	122.8 ± 45.5 n=12
Lung	42.2 ± 9.0 n=7	45.0 ± 13.9 n=7

[a]n = number of animals

Fahrig (16) also studied the effects of dimethylnitrosamine on strain D4 injected into the peritoneal cavity of random bred C_3H male and female mice and inbred BD rats (Druckrey, Freiburg).

The substance was administered as above, and cells recovered after 4 hr. The results are presented in Table 8.

Table 8. Dimethylnitrosamine[a] and D4

	n[b]	ADE conversion/10^6 recovered cells[b]	TRP conversion/10^6 recovered cells[b]
Control mice	3	4.89 ± 0.94	3.89 ± 0.78
DMN mice	7	10.05 ± 2.65	12.51 ± 2.21
Control rats	5	5.05 ± 0.82	5.10 ± 0.84
DMN rats	6	15.07 ± 2.32	17.11 ± 3.98
DMN rat testes	6	7.30 ± 1.84	7.31 ± 1.43

[a]10 mmole/kg body weight. [b]n = number of animals.

Ethylenimine

Zimmermann and von Laer (17) studied the induction of mitotic crossing-over in strain D1-RD, a respiratory deficient derivative of a strain with the same relevant markers as commonly used D3. Ethylenimine was redistilled and applied to a suspension of stationary phase cells in a 0.1 M potassium phosphate buffer, pH 7.0, at 20°C for 30 min. The treatment was stopped by diluting 1:100 into a 1% solution of $Na_2S_2O_3$. Mitotic crossing-over was scored on the basis of red, histidine-requiring sectors or whole colonies. Representative samples of red and white sectored colonies were also tested for the presence of the cycloheximide-sensitive phenotype in the white sector which should arise along with a red sector in the course of reciprocal mitotic crossing-over. Such types were observed. A 16-fold increase of recombinants was observed with a 10 mM solution, and also a slight effect, though based on a small sample of colonies, with 4 mM, killing being 10% at a 10 mM solution. Concentrations of 20, 40, 80, and 100 mM were also used. Induction reached a plateau at 40 mM with killing between 30 and 37%. The dose response curve below that concentration is probably sigmoidal.

The same batch of ethylenimine was used three years later (after storage at -25°C) to study induction of mitotic gene conversion in stationary phase cells of D4 in a potassium phosphate buffer, pH 7.0, at 25°C for 30 min. Treatment was terminated by diluting 1:10 and two washings. At a 20 mM concentration there was an increase by a factor 11 for conversion at the ade2- and ade-15 for the trp5 gene locus with no concomitant killing. The effect increased with 50 and 100 mM concentration, the dose response curve could be roughly linear. Killing at 100 mM was 20% (5).

Ethyl Methanesulfonate

Induction of both types of mitotic recombination, mitotic crossing-over and mitotic gene conversion was studied in strain D7 by Zimmermann et al. (3), where the same population can be screened for both end points. Gene conversion was monitored by the generation of tryptophan prototrophs arising from gene conversion between the alleles trp5-12 and trp5-27 (also present in D4) and mitotic crossing-over by the formation of colonies with red and pink twin spotted colonies. The two alleles were ade2-40 and ade2-119 (present in strain D5). The dose response curves are shown in Figure 2. The data clearly showed the more sensitive response of mitotic crossing-over compared to that of mitotic gene conversion.

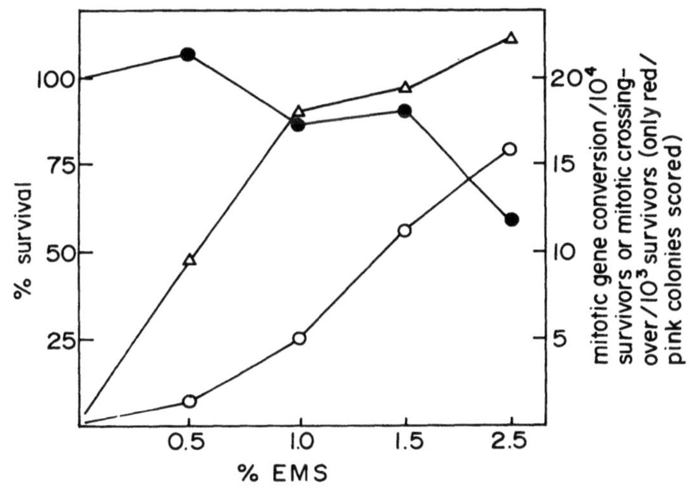

Figure 2. EMS and D7: (•) percent survivors; (o) mitotic gene conversion; (Δ) mitotic crossing-over. Stationary phase cells of D7 were treated with EMS in 0.1 M potassium phosphate buffer, pH 7.0, at 28°C for 4 hr. Treatment was terminated by centrifugation and two washings with distilled water.

A detailed study by Loprieno et al. (18) using mitotic gene conversion in strain D4 in 0.1 M Tris buffer, pH 7.5, at 37°C showed a nonlinear dose response. Concentrations were varied as well as times of treatment. The lowest concentration used was 0.025 M and the shortest time 30 min. This led to an almost 4-fold increase in conversion frequencies at the trp5 and ade2 gene loci. The conclusion was that at higher dose ranges, the increase in gene conversion proceeds at a more than linear rate for the trp5 gene locus whereas a continuous linear relationship obtains with the ade2 gene locus. A microsomal activation fraction was prepared from CF1 or Swiss male mice and supplemented with NADPH and incubated along with EMS at pH 7.5. This led to an about 50% enhancement of the effects of EMS. A host-mediated assay in Swiss mice peritoneal cavities with 10 mM EMS (concentration estimated on the assumption of equal distribution throughout the body) gave no effects. Higher concentrations could not be used because of the toxicity. The same concentration was inactive in vitro with incubations of up to 3 hr.

McGregor (19) studied the effects of rat liver activation preparations on EMS-induced mitotic gene conversion in D4 and found an about 50-60% enhancement in strain D4 stationary phase cells.

The first report on EMS-induced mitotic gene conversion has been provided by Yost et al. (20). These authors used hetero-allelic diploids of the his4 gene locus. Cells were treated in the stationary phase with 3% EMS at 30°C in a buffered solution (pH not indicated) for up to 35 min. The dose response was linear. Cells were washed twice with 6% $Na_2S_2O_3$ in water and once with distilled water.

ICR-170

Brusick and Andrews (14) studied induction of mitotic crossing-over in strains D3 and D5 (Table 9) and mitotic gene conversion in strain D4 (Table 10). Stationary phase cells were used in all cases. ICR-170 was applied as a solution in distilled water. In the case of D3 and D5, treatment was terminated by diluting into cold saline (4°C). The treatment temperature was 30°C and the time 3 hr.

Methyl Methanesulfonate

Snow and Korch (21) studied the induction of mitotic gene conversion in a series of diploids heteroallelic for various alleles of the his4 gene locus in a 0.067 M phosphate buffer, pH 6.8, at room temperature (about 25°C) with 0.5 % MMS (Eastman Kodak Co.). Treatments were terminated by a 10-fold dilution into a 4% solution of $Na_2S_2O_3$. A doubling of the spontaneous frequency of mitotic gene conversion was observed after 5 min incubation. The dose response curves for 5, 10, 15, and 20 min periods of treatment showed a correlation between the induced frequencies and the square of the treatment time. The dose-response curves for all the heteroallelic combinations were standardized as 1 convertant per 1×10^8 surviving cells per min^2 and compared to the effects of x-rays on the basis of 1 convertant per 1×10^8 surviving cells per rad (50 kV, 160 r/sec from a Machlett OEG-60 tube). On the average, a MMS unit induced 10 times more convertants than an x-ray unit.

Table 9. ICR-170 and D3 and D5

ICR-170 concn (μg/ml)	D3 Survival (%)	D3 Events/10^5[a]	D5 Survival (%)	D5 Events/10^4[a]
Control	100	23.1	100	4.4
0.1	100	16.7	100	3.2
1.0	100	18.8	100	7.8
10.0	53	1 362.3	3	250.0
100	0		0	

[a]Events refer to the sum of colonies with pigmented sectors and whole pigmented colonies per survivors.

Fahrig (22) studied the induction of mitotic gene conversion in strain D4 with MMS at five concentrations in the range between 1 and 4 mM in a 0.1 M potassium phosphate buffer, pH 7.0, at 25°C for a uniform period of 210 min. The dose response was linear when the logarithm of the conversion frequencies was plotted against the logarithm of the concentration. A slight effect was observed at the lowest concentration.

N-Methyl-N'-nitro-N-nitrosoguanidine

Induction of mitotic crossing-over was studied by Zimmerman et al. (1) in strain D1, which has the same relevant marker arrangement as the more commonly used strain D3. Stationary phase cells were incubated at 20°C for 30 min in 0.1 mM solution of MNNG in a 0.067 M potassium phosphate buffer, pH 7.0. Treatment was terminated by a 1:100 dilution into a 1% solution of $Na_2S_2O_3$.

Mitotic crossing-over was assessed by the appearance of red and histidine-requiring sectors or whole colonies. True homozygosis was also verified by tetrad analysis of one segregant. The control frequency was 0.063×10^{-3}. In the experiment the frequency was 9.23×10^{-3}. In another experiment with 0.2 mM MNNG, the frequency was 15.8×10^{-3} with 20% killing.

Table 10. ICR-170 and D4[a]

ICR-170 concn (μg/ml)	Survival (%)	ADE convertants/10^5 survivors	TRP convertants/10^5 survivors
Control	100	1.0	1.9
0.1	79	1.8	2.4
1.0	83	3.0	5.8
10.0	73	35.8	43.9
100.0	8	61.3	91.8

[a]Platings for convertants were done directly from the treatment solution, 0.1 ml on a plate with about 25 ml of medium.

Induction of mitotic gene conversion was studied by Zimmerman and Schwaier (4) with stationary phase cells of strain D4. Treatment conditions were 0.067 M potassium phosphate buffer pH 7.0 at 20°C for 30 min. Treatment was terminated by diluting into a 1% solution of $Na_2S_2O_3$ and two washings. The results are shown in Table 11.

The lower dose response for TRP conversion could not be extrapolated to the origin, whereas for ADE conversion, there was a linear relation. There were too few concentrations tested to permit any firm conclusions to be drawn as to the definit dose-response curves.

Mitomycin

Holliday (23) studied the induction of mitotic gene conversion at the trp5 gene locus with alleles trp5-1 and trp5-2 as well as the expression of a recessive resistance to cycloheximide, gene cyh2, from an originally heterozygous condition. Appearance of colonies on a medium with 2 ppm cycloheximide was taken as an indication for the occurrence of mitotic crossing-over. Cells were incubated at 30°C in a synthetic complete medium supplemented with 0.4 mg mitomycin C/ml. No killing was observed during 580 min of incubation. Cells were plated to detect mitotic gene conversion. The frequency increased from 6 × 10^{-5} cells/ml in the control linearly for about 2 hr and remained constant after that. Mitotic crossing-over was monitored by the appearance of cycloheximide resistant colonies. With direct plating on cycloheximide medium, there was only a slight effect: 3.7 × 10^{-4} resistants in the control versus 9.0 × 10^{-4} after 580 min. However, when cells were removed after 4 hr and incubated in medium without cycloheximide for 350 min, the frequency of resistant cells was increased to 21.0 × 10^{-4}, a 5.7-fold increase.

Table 11. N-Methyl-N'-nitro-N-nitrosoguanidine and D4

MNNG concn (μg/ml)	Survival (%)	ADE convertants/10^5 survivors	TRP convertants/10^5 survivors
Control	100	0.92	0.87
0.1	92	29.4	20.4
0.2	80	61.8	56.4
0.5	64	123.0	226.0
1.0	14.8	172.0	344.0

TEM

Fahrig (22) studied the induction of mitotic gene conversion in stationary phase cells of strain D4 in a 0.1 M potassium phosphate buffer, pH 7.0, at 25°C for a uniform period of 210 min. The cell concentration in the treatment solution was 5 × 10^7 cells/ml. Treatment was terminated by diluting 1:10 and plating of 0.1 ml samples onto medium in plates (about 25 ml per plate). Eight different concentrations were used, the lowest being 5 × 10^{-6}, the highest 1 × 10^{-3} M. The logarithm of the conversion frequency per surviving cells was plotted against the logarithm of the concentration of TEM. In this plot, the dose-response relationship was linear. The conversion frequency at the lowest concentration was slightly above the spontaneous background, at the highest concentration it was about 100 times the spontaneous level. No data on cell killing were presented.

Trenimon

Induction of mitotic gene conversion in strain D4 was studied by Zimmermann (5) by using stationary phase cells in a 0.1 M potassium phosphate buffer, pH 7.0, at 25°C for a period of 24 hr. Treatment was terminated by diluting 1:10 and two washings. The data are shown in Table 12.

Table 12. Trenimon and D4

Trenimon concn. (M)	Survival (%)	ADE convertants/10^5 survivors	TRP convertants/10^5 survivors
Control	100	0.63	0.68
4.32×10^{-9}	111	1.82	2.12
8.65×10^{-9}	117	3.92	3.74
2.16×10^{-7}	89	61.4	77.1

A detailed dose-response curve was obtained by Zimmermann (24), for D4 by using the same protocol except for a shorter time of treatment, which was 12 hr. The dose response curve was linear in the range from 1 to 50 × 10^{-9} g/ml for the absolute number of convertants and also for the conversion frequency per 10^5 survivors. There was up to 10% killing to be observed. The line extrapolated directly to the control values.

Vinyl Chloride

Loprieno et al. (25) studied the induction of mitotic gene conversion in stationary phase cells of strain D4 at pH 7.5 at 37°C with and without an activating fraction prepared from Swiss albino mice livers. The incubation mixture was enriched with vinyl chloride by bubbling a mixture of 50% vinyl chloride and air through the reaction mixture. This resulted in a final concentration of 48 mM. The effects are presented in Table 13.

Table 13. Vinyl chloride and strain D4

	Survival (%)	ADE convertants/10^5 survivors	TRP convertants/10^5 survivors
Control	100	0.49	0.89
180 min + microsomes	100	8.06	4.74
300 min + microsomes	100	8.47	4.36
240 min, no microsomes	100	0.71	1.51
360 min, no microsomes	100	1.05	1.25

REFERENCES

1. Zimmermann, F. K., Schwaier, F., and von Laer, U. Z. Verebungsl. 98: 230-246 (1966).

2. Zimmermann, F. K. Mutat. Res. 21; 263-269 (1973).

3. Zimmermann, F. K., Kern, R., and Rasenberger, H. Mutat. Res. 28: 381-388 (1975).

4. Zimmermann, F. K., and Schwaier, R. Mol. Gen. Genet. 100: 63-76 (1967).

5. Zimmermann, F. K. Mutat. Res. 11: 327-337 (1971).

6. Parry, J. M., and Zimmermann, F. K. Mutat. Res. 36: 49-66 (1976).

7. Manney, T. R., and Mortimer, R. K. Science 143: 581-582 (1964).

8. Schwaier, R. Mol. Gen. Genet. 101: 203-211 (1968).

9. Mayer, V. W., Hybner, C. J., and Brusick, D. J. Mutat. Res. 37: 201-212 (1976).

10. Siebert, D. Mutat. Res. 17: 307-314 (1973).

11. Siebert, D. Mutat. Res. 28: 57-61 (1975).

12. Siebert, D., and Simon, U. Mutat. Res. 21: 275-262 (1973).

13. Mayer, V. W. Genetics 74: 433-442 (1973).

14. Brusick, D., and Andrews, H. Mutat. Res. 26: 491-500 (1974).

15. Fahrig, R. Mutat. Res. 31: 381-394 (1975).

16. Fahrig, R. Mutat. Res. 26: 29-36 (1974).

17. Zimmermann, F. K., and von Laer, U. Mutat. Res. 4: 377-379 (1967).

18. Loprieno, N., Barale, R., Bauer, C., Baroncelli, S., Bronzetti, G., Cammellini, A., Cinci, A., Corsi, G., Leporini, C., Neri, R., Nozzolini, M., and Serra, C. Mutat. Res. 25: 197-217 (1974).

19. McGregor, D. B. Mutat. Res. 23: 395-398 (1974).

20. Yost, H. T., Jr., Chaleff, R. S., and Finerty, J. P. Nature 215: 660-661 (1967).

21. Snow, R., and Korch, C. T. Mol. Gen. Genet. 107: 201-208 (1970).

22. Fahrig, R. Mol. Gen. Genet. 144: 131-140 (1976).

23. Holliday, R. Genetics 50: 323-335 (1964).

24. Zimmermann, F. K. In: Chemical Mutagens: Principles and Methods for their Detection, A. Hollaender, Ed., Vol. 3, Plenum Press, New York, 1973, pp. 209-239.

25. Loprieno, N., Barale, R., Baroncelli, S., Bauer, C., Bronzetti, G., Cammellini, A., Cercignani, G., Corsi, C., Gervasi, G., Leporini, C., Nieri, R., Rossi, A. M., Stretti, G., and Turchi, G. Mutat. Res. 40: 85-96 (1976).

CHAPTER 9

MUTAGENICITY OF SELECTED CHEMICALS IN DROSOPHILA

E. Vogel and A. Schalet

Department of Radiation Genetics and Chemical Mutagenesis, State University of Leiden, Leiden, The Netherlands

W. R. Lee

Department of Zoology and Physiology, Louisiana State University, Baton Rouge, Louisiana, U.S.A.

F. Würgler

Institute für Toxikologie der ETH und der Universität Zürich, Schwerzenbach bei Zürich, Switzerland

INTRODUCTION

This review of mutagenicity data involving the 20 reference mutagens in Drosophila is restricted to data that were considered relevant for the purpose of the comparative evaluation. Although an effort was made to include information from as many sources as possible (approximately 500 papers were checked and representative material from about 140 papers was selected), the reports are not intended to review all studies reported in the literature. Each of the 20 compounds is reviewed separately in a small report containing three sections (treatment conditions; biological effects; evaluation). References are given in a single bibliography at the end. In the following section we will consider briefly the genetic endpoints which were preferentially used for assaying the chemicals under consideration.

Types of Genetic Damage Evaluated

Recessive Lethal Mutations. It will be realized that the bulk of mutation work with Drosophila has been carried out by means of tests for recessive lethals (mostly with treated males because the use of females for treatment is far less convenient), because this multilocus test is considered the most reliable and sensitive procedure available at present to score for heritable mutations in Drosophila. The reasons for the preferential use of this assay are as relevant today as they were 50 years ago. Recessive lethal mutations represent a heterogeneous class of genetic damage, consisting of gene mutations, deletions and other structural rearrangements, and the proportion with which the various types occur depends strongly on the mutagen tested. Induction of recessive lethal mutations in Drosophila can be tested for on any chromosome, but the detection of such effects induced on the X-chromosome needs only two generations whereas autosomal recessive lethal mutations require 3 generations for detection. The euchromatic X-chromosome corresponds to 20% of the Drosophila genome, about 1000 bands in salivary gland giant chromosomes, and it can be estimated that 600-900 loci can mutate to produce a recessive lethal. The low and very constant spontaneous mutation frequency (ca. 0.2% recessive lethals/X-chromosome/generation) in strains used for this test, the simple protocol of the test which is free from personal bias, and the fact that gene mutations as well as damage resulting from rearrangements are registered by this method, make the sex-linked recessive lethal assay the most sensitive technique available to detect mutagenicity in Drosophila. Quite a number of mutagens exist which are only active in the recessive lethal test but do not induce dominant lethals or chromosome aberrations (1,2). The high resolving power of the recessive lethal test is also evident from the fact that 19 of the 20 reference mutagens were found to be active.

Dominant Lethals. It seems generally accepted that chromosome breakage events of various types are the predominant type of event which results in dominant lethality. The widely used method for measuring dominant lethality in Drosophila consists of treating males and mating them to untreated females. The percentage of larvae hatching from eggs or adults emerging from the puparium is determined in the treated (ht) and a control (hc) series. The relative dominant lethal frequency (RDL) is then calculated from the simple equation, % RDL = $[100 - ht/hc] \times 100$, where ht is percent hatchability for the treated series and hc denotes percent hatchability for the control series.

Dominant lethal tests are quick and easy to perform in Drosophila, but their meaning is very doubtful for three reasons. The first problem is that nongenetic damage, even after male treatment, may contribute to what would be scored as "dominant lethals" as a consequence of inadequate insemination, inactivation or malfunction of sperm, or blocking of spermatogenesis (3). These problems could partly be overcome by analyzing unhatched eggs cytologically according to the method developed by Würgler and Ulrich (4). The second problem is that there is considerable variation between control experiments, the rates of unhatched eggs obtained from untreated flies usually varying between 3% and 10%. The third and major drawback of the dominant lethal test is its basic insensitivity, which makes it of only limited practical value in the detection of genetic damage caused by chemicals in Drosophila. This has become evident from our comparative work in Drosophila with a wide range of reference mutagens of different types and structure. Furthermore, it will be noted from Table 2 that 3 out of 11 compounds assayed in the dominant lethal test were inactive, although all 11 are clearly active in the recessive lethal test.

Translocations. The induction of a translocation leads to the production of gametes with duplications and deficiencies for parts of the chromosomes which result in disturbance of genetic balance which are usually lethal to the zygote. This phenomenon is used as a test criterion when testing for translocations in Drosophila, e.g., for translocations between the two large autosomes, the chromosomes 2 and 3. Such a 2;3 translocation is recognized by absence of two out of four expected phenotypes in the second generation after chemical treatment. Presumptive translocations can be verified by genetic retesting. Since chemically induced breaks mature only slowly into open breaks (maturation of breaks can take up to 3 or 4 weeks) tests on translocations in Drosophila are meaningful only when storage experiments are performed. Results with EMS and MMS strongly support the view that storage is an absolute essential for detecting translocations by monofunctional alkylating agents. Thus, the conditions of tests (stored or unstored) will be mentioned when reporting on such data.

Spontaneous translocations (5) occur with a frequency of about 0.04%, which is why control tests are usually unnecessary and why this method is considered a reliable and sensitive tool for detecting rejoinable chromosome breakage.

Entire and Partial Chromosome Loss. Methods for detecting these effects are F_1-generation tests, and they are taken as a measure of the frequency of chromosome breakage.

The common practice is to score for losses of marked ring-X (spontaneous frequency roughly 0.5-1.5%) or rod-X and Y-chromosomes (combined spontaneous frequency 0.03-0.10%) one generation after chemical treatment. (Complete X- and Y-losses are usually indistinguishable from each other.) Partial chromosome loss (sponta-

Table 1. Results with 20 reference mutagens in *Drosophila melanogaster*. Comparison of LEC on basis of X-linked recessive lethal mutations

Rank[a]	Compound	Exptl. LEC or LAEC[b] (mmole/l.)		Exptl. LEC or LAEC[b] (μl/g)	Conditions for LEC only	LEC (rough estimate) (mmole/l.)	Concentration range with mutagenic effects: Feeding (mmole/l.)	Concentration range with mutagenic effects: Injection (mmole/l.)	Activation required/ functionality
1	Trenimon	LAEC	0.01	0.65		< 0.005	0.01-0.02	0.02	No; trifunctional
1	TEM	LAEC	0.03	2.6		< 0.005		0.03-0.34	No; trifunctional
1	TEPA	LAEC	0.10			< 0.005		0.1-0.3	No; trifunctional
1	ThioTEPA	LAEC	0.012			< 0.005	0.01-0.12		No; trifunctional
2	DMN	LAEC	0.068			< 0.01	0.07-10.0	1.0-10.0	Yes; monofunctional
2	Aflatoxin B1	LAEC	0.064	5.6		< 0.01	0.064		Yes; monofunctional

3	MMS	LEC ∿ 0.05		Feeding, 2d		0.05-6.8	2.5-7.5	No; mono-functional
3	Mito-mycin C	LAEC 0.3			< 0.01	0.3-1.8	0.3	Yes; mono- & bifunctional
3	EMS	LEC ∿ 0.1		Feeding, 24 hr		0.1-25.0	1.0-50.0	No; mono-functional
3	Myleran	LEC ∿ 0.1		Feeding, 3 da		0.09-0.2		No; bifunc-tional
3	DEN	LEC ∿ 0.2		Feeding, 2 da		0.02-10.0	7.0-30.0	Yes; mono-functional
3	ICR-170	LAEC 2.5	285		< 0.25		2.5	No; mono-functional
3	Procarba-zine	LEC ∿ 0.5		Feeding, 2 da		0.5-30.0		Yes; mono-functional
3	MNNG	LAEC 3.4			< 0.5	3.4-6.8	2.5-5.0	No; mono-functional

(continued)

Table 1. (continued)

Rank[a]	Compound	Exptl. LEC or LAEC[b] (mmole/l.)	Exptl. LEC or LAEC[b] (µl/g)	Conditions for LEC only	LEC (rough estimate) (mmole/l.)	Concentration range with mutagenic effects: Feeding (mmole/l.)	Concentration range with mutagenic effects: Injection (mmole/l.)	Activation required/ functionality
4	Ethylenimine	LAEC 5.0			< 1.0		5.0-20.0	No; monofunctional
4	MeTEPA	LAEC 16.7	285		< 1.0		16.7	No; trifunctional
4	Cyclophosphamide	LEC ∿ 1.0		Feeding, 24 hr		1.0-20.0	19.2-38.2	Yes; mono- & bifunctional
4	Epichlorohydrin	LEC ∿ 2.5		Injection			2.5-25.0	No; monofunctional
4	Vinyl chloride	LEC ∿13.6 (850 ppm)		Inhalation, 2 da		850-50,000 ppm		Yes; monofunctional
	Cadmium	Nonmutagenic (preliminary information)						

[a]Ranking: Group 1, LEC < 0.005 mM; 2, LEC 0.005-0.05 mM; 3, LEC 0.05-0.5 mM; 4, LEC < 0.5 mM.

[b]LEC = lowest effective concentration; LAEC = lowest adverse effective concentration; mmole/l. refers to mutagen concentration as mmole per liter; µg/g refers to mutagen concentration as µg per g Drosophila (adult males).

neous frequency roughly 0.01-0.03%), also referred to as Y-chromosome deletions or Y-marker losses, are most effectively carried out with a doubly marked Y, carrying suitable markers at the end or tip of each arm of the chromosomes (YL, long arm; YS, short arm of the Y-chromosome). Some variations in spontaneous frequencies, this being particularly true when ring-X loss is assayed, makes concurrent controls a stringent necessity.

Nondisjunction. Incidence of nondisjunction of sex-chromosomes was the criterion used in the very few attempts which have been made to detect such effects. Information on nondisjunction by the 20 reference mutagens was scanty and was found only for DEN, MMS, and MNNG. Although spontaneous nondisjunction is easily detectable in Drosophila and is inducible by x-rays and by changes in temperature, the effects obtained with chemical mutagens, as will be seen in this review, are negligible or even contradictory when studies by several investigators were compared.

For more detailed descriptions and discussions of the various Drosophila techniques the reader is referred to the review by Abrahamson and Lewis (6), Auerbach (3), Muller and Oster (7), Vogel and Sobels (8), and Würgler et al. (9).

Metabolic Activation Systems in Drosophila

The fact that some 50 precarcinogens (10) show mutagenic activity in Drosophila provides clear-cut evidence of the presence of activating enzymes in this system. Very recently, Baars et al. (11) have isolated cytochrome P-450 from Drosophila microsomes and have demonstrated considerable AHH activity (aryl hydrocarbon hydroxylase) in such preparations.

Further Information on the Effects of the Reference Mutagens

Besides the reporting on mutagenic effects in one of the tests listed in the above sections, regular mention was made, whenever information was available, of data about the genetic nature of recessive lethals, the ratio of mosaic to complete changes in various stages of germ cells, and the effect of storage on the proportion of recessive lethal mutations vs. translocations. Particular attention was paid to the relation of concentration and the types of genetic damage as a function of exposure. The term "dose" was studiously avoided in all cases except with EMS for which an exact dosimetry has been developed.

The results with the 20 reference mutagens were summarized in two tables, listing the experimental LEC values (lowest effective concentration) or, if not available, the LAEC (lowest adverse effect level). However, in certain cases, for example

Table 2. Results with 20 reference mutagens in *Drosophila melanogaster*. Induction of dominant lethals (DL), translocations (TR), partial chromosome loss (PCL), chromosome loss (CL), and nondisjunction (ND)

Compound	Genetic effects observed[a]					Concentration range with positive effects (mM)		Rank (on basis of rec. lethals)	Activation required
	DL	TR	PCL	CL	ND	Feeding	Injection		
Trenimon	+	+	+	+		0.01-0.02	0.02	1	No
TEM	+	+	+	+			0.03-0.34	1	No
TEPA	+	+				0.1-0.3		1	No
ThioTEPA								1	
DMN				+		10.0		2	Yes
Aflatoxin B1								2	
MMS	+	+	+	+	0	0.5-2.0	2.5-7.5	2	No
Mitomycin C		+			0	0.3-1.8		3	Yes
EMS	+	+	+	+		10.0-25.0	10.0-50.0	3	No
Myleran								3	
DEN	0	0		0	?			3	Yes

ICR-170		+	+	+				3	No
Procarbazine	+	0		+		20.0-30.0		3	Yes
MNNG	+	+	+	+	+	Suspensions were used		3	No
Ethylenimine	+	+	+	+			5.0-20.0	4	No
MeTEPA								4	
Cyclophosphamide	0			0				4	Yes
Epichlorohydrin								4	
Vinyl chloride	0	0		0				4	Yes
Cadmium									

[a] +, mutagenic; 0, nonmutagenic.

with ICR-170, mutational yield at the lowest concentration tested was far (50- to 100-fold) above spontaneous frequency, and such a concentration as a LAEC was not considered to provide any meaningful information. Thus an attempt was made to estimate the LED on basis of experience with other mutagens and by extrapolating from data obtained with structurally related compounds.

Finally, the 20 reference mutagens were arranged to their effectiveness in the sex-linked recessive lethal test. The chemicals are grouped on basis of their LEC, forming four groups. This grouping was performed in order to compensate for minor errors in the calculation of the LEC.

DATA ON INDIVIDUAL CHEMICALS

Aflatoxin B1 (CAS 1162-65-8)

Aflatoxin mutagenesis has not been studied extensively in Drosophila (Table 3). There exists only one study demonstrating incidence of IInd chromosome recessive lethal mutations after adult feeding (12,13).

Table 3. Genetic damage by aflatoxin B1 in Drosophila

Type of damage	Sperm	Spermatids	Spermatocytes	Spermatogonia	Reference
Recessive lethals (IInd chromosomes)	+	+	+	+	12,13
Dominant lethals	Nt[a]	Nt	Nt	Nt	
Translocations	Nt	Nt	Nt	Nt	
Chromosome losses	Nt	Nt	Nt	Nt	
Nondisjunction	Nt	Nt	Nt	Nt	

[a]Nt = no test.

Treatment Conditions. Males from an Oregon-R stock of *D. melanogaster* were weighed, allowed to feed on a drop of aflatoxin-DMSO solution (0.5% DMSO and 0.02 mg/ml aflatoxin) and then reweighed. The average aflatoxin uptake was 4×10^{-6} mg per fly (about 5.6 μg/g). This dose was not toxic to the flies.

Biological Effects. The induction of IInd chromosome recessive lethals was determined in successive stages of spermatogenesis, using a female-male ratio of 2 Cy/BilL ♀♀ : 1 Oregon-R ♂, and a breeding interval of 3 days (Table 4). It will be seen that aflatoxin B1 is effective in the production of IInd chromosome recessive lethal mutations in Drosophila. In the treated males, mutation frequency was lowest in sperm (brood I) and spermatogonia (brood VI), highest in the spermatocytes (or early spermatids?). It is interesting to note that the mutational pattern of aflatoxin B1 resembles that obtained with other precarcinogens in Drosophila, e.g., low mutagenic response of the metabolically inert sperm in which activating enzymes are not present.

Evaluation. There can be no doubt that aflatoxin B1 is a mutagen in Drosophila, and from this point of view the analysis is complete. However, the relatively rough brood pattern used by Lamb and Lilly (12) does not permit clear identification of which germ cell stage is most sensitive to aflatoxin. Since aflatoxins represent an interesting class of naturally occurring precarcinogens, a more detailed analysis of their mutagenic properties in Drosophila is recommended. Such an analysis should include data on the LEC for several genetic end points and the LD_{50}.

Cadmium (CAS 7440-43-9)

Information on mutagenic effects of cadmium in Drosophila is very scanty. Preliminary results by Sorsa and Pfeifer (14) did not reveal any mutagenic activity in the recessive lethal test after exposing males to medium which contained $CdCl_2$, 50.0 mg/l. (0.27 mM). No further details were given.

Cyclophosphamide (CAS 50-18-0)

Cyclophosphamide produces recessive lethal mutations in sperm, spermatids and spermatocytes but not in spermatogonia. Cyclophosphamide mutagenesis in Drosophila shows a saturation effect, that is, the mutational yield becomes concentration-independent at concentrations which do not cause mutation rates higher than 3-5%. Dominant lethals and transmissible chromosome aberrations cannot be detected with this mutagen in Drosophila (Table 5).

Treatment Procedure. The LD_{50} of cyclophosphamide is roughly 15 mM after 3-day exposure of adult males (17). There are three studies (2,15,17) which demonstrate mutagenic effects with cyclophosphamide in Drosophila. Although the experimental conditions used in those studies varied considerably, the recessive lethal yield in no single case (brood) exceeded 5.3% (Table 6), mostly ranging between limits of 0.7 and 3.5%. Such a consistency in results was entirely unexpected, since the cyclophosphamide concentrations applied ranged from 1.0 to 38.3 mM. Vogel (2)

Table 4. IInd chromosome recessive lethals found after feeding Drosophila males with aflatoxin B1[a]

Brood	Time after treatment (days)	Fed 0.5% DMSO		Fed aflatoxin in 0.5% DMSO		P^b
		Tests	% lethals	Tests	% lethals	
I	1-3	596	0.34 ± 0.24	597	0.67 ± 0.33	0.69
II	4-6	592	0.17 ± 0.17	567	0.53 ± 0.30	0.59
III	7-9	604	0	601	3.66 ± 0.77	0.0001
IV	10-12	578	0.69 ± 0.34	572	2.10 ± 0.60	0.071
V	13-15	560	0.54 ± 0.31	558	1.97 ± 0.59	0.055
VI	16-18	518	0.19 ± 0.19	530	1.13 ± 0.46	0.13

[a]Data of Lamb and Lilly (12).

[b]P is the probability obtained in a two-tailed Fisher's exact test.

demonstrated a lack of a relationship between concentration (1.0-10.0 mM), exposure time and mutation yield, in that above 1.0 mM cyclophosphamide the mutation frequency became independent of exposure. Such "saturation effects" are also characteristic for certain triazenes, trofosfamide (2), ifosfamide (2), and vinyl chloride (18,19), all of which need enzymatically mediated activation to manifest mutagenicity, but have not yet been observed with directly acting agents. Hence, failure to observe concentration-dependence of recessive lethal frequencies can best be explained as resulting from an enzyme saturation in the bioactivation of cyclophosphamide.

Table 5. Incidence of genetic damage by cyclophosphamide in Drosophila

Type of damage	Sperm	Spermatids	Spermatocytes	Spermatogonia	References
Recessive lethals	+	+	+	0	2,15-17
Dominant lethals	0	Nt[a]	Nt	Nt	2
Chromosome losses	0	Nt	Nt	Nt	2
Translocations	Nt	Nt	Nt	Nt	
Nondisjunction	Nt	Nt	Nt	Nt	

[a]Nt = no test.

There seems also general agreement that mutation frequencies are highest in spermatids and spermatocytes (2,17) (these cell stages possess a highly developed endoplasmic reticulum which is generally considered the location of the activating enzymes) and lowest in mature sperm (Table 7). On the other hand, there is no evidence suggesting that cyclophosphamide is capable of causing mutations in Drosophila spermatogonia (16,17). Furthermore, the agent has been consistently negative when used to treat mature sperm for incidence of dominant lethals, entire loss of the Y- (or X-) chromosome, and partial Y-chromosome loss (2). The reason for this discrepancy can be explained on the basis of quantitative differences, in that in Drosophila lower concentrations are required to induce point mutations, as compared to the substantially higher levels needed for the incidence of chromosome breakage (20).

Table 6. Incidence of genetic damage in Drosophila males by cyclophosphamide

Route of application	Treatment (mM)/time	Type of damage	Activity %	Activity Stage[a]	Peak activity	References
Adult injection	19.2-38.3 in 0.75% NaCl	Recessive lethals	1.6-5.3	Not specified	-	15
Larvae feeding	11.0,13.0,14.0[b]	Recessive lethals	0	sg		16
Adult feeding	10.0/3 da	Recessive lethals	0.7-2.4	sp,st,sc	st	
	20.0/3 da	" "	0.9-4.9	sp,st,sc	st	17
Adult feeding	1.0-10.0/24 hr	Recessive lethals	∿ 0.7	sp	-	
	3.0/3 da	" "	0.7-3.6	sp,st,sc	st	2
	5.0/3 da	" "	0.5-1.9	sp,st,sc	st	
	1.0-20.0/24 hr	Dominant lethals	0	sp		
	10.0/24 hr	Chromosome loss	0	sp		

[a] sp, sperm; st, spermatids; sc, spermatocytes; sg, spermatogonia.

[b] Exposure time not specified.

Table 7. Concentration-independent induction of recessive lethals by cyclophosphamide in Drosophila males

Treatment (3 days) (mM)	X-Chromosomes	Lethals	Reference
3	1957	1.5 ± 0.3	2
5	2219	0.90 ± 0.2	2
10	1964	1.1 ± 0.2	17
20	1543	1.8 ± 0.3	17

Evaluation. The most striking characteristics of cyclophosphamide mutagenesis in Drosophila is the lack of a relation between exposure (concentration) and recessive lethal induction. The LEC for point mutations is about 1.0 mM for mature sperm and may be somewhat lower for spermatids and spermatocytes. Since in the target cells (mature sperm) the concentration was apparently below the level needed for the production of breakage-associated damage, chromosome aberrations and dominant lethals could not be detected. These results seem the consequence of an enzyme saturation. No further analysis is indicated with this agent.

N-Nitrosodiethylamine, DEN (CAS 55-18-5)

DEN is a mutagen with pronounced specific actions in Drosophila. It causes complete and mosaic point mutations in all stages of the spermatogenesis cycle (Table 8). When fed to adult males, DEN produces mutations at concentrations ranging from 2×10^{-1} mM to 10 mM. A lower LEC of 5×10^{-3} mM is obtained if larvae prior to their treatment as adults were additionally exposed to DEN. The agent is less mutagenic when administered by microinjection into the haemocoel of adult males. The ability of DEN to produce up to 25% recessive lethals stands in sharp contrast to the lack of breakage-type damage (dominant lethals, 2;3 translocation and chromosome loss) below toxic (LD_{60}-LD_{90}) concentrations. This finding makes DEN an extremely interesting model compound for elucidating the underlying mechanisms of certain cases of extreme mutagen specificity.

Treatment Conditions. The procedures adopted for DEN were to either inject adult males or feed them for 1-3 days, by use of DEN concentrations which varied between 5×10^{-2} and 10 mM in the feeding experiments and 7.0 to 30 mM in the injection studies. Additionally, one study on larvae was carried out at DEN concentrations which ranged from 5×10^{-3} mM to 1.0 mM.

Table 8. Genetic damage by DEN in Drosophila

Type of damage	Sperm	Spermatids	Spermatocytes	Spermatogonia	References
Recessive lethals (complete and mos.)	+	+	+	+	1,21-24
Dominant lethals	0	Nt[a]	Nt	Nt	1
2;3-translocations	0	0,(+)[b]	0	0	1,21,23
Chromosome loss					
Y, rod-X	0	0	0	0	1,21
Y, ring-X	0	0	Nt	Nt	25
Partial loss (Y^L,Y^S)	0	0	?[c]	?[c]	21
Nondisjunction	-	-	?[c]	?[c]	21

[a]Nt = no test.

[b]Low incidence of translocations at LD_{60}-LD_{90}.

[c]See the objections against experimental design in text.

Biological Effects. After adult feeding of males with 5.0 mM DEN for 3 days, mortality rate is about 50% (LD_{50}). Mortality rate increases to 90% at 10 mM DEN. Intensive studies have been conducted to establish the relation between exposure (time x concentration) and mutagenic response (Table 9). It will be seen from Table 9 that DEN is apparently more active when fed to adult males, as compared with adult injection. The spermatid seems to be the cell stage in which mutation induction is highest, but spermatocytes are most susceptible to killing by DEN. When adult males are exposed to various DEN concentrations at or near the LEC level for 2 days, genetic activity is not observed at concentrations below 2×10^{-1} mM (which represents the LEC under this experimental condition). A lower LEC of roughly 5×10^{-3} mM is determined, however, when male larvae (prior to adult treatment) are also exposed to DEN. In this combination treatment, a DEN concentration of

5×10^{-3} mM which gives decidedly negative results with adult feeding, is clearly mutagenic. This shows the presence already in larvae of the enzymatic machinery needed to activate precarcinogens.

Although large-scale experiments have been conducted in an attempt to produce chromosome aberrations and dominant lethals, at high and low concentration levels of DEN, there is yet no convincing evidence indicating this type of damage by DEN in Drosophila (Table 10). Tests on dominant lethals, 2;3-translocations, entire and partial chromosome loss proved decidedly negative at or below the LD_{50} (5.0 mM DEN, 3 days exposure). One 2L;3L translocation was found among 1643 tested gametes at the LD_{60} (7.5 mM DEN), and one 2R;3L translocation (in 629 gametes) at the LD_{90} (10 mM) (25). Since the spontaneous frequency of 2;3 translocations is about 0.4%, this is no convincing evidence that DEN is capable of producing translocations in Drosophila. Fahmy et al. (21) reported elimination of $\underline{y}^+$ and $\underline{B}$ markers of a doubly marked Y-chromosome (sc^8YB^S) and a low incidence of nondisjunction in premeiotic stages. These authors concluded that the majority, if not all, of the marker losses from the Y chromosome with DEN were genuine deletions. However, the method adopted for the simultaneous assay of nondisjunction and loss of the sex-chromosomes, was based on the utilization of marker chromosomes of the constitution: In(1) $sc^8y^{31d}w^a$ in females, and sc^8YB^S in males. This experimental design does not allow the distinction between X-Y^S exchange, deletions (loss of B^S), and exchange between the arms of the Y-chromosome at the four-strand stage, all events giving rise to males with the phenotype w^a. The same objection holds true for deletions of $\underline{y}^{31d}$ (w^a B^S ♂♂), since loss of y^+ could also be caused by X-Y^L exchange and again an exchange between the two arms of the Y-chromosome. Finally, B^S - ♀♀, which are thought to indicate nondisjunction in males (21), could also arise from the reciprocal product of X-Y^S exchange. On the other hand, incidence of crossovers and translocations, at concentrations which were as high as those used by Fahmy et al. (21), demonstrates that DEN is capable of producing breaks at extraordinarily high concentrations.

Evaluation. DEN is considered a very important compound both from an environmental point of view and with regard to its molecular mode of action. It shows extreme specificity by nearly exclusively producing point mutations, with LEC's of roughly 2×10^{-1} mM after 2 days exposure of adults, and 5×10^{-3} mM after treating male larvae. Induction of point mutations is a function of concentration over more than two orders of magnitude. In sharp contrast, genetic damage associated with chromosome breakage is not detectable unless very high doses or exposure levels (LD_{60}-LD_{90}) are applied. Such exposures may lead to nonspecific effects which would not occur with subtoxic concentrations. From this point of view one still hesitates to call DEN a "chromosome breaker." Further analysis of DEN mutagenesis should focus most attention on

Table 9. Incidence of sex-linked recessive lethal mutations by DEN in Drosophila males

Route of application	Treatment (mM)	Stages tested[a]	No. of gametes tested	Activity %	Activity + or 0	Peak activity	References
Adult injection	7.0 in 0.4% NaCl	sp-sg	3260	1.9 ± 0.8	+	st,sg	21
	15.0 in 0.4% NaCl	sp-sg	3460	1.3 ± 0.2	+	st	21,22
	30.0 in 0.4% NaCl	sp-sc	2703	6.1 ± 0.5	+	post-meiotic[b]	22
Adult feeding for 3 days	1.0	sp-sg	3287	5.0 ± 0.38	+	c	23,24
Adult feeding for 24 h	1.0×10^{-2}	sp	613	0.16	0	-	1
	1.0×10^{-1}	sp	610	0.49 ± 0.89	?) LEC		
	1.0	sp-sg	4425	2.3 ± 0.23	+) LEC	st	
	2.0	sp	562	2.0 ± 0.59	+		
	5.0	sp	572	3.5 ± 0.77	+		
Adult feeding for 2 days	5.0×10^{-2}	sp-sc	2484	0.16 ± 0.08	0		10
	1.0×10^{-1}	sp	360	0.28 ± 0.28	0		
	2.0×10^{-1}	sp-sc	2536	0.43 ± 0.13	+	3% in st	
	5.0×10^{-1}	sp-sc	1198	2.42 ± 0.44	+	st	
Adult feeding for 3 days	5.0	sp-sc(sc†)[d]	1134	17.6 ± 1.13	+	st	
	10.0	sp	703	19.6 ± 1.50	+		

Larvae feeding for 6 days	1.0	sc,sg	1388	8.72 ± 3.95[e]	+	
Larvae feeding for 6 days and adult feeding for 2-3 days	5.0×10^{-3}	sc,sg	2150	0.37 ± 0.13	(+)	10
	1.0×10^{-2}	sc,sg	2127	0.75 ± 0.53[d]	+	
	5.0×10^{-2}	sc,sg	3687	1.46 ± 0.20	+	
	1.0×10^{-1}	sc,sg	1616	2.23 ± 0.37	+	
	5.0×10^{-1}	sc,sg	1198	9.7 ± 2.0[e]	+	

[a]Stages: sp, sperm; st, spermatids; sc, spermatocytes; sg, spermatogonia.

[b]Not further specified.

[c]Clear-cut evaluation not possible; a 1:1 female:male ratio was used in the breeding experiments.

[d]†, treatment toxic to sc.

[e]Corrected for mutants of common origin.

Table 10. Failure of DEN to induce in Drosophila chromosome aberrations

Route of application	Treatment (mM)	Stage tested[a]	Type of damage	Activity	Reference
Adult injection	15	sp-sg	Entire X-(Y-)loss	0	21
		sc-sg	Nondisjunction (XXY)	?	21
		sp-sg	Deletions (Loss Y^L;Y^S)	?	21
		sp-sg	2;3-translocations	0	21
Adult feeding for 24 hr	0.1-10.0	sp	Loss of Y^L and Y^S	0	1
	0.1- 5.0	sp	Dominant lethals	0	1
Adult feeding for 3 days	1.0-10.0	sp	Dominant lethals	0	1
	5.0		2;3-translocations	0	1
	7.5,10.0		2;3-translocations	(+)?[c]	25

[a]Stage: sp, sperm; st, spermatids; sc, spermatocytes; sg, spermatogonia.

[b]0.09%, 2 translocations in 2272 tests.

the nature of changes at the DNA level. It may be also very instructive to know the sites of DNA and proteins alkylated by DEN, and to compare this with a monofunctional chromosome-breaking agent, such as MMS.

N-Nitrosodimethylamine, DMN (CAS 62-75-9)

Dimethylnitrosamine represents one of the most potent environmental mutagens known for Drosophila, with LED below 10^{-2} mM after adult feeding. Extensive studies were carried out to establish its activity in the recessive lethal assay, by feeding or injecting adult males (Table 11). DMN attacks all stages of the spermatogenesis cycle. Its low activity when tested for the production of breakage-type damage, such as loss of the ring-X chromosome, indicates that most recessive lethals are point mutations.

Table 11. Genetic damage by DMN in Drosophila

Type of damage	Sperm	Sperm-atimds	Sperma-tocytes	Sperma-togonia	Refer-ences
Recessive lethals	+	+	+	+	24,26-29
Dominant lethals	Nt[a]	Nt	Nt	Nt	
2;3-Translocations	Nt	Nt	Nt	Nt	
Chromosome loss (ring-X chromosome)	(+)	(+)	Nt	Nt	25
Partial chromosome loss	Nt	Nt	Nt	Nt	
Nondisjunction	Nt	Nt	Nt	Nt	

[a]Nt = no test.

Treatment Conditions. The mutagenic effects of DMN on Drosophila, with regard to the types of genetic end points discussed in this survey, have received only moderate attention. DMN nevertheless constitutes one of the first precarcinogens for which mutagenic activity was established in this system (24,29). A number of reports have been published describing substantial activity of DMN in the recessive lethal test, with regard to induction of crossingover, induction, and the incidence of bobbed-mutations and minutes (26,27), but scanty information exists about the extent

Table 12. Mutagenicity of DMN in Drosophila males

Route of application	Treatment (mM)	Type of damage
Adult injection	1.0 in 0.4% NaCl	Recessive lethals+visibles[b]
	2.5 in 0.4% NaCl	" " "
	5.0 in 0.4% NaCl	" " "
	10.0 in 0.4% NaCl	" " "
Adult injection	5.0 in 0.4% NaCl	Recessive lethals
Adult feeding for 3 days	6.8×10^{-2} 6.8×10^{-1}	Recessive lethals
Adult feeding for 2 days	1.0	Recessive lethals
Adult feeding for 27 hr	10.0	Recessive lethals
Adult feeding for 27 hr	10.0	Ring-X loss

[a]Stages: sp, sperm; st, spermatids; sc, spermatocytes; sg, spermatogon
sg, spermatogonia

[b]Recessive lethals and visibles were summed up.

[c]Not yet clear, brood patterns were too rough.

[d]†. toxic to sc.

No. of gametes tested	Stages tested[a]	% (mean values)	Peak activity[c]	References
2,046	sp-sg	3.0	st ?	26,27
2,006	sp-sg	4.0	sc	
1,867	sp-sg	9.4	st,sc or sg?	
2,028	sp-sg	14.4	sc	
1,582	sp-sg	4.2	?	28
3,569	sp-sg	1.6	?	29
1,654	sp-sg	4.2	st	
1,319	sp-sc	8.6	st (but sc†)[d]	25
301	sp	18.9		
4,102	sp	1.72 (control 1.03%)		25

to which DMN can cause chromosome aberrations in Drosophila. Both adult feeding and injection of males were proven to be valuable for studying DMN mutagenesis. The concentrations used in feeding experiments (24,29) varied between 6.8×10^{-2} mM and 10.0 mM, and between 1.0 mM and 10.0 mM in injection studies (26-28,30).

Biological Effects. LEC and LD_{50} values have not yet been determined for DMN. In experiments on induction of recessive lethals, substantial activity (3.9%) was still noted when males were exposed to concentrations as low as 6.8×10^{-2} mM for 3 days (24,29). This suggests that the LEC must be somewhere in between 10^{-2} mM and 5×10^{-3} mM. DMN is clearly more active when fed, as compared with adult injection (Table 12). Injection of 0.25 μl of 2.5 mM DMN (∿ 66 μg/g), gives rise to 4% lethals. A lower concentration of 6.8×10^{-1} mM, fed to adult males for 3 days, is sufficient to produce the same number of lethals (26,27). All published data show consistency in the sense that the mutational yield is lowest in mature sperm (which are nonmetabolizing cells), yet, it is not possible to define the stage which is most susceptible to mutation induction by DMN (see Table 12). This is so because fine brood pattern analysis has not been conducted with DMN in Drosophila.

A small amount of data exists demonstrating ring-X loss (25) and activity in the specific locus test with DMN (26). In a total of 106013 specific locus tests for the response of any of three loci, w, f and mal^{b7}, over the investigated range of DMN (1-10 mM), 16 mutants were recovered (11 f, 5 w), 8 of which (7 f, 1 w) were fertile and transmissible. Preliminary experiments (25) with 10 mM DMN revealed in sperm 1.72% ring-X losses (control frequency 1.03%; the recessive lethal yield was 18.9%), this indicating a low capacity of DMN for inducing chromosome breaks.

Evaluation. DMN is highly mutagenic in Drosophila, with an LEC for recessive lethals which is thought to be located between 10^{-2} mM and 5×10^{-3} mM after adult male feeding for 3 days. This concentration is much lower than the LEC for DEN, for which mutagenic activity was not observed below 2×10^{-1} mM after 2 days exposure.

Previous reports have shown that DEN produces nearly exclusively point mutations in Drosophila, which coincides well with our observation that chromosome breakage by DEN has not been detected. One is faced with a similar situation with DMN. However, a far more thorough study of its mutagenic potential and properties is needed before a meaningful evaluation is possible.

Epichlorohydrin (CAS 106-89-8)

There have been two reports on mutagenicity of epichlorohydrin in Drosophila (Table 13). A low incidence by epichlorohydrin of 0.7% recessive lethal mutation (4 lethals in 526 gametes tested; control 0/887) was obtained by Rapoport (31); no information was given about the experimental conditions or the concentration used. The other paper is a very recent study by Kramers (32), who used both injection and feeding procedures. His data clearly show that there can be no doubt that epichlorohydrin is mutagenic in Drosophila (Table 14). In injection experiments, concentrations of 5.1 mM and 25.5 mM epichlorohydrin produced recessive lethal mutations in all cell stages tested; these were post-meiotic (broods A, B, and C) and meiotic cell stages (D + E). A level of 2.6 mM seems to be the lowest effective concentration, for mutation yield was still observed at that concentration. However, no mutagenic activity was found in males fed 2.6 mM or 5.1 mM epichlorohydrin for 24 hr. In the feeding experiments with 5.1 mM 70% of the males died within 24 hr.

Table 13. Genetic damage by epichlorohydrin in Drosophila

Type of damage	Sperm	Spermatids	Spermatocytes	Spermatogonia	References
Recessive lethals	+	+	+	Nt	31,32
Dominant lethals	Nt[a]	Nt	Nt	Nt	
Translocations	Nt	Nt	Nt	Nt	
Chromosome losses	Nt	Nt	Nt	Nt	
Nondisjunction	Nt	Nt	Nt	Nt	

[a]Nt = no test.

Ethylenimine (CAS 151-56-4)

A number of points about the mutagenic effects of ethylenimine (EI) in Drosophila seem relatively clear: EI causes point mutations in all stages of male germ cells, while translocation induction was observed in meiotic and post-meiotic stages but not in premeiotic cells (Table 15). The most striking feature of EI is the very high proportion of translocations to sex-linked recessive lethals and of interchromosomal to intrachromosomal

Table 14. Induction of sex-linked recessive lethals in Drosophila

Expt.	Concn (mM)	Administration	Breeding scheme	Brood A No. chrom.	Brood A % lethal
1	2.6	Injection	2-2-2-2-2	526	0
2	5.1	Injection	2-2-2-2-2	958	0.52
3	25.5	Injection	2-2-2-2-2	448	0.67
4	25.5	Injection	2-2-2-2-2	992	1.74
5	2.6	24 hr feeding	2-2-2-2	628	0
6	5.1	24 hr feeding	2-2-2-2-2	773	0.13

[a]The 4-day old males were treated (Oregon-K) and individually mated was added in injection and feeding solutions, respectively. After E; after feeding of 2.6 mM this was the same; during feeding of was 100% lethal; here a higher stickiness of the solution may auxiliary solvent at final concentrations of 0.5% or lower water). Data from Kramers (32).

by epichlorohydrin[a]

Brood B		Brood C		Brood D		Brood E	
No. chrom.	% lethal	No. chrom.	% lethal	No. chrom.	% lethal	No. chrom.	% lethal
537	0.19	521	0.19	525	0	465	0
1057	0.47	957	0.31	1018	0.10	1108	0.45
528	1.33	412	0.49	450	0.22	252	0.79
836	0.96	801	0.37	844	0.47	817	0.12
772	0	580	0	704	0.14		
701	0	702	0.28	603	0	811	0

to 3 Basc females (virgins) per brood. 0.7% NaCl or 5% sucrose
injection of 25.5 mM, 75% of the flies were fertile through brood
5.1 mM, however, 70% of the flies died within 24 hr, while 25.5 mM
have played a role. DMSO was used (except in expt. 6) as an
(although it was not necessary to get perfect mixing of ECH in

changes (0.78), as compared with other mutagens. EI induces a pattern of damage like that of x-rays.

Table 15. Genetic damage by EI in Drosophila males

Type of damage	Sperm	Spermatids	Spermatocytes	Spermatogonia	References
Recessive lethals	+	+	+	+	33-39
Dominant lethals	+	Nt[a]	Nt	Nt	
2;3 Translocations	+	+	+	0	33-37, 40
Chromosome loss	Nt	Nt	Nt	Nt	
Partial chromosome loss	Nt	Nt	Nt	Nt	
Nondisjunction	Nt	Nt	Nt	Nt	

[a]Nt = no test.

Treatment Conditions. Adult injection has been the only technique for investigating mutagenicity of ethylenimine in Drosophila. Males were injected 0.2-0.35 μl EI solutions at concentrations of 5.0, 10.0, 15.0, or 20.0 mM.

Biological Effects. Several investigators have studied the mutagenic activity of EI and all studies have yielded positive results. EI produces in Drosophila sex-linked and autosomal recessive lethal mutations, dominant lethals, and translocations. LD_{50} and LEC's were not determined for ethylenimine, nor were there any attempts to study the relation of concentration and mutagenic effectiveness. Major emphasis was paid to genetic damage by EI in mature sperm and little attention has been given to the response to EI of younger germ cells. The study of Alexander and Glanges (34) requires particular mention, since this was a most comprehensive study with the entire spermatogenesis cycle (Table 16). Alexander and Glanges found the frequencies of sex-linked and autosomal recessive lethal mutations as well as the rate of translocations to increase in all stages except the premeiotic cells in which no translocations were discovered (Table 16). The response was generally lower in younger germ cells but there was no indication of any stage-specificity with ethylenimine.

Table 16. Genetic damage induced with 10.0 mM EI in Drosophila males[a]

Type of damage	Mating periods (days after treatment)						
	1-3	3-6	6-9	6-12	12-15	15-18	Average
Sex-linked recessive lethals							
Total sample	615	579	579	582	581	567	3503
% lethals	3.6	2.4	2.8	3.3	1.2	0.35	2.3
Translocations							
Total sample	579	569	528	512	603	592	3383
% translocations	2.1	1.9	3.2	1.2	0.33	0	1.42

[a]Data from Alexander and Glanges (34).

There is general agreement that ethyleneimine is very efficient in producing chromosomal abnormalities since translocations are nearly as frequent as sex-linked recessive lethals. A nearly 1:1 ratio for post-meiotic stages was found by Alexander and Glanges (34) and also by Lim and Snyder (36), this resembling the action of x-rays rather than that observed for other monofunctional alkylating agents (Table 17). Unlike x-rays, however, ethylenimine produced delayed lethals and translocations (33,34). Ethylenimine constitutes one of the chemicals which was used extensively for studying delayed opening of breaks (storage effect) after treatment with chemical mutagens. The results of the EI studies appear to be discordant, this largely being attributable to the findings of Ratnayake (38) and Watson (37), who failed to find evidence of storage effects by EI (Table 17). However, it can be seen from Table 17 that storage effects were observed for translocations (40,41) and for dominant lethals when there was a 12-day to 3-week storage period. This helps to explain the negative result obtained by Watson and Ratnayake using a total period of storage of only six days.

Slizynska (41,42) investigated mutagenic effects of EI both before and after storage by salivary gland chromosome analysis. Slizynska showed that most of the chracteristic features that distinguish the effects of most chemical mutagens (TEM and formaldehyde were used) from those of x-rays were also present after treatment with EI; but the differences between EI and x-rays were, on the whole, smaller than between x-rays and either TEM or formaldehyde. The pattern of EI-induced genetic damage in unstored spermatozoa differed from that caused by x-rays by a higher proportion of mosaics, repeats and deficiencies and by a lower one of

Table 17. Genetic damage induced with ethylenimine in Drosophila sperm

Treatement (injection)	Concentration (mM)	Test	Pre-stored No. tested	Pre-stored %	Storage effect increase	Storage period total	References
	10.0	X-chromos.rec.lethals	1225	2.7	Nt[a]		33,34
0.4 µl in 0.4% saline		Translocations	1287	1.0	Nt		33,34
0.3 µl in 0.4% saline	10.0	Translocations (Y-2-3)	737	4.2	3.8 fold	12 days	40
0.35 µl in 0.7% saline	5.0	X-chromos.rec.lethals	1607	5.9	Nt		36
		Translocations (Y-2-3)	1306	4.5	Nt		36
	10.0	X-chromos.rec.lethals	1077	11.5	Nt		36
		Translocations (Y-2-3)	911	7.1	Nt		36
	15.0	X-chromos.rec.lethals	531	13.4	Nt		36
		Translocations (Y-2-3)	420	12.6	Nt		36
3 µl in 0.7% saline	10.0	Dominant lethals	367	17.8	No effect	6 days	38
	20.0	" "	521	71.7	No effect	6 days	38

0.2 μl in 0.7% saline	5.0	X-chromos.rec.lethals	660	3.3	Questionable	3 weeks	35
	5.0	Translocations (Y-2-3)	547	0.36	9.7 fold	3 weeks	35
0.2 μl in 0.7% saline	10.0	Dominant lethals	1393	21.8	2.7 fold	12 days	39
2 μl in saline (?)	10.0	X-chromos.rec.lethals	641	3.4	No effect	6 days	37
	10.0	" " "	715	7.3	No effect	6 days	37
	10.0	Translocations (Y-2-3)	926	0.65	No effect	6 days	37
	20.0		589	2.5	No effect	6 days	37

[a] Nt = no test.

Table 18. Comparison of cytological effects induced by EI and x-rays[a]

Samples[b]	Time after treatment (days)	Changes in 100 spermatozoa[c]		M among changes (%)	Ratio[d] T/In,Rp,Df	Breaks in the heterochromatin (%)
		C	M			
I - U	2-3	2.45	0.35	12.5	0.78	11.1
II - VI	4-12	4.1	1.6	28.1	0.84	16.9
VII - VIII	13-16	11.9	4.4	27.0	0.90	14.9
IX - XII	17-24	8.2	4.2	33.9	0.80	15.0
X-rays	2-5	26.5	3.5	11.7	1.1	19.0

[a]Data from Slizynska (41,42).

[b]U = unstored, first sample after treatment; II - XII = stored samples in EI experiments.

[c]C = complete change; M = mosaic change.

[d]Df = deficiencies; In = inversions; Rp = repeats; T = translocations.

translocations, inversions and interchromosomal changes (Table 18). In spermatozoa stored following EI treatment the whole pattern of changes become more like that found after irradiation, which indicates the high proportion of prebreakage lesions which require more than 6 days to mature into open breaks.

Evaluation. There is a considerable literature on the mutagenesis of EI in Drosophila demonstrating inductions of point mutation (recessive lethals), structural changes in the chromosomes (translocations, duplications, inversions and deletions) as well as dominant lethals. Further investigation is suggested concerning the proportion of sex-linked to autosomal recessive lethal mutations as well as the qualitative and quantitative relation between exposure and mutational spectrum. This is of particular importance since EI, more than any other monofunctional alkylating agent, seems to induce a pattern of mutational damage that resembles that produced by x-rays.

Ethyl Methanesulfonate (EMS) (CAS 62-50-0)

EMS is a very potent mutagen in Drosophila, inducing recessive lethal mutations, specific locus mutations, dominant lethals, translocations, entire chromosome loss, and deletions (Table 19). EMS acts primarily on postmeiotic stages, meiotic and premeiotic effects being relatively immune to its effects. Deletions by EMS are extremely rare among recessive lethal mutations, as proved from cytological analysis of salivary gland giant chromosomes and complementation tests. The exact dosimetry developed with ^{3}H-EMS now allows one to relate mutation induction to ethylation/nucleotide. Adult males were fed on a concentration of 0.1 mM EMS for 24 hr. This treatment actually represents a dose of 2.1×10^{-4} ethylations/nucleotide and was found to give rise to $0.55 \pm 0.08\%$ X-linked recessive lethals; up until now 2.1×10^{-4} alkylation/nucleotide represents the lowest adverse effect level for mutation induction by EMS in Drosophila.

Treatment Conditions: Dosimetry. Considering the amount of attention the effects of chemical mutagens on Drosophila have received over the past 30 years, remarkably little effort has been spent on developing an exact dosimetry. The bulk of mutation work carried out on EMS mutagenesis in Drosophila gives exposure of the whole organism to the mutagen and not the dose which is defined as the amount of the mutagen that reacts with the genetic targe molecule (61). A level of 25.0 mM EMS seems to represent a standard concentration for this compound, because many workers have followed the very convenient feeding procedure of Lewis and Bacher (48) which used this EMS concentration. The actual concentration must have been 22.1 mM, because 0.24 ml were dissolved in 100 ml sucrose solution according to Lewis and Bacher (48). On the other hand, EMS is the only chemical mutagen for which an exact dosimetry

Table 19. Genetic damage by EMS in Drosophila

Type of damage	Sperm	Sperm-atids	Sperma-atocytes	Sperma-atogonia	Refer-ences
Recessive lethal mutations	+	+	+	+	36,43-49
Dominant lethals	+	+	+	+	39,47
Translocations	+	+?	Nt[a]	Nt	36,46 49,50
Chromosome loss	+				47,51
Partial chromosome loss	+	+?	Nt	Nt	51,52
Specific locus mutations	+	+	+	+	53-60
Nondisjunction	Nt	Nt	Nt	Nt	

[a]Nt = no test.

has been developed for Drosophila. In their pioneering work, Lee and co-workers (43,47,54,56,61-63) have used ^{3}H-EMS to determine a dose-response relationship in Drosophila sperm cells. Sex-linked recessive lethal mutations were detected in conjunction with determination of DNA ethylation per nucleotide after male (sperm) treatment with labeled EMS. Strong evidence was obtained of a linear relation between sex-linked recessive lethals and dose as alkylation per nucleotide (43). Recessive lethal frequency was found to increase linearly with the dose over a range in dose of 2.1×10^{-4} to 1.4×10^{-2} ethylations per nucleotide. Thus suggesting that there were no change in mechanism of mutagenesis from low to high doses in Drosophila. Therefore, relative frequency of sex-linked recessive lethals can be used as a biological dosimeter for comparing different studies on EMS mutagenesis in Drosophila (47). Accordingly, recessive lethal frequencies were found to increase linearly over a range of exposures from 0.1 mM (= 2.1×10^{-4} ethylations per nucleotide) to 1 mM, but a further 25-fold increase in exposure, i.e., 25.0 mM (1.4×10^{-2} ethylations per nucleotide) only increased the recessive lethal frequency 11-fold. This departure below linearity at higher exposures can be seen in the data of Huang and Baker (60) in the range of concentrations from 2.0 mM to 12.0 mM EMS. On the other hand, when EMS is injected into adult males and recessive lethal frequencies are scored, the data of Epler (64) indicate a linear exposure-response in the range from 10.0 mM to 40.0 mM EMS. Aaron and Lee (65) showed that it can be very misleading if conclusions are based on exposure because this showed a feeding rejection behavior at EMS concentrations above 1.0 mM (this leading to considerable variance among repeat experiment) (63). Legator et al. (66) have suggested that the mutagenic effect of EMS in Drosophila "feeding" experiments may result from penetration of the mutagen to the tests via the spiracular-tracheal system rather than the digestive tract.

Biological Effects. With respect to point mutations, there is general agreement in the literature that EMS acts primarily on postmeiotic cells, the premeiotic and meiotic cells remaining relatively immune to its effect. Browning (44) reported a 4- to 5-fold higher yield in postmeiotic cells of recessive lethals compared with the yield in meiotic and premeiotic cells, whereas in experiments done by Fahmy and Fahmy (45) the difference between postmeiotic and premeiotic cells was even more than one order of magnitude. A similarly low mutagenic response to EMS of younger stages of germ cells was observed by Jenkins (53,55,57), who found that 90% of all EMS-induced dumpy mutations accounted in the broods from postmeiotic germ cells. These observations are consistent with the findings by Aaron (63) that mature sperm and later spermatids retain a high level of ethylation of DNA; much lower levels of alkylation are retained in cells which were treated in early spermatid stages or as spermatogonia.

LD_{50} values for EMS have not been determined in Drosophila. A concentration of 0.1 mM (dose 2.1×10^{-4} ethylations per nucleotide) is the lowest adverse effect level for EMS in Drosophila sperm (43), this yielding a recessive lethal frequency of 0.55% (Table 20). The frequency of sex-linked recessive lethals was found linear over a wide range in dose (factor 67 in the study by Aaron and Lee) (43). Therefore, in those experiments with EMS where exposure-effect relationships were determined for recessive lethals and, for example, for another kind of damage like Minutes mutants, as the case in the study by Huang and Baker (60), recessive lethal frequencies can serve as secondary dosimeter (47) to establish dose (alkylation/nucleotide). The linear relation between sex-linked recessive lethal mutations and Minute mutations observed by Huang and Baker (60) shows that a linear relationship must also exist between dose and Minute mutations (47).

EMS is very efficient in both multiple and specific locus tests, and there is good reason to believe that those mutations are predominantly the result of intralocus mutations. This conclusion is supported by complementation tests with EMS-induced recessive lethals in cytogenetically well defined regions that have been "saturated" for lethal producing loci. Hochman (67) found that

Table 20. Dose and relative frequency of sex-linked recessive lethals for four concentration levels of EMS fed to *Drosophila melanogaster* males[a]

EMS exposure concn. (mM)	Dose (ethylations per nucleotide in sperm DNA)[b]	Lethals/chromosomes tested	% Lethals[c]
25	$(1.4 \pm 0.2) \times 10^{-2}$	1028 / 2390	43 ± 1[d]
1.0	$(1.4 \pm 0.2) \times 10^{-3}$	102 / 2540	3.9 ± 0.4
0.5	$(7.8 \pm 0.3) \times 10^{-4}$	113 / 5549	1.9 ± 0.2
0.1	$(2.1 \pm 0.1) \times 10^{-4}$	42 / 6296	0.55 ± 0.08
Concurrent controls	-----	7 / 5347	0.13 ± 0.05
Accumulated controls	-----	34 / 27064	0.12 ± 0.02

[a]Data from Aaron and Lee (43).

[b]Confidence limits represent one standard deviation computed from repeatability within exposure level.

[c]Experimental values corrected for control relative frequency.

[d]Confidence limits represent one standard deviation computed from sampling error.

74/75 lethals on chromosome 4 affected only a single complementation unit, and Lim and Snyder (59) found that all 83 lethals in the zeste-white region of the X-chromosome affected only a single complementation unit. Similarly, 83 mutations on the X-chromosome at the yellow or white loci were intragenic changes (51). On the other hand, a few lethals affecting more than one complementation unit, some of which have been confirmed cytologically as interstitial deletions involving the loss of at least 10 salivary chromosome bands have been reported to occur at low frequencies among recessive lethals induced by EMS in the proximal regions of the X-chromosome and the left arm of chromosome 2 (68-70). Of the proximal X deletions, four have a proximal break in heterochromatin. In summary, all the data show that deletions and gross structural rearrangements are extremely rare among EMS-induced recessive lethal mutations.

With respect to chromosome aberrations, 10.0 mM is the lowest concentration for which translocation induction has been demonstrated. This concentration yields about 5-6% recessive lethals when injected and 15-16% when fed for 18 hr (Table 21). A higher translocation frequency of 2.5% was observed after feeding adult males for 24 hr with a 25.0 mM EMS solution. Recessive lethal yield reached 53.6% (dose 2×10^{-2} alkylations per nucleotide) under these conditions. At a lower dose of 9.3×10^{-4} alkylations per nucleotide, estimated from the sex-linked recessive lethal frequency by using the data of Aaron and Lee (43), no translocations were detected. Furthermore, no increase in dominant lethals was observed in sperm with an exposure of 2.5 mM EMS (2.3×10^{-3} alkylations/nucleotide) which produced 6.1% sex-linked recessive lethals (43). A higher exposure of 25.0 mM EMS gave 1.4×10^{-2} alkylations per nucleotide and a recessive lethal frequency of 38% but only a slight increase in dominant lethals. A high rate of 60.5% dominant lethals was obtained in injection experiments (39) with the extraordinarily high concentration of 50.0 mM EMS. It is evident that mature spermatozoa in Drosophila are very resistant to induction of dominant lethals and viable chromosome breakage effects, unless high concentrations are used. This is also evident from the study by Bishop and Lee (51) who obtained substantial induction of sex-chromosome losses at high concentration.

Using high EMS-concentrations in the range of 10.0 mM to 50.0 mM EMS, Sram (39) for dominant lethals, Abrahamson et al. (50) for translocations, and Schalet (52) for sex-chromosome marker loss obtained experimental support of a storage effect with EMS. Abrahamsom et al. (50) discovered one Y-2 translocation from 271 unstored sperm, but three from 20 stored gametes carried a 2-3 translocation, giving a translocation frequency of 0.15 (3 translocations in 20 cells).

Although only a few cultures were tested, due to sharp decrease

Table 21. Induction of recessive lethal mutations vs. translocations by EMS in Drosophila sperm

Treatment	X-chromosome recessive lethals		Dose	Translocations		References
	N gametes	% lethals	alkyl./nucl.	N gametes	% translocations	
Feeding; 12.5 mM EMS for 24 hr	Nt[a]		Unknown	271[b] 20[c]	0.4 (15%)	50
Injection (0.15 μl); 25.0 mM EMS in 1% sucrose; pH 9.5	508	5.7	9.3×10^{-4}	958[b]	0	46
Feeding; 25.0 mM EMS for 24 hr; pH 9.5	612	53.6	2.0×10^{-2}	760[b]	2.5	46
Injection; 10.0 mM EMS; pH 7.2	1605	4.1	$\sim 6.7 \times 10^{-4}$	1459[b]	0	36,58
Injection; 10.0 mM EMS; pH 9.5	725	5.5	$\sim 9.0 \times 10^{-4}$	671[b]	0.2	36,58
Feeding; 10.0 mM EMS for 18 hr	1616	15.9		1061[b]	0.38	49

[a]Nt = no test.

[b]Unstored cells.

[c]Stored in females for 10 days.

in fertility, it is clear that there is a storage effect of EMS on translocation. The data of Sram (39) show the same tendency for dominant lethals. A strong increase of translocations with storage was found in 3871 tests from untreated males.

Spermatids, meiotic and spermatocyte stages seem to be more susceptible to dominant lethal induction by EMS (no information exists on translocation induction for those stages). However, Aaron and Lee (43) and Lee (47) suggest that it was male sterility (cells were destroyed) and not true dominant lethality which causes these effects.

Evaluation. Analysis of EMS mutagenesis is fairly complete for Drosophila. 2.1×10^{-4} ethylations per nucleotide (24 hr adult feeding with 0.1 mM EMS solution) constitutes the lowest adverse effect level, this dose gives rise to $0.55 \pm 0.08\%$ X-linked recessive lethal mutations (43) in sperm, this frequency being 4-fold above that found in concurrent controls. Much higher doses are needed to significantly increase genetic damage associated with chromosome breakage. Until now, such effects were not found at doses below 9×10^{-4} ethylations per nucleotide. However, since there are indications of a strong storage effect with EMS at high EMS concentration (10.0 mM and higher), and since no efforts were made to check at lower EMS levels for such effects, the question remains open as to what extent EMS is capable of producing breakage-type damage at and below 9×10^{-4} alkylations/nucleotide.

Since the mature spermatozoa seems the stage most resistant to incidence of chromosome breakage by EMS, it would be interesting to know the relation of point mutations to the various types of chromosome rearrangements in early spermatids because this stage is less immune to chromosome breakage induction.

ICR-170; Acridine Mustard (CAS 146-59-8)

ICR-170 is highly mutagenic in Drosophila (Table 22). Attention has been focused on the mutational spectrum by ICR-170 in mature sperm, the recessive lethal induction in all stages of spermatogenesis, the nature of induced recessive lethals, and the ratio of completes vs. mosaically expressed mutations.

Treatment Conditions. ICR-170 has been almost exclusively administered by abdominal injection of adult males. In all studies with adult males a concentration of 0.1% ICR-170 (about 2.5×10^{-3} M; 2.5 mM or 285 μg/g) in 0.4 or 0.7% saline was used.

Biological Effects. Generally, very low frequencies of large chromosome rearrangements (translocations and deletions) are recovered following adult treatments with ICR-170. On the other hand, a substantial amount of inversions (10.4%) was claimed in

Table 22. Genetic damage by ICR-170 in Drosophila

Type of damage	Sperm	Sperm-atids	Sperma-atocytes	Sperma-togonia	Refer-ences
Recessive lethals	+	+	+	+	71-74
Dominant lethals	Nt[a]	Nt	Nt	Nt	
2;3-Translocations	+	Nt	Nt	Nt	72
Chromosome loss	+	Nt	Nt	Nt	72
Nondisjunction	Nt	Nt	Nt	Nt	

[a]Nt = no test.

salivary gland chromosomes from ICR-170 treated larvae [control frequency 0.8%; treatment 0.25 mM × 4(?) days] (75). Table 23 shows data from Snyder and Oster (72) on the incidence of losses of the entire X- (or Y-) chromosome, partial loss of the Y (as indicated by loss of one or the other of the terminal markers of the doubly marked Y-chromosome), sex-linked recessive lethals, and autosomal translocations. The frequencies of sex-linked recessive lethals obtained by treatment of mature spermatozoa represents a substantial underestimate, since the data given in Table 23 do not include the considerable number of mosaic lethals produced by ICR-170. ICR-170 does not exhibit any stage-specificity. Point mutation induction is observed in spermatozoa, spermatids, and the earlier stages of meiosis, including spermatogonia (71,72). Studies using sex-linked lethal tests and the incidence of mutations at the dumpy locus have revealed a high degree of both gonadal and somatic mosaicism (71,76-78).

Treatment of postmeiotic germ cells produces more mosaic mutations and fewer complete mutations than treatment of the premeiotic cells. Thus, if the recessive lethal test is carried out in the common way as a two-generation test, mosaically expressed lethals are then missed and mutagenic activity seems highest in premeiotic stages (Table 24). An F_3 sex-linked lethal test detects F_1 gonadal mosaic mutations, and this procedure reveals that postmeiotic and premeiotic germ cells are about equally sensitive to ICR-170.

Carlson et al. (79) found that of 98 postmeiotic lethals and of 25 premeiotic and meiotic ones, which were analyzed by localization experiments, none involved concomitant gross structural

Table 23. Disarrangement of the X and Y chromosomes, sex-linked recessive lethals, and translocations between chromosomes 2 and 3 by ICR-170 in mature sperm

Treatment	Loss of X or Y		Partial loss of Y^L, Y^S		X-chrom. rec.l.		2;3-trans-cations	
	Tested	%	Tested	%	Tested	%	Tested	%
Control	5161	0.04	5161	0.06	829	0.39	-	-
ICR-170	3305	0.18	3305	0.27	3017	12.26	1253	0.24

changes. The average postmeiotic frequency for sex-linked lethals was 12.5% (71). In contrast, an x-ray dose which produced a similar percentage of F_2 sex-linked lethals would have produced 1/4-1/3 lethals associated with gross structural rearrangements (79). Thus, the localization of ICR-170 induced mutations is strong evidence for their point mutational nature.

Table 24. Distribution by brood of F_2 and F_3 mutation frequencies[a]

Brood (in cells)	Cell stage	F_2 lethals induced (%)	F_3 lethals induced (%)	Index of F_2: F_3 lethal frequency
1	Postmeiotic	12.4	3.6	3.4
3		12.9	2.6	4.9
5	Meiotic	8.8	1.7	5.2
9	Premeiotic	4.6	2.3	2.0
12		15.4	0	(15.4)[b]
Total		11.2	2.7	4.2

[a]Data of Carlson and Southin (71).

[b]Actual infinity, but one F_3 lethal was assumed to permit a comparison.

Conflicting results exists as to the ability of ICR-170 to produce ts (temperature-sensitive) recessive sex-linked mutations in Drosophila. Carlson et al. (79) have reported that the compound does not induce ts mutations, and they concluded that ICR-170 acts as an acridine in Drosophila by producing frameshift mutation events. This is contrary to the observation of Woodruff and Gander (74) and Woodruff and Johnson (73), who recovered 7 ts mutations out of a total of 122 lethals tested. This observation, together with the recovery of some leaky ts mutations, seems to indicate that in Drosophila ICR-170 may function by inducing some base-pair substitution mutations. The discrepancy between the two studies may result, at least in part, from a difference in the temperature ranges used. Carlson et al. (79) isolated lethal mutations at 24-28°C and tested them for survival at 16°C. In the other studies (73,74) lethals were recovered either at 29°C and tested for survival at 18 and 24°C, or lethals were recovered at 24°C and tested for survival at 18 and 29°C.

Evaluation. A moderate concentration of 2.5 mM ICR-170 produced a substantial amount of point mutations, but only few translocations and chromosome losses. The relation between concentration and mutagenic effects has yet to be studied with this agent. However, since ICR-170 may act both as an alkylating agent and a frameshift mutagen in Drosophila, we have some doubts about the usefulness of such a study. There is no information on LD50 and LEC values for ICR-170.

Methyl Methanesulfonate, MMS (CAS 66-27-3)

MMS causes in Drosophila all categories of genetic damage (except nondisjunction) including point mutations, dominant lethals, translocations, entire sex-chromosome loss, and partial chromosome loss (deletions) (Table 25). Concentration-response studies with MMS showed qualitative differences in the mutational spectrum as function of the concentration: genetic damage associated with structural changes is not observed at nontoxic concentrations (∿ 1/100 of the LD_{50}) while mutations are still detectable. LEC's for the production of breakage-type damage are 10- to 20-fold higher than those needed to significantly increase the amount of point mutations.

Treatment Conditions. The mutagenic action of MMS in the male of Drosophila has been extensively studied. Adult feeding (1,83) is clearly the most effective route of application compared with inhalation (80,82) or injection techniques (45,81,89). MMS undergoes rapid hydrolysis in aqueous solution, and this may affect the pH of the external solution. Thus, Vogel and Leigh (1) dissolved MMS in m/30 phosphate buffer, pH 6.8. It is well known for MMS and other monofunctional alkylating agents, that changes in pH lead to considerable variation between repeat expteriments. In the study by Smith (85), the recessive lethal frequency varied between 2.4 and 12.8%, and this variability appears to be due to the high instability of unbuffered MMS solutions. Variation between repeat experiments is far smaller when a buffered system is used (1). The effect of pH may be a feeding behavior due to a rejection type mechanism.

Table 25. Genetic damage by MMS in Drosophila

Type of damage	Sperm	Sperm-atids	Sperma-tocytes	Sperma-togonia	Refer-ences
Recessive lethals	+	+	+	+	1,45,59, 80-85
Dominant lethals	+	+	+	+	1,86,87
Translocations	+	Nt[a]	Nt	Nt	84
X-, Y-loss	+	Nt	Nt	Nt	1
Partial loss (Y^L,Y^S)	+	Nt	Nt	Nt	1
Nondisjunction			0	0	88

[a]Nt = no test.

Biological Effects. The LD_{50} of MMS is roughly 5.0 mM after adult feeding treatment of Berlin K males for 3 days (1). It will be seen from Table 26 that adult feeding provides the most efficient procedure for studying the mutagenic effects of MMS. In sperm, the recessive lethal yield increases with increasing concentration, with an LEC between 1.0×10^{-2} mM and 5.0×10^{-2} mM MMS (1). The sensitivity to recessive lethal induction of successive stages in spermatogenesis was analyzed by several investigators (45, 80-82). There seems general agreement that the mature spermatozoa is the stage most susceptible to MMS, while spermatogonia did not reach more than 1/10 of the percentage determined in post-meiotic stages (45).

MMS causes (in sperm) dominant lethals, Y-marker loss ($B^S.Y.y^+$), and entire chromosome loss (Table 27). Vogel and Leigh (1) have calculated the LEC:LD_{50} ratio for recessive lethals and various categories of chromosome aberrations. There was a remarkable difference between the LEC:LD_{50} proportions. The ratio was smallest

(1:100) for recessive lethals (rough estimate), but increased considerably for dominant lethals and X(Y)-loss (1:10) and even more so for deletions of the Y-chromosome marker for which a 1:5 proportion was established.

Table 26. Incidence of sex-linked recessive lethals in mature sperm.

Route of application	Treatment (mM)	pH	Lethals (%)	References
Inhalation (adult males)	930 × 7 hr		2.3	80,82
Adult injection in 0.4% NaCl	2.5		6.3	45,81, 89
	4.5		11.6	45,81, 89
Adult injection in 0.4% NaCl	4.0		2.8	59
	7.5		11.6	59
Adult feeding	5.0 × 22 hr		11.6	90
	2.3 × 24 hr		19.2	83
	4.5 × 24 hr		29.9	83
	6.8 × 24 hr		30.0	83
Adult feeding	1.0×10^{-2} × 2 da[a]	6.8	0.17 (control level)	1
	5.0×10^{-2} × 2 da	6.8	0.80 LEC[a]	1
	1.0×10^{-1} × 2 da	6.8	0.91	1
	5.0×10^{-1} × 2 da	6.8	14.5	1
	1.0 × 2 da	6.8	27.0	1

[a]The MMS solution was replaced by a fresh one at the end of the first 24 hr.

These comparative experiments proved the existence of different effective concentrations for mutations vs. structural aberrations. Previous findings by Lim and Snyder (84) can be interpreted in the same way. At an MMS concentration of 7.5 mM, various types of genetic damage were found: recessive lethal mutations (11.6%), translocations (0.73%), inversions (1.8%) and deletions (1.8%), but at 4.0 mM MMS recessive lethals (2.8%) were the only category of damage observed. From Table 27 one might get the impression that MMS is not very effective as a chromosome breaker in Drosophila.

Table 27. Yield of recessive lethals, dominant lethals, entire sex-chromosome loss, and loss of B^S (Y^L) or y^+ (Y^S) in sperm treated with MMS for 2 days[a]

Concentration (mM)	Recessive lethals		Dominant lethals		Entire loss		Partial loss	
	%	Activity	%	Activity	%	Activity	%	Activity
Control	0.17		0		0.08		0.04	
1.0×10^{-1}	0.91	+	2.0	0	0.09	0	0.04	0
5.0×10^{-1}	14.5	+	44.6	+	0.44	+	0.08	(+)?
1.0	27.0	+	95.9	+	1.14	+	0.21	+

[a] Data from Vogel and Leigh (1).

But this is not the case. In sperm (Table 28) exposed to MMS concentrations of 5.0×10^{-1} mM, 1.0 mM, and 2.0 mM, only one translocation was detected in unstored cells but up to 25% 2;3-translocations were found in samples from sperm stored in females for up to 31 days (25). This shows that performance of storage experiments is an absolute necessity for the detection of this type of rearrangements by a monofunctional alkylating agent like MMS.

The ability of MMS to produce deletions in Drosophila has received considerable attention. Lim and Snyder (84) analyzed salivary gland chromosomes from 22 lines carrying an MMS-induced X-chromosomal lethal and found no evidence of small deficiencies. Fahmy and Fahmy (89) found 17% of their MMS-induced X-linked recessive lethals to be associated with cytologically detectable deletions. On the contrary, Lim and Snyder (59) detected only one deletion among 54 MMS-induced recessive lethals in the X-chromosome. This discrepancy is rather great, since the concentration of MMS used by Fahmy and Fahmy was 4.5 mM, whereas that employed by Lim and Snyder (59) was 7.5 mM. The results from complementation analysis by Liu and Lim (90,91) are in good agreement with the observation made by Lim and Snyder. Liu and Lim analyzed 106 recessive lethal mutations in the 3A1 to 3C2 region of the X-chromosome. With only one exception, these mutants were found to be lesions restricted to single loci.

Obviously, two aspects are important when considering the ability of monofunctional alkylating chemicals to produce deletion-associated mutations. The first of these concerns the target material. Gatti et al. (92) demonstrated chromatid-type aberrations in somatic cells from MMS-treated third instar larvae. The distribution of the breaks between and within chromosomes showed that they were not localized at random but were clustered in the heterochromatic centromere region of the X-chromosome and the autosomes. The second point which seems of particular significance is the concentration, as demonstrated in the study by Vogel and Leigh (1). Below certain MMS concentrations, breakage-type damage seemed absent although mutation induction provided clear-cut evidence that there was still interaction with DNA.

Evaluation. MMS causes all categories of genetic damage in Drosophila, but the LEC values are lower for recessive lethals (1.0×10^{-2} - 5.0×10^{-2} mM), as compared with those for dominant lethals (between 10^{-1} and 5×10^{-1} mM), loss of X or Y ($\sim 5 \times 10^{-1}$ mM), and Y-marker loss ($\sim$ 1.0 mM). Recessive lethals are produced in the entire cycle of spermatogenesis and a high degree of dominant lethality is found after treatment of sperm, spermatids and spermatocytes, this reaching up to 100% (1,86). MMS was not very efficient as a chromosome breaker in chromosome loss tests (1), whereas high yields of translocations were observed in storage experiments. This suggests that performance of storage is essential.

There exists some disagreement concerning the ability of MMS to produce deletions. Analysis of the data reveals that it is quite likely that the differences noted are associated with (a) the target material (euchromatin vs. heterochromatin) in the genome, and (b) the concentration. The situation seems similar with other organisms. In Neurospora, for example, the proportion of $\underline{ad\text{-}3}^{1R}$ among the ad-3 mutants increased from 2.3% at 5.0 mM MMS to 15.7% at 20.0 mM MMS (93).

MeTEPA (CAS 57-39-6)

MeTEPA has not received so far much attention in mutagenicity experiments with Drosophila (Table 29). A paper by Benes and Sram (94) is the only report. In this study MeTEPA was shown to exhibit mutagenic activity by the induction of recessive lethal mutations, following injection of 5.0 mM (about 285 μg/g) into the body cavity of adult males. Under the conditions of testing, MeTEPA was clearly mutagenic, giving rise to 4.6% (31/668) mutations in brood 1 (postmeiotic cells) and 4.4 (6/138) in brood 2 (presumably meiotic stages).

N-Methyl-N'-nitro-N-nitrosoguanidine, MNNG, MNG (CAS 70-25-7)

When injected or fed Drosophila males, MNNG causes a broad spectrum of genetic damage at meiotic and postmeiotic germ cell stages, as indicated by its effectiveness in producing recessive lethal mutations in the X-chromosome, loss of one or both markers of a doubly marked Y-chromosome (deletions), entire loss of the Y- (or X-) chromosome (breakage or nondisjunction), translocations between major autosomes, and nondisjunction (Table 30). In spermatogonia, the pattern of genetic response is far smaller. No translocations and only a few chromosome losses are found in premeiotic cells.

Treatment Conditions. The application procedures used in treatment of adult Drosophila males with MNNG were 18-24 hr feeding (97,101) and abdominal injection (95,96,98-100). The concentrations used were rather high in all studies, ranging from 2.5 mM to 6.8 mM. There is only one study (96) with a fairly complete analysis of the mutational pattern produced by MNNG after it had been injected into adult males. However, since this work was carried out with a saturated solution of MNNG, there is no information about the concentration the flies were exposed to.

Mutational Spectrum. MNNG attacks all germ cell stages in Drosophila testes, as indicated by the occurrence of high frequencies of recessive lethal mutations (Table 31). Similarly, MNNG is capable of causing deletions (marker loss of a $sc^8.Y.B^S$ chromosome) in both postmeiotic and premeiotic cells, with the B deletion about three times as numerous as the $\underline{y}^+$ deletion, as would be expected

Table 28. Occurrence of X-chromosome recessive lethals and of translocations between chromosome 2 and 3 in sperm exposed to MMS for 24 hr[a]

Concentration (mM)	Recessive lethal mutations: N gametes tested	Recessive lethal mutations: % lethals	Translocations: N gametes tested	Translocations: Translocations N	Translocations: %	Storage period in females (days)
5.0×10^{-1}	588	6.5	471	0		0
			365	0		1-3
			352	1	(0.28)	3-6
			180	1	(0.56)	6-8
			28	1		8-10
			9	0		10-13
1.0	568	8.1	346	0		0
			336	1	0.30	1-3
			341	4	1.17	3-6
			210	2	0.95	6-8
			29	0		8-10
			14	2	14.3	10-13
2.0	561	14.3	325	0		0
			346	0		1-3
			334	13	3.89	3-6
			115	29	25.2	6-8
			3	0		8-10
4.0[b]	497	19.5	63	1	1.59	0
			302	1	0.33	1-3
			339	12	3.54	3-6
			17	2	11.76	6-8
			37	1	2.7	8-10

[a] Data of Vogel (25).
[b] About the LD_{50}.

Table 29. Genetic damage by MeTEPA in Drosophila

Type of damage	Sperm	Sperm-atids	Sperma-tocytes	Sperma-togonia
Recessive lethals	+	+	+[a]	[a]
Dominant lethals	Nt[b]	Nt	Nt	Nt
Translocations	Nt	Nt	Nt	Nt
Chromosome loss	Nt	Nt	Nt	Nt
Nondisjunction	Nt	Nt	Nt	Nt

[a]Sampling procedure not clearly indicated in the paper.
[b]Nt = no test.

because B is attached to the long arm of the Y-chromosome and y^+ to the short arm (96). A less uniform picture in mutagenic response of germ cells is obtained when considering the incidence of gross structural aberrations with MNNG. Males showing entire loss of the X- or Y-chromosome appeared only in brood 1 or 2 (1-6 days after termination of treatment), this indicating that the mutations had arisen in post-meiotic and meiotic stages (96). Browning discussed that they could have been the result either of chromosome breakage or nondisjunction.

Another characteristics of MNNG mutagenesis in Drosophila is high incidence of lethal mutations expressed as mosaics (97,99). Browning (99) reported that when virgin males were injected abdominally with MNNG, the compound produced only fractional mutations (F_3 tests for recessive lethals) while the F_2 tests for completes were negative.

The general tendency with translocations is that their number decreased rapidly the younger the tested germ cell (at the time of treatment). In a detailed brood pattern analysis with ten 3-day broods, 3.4% translocations (14 in 412 gametes) were detected in brood 1, 2.5% (5 in 201) in brood 2, 0.5% (1/211) in brood 3, but none thereafter in about 1400 tests (96). Failure of MNNG to produce in premeiotic cells genetic damage associated with gross structural aberrations indicates that the recessive lethals, which were detected in substantial numbers (up to 5%) in spermatogonia, were predominantly point mutations.

Table 30. Mutational spectrum by MNNG in Drosophila

Type of damage	Sperm	Spermatids	Spermatocytes	Spermatogonia	References
Recessive lethals	+	+	+	+	95-101
Dominant lethals	Nt[a]	Nt	Nt	Nt	
Translocations	+	+	(+)?	0	95,96
Chromosome loss (X,Y)	+	+	(+)?	0	95,96
Partial loss (Y^L; Y^S)	+	+	(+)	(+)	95,96
Nondisjunction			+	+	95,96

[a]Nt = no test.

Evaluation. MNNG when injected or fed to Drosophila males causes high frequencies of point mutations in all cells of the spermatogenesis cycle. Gross structural aberrations (translocations and chromosome loss) are produced in post-meiotic and meiotic male germ cells but not in spermatogonia. Analysis is fairly complete in terms of the types of genetic damage caused by MNNG, but information is absent on LEC's for various end points and LD_{50} values and the interrelationship of concentration and mutagenic response. Since detection of mutagenicity with MNNG has created problems with all mammalian tests relying on the induction of chromosome breakage, a more in-depth study of its mutagenic potential in Drosophila is strongly recommended.

1,4-Butanediol, Myleran, Busulfan (CAS 55-98-1)

Only few papers on mutagenicity of Myleran in Drosophila were available (Table 32). Frequencies of up to 3% X-linked recessive lethal mutations were found in post-meiotic cells when adult males were injected with 0.3-0.4 μl Myleran solution of concentration 2.0 mM, 6.0 mM, or 10.0 mM (102-104). It is not clear from the literature whether Myleran could be dissolved at these high concentrations. Given per os to adult males for 3 days, Myleran produced 0.3-0.5% recessive lethal mutations at 9.4×10^{-2} mM (lowest effective concentration), and 0.3-1.6% at 2.0×10^{-1} mM (105). The few mutations at 9.4×10^{-2} mM Myleran all occurred in post-meiotic cells while not a single mutation was found in younger stages in

Table 31. MNNG mutagenicity in Drosophila

Treatment	Sex	Response	Activity (%)	Cell stage[a]	References
24 da, adult feeding, 3.4 mM	M	Recessive lethals +	3.5	sp	97
24 da, adult feeding, 6.8 mM	M	Recessive lethals +	7.1	sp	97
0.3 μl injection of 2.5 mM	M	Recessive lethals +	10.0	Not specified[c]	98,100
0.3 μl injection of 5.0 mM	M	Recessive lethals +	15.5	Not specified[c]	98,100
Injection, saturated soln	M	Recessive lethals +	5.3	sp,st,sc,sg	95,96
Injection, saturated soln	M	Entire chromosome loss (X,Y) +	0.12 (0 control)	sp,st,sc?, 0 in sg	96
Injection, saturated soln	M	Partial loss (Y^L + Y^S) +	0.50 (0.12 control)	sp,st,sc?, 0 in sg	96,99
Injection, saturated soln	M	Translocations (2,3,Y) +	0.5 - 3.5	sp,st,sc?, 0 in sg	96

[a]Mean mutation frequency.

[b]Stages: sp, spermatozoa, st, spermatids; sc, spermatocytes; sg, spermatogonia.

[c]Apparently all stages of spermatogenesis were tested.

about 2000 tests. Since meiotic and premeiotic cells were not sampled from males treated with 2.0×10^{-1} mM Myleran, the question remains open as to whether this mutagen produces genetic damage in early germ cells. Browning and Altenburg (103) investigated the ability of Myleran to produce visible mutations at 14 selected loci (X-chromosome) by the Maxi technique. In all, 39 visibles (13 completes and 26 mosaics) were found in 17,000 chromosomes tested; this gave a mutation rate of 0.23% at a recessive lethal frequency of 2.0% (40 lethals in 1987 chromosomes).

Table 32. Genetic damage by Myleran in Drosophila

Type of damage	Sperm	Sperm-atids	Sperma-tocytes	Sperma-togonia	Refer-ences
Recessive lethals	+	+	?	?	102-105
Dominant lethals	Nt[a]	Nt	Nt	Nt	
Translocations	Nt	Nt	Nt	Nt	
Chromosome losses	Nt	Nt	Nt	Nt	
Specific locus mutations	+	Nt	Nt	Nt	103
Nondisjunction	Nt	Nt	Nt	Nt	

[a]Nt = no test.

Mitomycin C (CAS 50-07-7)

Mitomycin C (MC) represents one of the few mutagens which was extensively tested in both Drosophila sexes (Tables 33 and 34). In immature oocytes and in oogonia it causes sex-linked recessive lethal mutations, X-chromosome loss, detachments (of a compound X-chromosome), Y-fragmentations, and various types of interchanges between nonhomologous chromosomes (X-Y, X-4R, Y-4R), but no effect of nondisjunction (XXY ♀♀) was observed. After treatment of the male, sex-linked recessive lethals are found in all cell stages, interchanges occurred in spermatocytes and spermatogonia (X-Y, Y-Y), and the rate of nondisjunction remained unaffected. Chromosome rearrangements are already produced at concentrations which give 1-5% lethals, a striking similarity with the mutational pattern of radiation. The dependence of the categories of genetic damage on concentration of this agent has so far received very little attention. The few data on this point (0.3 mM-1.8 mM) are too limited to permit any valid conclusion. LEC and LD_{50} values have yet to be determined.

Table 33. Genetic damage by mitomycin C in Drosophila males

Type of damage	Sperm	Sperm-atids	Sperma-tocytes	Sperma-togonia	Refer-ences
Recessive lethals	+	+	+	+	106-108
Dominant lethals	Nt[a]	Nt	Nt	Nt	
2;3-Translocations	Nt	Nt	Nt	Nt	
Chromosome loss	Nt	Nt	Nt	Nt	
Partial loss (deletions)	Nt	Nt	Nt	Nt	
Nondisjunction			0	0	109
Interchanges (X-Y; Y-Y)			+	+	109

[a]Nt = no test.

Table 34. Genetic damage by mitomycin C in Drosophila females

Type of damage	Oocytes (immature)	Oogonia	References
Recessive lethals	+	+	109,110
Interchanges (X-Y; Y-Y; X-4R; Y-4R)	+	+	109,111,112
X-nondisjunction	0	0,(+)?	106,109,110,112
Detachments (compound X)	+		112
X-loss and Y-fragments	+		112

Treatment Conditions. The two methods by which mitomycin C was most commonly administered to the flies were either by injection into the body cavity or by feeding. In the latter case adult or larval stages (Table 35) were treated. The concentrations used varied between 0.30 mM and 1.8 mM.

Table 35. Treatment procedures with mitomycin C in Drosophila

Route of application	Sex	Concentration (mM)	References
Adult injection	M	0.3 MC in 0.7 N NaCl	113
Adult feeding for 2 da	M	0.30; 0.75; 1.50 MC	107,108
Adult feeding for 20 hr	F	0.37 MC	109,111
Adult feeding for 24 hr	F	1.8 MC	112
Feeding of third instar larvae for 1 hr	F	1.8 MC	110

Biological Effects. LD_{50} and LEC's in males were not determined for mitomycin C. Mukherjee (113) has conducted the most extensive study for sex-linked recessive lethals, using daily broods and a female-male proportion of 6:1 (Table 36). MC is mutagenic in all stages of the tests of Drosophila, with a peak of 5% recessive lethals in the third brood, which presumably represents spermatid stages at the time of treatment. The relative frequencies of lethal mutation induction by feeding MC (107) were similar to those obtained after injection (113) and indicated that a concentration of 0.75 mM (48 hr feeding of males) yields a maximum number of 6.6% lethals in spermatids. This frequency was not significantly different from the 4.1% and 3.6% mutations obtained at 0.30 mM and 1.5 mM MC, respectively. Unfortunately, the concentration range in the studies with MC was too small for evaluating concentration-effect relationships.

MC produces nonhomologous interchanges (X-Y) in gonial cells of Drosophila males [adult feeding with 0.37 mM for 21 hr (109,111)], but yet did not induce crossing-over in males (it does so in females). Obviously the induction of X-Y and Y-Y interchanges by MC does not require the normal meiotic apparatus for crossing over as indicated by their recovery from male gonial cells (109,111) (this conclusion was also supported by the recovery of interchanges from C3G homozygotes and of clusters from oogonial cells after female treatment).

Table 36. Frequencies of recessive sex-linked lethal mutations in successive daily broods with adult injection of a 0.30 mM MC solution[a]

Time after treatment (days)	Number tested	Lethals (%)
1	1080	1.2
2	1239	2.0
3	1349	5.0
4	1051	4.1
5	935	3.0
6	1119	3.3
7	1035	2.5
8	1160	1.2
Totals	8978	2.8

[a]Data from Mukherjee (113).

In females, mitomycin C was found to produce sex-linked recessive lethal mutations in all stages of oogenesis with the highest frequency (4.2%) occurring in meiotic broods (in those cells presumed to be near completion of the first meiotic prophase at the time of MC treatment), followed by a decline in mutation rate (down to 0.4-0.6% in gonial cells) (109).

Conflicting reports have been published with respect to nondisjunction induction by MC. Ostertag and Haake (110), Schewe et al. (109), and Walker and Williamson (112) found no differences between treated and control groups. These observations are in contrast to the result of Hayashi and Suzuki (106), who found that MC increased primary nondisjunction in brood presumed to represent gonial stages at the time of treatment.

As in males, MC was found to induce a wide range of rearrangements in females (109,111,112). These are interchanges (X-Y, Y-Y, X-4R, Y-4R), detachments of a compound-X-chromosome, X-loss, and fragmentation of the Y-chromosome. Thus it induces the types of aberrations recovered after x-irradiation. Dose-response experiments with immature oocytes conducted by Parker and Hammond (114) indicate that the frequency of detachment-bearing progeny induced with a 1.8 mM solution of MC corresponds to a dose of 250-500 R of x-rays.

Evaluation. Walker and Williamson (112) discuss two apparent modes of action in MC mutagenesis. In the first place, detachment and Y-chromosome classes, and the Y-breakpoint distribution in both X-Y and Y-fragments are similar to those observed in progeny of irradiated males. Furthermore, MC is a potent chromosome-breaking agent in Drosophila, when relative frequency of chromosome breakage is compared to recessive lethal frequency. Secondly, the production of gynandromorphs and mosaics (112) shows a delayed effect which is often observed with chemical mutagens. Hence, it is believed that MC can act either as a monofunctional or as a bifunctional alkylating agent (115). MC produces point mutations and rearrangements in both males and females, but it is still an open question whether MC is capable of inducing nondisjunction.

Comparative studies on the relation of concentration and the type of genetic damage were not conducted with MC. In addition, no experiments to detect dominant lethals, 2;3-translocations, and deletions (Y^S; Y^L) were conducted. Such studies should be carried out because MC represents one of the few cases for which extensive data exist for both males and females.

Procarbazine (CAS 366-70-1)

The genetic activity of procarbazine has been demonstrated in the whole spermatogenesis cycle of Drosophila (Table 37). Procarbazine induced preferentially point mutations (recessive lethals) in metabolically active cells (spermatids and spermatocytes) in comparison with lower activities in metabolically less active sperm and spermatogonia (which have a poorly developed endoplasmic reticulum). The mutagenic response to procarbazine in Drosophila confirms that insect metabolism precedes the formation of mutagenic metabolites in the intact body. The well-documented genetic activity of procarbazine in the recessive lethal test stands in sharp contrast to its failure to produce chromosome breakage events, such as partial sex-chromosome loss and 2;3-translocations. A low incidence of total sex-chromosome loss (Y,X), translocations, and of dominant lethals was observed in spermatids.

Treatment Conditions. Procarbazine was tested extensively in Drosophila by Blijleven and Vogel (116) in feeding experiments at concentrations which ranged from 0.5 mM to 30 mM procarbazine. Under the conditions of test, the mutagen did not affect the survival rate after male exposure for 2 days, and an LD_{50} was not determined.

Table 37. Genetic damage by procarbazine in Drosophila

Type of damage	Sperm	Sperm- atids	Sperma- tocytes	Sperma- togonia
Recessive lethals	+	+	+	+
Dominant lethals	0	+	Nt[a]	Nt
2;3-Translocations	0	0(+)	Nt	Nt
Entire X-(Y-)loss	0	+	Nt	Nt
Partial loss (Y^L,Y^S)	0	0	Nt	Nt
Nondisjunction	Nt	Nt	Nt	Nt

[a]Nt = no test.

Mutational Spectrum. Procarbazine at concentrations ranging from 0.5 to 30 mM is highly efficient in the production of recessive lethals (Table 38). The LEC is roughly 0.5 mM after 2 days exposure. The response of various cell types of spermatogenesis is considered of particular interest for the understanding of procarbazine mutagenesis in Drosophila. In metabolically active spermatids, a steep increase is obtained for mutation induction between 0.5 and 10.0 mM, with a plateau above 10.0 mM. In sperm, a clear-cut concentration-effect relationship is not apparent. The mutational pattern described here for procarbazine resembles that given by most precarcinogens in Drosophila. This is consistent with the important role in this system of intragonadal metabolism (activation). Though very extensive experiments were carried out on the incidence of dominant lethals, 2;3-translocations, and chromosome loss in both sperm and spermatids, genetic damage by procarbazine was not easy to detect there. Low relative frequencies of dominant lethals and total chromosome loss was the only damage observed at the extraordinarily high concentration of 30.0 mM. Even well-defined breakage events such as partial sex-chromosome loss and 2;3-translocations could not be detected in the first study. In further experiments (117), 0.16% 2;3-translocations (2/1220) were obtained from a storage experiment with spermatids. Comparison of the LEC: LD_{50} proportion demonstrates the enormous discrepancy in sensitivity between the recessive lethal test and those relying on the production of chromosome breakage.

Evaluation. It is somewhat surprising that the cytostatic procarbazine does not cause chromosome breakage in Drosophila. The

Table 38. Relation between LD_{50} and lowest genetically active concentration (LEC) for different genetic end points in postmeiotic male germ cells of Drosophila exposed to procarbazine for 2 days

Germ cell stage	Concn. Range (mM)	LD_{50} (mM)	Recessive lethals		Dominant lethals		Total chromosome loss		Partial chromosome loss		Trans- location	
			Activity	LEC/LD_{50}	Activity	LEC/LD_{50}	Activity	LEC/LD_{50}	Activity	LEC/LD_{50}	Activity	LEC/LD_{50}
Sperm	0.5–30	>30[b]	+	1:60	(+)	1:1	(+)	1:1	0		0	
Spermatids	0.5–30	>30[b]	+	1:60	+	1:1	+	1:1	0		+	1:1

[a]Activity: +, mutagenic; (+), activity questionable; 0, nonmutagenic.

[b]Not toxic in the tested concentration range.

role of metabolism and possible mechanism of biological action of procarbazine have been reviewed (118). Azoprocarbazine, the initial product in the metabolism of procarbazine, has at least three alternate pathways of metabolism. This fact should be taken into consideration when comparing the biological effects of procarbazine in various systems. Formation of methyl dizene, which readily yields the corresponding methyl group, is thought to be one of the products of enzymatic activity. Such active compounds are typified by their extremely weak ability to induce chromosome breaks. Comparison with other agents which are thought to yield similar reactive intermediates after activation, is given in the section on mutational spectrum of nitrosamines.

Triethylenemelamine, TEM (CAS 51-18-3)

Analysis of the experimental data shows that TEM is capable of inducing genetic damage in all stages of the spermatogenesis cycle (Table 39).

Table 39. Genetic damage by TEM in Drosophila

Type of damage	Sperm	Spermatids	Spermatocytes	Spermatogonia	References
Recessive lethals	+	+	+	+	37,59, 119-125
Dominant lethals	+	+	+	+	38,124
Translocations	+	+	Nt[a]	Nt	37,38,59, 119,120, 122,126, 127
Entire chromosome loss	+	+	Nt	Nt	125
Partial loss (deletion)	+	+	+?	Nt	125
Nondisjunction	Nt	Nt	Nt	Nt	

[a]Nt = no test.

Information on mutagenicity by TEM is fairly complete for mature spermatozoa, for which incidence of recessive lethal mutations, translocations, dominant lethals, entire chromosome loss (X,Y), deletions of Y-chromosome markers, and specific locus mutations has been demonstrated. In addition, about 70-85% of all chromosome breaks and dumpy mutations are expressed as mosaics in unstored sperm, while in stored cells the major part (up to 97%) of induced breaks shows up as complete changes. It has also been found that X-linked recessive lethal mutations are most frequent in postmeiotic stages, the mutational yield being 5-10 times lower in meiotic and premeiotic cells.

Treatment Conditions. Most mutagenicity studies with TEM have concentrated on its mode of action and the kind of genetic changes which TEM produces in comparison with that caused by x-rays. Therefore, only a small range of concentrations was studied (3.0×10^{-2} mM to 3.4×10^{-1} mM), and no efforts were made to determine LEC and LD_{50} values. Adult injection has been the treatment procedure which was preferentially used with TEM.

Biological Effects. As would be expected from a trifunctional alkylating agent like TEM which causes crosslinks, a high degree of cytotoxic activity and dominant lethality is observed with this agent. Analysis of the various stages of spermatogenesis revealed a reaction pattern which is typical of alkylating agents with more than one reactive site: at high concentrations ($>2.0 \times 10^{-1}$ mM) postmeiotic stages are the only ones which can survive the treatment, showing a high rate of recessive and dominant lethal mutations, while meiotic and spermatocyte stages are destroyed by TEM as indicated by complete sterility of the later broods. In the concentration range from 1.0×10^{-1} to 3.0×10^{-2} mM, TEM younger cells survive, but recessive lethal yield was 5-10 times below that determined for postmeiotic cells similarly treated. A concentration of 3.0×10^{-2} mM (2.6 μg/g Drosophila) represents the lowest TEM concentration used in mutagenicity experiments (123) with Drosophila, and since at this level induced dominant lethal frequency was still 24% (control rate zero) and the recessive lethal yield was 2%, the LEC for both genetic endpoints must be considerably lower. A LEC between 1.0×10^{-3} to 5.0×10^{-3} mM (∿ 0.25-1.25 μg/g), seems a realistic estimate based on comparable experimental data with the closely related Trenimon.

Like other studies of genetic damage with chemical mutagens, examples of adverse effects of TEM on Drosophila have come primarily from experiments with sperm. Most of those investigations have been directed to the phenomenon of delayed opening of TEM-induced breaks (storage effect), the nature of chromosome rearrangements and recessive lethal mutations induced by TEM, and the proportion of complete vs. mosaic changes in both unstored and stored spermatozoa. As can be seen from Table 40, TEM produces all categories

Table 40. Pattern of genetic changes by TEM in sperm

Treatment	Concentration (mM)	Genetic endpoint	Pre-stored No. tested	Pre-stored %	Pre-stored Storage effect	Pre-stored Period of storage	References
Injection; 0.3 μl in 0.7% saline	2.0×10^{-1}	Loss of X (or Y)	2342	0.18[a]	Nt[b]	---	122
		Loss of $Y^L + Y^S$	2342	0.90[b]	Nt	---	122
		X-linked recessive lethals	824	10.2	Nt	---	122
		Translocations (Y-II-III)	889	2.59	Nt	---	122
Injection; 0.3 μl in 0.7% saline	1.5×10^{-1}	X-linked recessive lethals	1037	10.0	1-2 fold	10-12 days	128
		Translocations	834	3.0	12 fold	10-12 days	128
Injection; 3 μl in 0.7% saline	5.0×10^{-2}	X-linked recessive lethals	617	3.4	?	6 days	38
		Translocations	1095[d]	0.55[d]	∿8-11 fold	6 days	38
	1.0×10^{-1}	X-linked recessive lethals	199	9.5	?	6 days	38
		Translocations	274	1.1	∿10 fold	6 days	38
Injection; 2 μl	2.0×10^{-1}	X-linked recessive lethals	1816[e]	6.6[e]	1.3-1.9 fold	6 days	37
		Translocations	3084[e]	1.1[e]	8-13 fold	6 days	37

(continued)

Table 40. (continued)

Treatment	Concentration (mM)	Genetic endpoint	Pre-stored		Pre-stored		References
			No. tested	%	Storage effect	Period of storage	
Injection; 0.3 µl in 0.4% saline	1.5×10^{-1}	X-linked recessive lethals	867	9.7	Nt	---	59
24 hr adult feeding	1.1×10^{-1}	X-linked recessive lethals	612	5.2	Nt	---	121
Injection; 0.3 µl in 0.4% saline	3.0×10^{-2}	X-linked recessive lethals	458	2.2	Nt	---	123, 125
	6.0×10^{-2}	X-linked recessive lethals	442	2.5	Nt	---	125
	1.0×10^{-1}	X-linked recessive lethals	764	7.6	Nt	---	125
	2.0×10^{-1}	X-linked recessive lethals	243	9.5	Nt	---	125
	2.5×10^{-1}	X-linked recessive lethals	570	14.0	Nt	---	125
	3.4×10^{-1}	X-linked recessive lethals	108	14.8	Nt	---	125

[a] Control rate 0.04% (5161 tests).

[b] Nt = no test.

[c] Control rate 0.06% (5161 tests); treated group; 0.47% losses of Y^L + 0.43% losses of Y^S.

[d] Average of 2 experiments.

[e] Average of 3 experiments.

of genetic damage. Concentration-dependence was found between 3.0×10^{-2} mM and 3.4×10^{-1} mM for recessive lethal mutations (125).

A very characteristic feature of TEM mutagenesis is the very high proportion of mosaic changes. Snyder and Oster (120) found that no more than 27.5% of mutations at the dumpy locus were expressed as completes (Table 41). TEM constitutes one of the few mutagens which have been used extensively to analyze the delayed opening of chemically induced breaks and consequences of this delay for the translocation/recessive lethal ratio. By treating spermatozoa in inseminated females with "vaginal douches" of TEM, Herskowitz (119,120) found a many-fold increase in the frequency of translocations when the spermatozoa were stored in the seminal receptacles for days or weeks, whereas there was no clear tendency in this direction detectable for chromosome losses or recessive lethals. Also a storage effect for recessive lethals was not obvious from experiments carried out by Ratnayake (38), while a small increase (one- to twofold) seemed to exist in other studies (37,128). The increase in the frequency of lethals is probably entirely due to the fact that, particularly at higher concentrations, a certain proportion of lethals is associated with structural changes and that these are subject to a storage effect.

Table 41. Incidence of dumpy mutations and the frequency of mosaics induced by TEM[a]

Treatment	Concentration (mM)	Dumpy mutations Tested	Dumpy mutations %	Frequency of mosaics
Injection; 0.3 μl in 0.7% saline	2.0×10^{-1}	8378	1.64	72.5
	control	11137	0.03	100.0

[a]Data from Snyder and Oster (122).

Data are available on the increase of interchromosomal changes in stored spermatozoa (37,38,119,120,128), suggesting an approximate 6- to 13-fold increase in translocation frequencies if the time interval between termination of treatment and use of sperm is delayed for 6-12 days. In a very comprehensive in-depth study with salivary gland giant chromosomes, Slizynska (127,129) analyzed the pattern of genetic changes produced by TEM in comparison to the effects by x-rays. The proportion of latent to immediate breaks was 90-99% in unstored sperm, which she used to explain the higher proportion of mosaics, repeats and deficiencies and the lower percentage of translocations and inversions as compared with the effects given by x-rays (Table 42). Storing of treated spermatozoa reduced the proportion of latent breaks and thus the whole pattern of damage become more like that observed after irradiation. It may be of interest to note that a high proportion of recessive lethals by TEM, like those induced by x-rays but in contrast to those induced by monofunctional agents, are associated with structural rearrangements. Salivary gland chromosomes of 50 TEM-induced X-chromosome recessive lethals were analyzed by Lim and Snyder (59), and 11 (22%) were found to be associated with detectable aberrations.

Table 42. Frequencies of immediate and latent breaks induced by TEM (1.5×10^{-1} mM) and x-rays (2500 R).[a]

Mutagen and samples	Time after treatment (days)	Changes in 100 spermatozoa		Mosaic changes (%)	Ratio T/In,Rp,Df[b]
		Complete	Mosaic		
TEM, Expt. I					
Unstored	2-3	0.6	3.5	85.4	0.6
Stored I	13-17	26.7	1.8	6.3	0.7
Stored II	18-20	40.4	7.0	14.8	1.4
TEM, Expt. II					
Unstored	2-4	5.4	11.4	67.9	0.4
Stored	11-13	47.4	1.3	2.7	1.5
x-Rays	2-5	26.5	3.5	11.7	1.1

[a]Data of Slizynska (127,129).

[b]Abbreviations: Df, deficiencies; H, stored samples; In, inversions; Rp, repeats; T, translocations.

Evaluation. Even though relevant toxicological data like LEC and LD_{50} have yet to be collected for TEM in Drosophila, no further analysis of this compound is recommended. One must realize that TEM as a polyfunctional alkylating agent in its mutagenic properties is very unlike the kind of mutagens and thus the problems one encounters when dealing with environmental mutagens (which are almost always monofunctional compounds).

TEPA (CAS 545-55-1)

In Drosophila melanogaster, TEPA causes both point mutations (recessive lethals) and chromosome aberrations in all cells of spermatogenesis (Table 43). The compound is most active in mature sperm, with LEC estimated to be around 10^{-2} mM (0.5 μg/g) after injection, and ∿ 10^{-3} mM TEPA after adult feeding.

Table 43. Genetic damage by TEPA in Drosophila males

Type of damage	Sperm	Sperm-atids	Sperma-tocytes	Sperma-togonia	Refer-ences
Recessive lethals	+	+	+	+	94,130, 131
Dominant lethals	+	Nt[a]	Nt	Nt	130
Translocations	+	+	+	+	94,130, 131
Chromosome loss	Nt	Nt	Nt	Nt	
Nondisjunction	Nt	Nt	Nt	Nt	

[a]Nt = no test.

Treatment Conditions. The ability of TEPA to produce in male germ cells sex-linked recessive lethal mutations, dominant lethals and translocations (involving the two major autosomes and the Y-chromosome) was investigated by Sram and his collaborators (94,130, 131). The procedure followed for all experiments was to inject Oregon K males 0.2 μl of a solution which contained 0.7% NaCl and TEPA at concentrations varying between 1.0 × 10^{-1} and 3 × 10^{-1} mM. When calculated on a μg/g basis (the weight of a Drosophila male is about 0.7 mg) the injected doses, very roughly, were 5, 10, and 15 μg/g, respectively.

Biological Effects. The literature gives no information about LEC and LD_{50} values. The summarized results of all reported exposures with TEPA are recorded in Tables 44 and 45. Although the sample sizes were rather small in all experiments, the mutation frequencies resulting at various concentrations show concentration-dependence (Table 44). From these data it can be seen TEPA is active both in the production of point mutations and breakage-type damage in mature sperm of Drosophila. The results with 3-day broods, beginning on the second day, also show clearly that treatment with 2×10^{-1} mM TEPA induced recessive lethal mutations and translocations in all male germ cells, with a persistent decline of mutational yield from mature sperm to spermatogonia - the stage which proved most resistant to TEPA. As would be expected with a trifunctional mutagen, translocation frequency could be drastically enhanced from 0.24% to 3.2% (Table 45) when treated sperm were kept in untreated females for a period of 14-16 days.

Table 44. Concentration-effect relationship for dominant lethals, recessive lethal mutations, and translocations in Drosophila sperm by TEPA[a]

	Dominant lethals		Recessive lethals		Translocations	
Concn. (mM)	Total number of eggs	% RDL[b]	Gametes tested	%	Gametes tested	%
Control	684	0				
1.0×10^{-1}	442	2.1	213	2.8	424	0
2.0×10^{-1}	312	10.9	315	6.0	410	0.24
3.0×10^{-1}	741	36.3	158	8.2	365	0.82

[a]Data of Benes and Šram (94).

Evaluation. TEPA was found to produce all categories of genetic damage in Drosophila. It seems as mutagenic as the structurally related ThioTEPA: 2×10^{-1} mM TEPA (10 μg/g) yield mutation rates that are similar to those obtained after treating sperm with 1.2×10^{-1} mM ThioTEPA for 6 hr. Mutational yield still remains the same (ca. 6-8%) with a lower ThioTEPA concentration of 1.2×10^{-2} mM but extended exposure time of three days. The exact LEC's are unknown for either compound, but from results with other polyfunctional alkylating agents (Trenimon), presumably the LEC for mutation induction with TEPA and ThioTEPA are in the order of 10^{-2} mM (0.5 μg/g) after injection. Lower LEC values of roughly 10^{-3} (about 0.05 μg/g) may be expected with the feeding procedure, when adult males are exposed to these compounds for 3 days.

Although analysis of the mutagenic properties of TEPA is incomplete, we do not suggest further studies with Drosophila.

Table 45. Frequencies of recessive lethals and translocations in successive germ cells of the spermatogenesis cycle after injection with TEPA (2.0×10^{-1} mM or 10 μg/g); female/male ratio = 3:1[a]

Brood	Storage[b]	Time after treatment (days)	Recessive lethals Gametes tested	Recessive lethals %	Translocations[c] Gametes tested	Translocations[c] %
I	U	2-4	315	6.0	410	0.24
I	S	2-4	234	8.6	409	3.2
II	U	5-7	380	3.7	444	0.45
III	U	8-10	447	2.2	485	0.24
IV	U	11-13	448	0.45	368	0.27
V	U	14-16	424	0.47	509	0.39

[a]Data of Benes and Šram (94).

[b]U = unstored; S = storage in females for 14-16 days.

[c]Translocations between chromosomes 2, 3 and/or the Y.

ThioTEPA (CAS 52-24-4)

Incidence of sex-linked recessive lethals is the only type of genetic damage studied with ThioTEPA treatments in Drosophila (Table 46). Mutations were induced by exposing Berlin K males to a 1.2×10^{-2} mM solution of ThioTEPA for 3 days or by treating them with a higher concentration of 1.2×10^{-1} mM for a reduced period of only 6 hr (132,133). An attempt was made to analyze the mutagenic response to ThioTEPA of successive stages in spermatogenesis, by mating treated males in a 1:1 ratio to virgin females at 3-day intervals. This rough breeding scheme, however, gives little indication about the cell stages sampled in broods 2 and 3 (even in brood 3, they were presumably in spermatid stages at the time of treatment). After both treatment for 3 days and 6 hr, the frequency of mutations was highest in brood 1 which suggests that mature sperm is the stage most susceptible to the compound (Table 47). The trifunctional alkylating ThioTEPA is considered not interesting enough to justify further analysis in Drosophila.

Table 46. Genetic damage by ThioTEPA in Drosophila males

Type of damage	Sperm	Spermatids	Spermatocytes	Spermatogonia	References
Recessive lethals	+	+	+ ?	Nt[a]	132-134
Dominant lethals	Nt	Nt	Nt	Nt	
Translocations	Nt	Nt	Nt	Nt	
Chromosome loss	Nt	Nt	Nt	Nt	
Nondisjunction	Nt	Nt	Nt	Nt	

[a]Nt = no test.

Table 47. Frequencies of recessive lethal mutations after male treatment with ThioTEPA

Time after treatment	1.2×10^{-2} mM for 3 days: Gametes tested	1.2×10^{-2} mM for 3 days: Lethals (%)	1.2×10^{-1} mM for 6 hr: Stages[a]	1.2×10^{-1} mM for 6 hr: Gametes tested	1.2×10^{-1} mM for 6 hr: Lethals (%)
0-3	1227	7.5	sp	421	11.4
4-6	843	6.8	sp,st ?	581	10.8
7-9	1105	3.0	sp,st ?	972	6.0

[a]Stages: sp, sperm; st, spermatids; sc, spermatocytes.

Trenimon (CAS 68-76-8)

The main emphasis has been on the incidence of genetic damage in mature sperm. In sperm, Trenimon produces recessive lethal mutations, 2;3-translocations, entire chromosome loss (rod-X and ring-X), partial chromosome loss Y^L, Y^S), and dominant lethals (Table 48). In addition, Trenimon shows high cytotoxic activity to both mature and immature oocytes after female treatment, but the mutagen remains virtually ineffective with respect to the induction of X-chromosomal loss and nondisjunction (NDJ) in oocytes.

Table 48. Incidence of genetic damage by Trenimon in Drosophila

Type of damage	Sperm	Spermatids	Spermatocytes	Spermatogonia	References
Recessive lethals	+	+	+	+	132,133, 135-138
Dominant lethals	+	Nt[a]	Nt	Nt	137,139
Chromosome loss	+	Nt	Nt	Nt	137,139, 140
2;3-Translocations	+	Nt	Nt	Nt	25
Nondisjunction	Nt	Nt	Nt	Nt	

[a]Nt = no test.

Treatment Procedure. A level of 1.0×10^{-2} mM Trenimon constitutes the standard concentration employed in most studies with either injection or adult feeding. Comparisons of both application procedures revealed that injection of 0.2 μl of a 10^{-2} mM Trenimon solution is equally mutagenic as 17 hours adult feeding of the same concentration. Since the weight of a Drosophila male is roughly 0.7 mg, the Trenimon concentration administered by injection is 0.65 μg/g.

Biological Effects. The LD_{50} of Trenimon is unknown for Drosophila. Males can tolerate treatment with 8×10^{-2} mM for 24 hr but become sterile at this concentration (95-100% dominant lethals) (25). Various types of genetic damage were found in mature sperm which had been exposed to 1.0×10^{-2} mM Trenimon (Table 49). The discrepancy in the results on recessive lethals may be mainly due to the small size in some studies and to the low stability of Trenimon in aqueous solution.

In translocation experiments with stored and unstored sperm which had been exposed to 1.0×10^{-2} mM Trenimon for 24 hr, a low frequency of 0.35% (2 translocations in 577 gametes) translocations between the two large autosomes (25) was found. A concentration increase from 1.0×10^{-2} mM to 2.0×10^{-2} mM was accompanied by a 4-fold increase in the translocation frequency (Table 50). However, the recessive lethal frequency obtained at 2.0×10^{-2} mM did not differ significantly from that produced at the lower concentration. The findings on translocations are in agreement with previous observations that translocation frequency increases as the square of

Table 49. Occurrence of different types of genetic damage in sperm treated with Trenimon

Treatment	Recessive lethals (%) (X-chrom.)	Induced dominant lethals (%)	Chromosome loss (%)			Transloca-tions (%)	Refer-ences
			Ring-X	Rod X(Y)	$Y^L + Y^S$		
None (spontaneous frequency)	0.2	0	0.6-1.4	0.04	0.02	-	
1.1×10^{-2} mM, 3 days; adult feeding	17-24						132, 133
1.0×10^{-2} mM; (injection)	4.7-9.5		4.8				135, 136,138 140
1.0×10^{-2} mM, 17 hr; adult feeding	6.6-13	25-30[a]	3.2	0.47	0.47	0.35	25, 137-140
2.0×10^{-2} mM, 24 hr; adult feeding	8-18					1.47	25

[a] 52% if the chromosome set contains a ring-X-chromosome.

Table 50. Induction of 2;3-translocations in stored and nonstored sperm after exposure to 2.0×10^{-2} mM Trenimon for 24 hr[a]

Storage period (days)	Gametes	Translocations No.	Translocations %
1-2	1300	15	1.15
2-5	2705	25	0.92
5-7	1023	21	2.05
7-9	425	10	2.35
9-12	107	11	10.28
0-12	5560	82	1.47 ± 0.16

[a]Pooled data from four experiments (25).

the exposure concentration (141). As will be seen further from Table 50, a very pronounced storage effect was observed with this agent [pooled data from four different runs (25)].

Contrary to extensive investigation on mature spermatozoa, little attention has been paid on the incidence of genetic damage in immature stages of spermatogenesis - spermatocytes, and spermatogonia. Recessive lethal yield and incidence of sex chromosome loss was found to be highest in sperm and spermatids (presumably late spermatids) which seem to be equally sensitive to Trenimon (132,135,136). The mutation frequency then drops substantially in spermatocytes, and in spermatogonia (132,135). The low relative frequencies of recessive lethal mutations in spermatocytes and spermatogonia coincides well with the resistance to X-loss induction and absence of NDJ in mature and immature oocytes, in spite of the extremely high lethal effects of Trenimon on mature oocytes. The mortality rates obtained for mature oocytes are 92% at 1.0×10^{-2} mM, and 78.3% at 5.0×10^{-3} mM (142). The implication is that the trifuctional agent Trenimon is capable of inducing a substantial amount of breakage events which cannot pass through meiosis.

<u>Evaluation</u>. Trenimon mutagenesis in Drosophila shows two characteristics: sperm and spermatids are the most sensitive germ

cell stages to induction of recessive lethal mutations and various kinds of chromosome aberrations; and recovery of transmissible genetic damage is low (recessive lethals) in spermatocytes and spermatogonia, and absent (nondisjunction and X-loss) in mature and immature oocytes. An LD_{50} has not yet been determined for Trenimon. Since exposure to 1.0×10^{-2} mM (0.65 μg/g) Trenimon produces up to 13% recessive lethals (1-day exposure), the LEC for recessive lethals is assumed in the range of 10^{-3} mM (0.065 μg/g) or even below that level. Experiments are currently in progress to determine the exact LEC for Trenimon. No further analysis is proposed with this agent.

Vinyl Chloride (CAS 75-01-4)

Vinyl chloride monomer (VCM) was found mutagenic in the recessive lethal test both after short-term and long-term exposure. The LEC was 850 ppm after 2-day exposure, and this figure could be lowered to 30 ppm by prolonging the exposure time to 17 days. In strong contrast to those clearly positive results, dominant lethals and transmissible chromosome aberrations in sperm and spermatids were not observed with VCM (Table 51).

Table 51. Mutagenicity by vinyl chloride monomer in Drosophila males

Type of damage	Sperm	Spermatids	Spermatocytes	Spermatogonia	References
Recessive lethals	+	+	+	+	18,19
Dominant lethals	0	0	Nt[a]	Nt	18
Chromosome loss	0	0	Nt	Nt	18
2;3-Translocations	0	0	Nt	Nt	18
Nondisjunction	Nt	Nt	Nt	Nt	

[a]Nt = no test.

Treatment Procedure. In inhalation experiments, males were exposed to a wide array of concentrations which ranges from 30 ppm (17 days exposure) up to extremely high concentrations as high as 200,000 ppm (3 hr exposure).

Table 52. Pattern of genetic damage by VCM in Drosophila males

Route of application	Treatment (ppm)	Type of damage	Activity: + or 0	Activity: Stage[a]	Peak Activity	References
Inhalation for 3 hr	10,000	Recessive lethals[b]	+[c]	?	?	19
	200,000	Recessive lethals[b]	+[c]	?	?	
Inhalation for 17 days	30	Recessive lethals	+[d]	sp,st,sc,sg	Unknown	
	850	Recessive lethals	+	sp,st,sc,sg	Unknown	18
Inhalation for 2 days	30	Recessive lethals	0	sp,st,sc,sg	-	
	200	Recessive lethals	0	sp,st,sc,sg	-	
	850	Recessive lethals	(+)	sp,st,sc,sg	st (0.72%)	
	10,000	Recessive lethals	+	sp,st,sc,sg	st (3.3%)	
	30,000	Recessive lethals	+	sp,st,sc,sg	st (3.5%)	
	50,000	Recessive lethals	+	sp,st,sc,sg	st (3.1%)	
	30,000	Dominant lethals	0	sp,st		
	30,000	Entire (X,Y) loss	0	sp,st		
	30,000	Partial (Y^L,Y^S) loss	0	sp,st		
	30,000	2;3-translocations	0	sp,st		

[a]Stages: sp, sperm; st, spermatids; sc, spermatocytes; sg, spermatogonia.

[b]Completes and mosaics.

[c]Frequency not given.

[d]Spontaneous rate, 0.10%.

Biological Effects. Two attempts were undertaken to demonstrate mutagenic effects in Drosophila with VCM (18,19). There are several conclusions which can be deduced from the experiments presented in these papers. VCM does not enhance mortality rates when males are exposed to concentrations from 850 ppm to 50,000 ppm for 2 days or to 30 ppm and 850 ppm for 17 days. An LD_{50} has not been determined from VCM in Drosophila. VCM is decidedly active in producing point mutations (recessive lethals) in all stages of the spermatogenesis cycle. The LEC for point mutation induction is about 30 ppm for a 17 days exposure, and about 850 ppm for a 2 days exposure. Peak mutagenic activity is found in spermatids (the endoplasmic reticulum is highly developed in early spermatids) which supports the hypothesis that intragonadal activation plays a major role in mutagenesis by indirectly acting mutagens in Drosophila (8). It is a further characteristic of VCM mutagenesis that mutation yields are not related to concentration at levels above 10,000 ppm (2 days exposure). This is most simply interpreted as an indication of enzyme saturation in the bioactivation of VCM. The lack of any concentration response relationship for recessive lethal induction (above certain concentration levels) helps to explain the failure of VCM to produce dominant lethals and various categories of viable chromosome aberrations (Table 52). There is now a substantial body of data which provides strong evidence that breakage-type damage is almost never detectable at mutagen concentrations which induce point mutation frequencies 30- to 50-fold above control frequencies (8). This phenomenon of two effective concentrations for mutations vs. chromosome breakage effects in the case of VCM would yield a "false negative," since the concentration of active material in the germ cells apparently does not reach a level which is sufficient for breakage induction.

Evaluation. A level of 30 ppm VCM constitutes the lowest concentration for which mutagenic activity has been suggested in Drosophila which demonstrates the sensitivity of the inhalation procedure and the high resolving power of recessive lethal assays. It would be informative to know whether prolongation of exposure time could shift the LEC towards even lower concentrations.

ACKNOWLEDGEMENTS

It is a pleasure to thank Mr. John S. Wassom, Director of the Environmental Mutagen Information Center, Oak Ridge, Tennessee, U.S.A., for assistance with print-outs of literature, and Dr. Charles S. Aaron for critical reading of the manuscript.

REFERENCES

1. Vogel, E., and Leigh, B. Mutat. Res. 29: 383-396 (1975).

2. Vogel, E. Mutat. Res. 33: 221-228 (1975).

3. Auerbach, C. In: Mutation - an Introduction to Research on Mutagenesis. Part I. Methods. C. Auerbach, Ed., Oliver and Boyd, Edinburgh and London, 1962, pp. 27-67.

4. Würgler, F. E., and Ulrich, H. In: The Genetics and Biology of Drosophila, Vol. 1c, M. Ashburner and E. Novitiski, Eds., Academic Press, London-New York-San Francisco, 1976, pp. 1269-1298.

5. Traut, H. Internat. J. Rad. Biol. 7: 401-403 (1963).

6. Abrahamson, S., and Lewis, E. G. In: Chemical Mutagens, Vol. 2, A. Hollaender, Ed., Plenum Press, New York, 1971, pp. 461-487.

7. Muller, H. J., and Oster, I. I. In: Methodology in Basic Genetics, W. J. Burdette, Ed., Holden-Day, San Francisco, 1963, pp. 240-278.

8. Vogel, E., and Sobels, F. H. In: Chemical Mutagens, Vol. 4, A. Hollaender, Ed., Plenum Press, New York, 1976, pp. 93-141.

9. Würgler, F. E., Sobels, F. H., Vogel, E. In: Handbook of Mutagenicity Test Procedures, B. J. Kilbey, M. Legator, W. Nichols and C. Ramel, Eds., Elsevier/North Holland Biomedical Press, 1977, pp. 335-373.

10. Vogel, E. In: The Origins of Human Cancer, H. H. Hiatt, J. D. Watson, and J. A. Winston, Eds., Cold Spring Harbor Lab., 1977, pp. 1483-1497.

11. Baars, A. J., Zijlstra, J. A., Vogel, E., and Breimer, D. D. Mutat. Res. 44: 257-268 (1977).

12. Lamb, M., and Lilly, J. L. Mutat. Res. 11: 430-433 (1971).

13. Lilly, J. L., and Lamb, M. J. EMS Newsletter 6: 26 (1972).

14. Sorsa, M., and Pfeifer, S. Hereditas 75: 273-277 (1973).

15. Bertram, C., Höhne, G. Strahlentherap. 43: 388-391 (1959).

16. Frye, S. Brit. Empire Cancer Campaign, 38th Ann. Rept., Pt. 2: 670 (1960).

17. Röhrborn, G. Mol. Gen. Genetics, 102: 50-68 (1968).

18. Verburgt, F. G., and Vogel, E. Mutat. Res. 48: 327-336 (1977).

19. Magnusson, J., and Ramel, C. Mutat. Res. 38: 115 (1976).

20. Vogel, E. In: Screening Tests in Chemical Carcinogenesis (IARC Scientif. Publ. No. 12), R. Monesano, H. Bartsch, and L. Tomatis, Eds., IARC, Lyon, 1976, pp. 117-132.

21. Fahmy, O. G., Fahmy, M. J., Massasso, J., and Ondrej, M. J. Mutat. Res. 3: 201-217 (1966).

22. Fahmy, O. G., and Fahmy, M. J. Mutat. Res. 6: 139-154 (1968).

23. Pasternak. L. Acta Biol. Med. Ger., 10: 436-438 (1963).

24. Pasternak, L. Arzneimittelforsch. 14: 802-804 (1964).

25. Vogel, E. Unpublished data.

26. Fahmy, O. G., and Fahmy, M. J. Chem. Biol. Interact. 11: 395-412 (1975).

27. Fahmy, O. G., and Fahmy, M. J. Cancer Res. 35: 3780-3784 (1975).

28. Ondrej, M. Folia Biol. (Prague), 15: 17-25 (1969).

29. Pasternak, L. Naturwiss. 49: 381 (1962).

30. Ondrej, M. Folia Biol. (Prague) 16: 225-229 (1970).

31. Rapoport, J. A. Genetika 60: 469-472 (1948).

32. Kramers, P. G. N. Unpublished data.

33. Alexander, M. L., and Glanges, L. Genetics 50: 231-232 (1964).

34. Alexander, M. L., and Glanges, L. Proc. Nat. Acad. Sci. (U.S.A.) 53: 282-288 (1965).

35. Sram, R. J. Mutat. Res. 9: 243-244 (1970).

36. Lim, J. K., and Snyder, L. A. Mutat. Res. 6: 129-137 (1968).

37. Watson, W. A. F. Z. Vererb. 95: 374-378 (1964).

38. Ratnayake, . W. Mutat. Res. 5: 271-278 (1971).

39. Sram, R. J. Mol. Gen. Genet. 106: 286-288 (1977).

40. Lim, J. K., and Snyder, L. A. Proc. Int. Congr. Genet., 12: 85 (1968).

41. Slizynska, H. Genet. Res. 4: 248-257 (1963).

42. Slizynska, H. Mutat. Res. 19: 199-213 (1973).

43. Aaron, C. S., and Lee, W. R. Dros. Inf. Service, 52: 64 (1977).

44. Browning, . S. Dros. Inf. Service, 45: 75-76 (1970).

45. Fahmy, O. G., and Fahmy, M. J. Genetics 46: 361 (1961).

46. Hotchkiss, S. K., and Lim, J. K. Dros. Inf. Service 43: 116 (1968).

47. Lee, W. R. Mutat. Res. 38: 311-316 (1976).

48. Lewis, E. B., and Bacher, F. Dros. Inf. Service 43: 193 (1968).

49. Watson, W. A. F. Mutat. Res. 14: 299-307 (1972).

50. Abrahamson, S., and Lewis, E. G. In: Chemical Mutagens, Vol. 2, A. Hollaender, Ed., Plenum Press, New York, 1971, pp. 461-487.

51. Bishop, J., and Lee, W. R. Genetics 61: S4-5 (1969).

52. Schalet, A. Genetics 86: 55 (1977).

53. Jenkins, J. B. Mutat. Res. 4: 90-92 (1967).

54. Lee, W. R., Sega, G. A., and Bishop, J. B. Genetics 60: 196 (1968).

55. Jenkins, J. B. Mutat. Res. 7: 487-489 (1969).

56. Lee, W. R., Sega, G. A., and Bishop, J. B. Mutat. Res. 9: 320-336 (1970).

57. Jenkins, J. B. Genetics 57: 783-793 (1967).

58. Lim, J. K., and Snyder, L. A. Genetics 74: S158-S159 (1973).

59. Lim, J. K., and Snyder, L. A. Genet. Res. 24: 1-10 (1974).

60. Huang, S. L., and Baker, B. S. Mutat. Res. 34: 407-414 (1976).

61. Lee, W. R. In: Radiation Research, Biomedical, Chemical and Physical Perspectives, O. F. Nijgaard, H. I. Adler, and W. K. Sinclair, Eds., Academic Press, New York, 1975, pp. 976-983.

62. Lee, W. R. In: The Genetics and Biology of Drosophila, M. Ashburner and E. Novitski, Eds., Academic Press, New York, 1976, pp. 1299-1341.

63. Aaron, C. S., Lee, W. R., Janca, F. C., and Seamster, P. M. Genetics 74: S1 (1973).

64. Epler, J. L. Genetics 48: 663-675 (1963).

65. Aaron, C. S., and Lee, W. R. Dros. Inf. Service 52: 64 (1977).

66. Legator, M. S., Zimmering, S., and Connor, T. H. In: Chemical Mutagens, Principles and Methods for Their Detection, A. Hollaender, Ed., Plenum Press, New York, 1976, pp. 171-190.

67. Hochman, B. Genetics 67: 235-252 (1971).

68. Lifschytz, E., and Falk, R. Mutat. Res. 8: 147-155 (1969).

69. Schalet, A., and Lefevre, Jr., G. Chromosoma 44: 183-202 (1973).

70. Wright, R. F., Hodgetts, R. B., and Sherald, A. F. Genetics 84: 267-285 (1976).

71. Carlson, E. A., and Southin, J. L. Genetics 48: 663-675 (1963).

72. Snyder, L. A., and Oster, I. I. Mutat. Res. 1: 437-445 (1964).

73. Woodruff, R. C., and Johnson, T. K. Genetics 74: S299 (1973).

74. Woodruff, R. C., and Ganer, R. M. Mutat. Res. 25: 337-345 (1974).

75. Kumar, S., and Sharma, R. P. Experientia 23: 266-267 (1967).

76. Browning, L. S., and Altenburg, E. Genetics 47: 945 (1962).

77. Carlson, E. A., and Oster, I. I. Genetics 47: 561-576 (1962).

78. Southin, J. L., Mutat. Res. 3: 54-65 (1966).

79. Carlson, E. A. Genetics 55: 295-313 (1967).

80. Burychenko, G. M. Sov. Genet. 5: 1494 (1969).

81. Fahmy, O. G., and Fahmy, M. J. Nature 180: 31 (1957).

82. Burychenko, G. M. Sov. Genet. 4: 1337 (1968).

83. Khan, A. H. Mutat. Res. 8: 165-175 (1969).

84. Lim, J. K., and Snyder, L. A. Genetics 61: S37 (1969).

85. Smith, P. D. Mutat. Res. 20: 215-220 (1973).

86. Bateman, A. J., and Chandley, A. C. Heredity 19: 711 (1964).

87. Würgler, F., personal communication.

88. Leigh, B., personal communication.

89. Fahmy, O. G., and Fahmy, M. J. Genetics 46: 447 (1961).

90. Liu, C. P., and Lim, J. K. Genetics 79: 601-611 (1975).

91. Liu, C. P., and Lim, J. K. Genetics 77: S40 (1974).

92. Gatti, M., Pimpinelli, S., de Marco, A., and Tanzarella, C. Mutat. Res. 33: 201-212 (1975).

93. Malling, H. V., and de Serres, F. J. Mutat. Res. 18: 1-14 (1973).

94. Benes, V., and Sram, R. Ind. Med. 38: 50-52 (1969).

95. Browning, L. S. Genetics 60: 165-155 (1968).

96. Browning, L. S. Mutat. Res. 8: 157-164 (1969).

97. Khan, A. H. Dros. Inform. Service 43: 112 (1968).

98. Fahmy, O. G., and Fahmy, M. J. Mutat. Res. 13: 19-34 (1971).

99. Browning, L. S. EMS Newsletter 3: 37 (1970).

100. Fahmy, O. G., and Fahmy, M. J. Cancer Res. 32: 550-557 (1972).

101. Judd, B. H., Shen, M. W., and Kaufman, T. C. Genetics 71: 130-156 (1972).

102. Fahmy, O. G., and Fahmy, M. J. J. Genetics 54: 146-164 (1956).

103. Browning, L. S., and Altenburg, E. Genetics 46: 1317-1321 (1961).

104. Fahmy, O. G., and Fahmy, M. J. Genetics 46: 1111-1123 (1961).

105. Röhrborn, G. Z. Vererb. 90: 116-131 (1959).

106. Hayashi, S., and Suzuki, D. T. Can. J. Gen. Cytol. 10: 276-282 (1968).

107. Suzuki, D. T., Piternick, L. K., Hayashi, S., Tarasoff, M., Baillie, D., and Erasmus, U. Proc. Nat. Acad. Sci. (U.S.) 57: 907-912 (1967).

108. Suzuki, D. T. Science 170: 695-706 (1970).

109. Schewe, M., Suzuki, D. T., and Erasmus, U. Mutat. Res. 12: 255-267 (1971).

110. Ostertag, W., and Haake, J. Z. Vererb. 98: 299-301 (1966).

111. Schewe, M., Suzuki, D. T., and Erasmus, U. Mutat. Res. 12: 269-279 (1971).

112. Walker, V. K., and Williamson, J. H. Mutat. Res. 28: 227-237 (1975).

113. Mukherjee, R. Genetics 51: 947-951 (1955).

114. Parker, D. R., and Hammond, A. E. Mutat. Res. 9: 273-286 (1970).

115. Szybalski, W., and Iyer, V. N. Microbiol. Genet. Bull. 21: 16 (1964).

116. Blijleven, W. G. H., and Vogel, E. Mutat. Res. 45: 47-59 (1977).

117. Blijleven, W. G. H., unpublished data.

118. Reed, D. J. In: Handbook of Experimental Pharmacology, A. C. Sartorelli and D. G. Johns, Eds., Vol. 38, Part 2, Springer Verlag, Berlin-Heidelberg-New York, 1975, pp. 747-765.

119. Herskowitz, I. H. Genetics 40: 574 (1955).

120. Herskowitz, I. H. Genetics 41: 605-609 (1956).

121. Lüers, H. Arch. Geschwulstforsch. 6: 77-83 (1953).

122. Snyder, L. A., and Oster, I. I. Mutat. Res. 1: 437-445 (1968).

123. Bird, M. J. J. Genet. 50: 480-485 (1952).

124. Fahmy, O. G., and Fahmy, M. J. J. Genet. 52: 603-619 (1954).

125. Fahmy, O. G., and Fahmy, M. J. J. Genet. 53: 566-584 (1955).

126. Slizynska, H. Mutat. Res. 19: 199-213 (1973).

127. Slizynska, H. Mutat. Res. 8: 165-175 (1969).

128. Snyder, I. A. Z. Vererb. 94: 182-189 (1963).

129. Slizynska, H. Genet. Res. 4: 248-257 (1963).

130. Sram, R. J. Folia Biol. (Prague) 18: 139-148 (1972).

131. Kocisova, J., and Sram, R. J. Folia
332 (1974).

132. Lüers, H., and Röhrborn, G. Mutat. Res. 2: 29-44 (1965).

133. Lüers, H., and Röhrborn, G. Genet. Today, Proc. Int. Congr. 11th, 1: 64-65 (1963).

134. Fahmy, O. G., and Fahmy, M. J. Cancer Res. 30: 195-205 (1970).

135. Mollet, P. Mutat. Res. 21: 137-148 (1973).

136. Mollet, P., Graf, U., and Würgler, F. E. Arch. Genetik 47: 184-190 (1974).

137. Vogel, E. Mutat. Res. 20: 330-352 (1973).

138. Vogel, E., and Lüers, H. Dros. Inform. Service 51: 113-114 (1974).

139. Büchei, R., and Bürki, K. Arch. Genetik 48: 59-67 (1975).

140. Mollet, P., and Büchi, R. Dros. Inform. Service 51: 96-97 (1974).

141. Nasrat, G. E., Kaplan, W. D., and Auerbach, C. Indukt. Abstammungsl. Vererb. 86: 240-262 (1964).

142. Vogel, E. Mutat. Res. 14: 250-253 (1972).

CHAPTER 10

MUTAGENICITY OF SELECTED CHEMICALS IN SOYBEAN TEST SYSTEMS

B. K. Vig

Nevada Mental Health Institute and Department of Biology

University of Nevada, Reno, Nevada 89557

and

Dr. Renee Sung

Department of Botany, University of California at Berkeley

Berkeley, California 94720

INTRODUCTION

General

The use of Glycine max (soybean) in assessment of genetic damage by mutagenesis is rather new. The organism has been employed in three types of studies, viz., leaf mosaicism, tissue culture, and traditional progeny testing. Most of the work has been confined to the first type of system which has proven a very sensitive indicator of several types of genetic alterations produced by mutagens. The system makes use of spots induced on heterozygous or homozygous recessive genotypes which express damage in the form of spots of apparently homozygous or hemizygous origin on heterozygous background and heterozygous spots on homozygous leaves. As discussed later, distinction is possible between the induction of somatic crossing over, duplications, deletions, nondisjunction, and specific point mutations by a single treatment of the seed of selfed heterozygous parentage. Since the investigator uses only the colonies of leaf cells whose primordial ancestors have been treated with the mutagen, and there is minimal competition for these cells for expanding in one or a few layers, the system offers advantages over the progeny test method. Additionally, the system is rapid, inexpensive and eukaryotic in nature with the added advantage that mosaicism is being studied in an in vivo situation where the background tissue is genetically identical except for the induced changes. These and other features of the system have been reviewed in the following pages and also discussed in detail in a recent review by Nilan and Vig (1).

Tissue culture provides an ideal system for experiments at the cellular level. The effect of any test compound can be judged solely from the response of the cultured cells without correcting interfering activities from neighboring tissues, as required with organs or whole organisms. Furthermore, liquid suspension cultures are comprised of a large, homogenous population of single cells or clumps of cells. This offers much of the advantages of a microbial system and it is also suitable for the selection of a rare event, e.g., isolation of variants. In plants, the heritability of a somatic variant and the mode of inheritance of a mutation can be studied via the progenies of the regenerated plants. Unfortunately, soybean system has not been vulnerable to such manipulations.

Recent technical advances in plant tissue culture offer alternatives of culturing haploid cells and protoplasts. The latter give rise to a population originating from single cells. Thus, spontaneous mutations have been isolated from such cultures (2). Also chemical mutagens can be employed to induce mutations (2-6). However, in soybean, so far only one systematic mutagenesis has been reported (4).

Routine progeny testing for detecting induced mutations in soybean has been tried for characters like yield (7) and leaflet number (8). However, such studies, in spite of being the surest index of mutation induction, suffer the disadvantage of long time lag required between the treatment and availability of results.

The Systems and Methodologies

Leaf Mosaicism. The occurrence of somatic crossing over and other chromosomal processes leading to mosaicism on the leaves of *Glycine max* have been reported (1,9-11). The gene combination $Y_{11}y_{11}$ have been used as a marker system. In varieties L65-1237 and T 219, the $Y_{11}Y_{11}$ plants are dark green in color, $Y_{11}y_{11}$ are light green, and $y_{11}y_{11}$ are golden yellow. The two simple leaves and, occasionally, the first compound leaf of the $Y_{11}y_{11}$ plants may exhibit some dark green (resembling the phenotype of $Y_{11}Y_{11}$ leaves), yellow (like $y_{11}y_{11}$ leaves) or twin (double) spots. The latter are composed of a dark green component placed adjacent to, and almost mirror image of, a yellow component. These double spots are inferred to originate by somatic crossing over leading to $Y_{11}Y_{11}$-$y_{11}y_{11}$ type of sectoring on the $Y_{11}y_{11}$ background. Some of the single (dark green or yellow) spots may have their origin in the failure of one component of the $Y_{11}Y_{11}$-$y_{11}y_{11}$ sectors. Other spots may be due to chromosomal segmental losses or numerical inequalities in chromosome distribution during mitosis (12) or simple point mutations (1,13). Thus an increase in the frequency of double or single spots on the $Y_{11}y_{11}$ leaves reflects the capability of an agent for inducing somatic crossing over or chromosomal inequalities in the daughter cells during mitosis, respectively. An increase in the frequency of light green spots on the $y_{11}y_{11}$ leaves will point to the induction of mutation at the y_{11} locus. The system, therefore, has the advantage for testing new mutagens and, also, for detecting the various modes of action of these mutagens, all by one treatment.

Methodology. All experiments discussed in here were carried out by soaking the seeds of variety L65-1237 or T 219 in aqueous solutions of the chemicals at room temperature. Controls were submerged for equivalent times in distilled water. In one series of experiments with Mitomycin C (MC), the chemical was applied in lanolin by camel hair brush to the apical portion of plants in which the first compound leaf has just started unfolding.

In all experiments the seed was thoroughly washed after treatment in running tap water and sown in steam-treated soil beds (for only the first series of experiments with MC) or in prewashed sand of nonnutritive value in galvanized metal flats in the greenhouse kept at ambiant temperatures. The seedlings emerge in about three days after which the plants were watered as desired. In about 4-5 weeks the two simple leaves and the first compound leaf are properly developed and expanded to permit analysis for the types and frequency of spots. Most of the data reported in the following pages were obtained from these leaves. The data on spot frequency was obtained by adding the total spots on the two simple leaves and the three leaflets of the first compound leaf of every plant used in analysis and divided by the number of total leaves, considering each leaflet of the compound leaf equivalent of a simple leaf.

Cell Cultures

Suspension cultures were initiated from stems of diploid soybean [*Glycine max* (L.) Merr. cv. Kanrich]. They were maintained in Linsmeir and Skoog medium (3) supplemented with 3 mg/l. naphthaleneacetic acid and 0.5 mg/l. of 2-esopentenyladenine.

Growth Measurements. The sidearm-turbidity method was employed to measure cell culture growth. The increase in cell density measured by turbidity generally corresponds with that of the dry weight and cell number of the culture. Thus, cell number at any point of the growth stage can be estimated from the turbidity expressed in arbitrary Klett units (3,4).

Table 1. Summary of inferred effect of some chemical mutagens in

Mutagen		Treatment			
CAS	Name	LCT (ppm)[a]	Time (hr)[b]	pH	Co-mutagen
50-07-7	MC	3.25	0-12	7	--
		25.0	0-4	7	--
		25.0	0-4	7	Colchicine
		5.0	0-18	7	NaN_3
		12.5	0-5	7	d-Nucleosides
151-56-4	EI	80	0-18	--	--
62-50-0	EMS	1000	0-24	--	--
		1000	2	5-7	--
		500	0-6	7	--
66-27-3	MMS	60	0-4	--	--
70-25-7	MNNG	6	24-48	--	--
62-75-9	DMN	60	0-4	--	--
68-76-8	TM	0.25	0-24	--	--

[a]Lowest concentration tested.

[b]Time for which cells or seeds were treated.

[c]+, effective; -, not effective; Nt, not tested for.

Mutagenesis Procedures. Late log phase cells at about 10^7 cells/ml [reading of 500 on a Kleet-Summerson colorimeter (3,4)] were resuspended in fresh medium and incubated with mutagen on a rotary shaker at 27°C. The mutagen, EMS, was dissolved in MNNG growth medium at pH 5.7. After treatment, cells were washed 3 times and resuspended in fresh medium (6). Dilutions of the mutagenized culture were then plated nonselectively on agar medium for counting of surviving colonies. The culture was then shaken in a sidearm flask, and a growth curve was obtained by monitoring turbidity and survival curves were obtained from growth arrest.

various test systems using Glycine max (soybean)

System studied	Possible effect/genetic end point[c] SCO	Segmental deletions	Locus-specific mutations	Reference
Leaf mosaicism	+	-	-	14
Leaf mosaicism	+	-	-	14
Leaf mosaicism	Some synergistic effect with MC			15
Leaf mosaicism	Only additive, no synergism			16
Leaf mosaicism	Synergistic effect, especially with d-C			17
Progeny testing	Nt	Nt	+	7
Leaf mosaicism	+	+	+	18
Cell culture	Nt	Nt	+	4
Progeny testing	Nt	Nt	+	8,7
Leaf mosaicism	+	+	+	11
Leaf mosaicism	+	-	-	19
Leaf mosaicism	+	+	-	19
Leaf mosaicism	+	+	+	20

Progeny Testing

For this type of study the seed is treated for varying periods of time, at different concentrations of the chemical. Also pH has been adjusted in some studies. The plants raised from this seed are observed for any phenotypic changes, and a sample of the progeny seed from naturally cleistogamous, selfed flowers is raised again for further study. Any variants are propagated further in order to establish heritable variations. These techniques are the routine practice with most studying mutagenesis by traditional progeny testing.

RESULTS

Before considering the details of data available on all the above discussed aspects of soybean mutagenesis a summary table (Table 1) is presented.

Of the chemicals used for induction of leaf mosaicism Mitomycin C (MC) has been studied most extensively. Other relevant chemicals tested for induction of genetic alterations in soybean include: dimethylnitrosamine (DMN), ethyl methanesulfonate (EMS), methyl methanesulfonate (MMS), methyl N-nitro-N'-nitrosoguanidine (MNNG), Trenimon, and ethylenimine.

Mitomycin C

The first series of experiments conducted with MC showed beyond doubt the capability of this chemical to induce somatic sectors. The concentrations used in these studies were as low as 0.000325% for 24 hr and 0.0025% for 12 hr. The highest concentration used was 0.005% for 24 hr. In these cases a significant increase in the frequency of spots was observed. It is of interest that the highest relative increase was that for double spots as if it were the major effect of MC treatment. Thus, it was concluded (1,9,13,14,16,21) that MC brings about complimentary segregation for the gene pair $Y_{11}y_{11}$ through the induction of somatic crossing over as its primary effect. The failure of one component or the other of the products of somatic crossing over would produce a single spot--dark green or yellow--and the two types should be nearly equal in frequency. This was generally true and lead us to believe that single spots originate from such failures or from chromosome aberrations like exchanges between nonhomologs, deletions, etc. The idea of somatic crossing overs having occurred is supported strongly not only by the type of effect on Glycine but also for similar results obtained with Ustilago and Saccharomyces by Holliday (22) as well as the quadriradials formed in human (10, 23) and plant (10,24) cells after MC treatment.

A summary of representative results obtained with MC is given in Table 2. The right part of the table gives the relative frequency between various types of spots. An examination of these calculations shows that the double spots generally constitute more than one third of the total spot frequency whereas the singles, yellows and dark greens are approximately equal in frequency. That double spots express the most increase can be better supported by considering the absolute increase in relation to their frequency in the untreated control.

It is a common observation that the two simple leaves have a far higher frequency of spots than the compound leaves. This has been attributed to the number of initial cells present in the embryonic leaves in the seed. This is confirmed by the virtual absence of spots on second and subsequent compound leaves and the fact that the spots on the first compound leaf are larger than those on the simple leaves. This points to the fact that MC has no systemic or much residual effect in this system and that the frequency of spots can be used as a good indicator of the rate at which the changes can be induced. This system, thus, is a beautiful eukeryotic equivalent of the bacterial system in that the treated cells on the leaves do not undergo a strong competition or selection for expression of the genetic alterations induced because cells from a given layer only undergo division and expansion to constitute more cells within that layer.

This system starts responding to concentrations as low as 0.0003% when treatment is given for at least 24 hr. Treatments at concentrations as high as 0.005% for 12 hr are tolerated but leaf expansion and development is adversely effected consequently, resulting in an overall reduction in the frequency of spots (Table 2).

That MC effects the cells which come in contact with it, and has no systemic effect, was further tested by application of 0.005% MC in lanolin paste to the few cells of the third and fourth compound leaves after the first compound leaf had partially unfolded. The results were clear cut: a frequency of 0.46 dark green spots per leaf in the treated material compared with 0.014 spots in the control, 0.33 yellow spots vs. 0.005 in the control and 0.27 double spots/leaf compared to the low frequency of 0.005 in the control (14).

The effect of MC is known to vary with the physiological age of the seed as studied by treating the seed at different times after initiation of germination (21). In these studies the seeds were treated with aqueous solution of 0.0025% Mitomycin C during 0-36 hr for a period of only 4 hr in successive but nonoverlapping sequences, e.g. 0-4 hr, 4-8 hr, 8-12 hr, etc. The general pattern that emerged was interesting in that a so-called cyclic pattern of increase in the frequency of spots developed from these data. Thus the frequency of spots in the material treated during 0-4 hr was always the lowest and increased in treatments lasting 16 hr post-germination age of the seed, when a decline in the frequency was observed, which was followed by another sharp increase at about 24-28 hr. This was followed by another drop. Whereas it is not convenient to reproduce the entire data, the results from one set of such studies are reproduced in Table 3.

These data indicate only a loose correlation between the frequency of spots induced by MC and the synthetic activity of the DNA molecule. DNA, as measured by ^{3}H-TdR incorporation synthesis in soybean embryos appears to start at about 24 hr after the ger-

Table 2. Types and frequencies of spots on the two simple leaves and the first compound leaf of the population obtained from seed treated with Mitomycin C (variety T219)[a]

Treatment	Number of leaves analyzed	Spot frequency per leaf[b]				Proportion of spots			
		DG	Y1	Db	T	DG/Db	Y1/Db	DG/Y1	T/Db
Control	230	0.35	0.15	0.06	0.56	5.83	2.50	2.33	9.33
Mitomycin C									
0.000325%, 24 hr	150	0.75	0.50	0.65	1.90	1.13	1.07	1.14	2.91
0.00065%, 24 hr	130	1.25	1.39	0.89	3.53	1.40	1.56	0.90	3.98
0.00125%, 24 hr	185	1.62	1.40	1.79	4.81	0.91	0.78	1.16	2.96
0.0025%, 24 hr	175	1.15	1.23	1.90	4.28	0.61	0.65	0.93	2.24
Control	125	0.20	0.02	0.04	0.26	5.00	0.50	10.00	6.50
Mitomycin C									
0.000625%, 24 hr	175	1.01	0.46	1.61	3.08	0.63	0.29	2.20	1.91
0.00125%, 24 hr	105	2.44	1.09	2.17	5.70	1.12	0.50	2.24	2.62
0.0025%, 24 hr	125	1.93	0.84	2.37	5.14	0.81	0.35	0.81	2.17
Control	90	0.02	0.19	0.06	0.53	4.67	1.67	2.80	8.67
Mitomycin C									
0.0025%, 12 hr	115	2.07	1.91	1.72	5.70	1.20	1.11	1.09	3.31
0.0025%, 24 hr	110	4.43	3.85	6.65	14.93	0.67	0.58	1.15	2.24
0.005%, 12 hr	145	3.59	2.72	3.04	9.35	1.11	0.89	1.32	3.72
0.005%, 24 hr	110	3.82	3.09	3.12	10.03	1.22	0.99	1.24	3.21

[a]Sources: Vig and Paddock (14).

[b]DG = dark green; Y1 = yellow; Db = double or twin; T = total spots.

Table 3. Data on the frequency and proportions of different types of spots on the leaves of $Y_{11}y_{11}$ Glycine max var. L65-1237 treated with 0.0025% of Mitomycin C during different periods of germination[a]

Treatment period (hr)[b]	Number of leaves analyzed	Spot frequency and type per leaf[c] DG	Y1	Db	T	t-Test[d]
--	135	0.33	0.18	0.22	0.73	
8 hr						
H_2O	85	0.19	0.29	0.19	0.67	
0-4	80	0.35	0.36	0.33	1.04	1
4-8	110	0.56	0.49	0.83	1.88	1,2,3
8-12	140	0.67	0.95	1.41	3.04	1,2,3
12-16	145	1.03	0.95	1.34	3.32	1,2,3
16-20	105	0.64	0.50	0.78	1.92	1,2
20-24	140	1.51	1.51	1.72	4.76	1,2,3,4,7
24-28	150	0.73	0.92	0.95	2.60	1,2,3
28-32	115	0.78	1.35	1.30	3.43	1,2,3
32-36	140	1.01	1.56	1.26	3.83	1,2,3,4

[a]Source: Vig (21).

[b]Time of initiation of soaking the dry seed = 0. Treatment period refers to the period during which seeds were in contact with mitomycin C solution. Seeds were kept in H_2O before and after this treatment for a total of 36 hours before sowing.

[c]DG = dark green, Y1 = yellow, DB = double or twin and T = total spots.

[d]Numbers refer to the treatments which differ significantly from the treatment in question (at 5% level).

mination period and continues to increase until about 42 hr, then reaches a plateau for about 20 hr. Then, there is again a burst of activity of this synthesis up until at least 80 hr into the post-germination period (21). It is possible that an increase in the frequency of spots depends upon the duplication of the chromosome segment carrying the gene Y_{11} and/or DNA synthesis required for completing the task of reciprocal rejoining. There is need for study of DNA synthesis in materials treated with MC.

One may also present a case of protein synthesis required for recombination. As a matter of fact, there are more drastic fluctuations in the uptake of ^{3}H-argenine by the embryonic cells, and variations continue until about 48 hr or so post-germination. The data therefore do not present a definite answer in terms of such activity of the cells as a cause of induction of spots by MC (21) even though it is known that the crosslinking efficiency of MMC is dependent upon the presence or absence of proteins associated with DNA (25).

When MC treatment is given to the seeds, which have been previously soaked in colchicine solution, a mild synergistic effect is observed (15). In one experiment wherein 0.01% colchicine treatment was applied for 4 hr followed by soaking of the seed in 0.0025% solution of MC for another 4 hr, an increase in the frequency of spots was observed which was slightly more than that calculated due to the additive effects. A representative set of data have been given in Table 4. Thus, when one adds up, for instance, the increase in the frequency of spots on the leaves of colchicine-treated material (spots in the colchicine-treated minus the spots in the control) and the increase induced by MC, the total frequency is lower than the increase in frequency of spots observed for colchicine + MC treatment [(0.72 - 0.50) + (0.84 - 0.50) total spots vs. (1.51 - 0.50) spots, i.e., 0.56 vs. 1.01]. Similar increases are evident for the various spot types. It is not known if these increases have anything to do with the expected effect of colchicine on proteins, or microtubules which have been found associated with chromosomes in the interphase (26).

One wonders if the action of MC is dependent upon the concentration of ATP in the cell. The inhibitor of oxidative phosphorylation, NaN_3 (27), which interferes with the mechanism of chromosome repair (28) was used in combination with MC treatments (16). NaN_3 does not cause chromosome breakage, per se (29). The idea was that an inhibition of SCO induced by NaN_3 would leave the free fragments unjoined thus creating single spots, if the phenomenon utilized breakage (first) rejoining sequence. However, if complementary exchanges are brought about by the crosslinking or alkylating property of the MC molecule, the NaN_3-induced changes may not interfere in the process required for DNA synthesis.

The experiments were carried out by treating the seed with 0.01 or 0.02% aqueous solution of NaN_3 with or without MC in final concentrations of 0.0005 and 0.001%. NaN_3 alone increased the frequency of dark green and yellow spots 6- to 9-fold but only 2- to 3-fold of the doubles (Table 5). Mitomycin C, on the other hand, increased the frequencies of all three types of spots equally. These are approximately 2- to 2.5-fold in case of 0.0005% treatment and 4- to 5-fold for 0.001% treatment. The relative frequencies between different types of spots are very similar to those observed in the control. The frequency of spots in case of combined Mitomycin C + NaN_3 treatment is higher than in case of either chemical alone. However, as seen by comparing the data in rows 5 vs. 6, 7 vs. 8, 10 vs. 11 and 12 vs. 13, it is clear that the effect of the two chemicals is merely additive and no synergism is is operative. The spot frequency in one case (0.001% Mitomycin C + 0.02% NaN_3 treatment) is approximately 75% of what is expected due to additive effect. This has been attributed to reduced leaf expansion, slow growth of the treated material and other physical effects of the combined treatment. The relative frequencies among different types of spots are also in between those for Mitomycin C or NaN_3 alone. The only exception, as expected, is the ratio between dark green and yellow spots because of the two increasing parallel to each other in case of both Mitomycin C or NaN_3 treatments. The data, therefore, permit the conclusion that, whereas MC may act synergistically with colchicine, such is not the case with NaN_3.

In the field of mutagenesis, whereas the problem of comutagenesis is generally a difficult one to assess, it can be heartening to find a situation wherein the mutagenic efficiency of a chemical can be reduced by the simultaneous or near simultaneous administration of another agent. One such case known in Glycine leaf mosaicism is the partial reversal or inhibition of the biological effect of MC by nucleosides. The incidence of somatic crossing over in the variety L65-1237 is reduced when seeds are treated with 1 to 2 × 10^{-4} M solutions of d-thymidine, d-adenosine, d-cytidine, d-guanidine, or d-uridine for varying lengths of time. This reduction in the frequency of spots induced by nucleosides is best evidenced in the material treated with d-C and is maintained even when d-C treated seeds are post-treated with 0.00125% solution of MC for 4-5 hr or post-treated after MC treatment. However, in these experiments the frequency of yellows is slightly higher than those of dark greens; but twins, in general, constitute about one-third of total spots. Examples of data on these studies are provided in Table 6. It is suggested that d-C interferes with the process of interphase chromosome arrangement as it affects the pairing of corresponding segments of the heterochromatic zones of homologous chromosomes (17).

Table 4. Types and frequency of spots in three sets of materials treated with water or 0.005% acqueous colchicine for different time periods and with or without post-treatment with Mitomycin C (0.0025%)[a]

Seed treatment				Type and frequency of spots per leaf				
Time (hr)	Agent	Time of post-treatment with Mitomycin C (hr)	No. of leaves	DG	Y1	Db	T	t (calculated)
4	Water	None	100	0.20	0.20	0.19	0.59	0.20 > p > 0.40
4	Colchicine	None	125	0.35	0.23	0.21	0.79	
4	Water	4	135	0.40	0.41	0.41	1.22	0.025 > p > 0.05
4	Colchicine	4	105	0.62	0.69	0.69	2.00	
8	Water	None	170	0.12	0.14	0.16	0.42	0.05 > p > 0.10
8	Colchicine	None	180	0.29	0.23	0.24	0.76	
8	Water	4	130	0.45	0.64	0.62	1.71	
8	Colchicine	4	105	0.52	0.50	0.63	1.65	
16	Water	None	150	0.18	0.11	0.14	0.43	0.05 > p > 0.10
16	Colchicine	None	125	0.52	0.38	0.32	1.22	
16	Water	4	150	0.51	0.51	0.75	1.77	0.20 > p > 0.40
16	Colchicine	4	150	0.91	0.75	0.95	2.61	
4	Water	None	105	0.19	0.17	0.14	0.50	0.20 > p > 0.40
4	Colchicine	None	170	0.29	0.18	0.25	0.72	
4	Water	4	95	0.31	0.31	0.23	0.84	0.025 > p > 0.05
4	Colchicine	4	110	0.45	0.45	0.50	1.51	
8	Water	None	135	0.25	0.22	0.13	0.61	0.40 > p > 0.50
8	Colchicine	None	120	0.35	0.23	0.13	0.72	
8	Water	4	125	0.40	0.47	0.26	1.12	p > 0.001
8	Colchicine	4	105	0.66	0.66	0.59	1.90	

[a]Source: Vig (15).

Table 5. Spot frequency on the leaves of $Y_{11}y_{11}$ plants from the seed treated with NaN_3 and/or Mitomycin C for the first 18 hr of germination[a]

No.	Treatment (% solution)		No. of leaves analyzed	Type and frequency of spots per leaf				Relative frequency between spots			
	NaN_3	Mitomycin C		DG	T1	Db	T	DG/Db	Y1/Db	DG/Y1	T/Db
1	H_2O control		170	0.62	0.52	0.54	1.67	1.15	0.96	1.19	3.09
2	0.01	--	120	3.62	3.49	1.04	8.17[b]	3.49	3.36	1.04	7.89
3	0.02	--	95	4.62	4.62	1.67	10.72[b]	2.65	2.77	0.96	6.42
4	--	0.0005	170	1.39	1.43	1.32	4.14[b]	1.05	1.08	0.97	3.14
5	0.01	0.0005	130	3.95	4.55	1.99	10.49[b]	1.98	2.29	0.87	5.27
6	Expected (additive) for 5[c]			4.39	4.40	1.82	10.64[d]	2.41	2.42	1.00	5.85
7	0.02	0.0005	70	5.24	5.54	2.33	13.24[b]	2.25	2.38	0.95	5.68
8	Expected for 7			5.39	5.53	2.45	13.19[d]	2.20	2.26	0.97	5.38
9	--	0.001	170	2.52	2.54	2.78	7.85[b]	0.90	0.91	0.99	2.82

(continued)

Table 5. (continued)

No.	Treatment (% solution)		No. of leaves analyzed	Type and frequency of spots per leaf				Relative frequency between spots			
	NaN_3	Mitomycin C		DG	Tl	Db	T	DG/Db	Yl/Db	DG/Yl	T/Db
10	0.01	0.001	125	5.56	5.70	3.36	14.64[b]	1.65	1.69	0.97	4.35
11	Expected for 10[c]			5.52	5.51	3.28	14.35[d]	1.68	1.68	1.00	4.38
12	0.02	0.001	85	4.45	5.72	2.47	12.64[b]	1.80	2.32	0.78	5.12
13	Expected for 12[c]			6.52	6.64	3.91	16.90	1.67	1.70	0.98	4.32

[a]Aqueous solution (25 cc) of the chemical(s) used on 10-g sample of L65-1237 seed. Source: Vig (16).

[b]Significantly different from the control (χ^2, $p < 0.01$).

[c]Expected spot frequency, in the event of additive effect of NaN_3 and Mitomycin C, is calculated by additing the spot frequency with NaN_3 alone to spot frequency observed with Mitomycin C alone less spot frequency in the control treated with neither chemical. For example, spot frequency expected for 0.01% NaN_3 + 0.0005%, Mitomycin C = spot frequency in (2) + spot frequency in (4) minus spot frequency in (1).

[d]Does not differ significantly from respective experimental data.

Table 6. Type and frequency of spots on the leaves of $Y_{11}y_{11}$ plants treated with 24 cc of 2×10^{-4} M nucleosides/water followed by or preceeded by 0.00125% MC[a]

Treatment	d-Nucleoside	No. of leaves	Types and frequency of spots/leaf: DG	Y1	Db	T	Proportion of spots: Dg/Db	Y1/Db	T/Db	Db/Y1
Nucleoside 0-16 hr, followed by MC for 5 hr	None (H_2O)	155	0.59	0.50	0.48	1.57	1.23	1.04	3.27	1.18
	d-A	185	0.39	0.69	0.37	1.45	1.05	1.76	3.84	0.60
	d-C	170	0.42	0.49	0.42	1.32[b]	1.00	1.17	3.14	0.86
	d-G	195	0.41	0.58	0.24	1.24	1.70	2.42	5.17	0.71
	d-T	165	0.38	0.62	0.42	1.42	0.90	1.48	3.38	0.61
	d-U	145	0.27	0.61	0.41	1.29	0.66	1.50	3.15	0.44
Water 0-16 hr, nucleoside 16-22 hr, MC 22-26 hr	None (H_2O)	255	0.72	0.76	0.90	2.38	0.80	0.84	2.64	0.95
	d-A	100	0.68	0.55	0.73	1.96	0.93	0.75	2.68	1.24
	d-C	90	0.22	0.32	0.36	0.90[b]	0.61	0.89	2.50	0.69
	d-G	90	0.67	0.62	0.68	1.97	0.99	0.91	2.90	1.08
	d-T	125	0.52	0.42	0.57	1.51	0.91	0.74	2.65	1.24
Nucleoside 0-24 hr, MC 24-29 hr	H_2O	150	0.74	0.82	0.66	2.22	1.13	1.24	3.36	0.90
	d-C	150	0.35	0.42	0.37	1.14[b]	0.95	1.14	2.71	0.83
	d-G	135	0.27	0.51	0.36	1.13[b]	0.75	1.42	3.14	0.53
	d-T	50	0.42	0.80	0.68	1.90	0.62	1.18	2.79	0.53
	d-U	125	0.50	0.58	0.51	1.59[b]	0.98	1.14	2.59	0.86
MC (16-20 hr) and followed by 2×10^{-4} M deoxynucleosides (20-25 hr)	None (H_2O)	175	0.63	0.73	0.67	2.03	0.94	1.10	3.05	0.90
	d-A	115	0.48	0.54	0.50	1.51	0.96	1.80	3.20	0.85
	d-C	115	0.29	0.29	0.24	0.82[b]	1.20	1.20	3.47	1.00
	d-G	90	0.29	0.24	0.16	0.68[b]	1.81	1.50	4.25	1.20
	d-T	90	0.53	0.51	0.41	1.44	1.29	1.24	3.51	1.04

[a]Source: Vig (17).

[b]Significantly different from the control as shown by analysis of variance and t-test (5%).

The data with MC, therefore, permit the conclusions that MC is capable of inducing somatic crossing over in the Glycine system, shows additive effects with NaN_3, a synergistic effect with colchidine, and is inhibited in its effect by deoxyribose cytidine (d-C). Thus, there is a strong correlation between the capability of MC in producing quadriradials in plant and animal cells (23, 24) and the induction of somatic recombination (1,10,14).

Even though a weak mutagenic effect of MC has been reported in E. coli, S. typhimurium, Drosophila, and mouse (10), the data from Glycine max $y_{11}y_{11}$ leaves do not permit such conclusions regarding specific locus mutations in this system. It has been postulated that MC, unlike diepoxybutane (DEB), MMS, or Trenimon (TM), is incapable of inducing minute deficiencies in the DNA molecule which eventually lead to the appearance of $y_{11}y_{11}$ type spots on the $y_{11}y_{11}$ background (1).

EMS

The T219 soybean seed treated with EMS produced mosaicism which was qualitatively quite different from that observed for MC. In a preliminary test (18), EMS at concentrations of 0.125% and 0.25%, when applied for 24 hr, produced a substantial increase in the frequency of spots of all three types. Contrary to what was observed for MC, the increase produced by EMS was the least in the frequency of twin spots and the greatest for the dark greens. The doubles constituted less than 1/7 of the total spots, about a fourth that of the dark greens and less than one half of those of yellows. The data are shown in Table 7.

The data from other experiments carried out with variety L65-1237 confirm these findings. One set of data (11), also reproduced in Table 7 indicate the poor potential of EMS in inducing twin spots in comparison to its capability to produce single spots. In this case, as in several other sets of data not reprinted here, an increase in the frequency of dark green spots was somewhat lower than those in the frequency of yellows. The data indicate to the capability of EMS in inducing a few homologous exchanges but primary effect being the induction of fragments and their subsequent loss resulting in production of single spots.

A finding of interest is the capability of EMS in inducing point mutations. In one set of leaves obtained from $y_{11}y_{11}$ seed treated with 0.025% EMS solution for 24 hr, as many as 13.46 light green spots per leaf were found in an analysis of 26 simple leaves. The control in this study showed a lack of any light green sectors on the $y_{11}y_{11}$ leaves. In another study where seeds were treated with 0.1, 0.05, and 0.025% solution of EMS, a mutation frequency of 45.78, 17.06, and 0.42 $y_{11} \rightarrow Y_{11}$ events per leaf was observed in respective samples of 40, 32, and 24 leaves on the $y_{11}y_{11}$ plants.

Table 7. Data on the frequencies of three types of spots observed on the two simple leaves and first trifoliate leaf of $Y_{11}y_{11}$ plants obtained by treating the seed (T219 or L65-1237) in solutions of ethyl methanesulfonate (20 ml) for the first 24 hr of germination[a]

EMS (concentration)	No. of leaves analyzed	Spot frequency per leaf				Relative frequencies of spots			
		DG	Y1	Db	T	Dg/Y1	DG/Db	T1/Db	T/Db
0 (control)	275	0.15	0.09	0.05	0.29	1.67	3.00	1.80	5.00
0.125%	140	1.19	0.59	0.27	2.06	2.02	4.41	2.19	7.63
0.25%	100	3.27	2.30	0.83	6.40	1.42	3.94	2.77	7.71
0 (control)	95	0.07	0.06	0.05	0.19	1.17	1.40	1.20	3.80
0.1%	95	8.73	11.18	3.54	23.44	0.78	2.47	3.16	6.62
0.25%	40	8.65	14.23	5.80	28.68	0.61	1.49	2.45	4.94

[a]Sources: Vig and Paddock (18); Vig et al. (11).

The control in this case gave a frequency of 0.21 spots per leaf. The data indicate to the capability of the system in permitting analysis of the effect of a chemical in induction of locus specific mutations. The findings that EMS induces very few double spots, produces a preponderance of yellow spots and dark green spots on the $Y_{11}y_{11}$ leaves, causes change of $y_{11} \rightarrow Y_{11}$, and that these effects are different from those of MC show that Glycine systems can distinguish between various effects of a given chemical.

An attempt was made to obtain survival curves and score the appearance of mutants from cultured soybean cells after EMS mutagenesis. Details of the former aspect have been reproduced elsewhere (3,4) and are not of significance for present discussions.

Drug resistance is generally used as markers to score for the appearance of mutants arising in cell cultures. Comparison of spontaneous and mutagen-induced mutation frequencies often reveals, on an average, a 10-fold increase in mutation frequency (4). When 5-methyl tryptophan was used as a marker to score for mutation frequency in soybean culture, resistant colonies appeared spontaneously at frequencies of the order of, or below 10^{-8}, whereas, in a total of three experiments, a treatment of 0.2-0.3% EMS for 2 hr at 28°C produced resistant colonies at a frequency of about 10^{-7}. A tenfold increase in resistant colonies is not enormous. Nevertheless, since it is difficult to plate more than 10^{-7} cells per plate, resistant colonies appearing spontaneously at 10^{-8} are at the limit of convenient detection, hence a 10-fold increase due to mutagenesis may be quite useful.

Kiang and Halloran (8) reported an interesting mutation induced in Glycine with 0.05 M EMS solution applied to seed. They found, instead of trifoliate compound leaves, multiple-leaflet mutants which bred true in M_1 and M_2. Several types of chlorophyll mutations were also reported. These mutations were induced at pH 7 when seeds were treated for 6 and 12 hr at 21°C. The authors agree with Blixt's idea that there exists a high correlation between sector formation and the frequency of mutations in M_2. In case of hydroxylamine, leaf sectors are pH-dependent, appearing at pH 8 but not at pH 6 or 7 (8).

MMS

MMS is effective in inducing mosaicism on soybean leaves, even at concentrations as low as 0.00625% applied to seed for 24 hr. The distribution of spots is similar to that for MC but differs from those found after EMS treatment in that the doubles constitute about 1/3 of all spots. Also, the frequency of spots on the upper surface of the leaf is about three times as much as on the lower surface (11). The representative data for the upper surface are given in Table 8. It is found that at equivalent concentration MMS is far more effective in producing total spots or twin spots than EMS.

The effect of 0.02% MMS on mosaicism was studied during early periods of germination by treating the seed for the first 6 hr of germination and also for 6-12, 12-18, and 18-24 hr treatments. The relative frequencies of spots were, in all cases, similar to those obtained for a continuous 24 hr treatment. However, a total spot frequency of 29.49, 24.48, 19.34, and 19.04 spots/$Y_{11}y_{11}$ leaf was observed for 0-6,6-12, 12-18, and 18-24 hr treatments, respectively (control 0.16 spots/leaf). Similar results were obtained from other experiments in which seeds were thoroughly washed after EMS treatment period was over. However, when seeds were not washed until the 24th hr and only water was added to the germinating seed at the end of the treatment, the differences disappeared. Thus, in one such study where 0.02% MMS was added at 0, 6, 12, or 18 hr, the frequency of spots was 47.55, 47.73, 59.57, and 54.88 spots/$y_{11}y_{11}$ leaf. This indicates that the total absorption of MMS in all these cases was similar and that the embryo is uniformly sensitive to the chemical during the first 24 hr of germination (8).

Like EMS, MMS has the capability of producing mutations of the y_{11} locus. The data in Table 9 clearly demonstrate the effectiveness of MMS in producing light green sectors on the yellow plants. Also, as pointed out by the data in Table 10, at equivalent concentrations, MMS appears to be as much as 50 times more effective in producing point mutations than EMS. However, the effectiveness of MMS in producing all types of spots on the $Y_{11}y_{11}$ plants is only about tenfold, indicating that the spots on $y_{11}y_{11}$ plants do not utilize the same mechanism of origin as those on $Y_{11}y_{11}$ leaves (8).

The data in Tables 8-10, thus, indicate the effectiveness of MMS in inducing somatic recombination and point mutations involving the Y_{11}-y_{11} locus. The frequency of single spots on the heterozygous leaves is such that events like the failure of a component of the twin spot, point mutations of the locus, nondisjunction or loss/gain of the chromosome segment carrying the Y_{11} locus must occur in a manner as to produce both dark green and yellow spots in about equal frequency. From the information available with other chemicals, like NaN_3 (16), the tempting conclusions are that some of these single spots result from failure of components of double spots while others may result from loss/gain of Y_{11}/y_{11} through simple breakge or reciprocal nonhomologous exchanges (13).

When seed samples were treated with 250 ppm MMS for 4 hr at staggered intervals, viz., 0-4, 4-8, 8-12, etc., up to 44-48 hr, washed and sown in two experiments, the frequency of total spots, even though higher than the control in every case, was not uniform.

Table 8. Frequency and relative proportion of the different types of spots on the leaves of $Y_{11}y_{11}$ *Glycine max* var. L65-1237 treated with various concentrations of methyl methanesulfonate[a]

Treatment	Concentration (%)	Number of leaves analyzed	Frequency of spots/leaf				Relative proportion of spots			
			DG	Y1	Db	T	DG/Db	Y1/Db	DG/Y1	T/Db
MMS (0-24 hr)	Control	125	0.7	0.6	0.5	1.8	1.4	1.2	1.2	3.6
	0.006	105	2.9	2.6	3.3[b]	8.8[b]	0.9	0.8	1.1	2.7
	0.0125	115	4.6	3.8	5.8[b]	14.2[b]	0.8	0.7	1.2	2.4
	0.025	120	7.6	7.3	9.8[b]	24.7[b]	0.8	0.7	1.0	2.5
	0.05	115	12.2	10.2	14.1[b]	36.6[b]	0.9	0.7	1.2	2.6
	Control	120	0.3	0.4	0.1	0.8	3.0	4.0	0.8	8.0
	0.006	110	3.8	5.0	6.3[b]	15.0[b]	0.6	0.8	0.8	2.4
	0.0125	105	4.4	7.3	9.4[b]	21.1[b]	0.5	0.8	0.6	2.2
	0.025	105	5.4	8.7	11.6[b]	25.7[b]	0.5	0.8	0.6	2.2
MMS (24-48 hr)	Control	105	0.5	0.6	0.3	1.4	1.7	2.0	0.8	4.7
	0.006	105	1.1	1.0	1.1[b]	3.2[b]	1.0	0.9	1.1	2.9
	0.0125	105	3.4	3.8	4.6[b]	11.8[b]	0.7	0.8	0.9	2.6
	0.025	110	5.8	6.7	9.7[b]	21.6[b]	0.6	0.7	0.9	2.2
	0.05	90	6.2	7.2	9.8[b]	23.2[b]	0.6	0.7	0.9	2.4
	Control	110	0.4	0.4	0.3	1.1	1.3	1.3	1.0	3.7
	0.006	110	1.7	2.2	2.3[b]	6.2[b]	0.7	0.9	0.8	2.7
	0.0125	110	3.2	4.4	5.6[b]	13.2[b]	0.6	0.8	0.7	2.4
	0.025	105	5.0	7.2	9.3[b]	21.5[b]	0.5	0.8	0.7	2.3
	0.05	75	9.8	11.3	16.1[b]	37.2[b]	0.6	0.7	0.9	2.3

[a]Source: Arenaz (19).

[b]Differs significantly from control, $\times^2$: $p > 0.05$.

Table 9. Data on the frequency of light green sectors per leaf on $y_{11}y_{11}$ plants from seed treated with methyl methanesulfonate[a]

MMS treatment		Number of	
Concentration (%)	Time (hr)	leaves analyzed	Light green sectors/leaf
0 (Control)		36	0.33
0.00625	0-24	40	
0.0125	0-24	32	14.81
0.025	0-24	38	44.13
Control		28	0.0
0.006	0-24	36	1.5
0.0125	0-24	40	3.7
0.025	0-24	30	18.6
0.05	0-24	34	68.3
Control		6	0.6
0.006	24-48	28	3.3
0.0125	24-48	30	21.7
0.025	24-48	26	51.4
0.05	24-48	6	42.7
Control		30	0.2
0.006	0-24	24	10.8
0.0125	0-24	30	33.6
0.025	0-24	32	91.7
Control		26	0.5
0.006	24-48	22	1.3
0.0125	24-48	10	10.5
0.025	24-48	22	27.3
0.05	24-48	28	50.9

[a]Source: Vig et al. (20).

Thus the first peak was observed in the material treated for 4-8 hr, followed by a decrease. Other increases were observed again around 20, 28, and 36 hr, with fluctuations within 4 hr between experiments. Uniformity was observed for increase/decrease in spot frequency for all three types of spots in every population analyzed.

MNNG

Of the several experiments reported for MNNG by Arenaz (19), the successful ones are those in which seeds were treated in the

Table 10. Comparison of frequency of light green sectors per leaf on $y_{11}y_{11}$ plants from seed treated with methyl methanesulfonate and ethyl methanesulfonate[a]

Agent	Concentration (%)	Number of leaves analyzed	Spot frequency per leaf	Range of spots per leaf
EMS	0.05	8	21.13	15-26
EMS	0.02	14	2.71	0-11
MMS	0.02	4	117.0	80-154
H_2O		18	0.56	0-2
EMS	0.02	24	1.41	0-6
MMS	0.02	20	107.00	46-158
EMS	0.01	26	0.65	0-4
MMS	0.01	26	30.03	1-76
H_2O		35	0.20	0-2

[a]Source: Vig et al. (20).

MNNG solution for 24-48 hr after 24 hr of initial soaking in water. The treatments lasting the first 24 hr did not produce a significant increase in the frequency of spots on either yellow or light green plants. However, a 24-48 hr treatment with 6 to 50 ppm MNNG solution increased the spot frequency significantly for all types produced on the $Y_{11}y_{11}$ leaves, even though these spots did not show an exactly equal distribution among those classes, and dark greens generally were over represented, the general patterns are more like those produced by MC or MMS treatments rather than those produced by EMS treatments. The relative increase in the frequency of doubles is always greater than such increase in the frequency of dark greens or yellows when compared with the controls. The data from a series of four experiments are presented in Table 11.

MNNG is a potent mutagen and can induce gene conversion, chromosome breakage in animals, point mutation, somatic crossing over, and nondisjunction, with apparently poor chromosome breakage capacity in higher plants (19). It has previously been observed (13) that chemicals capable of inducing exchanges in plants are the ones most successful in creating reciprocal homologous exchanges and, hence, twin spots. This may explain the poor performance of MNNG in inducing spots in this system. Since the chemical is known to be specific for replication fork (in bacteria) (30), it is not surprising that no effect, or very poor effect was seen when seeds were treated for 0-24 hr. It is, however, surprising that y_{11} gene did not mutate to Y_{11} under the influence of MNNG in either 0-24 or 24-48 hr treatments. Perhaps this indicates

Table 11. Frequency and relative proportion of the different types of spots on the leaves of $Y_{11}y_{11}$ *Glycine max* var. L65-1237 treated with various concentrations of N-Methyl-N-nitrosoguanidine for 24-48 hr[a]

MNNG Concentration (ppm)	Number of leaves analyzed	Frequency of spots/leaf				Relative proportion of spots			
		DG	Y1	Db	T	DG/Db	Y1/Db	DG/Y1	T/Db
Control	125	0.7	0.6	0.5	1.8	1.4	1.2	1.2	3.6
6	120	1.9	1.6	1.9[b]	5.3[b]	1.0	0.8	1.2	2.8
12.5	120	2.3	2.4	3.0[b]	7.7[b]	0.8	0.8	1.0	2.6
25	120	2.1	1.7	2.5[b]	6.3[b]	0.8	0.7	1.2	2.5
50	110	1.5	1.8	1.8[b]	5.0[b]	0.8	1.0	0.8	2.8
Control	100	0.1	0.2	0.05	0.3	2.0	4.0	0.5	6.0
6	100	1.3	1.1	0.7[b]	3.6[b]	1.9	1.6	1.2	5.1
12.5	No growth	-	-	-	-	-	-	-	-
25	105	1.8	1.0	1.1[b]	3.9[b]	1.6	0.9	1.8	3.6
50	100	1.3	0.9	0.8[b]	3.0[b]	1.6	1.1	1.4	3.8
Control	85	0.7	0.5	0.3	1.5	2.3	1.7	1.4	5.0
6	35	1.9	1.2	1.1[b]	4.2[b]	1.7	1.1	1.6	3.8
12.5	105	1.3	1.1	0.9[b]	3.3[b]	1.4	1.2	1.2	3.7
25	105	1.8	1.0	1.1[b]	3.9[b]	1.6	0.9	1.8	3.5
50	110	1.3	0.8	0.8[b]	2.9[b]	1.6	1.0	1.6	3.6
Control	120	0.3	0.4	0.1	0.8	3.0	4.0	0.8	8.0
6	105	0.8	0.4	0.5[b]	1.7[b]	1.6	0.8	2.0	3.4
12.5	105	1.2	0.5	0.8[b]	2.4[b]	1.5	0.6	2.4	3.0
25	105	0.9	0.8	0.8[b]	2.5[b]	1.1	1.0	1.1	3.1
50	40	5.2	4.0	3.2[b]	12.3[b]	1.6	1.3	1.3	3.8

[a]Source: Arenaz (19).

[b]Differs significantly from control, χ^2: $p > 0.05$.

the differences of action between MNNG and MMS on the Glycine system. Also since MNNG is generally a far more potent mutagen than MMS (31), the data in Glycine might suggest that the system may lack some essential factor which may be present in a mammalian system or in Drosophila. The mutagenic potential of MNNG as found in barley (32,33) may not mean that the chemical is capable of inducing somatic recombination in plants.

Table 11 reveals that, with one exception, where MNNG-treated material showed very poor germination, the system shows a saturation type of effect even at low doses. There, data, as also for DMN discussed below, perhaps are compatible with the suggestion that effects of MNNG are the result of some active metabolite rather than direct alkylation (34). If a pH-dependent intermediate like diazomethane or nitrous acid is involved, then the enzymatic system in soybean may not be efficient enough to convert all the intracellular MNNG to the active agent which would be capable of inducing somatic recombination.

Also, in order to dissolve MNNG in water, the solvent was heated to 65°C and treatments carried out in light. Both these factors may be responsible for break down of MNNG and, hence, its low efficiency. This, however, cannot be the reason for nonmutability of y_{11} which perhaps is due to some factor inherent in the structures of y_{11} and Y_{11} (1).

In spite of low frequency of spots in MNNG-treated material, a cyclic-type effect was found when seeds were treated for 0-4, 4-8, 8-12, ... 44-48 hr with periods of high response coinciding with treatments ending at 24-28 and 40 hr. A relationship between cell cycle activity, conversion of MNNG to active metabolite or availability of target molecule is not yet known.

DMN

Of the chemicals studied on the Glycine system, DMN turned out to be one of the most potent in inducing somatic mosaicism. Thus, in treatment with doses as low as 60 ppm for 0-24 hr, as high as a 20-fold increase in the frequency of total spots was observed (Table 12). However, two points are of interest in the data reported from three experiments carried out with 0-24 hr treatment. First, the frequency of double spots is much lower than the frequency of single spots and the frequency of dark greens is lower than those of yellows. Secondly, the frequency of spots does not increase with an increase in the concentration of DMN. A typical saturation effect is observed at doses as low as 60 ppm. At lower doses, however, this saturation effect is not observed (Table 13) (29). The frequency of twin spots, however, increased manyfold in the treated material, indicating to the capability of DMN in inducing somatic recombination in this system. The greater increase in the frequency of single spots may indicate to some additional effects of this chemical. There can be the production of deletions or duplications created by breakage or/and exchanges involving nonhomologous chromosomes or, at noncorresponding points, even the homologs.

It is of interest that in mammalian systems in vitro, bacterial, and fungal systems, DMN is ineffective as a mutagen unless metabolized to an active intermediate, probably diazomethane (for metabolic activation see 35-37), but requires no such artificial activation in higher plants, e.g., barley (33) and Arabidopsis (38) and, now, Glycine. Does this mean that higher plants have enzymes similar to those in mammals which can convert DMN to the active form? Even then, it is difficult to explain the saturation effect at low concentrations of DMN. The idea that there is a maximal conversion of DMN to active mutagen, not dependent upon the concentration of DMN, is a weak one, at best. It would be of great interest if something like the mammalian liver S-9 fraction can be isolated from Glycine with capability of converting DMN outside the living cell.

When seed were treated with DMN at 4 hr intervals for 4 hr periods, e.g., for 0-4, 4-8, 8-12, ... 44-48 hr, the frequencies of spots on the resulting leaves were not uniform. The frequency usually increased from 0 to 8 hr, then decreased, sometimes to half of the highest around 12 hr, and increased again in a cyclic fashion. The highest frequencies were observed at around 8, 16, and 36 hr in a 0-48 hr treatment schedule. In general, these responses are similar to those observed earlier for MC, MMS, and MNNG. However, there is no definite conclusion about this uniformity of response to all these chemicals based on either the availability of enzymatic background or the specific phase of cell cycle. That this so-called cyclic effect is common to many chemicals indicates some intrinsic property of the embryonic cells.

The gene y_{11} showed only a few mutations to Y_{11} under the influence of DMN. It could, like MC or MNNG, be based on some property of the y_{11} and its structural relationship with its allele Y_{11}. It is intriguing in view of the fact that chlorophyl mutations have been observed in other plants treated with this chemical (33).

Trenimon

Seed samples which were treated for 24 hr with 100 ppm or more of TM failed to germinate. In later experiments, concentrations as low as 0.25 ppm, 5 ppm, and 50 ppm were used. Again, there was no growth in case of 50 ppm treatment, but spot frequencies on the $Y_{11}y_{11}$ leaves in case of 0.25 and 5 ppm treated materials were significantly higher than those in the controls. The frequency of

Table 12. Frequency and relative proportion of the different types of spots on the leaves of $Y_{11}y_{11}$ *Glycine max* var. L65-1237 treated with various concentrations of dimethylnitrosamine (DMN).[a]

Treatment	DMN Concentration (ppm)	Number of leaves analyzed	Frequency of spots/leaf				Relative proportion of spots			
			DG	Y1	Db	T	DG/Db	Y1/Db	DG/Y1	T/Db
0-24 hr	Control	120	0.3	0.4	0.1	0.8	3.0	4.0	0.8	8.0
	60	No growth	-	-	-	-	-	-	-	-
	125	110	5.6	8.1[b]	2.6	16.3[b]	2.2	3.1	0.7	6.3
	250	105	4.9	9.1[b]	3.1	17.1[b]	1.6	2.9	0.5	5.5
	500	115	5.7	9.3[b]	3.9	18.8[b]	1.5	2.4	0.6	4.8
	Control	125	0.7	0.6	0.5	1.8	1.4	1.2	1.2	3.6
	60	115	6.1	8.3[b]	2.9	17.3[b]	2.1	2.9	0.8	6.0
	125	115	5.5	7.6[b]	2.7	15.8[b]	2.0	2.8	0.7	5.9
	250	115	6.2	7.1[b]	2.6	15.8[b]	2.4	2.7	0.9	6.1
	500	110	4.9	8.6[b]	2.8	16.3[b]	1.8	3.1	0.6	5.8
	Control	115	0.6	0.8	0.7	2.1	0.9	1.1	0.8	3.0
	60	110	5.3	10.8[b]	2.5	15.5	2.1	4.3	0.5	6.2
	125	110	4.2	9.0[b]	2.6	15.8	1.6	3.5	0.5	6.1
	250	105	4.2	9.8[b]	2.2	16.2	1.9	4.5	0.4	7.4
	500	75	4.4	9.5	2.2	16.1	2.0	4.3	0.5	7.3

24-48 hr	Control	105	0.6	0.8	0.4	1.8	1.5	2.0	0.8	4.5
	60	110	5.4	8.5[b]	3.1	17.0[b]	1.7	2.7	0.6	5.5
	125	85	6.0	10.1[b]	2.8	18.9[b]	2.1	3.6	0.6	6.8
	250	75	6.0	9.8[b]	3.6	19.2[b]	1.7	2.7	0.6	5.3
	500	55	5.8	9.8[b]	2.6	18.1[b]	2.2	3.8	0.6	7.0
	Control	100	0.3	0.2	0.1	0.6	3.0	2.0	1.5	6.0
	60	110	5.9	8.3[b]	3.6	17.8[b]	1.6	2.3	0.7	4.9
	125	100	6.1	9.1[b]	3.4	18.6[b]	1.8	2.7	0.7	5.5
	250	65	6.7	11.4[b]	3.8	21.9[b]	1.8	3.0	0.6	5.8
	500	55	6.7	13.5[b]	3.5	23.2[b]	1.9	3.7	0.5	6.7

[a] Source: Arenaz (19).

[b] Differs significantly from control, χ^2: $p > 0.05$.

Table 13. Type and frequency of various types of spots induced on $Y_{11}y_{11}$ leaves with low concentrations of DMN[a]

Experiment	DMN Concentration (ppm)	Number of leaves analyzed	Types of spots				Relative frequency of spots			
			DG	Y1	Db	T	DG/Y1	DG/Db	Y1/Db	T/Db
Expt. A	Control	110	0.45	0.51	0.15	1.11	0.88	3.0	3.40	7.40
	12.5	110	2.15	2.32	0.65	5.12	0.93	3.31	3.57	7.88
	25	110	3.47	4.15	0.79	8.51	0.84	4.39	5.25	10.77
	50	90	3.94	5.22	1.13	10.30	0.75	3.49	4.62	9.12
Expt. B	Control	50	0.54	1.40	0.34	2.28	0.39	1.59	4.12	6.65
	1.25	50	1.40	2.36	0.94	4.70	0.56	1.49	2.51	5.00
	2.50	50	2.48	5.24	1.84	9.56	0.47	1.35	2.85	5.20
	5.00	50	3.24	6.76	1.38	11.38	0.48	2.35	4.90	8.25
	10.00	50	3.88	8.06	1.14	13.08	0.48	3.40	7.07	11.47

[a] Seeds were treated for 0-24 hr in Experiment A and for 0-24 hr in Experiment B; 12 g seed of L65-1237 was soaked in 30 ml of solution at room temperature. In B a sample of only 10 plants was used in analysis in each treatment. Source: Vig and Zimmermann (20).

Table 14. Type and frequency of spots of various types observed on the $Y_{11}y_{11}$ leaves after treatment of *Glycine max*. var. L65-1237 see with Trenimon (TM)[a,b]

TM Concentration (ppm)	Duration (hr)	No. of leaves analyzed	Frequency of spots/leaf				Relative frequencies			
			Dg	Y1	Db	T	DG/Y1	DG/Db	Y1/Db	T/Db
Water	0–24	95	0.14	0.30	0.13	0.57	0.47	1.08	2.31	4.38
0.25	0–24	85	0.38	0.49	0.32[c]	1.19[c]	0.78	1.19	1.53	3.72
5.0	0–24	75	0.81	0.91	0.84[c]	2.56[c]	0.89	0.96	1.08	3.05
50.0	0–24	--	No germination							
Water	0–18	100	0.11	0.18	0.14	0.43	0.61	0.78	1.29	3.07
12.5	0–18	55	0.35	0.96	0.75[c]	2.06[c]	0.36	0.47	1.28	2.75
25	0–18	--	No germination							
50	0–18	--								

[a]Source: Vig and Zimmermann (20).

[b]A 10 g portion of seed was soaked in 20 ml TM solution or water.

[c]Significantly different from control: χ^2 $p > 0.05$.

spots of all types, including twins, increased in this case, as also in another experiment using 12.5 ppm TM solution for only 0-18 hr (Table 14). The twins constitute about one third of the total spots observed. These few studies also showed a rather narrow range of effective treatments with TM, within which the present system could be worked out. Also, TM can induce spots when treatments are given prior to onset of DNA synthesis in these seeds.

The data obtained from the $y_{11}y_{11}$ leaves from above treatments, also show the efficiency of TM in inducing point mutations in the system under study. Thus, a frequency of 0.58, 1.89, and 9.89 spots/leaf was observed in materials treated for 24 hr with H_2O, 0.25 ppm TM, and 5 ppm TM, respectively. The range of spots on individual leaves was between 0-2, 0-7, and 2-37, respectively. In another experiment a frequency of 15.14 spots/leaf analyzed from the material treated with 25 ppm was found against 0.10 spots/leaf on the control. These data leave no doubt that y_{11} gene mutates to Y_{11} rather frequently under the influence of TM.

Ethylenimine

The only data available for soybean are those dealing with yield and productivity in up to M_4 after 0.02, 0.01, and 0.008% ethylenimine (EI) treatment of the seed for 18 hr at 18-20°C. An increase in yield as reported, for example, by Melchenko (7) indicates the effect of the chemical on polygenes. Such effects on quantitative traits appear to be of special significance, in that such mutations in human populations will not be easily detectable or wiped out. It is logical to assume that many harmful effects on human genome may result from such minor, imperceptible alterations whose ultimate contribution can be drastically modified by extragenic environment. (Perhaps examples can be taken from increased incidence of spina bifida, anencephaly, hydrocephalus, cleft lip, and cleft palate in human populations exposed to environmental mutagens.)

DISCUSSION AND CONCLUSIONS

It is evident from the foregoing summaries on effects of various chemicals that somatic mosaicism and cell-culture mutagenesis in *Glycine max* are of value in determining rapidly the actions of various chemicals on different genetic parameters. The leaf mosaicism is far more useful in this regard since it has led to discrimination between genetic alterations induced by a given mutagen within a few weeks after a single experimental dose is given to one seed sample obtained from selfed $Y_{11}y_{11}$ plants. Thus, somatic crossing over appears to be the only genetic change giving rise to complementary homozygous sectors on the heterozygous background. Even though it has not yet been possible to regenerate plants from leaf discs in the soybean, such successful analysis with Nicotiana tabaccum by Carlson (39) and Deshayes and Dulieu (40) strengthens this proposal made for soybean several years before studies were conducted with tobacco. The failure of a component of the original twin spot would produce the two types of single spots in equal frequencies as would some other changes, e.g., nonreciprocal translocations or nondisjunction (both producing Y_{11}-$Y_{11}y_{11}y_{11}$ or y_{11}-$Y_{11}Y_{11}y_{11}$ components, i.e., near light green - near yellow, or yellow - near light green, respectively) or segmental losses (producing Y_{11}-$Y_{11}y_{11}$ or y_{11}-$Y_{11}y_{11}$). However, in the case of nondisjunction monosomic cells may not compete with normals or trisomics, thus producing mostly single spots. Additionally deletions could produce yellow spots, predominantly. The ideas of controlling elements, paramutations, conversion, cytoplasmic factors or point mutations have been discredited (13) as mechanisms for production of equivalent frequencies of dark green or yellow spots on the light green background.

The production of light green spots on yellow leaves can be attributed to the mutation of gene y_{11} to Y_{11} only. The alternative hypotheses, like gene activation/inactivation, with or without the help of controlling elements cannot explain the sectors on the $Y_{11}y_{11}$ leaves. It has, however, been proposed (1) that the gene y_{11} is composed of a number of tandem repeats of DNA represented fewer times in Y_{11}. Thus, the production of small intracistronic deletions would change y_{11} to Y_{11}. Similar results would be obtained if $y_{11}y_{11}$ segments undergo unequal somatic crossing over. However, for the point of discussion in this paper all changes result eventually into point mutation of y_{11}, producing light green sectors on the yellow leaves.

The genesis of rare light green sectors on the $Y_{11}Y_{11}$ genotype is not very clear. In some instances Y_{11} can mutate to y_{11}. This change, however, is rare, since y_{11} does not represent the simple loss of activity of Y_{11} but an alteration which produces yellow color. Thus, any random change in Y_{11} does not result in y_{11}. The very low frequency of light green sectors on the $Y_{11}Y_{11}$ leaves, but frequent change of y_{11} to Y_{11}, under the influence of certain chemicals (20) support the concept of differential mutation rate of y_{11} and Y_{11}. This also means that a preponderance of yellow spots on $Y_{11}y_{11}$ leaves in some treatments is not the result of point mutations but of the loss of Y_{11}.

These arguments may permit us to suggest the modes of action of various mutagens. A summary of main effects as concluded from the observations with the chemicals in question is given in Table 1.

Table 1 also provides a comparison with one of our reference agents, viz. EMS. Thus, one may conclude that the leaf mosaicism in Glycine offers a rapid, inexpensive eukaryotic, in vivo system which can be utilized for at least the preliminary screening of mutagens to get an idea about various effects produced by the agent on the genetic material. The data so far obtained are well in line with the information obtained from these chemicals when studied on long-term or short-term systems permitting test for only one genetic parameter.

There is need for regeneration of plants from leaf discs for a final analysis. There is need also to study in more detail the effects of pH, temperature at the time of treatment, and of growth, as well as a wider range of the concentration time interactions. Experiments should be designed to study cell cycle specificity in production of mutations, somatic recombination and deletions.

The data obtained with cell-culture mutagenesis is very limited but very promising for study of biochemical mutations, and with suitable markers, for study of somatic recombinations. The only conclusions to be drawn from these studies are that the chemicals tried produce mutations and that there is a strong need for further study in this system to correlate the information with that obtained for leaf mosaicism.

Whereas we do realize the importance of traditional breeding experiments for study of mutation induction, these efforts should be limited at this time to organisms other than soybean, e.g., barley, mouse, Drosophila, fungi, and bacteria. Work has been, however, carried out with successful modifications of quantitative traits, like yield.

ACKNOWLEDGEMENT

I am thankful to Dr. Z. R. Sung of University of California, Berkeley, California, for some help in writing the section dealing with Cell Culture Mutagenesis.

REFERENCES

1. Nilan, R. A., and Vig., B. K. In: Chemical Mutagens, Principles and Their Methods of Detection, A. Hollaender, Ed., Vol. 4, Plenum Press, New York, 1976, pp. 143-170.

2. Sung, Z. R. Plant Physiol. 56: 3-7 (1975).

3. Sung, Z. R. Plant Physiol. 57: 460-462 (1976).

4. Sung, Z. R. Genetics 84: 51-57 (1976).

5. Sung, Z. R., Pratt, M., and Signer, E. In: Proceedings of the First International Symposium on Haploids in Higher Plants--Advances and Potential, K. Kasha, Ed., University of Guelph, Guelph, Ontario, Canada, 1974, p. 392.

6. Sung, Z. R., Smith, J., and Signer, E. In: Genetic Manipulation with Plant Material, L. Ledoux, Ed., Plenum Press, New York, 1975, p. 547.

7. Melchenko, V. V. Soviet Genet. 6: 1462-1466 (1970).

8. Kiang, L. C., and Halloran, G. M. Mutat. Res. 33: 373-382 (1975).

9. Vig, B. K. Mutat. Res. 31: 49-56 (1975).

10. Vig, B. K. Mutat. Res. 49: 189-238 (1977).

11. Vig, B. K., Nilan, R. A., and Arenaz, P. Environ. Exptl. Botany 76: 223-239 (1976).

12. Vig, B. K. Can. J. Genet. Cytol. 11: 147-152 (1969).

13. Vig, B. K. Genetics 73: 583-596 (1973).

14. Vig, B. K., and Paddock, E. F. J. Hered. 59: 225-229 (1968).

15. Vig, B. K. Theor. Appl. Genet. 41: 145-149 (1971).

16. Vig, B. K. Genetics 75: 265-277 (1973).

17. Vig, B. K. Mol. Gen. Genet. 116: 158-165 (1972).

18. Vig, B. K., and Paddock, E. F. Theor. Appl. Genet. 40: 316-321 (1970).

19. Arenaz, P. M.S. Thesis, Univ. Nevada, 1976, 102 pp.

20. Vig, B. K., and Zimmermann, F. K. Environ. Exptl. Bot. 17: 113-120 (1977).

21. Vig, B. K. Theor. Appl. Genet. 43: 27-30 (1973).

22. Holliday, R. Genetics 50: 323-335 (1964).

23. German, J., and La Rock, J. Tex. Rep. Biol. Med. 27: 409-418 (1969).

24. Rao, R. N., and Natarajan, A. T. Genetics 57: 821-835 (1967).

25. Szybalski, W. Cold Spring Harbor Symp. Quant. Biol. 29: 151-159 (1965).

26. Vig, B. K. J. Theoret. Biol. 54: 191-199 (1975).

27. Loomis, W. F., and Lipmann, F. J. Biol. Chem. 179: 503-504 (1949).

28. Kihlman, B. A., Norton, K., Sturelid, S., Karlsson, M.-B., and Kronberg, D. Hereditas 69: 323-325 (1971).

29. Sander, C., Nilan, R. A., Kleinhofs, A., and Vig., B. K. Mutat. Res. 50: 67-75 (1978).

30. Hince, T. A., and Neale, S. Mutat. Res. 24: 383-387 (1974).

31. Roberts, J. J., Sturrock, J. E., and Ward, K. N. Mutat. Res. 26: 129-144 (1974).

32. Veleminsky, J., and Gichner, T. Mutat. Res. 5: 429-431 (1968).

33. Veleminsky, J., Zadrazil, J. S., Pokorny, V., Gichner, T., and Svachulova, J. Mutat. Res. 17: 49-58 (1973).

34. Delic, V., Hopwood, D. A., and Friend, E. J. Mutat. Res. 9: 167-182 (1970).

35. Ames, B. N., McCann, J., and Yamasaki, E. Mutat. Res. 31: 347-364 (1975).

36. Natarajan, A. T., Tates, A. D., von Buul, P. P. W., Meijers, N., and de Vogal, M. Mutat. Res. 37: 83-90 (1976).

37. Ong, T.-M., and Malling, H. Mutat. Res. 31: 195-196 (1975).

38. Veleminsky, J., and Gichner, T. Mutat. Res. 12: 65-70 (1971).

39. Carlson, P. Genet. Res. 24: 109-112 (1974).

40. Deshayes, A., and Dulieu, H. In: Polyploidy and Induced Mutations in Plant Breeding, IAEA, PL 503/14A, International Atomic Energy Agency, Vienna, 1974, pp. 85-99.

CHAPTER 11

MUTAGENICITY OF SELECTED CHEMICALS IN BARLEY TEST SYSTEMS

R. A. Nilan

Program in Genetics, Washington State University, Pullman

Washington 99164

and

J. Velemínský

Czechoslovak Academy of Sciences, Institute of Experimental

Botany, Department of Genetics, Flemingovo Nám. 2

160 00 Praha 6, Czechoslovakia

INTRODUCTION

Barley, Hordeum vulgare, has long been used as a model system in higher plants for studying the effects of a great number of physical and chemical mutagens. The advantages of this organism for analysis of mutagen effects include the following: (1) it is a true diploid; (2) it is a strict self-fertilizer; (3) it is easily grown in a relatively small space; (4) it has a low number ($2n = 14$) of large chromosomes whose alterations can be readily detected under a light microscope; and (5) it has a number of loci whose mutants are easily scored.

A wide variety of physiological and genetic endpoints can be scored following treatment of the seeds (caryopses), especially their embryos. Indeed, all of the genetic data presented in this review were obtained from seed treatment with the different chemicals. Following seed treatment, resulting seedlings and plants are of the M_1 generation while seedlings and plants produced from seed of the M_1 plants are of the M_2 generation.

Mutants that are scored in the M_2 generation originate as mutations in the meristematic cells of the treated embryo. The resting barley embryo consists of at least nine mutally exclusive meristems, each of which can develop one or more tillers and spikes that do not share a common mutation. Within each meristematic region three to four cells produce one primary tiller and spike. Each plant may produce four to five primary spikes.

Another unique factor of the seed and its embryo is that chemical treatments can be applied over a wide range of environmental conditions, such as water content, metabolic activity, temperature, pH, etc. This also means that chemical treatments can be highly controlled with respect to these conditions and highly reproducible experiments and results are possible.

Unfortunately, the experiments reviewed herein have been conducted with a considerable number of different varieties. Since it has been clearly demonstrated that varieties differ in response to different mutagens, this factor must be considered when assessing the data.

The major biological endpoints employed to described the effects of the chemicals of this study include: (1) chromosome and chromatid aberrations, scored during mitoses of cells of root tips or primary meristematic cells of the embryos of treated seeds (scored at anaphase or metaphase and calculated as % aberrations or % aberrated cells, usually dicentrics, or bridges, and fragments); (2) chromosome aberrations, mainly interchanges, during meiosis of M_1 plants and calculated as % of M_1 spikes; (3) gene mutations which produce chlorophyll-deficient M_2 seedlings (recessive mutants) and calculated as % of M_1 spikes or % of M_2 seedlings; and (4) lethals, usually egg or zygote lethals, as measured by % of non-surviving M_2 seedlings.

When several endpoints, including mutation and chromosome damage, are measured after mutagen treatment, it is possible to determine mutagen efficiency. In the barley mutagenesis literature this term usually refers to the ratio of mutations to chromosome or chromatid aberrations frequencies but also can be determined from ratios of mutation frequencies to M_1 plant survival and lethality. This term is distinct from mutagen effectiveness, which is the measure of mutagen effects in relation to concentration.

Scoring for chlorophyll-deficient mutants represents a reliable and sensitive multi-locus test for detecting spontaneous and induced heritable mutations in barley. Most of the mutants are caused by a range of nucleus damage including gene mutations and a variety of more gross chromosome changes, including deletions. Some are inherited cytoplasmically and are presumably due to changes in chloroplast DNA. Many of the nuclear chlorophyll-deficient mutants segregate in a 3:1 ratio, indicating they are due to gene changes or to minute chromosome changes that simulate single gene changes. Recently, it has been determined that at least 700 loci can mutate to produce a chlorophyll-deficient mutant. This estimate was calculated from the ratio of the number of mutants induced at the single locus controlling vine (gigas) seedling to the number of chlorophyll-deficient mutant events over several physical and chemical mutagen experiments (Nilan, unpublished).

The spontaneous or background chlorophyll-deficient mutation and chromosome aberration frequencies in barley are variable because different varieties are used in different experiments. However, in summary, the background mutation frequency is between 0.3 and 0.5% and chromosome aberration frequency is about 0.25%. Among the chemicals under study, ethyl methanesulfonate (EMS), ethylenimine (EI), and methyl methanesulfonate (MMS) have been studied extensively, and chemical and environmental parameters for obtaining reproducible results from a given concentration are fairly well established. The environmental parameters include temperature, water content, and metabolic activity of the embryo, and temperature and pH of the treatment solution. Chemical parameters include half-life and number of functional groups of the chemical in solution. DMN, MNNG, MeTEPA, Mitomycin C, Myleran, Trenimon, triethylenemelamine, and cadmium have been superficially analyzed, and environmental and chemical parameters for obtaining maximum genetic effects and reproducible results have received little attention. There is insignificant or no known genetic data for barley with respect to the effects of aflatoxin B_1, cyclophosphamide, ICR-170, Natulan, TEPA, thioTEPA, vinyl chloride, epichlorhydrin, and DEN.

For the most part the data for each chemical used in this review have been selected as being representative and from experiments involving adequate controls of environmental and biological parameters and treatment replications for reproducible and valid results. Based on this selection criterion, the data from only about one quarter of the papers reviewed could be included in this study.

The objective of many papers not included is to induce mutants for plant breeding, genetic markers, genetic studies, etc., and for specific traits. Thus, little attention is given to the treatment conditions and concentrations.

Dose-effect curves of the different genetic endpoints for EMS, MMS, and EI are presented. However, it is not possible to determine the dose of chemical reaching the embryonic cells that produce the M_1 and M_2 generations.

The effects of the lowest concentrations used are indicated, but these are not necessarily the lowest effective concentration

Table 1. Effectiveness of 11 chemical mutagens on barley

Mutagen (CAS No.)	Number of reports	Lowest reported effective concentration (mM)	Biological effects (all concentrations)[a]				Concentration range reported (mM)
			Chlorophyll mutations	Chromosome changes	Lethality	Other	
EMS (62-50-0)	20	8	+	+	+	DNA single-strand breaks	8-393
MMS (66-27-3)	7	0.037	+	+	n.t.	DNA single-strand breaks	0.037-27
EI (151-56-4)	12	7.73	+	+	+		7.73-1500
MNNG (70-25-57)	7	0.07	+	+	n.t.	DNA single-breaks	0.07-17
Myleran (55-98-1)	6	0.07	+	+	+		0.07-8.97
DMN (62-75-9)	1	50	-	n.t.	n.t.		
TEM (51-18-3)	1	0.5	+	n.t.	n.t.		
MeTEPA (57-39-6)	1	0.537	+	+	n.t.		

Trenimon (68-76-8)	1	0.001	n.t.	+	n.t.		0.001-0.05
Cadmium compounds	1	0.01	n.t.	+	n.t.		
Mitomycin C (50-07-7)	1	--	n.t.	n.t.	n.t.	Recombinagen	1.81-76

[a] + = positive effect; - = negative effect; n.t. = not tested

or dose that could be effective. It is usually impossible to extrapolate to a "no detectable effect" dose for the chemicals under study.

The concentrations for all chemicals are reported as millimolar (mM). In many cases these have been converted from concentrations stated in percent (%), parts per million (ppm), or molar (M).

RESULTS AND DISCUSSION

Summary

The effectiveness of 11 of the reference chemical mutagens on barley is summarized in Table 1.

Data on Radiations and Individual Chemicals

X- and γ-rays. Table 2 presents some fairly representative mutation and chromosome aberration frequency data. Other data for these sparsely ionizing radiations are presented under Comments in the chemical mutagen tables.

It is well known that slight changes in water and oxygen content of the seeds during and after radiation significantly affect the subsequent frequencies of genetic changes. The data in Table 2 are from experiments in which water and oxygen content were fairly well controlled.

Ethyl Methanesulfonate (62-50-0). Ethyl methanesulfonate (EMS) is one of the most potent mutagens that has been used on barley (Table 3). Maximum chlorophyll-deficient mutation frequencies of 50% on an M_1 spike basis have been reported. These high mutation frequencies are induced without appreciable frequencies of chromosome changes and significant amounts of physiological injury--making EMS a very efficient mutagen. It does induce high frequencies of seedling lethality and pollen sterility (which can be attributed to embryo and female gametophyte sterility). These events cannot be due primarily to usual types of gross chromosome changes evaluated under the microscope.

Differences in mutation frequencies (even at the same concentration) as described here and in other reports may be attributed to the time and temperature of EMS treatments. EMS is a slow-reacting compound with a half-life at 20°C of 93 hr and 9 min. Thus, more effective EMS treatments require longer treatment times and higher concentrations than have been used by a number of experimenters, particularly at 20°C. Other major reasons for variability in mutation frequencies are uncontrolled post-treatment

Table 2. Barley: x or γ-rays

Cultivar	Treatment conditions	Dose (krads)	Biological effects: Mutation (%) M_1 spike	Biological effects: Mutation (%) M_2 seedling	Aberrant cells (%)[a]	Lethality (%)	References
Himalaya	Seeds 9.5% H_2O	10 (γ-rays)	5.3		26.7 A		1
		20 (γ-rays)	7.9		54.0		
Himalaya	Seeds 10.2% H_2O, oxygen-free	40 (γ-rays)	8.0		6.0 fragments and 0.30 bridges/cell A		2
	Seeds 10.2% H_2O, oxygenated	15 (γ-rays)	12.5		10 fragments and 0.60 bridges/cell A		
Atlas 57	Seeds 13% H_2O, oxygenated	8.7 (γ-rays)		0.4			3
		17.5 (γ-rays)		0.5			
		26.2 (γ-rays)		0.9			
		35.0 (γ-rays)		1.1	4.0 fragments and 0.5 bridges/cell A		
Himalaya	Seeds 14% H_2O, oxygen-free	30 (x-rays)	8.0	1.2			4
		45 (x-rays)	8.9	1.3		33	
		60 (x-rays)	11.9	1.8		40	
Himalaya	Seeds 13% H_2O	30 (x-rays)	3.1		2.8 A	36	5

[a] A = anaphase.

Table 3. Barley: EMS (62-50-0)

Cultivar	Treatment conditions	Concentrations (mM)	Biological effects: Mutations (%) M_1 spike	Biological effects: Mutations (%) M_2 seedling	Biological effects: Aberrant cells (%)[a]	Biological effects: Lethality (%)	Comments	References
CV 292	Seeds at 10% H_2O; treated 18 hr, pH 7.0, 24°C, post-washed then soaked 4 hr; for combination treatment of EMS and gamma radiation seeds were irradiated at a dose rate of 3 kR/min	12	9.36	1.34	13 A	13	10-30 kR gamma induced 13-18% aberrations, 15-21% lethality, 1.07-2.61% mutations/M_1 spike and 0.21-0.48 mutations/M_2 seedling	7
Montcalm	Seeds presoaked 2 hr, 20°C; treated 8 or 4 hr at 10, 20, 30°C; rinsed 4 times, then bathed either 12 or 24 hr at 10, 20, 30°C	40	7.2-68.7	0.3-11.9				8
Himalaya	Seeds presoaked 5 hr or 17 hr, treated 1.5 hr at 5°C, followed by 15 min at 35°C; pH 3 or pH 11	30 and 50	2.0-16.2	0.16-1.82			15 kR gamma induced 5.1% mutations/M_1 spike and 1.21% mutations/M_2 seedling	9

							20 kR gamma induced 8.6% mutations/M_1 spike and 1.82% mutations/M_2 seedling	
Volla	Seeds treated 3 hr at 25°C; post-washed 24 hr at 25°C, redried at 40°C to 5, 13, 20, 30% H_2O content; stored 1-6 weeks, 25°C	280		1.0-15.2	24.5-32.7 A (un-stored)		Report increase in damage due to storage (in terms of % aberrations) at 20 and 13% H_2O, decrease in damage at 30% and no change at 5%	10
Himalaya	Seeds presoaked 4 or 17 hr, 20°C; treated 0.5 hr, 35°C; post-washed 4 hr, some dried to 13 or 2% H_2O	30			11.2-56.4 A			11
Himalaya	Seeds 14% H_2O treated at 30°C and 10°C for 5, 3.5, 4.5, 1.75 or 24 hr, post-washed	20-120	28.0-32.4	3.0-8.4		90-55	% lethality control 26%; 45 kR x-ray induced 55% lethality, 8.9% mutations/M_1 spike, 1.3% mutations/M_2 seedling;	12

(continued)

Table 3. (continued)

Cultivar	Treatment conditions	Concentrations (mM)	Biological effects: Mutations (%) M_1 spike	Mutations (%) M_2 seedling	Aberrant cells (%)[a]	Lethality (%)	Comments	References
							60 kR x-ray induced 65% lethality, 11.9% mutations/M_1 spike, 1.8% mutations/M_2 seedling	
Ekonom	Seeds treated 24 hr	20, 26, 40			0.3-11.7 A		LEC = 20 mM	13
Volla and Enkonom	Seeds treated 3 hr, 25°C; post-washed 24 hr, 25°C; stored in air or nitrogen 0, 1, 2 weeks	280			26.3-53.7 A		% aberrations about the same for air or nitrogen	14
	Seeds treated 3 hr, 25°C; post-washed 24 hr, 25°C; stored 0-8 weeks at -20, 0, 25°C	280			24.3-75.5 A		Total inhibition of mitosis at 25°C with 4 and 8 weeks storage	14

CU 292	Seeds treated 18 hr, pH 7.0, 24°C; post-washed 4 hr, 24°C	8 and 12	4.44-10.55			12.0-36.4		15
Ekonom	Seeds treated 3 hr, 25°C; washed 24 hr; redried at 40°C to 30% H_2O; stored 14 days at 25°C	280		1.1-14.3	8.0-24.5 A	15.5-83.5		16
Piroline	Seeds treated 3 hr, post-washed	24			0.63-0.75 A			17, 18
Atlas 57	Seeds presoaked 0, 8, 16 hr; treated 6 hr, pH 7.5, 21-23°C; post-washed 4 hr	30	10.83-18.86				15 kR gamma induced 5.04% mutations/ M_1 spike 490 R N_f induced 9.09% mutations/ M_1 spike	19
Himalaya CI 620	Seeds presoaked 5 hr, 20°C; treated 15 or 25 hr; post-washed 1-5 hr	20	28.2-32.6			1-21	45 kR x-ray induced 7.5% mutations/ M_1 spike	20
	Seeds presoaked 0, 5.5, 13, 29 hr, 30°C; treated 5, 3, 4.25, 5 hr, 30°C	20-40	8.7-43.5	1.0-7.6		5-36.6		

(continued)

Table 3. (continued)

Cultivar	Treatment conditions	Concentrations (mM)	Biological effects: Mutations (%) M_1 spike	Mutations (%) M_2 seedling	Aberrant cells (%)[a]	Lethality (%)	Comments	References
Himalaya CI 620	Seeds presoaked 6 hr, 0°C; treated 1, 2 or 3 hr, 20°C	25–50	12–24					21
		280					DNA single-strand breaks	22
		294 and 393			+			23
		280	+	+	+	+		24
		0–90			+			25
		24			+			26

[a] A = anaphase

conditions, e.g., duration of post-treatment washing of seeds, and different water content of seeds at the onset of germination prior to the first DNA replication in the meristematic cells of the embryo.

Data included in Table 3 are only from those experiments where optimum treatments which recognize the importance of the half-life of this chemical have been used.

Dose response curves for frequencies of chlorophyll-deficient mutations, chromosome aberrations, and DNA single-strand breaks, and amounts of M_1 germination and sterility are presented in Figure 1. These data are from Gichner et al. (6) and from Gichner and Veleminský (unpublished).

Methyl Methanesulfonate (66-27-3). Methyl methanesulfonate (MMS) produces in barley patterns of biological damage more similar to x- or γ-rays than to EMS (Table 4). At similar concentrations it induces relatively higher frequencies of aberrations and lower frequencies of mutation than EMS. Therefore, its efficiency of mutation induction is lower than EMS and similar to x- or γ-rays.

Limited data indicate that pH may be influential in MMS action. Gichner et al. (27) have demonstrated that frequencies of MMS-induced aberrations are influenced by storage of treated seeds.

Dose-response curves for frequencies of chlorophyll-deficient mutations and single-strand breaks and amounts of M_1 germination and seedling growth are presented in Figure 2. These data are from Gichner and Veleminský (unpublished).

Ethylenimine (151-56-4). Compared to x- or γ-rays, ethylenimine (EI) is highly mutagenic, as it induces chlorophyll-deficient seedling mutants at frequencies of about 30 per 100 M_1 spikes (Table 5). This may be compared to x-rays and γ-rays, which induce maximum frequencies of around 17-20%. It is a more efficient mutagen than the sparsely ionizing radiation at concentrations giving 50% M_1 plant lethality. However, unlike EMS, EI induces higher frequencies of chromosome alterations, approaching those of x-rays. Thus, it is not as efficient as ethyl methanesulfonate. M_2 seedling lethality is also about similar to that of x-rays, indicating that much of this lethality may be due to chromosome changes.

Among several papers reviewed, only a relatively few are included in Table 5. These studies are relevant because they control pH, a very important modifying factor in EI effectiveness (32,33). Slight changes in pH of the EI solution produce considerable differences in mutations and chromosome aberration frequencies. Thus experiments that do not take into account the pH and in fact do not control the pH of the EI solution during seed treatment will

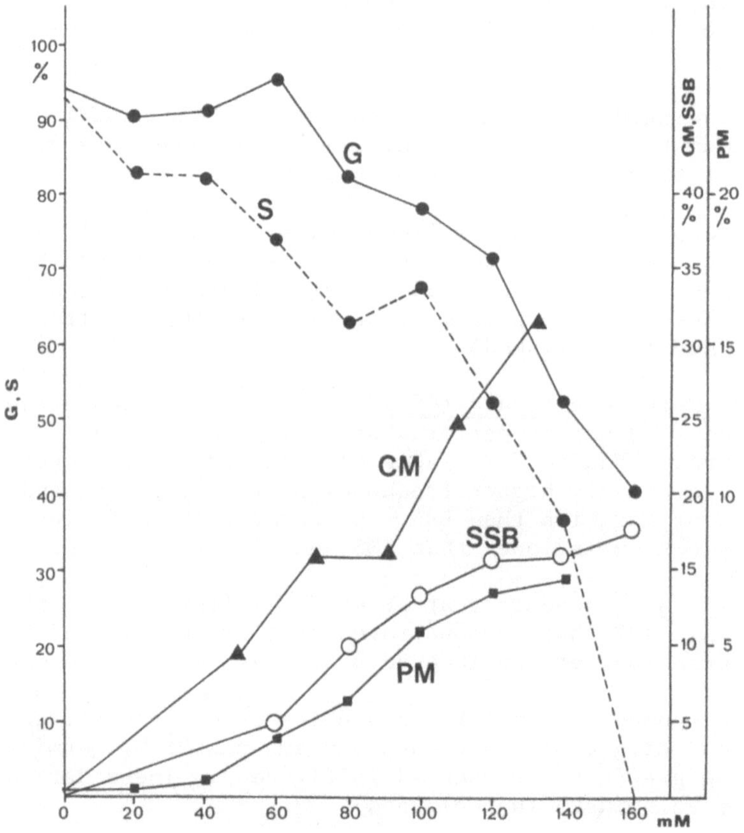

Figure 1. Dose response curve for EMS in barley.

S = sterility of spikes in the M_1 generation, G = germination of treated seeds, CM = chromosome aberrations in metaphases of the root tips, PM = recessive chlorophyll mutations segregating in M_2 generations (per total amount of M_2 seedlings), SSB = DNA single-strand breaks determined according to the sedimentation rate in alkaline sucrose gradients (5-20% sucrose, Beckman rotor SW 27.1) of DNA isolated from the barley embryos. SSB are expressed only relatively as a distance of central DNA peak of EMS-treated to untreated DNA in percentage of total gradient length. Treatment conditions: (PM, G, S) seeds soaked 5 hr in EMS solution at 25°C and washed in running water for 18 hr. (CM and SSB) 10 hr water presoaked seeds treated in EMS for 5 hr and washed for 1 hr.

Table 4. Barley: Methyl Methanesulfonate (66-27-3)

Cultivar	Treatment conditions	Concentrations (mM)	Biological effects: Mutations (%) M_1 spike	Biological effects: Mutations (%) M_2 seedling	Biological effects: Aberrant cells (%)	Biological effects: Lethality (%)	Comments	References
N.P. 13	Seeds at 12% content treated 12 hr at pH 6.5, 20°C, post-rinsed	23			6.68 (in microsporocytes)		45 kR x-ray induced 7.5 mutations/100 M_1 spikes	23
Himalaya	Seeds at 13.5% H_2O content treated 12 hr at pH 7.0, post-washed 5 hr at 20°C	2.3	6					20
Himalaya	Seeds at 14% H_2O content treated 12 hr, pH 7.0, 20°C, post-wash 5 hr at 20°C	2.3	5				30 kR x-ray induced 8.0 mutations/100 M_1 spikes and 53 bridges and 309 fragments/100 cells	28

(continued)

Table 4. (continued)

Cultivar	Treatment conditions	Concen-trations (mM)	Biological effects: Mutations (%) M_1 spike	M_2 seedling	Aberrant cells (%)	Lethal-ity (%)	Comments	Refer-ences
		0.037 and 0.045					Single-strand breaks in DNA	22
		8					Single-strand breaks	29, 30
		6, 10, 15			+		Single-strand breaks	27
		27	+		+			31

Table 5. Barley: Ethylenimine (151-56-4)

Cultivar	Treatment conditions	Concentrations (mM)	Biological effects: Mutations (%) M_1 spike	Mutations (%) M_2 seedling	Aberrant cells (%)[a]	Lethality (%)	Comments	References
Himalaya	Seeds 14% H_2O; 4 hr at 20°C, pH 6.4 and 7.01; post-wash 2 hr	7.73-25.1		1.09-4.36	5.59 A	14.9-51		32
	Dry seeds, 2 hr at 20°C, pH 7.7	9.29			70.7 A			34
Foma	Dry seeds, 5 hr at 21°C, pH 6.7 and 7.7, post-wash 1 min	12-1500	1-9.5			2-62		33
Bonus	Dry and presoaked (24 hr) seeds, 2 hr at 20°C, pH 7.5	9.29-21.4	5-30			5-45	10.5 kR x-ray induced ca. 10 mutations/100 M_1 spikes and 55% lethality	35
Freja	Dry seeds treated 90 min, 24°C; rinsed 15 min, pH 7	7.7			31.0-54.5 M		18 kR gamma induced 20.0-50.0% M aberrant cells	36

(continued)

Table 5. (continued)

Cultivar	Treatment conditions	Concentrations (mM)	Biological effects: Mutations (%) M_1 spike	M_2 seedling	Aberrant cells (%)[a]	Lethality (%)	Comments	References
		2-14	+		+			37
		2			+			38
		2 and 10	+		+	+		39
		10	+		+			40
		2	+		+			41
		2 and 4	+					42

[a] A = anaphase, M = metaphase.

not produce valid or reproducible results. The role of pH in modifying EI effectiveness is well described by Wagner et al. (32) and Konstantinov et al. (33).

No published experiments describe the least effect or the most effective doses of EI or concentrations that produce a doubling effect over spontaneous mutation frequencies.

Dose-response curves for frequencies, chlorophyll-deficient mutations and chromosome aberrations are presented in Figure 3. These data are from Wagner et al. (32) and are based on actual EI as well as imine form concentrations.

N-Methyl-N-nitro-N-nitrosoguanidine (MNNG) (70-25-57). Table 6 presents representative mutation and chromosome aberration frequency data for MNNG. At pH 5 to 6 this chemical induces chromatid aberrations (similar to those induced by other alkylating agents) in the root tips of treated, dry, presoaked and germinating seeds. The effects are dose-dependent but from the given data the shape of the dose-response curves and the lowest efficient dose (LED) cannot be accurately extrapolated. It appears MNNG is less capable of inducing aberrations in dry and presoaked seeds than EMS or methylnitrosourea (MNU). It may be more effective when applied to germinating seeds (S-phase cells), but confirmation requires more experiments.

Published data indicate that MNNG, even at sublethal doses, is much less effective in inducing recessive chlorophyll-deficient mutations than other alkylating agents. This applies to treatments of dry, soaked or germinating seeds with partially synchronized cell division (reached by cooling at 3°C for 40 hr). Even treatment of S-phase cells in the shoot apex (72 hours presoaking or germinating seeds) was not sufficient for inducing mutations. It should be pointed out that MNNG is a strong mutagen in certain other plants (e.g., *Arabidopsis thaliana*) when applied to dry seeds.

Myleran (55-98-1). Myleran induces very low frequencies of chlorophyll-deficient mutations (Table 7). Maximum reported frequency is 4% on an M_1 spike basis (31). It does, however, induce relatively high frequencies of chromosome changes. No differences in effectiveness were found in the concentrations ranging from 0.81 to 3.24 mM. Dose response was achieved by changing the treatment time (51).

DMN (62-75-9). DMN is very weakly mutagenic in barley (Table 8) even at sublethal doses which cause distinct plant injury (depression of germination, seedling growth, etc.). It should be pointed out that in another higher plant, *Arabidopsis thaliana*, DMN was mutagenic with relatively high activity (55,56).

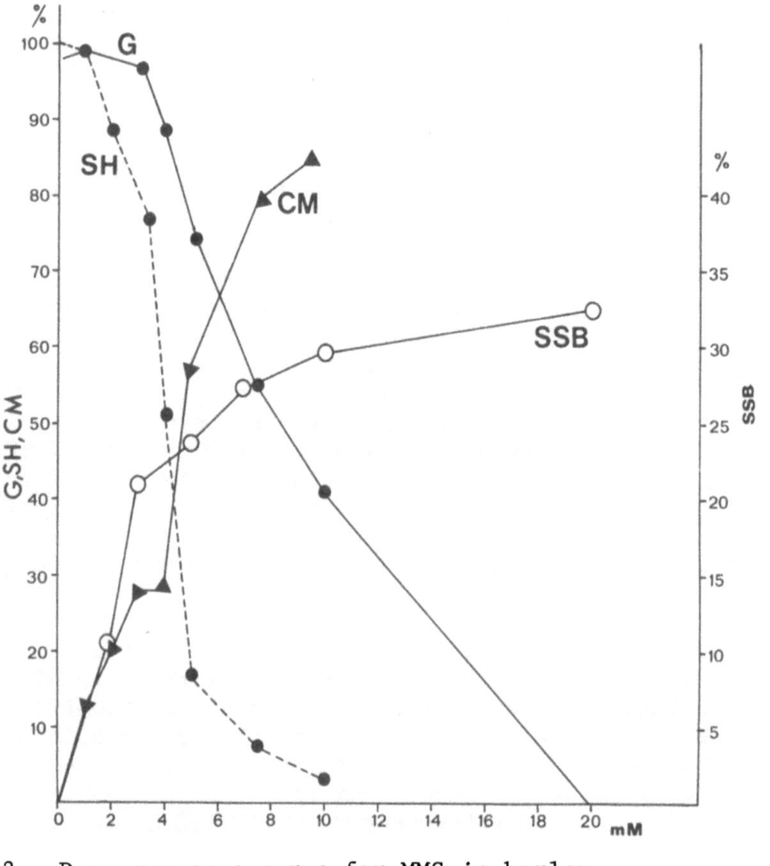

Figure 2. Dose response curve for MMS in barley.

SH = height of M_1 seedlings, G = germination of treated seeds, CM = chromosome aberrations in metaphases of the root tips, SSB = DNA single-strand breaks determined according to the sedimentation rate in alkaline sucrose gradients (5-20% sucrose, Beckman rotor SW 27.1) of DNA isolated from the barley embryos. SSB are expressed only relatively as a distance of central DNA peak of MMS-treated to untreated DNA in percentage of total gradient length. Treatment conditions: 10 hr water presoaked seeds soaked 5 hr in MMS solution and washed for 1/2 hr.

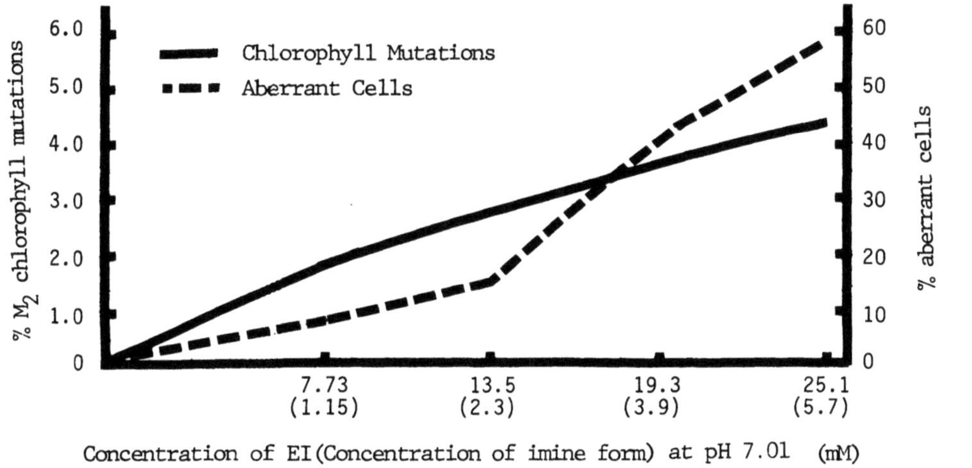

Figure 3. Dose response curves for frequencies of mutations and chromosome aberrations following EI treatment of barley seeds.

Seeds of the variety Himalaya at 14% water content were treated with solutions consisting of 0.1M phosphate buffers and to which ethylenimine (EI) was added shortly before treatment. Treatments were at pH 7.01. The graph is based on concentrations of ethylenimine as well as the imine form. Chromosome aberrations were scored at anaphase of the first division in embryonic shoots of treated seeds. Chlorophyll-deficient mutations were scored among M_2 seedling populations.

MeTEPA (57-39-6) and TEM (51-18-3). Available data are insufficient to measure the true effectiveness of these mutagens. Both induce chromosome aberrations but do not appear to induce significant frequencies of chlorophyll-deficient mutations. Besides the data in Table 8, Ehrenberg (57) has observed a very low frequency of mutations (1.5% of spikes segregating chlorophyll mutants in M_2 generations) after TEM treatment. No details of treatment conditions, concentrations, etc. were presented.

Trenimon (68-76-8). The ability of Trenimon to induce mutations in barley has not been tested. In *Arabidopsis thaliana* it was non-mutagenic when applied to dry seeds but mutagenic when applied to the apex of the dividing shoot (58).

Trenimon induces chromosome aberrations in the root tips of either dry treated seeds or barley seeds presoaked 16 hr and treated

Table 6. Barley: MNNG (70-25-57)

Cultivar	Treatment conditions	Concentrations (mM)	Biological effects: Mutations (%) M_1 spike	Biological effects: Mutations (%) M_2 seedling	Biological effects: Aberrant cells (%)[a]	Biological effects: Lethality (%)	Comments	References
Foma	Dry seeds treated 24 hr, pH 5.5-6 at 18°C	1-1.7			12.5 A			43
	Dry seeds treated 24 hr, pH 5.5-6 at 24°C	0.7-1.4			8.2 A			43
NP 113	Germinated seeds treated 30 min, pH 5.4, 20°C	6.8			90.5 M		S-phase cells treated (indicated autoradiographically)	44
Himalaya	Seeds presoaked 12 hr, treated 8 hr, 20°C, pH 5 and pH 6	0.07-0.7		0	23.2 A at pH 5 20.0 A at pH 6			45
XP 13	Seeds presoaked 4 hr, treated 12 hr, 26°C	0.34-0.7		1.24 (0.34 mM) 0.7 (0.7 mM)				46

Ekonom	Dry seeds and pre-soaked seeds treated either 3 hr or 24 hr at 25°C, pH 5.5	0.2-10	0.23 (24 hr dry, 1.2 mM) 0.02 (3 hr dry, 10 mM) 0 (3 hr, 72 hr pre-soak, 3 mM)		47
Ekonom	Dry seeds treated 24 hr, 25°C, pH 5.5-6	0.6-9.6		Induced single-strand breaks in DNA of non-germinating embryos	48
		17		17 mM MNNG unprobable due to insolubility	49

[a] A = anaphase, M = metaphase.

Table 7. Barley: Myleran (55-98-1)

Cultivar	Treatment conditions	Concentrations (mM)	Biological effects: Mutations (%) M_1 spike	Biological effects: Mutations (%) M_2 seedling	Biological effects: Aberrant cells (%)[a]	Biological effects: Lethality (%)	Comments	References
Rika	Dry seeds treated both 24 hr, 24°C, and and 72 hr, 3°C	0.81-1.62	1.6-0.17					50
Bonus	Dry seeds treated 45 min to 24 hr, 20-22°C	0.81-3.24			80 M, 54.5 A		LEC 45-90 min treatment	51
		8.97-4.49			+			52
		0.07-2.2			+			53
		1.62	+		+			54
		0.8-3.24	+		+			31

[a]A = anaphase, M = metaphase.

Table 8. Barley: DMN, TEM, MeTEPA, and Trenimon

Mutagen	Cultivar	Treatment conditions	Concentrations (mM)	Biological effects: Mutations (%) M_1 spike	Biological effects: Mutations (%) M_2 seedling	Biological effects: Aberrant cells (%)[a]	Comments	References
DMN (62-75-9)	Ekonom	Dry seeds treated 24 hr, 25°C	50-1000		1.0		1000 mM lethal	47
TEM (51-18-3)	Montcalm	Dry seeds treated 24 hr	0.5	0.64				59
MeTEPA (57-39-6)	Montcalm	Dry seeds treated 12 hr, pH 7, 24°C	0.537		0.42	1.42 in micro-sporo-cytes	5500 R x-ray induced 3.91% aberrations, 0.8 mM EMS induced 1.57% aberrations in microsporocytes	60
Trenimon (68-76-8)	Freja	Dry seeds treated 24 hr, 24°C	0.001-0.003			12.6 M		61
		6 hr presoaked seeds treated 3 hr, 24°C	0.02-0.05			25.2 M		62

[a] M = metaphase.

Table 9. Cadmium compounds

Compound	Cultivar	Treatment conditions	Concentration (mM)	Aberrant cells (%)[a]	References
Cadmium chloride (10108-64-2)	Piroline	Treated 2 hr, post-washed	10^{-2}	0.5 A	(18)
Cadmium bromide (7789-42-6)				0.3 A	(18)
Cadmium iodide (7790-80-9)				0.3 A	(18)
Cadmium nitrate (10325-94-7)				0.3 A	(18)
Cadmium sulfate (10124-36-4)				0.5 A	(18)
Cadmium acetate (549-90-8)				0.3 A	(18)

[a] A = anaphase.

(probably onset of DNA synthesis) (Table 8). It induces chromatid aberrations, preferentially isolocus breaks, localized mostly in regions of primary and secondary constrictions. Terminal deletions are very rare.

Mitomycin C (50-07-7). Dishler et al. (63) have published results that Mitomycin C is a recombinagen in barley. Its ability to induce mutations and chromosome aberrations in this plant has not been tested.

Cadmium Compounds. Cadmium compounds appear to be very weak chromosome breakers in barley at the concentrations used (Table 9). There are no published data about their abilities to produce chlorophyll-deficient mutations.

ACKNOWLEDGEMENT

The authors gratefully acknowledge the valuable assistance of Mr. Jeff Rosichan.

REFERENCES

1. Sideris, E. G., Nilan, R. A., and Bogyo, T. P. Rad. Bot. 13: 315-322 (1973).

2. Conger, B. V. Ph.D. Thesis, Program in Genetics, Washington State University, Pullman, Washington, 1967, 91 pp.

3. Conger, B. V., and Constantin, M. J. In: Biological Effects of Neutron Irradiation, IAEA, Vienna, 1974, pp. 417-432.

4. Froese-Gertzen, E. E., Konzak, C. F., Foster, R., and Nilan, R. A. Nature 198: 447-448 (1963).

5. Nawar, M. M., Konzak, C. F., and Nilan, R. A. Mutat. Res. 11: 339-346 (1971).

6. Gichner, T., Velemínský, J., Švachulová, J., and Pokorný, V. Biol. Plant. 19: 284-291 (1977).

7. Khalatkar, A. S., and Bhatia, C. R. Rad. Bot. 15: 223-229 (1975).

8. Satpathy, D., and Arnason, T. J. Can. J. Genet. Cytol. 11: 522-530 (1969).

9. Mikaelsen, K., Ahnström, G., and Li, W. C. Hereditas 59: 353-374 (1968).

10. Gichner, T., and Gaul, H. Rad. Bot. 11: 53-58 (1971).

11. Mikaelsen, K., and Li, W. C. In: Polyploidy and Induced Mutations in Plant Breeding, IAEA, Vienna, 1974, pp. 401-404.

12. Froese-Gertzen, E. E., Konzak, C. F., Nilan R. A., and Heiner, R. E. Rad. Bot. 4: 61-69 (1964).

13. Gichner, T., Ehrenberg, L., and Wachtmeister, C. A. Hereditas 59: 253-262 (1968).

14. Gichner, T., and Velemínský, J. Biol. Plant. 15: 72-79 (1973).

15. Khalatkar, A. S., Gopal-Ayengar, A. R., and Bhatia, C. R. Mutat. Res. 12: 331-334 (1971).

16. Velemínský, J., Zadražil, S., and Gichner, T. Mutat. Res. 14: 259-261 (1972).

17. Degraeve, N. Rev. Cytol. Biol. Vég. 34: 223-244 (1971).

18. Degraeve, N. Rev. Cytol. Biol. Vég. 34: 245-256 (1971).

19. Constantin, M. J., Conger, B. V., Chowdhury, J. B., and Ramage, R. T. In: Polyploidy and Induced Mutations in Plant Breeding, IAEA, Vienna, 1974, pp. 53-62.

20. Konzak, C. F., Nilan, R. A., Froese-Gertzen, E. E., and Foster, R. J. In: Induction of Mutations and the Mutation Process, Publishing House of the Czechoslovak Academy of Sciences, Prague, 1965, pp. 123-132.

21. Narayanan, K. R., and Konzak, C. F. In: Induced Mutations in Plants, IAEA, Vienna, 1969, pp. 281-304.

22. Velemínský, J., Zadražil, S., Pokorný, V., Gichner, T., and Švachulová, J. Mutat. Res. 19: 73-81 (1973).

23. Rao, R., and Natarajan, A. T. Mutat. Res. 2: 132-148 (1965).

24. Gichner, T., and Omura, T. Biol. Plant. 14: 155-163 (1972).

25. Kuenzel, G. Mutat. Res. 12: 397-409 (1971).

26. Moutschen, J., Moehs, A., and Gilot, J. Experientia 20: 494-495 (1964).

27. Gichner, T., Velemínský, J., Pokorný, V., and Zadražil, S. In: Polyploidy and Induced Mutation in Plant Breeding, IAEA, Vienna, 1974, pp. 143-159.

28. Froese-Gertzen, E. E., Nilan, R. A., Konzak, C. F., and Legault, R. R. Nature 200: 714-715 (1963).

29. Velemínský, J., Zadražil, S., Pokorný, V., and Gichner, T. In: Polyploidy and Induced Mutations in Plant Breeding, IAEA, Vienna, 1974, pp. 151-159.

30. Velemínský, J., Gichner, T., and Pokorný, V. Mutat. Res. 28: 79-86 (1975).

31. Natarajan, A. T., and Ramanna, M. S. Nature 211: 1099-1100 (1966).

32. Wagner, J. H., Nawar, M. M., Konzak, C. F., and Nilan, R. A. Mutat. Res. 5: 57-64 (1968).

33. Konstantinov, K., Harms-Ringdahl, M., Ehrenberg, L., Dumanović, J., and Bozović, I. Rad. Bot. 10: 499-509 (1970).

34. Moutschen-Dahmen, J. and M., and Ehrenberg, L. Hereditas 60: 267-269 (1968).

35. Ehrenberg, L., Lundqvist, U., and Ström, G. Hereditas 44: 330-336 (1958).

36. Nicoloff, H., and Gecheff, K. Mutat. Res. 34: 223-244 (1976).

37. Avetisov, V. A., and Valeva, S. A. Sov. Genet. 11: 281-287 (1975).

38. Avetisov, V. A., and Valeva, S. A. Sov. Genet. 4: 580-586 (1968).

39. Batygin, N. F., and Pitirimova, M. A. Sov. Genet. 6: 165-169 (1970).

40. Gechev, K. Sov. Genet. 9: 825-831 (1973).

41. Olimpienko, G. S., Vorob'eva, E. A., And Mitrofanov, Yu. A. Sov. Genet. 7: 1005-1012 (1971).

42. Sharma, B., and Bansal, H. C. Theor. Appl. Genet. 42: 25-31 (1972).

43. Ehrenberg, L., and Gichner, T. Biol. Zentrbl. 86: (Suppl.) 107-118 (1967).

44. Katiyar, R. K., Kalia, C. S., and Singh, M. P. Experientia 26: 645-646 (1970).

45. Kamra, O. P. Mutat. Res. 13: 327-335 (1971).

46. Prasad, M. V. R., Krishnaswami, R., and Swaminathan, M. S. Curr. Sci. 36: 438-439 (1967).

47. Gichner, T. Unpublished data.

48. Velemínský, J. Unpublished data.

49. Tarasenko, I. O. Dokl. Akad. Nauk. SSSR 196: 1456-1459 (1971).

50. Heslot, H., Ferrary, R., Lévy, R., and Monard, C. C. R. Acad. Sci. (Paris) 248: 729-732 (1959).

51. Moutschen-Dahmen, J. and M. Hereditas 44: 415-446 (1958).

52. Moutschen, J., and Reekmans, M. Caryologia 17: 495-508 (1964).

53. Wachtmeister, C. A., Moutschen-Dahmen, M., Moutschen, J., and Ehrenberg, L. Rad. Bot. 3: 187-192 (1963).

54. Ramanna, M. S., and Natarajan, A. T. Indian J. Genet. Plant Breeding 25: 24-45 (1965).

55. Velemínský, J., and Gichner, T. Mutat. Res. 5: 429-431 (1968).

56. Velemínský, J., and Gichner, T. Mutat. Res. 12: 65-70 (1971).

57. Ehrenberg, L. Abh. Akad. Wiss. Berlin (Med.) 1: 124-136 (1960).

58. Müller, A. J. Biol. Zentrbl. 91: 31-48 (1972).

59. Arnason, T. J., and Wakonig, R. Can. J. Genet. Cytol. 1: 16-20 (1959).

60. Wuu, K. O., and Grant, W. G. Can. J. Genet. Cytol. 8: 481-501 (1966).

61. Nicoloff, H. R., and Gecheff, K. Mutat. Res. 6: 257-261 (1968).

62. Nicoloff, H. R., Riefer, R., Künzel, G., and Michaelis, A. Mutat. Res. 30: 149-152 (1975).

63. Dishler, V. Ya., Nagle, E. F., Filipeka, V. F., and Rashal, I. D. Latv. Psr. Zinat. Akad. Vestis. 7: 49-52 (1975).

CHAPTER 12

MUTAGENICITY OF SELECTED CHEMICALS IN BARLEY: REPAIR AND RECOVERY

J. Velemínský and T. Gichner

Department of Genetics, Institute of Experimental Botany

Czechoslovak Academy of Sciences, Flemingovo Nám. 2

160 00 Praha 6, Czechoslovakia

CELLULAR AND ORGANISMAL LEVEL

At this level repair may be detected either directly by the recovery from the mutagenic induced damage achieved by various post-treatment or pretreatment modulations or indirectly by sensitization of the action of mutagens by various inhibitors known or supposed to interact with DNA repair in other systems.

Both possible courses of repair were observed in barley damaged by monofunctional alkylating agents and by ionizing radiation.

Storage Recovery

Several authors have demonstrated that storage of seeds which were treated with alkylating agents like EMS, MMS, or EI showed a different (increased or decreased) yield of mutations, aberrations, etc. compared to unstored seeds otherwise treated identically (1-8). Whether the degree of damage in the course of storage increases, decreases, or remains constant depends on the water content in the seeds during storage.

At 30% water content, the yield of mutations, chromosome aberrations, etc. induced by EMS, MMS, (and also by PMS, iPMS, DES, MNU) decreased due to storage. At 20% and 15% water content of the stored seeds, the degree of injury increased when seeds were treated with EMS and MMS (and with PMS or DES, too), whereas weak recovery was observed in seeds treated with iPMS and MNU stored at 15% water content (9,10). At 5% water content, no changes in the yield of mutations of aberrations were observed, even after very long seed storage (10-13). Changes obtained due to the storage of seed at 30% or 20% water content were reversible (14). Once achieved only partially, they cannot be reverted by changing the water content of the stored seeds.

The rate of the recovery at storage with 30% seed water content is positively correlated with the temperature (15), duration of storage (11-13), and duration of post-treatment seed washing before the storage (12). It is inversely related to the dose of mutagen. At extremely high doses of EMS (e.g., reducing the seed germination under 10%) (16) and of MMS (14), no recovery from induced damage takes place.

The recovery from EMS-induced injury due to storage is insensitive to hypoxia applied during the storage and to nontoxic doses of caffeine and sodium azide (pH 7) applied before the storage (15, 17,18). Besides the usually measured characters like chromosome aberrations in the root tips, recessive chlorophyll-deficient mutation, etc., the recovery of the activity of cytochrome oxidase and catalase in the seed, partially inhibited by MMS before the storage, was also observed (19).

No comparable storage recovery was observed when seeds were damaged by ionizing radiation (10,16).

Washing Recovery

Washing of EMS or MMS-treated seeds in tap water or in buffers led to the decrease of seedling injury and to a decrease of the amount of chromosome aberrations and chlorophyll mutations. The degree of recovery was related to the duration of post-treatment washing (7,13,20) and to the temperature of the washing medium (1, 2,7,10,12,13,20-22). This effect, especially for seedling injury, was strengthened by washing with water solution of thiourea, cysteine, thioacetamide, sodium thiosulfate etc. (23). The highest degree of recovery due to the postwashing, comparable to the storage recovery, was achieved in seeds treated with MNU or PMS and iPMS (9,24). In the case of MMS and EMS the washing recovery was less efficient.

Recovery Caused by Exogenous DNA

The amount of x-ray-induced chromosome aberrations in the root tips was lowered when exogenous DNA (calf thymus) was applied before the irradiation of seeds (25).

Sensitization of Mutagens by Inhibitors

The repair inhibitor caffeine increased the yield of damage

induced in barley by MMS, MNU, etc., and ionizing radiation even when applied at doses which were ineffective (18,26-28). In γ-irradiated barley, the oxygen-independent component of damage was potentiated, whereas the oxygen-dependent damage was protected by caffeine under certain conditions (29-32).

The potentiation by caffeine was mostly measured by the depression of seedling height and frequency of chromosome aberrations in the root tips. The spike fertility and frequency of EMS induced recessive chlorophyll-deficient mutations were not influenced by caffeine post-treatment (18).

The sensitization of mutagen by other inhibitors, which may be explained (among other possible explanations) by the inhibition of repair pathways, was observed for several arsenicals, mercuric chloride, cadmium and beryllium salts, and EMS-induced chromosome aberrations (33-35), for 5-fluorodeoxyuridine, 2,4-dinitrophenol or sodium azide, and γ- or x-ray-induced chromosome aberrations and seedling growth injury (36-39). Similarly as with caffeine (38), sodium azide did not increase the frequency of γ-ray-induced chlorophyll-deficient mutations.

REPAIR ON DNA LEVEL

Alkylating Agents

MMS, EMS (and other alkylating agents, like DES, MNU, iPMS) induced single-strand breaks in DNA of barley embryonic cells (10, 13,40-43) and the shape of the dose-response curves resembles those obtained for chromosome aberrations (44). These breaks are repaired during the storage of mutagen-treated seeds having 30% water content (10,13,41,42) or increased in their amount during the storage of seeds having 15% water content (41). The repair takes place before the onset of DNA synthesis as this synthesis starts after the stored seeds are sown and germinate (10,42). The endonuclease specific for apurinic sites in DNA was isolated from barley embryos and leaves (45-47). This endonuclease is most probably responsible for the rapid formation of DNA breaks from apurinic sites induced by alkylating agents and takes part in the observed repair pathway.

Another step of this process represents the repair synthesis detected in barley by means of ^{3}H BUdR and isopycnic CsCl gradient centrifugation in seeds and embryos treated with methyl nitrosourea (48).

Caffeine at nontoxic doses did not inhibit the "storage" repair of DNA single-strand breaks induced by MMS and EMS, but inhibited the joining of smaller newly synthesized DNA pieces to the large molecules in EMS- or MMS-treated seeds. It is proposed that caffeine inhibits or delays a repair system which acts after the DNA synthesis (postreplication repair) (49).

Excision like repair of DNA single-strand breaks was observed also during the post-treatment washing of seeds (40,50), during submerged storage of seeds under anaerobic conditions (48), and in seeds treated with mutagen and sodium azide (51). In all three cases, however, methyl nitrosourea was used as a mutagen.

Ionizing Radiation

Dose-dependent induction of DNA single-strand breaks by γ-rays in barley was detected (52), and their repair has been observed (58). It was also observed in γ-irradiated carrot protoplasts (53). The prereplication repair of γ-ray-induced DNA lesions was detected by unscheduled DNA synthesis observed autoradiographically during the early stage of germination of γ-irradiated seeds (54) and by differential activity of DNA polymerase in gamma and nonradiated samples as well as in radiosensitive and resistant varieties (55-57).

REFERENCES

1. Bender, K., and Gaul, H. Rad. Bot. 7: 289-301 (1967).

2. Brunner, H., and Mikaelsen, K. Z. Pflanzenzüchtung. 66: 9-36 (1971).

3. Frose-Gertzen, E.E., Konzak, C.F., Nilan, R.A., and Heiner, R.E. Rad. Bot. 4: 61-69 (1964).

4. Garina, K.P. Tsitol. & Genet. 7: 42-45 (1973).

5. Garina, K.P., and Romanova, N.I. Mol. Gen. Genet. 106: 93-105 (1970).

6. Gichner, T., and Ehrenberg, L. Biol. Plant. 8: 256-259 (1966).

7. Krausse, G.W., Dumanović, J., and Ehrenberg, L. Rad. Bot. 11: 13-19 (1971).

8. Mikaelsen, K., Ahnström, G., and Li, W.C. Hereditas 59: 353-374 (1968).

9. Gichner, T., and Veleminský, J. Mutat. Res. 16: 35-40 (1972).

10. Gichner, T., Veleminský, J., Pokorný, V., and Zadražil, S. In: Polyploidy and Induced Mutations in Plant Breeding, IAEA, Vienna, 1974, pp. 143-159.

11. Gichner, T., and Gaul, H. Rad. Bot. 11: 53-58 (1971).

12. Gichner, T., and Omura, T. Biol. Plant. 14: 155-163 (1972).

13. Gichner, T., Velemínský, J., and Pokorný, V. Environ. Exptl. Bot. 17: 63-67 (1977).

14. Gichner, T. and Velemínský, J. Mutat. Res. 66: 135-142 (1979).

15. Gichner, T., and Velemínský, J. Mutat. Res. 24: 73-75 (1974).

16. Gichner, T., Velemínský, J., Švachulová, J., and Pokorný, V. Biol. Plant. 19: 284-291 (1977).

17. Gichner, T., Švachulová, J., Velemínský, J., and Pokorný, V. Biol. Plant. 17: 202-206 (1975).

18. Gichner, T., and Velemínský, J. Mutat. Res. 25: 305-310 (1974).

19. Švachulová, J., Velemínský, J., and Gichner, T. Biol. Plant. 17: 109-112 (1975).

20. Mikaelsen, K., Brunner, H., and Li, W.C. Heriditas 69: 15-18 (1971).

21. Bender, K., and Gaul, H. Rad. Bot. 6: 505-518 (1966).

22. Satpathy, D., and Arnason, T.J. Can. J. Genet. Cytol. 11: 522-530 (1969).

23. Narayanan, K.R., and Konzak, C.F. In: Induced Mutations in Plants, IAEA, Vienna, 1969, pp. 281-304.

24. Gichner, T., Gaul, H., and Omura, T. Rad. Bot. 8: 499-507 (1968).

25. Ondřej, M., and Lim, E.S. Biol. Plant. 18: 366-372 (1976).

26. Ahnström, G. Mutat. Res. 26: 99-103 (1974).

27. Ahnström, G., and Natarajan, A.T. Int. J. Rad. Biol. 19: 433-443 (1971).

28. Yamamoto, K., and Yamaguchi, H. Mutat. Res. 8: 428-430 (1969).

29. Kesavan, P.C. Rad. Bot. 13: 355-359 (1973).

30. Kesavan, P.C., and Ahmad, A. Mutat. Res. 23: 337-346 (1974).

31. Kesavan, P.C., Trasi, S., and Ahmad, A. Int. J. Rad. Biol. 24: 581-587 (1973).

32. Nadkarni, S., and Kesavan, P.C. Int. J. Rad. Biol. 27: 569-576 (1975).

33. Degraeve, N. Rev. Cytol. Biol. Vég. 34: 233-244 (1971).

34. Moutschen, J., and Degraeve, N. Rev. Cytol. Biol. Vég. 29: 173-189 (1966).

35. Moutschen, J., and Degraeve, N. Experientia 21: 729 (1965).

36. Gaur, B.K., Joshi, R.K., and Joshi, V.G. Rad. Bot. 10: 29-34 (1970).

37. Nishiyama, I., and Ohta, N. Japan. J. Breeding 14: 47-53 (1964).

38. Sideris, E.G., Nilan, R.A., and Bogyo, T.P. Rad. Bot. 13: 315-322 (1973).

39. Yamamoto, K., and Yamaguchi, H. Mutat. Res. 8: 424-427 (1969).

40. Veleminský, J., and Gichner, T. In: Barley Genetics III, Proc. Third Internatl. Barley Genetics Symp., Garching, 1975, pp. 146-154.

41. Veleminský, J., Zadražil, S., and Gichner, T. Mutat. Res. 14: 259-261 (1972).

42. Veleminský, J., Zadražil, S., Pokorný, V., Gichner, T., and Svachulová, J. Mutat. Res. 17: 49-58 (1973).

43. Veleminský, J., Zadražil, S., Pokorný, V., Gichner, T., and Svachulová, J. Mutat. Res. 19: 73-81 (1973).

44. Veleminský, J., and Gichner, T. Unpublished data.

45. Šatava, J., Švachulová, J., and Veleminský, J. Paper presented at IUB 10th Internatl. Congr. Biochem., Hamburg, 1976; Abstracts, No. 14-2-089, p. 639 (1976).

46. Švachulová, J., Šatava, J., and Veleminský, J. Eur. J. Biochem. 87: 215-220 (1978).

47. Veleminský, J., Švachulová, J., and Šatava, J. Biol. Plant. 19: 346-352 (1977).

48. Veleminský, J., Zadražil, S., Pokorný, V., and Gichner, T. Mutat. Res., in press.

49. Velemínský, J., Zadražil, S., and Gichner, T. Mutat. Res. 28: 79-86 (1975).

50. Zadražil, S., Pokorný, V., Velemínský, J., and Gichner, T. Biol. Plant. 16: 7-13 (1974).

51. Gichner, T., Velemínský, J., and Pokorný, V. Environ. Exptl. Bot. 18: 27-31 (1978).

52. El-Metainy, A., Takagi, M., Tano, S., and Yamaguchi, H. Mutat. Res. 13: 337-344 (1971)

53. Howland, G.P., Hart, R.W., and Yette, M.L. Mutat. Res. 27: 81-87 (1975).

54. Yamaguchi, H., Tatara, A., and Naito, T. Japan. J. Genet. 50: 307-318 (1975).

55. Yamaguchi, H., Naito, T., Tatara, A., and Kurobane, I. Japan. J. Genet. 51: 59-60 (1976).

56. Yamaguchi, H., Tatara, A., and Naito, T. J. Rad. Res. 17: 32-33 (1976).

57. Yamaguchi, H., Naito, T., and Tatara, A. J. Rad. Res. 17: 32 (1976).

58. Tano, S., and Yamaguchi, H. Mutat. Res. 42: 71-78 (1977).

CHAPTER 13

MUTAGENICITY OF ETHYL METHANESULFONATE, MITOMYCIN C, AND VINYL CHLORIDE IN THE TRADESCANTIA TEST SYSTEMS

C. H. Nauman

Biology Department, Brookhaven National Laboratory, Upton Long Island, New York 11973 *

and W. F. Grant

Genetics Laboratory, MacDonald Campus of McGill University Ste. Anne de Bellevue, Quebec, Canada HOA 1CO

INTRODUCTION

Tradescantia is a classical plant for mutation studies and was used for radiation cytology experiments by many early workers (1). It is especially sensitive to both ionizing radiation (2-4) and chemical mutagens (5-7), the effects of very low exposures being readily detectable. It has a cellular radiosensitivity similar to that of many mammalian cells. The plant is relatively easy to grow under a wide range of environmental conditions and blooms continuously throughout the year. Flowering axillary cuttings are readily obtained and easily rooted, thus providing both root tips and floral tissues for study.

Somatic cells of Tradescantia have only 12 large chromosomes, which are well suited for detailed cytological analyses. This small number of large chromosomes makes Tradescentia excellent material for chromosome scoring of both somatic and meiotic chromosome abnormalities (8).

Tradescantia plants lend themselves to somatic mutation studies. Special clones heterozygous for flower color provide a useful test

*Present address: U.S. EPA, RD-682, Washington, D.C. 20460.

system for physical or chemical mutagens. In heterozygotes the flower and stamen hair color is blue, and somatic mutations are scored as phenotypic changes to pink or colorless. Somatic mutations in both petals and stamen hair tissues can be readily detected by using only a dissecting microscope (9,10). Three heterozygous clones are in use. Clone 02 is thought to be an interspecific hybrid of _T. occidentalis_ and _T. ohiensis_ and has been described in detail (9-11). Clone 0106, a highly mutable clone having a high spontaneous mutation frequency of blue to pink sectors in petals and stamen hairs, is a radiation-induced mutant derived from the parent clone 02 and is thought to be a periclinal chimera, since progeny of self-fertilization do not show the mutable character (7,12). A third clone being used for somatic mutation studies, Clone 4430, is unrelated to the previous two and originated as an interspecific hybrid between _T. hirsutiflora_ and _T. subacaulis_ (6).

The stamen-hair system has been used by a number of investigators, notably by Sparrow and colleagues at Brookhaven National Laboratory and the Mericles at Michigan State University, to study both somatic mutation rates and loss of reproductive integrity (6,13-15). The hairs develop in acropetal succession, from the bottom of the filament to the top. They increase in length mostly by successive divisions of terminal and subterminal cells. Cell division in the hairs stops about four or five days prior to anthesis. At maturity each of the six stamens normally possesses about 50 hairs, decreasing in number in older inflorescences. In normal untreated plants the hairs are shorter at the apex of the filament than at the base, ranging from about 12 to 30 cells in length, but the number of cells per hair varies in irradiated material.

Tradescantia has the advantage that plants can be used as monitors of gaseous chemicals over long exposure periods (16,17). Inspection of the flowers for somatic mutations once or twice a week reveals quickly whether or not exposure to a mutagen has occurred. The genetic basis for the somatic mutations is not definitely known, but it may be associated with gene mutations, chromosome breakage, altered repair systems, or nondisjunction. Increasing evidence has been presented for somatic crossing over (9,18,19).

This report is a summary of the studies which have been carried out by use of the Tradescantia systems for testing the cytological and mutational effects of the chemicals ethyl methanesulfonate (EMS), mitomycin C (MC), and vinyl chloride (VC). The results are summarized in Table 1.

ETHYL METHANESULFONATE (EMS)

In the studies by Sparrow and associates (Table 1), three

clones of Tradescantia have been used (6,7,12,17,20-24). Results (exposure-response curves, Figs. 1 and 2) show that these three clones are differentially sensitive to EMS, with clone 4430 being 6 to 7 times more sensitive to the induction of pink mutations in stamen hair cells than clone 02. It is of interest that clones 4430 and 02 are equally sensitive to x-rays. Clone 0106 shows a sensitivity to EMS that is intermediate between that of clones 4430 and 02; clone 0106 is significantly more sensitive to x-rays than clones 4430 and 02 ($p < 0.01$). The LEC (lowest effective concentration) was not determined for any of these clones.

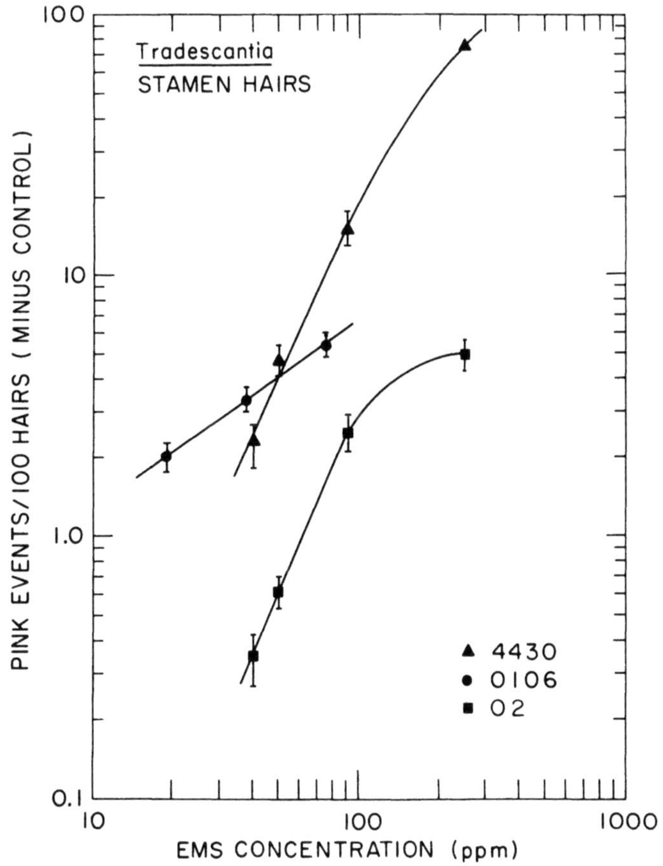

Figure 1. Dose-response curves for pink mutations in clones 02, mutable 0106 and 4430 following 6 hr exposures to EMS. Data of Nauman et al. (7).

Table 1. Summary of studies on ethyl methanesulfonate (EMS), mito-test systems.

Muta-gen	Stage treated	Target tissue	Clones treated	Chemical form
EMS	Cuttings bearing inflorescences	Developing stamen hairs	02[a] 0106[b] 4430[c]	Gaseous, flowed at 1-2 l./min
	Young rooted cuttings	Root tips	02	Aqueous soln, immersion of root tips
	Plants bearing inflorescences	Developing petals stamen hairs	02	Aqueous soln, in contact with inflorescence
MC	Plants with adventitious roots	Root tips	*T. paludosa*	Aqueous soln
VC	Cuttings bearing inflorescences	Developing stamen hairs	4430	Gaseous, flowed at 1-2 l./min

[a]Clone 0081, parentage unsure but thought to be interspecific hybrid, *T. occidentalis* × *T. ohiensis*.

[b]Radiation-induced mutant of clone 02.

[c]Interspecific hybrid of *T. hirsutiflora* × *T. subcaulis*.

The pink mutation endpoint scored following chemical exposure is thought to represent a mixture of chromosomal aberrations and point mutations. From Fig. 2, a determination can be made that the clone 4430 EMS-induced mutation response, for the range of exposures used, is equivalent to from 24 to over 160 rads of x-rays; values are from 16 to 33 rads for clone 0106, and from less than 10 rads to 49 rads for clone 02.

EMS is relatively highly mutagenic. The exposure of Tradescantia (clones 02, 0106, and 4430) inflorescences to 100 ppm of

mycin (MC), and vinyl chloride (VC) by use of the Tradescantia

Range of exposures	Conditions	Endpoint	Resulting data	References
19-250 ppm, 6 hr	19-26°C, 60-80% RH	Pink mutations in stamen hairs	Exposure-response curves	6,7, 12,20-24
200, 400 and 600 ppm, each for 3,6,12 hr	19-24°C, 70% RH	Chromosome aberrations in root tips	Increased chromosome aberrations	25
1500 ppm, 24 hr	20-24°C	Pink mutations in stamen hairs and petals	Pink mutation increase in petals	26
10 μg/ml, 1 hr	19-21°C	Chromosome aberrations in root tips	Increased chromosome aberrations	27
10-150 ppm, 6 hr	ca. 25°C, 60-80% RH	Pink mutations in stamen hairs	Exposure-response curve	17

gaseous EMS for 6 hr, that is, a 600 ppm-hr exposure, elicits pink mutation responses in stamen hairs equivalent to those induced by 32 rad of acutely delivered x-rays in clone 02, 37 rad in clone 0106, and 160 rad in clone 4430 (7).

In the study by Ahmed and Grant (25), EMS was used as a positive control for comparison with the production of chromosome aberrations by two pesticides using root tips of Tradescantia clone 02 as experimental material. Chromosome aberrations were relatively constant with exposure concentration for the 3 and 6 hr exposure periods but increased with exposure concentration for the 12 hr treatment time. The percentages of abnormal cells were all significant over control and the means for the 3, 6, and 12 hr treatment times were 4.55, 4.41, and 3.92%, respectively.

Tomkins and Grant (26) used EMS as a positive control for a comparison of the mutation events in stamen hairs and petals of Tradescantia with that produced by some pesticides. The increase in the frequency of pink mutations in the flower petals was highly significant. While there was an increase in frequency of pink mutations in stamen hairs, it was not significant over control.

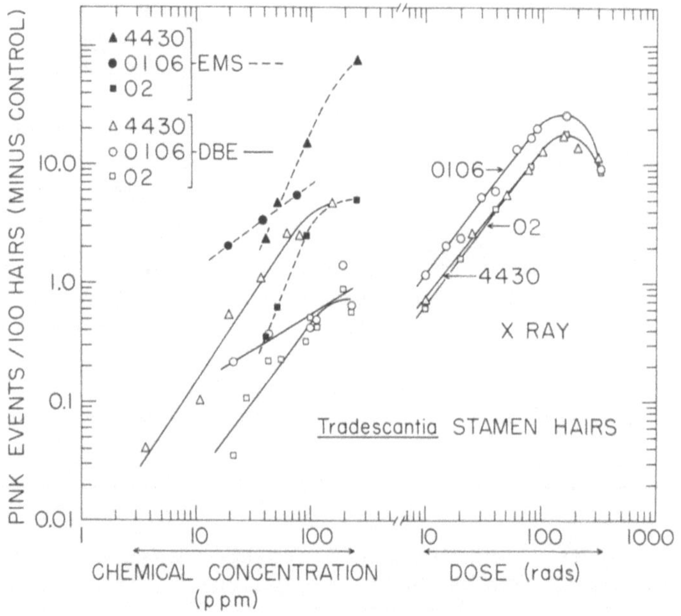

Figure 2. Comparison of dose-response curves for pink mutations in clones 02, mutable 0106, and 4430 following 6 hr exposure to EMS and 1,2-dibromoethane (DBE) and acute exposure to x-rays. Data of Nauman et al. (7).

MITOMYCIN C

In this single investigation by Utsumi (27), the cytological effects of MC were observed in root tip cells of *Tradescantia paludosa* and *Vicia faba* in a parallel study. Experimental procedures for Tradescantia are given in Table III of Utsumi (27). Roots were sampled at 24, 48 and 72 hr after treatment. The number of aberrations per 100 cells was found to be maximal (16.5) at 24 hr, decreased to 11.3 at 48 hr, and to 5.3 at 72 hr.

The structural changes scored in metaphase cells were almost all of the chromatid-type, that is, isochromatid breaks with sister reunion, isochromatid gaps, chromatid breaks and interchanges. Chromosome-type aberrations were so few they were omitted from the data presented.

In both species, isochromatid breaks were the most frequent type of aberration (54.5% for Tradescantia; 59.8% for Vicia). Yields of aberrations were higher in Vicia than in Tradescantia indicating differences in susceptibility between the two species.

Distribution of chromosome aberrations was nonrandom. Chromosome breakage was favored in the proximal and distal regions of the long arms of chromosomes of *T. paludosa* in contrast to *V. faba*, in which breaks occurred most frequently in the heterochromatic regions.

Differences in susceptibility between the two species were speculated to be due to differences in heterochromatin content. Chromosomes of *T. paludosa* have no visible segments of heterochromatin. The localization of MC-induced breaks in Tradescantia chromosomes suggested they may be a reflection of the presence of G-C rich segments in the chromosomes, since MC-induced breaks in the heterochromatic sites of Vicia chromosomes are considered to be due to the selective action of MC in G-C rich regions.

Another finding of the study was that MC caused mitotic delay.

VINYL CHLORIDE

In an unpublished study (Table 1) by Sparrow and associates (17), it may be seen in Figure 3 that the exposure-response curve for vinyl chloride has a very shallow slope and saturates at about 75 ppm for 6 hr (= 450 ppm-hr). No explanation is apparent for this type of exposure-response curve, but it might be speculated very tentatively that these kinetics are due to a saturation of the metabolic activation system by low exposures to vinyl chloride. In summary, it can be stated that vinyl chloride is weakly mutagenic in this system.

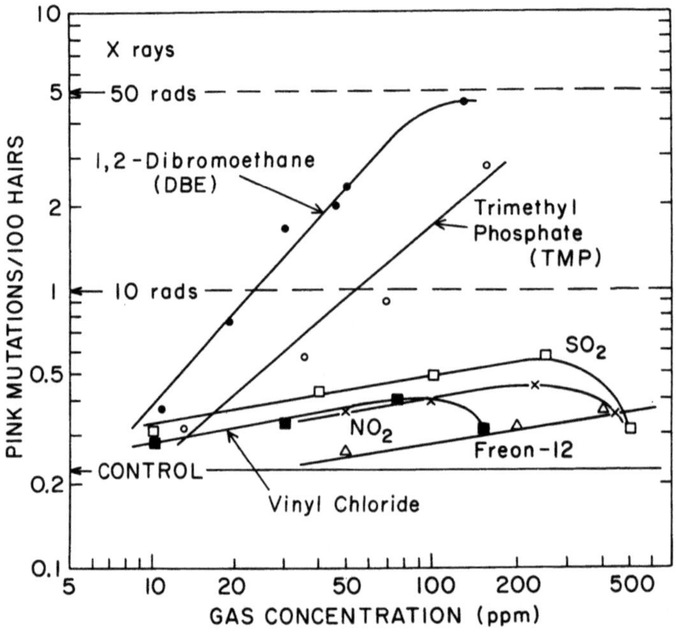

Figure 3. Comparison of exposure-response curves for pink mutations in Tradescantia clone 4430 following exposure to vinyl chloride, Freon-12, NO_2, SO_2, trimethyl phosphate, and 1,2-dibromoethane. All exposures were for 6 hr; mutation frequencies induced by acute 250-kVp x-rays are indicated on the ordinate for comparison. Data of Sparrow et al. (17).

REFERENCES

1. Swanson, C. P. Proc. Natl. Acad. Sci. (U.S.A.) 35: 237-244 (1949).

2. Mericle, L. W., and Mericle, R. P. Rad. Bot. 5: 475-492 (1965).

3. Sparrow, A. H., Baetcke, K. P., Shaver, D. L., and Pond, V. Genetics 59: 65-78 (1968).

4. Sparrow, A. H., Underbrink, A. G., and Rossi, H. H. Science 176: 916-918 (1972).

5. Kihlman, B. A. Hereditas 81: 384-404 (1975).

6. Sparrow, A. H., Schairer, L. A., and Villalobos-Pietrini, R. Mutat. Res. 26: 265-276 (1974).

7. Nauman, C. H., Sparrow, A. H., and Schairer, L. A. Mutat. Res. 38: 53-70 (1976).

8. Evans, H. J. Int. Rev. Cytol. 13: 221-231 (1962).

9. Mericle, L. W., and Mericle, R. P. Rad. Bot. 7: 449-464 (1967).

10. Underbrink, A. G., Schairer, L. A., and Sparrow, A. H. In: Chemical Mutagens, Principles and Methods for Their Detection, Vol. 3, A. Hollaender, Ed., Plenum Press, New York, 1973, pp. 171-207.

11. Sparrow, A. H., Cuany, R. L., Miksche, J. P., and Schairer, L. A. Rad. Bot. 1: 10-34 (1961).

12. Nauman, C. H., Sparrow, A. H., Schairer, L. A., and Sautkulis, R. C. Mutat. Res. 31: 318-319 (1975) (Abstr.).

13. Nayar, G. G., and Sparrow, A. H. Rad. Bot. 7: 257-267 (1967).

14. Mericle, L. W., and Mericle, R. P. Genetics 73: 575-782 (1973).

15. Sparrow, A. H., and Schairer, L. A. EMS Newsletter 5: 16-19 (1971).

16. Nauman, C. H., Sparrow, A. H., Underbrink, A. G., and Schairer, L. A. In: Radiological Protection, First European Symposium on Rad-equivalents, R. Chanet, Ed. Commission of the European Communities, Luxembourg, 5725e, 1977, pp. 13-23.

17. Sparrow, A. H. et al. Unpublished data, 1977.

18. Mericle, L. W., and Mericle, R. P. J. Hered. 62: 323-328 (1971).

19. Christianson, M. L. Mutat. Res. 28: 389-395 (1975).

20. Schairer, L. A., Sparrow, A. H., and Sautkulis, R. C. Rad. Res. 55: 599-600 (1973) (Abstr.).

21. Sparrow, A. H., Schairer, L. A., and Villalobos, R. Mutat. Res. 21: 238-239 (1973) (Abstr.).

22. Nauman, C. H., Villalobos-Pietrini, R., and Sautkulis, R. C. Mutat. Res. 26: 444 (1974) (Abstr.).

23. Sparrow, A. H., and Schairer, L. A. Mutat. Res. 26: 445 (1974) (Abstr.).

24. Villalobos-Pietrini, R., Sparrow, A. H., Schairer, L. A., and Sparrow, R. C. Rad. Res. 59: 153 (1974) (Abstr.).

25. Ahmed, M., and Grant, W. F. Can. J. Genet. Cytol. 14: 157-165 (1972).

26. Tomkins, D. J., and Grant, W. F. Can. J. Genet. Cytol. 14: 245-256 (1972).

27. Utsumi, S. Japan. J. Genet. 46: 125-134 (1971).

CHAPTER 14

MUTAGENICITY OF SELECTED CHEMICALS IN *VICIA FABA* AND *ALLIUM CEPA* TEST SYSTEMS

R. Rieger, A. Michaelis, I. Schubert, B. Kaina, and K. Heindorff

Zentralinstitut für Genetik und Kulturpflanzenforschung der Akademie der Wissenschaften der DDR, DDR-4325 Gatersleben German Democratic Republic

COMPARATIVE CHEMICAL MUTAGENESIS

VICIA FABA

The broad or horse bean, *Vicia faba* (2n = 12), has widely been used to study the capability of physical and chemical agents to induce chromosome and/or chromatid aberrations, to investigate the kinetics of aberration origination, and to study the influence of the metabolic state of the cells on the effect of mutagenic agents (1-6). The standard karyotype consists of five pairs of acrocentric, nearly equally long chromosomes and one pair of metacentric satellite chromosomes; a number of structurally reconstructed karyotypes have been developed whose chromosomes are all easily distinguishable by morphological criteria. These karyotypes are of special value for the investigation of the patterns of aberration distribution along individual chromosomes (7,8).

Seeds and main or lateral roots are used for treatments with the agents to be tested with respect to their capacity to induce chromosome structural changes. Prior to fixation (ethanol-acetic acid, 3:1) main or lateral roots are treated with colchicine, hydrolyzed (11 min, 1 N HCL, 60°C) and Feulgen-stained, squashed in 45% acetic acid, and made permanent by mounting in Euparal. Fixation is done after various recovery times in tap water to bring interphase cells into metaphase. Generally, several fixation times should be used, and fixation of the treated material should occur with short intervals and over a period long enough to cover at least one whole mitotic cycle. Agents inducing chromosome structural changes may show nondelayed or delayed effects (5). Nondelayed effects occur when the agent is able to produce aberrations in cells (interphases) that have completed DNA synthesis, whereas agents showing delayed effects are apparently unable to do so. In the latter case, aberrations are found in metaphases fixed 8 to 48 hr after the treatment; in the former case, the main effect is obtained 0 to 8 hr after exposure of the root tip meristem to the agent. Nondelayed and delayed effects differ in that delayed effects are produced in an earlier stage of interphase than nondelayed effects. Usually, metaphase analyses are being made. These allow the unequivocal scoring of the types and frequencies of induced chromatid and/or chromosome-type aberrations. The spontaneous frequency of metaphases with chromatid or chromosome aberrations is about 0.25% (9).

The chromosome-type aberrations scored are dicentrics, rings, and fragments; the chromatid-type aberrations represent isochromatid (isolocus breaks) breaks, with and without sister-chromatid reunion, chromatid translocations (complete and incomplete, symmetrical and asymmetrical), and triradials. "Subchromatid-aberrations" have also been reported but due to the uncertainty of their true character (they probably represent "masked" chromatid aberrations) data concerning them have been omitted from this report. The same is true with respect to "gaps". Recently, sister-chromatid exchanges (SCEs) have been found in V. faba and their frequency may be significantly increased by treatment with some mutagens. The SCE frequency is about 20 SCE per metaphase after BrdUrd incorporation (10,11). By means of the various banding techniques the positioning of constitutional heterochromatin has been described and correlated with clustering of induced chromatid aberrations (12).

Both energy-rich radiation and chemical agents have been found to result in patterns of nonrandom distribution of chromatid aberrations which are, in part at least, mutagen-specific. Generally, agents giving rise to "nondelayed" chromatid aberrations result in less pronounced aberration hotspots than agents resulting in delayed chromatid aberrations (which appear after recovery times of more than 8-10 hr)(13). Chromosome regions containing constitutional heterochromatin (Giemsa bands, fluorescence bands, late replicating DNA) have been found to become preferentially involved in induced chromatid aberrations, though it is presently not clear whether aberration clustering in fact involves the heterochromatin proper or the immediately adjacent euchromatin. Data about numerical chromosome aberrations (polysomy and polyploidy) are rare but indicate that these are of very seldom spontaneous occurrence. Mutagen-induced point and chlorophyll mutations have only been studied by Sjödin (14).

Due to the low number of large chromosomes and the relatively high frequency of dividing cells, V. faba root tip meristems provide a useful and easily handable material for testing chemical and physical agents with respect to the induction of chromosome-type and chromatid-type aberrations. The material is especially well suited for testing clastogenic effects of pesticides and for investigation of basic problems connected with mutagenesis. The reconstructed karyotypes are of special value for studies of differential sensitivities of individual chromosomes and chromosome segments after treatment with different mutagens. They furthermore allow to look for influences of differential segment positioning in different karyotypes on their involvement in chromatid aberrations.

The data presently available for the selected mutagens, including x-rays as reference mutagen, are compiled as references (1-111). Table 1 provides a short presentation of the results available for the selected compounds after treatment of V. faba.

COMPARATIVE CHEMICAL MUTAGENESIS

ALLIUM CEPA

The common onion, Allium cepa (2n = 16), and some other Allium species have been used to test the effects of chemicals on chromosomes and cell division since Levan (112) introduced standardized tests (6). From the selected agents dealt with in this report six have been used for treatment of Allium roots. In most cases the data are based on anaphase analyses for induced chromosome structural changes (scoring of anaphase bridges and fragments). Under these circumstances it is mainly the time of appearance of chromosomal damage which allows one to draw indirect conclusions with respect to the presence of chromatid- or chromosome-type aberrations. In the very rare cases of metaphase evaluation, isochromatid breaks, chromatid translocations, and chromatid breaks proved to be the aberration types most frequently observed after treatment with agents showing "delayed" effects. The spontaneous frequency of cells with chromosome structural changes has been reported to be 0.12 to 0.73% (6). No papers have been found which supply sufficient data with respect to x-rays as reference mutagen. Some limited possibilities for comparison with the clastogenic activities of chemical mutagens are provided by the γ-ray data of Sax and Sax (113).

In most experiments seeds were treated; more rarely were roots treated. The testing procedure usually consisted of germination of seeds on paper moistened with a solution of the agent in question. After various times of treatment, roots were immersed in colchicine or immediately fixed, hydrolyzed, Feulgen-stained, and squashed.

As compared to Vicia faba, the Allium system provides some

Table 1. Available results for the selected compounds after treatment of *Vicia faba*

Mutagen	Effect	Type of damage observed[a]	Dose range ($M \times 10^{-3}$)
Aflatoxin B_1	+	Anaphase aberrations, nd	0.67
Cadmium	+	Chromatid aberrations	0.01-10
Cyclophos-phamid (Endoxan)	+	Chromatid aberrations, d	0.125-100
DEN	-		1-2
DMN	-		1-<100
Epichlorhydrin	+	Chromatid aberrations, d	0.6-20
Ethylenimine	+	Chromatid aberrations, mutants	0.5-20
Ethyl methane-sulfonate	+	Chromatid aberrations, d, SCE, mutants	10-100
ICR-170	+	Chromatid aberrations, d, nd	0.001-0.25
MeTEPA	+	Anaphase aberrations, d	2.33
Methyl methane-sulfonate	+	Chromatid aberrations, d, SCE, mutants	3
Mitomycin C	+	Chromatid aberrations, d, SCE	0.0022-3.5
MNNG	+	Chromatid aberrations, d	0.0075-0.5
Myleran	+	Chromatid aberrations, d	0.1-50
Natulan	-		0.2-5
TEPA	+	Chromatid aberrations, d	0.75-1
ThioTEPA	+	Chromatid aberrations, d, SCE, mutants	0.5-2.5

(continued)

Mutagen	Effect	Type of damage observed[a]	Dose range (M x 10^{-3})
Trenimon	+	Chromatid aberrations, d	0.0033-0.002
Triethylene-melamine	+	Chromatid aberrations, d	0.001-20
Vinyl chloride	Nt		
X-rays	+	Chromatid aberrations, nd; chromosome aberrations, mutants, SCE	10-9000 R

[a]d = delayed, nd = nondelayed.

[b]Nt = not tested.

advantages which are, however, more or less compensated by the inferior cytological possibilities: (1) results concerning an agent's capability to induce cytological damage are rather quickly obtainable; (2) only relatively small amounts of the test substance are needed to reach definitive conclusions with respect to its ability to induce chromosome structural changes; (3) the test procedure is even less expensive than testing on the basis of Vicia faba. The disadvantages evident from the data available are: (1) use of several Allium species and, therefore, difficulties in comparing the data; (2) due to similar morphology of most chromosomes, Allium is not well suited for obtaining data on intrachromosomal aberration distribution and aberration clustering; (3) the data available provide a much less comprehensive body of literature for comparison; (4) most data are obtained from anaphase analyses.

The results presently available for the selected agents are sparse (21,22,29,44,55,86,112-118). Table 2 provides a short summary of the data. The majority of the selected agents have in fact not been tested for effects in Allium.

Table 2. Available results for the selected compounds after treatment of Allium

Mutagen	Effect[a]	Type of damage observed	Dose range (M x 10^{-3})
Aflatoxin B_1	+	Anaphase aberrations (nondelayed)	0.0032-0.64
Cadmium	Nt		
Cyclophos-phamid (Endoxan)	Nt		
DEN	Nt		
DMN	Nt		
Epichlorhydrin	Nt		
Ethylenimine	Nt		
Ethyl methane-sulfonate	+	Anaphase aberrations (delayed)	1-20
ICR-170	Nt		
MeTEPA	Nt		
Methyl methane-sulfonate	Nt		
Mitomycin C	+	Chromatid type aberrations (delayed)	0.0013
MNNG	+	Chromatid type aberrations (delayed)	0.01-3.5
Myleran	+	Anaphase aberrations (delayed)	0.01-0.75
Natulan	Nt		
TEPA	Nt		
ThioTEPA	+	Chromatid type aberrations (delayed)	0.5

(continued)

Mutagen	Effect[a]	Type of damage observed	Dose range ($M \times 10^{-3}$)
Trenimon	Nt		
Triethylene-melamine	+	Chromatid type aberrations (delayed)	0.01-1
Vinyl chloride	Nt		

[a]Nt = not tested.

REFERENCES

1. Ford, C. E. Proc. 8. Int. Congr. Genetics, Hereditas, Lund (Suppl.) 570-574 (1949).

2. Wolff, S. In: Advances in Radiobiology, G. C. de Hevesy et al., Eds., Oliver & Boyd, Edinburgh, 1957, pp. 463-469.

3. Revell, S. H. Mutat. Res. 3: 34-53 (1966).

4. Evans, H. J. Int. Rev. Cytol. 13: 221-321 (1962).

5. Kihlman, B. A. Actions of Chemicals on Dividing Cells. Prentice-Hall, Englewood Cliffs, N. J., 1966.

6. Rieger, R., and Michaelis, A. Die Chromosomenmutationen. VEB Gustav Fischer Verlag, Jena, 1967.

7. Michaelis, A., and Rieger, R. Chromosoma 35: 1-8 (1971).

8. Döbel, P., Rieger, R., and Michaelis, A. Chromosoma 43: 409-422 (1973).

9. Rieger, R., and Michaelis, A. Kulturpflanze 8: 230-243 (1960).

10. Kihlman, B. A. Chromosoma 51: 11-18 (1975).

11. Schubert, I., Sturelid, S., Döbel, P., and Rieger, R. Mutat. Res. 59: 27-38 (1979).

12. Rieger, R., Michaelis, A., Schubert, I., Döbel, P., and Jank, I.-W. Mutat. Res. 27: 69-79 (1975).

13. Schubert, I., and Rieger, R. Mutat. Res. 44: 337-344 (1977).

14. Sjödin, J. Hereditas 66: 215-232 (1970).

15. Ahmed, M., and Grant, W. F. J. Genet. Cytol 14: 157-165 (1972).

16. Asp, B. Mutat. Res. 21: 22 (1973).

17. Bempong, M. A. Bull. Torr. Bot. Club 99: 113-118 (1972).

18. Bempong, M. A., and Newsone, Y. L. Can. J. Genet. Cytol. 14: 655-666 (1972).

19. Buiatti, M., and Nuti-Ronchi, V. Caryologia 16: 397-403 (1963).

20. Caspersson, T., Zech, L., Modest, E. J., Foley, G. E., Wagh, U., and Simonsson, E. Exptl. Cell Res. 58: 128-140 (1969).

21. Deysson, G. C. R. Soc. Biol. (Paris) 168: 687-693 (1974).

22. Deysson, G., and Truhaut, R. C. R. Acad. Sci. (Paris) Ser. D 268: 83-85 (1969).

23. Döbel, P. Biol. Zbl. 89: 481-495 (1970).

24. Fucik, V., Michaelis, A., and Rieger, R. Biochem. Biophys. Res. Commun. 13: 366-371 (1963).

25. Gichner, T., Michaelis, A., and Rieger, R. Biochem. Biophys. Res. Commun. 11: 120-124 (1963).

26. Gläss, E. Zeitschr. Botanik 43: 359 (1955).

27. Gläss, E. Zeitschr. Botanik 44: 1-58 (1956).

28. Gläss, E. Chromosoma 8: 260-284 (1956).

29. Hartley-Asp, B. Hereditas 83: 223-236 (1976).

30. Hussein, H. A. S., and Abdalla, M. M. F. Egypt. J. Genet. Cytol. 3: 246-258 (1974).

31. Hussein, H. A. S., Heakel, M. Y., and Fayed, M. A. Egypt. J. Genet. Cytol. 3: 299 (1974).

32. Izard, C. C. R. Acad. Sci. (Paris) Ser. D 274: 1660-1662 (1972).

33. Jacob, K. M., and Wolff, S. Int. Rad. Biol. 15: 519-523 (1969).

34. Kaul, B. L. Mutat. Res. 7: 43-49 (1969).

35. Kaul, B. L. Chromosoma 26: 469-474 (1969).

36. Kaul, B. L. Naturwiss. 57: 455-456 (1970).

37. Keller, R. Arch. Julius Klaus-Stift. Vererbungsforsch. Sozial-anthropol. Rassenhyg. 43 (Suppl.): 105-112 (1968).

38. Kihlman, B. A. Rad. Bot. 1: 35-41 (1961).

39. Kihlman, B. A. Rad. Bot. 1: 43-50 (1961).

40. Kihlman, B. A. In: Chemical Mutagens - Principles and Methods for Their Detection, Vol. 2, A. Hollaender, Ed., New York, 1971, pp. 489-515.

41. Kihlman, B. A., Hartley-Asp, B., Nilsson, K., and Sturelid, S. Mutat. Res. 21: 191-192 (1973).

42. Kihlman, B. A., and Sturelid, S. Hereditas 88: 35-41 (1978).

43. Kihlman, B. A., Sturelid, S., Hartley-Asp, B., and Nilsson, K. Mutat. Res. 17: 271-275 (1973).

44. Kihlman, B. A., Sturelid, S., Hartley-Asp, B., and Nilsson, K. Mutat. Res. 26: 105-122 (1974).

45. Kihlman, B. A., Sturelid, S., Palitti, F., and Becchetti, A. Mutat. Res. 46: 130-131 (1977).

46. Kumar, S., Aggarwal, U. U., and Swaminathan, M. S. Mutat. Res. 4: 155-162 (1967).

47. Lilly, L. J. Nature 207: 433-434 (1965).

48. Loveless, A. Nature 167: 338-342 (1951).

49. Merz, T. Science 133: 329-330 (1961).

50. Michaelis, A., and Rieger, R. Züchter 30: 150-163 (1960).

51. Michaelis, A., and Rieger, R. Nature: 199: 1014-1015 (1963).

52. Michaelis, A., and Rieger, R. Kulturpflanze 11: 403-415 (1963).

53. Michaelis, A., and Rieger, R. In: Induction of Mutations and the Mutation Process, Czechoslov. Akad. Sci., Praha, 1965, pp. 101-106.

54. Michaelis, A., Nicoloff, H., and Rieger, R. Biochem. Biophys. Res. Commun. 9: 280-284 (1962).

55. Mohandas, R., and Grant, W. F. Can. J. Genet. Cytol. 14: 773-783 (1972).

56. Moutschen, J. Hereditas 46: 471-480 (1960).

57. Moutschen-Dahmen, J., and Degraeve, N. Experientia 21: 200-202 (1965).

58. Moutschen-Dahmen, J., and Moutschen-Dahmen, M. Hereditas 44: 415-446 (1958).

59. Moutschen-Dahmen, J., and Moutschen-Dahmen, M. Experientia 15: 310-311 (1959).

60. Moutschen-Dahmen, J., and Moutschen-Dahmen, M. Experientia 19: 144-145 (1963).

61. Moutschen-Dahmen, J., and Moutschen-Dahmen, M. Rad. Bot. 3: 297-310 (1963).

62. Moutschen-Dahmen, M., Ehrenberg, L., and Moutschen, J. Rad. Bot. 5: 271 (1965).

63. Natarajan, A. T. EMS Newsletter 6: 22 (1972).

64. Natarajan, A. T., and Ahnström, G. Chromosoma 28: 48-61 (1969).

65. Natarajan, A. T., and Upadhya, M. D. Chromosoma 15: 156-169 (1964).

66. Nicoloff, H., Rieger, R., and Michaelis, A. Genetics and Plant Breeding (Sofia) 3: 325-332 (1970).

67. Ninan, T., and Wilson, B. B. Genetica 40: 103-119 (1969).

68. Nuti-Ronchi, V. Buiatti, M., and Ipata, P. L. Mutat. Res. 4: 315-321 (1967).

69. Obe, G. Mutat. Res. 13: 421-424 (1971).

70. Obe, G. Umschau 11: 359-360 (1972).

71. Ockey, C. H. J. Genetics 55: 525-549 (1957).

72. Ockey, C. H. In: Chemische Mutagenese, Deut. Akad., 1960, pp. 47-53.

73. Panosyan, G. A., and Tamrazyan, E. E. Sov. Genet. 9: 561-565 (1973); Genetika 9 (5): 36-42 (1973).

74. Revell, S. H. In: Effect of Ionizing Radiations on Seeds, IAEA, Vienna, 1961, pp. 229-242.

75. Rieger, R. In: Induction of Mutations and the Mutation Process, J. Velemínský and T. Gichner, Eds., Czechoslovak Acad. Sci., Praha, 1965, pp. 85-100.

76. Rieger, R. Mutat. Res. 21: 232 (1973).

77. Rieger, R., and Michaelis, A. Chromosoma 10: 163-178 (1959).

78. Rieger, R., and Michaelis, A. Chromosoma 11: 673-581 (1961).

79. Rieger, R., and Michaelis, A. Biol. Zbl. 80: 301-317 (1961).

80. Rieger, R., and Michaelis, A. Kulturpflanze 10: 212-292 (1962).

81. Rieger, R., and Michaelis, A. Exptl. Cell Res. 31: 202-205 (1963).

82. Rieger, R., and Michaelis, A. Mutat. Res. 1: 109-112 (1964).

83. Rieger, R., and Michaelis, A. Nature 206: 741-742 (1965).

84. Rieger, R., Nicoloff, H., and Michaelis, A. Biol. Zbl. 82: 393-412 (1963).

85. Rieger, Michaelis, A., Schubert, I., and Meister, A. Mutat. Res. 20: 295-298 (1963).

86. Roy, S. C. Indian J. Exptl. Biol. 10: 244-246 (1972).

87. Scalera, S. E., and Ward, O. G. Mutat. Res. 12: 71-79 (1971).

88. Schubert, I., and Rieger, R. Experientia 32: 854-855 (1976).

89. Schubert, I., and Rieger, R. Mutat. Res. 35: 79-90 (1976).

90. Shan, V. C., Rao, S. R. V., and Arora, O. P. Nucleus (Calcutta) 25 92-96 (1972).

91. Shan, V. C., Rao, S. R. V., and Arora, O. P. Indian J. Exptl. Biol. 10: 431-435 (1972).

92. Shan, V. C., Rao, S. R. V., and Arora, O. P. Indian J. Biochem. Biophys. 9: 251-253 (1972).

93. Sidorov, B. N., Sokolov, N. N., and Andreev, V. S. Sov. Genet. 2 (7): 81-87 (1966); Genetika 2 (7): 124-122 (1966).

94. Singh, M. P., Kalia, C. S., and Gupta, M. Proc. Int. Congr. Genet. (Tokyo) 12: 116 (1968).

95. Sjödin, J. Hereditas 67: 155-180 (1970).

96. Sjödin, J. Hereditas 68: 1-34 (1971).

97. Slotova, J., Karpfel, Z., and Kubickova, D. Biol. Plant. 16: 21-27 (1974).

98. Sturelid, S. Hereditas 68: 255-276 (1971).

99. Sturelid, S., and Kihlman, B. A. EMS Newsletter 3: 15-17 (1970).

100. Sturelid, S., and Kihlman, B. A. Hereditas 79: 29-42 (1975).

101. Sturelid, S., and Kihlman, B. A. Hereditas 80: 233-246 (1975).

102. Swietlinska, Z., and Zuk, J. Mutat. Res. 26: 89-97 (1974).

103. Utsumi, S. Japan. J. Genet. 46: 125-134 (1971).

104. Wachtmeister, C. A., Moutschen-Dahmen, M., Moutschen, J., and Ehrenberg, L. Rad. Bot. 3: 187-192 (1963).

105. Wakonig, R., and Arnason, T. J. Abstracts X Int. Congr. Genet. Vol. II, 1958, p. 305.

106. Wakonig, R., and Arnason, T. J. Can. J. Bot. 37: 403-411 (1959).

107. Ward, O. G., and Glover, D. V. Abstracts XI Int. Bot. Congr., 1969, p. 232.

108. Wolff, S. Mutat. Res. 21: 349 (1973).

109. Wolff, S., and Cleaver, J. E. Mutat. Res. 20: 71-76 (1973).

110. Wu, T.-P. Taiwania 17: 248-254 (1972).

111. Wuu, K. D., and Grant, W. F. Nucleus (Calcutta) 10: 37-46 (1967).

112. Levan, A. Proc. 8th Int. Congr. Genet., Hereditas (Suppl.), 1949, pp. 325-337.

113. Sax, K., and Sax, H. J. Proc. Natl. Acad. Sci. (U.S.) 55: 1431-1435 (1966).

114. Benbadis, M.-C. C. R. Acad. Sci. (Paris) Ser. D 260: 268-270 (1965).

115. Biesele, J. J., Berger, R. E., Clarke, M., and Weiss, L. Exptl. Cell Res. (Suppl. 2): 279-303 (1952).

116. Grant, W. F. Can. J. Genet. Cytol. 15: 658 (1973).

117. Kalia, C. S., and Singh, M. P. Caryologia 26: 347-355 (1973).

118. Reiss, J. Experientia 27: 971-972 (1971).

CHAPTER 15

MUTAGENICITY OF SELECTED CHEMICALS IN THE HOST-MEDIATED ASSAY

R. Braun and J. Schöneich

Akademie der Wissenschaften der DDR, Zentral Institut für Genetik und Kulturpflanzenforschung
4325 Gatersleben, DDR

M. S. Legator

Department of Preventive Medicine and Community Health
University of Texas Medical Branch, Galveston
Texas 77550 (U.S.A.)

D. B. McGregor

Inveresk Research International Musselburgh,
Midlothian EH21 7UB, United Kingdom

and

G. R. Mohn

Department of Genetics and Chemical Mutagenesis, Wassenaarsweg 72, State University of Leiden, Leiden, The Netherlands

INTRODUCTION

Each test for mutagenicity comprises two components, a metabolic activation and a response system. An effective metabolizing system is especially necessary, since the majority of mutagens are promutagens and require biotransformation to the reactive metabolites or ultimate mutagens. The development of well characterized and highly sensitive genetic indicators (bacteriophages, bacteria, yeasts, fungi, mammalian cell cultures) which themselves do not possess the capacity for metabolic activation as do living mammals, led to the development of two-component tests later called by Garbridge and Legator host-mediated assay. The introduction of foreign mammalian cells into laboratory animals for mutagenicity testing was established about 20 years ago, tumors being used as genetic indicators for the detection of chromosomal damage. Ten years later, this method was extended to the use of microbial indicators and mammalian cell cultures. Today, practically all types of genetic changes may be detected, depending on the indicator component used, such as gene mutations, gene conversions, and recombinations. As compared with in vitro tests, in which the metabolic activation is supplied by mammalian livers, the pharmacokinetics of a given compound is taken into consideration in the host-mediated assay (Fig. 1). The genetic indicators can be recovered from all sites of the body, depending on the site of inoculation, which offers the possibility for studying organospecific mutagenesis. Furthermore, complex interactions between chemicals which influence mutagenic activity are taken into consideration. Besides such drug interactions, the formation of mutagenic compounds from nonmutagenic precursors, as exemplified by nitrosation reactions, may be proven under in vivo conditions. Depending on the main goal of the studies, several modifications of this method have been published.

METHODS

In the standard procedure the indicator organisms are introduced into the peritoneal cavity of treated animals, generally rats or mice. Since the peritoneal cavity represents--in pharmacokinetic terms--a deep compartment, and promutagens giving rise to short-lived ultimate reactive metabolites are not detected, other inoculation sites have been proven. The variations of the host-mediated assay technique are summarized in Table 1. The injection of indicators into the lateral tail vein offers the possibility of bringing them in close contact with the metabolizing system in the liver and other organs since they accumulate predominantly in liver, spleen, and lungs. Injection directly into the testes allows estimation of mutagenic substances in the gonadal region. Bacteriophages injected intravenously do not accumulate in any specific organ and circulate for many hours in the blood stream.

The compounds to be tested are given a route other than indicators, in general simultaneously. Application of the test substances is also possible prior to or later than injection of genetic systems, one time or several times. The exact procedure depends on the incubation times of indicator organisms, which vary from several minutes to many hours. After a sufficiently long period of time the indicators are withdrawn from the peritoneal cavity, blood stream, or the organ chosen for testing, and induced genetic changes are determined by means of standard procedures.

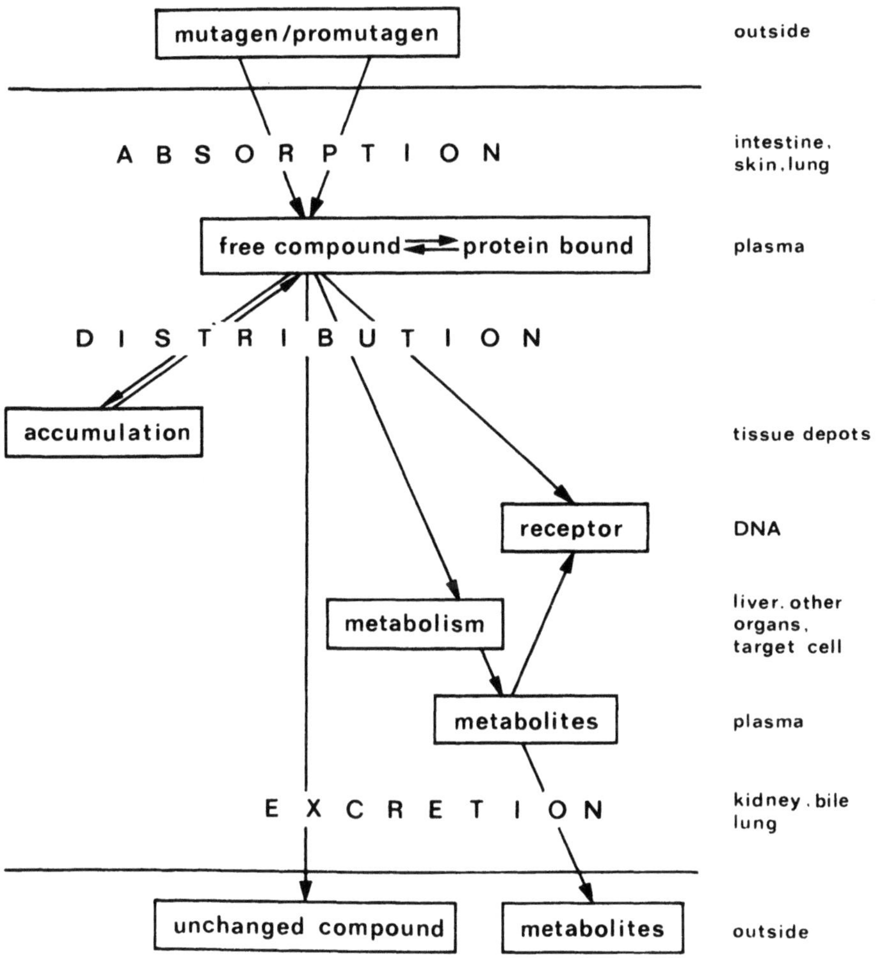

Fig. 1. Pharmacological parameters which determine the mutagenic response in mammalian organisms.

Table 1. Variations in the procedure of the host-mediated assay

Site of inoculation of genetic indicators	Site of recovery of genetic indicators	Suitable indicator organisms
Peritoneal cavity	Peritoneal cavity	Mammalian cells Fungal spores Yeasts Bacteria Bacteriophages
Testes	Testes	Yeasts Fungal spores
Blood	Liver Spleen Lung	Fungal spores Yeasts Bacteria
Blood	Blood	Bacteriophages
Intestinal tract	Intestinal tract	Bacteria

On reviewing the literature concerning mutagenic activity of chemicals in the host-mediated assay it has been found that in several papers the methodological aspects as well as basic questions for the interpretation of the results obtained were only poorly described. In general, more attention should be paid to the description of the host animals and the methods used for the passage of the indicator organisms through the host. We feel that it is necessary to adapt further publications of results obtained in host-mediated assay to the scheme given below. This procedure will not only facilitate comparisons of different papers but will also avoid any false interpretation of the results by the reader. The following points should be mentioned exactly in any publication:

1. Name of compound (trade name, chemical exact name, abbreviation used)

2. Formula

3. Constitution of the compound

   a. Purity
   b. Producer
   c. Impurities known to the author

4. Genetic indicator system

    a. Species
    b. Strain
    c. Mutation type analyzed

5. Host animal

    a. Species
    b. Strain
    c. Breeding conditions (SPF, barrier conditions, conventional, others)
    d. Genetic constitution (inbred, outbred, hybrids)
    e. Sex
    f. Age (weight)

6. Application site of indicator organisms (blood, peritoneal cavity, testes)

7. Incubation compartment (testes, liver, lung, spleen, peritoneal cavity)

8. Incubation time of indicator organisms within the host animal

9. Treatment conditions

    a. Solvent
    b. Application volume
    c. Application site
    d. Dose (dose range)
    e. Animals per dose
    f. Time relationship between application of indicator strains and compounds (simultaneously, prior, later, others)

10. Effects observed

    a. Inactivation
    b. Induction of genetic events at different times and concentrations
    c. Lowest effective dose
    d. Mutant frequency at the lowest effective dose
    e. Toxicity of compounds to the host animal preventing higher dosage
    f. Dose-effect relationships

11. Reproducibility (how many tests)

12. Statistic test used, probability of a mutagenic event

13. Spontaneous mutant frequency

    a. Typical for the test organism
    b. Found for negative or solvent control

14. Positive control

    a. Compound
    b. All points mentioned also for the test compound

15. Clear interpretation of the results obtained

RESULTS AND DISCUSSION

Detailed descriptions or reviews have been given by Fahrig (1), Grafe et al. (2), Legator (3), and Mohn and Ellenberger (4).

Indicator organisms currently or occasionally used in host-mediated assays are listed in Table 2. Table 3 surveys the compounds reviewed and results obtained.

Vinyl Chloride

The genetic activity of vinyl chloride is not well documented in the host-mediated assay (Table 4). Only on paper gives evidence for the induction of forward mutations in high doses.

Epichlorhydrin

The mutagenic effectivity of epichlorhydrin has been demonstrated in the host-mediated assay with *S. typhimurium* strains G 46, TA 100, and TA 1950 by Šràm et al. (6). The results of these experiments (Table 4) are ambiguous since no statistical evaluation has been presented and the spontaneous mutant frequency is not stable in different experiments and seems to be very high.

Mitomycin C

The mutagenic activity of Mitomycin C is not well established in the host-mediated assay by using microbial indicator organisms. It was negative with *S. typhimurium* and *E. coli* but showed preferential inhibition of the repair-defective strain in a host-mediated rec-type repair test (Table 4). It was mutagenic in a not further characterized strain of *E. coli* (K 12) at a dose of 10 mg/kg.

Table 2. Indicator organisms currently or occasionally used in host-mediated assays

Species/strain/cell type	Major genetic effect
Haploid and diploid microorganisms	
Neurospora crassa H-12 conidia	Recessive lethal mutations Deletion Forward and back mutations
Saccharomyces cerevisiae	Mitotic gene conversions Mitotic recombinations Forward and back mutations
Schizosaccharomyces pombe	Forward and back mutations
Salmonella typhimurium LT 2	Back mutations
Escherichia coli B/r and K-12	Forward and back mutations
Serratia marcescens HY	Back mutations
Bacillus subtilis 17A and 45T	Differential DNA repair
Listeria monocytogenes 3-54	Cellular and colony morphology
Bacteriophages, T4, λ, and K	Forward and back mutations Differential DNA repair
Mammalian cell cultures	
Walker carcinoma 256	Chromosomal aberrations
Ehrlich-Lettre ascites cells	Chromosomal aberrations DNA single-strand breaks
L5178Y murine leukemia (and other substrains)	Chromosomal aberrations Forward mutations Back mutations
Human and animal peripheral blood cells	Chromosomal aberrations
Chinese hamster cell lines	Forward mutations

Table 3. Compounds reviewed and summary of results obtained

Compound	Mutations and/or recombinations: Prokaryotic cells	Mutations and/or recombinations: Eukaryotic cells	Mutations and/or recombinations: Mammalian cells	Chromosomal aberrations in mammalian cells
Vinyl chloride		+		
Epichlorhydrin	+			
Mitomycin C	+			+
ThioTEPA	+		+	
DEN	+	+	(+)	
DMN	+	+	+	
Cyclophos-phamide	+	+	+	+
Trenimon	+			+
MNNG	+	+	+	
EMS	+	+	+	+
MMS	+	+		+
Myleran	(+)			
TEPA	(+)			
ICR-170		-		
Cadmium				
Natulan				
Ethylenimine				
Aflatoxin B1				
MeTEPA				

The induction of chromatid aberrations has been demonstrated in ascites tumor cells after intraperitoneal and subcutaneous application of Mitomycin C.

ThioTEPA

ThioTEPA is mutagenic both in *S. typhimurium* and the leukemic cell line L 5178 Y (Table 5).

Diethylnitrosamine

There are ambiguous results published in the literature concerning the genetic activity of diethylnitrosamine in the host-mediated assay (Table 6). It seems that genetic activity in the intraperitoneal host-mediated assay is difficultly reproducible, perhaps due to the instability of the ultimate mutagens produced. In other cases, by incubating the indicator organisms close to the metabolizing sites in the liver, a mutagenic activity may be demonstrated in suitable markers.

Dimethylnitrosamine

Dimethylnitrosamine is one of the best analyzed promutagenic compound (Table 7). It was found to be mutagenic in the host-mediated assay after oral, intramuscular, subcutaneous application with *S. typhimurium*, *S. marcescens*, *Neurospora crassa*, *E. coli* (different forward and reversion systems), and induced both gene conversion and mitotic crossing over in *Saccharomyces cerevisiae*, incubated in liver, lung, and testes. Furthermore, it induced point mutations in the host-mediated assay with leukemic L5178Y cells in a very low dose of 2.5 mg/kg.

Cyclophosphamide

The mutagenic and clastogenic activity of cyclophosphamide has been tested in several experiments with mice and rats as host animals and *S. typhimurium*, *S. marcescens*, *E. coli*, *S. cerevisiae*, and ascites tumor cells as indicator organisms (Table 8). Cyclophosphamide induced gene mutations in the host-mediated assay with *S. typhimurium* in doses of 200 mg/kg and higher (Table 9). It is convertogenic with *S. cerevisiae* D-4 in doses of 500 mg/kg and induced chromatid aberrations in ascites tumor cells in vivo.

Table 4. Vinyl chloride, epichlorhydrin, and mitomycin C

Compound	Genetic indicator	Genetic event	Host animal	Strain and sex
Vinyl chloride	Saccharomyces pombe, haploid SP 198, ade 6-60/rad 10-198	Forward mutation	Mouse	Swiss albino
Epichlorhydrin	S. typhimurium G 46, TA 100, TA 1950	Base substitution	Mouse	ICR
Mitomycin C	S. typhimurium; E. coli	Gene mutations	Mouse	Swiss albino
	B. subtilis 45T, 17A	DNA damage	Mouse	CBA
	E. coli K 12	Gene mutations	Mouse	COBS
	Ehrlich ascites tumor cells	Chromatid aberrations	Mouse	AB-Jena

Table 5. ThioTEPA

Genetic indicator	Genetic event	Host animal	Strain and sex	Assay type
S. typhimurium TA 1950	Base substitutions	Mouse	NMRI/male	IP
S. typhimurium TA 1535	Base substitutions	Mouse	NMRI/male	Intrasanguine (liver)
Leukemic cell line L 5178 Y	Gene mutations	-	-	IP

Assay type	Exposure range (mg/kg)	Lowest effective dose (mg/kg)	Comment	References
IP	700, PO, 3-12 hr	700, PO, 6 hr	Induction of forward mutations	5,62
IP	-	100, SC and IM, 3 hr	Induction of gene mutations	6
IP	15, IM, ×3	-	Negative	7
IP	3, IM	-	Induction of DNA damage	8
IP	10, SC, 4 hr	-	Induction of gene mutations in a not further characterized strain	9
IP	-	2.3, SC 0.11, IP	Induction of chromatid-type aberrations	10

Exposure or range	Lowest effective dose (mg/kg)	Comment	References
-	25, SC	Mutagenic	11
-	12.4, IP	Mutagenic in acute experiments but not in subacute treatments (5 days)	12
-	7.5, SC	Significantly mutagenic	48

Table 6. Diethylnitrosamine

Genetic indicator	Genetic event	Host animal	Strain and sex	Assay type
S. typhimurium G46	Base substitution	Mouse	C57BL/6 × C3H/male	IP
S. typhimurium G46	Base substitution	Mouse	-	IP
S. typhimurium TA 1950	Base substitution	Mouse	NMRI/male	IP
S. typhimurium TA 1535	Base substitution	Mouse	NMRI/male	Intra-sanguine (liver)
E. coli 343/113	Base substitution	Mouse	NMRI/female	Intra-sanguine (liver)
N. crassa ade_3A/B	Forward mutation	Mouse	-	Intra-sanguine (liver)
S. cerevisiae	Mitotic crossing over	Mouse	Swiss albino albino	IP
S. typhimurium G46	Base substitution	Mouse	ICR/male	IP
S. typhimurium G46	Base substitution	Mouse	NMRI/male	IP
S. marcescens a13	Base substitution	Mouse	NMRI/male	IP
S. marcescens a21	Base substitution	Mouse	NMRI/male	IP
E. coli 343/113	Forward mutation	Mouse	NMRI/male	IP

Exposure or range (mg/kg)	Lowest effective dose (mg/kg)	Comment	References
400, IM	-	Weakly mutagenic	13
400, IM	-	Weakly mutagenic	14
250, SC and IM	-	Weakly mutagenic	15
150, PO	60, SC	Significantly mutagenic in both cases	12
-	66.66, SC	Significant mutagenic activity	16
400, IM	-	Weakly mutagenic	14
1900	-	Weakly recombinogenic	17
-	-	Nonmutagenic at 500 mg/kg IM	18
-	-	Nonmutagenic at 210 mg/kg SC, IM, PO	15
-	-	Nonmutagenic at 1500 mg/kg SC	19
-	-	Nonmutagenic at 1500 mg/kg SC	19
-	-	Nonmutagenic at 1500 mg/kg SC	19

(continued)

Table 6. (continued)

Genetic indicator	Genetic event	Host animal	Strain and sex	Assay type
E. coli 343/113	Forward mutation	Mouse	NMRI/female	Intra-sanguine (liver)
N. crassa ade_3A/B	Forward mutation	Mouse	-	IP
S. typhimurium G46	Base substitution	Rat	Sprague/Dawley	IP
S. typhimurium G46	Base substitution	Rat		IP

Table 7. Dimethylnitrosamine

Genetic indicator	Genetic event	Host animal	Strain and sex	Assay type
S. typhimurium G46	Base substitution	Mouse	C57Bl × C3H male	IP
S. typhimurium G46	Base substitution	Mouse	ICR	IP
S. typhimurium G46	Base substitution	Mouse	NMRI/male	IP
S. typhimurium G46	Base substitution	Mouse	-	IP
S. typhimurium G46	Base substitution	Mouse	-	IP
S. typhimurium G46	Base substitution	Mouse	-	IP

Exposure or range (mg/kg)	Lowest effective dose (mg/kg)	Comment	References
-	-	Nonmutagenic at 66.66 mg/kg SC	16
-	-	Nonmutagenic at 200 mg/kg IM	14
-	-	Nonmutagenic at 400 mg/kg IM	13
-	-	Nonmutagenic at 400 mg/kg IM	14

Exposure or range (mg/kg)	Lowest effective dose (mg/kg)	Comment	References
25, IM, 3 hr	-	Mutagenic	13
65, IM, 3 hr	-	Mutagenic	18
182, PO, SC, IM, 3 hr	-	Mutagenic	15
20, IM, 5 min	-	Mutagenic	20
25, IM	-	Mutagenic	14
-	20, SC 100, PO	Stronger mutagenic after SC application	21

(continued)

Table 7. (continued)

Genetic indicator	Genetic event	Host animal	Strain and sex	Assay type
S. typhimurium G46	Base substitution	Mouse	Swiss albino	IP
S. typhimurium G46	Base substitution	Mouse	NMRI	IP
S. typhimurium G46	Base substitution	Mouse	NMRI/male	IP
S. typhimurium G46	Base substitution	Mouse	Swiss albino	IP
S. typhimurium G46	Base substitution	Mouse	ICR/male	IP
S. typhimurium TA 1950	Base substitution	Mouse	NMRI/male	IP
S. typhimurium G46	Base substitution	Rat	-	IP
S. typhimurium G46	Base substitution	Rat	Sprague-Dawley	IP
S. marcescens a21 leu	Base substitution	Mouse	-	IP
E. coli K-12/C600	$lac^- \rightarrow lac^+$	Mouse	COBS/CD	IP
E. coli K-12/C600	$lac^- \rightarrow lac^+$	Papio cynoceph.	-	IP
E. coli K-12	Forward mutation $MTR^s \rightarrow MTR^r$	Mouse	-	Intrasanguine (liver)
E. coli K-12	Forward mutation	Mouse	NMRI/female	Intrasanguine (liver)

Exposure or range (mg/kg)	Lowest effective dose (mg/kg)	Comment	References
3 × 0.1 ml 1% solution	-	Strongly mutagenic	7
100, PO, 3 hr	-	Mutagenic	22
500, PO and SC, 3 hr	-	Mutagenic	23
20, IM, 5 min	-	Mutagenic	23
16.5, IM, 3 hr	-	Mutagenic	24
-	30, PO, 5 hr	Mutagenic	15
100, IM	-	Mutagenic	14
25, IM, 3 hr	-	Mutagenic	13
3 × 100, SC	-	Mutagenic	21
100, PO, 4 hr	-	Mutagenic	9
100, PO, 4 hr	-	Mutagenic	9
-	30, SC, 210 min	Mutagenic	26
0.1 ml 1% solution SC, 15 min	-	Induction of forward muta- mutation $MTR^s \rightarrow MTR^r$	27

(continued)

Table 7. (continued)

Genetic indicator	Genetic event	Host animal	Strain and sex	Assay type
E. coli K-12	Base substitution	Mouse	NMRI/female	Intra-sanguine (liver)
N. crassa ade 3A/ade 3B	Forward mutation	Mouse	-	Blood mediated
Saccharomyces cerevisiae D-3	Mitotic crossover	Mouse	Swiss albino	IP
Saccharomyces cerevisiae D-4	Gene conversion	Mouse	C3H	IP
Saccharomyces cerevisiae D-4	Gene conversion	Mouse, rat	C3H/male, female BD male	IP and testes
Saccharomyces cerevisiae D-4	Gene conversion	Rat	Wistar/ male	Intra-sanguine (liver, lung) and testes
Leukemic cell line L5178Y	Gene mutation	Mouse	$BD2F_1$	IP

Exposure or range (mg/kg)	Lowest effective dose (mg/kg)	Comment	References
0.1 ml 1% solution SC, 15 min	-	Induction of back mutation $arg^- \rightarrow arg^+$	27
75, IM	-	Mutagenic	14
1300, 3 hr	-	Recombinogenic	17
740	-	Convertogenic	28
740, SC 4 hr	-	Effectivity in this comparative study: rat peritoneum > mouse peritoneum > rat testes	29
740, SC, 4 hr	-	Effectivity in this comparative study: liver > lung > testes	30
2.5, SC, 6 hr	-	Induction of gene mutation $asn^- \rightarrow asn^+$	31

Table 8. Cyclophosphamide

Genetic indicator	Genetic event	Host animal	Strain and sex	Assay type
S. typhimurium G46	Base substitution	Mouse	NMRI/male	IP
S. typhimurium G46 and TA 1950	Base substitution	Mouse	NMRI/male	IP
S. marcescens a13	Back mutation	Mouse	NMRI/male	IP
S. marcescens a21	Back mutation	Mouse	NMRI/male	IP
E. coli 343/113	Forward mutation	Mouse	NMRI/male	IP
S. typhimurium TA 1535	Base substitution	Mouse	NMRI/male	Intra-sanguine (liver)
S. cerevisiae D4	Gene conversion	Rat	BD/male	Testes
S. cerevisiae D4	Gene conversion	Rat	Wistar/male	Testes
S. cerevisiae D4	Gene conversion	Rat	Wistar/male	Intra-sanguine (liver, lung)
S. cerevisiae D3	Mitotic crossing over	Mouse	Swiss albino/male	IP

Exposure or range (mg/kg)	Lowest effective dose (mg/kg)	Comment	References
870, SC	-	Weakly mutagenic	19
-	870, SC (G46) 200, SC (TA1950)	-	33
870, SC	-	Nonmutagenic	19
870, SC	-	Mutagenic	19
870, SC	-	Nonmutagenic	19
117-900, PO	390, PO	Strongly mutagenic in intermediate dose ranges, but no or only slight activity in low (up to 200 mg/kg) and high (higher than 750 mg/kg) doses	12
760, SC	-	Convertogenic	29
380, SC	-	Convertogenic	30
380, SC	-	Convertogenic both in liver and lung	30
400	-	Recombinogenic	17

(continued)

Table 8. (continued)

Genetic indicator	Genetic event	Host animal	Strain and sex	Assay type
S. cerevisiae D4	Gene conversion	Mouse	AB	IP
S. cerevisiae D4	Gene conversion	Mouse	C3H	IP
S. cerevisiae D4	Gene conversion	Mouse	C3H/male and female	IP
S. cerevisiae D4	Gene conversion	Mouse	Swiss albino/male	IP
S. cerevisiae D4	Gene conversion	Rat	BDII/male	IP
S. cerevisiae D4	Gene conversion	Rat	BD/male	IP
S. cerevisiae D4	Gene conversion	Rat	BDIX/male	IP
S. cerevisiae D4	Gene conversion	Rat	BDII/male	IP
S. cerevisiae D4	Gene conversion	Mouse	AB	In vitro urine
S. typhimurium TA 1535	Base substitution reverse mutation	Mouse	NMRI/male	In vitro, serum-mediated
S. cerevisiae D4	Gene conversion	Rat	-	In vitro, urine
S. cerevisiae D4	Gene conversion	Rat	BDIX	In vitro, urine
S. cerevisiae D4	Gene conversion	Rat	BDII	In vitro, urine
S. cerevisiae D4	Gene conversion	Man/ female	-	In vitro, urine

Exposure or range (mg/kg)	Lowest effective dose (mg/kg)	Comment	References
500, SC negative	-	Nonconvertogenic	32
760, SC	-	Convertogenic	28
760, SC	-	Convertogenic	29
800, negative	-	Nonconvertogenic	17
-	500, PO	Convertogenic	34
-	750, SC	Convertogenic	29
500, PO	-	Convertogenic	35
500, PO	-	Convertogenic	36
500	-	Convertogenic	32
200	-	Serum from treated mice is mutagenic	11
500	-	Convertogenic	37
500	-	Convertogenic	36
340, PO	-	Convertogenic	35
3.1, IV	-	Urine is convertogenic 24 hr after treatment	38

Table 9. Mutagenic activity of cyclophosphamide[a]

S. typhimurium strain	Dose (mg/kg)	Survival (%)	Mutant frequency × 10^{-9}	C	P
G 46	200	108	2.45	1.07	>>0.01
	400	96	5.43	1.57	>>0.01
	3×200[b]	94	3.97	1.51	>>0.01
	3×290[b]	86	6.33	2.68	<0.01
	3×400[b]	66	10.91	3.40	<0.01
TA 1950	40	123	19.08	2.10	>0.01 <0.05
	80	127	24.69	2.72	<0.05
	200	73	51.04	7.64	<0.01
	400	73	63.09	9.41	<0.01

[a]Host-mediated assay with Salmonella typhimurium as genetic system and male NMRI mice as host animals; 3 hr incubation of bacteria within the peritoneal cavity (11).

[b]Test compound given subcutaneously simultaneous with and 1 and 2 hr after injection of bacteria.

There are also some negative results with S. marcescens and E. coli as indicators but also with S. cerevisiae D-4. As compared with the intraperitoneal host-mediated assay, the convertogenic activity of this compound is well documented in the urine assay using the same host animals as well as the same indicator strain of S. cerevisiae (32). The mutagenic activity of cyclophosphamide is influenced by hepatotoxic agents.

It has been shown that convertogenic metabolites of cyclophosphamide are present in the urine of treated patients.

Trenimon

In dose-response studies with Trenimon, mutagenic activity has been demonstrated in doses of 1.5 mg/kg and higher by using S. typhimurium TA 1950 (11), while the corresponding wild type strain G46 revealed no mutagenic activity (19). In the host-mediated assay with S. marcescens and E. coli as indicator organisms, Trenimon induced base substitutions and forward mutations (Table 10). In experiments with ascites tumor cells in vivo, Trenimon induced chromatid aberrations in doses of 0.25 mg/kg and higher (39).

Induction of chromatid aberrations in S_2 sarcoma of mice is shown in Table 11, and mutagenic activity of Trenimon with S. typhimurium after subcutaneous application is detailed in Table 12.

N-Methyl-N'-nitro-N-nitrosoguanidine

MNNG induced point mutations in the host-mediated assay with various bacterial indicator strains and leukemic L5178Y cells. Mitotic gene conversion and mitotic crossing over were produced in S. cerevisiae but chromosome deletion or forward mutation in the fungus Neurospora crassa were not (Table 13). These findings may be connected with the impermeability of the conidia cell wall for MNNG.

Ethyl Methanesulfonate

The mutagenic and recombinogenic activity of EMS in the host-mediated assay is well documented (Table 14). Very recently a summary of the EMS results has been published (61).

Methyl Methanesulfonate

MMS was mutagenic in the host-mediated assay with mice as host animals and E. coli (27), S. marcescens (56), and S. typhimurium (21) as indicator organisms. In contrast to this single result of Propping et al., Fahrig et al., reported negative findings using the same strain of S. typhimurium (56).

Otherwise, the convertogenic effectivity of MMS is very well established with S. cerevisiae (Table 15). There are also clear results concerning the induction of genetic recombination and forward mutations in the host-mediated assay with Schizosaccharomyces pombe and forward-mutation induction in N. crassa.

Table 10. Trenimon

Genetic indicator	Genetic event	Host animal	Strain and sex	Assay type	Exposure or range (mg/kg)	Lowest effective dose (mg/kg)	Comment	References
S. typhimurium G46	Base substitution	Mouse	NMRI	IP	1, SC, ×3	-	Nonmutagenic	19
S. marcescens his a 13	Base substitution	Mouse	NMRI	IP	4, SC, ×3	-	Mutagenic	19
S. marcescens leu a 21	Base substitution	Mouse	NMRI	IP	4, SC, ×3	-	Mutagenic	19
E. coli 343 Gal R^{S}_{18}/ B_1/F^-	Forward mutation	Mouse	NMRI	IP	16, SC, ×3	-	Mutagenic	19
S. typhimurium G46 and TA 1950	Base substitution	Mouse	NMRI/ male	IP	4 SC (G46)	1.5, SC (TA 1950)	Nonmutagenic in G46, linear dose-response in TA 1950	11
Ascites tumor cells S_2	Chromatid aberrations	Mouse	AB-Jena	IP	-	0.25, IP 12.50, PO	Induction of chromatid aberrations	39

Table 11. Induction by Trenimon of chromatid aberrations in S_2 sarcoma of mice, and its dependence on Trenimon concentration (injected volume 1 ml in all cases)[a]

Concentration (M)	No. of animals	No. of cells scored	Metaphases with aberrations No.	%
2×10^{-7}	4	600	10	1.7 ± 2.2
1×10^{-6}	4	650	116	17.8 ± 13.7
2×10^{-6}	4	700	314	44.8 ± 4.6
5×10^{-6}	2	300	186	62.0 ± 8.0
1×10^{-5}	4	650	471	72.5 ± 6.9
Control	5	1000	2	0.2

[a]Data of Schöneich (39).

Table 12. Mutagenic activity of Trenimon after subcutaneous application in the host-mediated assay with S. typhimurium TA 1950 as indicator system and male NMRI mice as host animals with an incubation time of 3 hr[a]

Dose (mg/kg)	Survival (%)	Mutant frequency $\times 10^{-9}$	C	P
0.5	0.70	7.18	1.06	>>0.01
1.5	0.66	16.04	3.06	<0.01
3.5	0.53	20.76	3.07	<0.01
4.0	0.71	23.17	5.76	<0.01
5.0	0.68	25.69	6.39	<0.01

[a]Data of Braun (11).

Table 13. N-Methyl-N'-nitro-N-nitrosoguanidine

Genetic indicator	Genetic event	Host animal	Strain and sex	Assay type
S. typhimurium G46	Base substitution	Mouse	NMRI	IP
S. typhimurium G46	Base substitution	Mouse	Flander's	IP
S. typhimurium G46	Base substitution	Mouse	-	IP
S. typhimurium G46	Base substitution	Mouse	NMRI/male	IP
S. typhimurium G46 and TA 1530	Base substitution	Mouse	Swiss-Webster	IP
E. coli K12 343	Forward	Mouse	NMRI	IP
S. marcescens a13 his$^-$ and a21 leu$^-$	Base substitution	Mouse	NMRI	IP
S. cerevisiae D4	Mitotic gene conversion	Mouse	C3H	IP
S. cerevisiae D3	Mitotic cross over	Mouse	Swiss-Webster	IP
S. cerevisiae D5	Mitotic recombination	Mouse	NMRI/male	IP
N. crassa ade 3A/ade 3B	Forward mutation or chromosome deletion	Mouse and Chinese hamster	-	-

Exposure or range (mg/kg)	Lowest effective dose (mg/kg)	Comment	References
20, SC, 3 hr	-	Mutagenic	19
30	-	Mutagenic	7
-	-	Mutagenic	40
10, SC, 3 hr	-	Mutagenic with SC injection; nonmutagenic with PO intubation	23
-	-	Mutagenic in both strains	41
60, SC, 3 hr	-	Mutagenic	19
60, SC, 3 hr	-	Mutagenic in both strains	19
2	-	Convertogenic	42
-	-	Recombinogenic	41
25	-	Very low activity at the dose given	43
-	-	Nonmutagenic in both species	44

(continued)

Table 13. (continued)

Genetic indicator	Genetic event	Host animal	Strain and sex	Assay type
N. crassa ade 3A/ade 3B	Forward mutation or chromosome deletion	Mouse	101×C3H	-
Mouse cell line L51784 asn^-	Point mutation	Mouse	BD2F/male	IP

Table 14. Comparison of lowest effective or lowest tested doses of ethyl methanesulfonate (EMS) that induce genetic effects in host-mediated assays with various indicators

Indicator species and/or strain	Induced genetic event	EMS concn (mg/kg)	References
ELD ascites tumor	Chromatid aberrations	250 (IV)	46
L5178Y murine leukemia	Various forward mutations (MTX, TdR, ara-C)	100 (SC)	47,48
N. crassa, heterokaryon 12	ade-3 forward mutations	300	49,50
S. cerevisiae, strain D4	ade-2 and trp-5, mitotic gene conversion	400	51
		1500	52
		1200	53
S. cerevisiae, strain D3 and others	Mitotic recombination	800	17,54
	canR forward mutations, met^+ reversions	150	
Schizosaccharomyces pombe, strain SP 168	ade^-, forward mutations	300	53
S. typhimurium G 46	his^+ reversions	35 (IM)	55

Exposure or range (mg/kg)	Lowest effective dose (mg/kg)	Comment	References
-	-	Nonmutagenic	45
25	-	Significantly mutagenic	31

Myleran

Serum of Myleran-treated rats was found to be mutagenic towards *S. typhimurium* (Table 16).

TEPA

Serum of TEPA-treated rats was found to be mutagenic towards *S. typhimurium* (Table 16).

ICR-170

No evidence for genetic activity of ICR-170 in the host-mediated assay has appeared in the literature to date (Table 16).

Table 15. Methyl methanesulfonate

Genetic indicator	Genetic event	Host animal	Strain and sex	Assay type
S. typhimurium his G46	Base substitution	Mouse	NMRI C3H/ male and female	IP
S. typhimurium his G46	Base substitution	Mouse	NMRI	IP
E. coli K-12 343/113/gal R^{s}18/arg 56/ nad 113	Forward mutations, reversions	Mouse	NMRI/male	Intra- sanguine (liver)
S. marcescens a 21 leu^{-}	Base substitution	Mouse	NMRI, C3H, male and female	IP
Schizosaccharo- myces pombe ade 7-50	Genetic recom- binations (ade 7-50)	Mouse	-	IP
Schizosaccharo- myces pombe ade 6-60 and ade 7-152	Forward mutations (ade 6-60 and genetic genetic recom- bination (ade 7-152)	Mouse	-	IP
Shizosaccharo- myces pombe	Forward mutation	Mouse	-	IP
N. crassa	Not specified	Mouse	-	Not specified
N. crassa	Forward mutation in ad-3A/ad-3B or deletions	Mouse	101×C3H male	IP

Exposure or range (mg/kg)	Lowest effective dose (mg/kg)	Comment	References
0.1-1 mmole/kg SC, 1 hr negative	-	Nonmutagenic	56
150, SC	-	Mutagenic in SC application, nonmutagenic after PO intubation	21
-	132, SC, 30 min	Both mutagenic in the reversion system $arg^- \rightarrow arg^+$ and forward mutation system $gal^- \rightarrow gal^+$	27
-	0.15 mmole/kg SC, 4 hr	Mutagenic	56
5 mmole/kg, 2 hr	-	Recombinogenic	5
5.4 mmole/kg, 2 hr	-	Mutagenic and recombinogenic	57
-	-	Mutagenic Mutation frequency 87.75×10^{-3}/locus	58
250, 8 hr	-	Genetically active	44
100, IV, 10 hr	-	Mutagenic	45

(continued)

Table 15. (continued)

Genetic indicator	Genetic event	Host animal	Strain and sex	Assay type
N. crassa ad-3	Forward mutation	Mouse	-	IP
S. cerevisiae D-4	Gene conversions	Mouse, rat	C3H, male and female; BD male	IP and testes
S. cerevisiae D-4	Gene conversions	Mouse	-	IP
S. cerevisiae D-4	Gene conversions	Mouse	Swiss/ male	IP
S. cerevisiae D-4	Gene conversions	Mouse	-	IP
S. cerevisiae D-4	Gene conversions	Mouse	C3H male and female	IP
S. cerevisiae D-4	Gene conversions	Mouse, rat	C3H male and female; Wistar male	IP, intra-sanguine (liver, lung), and testes

Exposure or range (mg/kg)	Lowest effective dose (mg/kg)	Comment	References
20 µl/kg IP	-	Mutagenic	59
2 mmole/kg, SC, 1 hr	-	Convertogenic effectivity: mouse peritoneum > rat testes > rat peritoneum	29
4.8 mmole/kg, 2 hr	-	Convertogenic	5
1.2 mmole/kg, PO, 21	-	Convertogenic	57
-	-	Convertogenic 1.30×10^{-5}/locus-ppm for ade 2 locus; 1.20×10^{-5}/locus - ppm for trp 5 locus	58
0.4 mmole/kg, SC, 1 hr	-	Convertogenic	30
2-4 mmole/kg	-	Convertogenic in mouse peritoneal cavity at 2 mM/kg SC, 4 hr; in intrasanguine tests with rat convertogenic at 4 mM/kg SC, 1 hr Effectivity: liver > lung > testes	30

Table 16. Myleran, TEPA, and ICR-170

Compound	Genetic indicator	Genetic event	Host animal	Strain and sex	Assay type	Exposure or range (mg/kg)	Comments	References
Myleran	S. typhimurium his- auxotrophs	Back mutations	Rat	-	In vitro testing using blood serum of treated animals	-	In vitro testing for 1 hr	60
TEPA	S. typhimurium his- auxotrophs	Back mutations	Rat	-	In vitro testing using blood serum of treated animals	-	In vitro testing for 1 hr	60
ICR-170	N. crassa ade $_3$A/B	Forward mutation	Mouse	101×C3H	IP	IV	IV, nonmutagenic	45
	N. crassa ade 3A/B	Forward mutation	Rat	-	IP	8 mg/kg SC	8 mg/kg SC, nonmutagenic	44

REFERENCES

1. Fahrig, R. Mutat. Res. 31: 381-394 (1975).

2. Grafe, A., Lorenz, R., and Vollmar, J. Mutat. Res. 31: 205 (1975).

3. Legator, M. S. Agents and Actions 3: 111 (1973).

4. Mohn, G. R., and Ellenberger, J. In: Handbook of Mutagenicity Test Procedures. B. J. Kilbey, M. Legator, W. Nichols, and C. Ramel, Eds., Elsevier, Amsterdam, 1977.

5. Loprieno, N., Barale, R., Baroncelli, S., Bauer, C., Bronzetti, G., Cammellini, A., Cerecignani, G., Corsi, C., Gervasi, G., Leporini, C., Nieri, R., Rossi, A. M., Stretti, G., and Turchi, G. Mutat. Res. 40: 85-96 (1976).

6. Šràm, R. J., Černa, M., and Kučerova, M. Biol. Zbl. 95: 451-462 (1976).

7. Gabridge, M. G., and Legator, M. S. Fed. Proc. 130: 831-834 (1969).

8. Kada, T., Tutikawa, K., and Sadaie, Y. Mutat. Res. 16: 165-174 (1972).

9. Mazue, G., Vindel, J. A., Landsmann, F., and Brunaud, M. Arzneim.-Forsch. 24: 1422-1425 (1975).

10. Wobus, A. M., Thieme, R., and Schöneich, J. Abstr. 6th Ann. Meeting EEMS, Gernrode (GDR), 1976, pp. 163-164.

11. Braun, R. Unpublished results.

12. Arni, P., Mantel, T., Deparade, E., and Müller, D. Mutat. Res. 45: 291-307 (1977).

13. Malling, H. V. Mutat. Res. 26: 465-472 (1974).

14. Malling, H. V., and Frantz, C. N. Environ. Health Perspect. 6: 71-82 (1973).

15. Braun, R., and Schöneich, J. Biol. Zentralbl. 94: 661-669 (1975).

16. Mohn, G., Ellenberger, J., McGregor, D., and Merker, H.-J. Mutat. Res. 29: 221-233 (1975).

17. Brusick, D., and Mayer, V. M. Environ. Health Perspect. 6: 83-96 (1973).

18. Couch, D. B., and Friedman, M. A. Mutat. Res. 38: 89-96 (1976).

19. Propping, P., Röhrborn, G., and Buselmaier, W. Mol. Gen. Genet. 117: 197-209 (1972).

20. Zeiger, E. Environ. Health Perspect. 6: 101-109 (1973).

21. Propping, P., Buselmaier, W., and Röhrborn, G. Arzneim.-Forsch. 23: 746-749 (1973).

22. Münzer, R., and Renner, H. W. Int. J. Rad. Biol. 27: 371-375 (1975).

23. Propping, P., and Buselmaier, W. Arch. Toxicol. 28: 129-134 (1971).

24. Zeiger, E., and Legator, M. S. Mutat. Res. 12: 469-471 (1971).

25. Couch, D. B., and Friedman, M. A. Mutat. Res. 26: 371-376 (1974).

26. Mohn, G., and Ellenberger, J. Mutat. Res. 19: 257-260 (1973).

27. Mohn, G., Ellenberger, J., and McGregor, D. Mutat. Res. 25: 187-196 (1974).

28. Fahrig, R. Mutat. Res. 21: 219-220 (1973).

29. Fahrig, R. Mutat. Res. 26: 29-36 (1974).

30. Fahrig, R. Mutat. Res. 31: 313 (1975).

31. Capizzi, R. L., Papirmeister, B., Millins, J. M., and Cheng, E. Cancer Res. 34: 3073-3082 (1974).

32. Schubert, A. Biol. Zbl. 94: 451-454 (1975).

33. Braun, R., and Schöneich, J. Mutat. Res. 31: 191-194 (1975).

34. Siebert, D. Mutat. Res. 28: 57-61 (1975).

35. Siebert, D. Z. Krebsforsch. 81: 261-267 (1974).

36. Siebert, D. Mutat. Res. 17: 307-314 (1973).

37. Marquardt, H., and Siebert, D. Naturwiss. 58: 568 (1971).

38. Siebert, D., and Simon, U. Mutat. Res. 21: 257-262 (1973).

39. Schöneich, J. Unpublished results.

40. Gabridge, M. G., Oswald, E. J., and Legator, M. S. Mutat. Res. 7: 117-119 (1969).

41. Maxwell, W. A., and Newell, G. W. In: Molecular and Environmental Aspects of Mutagenesis. Proc. Rochester Int. Conf. Environ. Toxicol., Springfield, 1974, pp. 223-252.

42. Fahrig, R. Agents and Actions 3: 99-110 (1973).

43. Ryttman, H., and Zetterberg, G. Mutat. Res. 34: 201-216 (1976).

44. Malling, H. V. In: Molecular and Environmental Aspects of Mutagenesis. Proc. Rochester Int. Conf. Environ. Toxicol., Springfield, 1974, pp. 159-177.

45. Malling, H. V. Int. Cancer Congr. Abstr. 10: 800-801 (1970).

46. Chu, E. H. Y. Genetics 64: 12-13 (1970).

47. Fischer, G. A., Lee, S. Y., and Calabresi, P. Mutat. Res. 26: 501-511 (1974).

48. Lee, S. Y. Environ. Health Perspect. 6: 145-149 (1973).

49. Legator, M. S., and Malling, H. V. In: Chemical Mutagens, A. Hollaender, Ed., Plenum Press, New York, 1971, pp. 569-589.

50. Malling, H. V. Mol. Gen. Genet. 116: 211-222 (1972).

51. Dean, B. J., Doak, S. M. A., and Funnell, J. Arch. Toxicol. 30: 61-66 (1972).

52. Fahrig, R. Mutat. Res. 21: 30 (1973).

53. Loprieno, N., Barale, R., Bauer, C., Baroncelli, S., Bronzetti, G., Cammellini, A., Cinci, A., Corsi, G., Leporini, C., Nieri, R., Nozzolini, M., and Serra, C. Mutat. Res. 25: 197-217 (1974).

54. Brusick, D., and Legator, M. S. EMS Newsletter 4: 31-32 (1971).

55. Ray, V. A., Holden, H. E., Salsburg, D. S., Ellis, J. H., Jr., Just, L. J., and Hyneck, L. M. Toxicol. Appl. Pharmacol. 30: 107-116 (1974).

56. Fahrig, R., Grafe, A., and Buselmaier, W. Arch. Toxicol. 34: 9-15 (1975).

57. Loprieno, N., Barale, R., Baroncelli, S., Bronzetti, G., Cammellini, A., Corsi, I. G., Leporini, C., Nieri, R., Nozzolini, M., and Rossi, A. M. Consig. Naz. Rec. Lab. Mutagen. Differenz. 109: 1-49 (1974).

58. Bronzetti, G., Barale, R., Baroncelli, S. Gammellini, A., Corsi, C., Leporini, C., Loprieno, N., Nieri, R., Nozzolini, M., and Rossi, A. M. Atti. Assoz. Genet. Ital. 20: 6-8 (1975).

59. Malling, H. V., and Cosgrove, G. E. In: Chemical Mutagenesis in Mammals and Man, F. Vogel and G. Röhrborn, Eds., Springer Verlag, Berlin, 1970.

60. Styles, J. A. Mutat. Res. 21: 50-51 (1973).

61. Mohn, G. R. Arch. Toxicol. 38: 109-133 (1977).

62. Loprieno, N., Barele, R., Baroncelli, C., Bronzetti, G., Cammellini, A., Corsi, G., Leporini, C., Nieri, R., and Rossi, A. M. IARC Sci. Publ. 12: 505-519 (1976).

CHAPTER 16

MUTAGENICITY OF SELECTED CHEMICALS IN SPECIFIC LOCI OF CULTURED MAMMALIAN CELLS

D. Clive, K. O. Johnson, A. G. Batson, and J. F. S. Spector

Genetic Toxicology Laboratory, The Wellcome Research Laboratories, Burroughs Wellcome Company, 3030 Cornwallis Road, Research Triangle Park, North Carolina (U.S.A.) 27709

INTRODUCTION

Specific locus mutations in mammalian cells in culture can be used to quantitate and qualitate genetic damage occurring in these higher eukaryotic cells. Qualitatively, mutations to ouabain-resistance (1-3), reversion from auxotrophy to prototrophy (4-6), or even some mutations to drug resistance (7-9) probably represent minute genomal damage (involving a small number of base pairs); at the other extreme, some thymidine kinase (TK)-deficient mutations induced in $TK^{+/-}$ heterozygous cells may represent extensive chromosomal damage involving other loci linked to the newly lost TK gene (10).

Various metabolic activation systems have been coupled with cultured mammalian cells, allowing for the detection of specific locus mutations induced by metabolites of promutagens and procarcinogens (10-15) (Table 1). This capability has led to the demonstration that, for a wide variety of chemical carcinogens, a quantitative relationship exists between in vitro mutagenic potency and in vivo oncogenicity (10,14).

Although mutations have been induced in cultured mammalian cells since 1968 (16,17), a full picture of the critical parameters involved in detecting such mutations has only recently started to emerge. For the most widely utilized locus--that specifying the enzyme hypoxanthine-guanine phosphoribosyl transferase (HGPRT)--the conditions necessary for complete expression of the mutant phenotype [resistance to 8-azaguanine (AG) or 6-thioguanine (TG)] are complex (18), the time required is usually 6 days or longer (19,20) and dispersion of the cells prior to selection is necessary for optimal recovery of the mutants (20). Controversy still exists as to the best selective agent for this marker (21,22). Prior to 1975, all of the mutagenesis experiments utilizing the HGPRT locus involved the standard protocol of Chu and Malling (16) wherein mutagen-treated cells attached to the culture dish were allowed to grow, divide and express in situ. The resulting microcolonies, probably containing a segregation-derived mixture of HGPRT-mutant and wild-type cells, were then exposed to 8-azaguanine (AG) to kill off the nonmutant cells. Typically an expression delay was observed immediately after treatment. Thereafter the AG-resistant presumptive mutant frequency would rise to a maximum, usually by 48 hr, then sharply decline. This decline is usually attributed to metabolic cooperation between wild-type and mutant cells (23,24), either within the same microcolony or among colonies.

When cells are subcultured (or maintained in suspension culture) no AG- or TG-resistant mutants are found at 48 hr expression time (19,25)--the optimal expression time for in situ selection (16) (without cell dispersion). The explanation for this is not clear. The implication, however, is a temporary reluctance to put in situ AG-selection data on a par with better defined mutagenesis experiments. This boycott will persist until it is convincingly demonstrated that in situ selection merely shortens the normally long AG-resistance expression time without either producing nonmutant phenocopies or seriously suppressing the optimal mutant frequency.

Thus, a majority of the specific locus mutagenicity literature on cultured mammalian cells must await re-evaluation. Since part of the purpose of this Comparative Chemical Mutagenesis Workshop is to evaluate the performance and potential of various mutagen assay systems, the complete exclusion of pre-1975 HGPRT data on the grounds given would grossly underestimate the potential of cultured mammalian cell systems. On the other hand, to place these HGPRT results in comparison with other experiments lacking these problems without proper identification and qualification might be misleading when quantitative comparisons are being made among systems.

A proper balance may be achieved in the following fashion. When more reliable data are available for a given chemical's mutagenicity at specific loci of cultured mammalian cells, the less reliable HGPRT data are completely excluded. Where in situ AG selection affords the best available data, they are used. In these instances, reference is made to the qualified nature of the data and to the fact that such data should reflect no more than a plus-minus sort of response for that chemical.

This dearth of reliable specific locus mutagenicity data has the effect of seriously but deceptively weakening the data base for these systems. The serious effect lies in the poor showing that these systems make relative to other mutagenicity assays; that this is misleading can be gleaned from the fact that most of the better quality data has been acquired in a space of only a few months and often as a direct outcome of the deficiencies in the published literature. This forced approach suffers from the deficiencies normally inherent in relying too heavily on one laboratory and one (or at best two) loci (26-28) for drawing general conclusions. All of these recent results will be published (or at least in manuscript from) by the time of the workshop this fall.

Other problems have been encountered in the literature which have been frustrating but hopefully will eventually prove beneficial. It is depressing, for example, to find one of the finest MNNG studies (29) rendered useless by the omission of the actual 16 doses of mutagen used, these being summarized as "(0.2-2.0 μg/ml)." All of the selection and expression conditions were carefuly worked out, 16 different mutant frequencies were plotted against the logarithim of 16 different survivals, and a correlation coefficient of 0.90 was calculated between mutant frequency and log survival. The extraction of the data relevant to this Workshop's objectives was impossible because of the absence of the most common item of all, a dose-response relationship!

As an outgrowth of such occurrences a number of criteria were established for the selection and evaluation of specific locus mutagenicity data in cultured mammalian cells. These are specified below.

MINIMAL ACCEPTABLE CRITERIA FOR MUTAGENICITY TESTING

Dose-Response Curve

1. A measure of the lowest effective concentration (LEC) should be extractable by interpolation of actual data points.

2. Predetermined upper-dose endpoint criteria should be established as a routine component of each system. These criteria should take into account the characteristics and quirks of each system and chemical; their application should have been empirically demonstrated as optimizing extrapolability to some relevant ideal (such as risk to man). For cultured mammalian cells these could be either (a) 90% kill or (b) the presence of a precipitate which persists throughout the entire treatment time.

Table 1. Ranking of mutagens at specific loci of cultured mammalian cells

Chemical[a]	Cells	Locus (marker)	Selection	S-9	LEC (μg/ml)	Ranking
Aflatoxin B1	V79	HGPRT	6-TG	+	0.07	3
Cyclophosphamide	L5178Y	$TK^{+/-}$	TFT	+	1.0	10
DEN	L5178Y	$TK^{+/-}$	BUdR	+	25	15
DMN	L5178Y	$TK^{+/-}$	BUdR	+	0.5	8
EMS	L5178Y	$TK^{+/-}$	BUdR	+ or -	2	13
ICR-170	V79	HGPRT	8-azg	-	<0.1	4
MMC[d]	L5178Y	$TK^{+/-}$	TFT	-	0.02	2
MMC[d]	L5178Y	$TK^{+/-}$	TFT	+	0.5	8
MMS[d]	L5178Y	$TK^{+/-}$	BUdR	-	1.0	10
MMS[d]	L5178Y	HGPRT	6-TG	-	1.0	10
MNNG[d]	L5178Y	$TK^{+/-}$	BUdR	-	0.004	1
MNNG[d]	L5178Y	$TK^{+/-}$	BUdR	+	0.15	5
Myleran	L5178Y	$TK^{+/-}$	TFT	+ or -	∿7.5	14
Natulan	L5178Y	$TK^{+/-}$	TFT	+	0.3	7
TEM	L5178Y	$TK^{+/-}$	BUdR	-	0.2	6

[a]Chemicals not tested: cadmium salts; epichlorohydrin; ethlenimine;

[b]Mutagenic potency = number of induced mutants/10^6 survivors/μg

[c]f^i at LD_{90} = number of induced mutants/10^6 survivors evaluated

[d]Because of large discrepancies two separate evaluations are

ID_{50} (μg/ml)	Ranking	Mutation potency ($\times 10^6$)[b]	Ranking	f^i at ID_{90} ($\times 10^6$)[c]	Ranking	Average ranking
0.1	2	200	4	170	13	6
4	10	20-33	10	1500	1	8
750	15	0.5	14	>475	12	14
3.0	9	33	9	1300	2	7
200	14	0.26	15	1100	5	12
<1.0	5	380	3	740	10	6
0.10	2	1088	2	800	8	4
1.0	7	45	7	500	11	8
∿5	11	10	11	800	8	10
∿5	11	1.3	13	70	15	12
<0.02	1	12,000	1	1000	6	2
0.8	6	60	5	1000	6	6
∿15	13	8	12	∿1200	3	10
2	8	60	5	∿1200	3	6
0.2	4	40	8	∿100	14	8

MeTEPA; TEPA; Thio-TEPA; Trenimon; and vinyl chloride.

mutagen/ml/hr treatment.

at approximately 10% survival dose.

tabulated.

3. There should be sufficient dose points between these extremes so as to provide kinetic information which might be relevant to interpreting mechanism of action.

4. Quantitation of survival should be determined over the full range of relevant doses studied.

System Analysis

1. The test system (cell line; selective agent; conditions of exposure, expression and cloning) should have been adequately analyzed.

2. During the developmental phase of each system the variants which serve as the mutational end point should have been characterized as to (a) spontaneous and induced "mutation" rates to the altered phenotypes; (b) persistence of the altered phenotype under non-selective conditions; (c) reversion rates of the variant to the original phenotype, either spontaneous or induced; (d) enzymatic characterization of a representative sampling of variants, where this is feasible.

3. Sufficient cell numbers should be treated and cloned for variant number and plating efficiency so that measured variant frequencies are statistically meaningful.

4. Any given mutagen assay should be run according to a well characterized protocol; or any and all significant alterations should be justified with actual data or references to published data.

AFLATOXIN B1 (CAS #1162-65-8)

LEC

1. Without S-9, aflatoxin B1 is cytotoxic above 10 μg/ml for a 3 hr exposure (LD_{90} = 75 μg/ml) in Chinese hamster cells, and the induced mutation frequency to 8-azaguanine-resistance (presumptive $HGPRT^+ \rightarrow HGPRT^-$ mutations) was at least 20-fold over the background (30).

2. With S-9, LD_{50} = 0.1 μg/ml in V79 Chinese hamster cells; this dose induced a 40-fold increase in thioguanine-resistant variant colonies (presumptive $HGPRT^+ \rightarrow HGPRT^-$ mutations) (15).

Neither of these studies meet the minimal acceptable mutagenicity standards referred to in the Introduction, and good quantitation of specific locus mutagenicity is not possible beyond the crude upper estimates of the LEC already mentioned. Neither abstract provides any protocol information, especially with regard to expression time or subculturing and cell resuspension prior to selection. All of these factors are critical to quantitating the HGPRT systems (19,20,25,31).

Aflatoxin B1 must simply be regarded as mutagenic at low (0.1 μg/ml) doses at the HGPRT locus in V79 Chinese hamster cells.

CADMIUM SALTS

There were no known data on the specific locus mutagenicity of cadmium salts in cultured mammalian cells at the time of this workshop (1977).

CYCLOPHOSPHAMIDE (CYTOXAN; ENDOXAN; CAS #50-18-0)

As of February 1977, there were no published data on the specific locus mutagenicity of cyclophosphamide (Cytoxan; Endoxan) in cultured mammalian cells. Subsequently published results from our laboratory (10) using the L5178Y/$TK^{+/-}$ mouse lymphoma system with or without S-9 indicate that this chemical is strongly mutagenic in the presence of S-9, but is neither toxic nor mutagenic in the absence of such metabolic activation (Table 2).

LEC

For a 4.0 hr exposure, 1.0 μg trifluorothymidine (TFT)/ml selection--which gives equivalent but cleaner results than 50 μg BUdR/ml--and 48 hr expression, the $TK^{+/-} \rightarrow TK^{-/-}$ assay appeared able to detect an LEC of 1.0 μg cyclophosphamide/ml in the presence of S-9 and an LEC of >1000 μg cyclophosphamide/ml in the absence of S-9.

Mutagenic Potency

Without S-9, the mutagenic potency (number of induced mutants/10^6 survivors/μg cyclophosphamide/ml/hr of exposure) of cyclophosphamide was 0.1; with S-9, it was 15-62.

DIETHYLNITROSAMINE (DEN; CAS #55-18-5)

As of February 1977, there were no published data on the specific locus mutagenicity of DEN in cultured mammalian cells. Subsequently published results from our laboratory (10) using the L5178Y/$TK^{+/-}$ mouse lymphoma system coupled with S-9 indicate that this chemical is mutagenic.

Table 2. Comparative specific locus mutagenicity of

Cells	Locus	Selection	Exposure (hr)	Metabolic activation[a]	Expression (hr)
L5178Y	$TK^{+/-}$	1 μg TFT/ml (same as 50 μg BUdR/ml)	4.0	None	48
L5178Y	$TK^{+/-}$	1 μg TFT/ml (same as 50 μg BUdR/ml	4.0	I/R	48
L5178Y	$TK^{+/-}$	1 μg TFT/ml (same as 50 μg BUdR/ml)	4.0	I/R	72

[a] I = induced; N = noninduced; /M = mouse liver; /R = rat liver.

[b] f ($\times 10^6$) = number of mutants/10^6 survivors.

[c] Number of induced mutants/10^6 survivors/μg/ml/hr of treatment.

LEC

In the presence of Aroclor-induced rat liver S-9, the lowest tested dose of 50 μg DEN/ml for a standard 4 hr exposure was significantly positive in the $TK^{+/-} \rightarrow TK^{-/-}$ assay (3 days' expression) using BUdR (50 μg/ml) as the selective agent (10). This dose gave an 88% survival and 86 induced mutants/10^6 survivors (after subtracting a background of 78). At the HGPRT locus, using 5 μg thioguanine/ml as the selective agent and 6 days' expression, an (interpolated) LEC of 25 μg/ml was observed. The S-9 capability for converting DEN to toxic and mutagenic form(s) appeared to be saturated above 500 μg DEN/ml, while deviations from linearity of the dose-response curve appeared a doses in excess of 100 μg/ml.

Mutagenic Potency

Over the range of concentrations tested in our laboratory (50-2500 μg/ml) the mutagenic potency (frequency of induced

cyclophosphamide (Endoxan; Cytoxan)

Mutagen concn (μg/ml)	Survivors (% of control)	f (×10⁶)[b]	Mutagenic potency (×10⁶)[c]	LEC (μg/ml)	References
0	100	99	---	>1000	10
500	94	109	0.01		
750	97	117	0.01		
1000	101	130	0.01		
0	100	93	---	1.0	10
10	6.3	1159	27		
20	0.09	1280	15		
0	100	163	---	1-2	10
2.0	55	320	20		
4.0	46	710	34		
6.0	28	1657	62		
8.0	21	1356	37		
10.0	18	1495	33		

S-9 unless otherwise indicated.

mutations per survivor/μg DEN/ml/hr of treatment) at the TK locus, 3 days' expression, decreased from 0.43×10^{-6} (at 50 μg/ml) and 0.36×10^{-6} (at 100 μg/ml) to 0.040×10^{-6} (at 2500 μg/ml) (10). This decline resulted from the above-mentioned saturation of the S-9 as reflected by a constant induced mutant frequency over the range of 500-2500 μg DEN/ml. Nearly identical values for mutagenic potency were observed at the HGPRT locus (6 days' expression).

DIMETHYLNITROSAMINE (DMN; CAS #62-75-9)

LEC

The lowest effective dose of DMN in L5178Y mouse lymphoma cells by the $TK^{+/-} \rightarrow TK^{-/-}$ assay (Table 4) was:

1. 0.5 μg/ml using noninduced rat liver S-9 and 4 hr exposure (10);

Table 3. Comparative specific locus mutagenicity of diethylnitrosamine

Cells	Locus	Selection	Exposure (hr)	Metabolic activation[a]	Expression (hr)
L5178Y	$TK^{+/-}$	50 μg BUdR/ml	4.0	I/R	72
L5178Y	HGPRT	5 μg TG/ml	4.0	I/R	144

[a]I = induced; N = noninduced; /M = mouse liver; /R = rat liver.

[b]f ($\times 10^6$) = number of mutants/10^6 survivors.

[c]Number of induced mutants/10^6 survivors/μg/ml/hr of treatment.

2. 4 μg/ml using Aroclor-induced rat liver S-9 and 4 hr exposure (10); and

3. 5 μg/ml using noninduced mouse liver $CaCl_2$-aggregated microsomes, high (1.6 mg/ml) NADPH concentration and 15 min exposure (11).

No other specific loci in mammalian cells in culture have adequately examined the mutagenicity of DMN.

Mutagenic Potency

The mutagenic potency (number of induced mutants/10^6 survivors/μg DMN/ml/hr) in L5178Y mouse lymphoma cells, $TK+/^- \rightarrow TK^-/^-$ assay was:

1. 85 ± 10 using noninduced mouse liver $CaCl_2$-aggregated microsomes, high (1.6 mg/ml) concentration of NADPH, and 15 min exposure (11);

Mutagen concn (μg/ml)	Survivors (% of control)	f (×10[6])[b]	Mutagenic potency (×10[6])[c]	LEC (μg/ml)	Remarks	References
0	100	78	-	50	S-9 saturated by 500 μg DEN per ml	10
50	88	163	0.43			
100	78	220	0.36			
250	86	292	0.21			
500	58	430	0.18			
1000	44	423	0.09			
2500	30	475	0.04			
0	100	20	-	25	S-9 saturated by 500 μg DEN per ml	10
50	88	118	0.49			
100	78	112	0.23			
250	86	198	0.18			
500	58	280	0.13			
1000	44	370	0.09			
2500	30	383	0.04			

S-9 unless otherwise indicated.

2. 33 ± 4 using noninduced rat liver S-9 and 4 hr exposure (10); and

3. 2.7 ± 0.6 using Aroclor-induced rat liver S-9 and 4 hr exposure (10).

No other specific loci in cultured mammalian cells have adequately explored the mutagenicity of DMN.

EPICHLOROHYDRIN (CAS #106-89-8)

There were no known data on the specific locus mutagenicity of epichlorohydrin in cultured mammalian cells at the time of this workshop (1977).

ETHYLENIMINE (CAS #151-56-4)

There were no known data on the specific locus mutagenicity

Table 4. Comparative specific locus mutagenicity of dimethylnitrosamine

Cells	Locus	Selection	Exposure (hr)	Metabolic activation[a]	Expression (hr)
L5178Y	$TK^{+/-}$	50 μg BUdR/ml	0.25	N/M $CaCl_2$-aggregated microsomes	48-60
L5178Y	$TK^{+/-}$	50 μg BUdR/ml	4.0	N/R	72
L5178Y	$TK^{+/-}$	50 μg BUdR/ml	4.0	I/R	72

[a]I = induced; N = noninduced; /M = mouse liver; /R = rat liver.

[b]f ($\times 10^6$) = number of mutants/10^6 survivors.

[c]Number of induced mutants/10^6 survivors/μg/ml/hr of treatment.

of ethylenimine in cultured mammalian cells at the time of this workshop (1977).

ETHYL METHANESULFONATE (EMS; CAS #62-50-0)

An abundance of published data exists for the mutagenicity of EMS at specific loci in cultured mammalian cells, and the problem has been one of representative and illustrative choices.

Linear vs. Nonlinear Dose-Response Curves

Hsie et al. (25) have shown that EMS-induced mutations are detected at the X-linked (and hence functionally monosomal) HGPRT locus of Chinese hamster ovary (CHO) cells with linear kinetics over a survival range of 0.3-100% of the untreated control, using

Mutagen concn (μg/ml)	Survivors (% of control)	f ($\times 10^6$)[b]	Mutagenic potency ($\times 10^6$)[c]		LEC (μg/ml)	Remarks	References
0	100	100	-		5	1.6 mg NADPH/ml in activation mixture	11
7.4	78	240	76)				
14.8	53	480	100)				
22.2	47	530	77)	85±10			
37	36	900	86)				
74	10	1720	88)				
0	100	65	-		0.5		10
1.0	66	218	37)				
3.0	44	436	31)	33±4			
10	7.8	1324	30)				
0	100	16	3.3)		4		10
9.2	45	137	2.9)	2.7±0.6			
18	33	233	2.6)				
27	20	402	2.0)				
74	10	618					

S-9 unless otherwise indicated.

6-thioguanine (TG, 1.7 μg/ml) as the selective agent and 168 hr expression (Table 5). Identical linear kinetics are also observed at both the HGPRT (5 μg TG/ml selection) and the autosomal, heterozygous TK (50 μg BUdR/ml selection) loci of L5178Y mouse lymphoma cells at 144 hr expression (10). (For other reasons, the cells were treated in the presence of S-9, but this had little or no effect on the results as compared to no-S-9 treatment conditions.) In these three systems (CHO/HGPRT; L5178Y/HGPRT; L5178Y/TK; all at 144-168 hr expression) the mutagenic potencies were, respectively, 0.26 ± 0.07; 0.23 ± 0.03; and 0.24 ± 0.04 induced mutants/ 10^6 survivors/μg EMS/ml/hr of treatment, averaged over the complete dose ranges studied (Table 5).

Expression is complete by 48 hr at the TK locus of L5178Y cells and a nonlinear dose-response relationship exists for this

Table 5. Comparative specific locus mutagenicity of EMS

Cells	Locus	Selection	Exposure (hr)	Metabolic activation[a]	Expression (hr)
CHO	HGPRT	1.7 μg TG/ml	16	None	168
L5178Y	HGPRT	30 μg TG/ml	2.0	None	144-192
L5178Y	HGPRT	5 μg TG/ml	4.0	I/R	144

Mutagen concn (μg/ml)	Survivors (% of control)	f (×10^6)[b]	Mutagenic potency (×10^6)[c]	LEC (μg/ml)	Remarks	References
0	100	4	--	2	Subcultured during expression	25
12.5	100	56	0.26			
25	96	147	0.36			
50	96	282	0.35			
100	85	310	0.19		Avg. mutagenic potency = (0.26 ± 0.07) × 10^{-6}; linear dose-response curve	
200	58	810	0.25			
400	8	1152	0.18			
800	0.3	2639	0.21			
0	100	3	--	∿100	Avg. mutagenic pogency = (0.055 ± 0.015) × 10^{-6}; TG unnecessarily high	2
508	93	42	0.038			
1016	75	134	0.064			
2046	14	256	0.063			
0	100	1	--	5	Avg. mutagenic potency = (0.23 ± 0.03) 10^{-6}; linear dose-response	10
100	63	81	0.20			
300	48	271	0.22			
700	6.4	730	0.26			
900	2.3	896	0.25			

(continued)

Table 5. (continued)

Cells	Locus	Selection	Exposure (hr)	Metabolic activation[a]	Expression (hr)
L5178Y	$TK^{+/-}$	50 μg BUdR/ml	2.0	None	72
L5178Y	$TK^{+/-}$	50 μg BUdR/ml	4.0	I/R	72
L5178Y	$TK^{+/-}$	50 μg BUdR/ml	4.0	I/R	144
L5178Y	$Oua^s \rightarrow Oua^r$	1.0 mM ouabain	2.0	None	24-96 (little effect on results)
V79	$Oua^s \rightarrow Oua^r$	1.0 mM ouabain	2.0	None	50h(?)

[a]I = induced; N = noninduced; /M = mouse liver; /R = rat liver.

[b]f ($\times 10^6$) = number of mutants/10^6 survivors.

[c]Number of induced mutants/10^6 survivors μg/ml/hr of treatment.

Mutagen concn (μg/ml)	Survivors (% of control)	f (×10⁶)[b]	Mutagenic potency (×10⁶)[c]	LEC (μg/ml)	Remarks	References
0	100	41	--	50	Nonlinear dose-response	10
248	60	313	0.55			
496	54	505	0.47			
744	35	642	0.40			
992	21	887	0.42			
1240	12	1212	0.47			
1488	5.5	1972	0.65			
1736	3.0	2640	0.65			
0	100	30	--	25	Nonlinear dose-response	10
100	63	153	0.31			
300	48	345	0.26			
700	6.4	1302	0.45			
900	2.3	1993	0.55			
0	100	50	--	30	Linear dose-response avg. mutagenic potency = 0.24 ± 0.04; same as HGPRT	10
100	63	158	0.27			
300	48	299	0.21			
700	6.4	701	0.23			
900	2.3	1040	0.28			
0	100	≦5	--	250	Low sensitivity marker in L5178Y	2
508	93	14	0.009			
1016	65	22	0.009			
2046	17	29	0.007			
0	100	0	--	∿20	Higher sensitivity than L5178Y	32
510	90-100	120-250	0.12-0.25			
1020	70-100	400-600	0.20-0.30			
1540	30-70	800-2000	0.26-0.65			
2050	10-25	1700-2300	0.42-0.56			

S-9 unless otherwise indicated.

locus at shorter (48 or 72 hr) expression times (Table 5). At these shorter expression times the mutagenic potencies are higher than at 144 hr, and the extent of this difference increases with dose. This has been interpreted as representing a time-dependent dilution of slower growing (chromosomal?) TK-deficient mutants (10), suggesting the possibility that rapidly expressing autosomal specific locus mutations may serve to detect chromosomal as well as gene mutations.

Effect of Concentration of Selective Agent

Cole and Arlett (2) have chosen 30 μg TG/ml as selection conditions for detecting $HGPRT^-$ mutants in L5178Y mouse lymphoma cells. The resulting mutagenic potency is 0.055 ± 0.015 induced mutations/10^6 survivors/μg EMS/ml/hr of treatment over the dose range studied. This same study indicates that, had 5 μg TG/ml been used their mutant frequencies--and hence mutagenic potency--would have increased 4-5 fold, bringing these values up to those observed by Hsie et al. (25) and Clive (10).

Sensitivity of Genetic Markers

From what has been already said it is clear that comparable quantitative results can be obtained at the HGPRT locus in two different cell types and in three laboratories. Further, the TK locus yields results which are quantitatively indistinguishable from those seen at the HGPRT locus when compared at similar expression times. These all represent indiscriminate forward mutations to a completely dispensable gene, and thus to a relatively large genetic target.

On the other hand, mutation to ouabain-resistance requires a fastidious alteration to a gene product (Na^+-K^+-ATPase) which has decreased sensitivity to the normally inhibitory ouabain, yet which still functions enzymatically to maintain viability of this mutated cell. The probability that both criteria are simultaneously met should depend on the amino acid sequence in the starting target enzyme and on how readily critical sites in the corresponding gene are appropriately mutated to yield a ouabain-resistant but otherwise functional structure.

With these factors in mind, the great differences seen in the mutagenic potencies and in LEC's of EMS at this marker in cells of two different rodent species becomes understandable. It also serves as a warning that the use of such mutagenically fastidious markers must be approached with caution, not only from cell type to cell type, but also from mutagen to mutagen. It is unlikely, for example, that frameshift mutations would be readily detected in such systems (although frameshift mutagens might produce detectable base-pair substitutions by affecting the fidelity of repair and in this manner make their secondary mutagenic potential known).

From Table 5 the following quantitative measures can be evoked.

LEC

1. CHO cells, $HGPRT^+ \rightarrow HGPRT^-$: ∿2 μg EMS/ml at 16 hr exposure, 1.7 μg TG/ml selection (25).

2. L5178Y cells, $HGPRT^+ \rightarrow HGPRT^-$:

   (i) 5 μg EMS/ml at 4 hr exposure, with or without S-9, 5 μg TG/ml selection (10).

   (ii) ∿100 μg EMS/ml at 2 hr exposure, 30 μg TG/ml selection (2).

3. L5178Y cells, $TK^{+/-} \rightarrow TK^{-/-}$: 25-30 μg EMS/ml at 2-4 hr exposure, with or without S-9, 72 or 144 hr expression, 50 μg BUdR/ml selection (26,10).

4. L5178Y cells, $Oua^s \rightarrow Oua^r$: 250 μg EMS/ml at 2 hr exposure, 1.0 mM ouabain selection (2).

5. V79 cells, $Oua^s \rightarrow Oua^r$: ∿20 μg EMS/ml at 2 hr exposure, 1.0 mM ouabain selection (32).

Mutagenic Potency

1. CHO cells, $HGPRT^+ \rightarrow HGPRT^-$: 0.26 ± 0.07 at 16 hr exposure, 1.7 μg TG/ml selection (25).

2. L5178Y cells, $HGPRT^+ \rightarrow HGPRT^-$:

   (i) 0.23 ± 0.03 at 4.0 hr exposure, with or without S-9, 5 μg TG/ml selection (10).

   (ii) 0.055 ± 0.015 at 2.0 hr exposure, 30 μg TG/ml selection (2).

3. L5178Y cells, $TK^{+/-} \rightarrow TK^{-/-}$:

   (i) 0.24 ± 0.04 at 4.0 hr exposure, with or without S-9, 144 hr expression, 50 μg BUdR/ml selection (26,36).

Table 6. Comparative specific locus mutagenicity of ICR-170

Cells	Locus	Selection	Exposure (hr)	Metabolic activation[a]	Expression (hr)
V79	HGPRT	30-50 μg AG/ml	2.0	None	42

[a]I = induced; N = noninduced; /M = mouse liver; /R = rat liver.

[b]f ($\times 10^6$) = number of mutants/10^6 survivors.

[c]Number of induced mutants/10^6 survivors μg/ml/hr of treatment.

(ii) 0.26-3.4 at 2 or 4 hr exposure, with or without S-9, 48-72 hr expression, 50 μg BUdR/ml selection. This is a decidedly non-linear dose response (10).

4. L5178Y cells, $Oua^s \rightarrow Oua^r$: 0.008 ± 0.001 induced mutants/10^6 survivors at 2.0 hr exposure, 1.0 mM ouabain selection (20).

5. V79 cells, $Oua^s \rightarrow Oua^r$: 0.12-0.56 at 2.0 hr exposure, 1.0 mM ouabain selection; nonlinear dose-response (32).

ICR-170 (CAS #146-59-8)

As of February 1977, there were only three published references to specific locus mutagenicity of ICR-170 in cultured mammalian cells; none of these satisfied the minimal acceptable mutagenicity standards referred to in the Introduction. As a consequence, it is possible to say only that ICR-170 appears mutagenic at the HGPRT locus in cultured V79 Chinese hamster cells (33-35), with only primitive approximations of quantitation (such as LEC or mutagenic potency) possible.

Mutagen concn (μg/ml)	Survivors (% of control)	f $(\times10^6)^b$	Mutagenic potency $(\times10^6)^c$	LEC (μg/ml)	Remarks	References
0	100	10.4	--	Probably	No sub-	35
0.96	8.9	742	∿380	<0.1	culturing or resuspension prior to selection	

S-9 unless otherwise indicated.

METEPA (CAS #57-39-6)

There were no known data on the specific locus mutagenicity of METEPA in cultured mammalian cells at the time of the workshop (1977).

METHYL METHANESULFONATE (MMS; CAS #66-27-3)

Although many references exist for MMS mutagenicity to specific loci in cultured mammalian cells, they suffer from remarkably diverse deficiencies for our present purposes of quantitation, such as only a single dose being studied; doses not explicitly given with the mutant frequencies and survivals; or lack of subculturing for the determination of azaguanine-resistant mutant frequencies. Consequently we have had to rely heavily on unpublished data for this mutagen.

LEC

1. L5178Y cells, $TK^{+/-} \rightarrow TK^{-/-}$ assay (Table 7):

   (i) 1.0 μg MMS/ml at 4.0 hr exposure, 50 μg BUdR/ml selection, 72 hr expression (10).

   (ii) 4.0 μg MMS/ml at 4.0 hr exposure, 50 μg BUdR/ml selection, 144 hr expression (10).

Table 7. Comparative specific locus mutagenicity of MMS

Cells	Locus	Selection	Exposure (hr)	Metabolic activation[a]	Expression (hr)
L5178Y	$TK^{+/-}$	50 μg BUdR/ml	4.0	I/R	72
L5178Y	$TK^{+/-}$	50 μg BUdR/ml	4.0	I/R	144
L5178Y	HGPRT	5 μg TG/ml	4.0	I/R	144
V79	HGPRT	10 μg AG/ml,	--	None	0
		dialyzed serum,	1.0	None	100
		reseeding	2.0	None	110
V79	$Oua^s \rightarrow Oua^r$	1.0 mM ouabain	0.5	None	50-60?

[a]I = induced; N = noninduced; /M = mouse liver; /R = rat liver.

[b]f ($\times 10^6$) = number of mutants/10^6 survivors.

[c]Number of induced mutants/10^6 survivors/μg/ml/hr of treatment.

Mutagen concn (μg/ml)	Survivors (% of control)	f (×10[6])[b]	Mutagenic potency (×10[6])[c]	LEC (μg/ml)	Remarks	References
0	100	41	--	1.0		10
7.5	35	354	10.5			
10	20	393	8.8			
15	7.4	863	13.8			
20	2.2	1467	18			
22.5	1.6	1491	16			
0	100	69	--	4.0		10
7.5	35	189	4.0			
10	20	296	5.8			
15	7.4	372	5.0			
20	2.2	478	5.0			
22.5	1.6	806	8.2			
0	100	7	--	1.0		10
7.5	35	52	1.5			
10	20	24	0.4			
15	7.4	77	1.2			
20	2.2	125	1.5			
22.5	1.6	133	1.4			
0	100	0-6	--	5-10		20
55	95	40	0.73			
110	6	230	1.05			
0	0	0	--	∿25	Nonlinear dose-response	32
132	80; 80	12; 20	0.18; 0.30			
198	--; 50	--; 60	--; 0.61			
264	8; 4	200; 700	1.52; 5.30			

S-9 unless otherwise indicated.

Table 8. Comparative specific locus mutagenicity of MNNG

Cells	Locus	Selection	Exposure (hr)	Metabolic activation[a]	Expression (hr)
V79	HGPRT	30 μg AG/ml	2.0	None	144 (dispersed at 72 hr)
L5178Y	$Asn^- \rightarrow Asn^+$	Asn-free medium	2.0	None	72
L5178Y	$TK^{+/-}$	50 μg BUdR/ml	2.0	None	72
L5178Y	$TK^{+/-}$	50 μg BUdR/ml	4.0	I/R* N/R†	72
L5178Y	HGPRT	5 μg TG/ml	4.0	I/R* N/R†	168

[a]I = induced; N = noninduced; /M = mouse liver; /R = rat liver.

[b]f ($\times 10^6$) = number of mutants/10^6 survivors.

[c]Number of induced mutants/10^6 survivors/μg/ml/hr of treatment.

Mutagen concn (μg/ml)	Survivors (% of control)	f ($\times 10^6$)[b]	Mutagenic potency ($\times 10^6$)[c]	LEC (μg/ml)	Remarks	References
0.0	100	175	--	0.02		36
0.0147	96	185	700			
0.044	83	660	11,000			
0.147	71	1300	7000			
0.44	59	3100	6600			
1.47	32	3400	2200			
0.0	100	3	--	0.01		4
0.037	85	31	360			
0.074	35	56	360			
0.0	100	100	--	0.004		10
0.026	14	717	12,000			
0.0.47	6.2	1316	13,000			
0.082	1.3	2006	11,500			
0.0	0	44	--	0.15	S-9 detoxifies MNNG; approx. 50× high doses required. Mimics in vivo situation	10
0.78†	60	230	60			
0.78*	50	99	18			
1.6†	42	425	60			
3.1*	14	890	68			
3.1†	6.8	1729	136			
0.0	0	3	--	0.1-0.2	S-9 detoxifies MNNG; approx. 50× high doses required. Mimics in vivo situation	10
0.78†	60	22	6.1			
0.78*	50	19	5.1			
1.6†	42	22	3.0			
3.1*	14	93	7.3			
3.1†	6.8	153	12.1			

S-9 unless otherwise indicated.

2. L5178Y cells, $HGPRT^{+} \rightarrow HGPRT^{-}$: 1.0 μg MMS/ml at 4.0 hr exposure, 5 μg TG/ml selection, 144 hr expression (10).

3. V79 cells, $HGPRT^{+} \rightarrow HGPRT^{-}$: 5-10 μg MMS/ml, at 1-2 hr exposure, 10 μg AG/ml (using dialyzed serum) and reseeding of cells during 100-110 hr expression (20).

4. V79 cells, $OUA^{s} \rightarrow OUA^{r}$: ∿25 μg MMS/ml, at 0.5 hr exposure, 1.0 mM ouabain selection, 50-60 (?) hr expression (32).

Mutagenic Potency

The mutagenic potency (number of induced mutants/10^6 survivors/μg MMS/ml/hr of exposure) was:

1. L5178Y cells, $TK^{+/-} \rightarrow TK^{-/-}$:

   (i) 9-18 at 4.0 hr exposure, 50 μg BUdR/ml selection, 72 hr expression, with a suggestion of a nonlinear dose-response (10).

   (ii) 4-8 at 4.0 hr exposure, 50 μg BUdR/ml selection, 144 hr expression, with a suggestion of a nonlinear dose-response (10). (This represents approximately 50% of the mutagenic potency seen at 72 hr expression.)

2. L5178Y cells, $HGPRT^{+} \rightarrow HGPRT^{-}$: 0.4-1.4 at 4.0 hr exposure, 5.0 μg TG/ml selection, 144 hr expression. There was no significant evidence of nonlinear dose-response (10).

3. V79 cells, $HGPRT^{+} \rightarrow HGPRT^{-}$: 0.7-1.1 at 1-2 hr exposure, 10 μg AG/ml (using dialyzed serum) and reseeding of cells during 100-110 hr expression (20).

4. V79 cells, $OUA^{s} \rightarrow OUA^{r}$: 0.2-5 at 0.5 hr exposure, 1.0 mM ouabain selection, 50-60 (?) hr expression, with markedly nonlinear dose-response (32).

N-METHYL-N'-NITRO-N-NITROSOGUANIDINE (MNNG; CAS #70-25-7)

LEC

1. L5178Y mouse lymphoma cells, $TK^{+/-} \rightarrow TK^{-/-}$:

   (i) 0.004 μg MNNG/ml without S-9; 4 hr exposure (10).

(ii) 0.15 μg MNNG/ml in the presence of either Aroclor-induced or noninduced rat liver S-9, 4 hr exposure (10).

2. L5178Y mouse lymphoma cells, $HGPRT^{+} \rightarrow HGPRT^{-}$:

0.1-0.2 μg MNNG/ml in the presence of either Aroclor-induced or noninduced rat liver S-9, 4 hr exposure (10);

3. L5178Y mouse lymphoma cells, $Asn^{-} \rightarrow Asn^{+}$:

0.01 μg MNNG/ml, in the absence of S-9, 2 hr exposure (4).

4. V79 Chinese hamster cells, $HGPRT^{+} \rightarrow HGPRT^{-}$:

0.02 μg MNNG/ml in the absence of S-9; 2 hr exposure (36).

Mutagenic Potency

The mutagenic potency (number of induced mutants/10^6 survivors/μg MNNG/ml/hr exposure) was:

1. L5178Y mouse lymphoma cells, $TK^{+/-}$ $TK^{-/-}$:

(i) 12,000 in the absence of S-9 (10).

(ii) ca. 60 in the presence of either Aroclor-induced or noninduced rat liver S-9 (10).

2. L5178Y mouse lymphoma cells, $HGPRT^{+} \rightarrow HGPRT^{-}$:

ca. 6 in the presence of either Aroclor-induced or non-induced rat liver S-9 (10).

3. L5178Y mouse lymphoma cells, $Asn^{-} \rightarrow Asn^{+}$:

360 in the absence of S-9 (4).

4. V79 Chinese hamster cells, $HGPRT^{+} \rightarrow HGPRT^{-}$:

ca. 5000 in the absence of S-9 (36).

General Comments

MNNG is rapidly and effectively detoxified by liver microsomal P-450 enzymes (37) to nonmutagenic forms. This is reflected here by the 40-fold difference in LEC's and 200-fold difference

Table 9. Comparative specific locus mutagenicity of mitomycin C

Cells	Locus	Selection	Exposure (hr)	Metabolic activation[a]	Expression (hr)
L5178Y	$TK^{+/-}$	1.0 μg TFT/ml	4.0	--	72
L5178Y	$TK^{+/-}$	1.0 μg TFT/ml	4.0	I/R	72

[a]I = induced; N = noninduced; /M = mouse liver; /R = rat liver.

[b]f ($\times 10^6$) = number of mutants/10^6 survivors.

[c]Number of induced mutants/10^6 survivors/μg/ml/hr of treatment.

in mutagenic potencies for MNNG when tested with and without S-9 (10) (Table 8).

Under identical treatment conditions (i.e., with S-9) the TK locus of L5178Y mouse lymphoma cells responds nearly 10 times more sensitively than does the HGPRT locus in these same cells (10). Under similar conditions (no S-9) the HGPRT locus of V79 Chinese hamster cells is nearly 20 times as sensitive (in terms of induced mutant frequencies and mutagenic potencies) as the asn marker in L5178Y mouse lymphoma cells.

Both of these quantitative differences point out possible differences in the cross section of the respective genetic targets. First, the asn marker probably measures reverse mutations at a single base pair, since the mutants acquire previously absent asparagine synthetase activity (38). The HGPRT locus measures forward mutational events, probably at a number of target base pairs (7,9). Thirdly, the autosomal TK locus may detect chromosomal as well as gene mutations resulting in loss or suppression of thymidine kinase activity (10).

MITOMYCIN C (CAS #50-07-7)

As of February 1977, there were no satisfactory published data on the specific locus mutagenicity of mitomycin C in cul-

Mutagen concn (μg/ml)	Survivors (% of control)	f $(\times 10^6)^b$	Mutagenic potency $(\times 10^6)^c$	LEC (μg/ml)	Remarks	References
0	100	61	--	0.02		10
0.10	24	496	1088			
0.30	1.0	932	726			
0	100	59	--	∿0.5		10
0.10	93	56	<0			
0.30	73	81	18			
1.00	44	234	44			
3.00	7.6	611	46			

S-9 unless otherwise indicated.

tured mammalian cells. Results from our laboratory published since this time (10), using the L5178Y/TK$^{+/-}$ mouse lymphoma system, indicate that this chemical is strongly mutagenic at low doses (Table 9).

LEC

In the absence of S-9, at a 4 hr exposure, 1.0 μg trifluorothymidine (TFT)/ml selection (equivalent to 50 μg BUdR/ml selection) and 72 hr expression, the TK$^{+/-}$→TK$^{-/-}$ assay appeared capable of detecting 0.02 μg mitomycin C/ml. In the presence of S-9 and otherwise similar conditions, the LEC was about 0.5 μg/ml, a 25-fold increase arising most likely from the detoxification of MMC by the S-9.

Mutagenic Potency

At the lowest dose tested (0.1 μg mitomycin C/ml) the mutagenic potency was at a maximum of 1088 induced mutants/10^6 survivors/μg mitomycin C/ml/hr of exposure) in the absence of S-9. The addition of S-9 to the reaction mixture lowered the mutagenic potency by a factor of nearly 50 (at comparable survivals) to about 45.

Table 10. Comparative specific locus mutagenicity of Myleran

Cells	Locus	Selection	Exposure (hr)	Metabolic activation[a]	Expression (hr)
L5178Y	$TK^{+/-}$	1 μg TFT/ml (same as 50 μg BUdR/ml)	4.0	None	72
L5178Y	$TK^{+/-}$	1 μg TFT/ml (same as 50 μg BUdR/ml)	4.0	I/R	72

[a]I = induced; N = noninduced; /M = mouse liver; /R = rat liver.

[b]f ($\times 10^6$) = number of mutants/10^6 survivors.

[c]Number of induced mutants/10^6 survivors/μg/ml/hr of treatment.

MYLERAN (BUSULFAN; CAS #55-98-1)

As of February 1977, there were no published data on the specific locus mutagenicity of Myleran (Busulfan) in cultured mammalian cells. Results from our laboratory published since this time (10), using the L5178Y/$TK^{+/-}$ mouse lymphoma system, with or without S-9, indicate that this chemical is strongly mutagenic, and is not significantly affected by such metabolic activation (Table 10).

LEC

With a 4.0 hr exposure, 1.0 μg trifluorothymidine (TFT)/ml selection (which gives equivalent but cleaner results than 50 μg BUdR/ml) and 72 hr expression, the $TK^{+/-} \rightarrow TK^{-/-}$ assay appeared able to detect an LEC of 3.0 μg Myleran/ml in the presence of S-9 and a LEC of 6.0 μg Myleran/ml in the absence of S-9. This difference in LEC's results from a nearly twofold lower background mutant frequency in the assay run with S-9 relative to that lacking

(Busulfan)

Mutagen concn (μg/ml)	Survivors (% of control)	f (×10⁶)[b]	Mutagenic potency (×10⁶)[c]	LEC (μg/ml)	Remarks	References
0	100	193	--	6.0	Probably no difference with and without S-9	10
30	21	1022	6.9			
45	23	1186	5.0			
60	∿6	∿1450	∿5.2			
75	0.92	3658	11.6			
0	100	119	--	3.0		10
30	15	1199	9.0			
45	5.9	1979	10.3			
60	2.7	1738	6.7			
75	0.51	3913	12.6			

S-9 unless otherwise indicated.

S-9, and is belied by the nearly identical mutagenic potencies of Myleran under these different metabolic conditions.

Mutagenic Potency

Without S-9, the mutagenic potency of Myleran was 7.2 ± 3.1 induced mutants/10^6 survivors; with S-9, it was 9.6 ± 2.5. There is no significant difference between these two values, and they can be averaged to give a value of 8.4 ± 2.9 for the mutagenic potency of Myleran, with or without S-9.

NATULAN (PROCARBAZINE; CAS #366-70-1)

As of February 1977, there were no published data on the specific locus mutagenicity of Natulan (procarbazine) in cultured mammalian cells. Results from our laboratory published since this time (10) using the L5178Y/$TK^{+/-}$ mouse lymphoma system, with or without S-9, indicate that this chemical is strongly mutagenic in the presence of S-9 and is weakly mutagenic without such metabolic activation (Table 11).

Table 11. Comparative specific locus mutagenicity of Natulan (Procarbazine)

Cells	Locus	Selection	Exposure (hr)	Metabolic activation[a]	Expression (hr)
L5178Y	$TK^{+/-}$	1 μg TFT/ml (same as 50 μg BUdR/ml)	4.0	I/R	72
L5178Y	$TK^{+/-}$	1 μg TFT/ml	4.0	I/R	72
L5178Y	$TK^{+/-}$	1 μg TFT/ml	4.0	None	72

[a] I = induced; N = noninduced; /M = mouse liver; /R = rat liver.

[b] f ($\times 10^6$) = number of mutants/10^6 survivors.

[c] Number of induced mutants/10^6 survivors/μg/ml/hr of treatment.

Mutagen concn (μg/ml)	Survivors (% of control)	f (×10⁶)[b]	Mutagenic potency (×10⁶)[c]	LEC (μg/ml)	Remarks	References
0	100	111	--	0.3		10
0.5	75	278	84			
1.0	79	311	50			
4.0	23	1096	62			
10	1.9	1834	43			
30	0.10	2253	18			
60	0.09	3000	12			
100	0.07	2351	5.6			
1500	7.4	1278	0.21			
0	100	73	--	<<100		10
100	0.05	2534	6.2			
200	0.07	3200	3.9			
600	1.02	2446	1.0			
1000	3.0	1757	0.42			
1500	7.5	1454	0.23			
1600	6.1	1580	0.24			
0	100	67	--	700	Negligible mutagenicity without metabolic activation (S-9)	10
100	101	81	0.04			
300	99	103	0.03			
500	104	110	0.02			
700	56	168	0.04			
800	26	208	0.04			
900	8.7	193	0.04			

S-9 unless otherwise indicated.

Table 12. Comparative specific locus mutagenicity of TEM

Cells	Locus	Selection	Exposure (hr)	Metabolic activation[a]	Expression (hr)
L5178Y	TK	50 μg BUdR/ml	3.0	None	72
L5178Y	TK	1.0 μg TFT/ml	4.0	None	72

[a] I = induced; N = noninduced; /M = mouse liver; /R = rat liver.

[b] f ($\times 10^6$) = number of mutants/10^6 survivors.

[c] Number of induced mutants/10^6 survivors/μg/ml/hr of treatment.

LEC

With a 4.0 hr exposure, 1.0 μg trifluorothymidine (TFT)/ml selection (which gives equivalent but cleaner results than 50 μg BUdR/ml) and 72 hr expression, the $TK^{+/-} \rightarrow TK^{-/-}$ assay appeared able to detect an LEC of 0.3 μg Natulan/ml in the presence of S-9 and an LEC of 700 μg Natulan/ml in the absence of S-9.

Mutagenic Potency

With S-9, the mutagenic potency of Natulan was ≥50 at low doses, then decreased with dose (see below); without S-9, it was 0.04.

Mutagen concn (μg/ml)	Survivors (% of control)	f (×10⁶)[b]	Mutagenic potency (×10⁶)[c]	LEC (μg/ml)	Remarks	References
0	100	18	--	0.2	Nonlinear dose-response	39
0.05	100	23	33			
0.10	75	32	47			
0.20	63	38	33			
0.25	4.8	104	115			
0	76	55	--	0.004		40
0	118	65	--			
0.010	42	375	8000			
0.010	37	503	10,900			
0.031	8.4	827	6200			
0.031	7.3	874	6500			

S-9 unless otherwise indicated.

Shape of Dose-Response Curve

In the presence of S-9, concentrations of Natulan ≤200 μg/ml resulted in a normal monotonic increase in induced mutant frequency and cytotoxicity with dose. At higher concentrations both cytotoxicity and mutagenicity declined with increasing dose. This effect was observed in two separate experiments (Table 11) with S-9, but not in the absence of S-9. (Low % survivals included in Table 11 are contrary to our usual procedure but were included here to demonstrate the unusual dose-response curve.)

TRIETHYLENEMELAMINE (TEM (CAS 51-18-3)

As of November 1977, there were no published data on the specific locus mutagenicity of TEM in cultured mammalian cells. Unpublished results from Dr. Dale Matheson's laboratory at Litton Bionetics (39) and from Dr. Anne Mitchell's laboratory at Stanford Research Institute International (40) on the L5178Y/TK$^{+/-}$ mouse lymphoma system indicate that this chemical is mutagenic. No data are available using S-9 for activation or detoxification.

LEC

In the absence of S-9, a 3 hr exposure to 0.2 μg TEM/ml was detected as mutagenic at the TK locus of L5178Y mouse lymphoma cells (39) using 3 days' expression and 50 μg BUdR/ml as the selective agent. This dose gave a 63% survival and 20 induced mutants/10^6 survivors (after subtracting a background of 18). A much lower dose of 0.010 μg/ml was readily detected in essentially the same system in another laboratory (40). This dose gave a 40% survival and 378 induced mutants/10^6 survivors. A linearly extrapolated LEC of 0.004 μg/ml is estimated (Table 12).

Mutagenic Potency

In the absence of S-9, the interlaboratory variation observed with LEC extends to mutagenic potencies. The mutagenic potencies range from 33 to 115 in Matheson's laboratory (39) and from 9500 to 6400 in Mitchell's laboratory (40). The reasons for these discrepancies between laboratories are not clear at present.

TEPA (CAS 545-55-1)

There were no known data on the specific locus mutagenicity of TEPA in cultured mammalian cells at the time of this Workshop (1977).

THIO-TEPA (CAS #52-24-4)

There were no known data on the specific locus mutagenicity of Thio-TEPA in cultured mammalian cells at the time of this Workshop (1977).

TRENIMON (CAS #68-76-8)

There were no known data on the specific locus mutagenicity of Trenimon in cultured mammalian cells at the time of this Workshop (1977).

VINYL CHLORIDE (CAS #75-01-4)

There were no known data on the specific locus mutagenicity of vinyl chloride in cultured mammalian cells at the time of this Workshop (1977).

ACKNOWLEDGMENT

The authors gratefully acknowledge the contributions of the following individuals: Dr. Dale Matheson of Litton Bionetics, Kensington, Md., and Dr. Anne Mitchell of SRI, International, Menlo Park, California, who graciously permitted us to include their unpublished TEM data in this report; and to Drs. C. F. Arlett (University of Sussex, England), E. H. Chu (University of Michigan, Ann Arbor, Michigan, U.S.A.), and J. W. I. M. Simons (State University of Leiden, The Netherlands) for critically reading the manuscript. The authors acknowledge the Editors' contributions in the textual revisions.

REFERENCES

1. Baker, R. M., Burnette, D. M., Mankovitz, R., Thompson, L. H., Whitmore, G. F., Siminovitch, L., and Till, J. E. Cell 1: 9-21 (1974).

2. Cole, J., and Arlett, C. F. Mutat. Res. 34: 507-526 (1976).

3. Mankovitz, R., Buchwald, M., and Baker, R. M. Cell 3: 221-226 (1974).

4. Capizzi, R. L., Papirmeister, B., Mullins, J. M., and Cheng, E. Cancer Res. 34: 3073-3082 (1974).

5. Suzuki, N., and Okada, S. Mutat. Res. 34: 489-506 (1976).

6. Uren, J. R., Summers, W. P., and Handschumacher, R. E. Cancer Res. 34: 2940-2945 (1974).

7. Beaudet, A. L., Roufa, D. J., and Caskey, C. T. Proc. Natl. Acad. Sci. (U.S.A.) 70: 320-324 (1973).

8. Clive, D., and Voytek, P. Mutat. Res. 44: 269-278 (1977).

9. Sharp, J. D., Capecchi, N. E., and Capecchi, M. R. Proc. Natl. Acad. Sci. (U.S.A.) 70: 3145-3149 (1973).

10. Clive, D., Johnson, K. O., Spector, J. F. S., Batson, A. G., Brown, M. M. M. Mutat. Res. 59: 61-108 (1979).

11. Frantz, C. N., and Malling, H. V. J. Toxicol. Environ. Health 2: 179-187 (1976).

12. Huberman, E. Mutat. Res. 29: 285-291 (1975).

13. Huberman, E., and Sachs, L. Int. J. Cancer 13: 326-333 (1974).

14. Huberman, E., and Sachs, L. Proc. Natl. Acad. Sci.(U.S.A.) 73: 188-192 (1976).

15. Krahn, D. F., and Heidelberger, C. Proc. Am. Assoc. Cancer Res. 16: 74 (1975) (Abstract).

16. Chu, E. H., and Malling, H. V. Proc. Natl. Acad. Sci. (U.S.A.) 61: 1306-1312 (1968).

17. Kao, F. T., and Puck, T. T. Proc. Natl. Acad. Sci. (U.S.A.) 60: 1275-1281 (1968).

18. Arlett, C. F., and Harcourt, S. A. Mutat. Res. 16: 301-306 (1972).

19. Knaap, A. G. A. C., and Simons, J. W. I. Mutat. Res. 30: 97-110 (1975).

20. Myhr, B. C., and DiPaolo, J. A. Genetics 80: 157-169 (1975).

21. Cox and Masson. Mutat. Res. 36: 93-104 (1976).

22. Nikaido, O., and Fox, M. Mutat. Res. 35: 279-288 (1976).

23. Cox, R. P., Krauss, M. R., Balis, M. E., and Dancis, J. Exptl. Cell Res. 74: 251-268 (1972).

24. Subak-Sharpe, J. H., Burk, R. R., and Pitts, J. D. J. Cell Sci. 4: 353-367 (1969).

25. Hsie, A. W., Brimer, P. A., Mitchell, T. J., and Gosslee, D. G. Somatic Cell Genet. 1: 247-261 (1975).

26. Clive, D. Environ. Health Perspect. 6: 119-125 (1973).

27. Clive, D., Flamm, W. G., Machesko, M. R., and Bernheim, N. J. Mutat. Res. 16: 77-87 (1972).

28. Clive, D., and Spector, J. F. S. Mutat. Res. 31: 17-29 (1975).

29. Peterson, A. R., Krahn, D. F., Peterson, H., Heidelberger, C., Bhuyan, B. K., and Li, L. H. Mutat. Res. 36: 345-356 (1976).

30. Velȧzquez, A., deNava, C., Coutiño, R., and Pulido, I. Mutat. Res. 21: 241-242 (1973).

31. Chasin, L. A. J. Cell. Physiol. 82: 299-308 (1973).

32. Arlett, C. F., Turnbull, D., Harcourt, S. A., Lehmann, A. R., and Colella, C. M. Mutat. Res. 33: 261-278 (1975).

33. Chu, E. H. U. In: Chemical Mutagens: Principles and Methods for Their Detection, A. Hollaender, Ed., Vol. 2, New York, 1971, pp. 411-444.

34. Chu, E. H. Y., Brimer, P., and Malling, H. V. Genetics 61: S10-11 (1969).

35. Chu, E. H. Y., Brimer, P. A., Schenley, C. K., Ho, T., and Malling, H. V. Molecular Environmental Aspects of Mutagenesis, Proceedings 6th Rochester International Conference on Environmental Toxicology, 1963, 1974, pp. 178-195.

36. Harris, M. Biochem. Dis. 4: 575-590 (1974).

37. Czygan, P., Greim, H., Garro, A. J., Hutterer, F., Rudick, J., Schaffner, F., and Popper, H. J. Natl. Cancer Inst. 51: 1761-1764 (1973).

38. Colofiore, J., Morrow, J., and Patterson, M. K., Jr. Genetics 75: 503-514 (1973).

39. Matheson, D. (Litton Bionetics), personal communication.

40. Mitchell, A. (Stanford Research Institute International), personal communication.

CHAPTER 17

MUTAGENICITY OF SELECTED CHEMICALS IN MAMMALIAN IN VITRO CYTOGENETIC SYSTEMS

J. G. Brewen

Corporate Medical Affairs, Allied Chemical Corporation
Morristown, New Jersey 07960 (U.S.A.)

A. T. Natarajan

Dept. of Radiation Genetics and Chemical Mutagenesis
State University of Leiden, Sylvius Laboratories
Leiden, The Netherlands

G. Obe

Institut für Genetik der Freien Universitat Berlin
1 Berlin 33, Arnimallee 5-7, W. Germany

INTRODUCTION

The study of structural chromosome aberrations in mammalian cells grown in tissue culture offers the opportunity of observing directly the mutagenic effects of various agents without the need of maintaining an animal colony. A myriad of cell lines exists offering the investigator a wide choice of mammalian species, and in many instances a variety of karyological types within a species, e.g., unique marker chromosomes.

The mutagenic endpoints under consideration are structural changes to the chromosome. There are three broad types of chromosomal aberrations, depending upon the organizational unit of the metaphase chromosome involved in the aberration. The least understood, and consequently least considered, are the so-called subchromatid aberrations. These appear to involve a fiber of the coiled chromatid and have often been used as evidence to support the multineme model of chromosome organization. The production of sub-, or half-, chromatid aberrations is usually restricted to the mitotic portion of the cell cycle. The other two types of aberrations, namely chromosome-type and chromatid-type, are the most commonly studied. Their names, e.g., subchromatid-type, are derived from their appearance. Chromosome-type aberrations involve both the chromatids of the metaphase chromosome and the position of the aberration is identical on both chromatids. Chromatid-type aberrations, for the most part, involve only one of the two chromatids of the metaphase chromosome.

The formation of chromosome- and chromatid-type aberrations is dictated by the stage of the cell cycle. The cell cycle is generally partitioned into four parts, G_1, S, G_2, and M. The chromosomes are functionally single threads in G_1, and any aberration formed in G_1 will be a chromosome-type. During S and G_2 the chromosomes are functionally double, and the aberrations formed in these two portions of the cell cycle are chromatid-type. The transition from chromosome- to chromatid-type aberration formation has been shown to occur just prior to the onset of replicative DNA synthesis. With the exception of chromosome and isochromatid deletions, the two major types of structural aberrations are unambiguous and easily differentiated. Paired double fragments can be either a chromosome-type deletion or an isochromatid deletion. If they occur along with other chromatid-type aberrations they should be considered isochromatid deletions.

Generally speaking, there are three major classes of aberrations. These are deletions, interchanges, and intrachanges. The names describe the aberrations quite adequately as deletion implies the loss of a chromosome segment, interchanges are simply rearrangements between two different chromosomes, and intrachanges are rearrangements within one chromosome. The rearrangements are either symmetrical or asymmetrical. The asymmetrical type result in loss of chromosome material via an acentric fragment, whereas the symmetrical type conserve all the chromosome material with a centromere. For example, one type of asymmetrical chromosome-type intrachange generates a centric ring with fragment and the corresponding symmetrical type generates a pericentric inversion. Of all the aberrations that are analyzable in the light microscope, only the symmetrical inter- and intrachanges result in genetically recoverable events; all the others are lethal at either the cell or organism level.

The probability of recovery is different depending on whether the aberration is of the chromosome-, or chromatid-type. A symmetrical chromosome-type interchange (reciprocal translocation) will be recovered in all of the progeny cells, whereas a symmetrical chromatid-type interchange will only be recovered in 25% of the progeny cells due to random segregation of the four chromatids of the complex at mitosis. Symmetrical chromatid intrachanges will be recovered in 50% of the daughter cells.

There is another structural anomaly that is frequently observed after mutagen treatment that has introduced considerable confusion into the literature on the efficacy of mutagens at inducing chromosome aberrations. This anomaly goes under the guise of various names, the most common being "achromatic lesion" or "gap." In reporting data many investigators have lumped breaks, fragments, and gaps under one heading. This is a very serious misrepresentation of the results. First, "gaps" are not understood in regards to what they really are in relation to chromosome structure. Second, they have never been shown to result in true structural aberrations at subsequent cell divisions. Thus, their importance is highly questionable. If they are to be recorded, they should be done so as a separate class.

The great majority of chemical mutagens have been shown to produce only chromatid-type aberrations if the precaution is taken to analyze only cells in their first post-treatment mitosis (Table 1). This observation does not mean that these agents are effective on only S and G_2 cells and not G_1 cells. The data do show, however, that the lesions induced by most chemical agents require at least one round of chromosome replication to produce a structural aberration. This conclusion has been based on the observation that when a heterogeneous cell population is treated the G_2 cells contain no, or a few at most, aberrations, S cells contain many chromatid-type aberrations, and G_1 cells contain only chromatid-type aberrations. The obvious interpretation is that the lesions are induced in cells in all stages of the cell cycle but that only those cells that undergo chromosome replication translate the lesions into chromatid aberrations.

In vitro cytogenetic studies offer many advantages for studying the effects of known and putative mutagens. A very obvious advantage is the precision available for treating with a known concentration of a chemical for a specific period of time. Second, synchronous populations of cells can be obtained enabling the investigator to treat during a specific portion of the cell cycle. A third advantage is the short generation times that allow for an experiment to be performed within a short period of time when a cell line is used. Finally, the techniques for producing high quality chromosome preparations have been perfected and are quite simple.

Most cell types used for in vitro cytogenetic tests are amenable to adaptations involving S-9 fractions for microsomal activation studies. Thus the researcher is not limited to studying only those compounds that themselves exert a direct effect on the chromatin material. This technique expands the cytogeneticist's ability to do quick assays on potential mutagens and carcinogens.

The most commonly used cell types for in vitro cytogenetic analysis are Chinese hamster cell lines and human peripheral leukocytes. The hamster cells have the advantage of a low chromosome number and rapid proliferation, whereas the peripheral leukocyte has

Table 1. Clastogenic properties of various chemical and physical agents

Test substance	Cell origin	Aberration type	Cell stage-specific	Lowest effective dose tested	Microsomal activation necessary
UV	Human	Chromatid	Yes	10 $ergs/mm^2$	No
	Hamster	Chromatid	Yes	25 $ergs/mm^2$	No
X-rays	Human	Chromosome and chromatid	No	5 R	No
	Hamster	Chromosome and chromatid	No	10 R	No
Aflatoxin B1	Human	Chromatid	Probably	1 μg/ml	No
	Hamster	Chromatid	Probably	1 μg/ml	No
Cyclophosphamide	Human	Chromatid	Yes	Unknown	Yes
DEN	Hamster	Chromatid	Probably	4 mM	Yes
DMN	Hamster	Chromatid	Probably	1.35 mM	Yes
	Rat	Chromatid	Probably	-	Yes
	Human	Chromatid	Probably	50 mM	Yes
Ethylenimine	Human	Chromatid	Probably	10^{-4} M	No
	Hamster	Chromatid	Probably	2×10^{-5} M	

EMS	Human (Fanconi's anemia)	Chromatid	Yes	10^{-6} M	No
MMS	Human	Chromatid	Yes	1 μg/ml	No
	Hamster	Chromatid	Yes	6×10^{-4} M	No
Mitomycin C	Human	Chromatid	Yes	0.01 μg/ml	No
	Hamster	Chromatid	Yes	2.5×10^{-8} M	No
MNNG	Human	Chromatid	Yes	0.1 μg/ml	No
	Hamster	Chromatid	Yes	1 μgM	No
Myleran	Human	Chromatid	Yes	1×10^{-5} M	No
TEPA	Human	Chromatid	Yes	4.4×10^{-5} M	No
	Hamster	Chromatid	Yes	3.3×10^{-5} M	No
ThioTEPA	Human	Chromatid	Yes	2.5×10^{-5} M	No
	Hamster	Chromatid	Yes	2×10^{-5} M	No
Trenimon	Human	Chromatid	Yes	1×10^{-7} M	No
	Hamster	Chromatid	Yes	1×10^{-5} M	No
TEM	Human	Chromosome and chromatid	To a degree	5×10^{-6} M	No
Epichlorohydrin	Human	Chromatid	Unknown	1×10^{-5} M	No

the quality of human derivation and being nontransformed. Both cell types have a low spontaneous frequency of chromosomal aberrations and grow well on simple tissue culture medium. The human peripheral leukocyte system possesses the added advantage that all the cells are in a prereplicative stage of the cell cycle at the time the blood sample is drawn and the culture initiated. This offers the investigator the opportunity of studying the effects of compounds on quasi-synchronous cells without the necessity of using complex synchronization techniques. The principal precaution that must be followed is to sample cells 44-50 hr after stimulation with phytohemagglutinin in order to insure examining cells in their first post-treatment mitosis.

Two of the major problems confronted by the reader of the literature of the in vitro cytogenetic effects of chemical mutagens are the method of data reporting and experimental protocols. As mentioned earlier, the various classes of aberrations should be delineated with respect to dose and treatment times. Secondly, treating a heterogeneous population of cells for 6-48 hr with a compound that may become inactive within 30 min after its addition, and then sampling mitotic cells does not allow the reader to discern if the compound is cell stage-specific. It is virtually impossible to define dose effects when data are collected on sensitive and insensitive cells in varying proportions. Hence many of the data considered herein have not served a very useful purpose toward establishing dose effect curves. It is suggested that future experimentation include an effort to delineate cell stage specificity and to account for this in experimental design. The analysis of first post-treatment mitoses is absolutely necessary for quantiation of the data.

RESULTS AND DISCUSSION

Aflatoxin (AFT)

This toxic mold metabolite binds to DNA and in some cases needs metabolic activation to become effective. Two available studies have used no metabolic activation system and used human leukocytes and Chinese hamster cells as target cells (Table 2). Very prolonged treatment times (8 to 48 hr) and varying fixation times (0 and 40 hr or 48 hr to several days) have been used; thus there is no possibility of making any statement on the type of dose response, except that there is a concentration/duration-dependent increase in the frequencies of chromosome aberrations. The minimum effective dose can be estimated as 1 μg/ml for 48 hr.

Cyclophosphamide (CP)

Cyclophosphamide is a bifunctional compound and needs metabolic activation to become effective. Five relevant papers have been pub-

Table 2. Aflatoxin B1.

Types of cells used	Concentration and duration	Recovery time	Comments	References
Human leukocyte	Aflatoxin mixture containing 15% B1	0 or 40 hr	Concentration/length dependent increase in aberration frequency.	1
	8 or 48 hr; 1 to 50 μg/ml		Lowest dose (1 μg/ml) for 48 hr and 5 μg/ml for 8 hr, induced significant increase in aberrations	
Chinese hamster	10-75 μg/ml; 3 hr	48 hr or days	Breaks at 48 hr and reunion later	2

lished (Table 3), in four of which some kind of activation system has been employed. In human lymphocytes, nonactivated CP does not produce any chromosome aberrations even following a 24 hr treatment whereas in Burkitt lymphoma cells as well as in HeLa cells, 22 to 48 hr treatment increases the frequencies of aberrations. These data indicate the two tumor lines can activate CP to some extent. Plasma from rats injected with CP when added to cultures of HeLa cells and human lymphocytes induced chromosome aberrations. Since CP needs activation to be effective, no statement can be made with regard to the minimum effective concentration and for the same reason a correct dose response curve cannot be constructed from the published data.

Ethylenimine (EI)

Ethylenimine is a monofunctional directly acting compound. Only two papers with quantitative data have been published (Table 4). Human embryonic lung fibroblasts, human leukocytes and Chinese hamster cells have been used as target cells. No dose response curve can be constructed from the published data, as these studies were not planned for this purpose. 10^{-5} M solution of EI for 1 hr should be an effective concentration for producing chromosome aberrations.

Diethylnitrosamine (DEN)

Diethylnitrosamine is a monofunctional alkylating agent which needs metabolic activation to be effective. Only one study (Table 5) using CHO cells as target cells and rat liver microsomes for activation is available; 4 mM to 200 mM solutions of DEN have been tested, and the lowest concentration increased significantly the frequency of aberrations, compared to the controls. This ethylating agent is less effective than the methylating DMN for the production of chromosome aberrations.

Table 3. Cyclophosphamide[a]

Types of cells used	Concentration and duration	Comments	References
Human leukocytes	0.2-0.4 mg/ml, 24 hr	No effect	3
	0.1-0.5 ml of plasma from rats injected into 500 or 1000 mg/kg (cyclophosphamide), 24 hr	0.32 to 2.58 breaks/cell, dose dependent with an indication of increased efficiency of IV injection to IP injection	
Human leukocytes	Activated over rat liver slices (100-600 mg)	Maximum aberration frequency after 24 hr	4
	0.028-2.8 mg/culture (in activated form), 12, 24, 48 hr	Dose-dependent increase in aberrant metaphases and aberrations/cell	
HeLa	Nonactivated form: 6-84 μg/ml, 24 or 48 48 hr	3- to 4-fold increase over the control at high doses	5
	Activated form: 0.35-1.0 ml of plasma from rat injected into 19 mg/kg cyclophosphamide 22 and 46 hr	More effective in non-activated form	
HeLa	Activated plasma 0.35 ml + 9 ml of medium; 22 hr with or without chloramphenicol	Chloramphenicol reduces the frequency of exchanges, but not breaks	6
Burkitt Lymphoma	0.005-0.01 μg/ml, nonactivated 22 hr	Aberrations of all types in most any dose dependence	7

[a]Minimum effective dose cannot be stated, as the compound has to be activated.

Table 4. Ethylenimine[a]

Types of cells used	Concentration and duration	Recovery time	Comments	References
Human embyronic lung fibroblast (WI 38)	10^{-6}–10^{-3} M, 1 hr	1,7,10, 14 days	Dose-dependent increase in frequency of aberration at 1×10^{-6} M doubles the spontaneous frequency	8
Human leukocytes	10^{-3} and 10^{-4} M, 1 hr	24 hr (last 8 hr in H^3 thymidine)	10^{-4} M produces both breaks and exchanges; no control value given	8 8
Chinese hamster (clone 451)	2×10^{-4} M; 2×10^{-5} M for 2 hr or 4 hr; post-treated with aminopterin or actinomycin D or glyoxal, 4 hr	0	Pretreatment of actinomycin D (0.02 μg/ml) increases the frequency of EI-induced aberrations	9 9
Human leukocytes	2×10^{-5} M, 2 hr	2 hr with or without actinomycin D	No profound effect except for gaps	9

[a] 10^{-5} M for 1 hr should be an effective concentration.

Dimethylnitrosamine (DMN)

Dimethylnitrosamine is a monofunctional alkylating agent requiring metabolic activation. 50 mM of DMN activated with mouse liver microsomes induced a twofold increase of chromatid aberrations in human leukocytes (Table 6). Only one dose was used, and a high frequency of dicentrics (a class of aberration which is not generally expected to occur in the first mitosis following treatment with an alkylating agent) was obtained in this study. Using rat liver microsomes for activation and CHO cells as target cell, a wide range of concentrations from 0.135 to 135 mM has been investigated. One treatment time (1 hr) and one recovery time (16 hr) have been used. 1.35 mM of DMN was effective in inducing a significantly high frequency of aberrations. There is a threshold at low concentrations, and the response at higher concentration is more than linear.

Table 5. Diethylnitrosamine

Types of cells used	Concentration and duration	Comments	Reference
CHO	4-200 mM + rat liver microsomes (Arocolor-treated) 1 hr	Concentration-dependent increase in aberration frequencies; significant increase obtained at 4 mM	10

Table 6. Dimethylnitrosamine

Types of cells used	Concentration and duration	Recovery time (hr)	Comments	References
Rat leukocyte	30 mg/kg body in vivo action; 6 hr in vitro (0.09 mM), 24 hr	48 0	3-fold increase in chromatid aberrations over the control	11
Human lymphocytes	50 mM, activated into mouse liver microsomes (PB), 45 min	24	2-fold increase in chromatid breaks and high frequency of dicentrics	12
CHO cells	8-135 mM + rat liver S-9 (Aroclor treated), 1 hr	16	Concentration-dependent increase in exchanges + breaks	10
CHO cells	0.135-13.5 mM + rat liver S-9 (Aroclor treated) 1 hr	16	Significant increase aberrations obtained at 1.35 mM	10

MNNG

MNNG induces chromatid aberrations in M1 giving rise to derived chromosome type aberrations in M2 (Table 7). Aberrations in M2 exhibit dose-effect relationships. Cells from Fanconi's anemia, Xeroderma pigmentosum and normal probands show the same sensitivity against the chromosome breaking activity of MNNG.

Myleran

Myleran has only been tested in human leukocyte cultures for chromosome breaking activities (Table 8). Chromatid aberrations are induced in M1; chromosome type aberrations seem to be at least partly of derived type in M2. The chromosome breaking activity is especially high when cells are treated in S-phase. The aberrations show dose-effect relationships. Independent of dose and fixation time the frequencies of chromatid translocations are very low. The frequencies of Myleran-induced aberrations are suppressed in the presence of L-cysteine. The inter- and intrachromosomal localizations of achromatic lesions and chromatid breaks induced by Myleran are not random.

Mitomycin C (MMC)

MMC induces chromatid aberrations in M1 which show dose-effect relationships and give rise to chromosome aberrations of the derived type in M2 (Table 9). Cells of Fanconi's anemia are more sensitive against the chromosome breaking activity of MMC as normal cells. The frequencies of MMC induced chromosomal aberrations are enhanced by post-treatment with caffeine. The MMC-induced aberrations are preferably localized in the constitutive heterochromatin and in the secondary constrictions of human chromosomes 1, 9, 16. In human cells, chromatid translocations between chromosomes 1, 9, 16 are frequently found between the homologs at homologous sites, i.e., the secondary constrictions: somatic crossing over.

In one report no chromatid translocations were found after 12 and 24 hr treatment of lymphocyte cultures with 0.01 μg/ml MMC.

Ethyl Methanesulfonate (EMS)

In two papers the ability of EMS to induce SCE is shown with human leukocytes and with CHO cells. Auerbach and Wolman (73,74) have shown that daily addition of 10^{-6} M EMS to fibroblasts for 6 days leads to a higher aberration frequency in Fanconi's anemia cells as compared to normal ones has also been reported by Latt (51) in leukocyte cultures treated 24 hr with EMS. One paper of Woman and Auerbach is in the form of an abstract with no exact data given. EMS is a weak chromosome breaking agent that induces

Table 7. MNNG

Cell type used	Concentration used	Length of treatment (hr)	Recovery time (hr)	Comments	References
Human peripheral lymphocytes, Syrian hamster embryo cultures	0.1 μg/ml	24	0	Induction of banding patterns and of chromatid aberrations	13
Chinese hamster fibroblasts	0.5, 1.0 μg/ml			Complex aberrations in 12-28% of mitoses	14
Chinese hamster cells derived from DON line	1, 4, 10, 50, 100 μM	1	24	Chromatid aberrations; more than aberrant mitoses with 100 μM in M2; dose-effect relationship of aberrations in M2	15
Human embryonic lung lung cells, heteroploid	0.5, 1.0 μg/ml	2	0-38 every 2 hr	Cells treated in S-phase showed the highest frequencies of chromatid aberrations with a few exchange figures; cells treated in G1 show chromatid breaks and complex exchange figures	16

Human embryonic lung cells, heteroploid	0.5,1.0 μg/ml	2	6,24,48	Chromatid aberrations at 24 and 48 hr recovery time, but not at 6 hr	17
CHO	10^{-7}, 10^{-6} M	24-36	0	Very few chromatid and chromosome aberrations in M2, but clear elevation of SCE frequencies 10^{-7} M. 3 aberrant mitoses in 20; 10^{-6} M, 12 aberrant mitoses in 20	18
Lymphocyte cultures of: normal adult and of patients with Fanconi's anemia	1 μg/ml	24	0	No difference in the frequencies of chromatid aberrations in the two cell types; cells with aberrations: FA, 63.5%; normal, 23.0%; chromatid aberrations per cell: FA, 1.165%; normal, 0.270%	19
Xeroderma pigmentosum skin fibroblasts and fibroblasts of normal proband	2×10^{-5} M	3	24	No difference in the frequencies of chromatid aberrations in the two cell types: XP, 68% breaks, 32% exchanges; normal, 64% breaks, 36% exchanges	20
Chinese hamsters DON cells	Not given			Chromatid aberrations	21
Chinese hamsters DON cells	2000, 1.5, 0.5 μg/ml	2	27	Very few endoreduplications: <0.01%	

Table 8. Myleran (Busulfan): human leukocyte chromosome

Cell type used	Concentrations used	Length of treatment	Recovery time (hr)	Comments	References
Human leukocytes	2.5,5,10, 15 μg/ml	5 days		Dose and time effect relationships of chromatid-type aberrations	23
	5.0 μg/ml	4,5,6,7 days			
Human leukocytes	2×10^{-5} M	6,12,16, 24,35,46 hr	0	Aberrant metaphases: 17, 28, 30, 35, 58 68%; no chromatid exchanges	24
Human leukocytes	2×10^{-5} M	24 hr	0,24,46	Aberrant metaphases: 35, 71, 85%; very few exchange-type aberrations	24
Human leukocytes	$0.5,1,2,4 \times 10^{-5}$ M	24 hr	24	Dose-effect relationships; only very few exchange-type aberrations	24

Human leukocytes	8×10^{-5} M	1 hr	3,6,12, 16,22,34, 48	Indication of activity in S-phase of the cell cycle; very few exchange-type	24
Human leukocytes	2×10^{-5} M	12,18,24, 36,48 hr	0	L-Cysteine leads to a suppression of chromosomal aberrations induced by Myleran	25
				Nonrandom inter- and intrachromosomal distribution of Myleran-induced achromatic lesions (n=1503) and chromatid breaks (n=1342)	26
Human leukocytes	$0.5,1,2,4 \times 10^{-5}$ M	24 hr	24	Dose-effect relationship of achromatic lesions and chromatid breaks	27

(continued)

Table 8. (continued)

Cell type used	Concentrations used	Length of treatment	Recovery time (hr)	Comments	References
	2×10^{-5} M	24 hr	24	Time-effect relationship of achromatic lesions and chromatid breaks	
	8×10^{-5} M	1 hr	3,6,12, 16,22, 34,48	Time-effect relationship of achormatic lesions and chromatid breaks	
	2×10^{-5} M	12,18,24, 36,48 hr	0	L-Cysteine leads to a slight suppression of the frequencies of achromatic lesions and of chromatid breaks	
Human leukocytes	8×10^{-5} M	1 hr	48	Inter- and intrachromosomal localization of achromatic lesions, chromatid breaks and breaks participating in exchange-type aberrations	28
Human leukocytes	0.01-1000 μg/ml	72 hr	0	At doses of 1, 10, and 20 μg/ml chromosomal aberrations are 1.7, 10.9, and 25.3%, respectively	29

Human leukocytes	0.01-1000 μg/ml	48 hr		At doses of 5, 10, and 20 μg/ml, chromosomal aberrations are 8.1, 12.3, and 27.6%, respectively	29
Human leukocytes	1×10^{-4} M	24 hr	0	32% aberrant metaphases, no chromatid translocations, 7 dicentric chromosomes in 100 mitoses (derived aberrations)	30
	2×10^{-4} M	1 hr	6	19% aberrant metaphases	
			16	12% aberrant metaphases	
				No chromatid translocations; 3 dicentric chromosomes in 100 mitoses analyzed at 16 hr recovery time	
	4×10^{-4} M	1 hr	16	34% aberrant metaphases	
			22	37.5% aberrant metaphases	
			29	31% aberrant metaphases	
				One chromatid translocation in each series; 2 dicentric chromosomes at 16 and 29 hr recovery time (n=100 each), 6 at 22 hr recovery time (n=200)	

Table 9. Mitomycin C

Cell type used	Concentrations used	Length of treatment (hr)	Recovery time (hr)	Comments	References
Human leukocytes	0.1, 1.0 μg/ml	24	0	Very few endoreduplications (<1%)	31
Human leukocytes	1.0 μg/ml	24	0	Analysis of aberrations with reversed banding; preferential location of aberaberrations in the centromeric regions of chromosomes 1,5,9,16,19,20	32
Human leukocytes	1.0 μg/ml	2	22	Analysis of intercellular distributions of aberrations; 1 aberration/cell	33
Human leukocytes	0.1, 1.0 μg/ml	24	0	Homologous and nonhomologous exchanges between chromosomes 1, 9, and 16 and between and with the acrocentric chromosomes were more frequent than expected	34
Human leukocytes	1.0 μg/ml	24	0	Electron microscopic analysis of aberrations	35

Human leukocytes	0.1, 0.5 μg/ml	Not given		Enhancement of chromosomal damage by caffeine	36
Human leukocytes	0.01,0.1, 1.0 μg/ml			Intrachromosomal distribution of aberrations; preferential location of aberrations in the secondary constriction of chromosomes 1, 9, and 16	37
Human leukocytes	0.1, 0.5 μg/ml	24	0	Enhancement of chromosomal damage by caffeine; at 0.5 μg: 0.15 lesions/cell, 0.03 breaks/cell, 0.05 chromatid tid translocations/cell	38
Human leukocytes	0.01,0.1, 1.0 μg/ml	Not given		Chromatid translocations between chromosomes 1-1, 1-9, and 9-9 most frequent	39
Human leukocytes	0.01 μg/ml	24,12,6	0	Induction of secondary constrictions after 12 hr treatment (48 hr cultures)	40
Human leukocytes	0.01 μg/ml	12,24,36, 42	0	Induction of secondary constrictions in chromosomes 9 at 12 and 24 hr treatment times; no chromatid translocations	41

(continued)

Table 9. (continued)

Cell type used	Concentrations used	Length of treatment (hr)	Recovery time (hr)	Comments	References
Human leukocytes	0.1,1.0 µg/ml	24	0	Breaks localized especially in chromosomes 1, 9, 16 (secondary constrictions); chromatid translocations between homologous chromosomes at homologous sites far in excess; chromosomes 1 and 9 are most frequently affected (somatic crossing-over)	42
Human leukocytes	1.0 µg/ml	24	0	Lysosomal stabilizing agents do not influence the aberration rate, 2.924 breaks/cell	43
Human leukocytes	1.0 µg/ml	24	0	2.924 breaks/cell	44
Human leukocytes	0.1,1.0, 5.0 µg/ml	24	0	Marked excess of breaks in the secondary constrictions of chromosomes 1, 9, 16; 60.2% of 118 chromatid translocations were between homologous elements (somatic crossing-over)	45
				5.0 µg/ml toxic; mean number of breaks per cell at 0.1 µg/ml, 0.75; at 1.0 µg/ml, 2.30; chromatid translocations per cell at 0.1 µg/ml, 0.10; at 1.0 µg/ml, 0.10; at 1.0 µg/ml, 0.67	
Human leukocytes	1.0 µg/ml	2	22	Excess of breaks in chromosome 1; hot spots in areas of secondary constrictions of chromosomes 1 and 9	46

Human leukocytes				Homologous exchanges were more frequent (58%) than nonhomologous ones (27%) especially involving chromosomes 1, 9, and 16	48
Human leukocytes	3,10,30, 100 μg/ml	73	0	Very low frequencies of aberrations: 0.06 breaks/cell with 100 μg/ml; clear elevation of SCE frequencies	49
Human leukocytes	Not given	72	24	Chromatid breaks and rearrangements Chromosomal breaks and rearrangements in Fanconi's anemia may reflect a defect in a form of repair of DNA damage	50
Human leukocytes	0.01,0.03 μg/ml	24	0	Higher frequencies of aberrations in Fanconi's anemia (FA) as compared to that in normal cells Clear elevation of SCE frequencies in both cell types. Breaks per cell at 0.01 μg/ml are 0.22-0.69 (FA) and 0.01-0.04 normal; at 0.03 μg/ml, 0.72-2.20 (FA) and 0.03-0.04 (normal)	51

(continued)

Table 9. (continued)

Cell type used	Concentrations used	Length of treatment (hr)	Recovery time (hr)	Comments	References
Human leukocytes	0.1,0.25, 1.0 μg/ml	6,48,72	0	Analysis with quinacrine banding; over representation of chromatid translocations involving chromosomes 1, 9, 16; especially in C-band regions	52
Human leukocytes	1.0 μg/ml	1	71,47, 19,9	Overrepresentation of chromatid translocations between the regions of secondary constrictions of chromosomes 1, 9, 16 (somatic crossing-over)	53
				Aberrant mitoses for recovery times of 71, 47, 19, and 9 hr were 62, 56, 20, and 6%, respectively	
		68-72 hr	None	For 68-72 hr, 2% aberrant mitoses	
Human leukocytes	0.6 μg/ml	24	0	Analysis with Hoechst stain 32258	54
				Chromosomal aberrations as positive control in a study concerning possible chromosome breaking activities of LSD in vitro	55

Human leukocytes	0.01 μg/ml	24	0	Lymphocytes from Fanconi's anemia are more sensitive than that from aplastic anemia patients Aplastic anemia: aberrant mitoses, 24.0%, aberrations per cell 0.282; Fanconi's anemia: aberrant mitoses, 93.0-100.0%, 8.45-28 aberrations per cell	19
Human leukocytes	Not given			Fanconi's anemia lymphocytes show surprisingly few effects	56
Human leukocytes	0.1, 1.0 μg/ml	24	0	Breaks and chromatid translocations occur primarily in the proximal heterochromatin of the chromosomes and in the secondary constrictions of chromosomes 1 and 9; 50% of the translocations seem to be somatic crossing-over	57

(continued)

Table 9. (continued)

Cell type used	Concentrations used	Length of treatment (hr)	Recovery time (hr)	Comments	References
Human leukocytes	0.02,0.05, 0.1,0.2, 0.5,1.0, 5.0 μg/ml	2,4,8, 12,24	0	Aberrations occur primarily in the secondary constrictions of chromosomes 1, 9, 16; for 1.0 μg/ml, 24 hr recovery time: 20% aberrant mitoses	59
HeLa cells	Not given			Electron microscopic analysis of chromatid gaps	60
HeLa cells		6	60–70	Bridges, chromosome and chromatid breaks	61
Human diploid	10,50,100, 500 μg/ml	18	0,6,24	6 from 14 chromatid interchanges were between homologous chromosomes (somatic crossing-over); 500 μg/ml is toxic	62
Primary embryonic fibroblasts	0.02,0.05, 0.1,0.2, 0.5,1.0, 5.0 μg/ml	2,4,8, 12,24	0 0	Aberrations occur primarily in the secondary constrictions of chromosomes 1, 9, 16; for 1.0 μg/ml, 24 hr recovery time, 50% aberrant mitoses found	59

Chinese hamster lung fibroblasts, diploid	Not given			Enhancement of frequencies of chromatid aberrations by caffeine	63
Chinese hamster lung fibroblasts diploid	0.1 μg/ml	1	16	Enhancement of frequencies of chromatid aberrations by post-treatment with caffeine; 24% aberrant metaphases	64
Chinese hamster fibroblasts, pseudodiploid	1,4,10,50, 100 μM	1	24	Chromatid and chromosome aberrations, the latter of derived type; with 2 μM, 77% aberrant mitoses of M2; dose-effect relationships of aberrations in M2	15
Chinese hamster fibroblasts, pseudodiploid	0.025,0.05, 0.1,0.25, 1.00,2.50, 5.00,10.00 $\times 10^{-7}$ M	8		Dose-effect relationship for different chromatid-type aberrations, starting at 0.25×10^{-7} M Aberrations in 200 mitoses at 0.25, 0.5, 1.0, 2.5, 5.0, and 10.0×10^{-7} M were 7, 16, 23, 41, 82, and 111 breaks and 0, 0, 3, 4, 19, and 16, translations, respectively	65

(continued)

Table 9. (continued)

Cell type used	Concentrations used	Length of treatment (hr)	Recovery time (hr)	Comments	References
Chinese hamster lung fibroblast, diploid	Not given			Enhancement of frequencies of chromatid aberrations by post-treatment with caffeine	66
Chinese hamster fibroblasts	0.1,0.75 μg/ml	3,5,7, 10,15,18, 21,25,33		Localization of chromatid aberrations in constitutive heterochromatin of the autosomes and the X chromosome	67
CHO	10^{7}, 10^{-8} M	24-36	0	Very few chromatid and chromosome aberrations in M2, but clear elevation of SCE frequencies	18
	10^{-7} M	24-36	0	0 aberrant mitoses in 20	
	10^{-8} M	24-36	0	6 aberrant mitoses in 20	
Rat kangaroo kidney fibroblasts	10 g/ml	8	40,64	Chromatid breaks; with 40 hr recovery time, 43% mitoses with lesions and breaks	69
Muntjak fibroblasts Mouse L cells	0.33-0.5 μg/ml	18-22	0	Chromatid exchanges between homologous chromosomes at homologous sites as revealed with Giemsa-banding; somatic crossing-over 1-2% quadriradials in muntjak 4-5% quadriradials in mouse	70

Microtus agrestis lung fibroblasts	0.6,0.75 μg/ml	12,15,17, 19,21,25	Preferentially localization of chromatid aberrations in the constitutive heterochromatin; with 15-21 hr recovery time, 47% aberrant mitoses	71
Fibroblasts of the lab mouse and lab mouse × tobacco mouse	Not given		Preferentially localization of chromatid aberrations in the constitutive heterochromatin	72

Table 10. Ethyl Methanesulfonate (EMS)

Cell type used	Concentrations used	Length of treatment	Recovery time (hr)	Comments	References
Fibroblasts of Fanconi's anemia, Normal controls	10^{-6}, 10^{-5}, 10^{-4} M	6 days, daily addition	0	With 10^{-6} M a significant increase of chromosome aberrations was found with Fanconi's anemia cells but not with normal controls 10^{-5} and 10^{-4} M are toxic concentrations Mitoses with breaks (at 10^{-6} M) are 5/100 for controls and 47/100 for Fanconi's anemia	73
Leukocytes of Fanconi's anemia, Normal controls	250.0 μg/ml	24 hr	0	Slight elevation of the frequencies of chromatid aberrations in Fanconi's anemia but not in normal cells: Normal cells, 0.00-0.03 breaks/cell; FA cells, 0.11-0.24 breaks/cell	51
CHO	3×10^{-3} M 3×10^{-4} M	24-36 hr	0	Very few chromatid and chromosome aberrations in M2, but clear elevation of SCE frequencies. At 3×10^{-4} M: 4 aberrant mitoses/20; at 3×10^{-3} M: 3 aberrant mitoses/20	18
Fibroblasts of Fanconi's anemia, xeroderma pigmentosum, ataxia telangiectasia	10^{-6} M	Up to 7 days	0	Chromosome damage	74

chromatid aberrations in M1, and these may lead to chromosome type aberrations of the derived type in M2.

Methyl Methanesulfonate (MMS)

MMS induces chromatid aberrations in M1 which show dose-effect relationships and give rise to chromosome aberrations of the derived type in M2. Cells from Fanconi's anemia, xeroderma pigmentosum, and normal probands exhibit the same sensitivities against MMS. A maximum of chromatid aberrations is obtained in cells treated in the S-phase. The MMS-induced aberration rates can be elevated by post-treatment with caffeine and with 8-chlorocaffeine. The chromatid aberrations seem to be preferentially located in the C-band regions of human chromosomes.

TEPA

TEPA [tris(1-aziridinyl)phosphine oxide, M.W. 186] is a trifunctional alkylating agent. The data that have been derived from in vitro cytogenetic tests (Table 12) show very clearly that it is a potent mutagen. It gives clear positive results at 3.3×10^{-5} M. As other alkylating agents, it requires an intervening round of DNA syntheses to produce its effect as treatment of G_2 does not give rise to chromatid aberrations at the next mitosis. The lesion it introduces into the chromatin is long-lived, as treatment of unstimulated leukocytes results in chromatid aberrations after stimulation. The dose response characteristics show the presence of a multi-hit component. Aberrations are not distributed randomly among the chromosomes. It is more effective than monofunctional alkylating compounds with similar structure.

Thio-TEPA

Thio-TEPA [thiophosphoramide, tris(1-aziridinyl)phosphine sulfide] is a trifunctional alkylating agent. Its effect on mammalian chromosomes in vitro has been methodically studied in human peripheral leukocytes by Bockkov and colleagues (33,46,85-92).

Thio-TEPA produces predominantly chromatid-type aberrations, regardless of when in the cell cycle the treatment occurs, thus implicating the necessity for an intervening round of DNA synthesis (Table 13). This is particularly clear in the study of Dubinina (94), where G_2 treatment with 5×10^{-5} M results in no aberrations at the subsequent mitosis but a significant level of chromatid aberrations at the second and third post-treatment mitosis. Very high (0.2-1.0 mg/ml) doses do cause some aberrations in G_2.

Table 11. Methyl Methanesulfonate (MMS)

Cell type used	Concentrations used	Length of treatment (hr)	Recovery time (hr)	Comments	References
Human hematopoietic cell	10,20,40,80 μg/ml	6,12,24, 48	0	Open chromatid breaks and pulverizations	75
			12	Mitoses with aberrations were 8, 11, and 25% at 10, 20, and 40 μg/ml	
			24	Mitoses with aberrations were 8, 18, and 35% at 10, 20, and 40 μg/ml	
CHO	3×10^{-4}, 3×10^{-5} M	24-36	0	Very few chromatid and chromosome aberrations in M2, but clear elevation of SCE frequencies: 1 aberrant mitosis/20 at both 3×10^{-4} and 3×10^{-5} M	18
Chinese hamster lung fibroblasts, diploid	6×10^{-4} M	1	18	Potentiation of the rate of chromatid aberrations by post-treatment with 8-chlorocaffeine; 13% mitoses with multiple aberrations	76

	0.2 mM	24	0	C-band regions seem to be frequently involved in the chromatid aberrations	
Human lymphocytes	2×10^{-5} 2×10^{-4} M	24	0	Enhancement of chromosomal damage by caffeine At 2×10^{-4} M: 0.3 lesions/cell; 0.05 breaks/cell; 0.03 chromadid translocations/cell	38
	4 μg/ml	4,8,12,16, 20,24,28, 32,36		Maximum of chromatid aberrations 8-12 hr after treatment No differences in the frequencies of aberration in xeroderma and normal cells With 4-12 hr recovery, 1 aberration/cell; with 20-32 hr recovery 0.1 aberration/cell	77
	Not given	Not given	Not given	Chromatid aberrations No differences in the frequencies of aberrations in xeroderma and normal cells	78

(continued)

Table 11. (continued)

Cell type used	Concentrations used	Length of treatment (hr)	Recovery time (hr)	Comments	References
Human lymphocytes	1 μg/ml	24	0	Lymphocytes from normal probands, probands with Fanconi's anemia and with aplastic anemia show the the same aberration frequencies	19
				Aberrant cells: 15.8 and 72% in normal and FA, respectively	
				Chromatid aberrations per cell: 0.178 and 1.480 in normal and FA	
	5 μg/ml	24	0	Aberrant cells: 100.0 and 100.0% both normal and FA lymphocytes	
				Chromatid aberrations 20.62 and 14.98 per cell in normal and FA cells, respectively	

The lesions induced in G_0 and G_1 chromosomes persist until DNA synthesis. Exchange aberrations are formed but are infrequent compared to deletions. The aberrations are distributed according to a binomial. The aberrations are not distributed randomly in the karyotype; there is an excess in chromosome 1 and the D group with concommittant shortage in the C and G groups.

Excepting one report (3), the dose-response kinetics appear to be multi-hit down to 1 μg/ml.

When radioprotecting agents are given at the time of DNA synthesis following G_1 treatment, a lower yield of aberrations, but not affected cells, is observed. This observation fits well with the observation that thio-TEPA lesion interacts with x-ray induced lesions to give a synergistic effect in dual treatments.

Caffeine and methylated oxypurines enhance the yield of thio-TEPA-induced aberrations suggesting a possible role of post-replication repair. This latter observation is restricted to Chinese hamster cells.

Trenimon

Trenimon (2,3,5-triethyleniminobenzoquinone-1,4) is a trifunctional alkylating agent. As a trifunctional alkylating agent it is efficient at inducing chromatid type aberrations that are presumably formed during the DNA replicative phase of the cell cycle (Table 14). The effect of Trenimon can be moderated by the radioprotector L-cysteine. The protective effects of L-cysteine seem to be related to the cell cycle (late S), and this may argue for the involvement of post-replication repair process as in the case of thio-TEPA. The aberrations induced by Trenimon are not randomly located throughout the karyotype. Exchange figures (symmetrical and asymmetrical) occur at equal frequencies.

In many of the studies gaps were included as aberrations, and this makes the dose-response data difficult to interpret.

Triethylenemelamine (TEM)

TEM (M.W. 204) is a trifunctional alkylating agent. The few studies on in vitro systems (Table 15) indicate that, as in the case of other alkylating agents, TEM requires DNA synthesis to produce structural chromosome aberrations. The lesions induced in the chromatin are obviously long-lived, as treatment of G_0 leukocytes leads to a significant level of chromatid aberrations at the next mitosis. TEM is much more potent than TEPA.

Table 12. TEPA

Cell type used	Concentrations used	Lowest Concn with positive effect	Time of treatment
Human leukocyte	4.4-26.4 × 10^{-5} M	4.4 × 10^{-5}	G_1
Chinese hamster	7.5 × 10^{-4} M 3.3 × 10^{-5} M	3.3 × 10^{-5} M	G,S,G_2
Human leukocytes	1 × 10^{-4} M	1 × 10^{-4} M	G_0 G_1,S
Chinese hamster	100 μg/ml	100 μg/ml	G_1
Human leukocytes	Not given	Not given	S
Chinese hamster	3.3 × 10^{-5} M	3.3 × 10^{-5} M	G_1-S
Human leukocytes	1 × 10^{-4} M	1 × 10^{-4} M	S-G_2

Length of treatment (hr)	Aberration type	Comments	References
1	Chromatid	Small multi-hit component present in the dose-response curve Deletions predominate but symmetrical and asymmetrical exchanges occur at equal frequency with 50% completeness	79
2	Chromatid exclusively	High frequency of deletions, and 50% incomplete exchanges Equally potent as bifunctional agents and more potent than monofunctional agents No effect on G_2 cells in X_1 Very effective on S and G_0 cells, the latter having only chromatid aberrations ^{3}H-TdR used for staging cells	80
1 24	Chromatid	Induced chromatid-type aberrations when G_0 treated G_1 and S more sensitive than G_0 Aberrations not distributed randomly among chromosomes (too few in D group)	81
2	Polyploidy	Negative results	22
24	Chromatid	Nonrandom distribution in karyotype	82
2	Chromatid	Chromatid deletions most common, exchanges were 50% incomplete	83
8	Chromatid	1/50 as effective as TEM	84

Table 13. Thio-TEPA

Cell type	Concentrations used	Lowest tested concn with positive effect	Time of treatment
Human leukocytes	20-350 μM	50 μM	S-G_2
Human	10,20,30 μg/ml	10 μg/ml	G_0, G_1, S, G_2
Human leukocytes	4-1200 μg/ml + temp	1 μg/ml	S
Human leukocytes	5.2-18.2 × 10^{-5} M	5.2 × 10^{-5} M	S
Human leukocytes	5.2-18.2 × 10^{-5} M	5.2 × 10^{-5} M	S
Human leukocytes	20, 30 μg/ml	20 μg/ml	G_0

Length of treatment	Aberration type	Comments	References
24 hr	Chromatid	Linear dose response	3
1 hr	Chromatid	G_2 insensitive residual lesions from G_0 and G_1 treatments give chromatid aberrations Binomial distribution of aberrations	33
1 hr	Chromatid	Aberrations not randomly distributed among chromosomes: excess in chromosome 1 and D group, too few in #3 and C and G groups	46
1 hr	Chromatid	Produced predominantly chromatid deletions at a ratio of 7:3 to exchanges. More (67%) symmetrical vs. asymmetrical exchanges, with 50% uncompleteness	79
1 hr	Chromatid	Multihit dose response curve. Thio-TEPA more effective than TEPA, dipine, and fotrin	85
10-140 min	Chromatid	Aberration yield increased with increasing exposure time at both concentrations	86

(continued)

Table 13. (continued)

Cell type	Concentrations used	Lowest tested concn with positive effect	Time of treatment
Human leukocytes	5-50 μg/ml	5	G
Human leukocytes	50-1000 μg/ml	50	G (?)
Human leukocytes	10,20,30,40 μg/ml	10	S
Human leukocytes	20 μg/ml	20 μg/ml	S
Human leukocytes	15-20 μg/ml + cysteine	15 μg/ml	G_0 and S
Human leukocytes	1,3,10 μg/ml	1 μg/ml	28 hr prior to fixation ($S-G_2$?)
Human leukocytes	1,3,10 μg/ml	1 μg/ml	28 hr prior to fixation ($S-G_2$?)
Human leukocytes	4-1200 μg/ml + temp.	8 μg/ml (37°C) 240 μg/ml (5°C)	G_0
Human embryo fibroblast	1×10^{-4} M + x-ray	1×10^{-4}	$S-G_2$

Length of treatment	Aberration type	Comments	References
1 hr	Not known	Multihit dose response curve	87
1 hr	Not known (presumably chromatid)	Multihit dose response curve. Very large differences in differences in sensitivity	87
1 hr	Chromatid	Nonrandom distribution among chromosomes	88
1 hr	Chromatid	Cysteine had no effect when added with Thio-TEPA during G_0 but protected both G_0 and S treated cells when added 28 hr after culture initiation	89
28 hr	Chromatid (all classes)	Older (> 60 yrs age) donors more sensitive, no sex dependency	90
1 hr	Chromatid (all classes)	Older (> 60 yrs age) donors more sensitive, no sex dependency	91
1 hr	Chromatid	Temperature dependence with reduction in effect with reduced temperature	92
1 hr	Chromatid	Synergistic effect between x-ray and Thio-TEPA with x-ray doses of 25, 50, 100 R	93

(continued)

Table 13. (continued)

Cell type	Concentrations used	Lowest tested concn with positive effect	Time of treatment
Human leukocytes	5×10^{-5} M	4×10^{-5} M	G_0, G_1, S, G_2
Human embryo fibroblasts	3×10^{-5} M, 1×10^{-4} M, 3×10^{-4} M + x-rays	3×10^{-4}	6-30 hr prior to fixation (S-G_2)
Human leukocytes	5×10^{-3} M	5×10^{-3} M	G_1, S, G_2
Human leukocytes	5×10^{-5}, 7.5×10^{-5}, 1×10^{-4} M	5×10^{-5} M	G_0
	2.5×10^{-5}, 5×10^{-5}, 1.5×10^{-4} M	2.5×10^{-5} M	G_1-S
Human leukocytes	5×10^{-5} 1×10^{-4} M	5×10^{-5}	G_1 thru S
Chinese hamster	2×10^{-5} M	2×10^{-5} M	S

Length of treatment	Aberration type	Comments	References
1 hr	Predominantly chromatid	Probably acts through DNA synthesis G_0 and G_1 treatment results in chromatid aberrations G_2 treatment gives no positive effect in X_1 (^{3}H-TdR tag conconfirms) G_2 treatment gives chromatid aberrations at X_2 and X_3 S less sensitive than G_1	94
1 hr	Anaphase fragments	Anaphase fragments scored; apparent synergistic effect	95
1 hr	Chromatid	Produces chromatid aberrations when G_1 is treated, yielding a a lingering lesion Treatment in S more effective G_2 very little effect	96
1 hr	Chromatid	Several radioprotectors and cysteine, 2-mercaptoethylamine hydrochloride, AET, and unithiol) reduced the aberration yield but not the number of aberrant cells	97
1 hr	Chromatid		
28 hr	Chromatid	Radioprotective agents (5-methoxytryptamine, aminopropylaminoethylthiophosphoric acid, and cystaphos) reduced aberration yield at high concentrations of the Thio-TEPA when added after treatment	98
ıknown	Chromatid	Treatment of thio-TEPA treated cells with caffeine or theophyine increased the aberration yield	66,76, 99

Table 14. Trenimon

Cell type	Concentrations used	Lowest concn with positive effect	Time of treatment
Human leukocytes	3.3×10^{-7} M	3.3×10^{-7} M	$S\text{-}G_2$
Human leukocytes	$1.25\text{-}5 \times 10^{-7}$ M	1.25×10^{-7} M	$S\text{-}G_2$
Human leukocytes	6.25×10^{-7} M	6.25×10^{-7} M	$S\text{-}G_2$
Human leukocytes	1×10^{-6} - 1×10^{-7} M	1×10^{-7} M	-
Human leukocytes	$0.5\text{-}1 \times 10^{-6}$	0.5×10^{-6} M	$S\text{-}G_2$
Human leukocytes	2.5×10^{-7} M	2.5×10^{-7} M	$S\text{-}G_2$

Length of treatment (hr)	Aberration type	Comments	References
24	Chromatid	Based on equal numbers of active groups per culture Trenimon much more effective than mono- or bi-functional agents, possibly due to its ability to crosslink	100
24	Chromatid	Some protection by l-cysteine afford some protection Dose response less than linear	101
1	Chromatid	L-Cystein protected against Trenimon-induced aberrations if present with the mutagen	102
-	Chromatid	Exchange figures analyzed Symmetrical and asymmetrical occur with equal frequency Homologous chromosomes participate more than expected Excess in C group	103
24	Chromatid gaps	Frequency of gaps and breaks dose dependent (approx. linear) and L-cysteine protected against both S the most sensitive stage; cysteine most protective 6-12 hr before mitosis	27
24	Chromatid	L-Asparagine and metheonine protected against Trenimon when added with the mutagen L-Alanine and L-arginine enhanced the effect of Trenimon particularly the frequency of interchanges	104

(continued)

Table 14. (continued)

Cell type	Concentrations used	Lowest concn with positive effect	Time of treatment
Human leukocytes	1×10^{-7} M	1×10^{-6} M	S-G_2
Human leukocytes	1×10^{-7} M	1×10^{-7} M	S-G_2
Human leukocytes	1.4 10^{-}	1.4×10^{-10} 1.4×10^{-9} M	S-G
Chinese hamster (CHO)	1×10^{-6} M	1×10^{-6} M	G_2
Chinese hamster	1.25×10^{-6} - 4×10^{-2} μg/ml	1×10^{-5} μg/ml	S-G_2
Human fibroblast	2.5×10^{-6} - 1×10^{-2} μg/ml	2×10^{-5} μg/ml	S-G_2
Human leukocytes	2×10^{-4} - 2×10^{-1} μg/ml	2×10^{-3} μg/ml	S-G_2

Length of treatment (hr)	Aberration type	Comments	References
24	Chromatid	Reducdyn (N-acetyl-homocysteine-thiolactone, cysteine, and fructose) decreased Trenimon effect when used as a pre- or co-treatment but not as post-treatment Reducdyn protection was dose dependent	105
24	Chromatid	NaF confers some protection	106
24	Chromatid	Level of aberrant metaphases is significant at 2.8×10^{-10} M (includes gaps)	107
	Chromatid	PCC of G_2 cells were examined and only gaps found	108
24	Chromatid	Human cells are less sensitive and the leukocytes are least of all	109
24	Chromatid		109
24	Chromatid		109

Table 15. TEM

Cell type	Concentrations used	Lowest concn with positive effect	Time of treatment
Human leukocytes	0.02-2 μg/ml	0.02 μg/ml	$S-G_2$
Human leukocytes	5×10^{-6} M	5×10^{-6} M	$S-G_2$
Human leukocytes	1×10^{-5} M	1×10^{-5} M	G
	1×10^{-5} M	1×10^{-5} M	G
	1×10^{-5} M	1×10^{-5} M	S
	1×10^{-5} M	1×10^{-5} M	G

Epichlorohydrin

Only one paper, by Kucerova et al. (111), is available. The data show epichlorohydrin (ECHH) to be a relatively mild clastogen when compared to TEPA, e.g., ECHH at 1×10^{-5} M for last 24 hr produced 9.7% breaks as compared to 51.5% for TEPA at the same concentration. A 1×10^{-4} M concentration of ECHH was toxic. Treatment of G_0 cells does not lead to aberrations, e.g., 1×10^{-4} M ECHH gives 4% and 1×10^{-4} M TEPA gives 40.5%. This is similar to other monofunctional alkylating agents.

Length of treatment (hr)	Aberration type	Comments	References
24	Chromatid	Dose-response curve shows multihit kinetics Aberrations distributed nonrandomly in karyotype Excess in distal regions of chromosomes	3
8	Chromatid	Very potent This dose gave average of 1.6 gaps + breaks/cell	84
2	Chromatid	Predominantly chromatid aberrations seen	110
2		Occasional dicentric from G_0 and G_1 treatment	
2		Lesions must persist until DNA synthesis	
4		Most sensitive stage was G_2 - S transition	

REFERENCES

1. Dolimpio, D. A., Jacobson, C., and Legator, M. Proc. Soc. Exptl. Biol. Med. 127: 559-562 (1968).

2. Velazquez, A., DeNava, C., Covtino, R., and Pulido, T. Mutat. Res. 21: 241-242 (1973).

3. Hampel, K. E., Kober, B., Roesch, D., Gerhartz, H., and Meinig, K. H. Blood 27: 816-823 (1966).

4. Hampel, K. E., Fritzache, M., and Stopik, D. Humangenetik 7: 28-37 (1969).

5. Vrba, M. Humangenetik 4: 362-370 (1976).

6. Vogel, F., and Vrba, M. Mutat. Res. 4: 874-875 (1967).

7. Bishun, N. P. Mutat. Res. 11: 258-260 (1971).

8. Chang, T.-H., and Elequin, F. T. Mutat. Res. 4: 83-89 (1967).

9. Seleznev, Yu. V., and Selezneva, T. G. Sov. Genet. 7: 879-883 (1971) Genetika 7(7): 71-76 (1971).

10. Natarajan, A. T., Tates, A. D., Van Buul, P. P. W., Meijers, M., and Vogel, N. Mutat. Res. 37: 83-90 (1976).

11. Lilly, L. J., Bahner, B., and Magee, P. N. Nature 258: 611-612 (1975).

12. Bimboes, D., and Greim, H. Mutat. Res. 35: 155-160 (1976).

13. Dipaolo, J. A., and Popescu, N. Brit. J. Cancer 30: 103-108 (1974).

14. Howard, P. N., Bloom, A. D., and Krooth, R. S. In Vitro 7: 359-365 (1972).

15. Kato, H. Exptl. Cell Res. 85: 239-247 (1974).

16. Kelly, F., and Legator, M. Mutat. Res. 10: 237-246 (1970).

17. Kelly, F., and Legator, M. Mutat. Res. 12: 183-190 (1971).

18. Perry, P., and Evans, H. J. Nature 258: 121-125 (1975).

19. Sasaki, M. S., and Tonomura, A. Cancer Res. 33: 1829-1836 (1973).

20. Stich, H. F., Stich, W., and San, R. H. C. Proc. Soc. Exptl. Biol. Med. 142: 1141-1144 (1973).

21. Sutou, S., Saito, F., Usuki, K., Arai, Y., and Tokurama, F. Mutat. Res. 26: 439-440 (1974).

22. Sutou, S., and Tokuyama, F. Cancer Res. 34: 2615-2623 (1974).

23. Fukushima, T., and Zeldis, L. J. Nippon Ketsueki Gakkai Zasshi 33: 321-328 (1970).

24. Gebhart, E. Humangenetik 7: 126-136 (1969).

25. Gebhart, E. Mutat. Res. 7: 254-257 (1969).

26. Gebhart, E. Chromosoma 29: 336-348 (1970).

27. Gebhart, E. Humangenetik 13: 98-107 (1971).

28. Gebhart, E. Humangenetik 21: 263-272 (1974).

29. Richmond, J. Y., and Kaufmann, B. N. Exptl. Cell. Res. 54: 377-380 (1969).

30. Schwanitz, G., and Gebhart, E. Aerztliche Praxis 20: 2491 (1968).

31. Amato, R. S., Mitra, J., Blinick, G., and Antopol, W. Cytologia 35: 584-592 (1970).

32. Ayraud, N., Cantrelle, C., and Lloyd, M. C. R. Seances Soc. Biol. Fil. 169(6): 1572-1578 (1975).

33. Bochkov, N. P., Yakovenko, K. N., Chebothrev, A. N., Cravioto, F. F., and Zhurkov, V. S. Sov. Genet. 8: 1595-1601 (1972); 8(12): 160-168 (1972).

34. Bourgeois, C. A. Chromosoma 48: 203-211 (1974).

35. Brinkley, B. R., and Shaw, M. W. In: Genetic Concepts and Neoplasia (Symp. Fundam. Cancer Res.) 23: 313-345 (1970).

36. Broegger, A. Genetics 74: 531 (1973).

37. Broegger, A. Hereditas 77: 205-208 (1974).

38. Broegger, A. Mutat. Res. 23: 353-360 (1974).

39. Broegger, A., and Johansen, J. Chromosoma 38: 95-104 (1972).

40. Brown, J., Palmer, C. G., and Yu, P. L. Proc. Am. Assoc. Cancer Res. 11: 13 (1970).

41. Brown, J. A., Palmer, C. G., and Yu, P. L. Can. J. Genet. Cytol. 14: 81-93 (1972).

42. Cohen, M. M. Can J. Genet. Cytol. 11: 1-24 (1969).

43. Cohen, M. M., and Hirschhorn, R. Exptl. Cell. Res. 64: 209-217 (1971).

44. Cohen, M. M., Hirschhorn, R., and Freeman, A. I. In: Genetic Concepts and Neoplasia (Symp. Fundam. Cancer Res.) 23: 228-252 (1970).

45. Cohen, M. M., and Shaw, M. W. J. Cell Biol. 23: 386-395 (1964).

46. Funes-Cravioto, F., Yakovienko, K. N., Kuleshov, N. P., and Zhurkov, V. S. Mutat. Res. 23: 87-105 (1974).

47. Honda, T. Acta Obstet. Gynaecol. Japan (Engl. Ed.) 18: 52 (1971).

48. Lakshmy, G. V., and Singh, S. J. Anat. Soc. India 23(3): 71-75 (1974).

49. Latt, S. A. Proc. Natl. Acad. Sci. (U.S.A.) 71: 3162-3166 (1974).

50. Latt, S. A., Juergens, L., Dubin, H. G., Buchanan, G. R., and Gerald, P. S. Pediatr. Res. 9: 314 (1975).

51. Latt, S. A., Stetten, G., Juergens, L. A., Buchanan, G. R., and Gerald, P. S. Proc. Natl. Acad. Sci. (U.S.A.) 72: 4066-4070 (1975).

52. Morad, M., Jonasson, J., and Lindsten, J. Hereditas 74: 273-282 (1973).

53. Nowell, P. C. Exptl. Cell Res. 33: 445-449 (1967).

54. Raposa, T., and Natarajan, A. T. Cytobiol. 11: 230-239 (1975).

55. Rendon, O. R. Cellule 70: 331-358 (1974).

56. Schroeder, T. M., and Mauge, C. Mutat. Res. 29: 283 (1975).

57. Shaw, M. W., and Cohen, M. M. Genetics 51: 181-190 (1965).

58. Singh, V., and Singh, S. J. Anat. Soc. India 22(1): 40 (1973).

59. Sinkus, A. G. Titologiya 11: 933-940 (1969).

60. Broegger, A. Exptl. Cell. Res. 67: 243 (1971).

61. Kosaka, Y. Mie Med. J. 13(3): 000 (1964).

62. German, J., and LaRock, J. Tex. Rep. Biol. Med. 27: 409-418 (1969).

63. Asp, B. Mutat. Res. 21: 22 (1973).

64. Hartley-Asp, B., and Kihlman, B. A. Hereditas 69: 326-328 (1971).

65. Kato, H., and Shimada, H. Mutat. Res. 28: 459-464 (1975).

66. Kihlman, B. A., Sturelid, S., Hartley-Asp, B., and Nilsson, K. Mutat. Res. 17: 271-275 (1973).

67. Natarajan, A. T., and Schmid, W. Chromosoma 33: 48-62 (1971).

68. Kihlman, B. A., Sturelid, S., Hartley-Asp, B., and Nilsson, K. Mutat. Res. 26: 105-122 (1974).

69. Green, S., Palmer, K. A., and Legator, M. S. Food Cosmet. Toxicol. 8: 617-623 (1970).

70. Huttner, K. M., and Ruddle, F. H. Chromosoma 56: 1-13 (1976).

71. Natarajan, A. T., Ahnstroem, G., and Sharma, R. P. Mutat. Res. 22: 73-79 (1974).

72. Natarajan, A. T., and Raposa, T. Mutat. Res. 29: 199-200 (1975).

73. Auerbach, A. D., and Wolman, S. R. Nature 261 494-496 (1976).

74. Wolman, S. R., and Auerbach, A. D. Proc. Am. Assoc. Cancer Res. 16: 69 (1975).

75. Huang, C. C. Proc. Soc. Exptl. Biol. Med. 142 36-40 (1973).

76. Sturelid, A., and Kihlman, B. A. Hereditas 80: 233-246 (1975).

77. Sasaki, M. S. Mutat. Res. 20: 291-293 (1973).

78. Sasaki, M. S., and Tonomura, A. Idengaku Zasshi 47: 370 (1972).

79. Azhaev, S. A. Sov. Genet. 10: 1580-1583 (1974); Genetika 10(12): 156-160 (1974).

80. Sturelid, S. Hereditas 68: 255-276 (1971).

81. Kucerova, M., and Polivkova, Z. Mutat. Res. 34: 279-290 (1976).

82. Hampel, K. E., and Stopik, D. Acta Haematol. 46: 136-141 (1971).

83. Sturelid, S., and Kihlman, B. EMS Newsletter 3: 15-17 (1970).

84. Chang, T.-H., and Klassen, W. Chromosoma 24: 314-323 (1968).

85. Yakovenko, K. N., Azhaev, S. A., and Bochkov, N. P. Sov. Genet. 10: 1299-1305 (1974); Genetika 10(10): 135-143 (1974).

86. Kirichenko, O. P. Sov. Genet. 10: 1172-1175 (1974); Genetika 10(9): 139-143 (1974).

87. Chebotarev, A. N., and Yakovenko, K. N. Sov. Genet. 10: 1048-1054 (1974); Genetika 10(8): 150-157 (1974).

88. Kravioto, F. F., and Yakovenko, K. N. Sov. Genet. 8: 783-787 (1972); 8(6): 131-137 (1972).

89. Arutyunayan, R. M., and Kuleshov, N. P. Sov. Genet. 8: 534-538 (1972); Genetika 8(4): 148-153 (1972).

90. Bochkov, N. P., and Kuleshov, N. P. Sov. Genet. 7: 372-376 (1971); Genetika 7(3): 132-138 (1971).

91. Bochkov, N. P., and Kuleshov, N. P. Mutat. Res. 14: 345-353 (1972).

92. Chebotarev, A. N. Sov. Genet. 10: 1178-1182 (1974); Genetika 10(9): 147-153 (1974).

93. Korotkikh, I.M., Tarasov, V. A., and Schvetzova, T. P. Sov. Genet. 10: 1577-1579 (1974); Genetika 10(12): 152-155 (1974).

94. Dubinina, L. G. Sov. Genet. 10: 1428-1436 (1974); Genetika 10(11): 129-137 (1974).

95. Korotkikh, I. M., and Tarasov, V. A. Radiobiology (USSR) 11(4): 59-65 (1971); Radiobiologiya 11: 528-523 (1971).

96. Dubinina, L. G. Dokl. Biol. Sci. 217 304-342 (1974); Dokl. Adak. Nauk USSR 217: 943-945 (1974).

97. Arutyunyan, R. M., and Egiazaryan, S. V. Cytol. Genet. (USSR) 9(4): 5-8 (1975); Tsitol. Genet. 9: 295-298 (1975).

98. Arutyunyan, R. M., and Uganyan, V. K. Cytol. Genet. (USSR) 9(5): 10-13 (1975); Tsitol. Genet. 9: 396-399 (1975).

99. Sturelid, S., and Kihlman, B. A. Hereditas 79: 29-42 (1975).

100. Obe, G. Mutat. Res. 6: 467-471 (1968).

101. Gebhart, E. Humangenetik 10: 115-126 (1970).

102. Gebhart, E. Humangenetik 11: 237-243 (1971).

103. Gebhart, E., and Bauer, D. Chromosoma 32: 152-161 (1970).

104. Gebhart, E. Humangenetik 18: 237-246 (1973).

105. Stosiek, M., and Gebhart, E. Humangenetik 25: 209-216 (1974).

106. Obe, G., and Slacik-Erben, R. Mutat. Res. 19: 369-371 (1973).

107. Kaufmann, W., Gebhart, E., and Horbach, L. Humangenetik 20: 1-8 (1973).

108. Hittleman, W. N., and Rao, P. N. Mutat. Res. 23: 259-266 (1974).

109. Arakaki, D. T., and Schmid, W. Humangenetik 11: 119-131 (1971).

110. Luippold, H. E., P. C. Gooch, and J. G. Brewen. Genetics 88: 317-326 (1978)

111. Kucerova, M., Polivkova, Z., Sram, R., and Matousek, V. Mutat. Res. 34: 271-278 (1976).

112. Rejniak, L. Patol Pol. 15: 549-551 (1969).

CHAPTER 18

MUTAGENICITY OF SELECTED CHEMICALS IN THE MAMMALIAN DOMINANT LETHAL ASSAY

B. J. Dean

Shell Research Ltd., Shell Toxicology Laboratory

Sittingbourne Research Centre (Tunstall), Sittingbourne

Kent ME9 8AG, England

D. Anderson

ICI Ltd., Central Toxicology Laboratory, Alderley Park

Nr. Macclesfield, Cheshire, SK10 7TJ, England

R. J. Šrám

Institute of Hygiene and Epidemiology, Srobarova 48

100 42 Praha 10, Czechoslovakia

INTRODUCTION

Compounds such as triethylenemelamine, EMS, cyclophosphamide, and Myleran are widely used model mutagens, and a wealth of information on their dominant lethal activity has been published. For other chemicals, such as Aflatoxin B1, DEN, and DMN, dominant lethal data are very sparse, perhaps because they have been regarded as model carcinogens rather than mammalian mutagens.

References were abstracted from the EMIC literature searches, or, for more recent publications, from other sources. Approximately, 300 publications were scrutinized, and data from 130 of these were selected for inclusion in the reviews. This report contains a concise tabulation of the most relevant data and a detailed review of each individual chemical including the references used to provide the data.

In spite of the extensive use of the dominant lethal assay it was surprisingly difficult to estimate "no detectable effect" doses for the majority of the compounds researched. It is unlikely that such doses have not been determined as a preliminary to some of the published experiments and the information might be available from unpublished sources.

While a comprehensive critique of the dominant lethal assay, taking in variation of methodology and statistical analysis, is not within the scope of this report, a brief description of the test might aid the interpretation of the compound reviews.

Dominant lethals in mammalian reproduction result in nonviable zygotes and fetal death and are associated with sterility and semisterility in F_1 progeny. Such effects are considered to be a result of structural or numerical changes in the chromosomes of the germinal cells but in some cases, particularly in female germ cells, nongenetic events might mimic dominant lethal mutations.

The assay in males involves sequential mating with one or more females at intervals during a period corresponding to the length of the spermatogenic cycle, i.e., 6-8 weeks in mice, 10-12 weeks in rats. Mating during the first 3 weeks after dosing mice measures effects induced in spermatids or spermatozoa while the results of later matings reflect the effect of the chemical on premeiotic cells (spermatogonial cells and spermatocytes). The pregnant females are dissected 12-14 days after mating and the uterus examined for dead embryos and for evidence of pre-implantation losses.

Most of the compounds so far assayed exert their most potent effect in postmeiotic stages, with a few exceptions, such as Mitomycin C, inducing a higher frequency of mutations in the premeiotic cells.

The dominant lethal assay in female mice is considered a less specific test of mutagenicity than the assay in males because of a variety of nongenetic factors which can influence ovulation, fertilization, and implantation. However, the resting oocyte normally develops into a mature oocyte without the selection and elimination associated with spermatogenesis and is considered by some workers to be particularly suitable for detecting genetic damage. By careful monitoring of ovulation and control of mating it is possible to dose females at precise stages of pre- and postcopulation oogenesis. In a typical assay in female mice, the dosed females are caged with fertile males at intervals of 0, 5, 10, and 15 days after treatment. The males are separated from the females either after observation of a mating plug or at the end of the five-day mating interval. Females are dissected 13 days after mating or presumed mating and the uterus examined for evidence of dominant lethality.

Detailed technical information on dominant lethal assays in mammals may be found in works by Röhrborn (1), Bateman and Epstein (2), and Generoso (3).

RESULTS

A summary of data on dominant lethals induced by single intra-preitoneal (IP) doses is given in Table 1. Individual results are tabulated in Table 2.

Aflatoxin B1 (1162-65-8)

Two dominant lethal assays have been performed in male mice given single IP doses of Aflatoxin B1 (4,5). Both studies produced results suggesting a dominant lethal effect, but only small numbers of mice were used and the data obtained were not conclusive. In one experiment (5) a mixture of Aflatoxin B1 and G1 was used.

A comprehensive study is necessary to elucidate the dominant lethality of Aflatoxin.

Vinyl Chloride (75-01-4)

Only one study has been published on the dominant lethality of vinyl chloride. This was an inhalation experiment in which male CD-1 mice were exposed to atmospheres containing 3,000, 10,000 and 30,000 ppm, 6 hr/day, for 5 days. No evidence of dominant lethal mutations was recorded (6).

MNNG (70-25-57)

Three dominant lethal assays with MNNG are reported. A single IP dose of 50 mg/kg to male mice gave results which were marginally positive in one study (7) and produced small differences in the results of the second study (8) which were not statistically significant. A dose of 100 mg/kg induced high degrees of dominant lethality, reducing fetal implants and increasing early fetal deaths in the first 4 days after dosing. The effect was confined to mature spermatozoa. The LD_{50} of MNNG to male mice under these conditions is 114 mg/kg, and, not surprisingly, some 50% of animals died during the second week after mating.

In female mice, single IP injections of 50 mg/kg induced pre- and postimplantation losses in the first 5-day mating period after dosing (9).

Table 1. Summary of dominant lethals induced by single IP doses

Compounds (CAS no.)	Species	Sex	Lowest dose to induced dominant lethals (mg/kg)	Highest dose with no significant dominant lethality (mg/kg)
Aflatoxin B1 (1162-65-8)	Mouse	Male	Results not conclusive	
Vinyl chloride (75-01-4)	Mouse	Male	Negative	30,000 ppm (in air)
MNNG (70-25-57)	Mouse	Male	50	Not determined
	Mouse	Female	50	Not determined
ICR-170 (146-59-8)	Mouse	Male	Negative	4
	Mouse	Female	Negative	4
Mitomycin C (50-07-7)	Mouse	Male	3.5	Not determined
Natulan (366-70-1)	Mouse	Male	200	Not determined
EMS (62-50-0)	Mouse	Male	150	100
	Mouse	Female	250	Not determined
	Rat	Male	100	Not determined
MMS (66-27-3)	Mouse	Male	20	Not determined
	Mouse	Female	100	Not determined
	Rat	Male	50	Not determined
DEN (55-18-5)	Mouse	Male	Results not conclusive	
DMN (62-75-9)	Mouse	Male	4.4 (SC)	9.0 (IP)
Cyclophosphamide (50-18-0)	Mouse	Male	60	48
	Mouse	Female	210	150 (oral)
Myleran (55-98-1)	Mouse	Male	10	Not determined
	Mouse	Female	20	Not determined
	Rat	Male	4	Not determined

(continued)

Table 1. (continued)

Compounds (CAS no.)	Species	Sex	Lowest dose to induced dominant lethals (mg/kg)	Highest dose with no significant dominant lethality (mg/kg)
TEPA (545-55-1)	Mouse	Male	1.0	0.625
METEPA (57-39-6)	Mouse	Male	12.5	6.25
ThioTEPA (52-24-4)	Mouse	Male	0.64	0.32
	Mouse	Female	2	Not determined
Cadmium (7440-43-9)	Mouse	Male	Negative	7.0
Epichlorhydrin	Mouse	Male	Negative	150
Ethylenimine (151-56-4)	Mouse	Male	5	Not determined
TEM (51-18-3)	Mouse	Male	0.05	0.035
	Mouse	Female	0.4	Not determined
	Rat	Male	0.025	Not determined
Trenimon (68-76-8)	Mouse	Male	0.125	Not determined
	Mouse	Female	0.125	Not determined

The highest dose showing no detectable effect was shown to be below 50 mg/kg for both sexes.

ICR-170 (146-59-8)

Dominant lethal assays have been carried out in both male and female mice after single IP doses of 4 mg ICR-170/kg body weight. In both cases no evidence of a dominant lethal effect was obtained (8-10).

In an experiment measuring the total reproductive capacity of female mice after IP dosing with 4 mg/kg ICR-170 (10), there was a slight reduction in the mean number of young produced per female when compared with untreated mice. The average interval between dosing and birth of the first litter, and the mean number of litters per female did not differ significantly from the control values. There was no suggestion of dominant lethality in these animals.

It is concluded that, under the experimental conditions described, ICR-170 does not induce dominant lethal mutations in mice.

Mitomycin C (50-07-7)

Mitomycin C is one of the few chemicals which have been shown to induce dominant lethal mutations in premeiotic cells in the male mouse. In a dose-response study, a single IP injection of 3.5 mg/kg Mitomycin C induced pre- and postimplantation dominant lethals in spermatocytes. IP doses of 5.25 and 7.0 mg/kg resulted in dominant lethals in spermatids, but as a lower frequency than those detected in premeiotic cells at the same doses (11).

A single IP dose of 1.75 mg/kg induced a low frequency of pre-implantation losses in one study (12) but in another experiment (4) a dose of 2.4 mg/kg did not produce effects significantly different from the control animals. The highest dose to induce no detectable dominant lethals in male mice requires confirmation.

In a study in which female mice were given single IP doses of 5.25 or 7.0 mg/kg Mitomycin C., a reduction in litter size was recorded in both dose levels (13).

Subcutaneous injections of 0.67 or 6.67 mg/kg in male rats also reduced litter size in subsequent matings (14).

Natulan (366-70-1)

Two studies have been carried out in male mice given IP injections of Natulan (procarbazine). The highest dose used in the first assay (4) showed some evidence of postimplantation losses in animals dosed with 110 mg/kg, but these were not statistically significant.

In the second study, pre-implantation losses occurred in the 200 mg/kg group, and early fetal deaths in males given 400 mg/kg Natulan. Both pre- and postmeiotic effects were detected. These were confined to spermatocytes and spermatids with no dominant lethals observed in mature spermatozoa, i.e., during the first week after dosing (16). It is concluded that the lowest dose to induce dominant lethality in male mice was 200 mg/kg Natulan, but a "no detectable effect" dose was not achieved with any certainty.

Ethyl Methanesulfonate (EMS) (62-50-0)

Ethyl methanesulfonate is one of the most widely tested chemicals in the dominant lethal assay and, at an IP or oral dose of

200 mg/kg, is often used as a positive control compound in male mice.

In experiments in which male mice were given single IP injections of EMS, a dose-response relationship was demonstrated at doses from 150 to 250 mg/kg with no detectable dominant lethals at 100 mg/kg (17-19). In another study of the dose response relationship, mice were given single IP doses of EMS ranging from 50 to 300 mg/kg. Dominant lethality was not detected below 200 mg/kg (20). No significant route-related differences were detected when male mice were dosed orally or IP with 200 mg/kg EMS (21). When the compound was injected directly into the testes at 200 mg/kg of testis weight, dominant lethality was not observed. (This may be due to rapid absorption and distribution of a very small total dose of EMS.) In the same series of experiments (21), no significant differences in the frequency of EMS-induced dominant lethals were detected between three strains of male mice, i.e., strain DBA/2J and two hybrid strains. In an earlier report (22), T-stock males showed a lower frequency of dominant lethals than two hybrid strains after IP dosing with 200 mg/kg EMS.

In an experiment in which male ICR mice were given five daily IP doses of 10, 20, 30, or 40 mg/kg EMS, dominant lethality was detected in the 20 mg/kg group during the first week after dosing (23). No dominant lethality was detected in mice given five daily doses of 10 mg/kg EMS.

Oral dosing of male mice on each of five successive days with 100 or 200 mg EMS/kg induced a high incidence of dominant lethals (6,24) and when male CDI mice were dosed orally with 100, 150 or 200 mg/kg EMS on each of five successive days a dose-related increase in postmeiotic dominant lethality was demonstrated (25).

In all experiments describing dominant lethal mutation induction by EMS in male mice, both pre- and postimplantation losses have been detected and were confined to spermatids and mature spermatozoa. Meiotic and premeiotic dominant lethality has not been recorded.

Experiments with female mice have shown that (a) much higher doses of EMS are required to induce dominant lethals and (b) the frequency of dominant lethality is greatly influenced by the strain of mouse used and the stage of oogenesis treated.

In contrast to the results in male mice (22), T-stock females were found to be relatively sensitive to the induction of dominant lethals by EMS showing a high frequency of pre- and postimplantation losses in pre-ovulatory oocytes after IP doses of 300 or 325 mg/kg (9,22). Two hybrid strains gave a very weak or negative response.

Table 2. Mammals: dominant lethals

Compound (CAS No.)	Species	Sex	Strain	Route[a]	Vehicles[b]
Aflatoxin B1 (1162-65-8)	Mouse	M	ICR/Ha	IP	-
	Mouse	M	CD-1	IP	DMSO
Vinyl chloride (75-01-4)	Mouse	M	CD-1	Inhal.	Air
MNNG (70-25-7)	Mouse	M	101/C_3H	IP	HBSS
	Mouse	M	$C_{57}BL$ x BALB/C	IP	PO Buff
	Mouse	F	T-stock	IP	HBSS
ICR-170	Mouse	M	101/C H	IP	PO Buff
	Mouse	F	T-stock	IP	PO Buff
	Mouse	F	SEC x $C_{57}BL$	IP	PO Buff
Mitomycin C (50-07-7)	Mouse	M	101/C_3H	IP	HBSS
	Mouse	M	101/C_3H	IP	Water
	Mouse	M	ICR/Ha	IP	Water
	Mouse	M	CF-1	IP	Saline

Dose (mg/kg)	Pre-implantation losses[c]	Post-implantation losses[c]	Dominant lethality[c]	Comments	References
50	+	+	+	Premeiotic inconclusive	4
70	-	-	-		
68	-	+	+	Pre- and postmeiotic	5
3,000 ppm	-	-	-		6
10,000 ppm	-	-	-		
30,000 ppm	-	-	-		
50	-	-	-		
50	-	+	+	Postmeiotic	7
70	-	+	+		
100	+	+	+		
70	+	+	+		9
4	-	-	-		8
4	-	-	-		9
4	-	-	-		10
3.50	+	+	+	Premeiotic most sensitive mutable stage	11
5.25	+	+	+		
7.00	+	+	+		
1.75	+	-	+	Marginal premeiotic effect only	12
2.4	-	-	-	Premeiotic	4
12.0	+	-	+		
5.0	+	+	+	Pre- and postmeiotic	15

(continued)

Table 2. (continued)

Compound (CAS No.)	Species	Sex	Strain	Route[a]	Vehicle[b]
Mitomycin C (cont'd.)	Mouse	F	$101xC_3H$	IP	-
	Rat	M	-	SC	-
Natulan (366-70-1)	Mouse	M	ICR/Ha	IP	Water
	Mouse	M	$101xC_3H$	IP	Water
EMS	Mouse	M	$101xC_3H$	IP	HBSS
	Mouse	M	$101xC_3H$	IP	HBSS
	Mouse	M	SEC x $C_{57}BL$	IP	HBSS
	Mouse	M	T-stock	IP	HBSS
	Mouse	M	$101xC_3H$	IP	HBSS
	Mouse	M	ICR/Ha	IP	Water

Dose (mg/kg)	Pre-implantation losses[c]	Post-implantation losses[c]	Dominant lethality[c]	Comments	References
5.25	-	-	-	Reduced litter size	13
7.00	-	-	-		
0.67	-	-	-	Reduced litter size	14
6.67	-	-	-		
22	-	-	-		4
100	-	-	-		
100	-	-	-	Post- and premeiotic	16
200	+	-	+		
400	+	+	+		
800	+	+	+		
100	-	-	-	Postmeiotic	10
200	+	+	+		
250	+	+	+		
200	+	+	+	Postmeiotic	22
300	+	+	+		
200	+	+	+	Postmeiotic	22
300	+	+	+		
200	+	+	+	Postmeiotic	22
300	+	+	+		
400	+	+	+	Postmeiotic	11
76	-	-	-	Postmeiotic (no pre-	4
380	-	+	+	implantation losses)	

(continued)

Table 2. (continued)

Compound (CAS No.)	Species	Sex	Strain	Route[a]	Vehicle[b]
EMS (cont'd.)	Mouse	M	101xC_3H	IP	-
	Mouse	M	CD-1	IP	Corn oil
	Mouse	M	ICR	IP	HBSS
	Mouse	M	101xC_3H	IP	HBSS
	Mouse	M	CD-1	IP	Saline
	Mouse	M	CD-1	Oral	Water
	Mouse	M	CD-1	Oral	Water

Dose (mg/kg)	Pre-implantation losses[c]	Post-implantation losses[c]	Dominant lethality[c]	Comments	References
50	-	-	-	Postmeiotic	17
100	-	-	-		
150	-	+	+	Dose response over 150, 200, 250	
200	+	+	+		
250	+	+	+		
50	-	-	-	Postmeiotic	20
100	-	-	-	Dose response over 200, 300, 400	
200	+	+	+		
300	+	+	+		
400	+	+	+		
10 (x5)	-	-	-	Postmeiotic	23
20 (x5)	+	+	+		
30 (x5)	+	+	+		
40 (x5)	+	+	+		
50	-	-	-	Postmeiotic	18
100	-	-	-	Dose response over 150, 200, 250	
150	-	+	+		
200	+	+	+		
250	+	+	+		
300	+	+	+		
30	-	-	-	Postmeiotic	19
40	-	-	-		
50	-	-	-		
100	-	-	-		
150	-	+	+		
200	+	+	+		
360	+	+	+		
100 (x5)	+	+	+	Postmeiotic	24
200 (x5)	+	+	+	Postmeiotic	6

(continued)

Table 2. (continued)

Compound (CAS No.)	Species	Sex	Strain	Route[a]	Vehicle[b]
EMS (cont'd.)	Mouse	M	CD-1	Oral	Water
	Mouse	M	DBA/2J DBA/2J C_3D_2 C_3D_2	Oral IP	HBSS
	Mouse	F	$SECxC_{57}BL$ $101xC_3H$ T-stock	IP IP	HBSS
	Mouse	F	T-stock	IP	HBSS
	Mouse	F	$SECxC_{57}BL$	IP	HBSS
	Mouse	F	C_3Hx101	IP	HBSS
			SBES	IP	
	Rat	M	-	IP	Arachis oil
	Rat	M	-	IP	-
MMS (66-27-3)	Mouse	M	$101xC_3H$	IP	HBSS

Dose (mg/kg)	Pre-implantation losses[c]	Post-implantation losses[c]	Dominant lethality[c]	Comments	References
100 (x5)	+	+	+	Postmeiotic	25
150 (x5)	+	+	+	Comparative study between two laboratories	
200 (x5)	+	+	+		
200	+	+	+	Postmeiotic	21
200	+	+	+	No route-related or strain-related differences in dominant lethality	
200	+	+	+		
200	+	+	+		
300	-	-	-	Pre-ovulatory oocyte	23
300	-	-	-	Strain differences in response	
300	+	+	+		
235	+	+	+	Pre-ovulatory oocytes	9
200	-	-	-		10
325	+	-	+		
250	+	+	+	Postmating, precleavage	26
300	+	+	+	No strain differences	
300	+	+	+		
100	+	+	+	Postmeiotic	27
200	+	+	+		
300	o	o	o	Reduced litter size in early weeks after dosing	27
20 (x15)	o	o	o		
50	+	+	+	Postmeiotic	8
100	+	+	+		
150	+	+	+		

(continued)

Table 2. (continued)

Compound (CAS No.)	Species	Sex	Strain	Route[a]	Vehicle[b]
MMS (cont'd.)	Mouse	M	$C_{57}BL$	IP	-
	Mouse	M	ICR/Ha	IP	TCP
	Mouse	M	-	IP	Saline
	Mouse	M	CFLP	IP	Saline
	Mouse	M	ICR/Ha	IP	-
	Mouse	M	CD-1	IP	Saline
	Mouse	M	Various	IP	-
	Mouse	M	CD-1	IP	Water
	Mouse	M	CD-1	IP	Corn oil

Dose (mg/kg)	Pre-implantation losses[c]	Post-implantation losses[c]	Dominant lethality[c]	Comments	References
30	-	+	+	Postmeiotic	33
60	+	+	+		
50	+	-	-	Postmeiotic	34
50	+	+	+	Postmeiotic	32
100	+	+	+		
50	+	+	+	Postmeiotic	35
100	+	+	+		
25	-	+	+	Postmeiotic	4
50	-	+	+		
100	+	+	+		
120	+	+	+		
50	+	+	+	Postmeiotic	36
20	-	+	+	Postmeiotic	28
40	+	+	+	Comparative study in 6 laboratories	
80	+	+	+		
20	-	+	+	Postmeiotic	29
40	+	+	+	Dose-related increase from 20 to 80	
80	+	+	+		
3.125	-	-	-	Postmeiotic	20
6.25	-	-	-	No effect dose not clearly defined	
12.5	-	+	+		
25	-	+	+		
50	-	+	+		
100	+	+	+		

(continued)

Table 2. (continued)

Compound (CAS No.)	Species	Sex	Strain	Route[a]	Vehicle[b]
	Mouse	F	T-stock 101xC_3H SECxC_{57}BL	IP	HBSS
	Mouse	F	-	IM	-
	Mouse	F	CF-1	IP	Water
	Rat	M	-	IP	Arachis oil
	Rat	M	-	IP	-
DEN (55-18-5)	Mouse	M	101xC_3H	IP	Saline
DMN (62-75-9)	Mouse	M	ICR/Ha	IP	Water
	Mouse	M	101xC_3H	SC	Saline
Cyclophosphamide (50-18-0)	Mouse	M	C_3H	IP	Saline
	Mouse	M	C_3H	IP	Saline

Dose (mg/kg)	Pre-implantation losses[c]	Post-implantation losses[c]	Dominant lethality[c]	Comments	References
150	-	+	+	First 10 days after dosing	9
150	-	-	-		
150	-	-	-	Strain differences in response	
100	-	+	+	Pro-estrus	31
100	+	+	+		30
50	+	+	+	Postmeiotic	27
100	+	+	+		
50	o	o	o	Postmeiotic fertility reduced at all doses	37
100	o	o	o		
2.5 (x25)	o	o	o		
5 (x25)	o	o	o		
10 (x25)	o	o	o		
13.5	-	-	-		38
8	-	-	-		4
9	-	-	-		
4.4	+	+	+	Postmeiotic	38
60	+	+	+	Postmeiotic	42
210	+	+	+		
60	+	+	+	Postmeiotic	39
210	+	+	+	(premeiotic losses at 210 mg/kg); 420 mg/kg group did not survive	
420	o	o	o		

(continued)

Table 2. (continued)

Compound (CAS No.)	Species	Sex	Strain	Route[a]	Vehicle[b]
Cyclophosphamide (cont'd.)	Mouse	M	101xC_3H	IP	-
	Mouse	M	CFLP	IP	Saline
	Mouse	M	ICR/Ha	IP	Water
	Mouse	M	CD-1	IP	Water
	Mouse	M	ICR	IP	-
	Mouse	M	ICR	IP	Saline
	Mouse	M	ICR	IP	Saline
	Mouse	M	ICR	IP	Saline
	Mouse	M	-	Oral	Water
	Mouse	M	CFI	Oral	Water
	Mouse	M	ICR	Oral	Saline

Dose (mg/kg)	Pre-implantation losses[c]	Post-implantation losses[c]	Dominant lethality[c]	Comments	References
210	+	+	+	Postmeiotic	43
60	-	-	-	Postmeiotic only	44
240	+	+	+	Less sensitive strain	
48	o	-	-	Postmeiotic	4
240	o	+	+		
200	+	+	+	Postmeiotic	24
100	+	+	+	Postmeiotic	23
2.5 (x20)	+	+	+		
5.0 (x20)	+	+	+		
10	-	-	-	Postmeiotic	41
20	-	-	-		
40	-	-	-		
80	+	+	+		
60	+	+	+	Postmeiotic	41
80	+	+	+		
100	+	+	+		
120	+	+	+		
160	-	+	+	Postmeiotic; marginal premeiotic losses	45
210	+	+	+		
100	+	+	+	Postmeiotic	46
60	-	-	-	Postmeiotic	41
80	-	+	+		
100	-	+	+		
120	+	+	+		
140	+	+	+		

(continued)

Table 2. (continued)

Compound (CAS No.)	Species	Sex	Strain	Route[a]	Vehicle[b]
Cyclophosphamide (cont'd.)	Mouse	M	ICR	Oral	-
	Mouse	M	ICR	Oral	Saline
	Mouse	M	NMRI	IV	Water
	Mouse	M	ICE	Oral	Water
	Mouse	M	-	Oral	Water
	Mouse	F	101xC_3H	IP	-
	Mouse	F	-	Oral	-
	Mouse	F	-	Oral	Water
Myleran (55-98-1)	Mouse	M	-	IP	-

Dose (mg/kg)	Pre-implantation losses[c]	Post-implantation losses[c]	Dominant lethality[c]	Comments	References
80	-	-	-	Postmeiotic	41
110	-	+	+		
140	-	+	+		
170	-	+	+		
200	+	+	+		
10 (x5)	-	-	-	Postmeiotic	41
20 (x5)	-	+	+		
30 (x5)	+	+	+		
40 (x5)	+	+	+		
100	+	+	+	Postmeiotic	40
0.005%	+	+	+	Administered in drinking water for 28 days	23
0.01%	+	+	+		
0.02%	+	+	+		
0.01%	+	+	+	Drinking water; maximum frequency after 2-4 weeks	47
210	+	+	+	Pre-ovulatory stages	43
150	-	-	-	Pro-estrus and 1 day postcoitum	31
200	-	+	+		
0.002%	-	-	-	Administered in drinking water	47
0.01%	+	-	+		
10	o	o	+	Pre- and postmeiotic	48
20	o	o	+		

(continued)

Table 2. (continued)

Compound (CAS No.)	Species	Sex	Strain	Route[a]	Vehicle[b]
	Mouse	M	ICR/Ha	IP	-
	Mouse	F	T-stock $101xC_3H$ $SECxC_{57}BL$	IP	Olive oil
	Mouse	F	$SECxC_{57}BL$	IP	Olive oil
	Rat	M	-	IP	Arachis oil
	Rat	M	-	IP	Arachis oil
TEPA (545-55-1)	Mouse	M	A/L	IP	Saline
	Mouse	M	ICR/Ha	IP	Water

Dose (mg/kg)	Pre-implantation losses[c]	Post-implantation losses[c]	Dominant lethality[c]	Comments	References
25	-	+	+	Postmeiotic and meiotic	4
50	+	+	+		
75	+	+	+		
100	+	+	+		
40	-	+	+	Detected in each mating period up to 3 weeks after dosing	49
40	-	+	+		
40	-	+	+	No strain differences	
10	o	o	+	All doses reduced fertility	10
20	o	o	+		
40	o	o	+		
80	o	o	+		
4	+	-	+	Premeiotic and meiotic	27
6	+	-	+	No early fetal deaths reported	
10	+	-	+		
4	-	-	-	Premeiotic and meiotic	37
6	+	-	+		
8	+	-	+		
10	+	-	+		
5 (x2)	+	-	+		
2 (x5)	+	-	+		
0.5 (x2)	-	-	-	Postmeiotic	51
1.0 (x2)	+	+	+		
2.5 (x2)	+	+	+		
5	-	-	+	Postmeiotic Translocations detected in F_1 and F_1 males	54

(continued)

Table 2. (continued)

Compound (CAS No.)	Species	Sex	Strain	Route[a]	Vehicle[b]
TEPA (cont'd.)	Mouse	M	ICR/Ha	IP	TCP
	Mouse	M	ICR	IP	Saline
	Mouse	M	ICR	IP	Saline
	Mouse	M	ICR/Ha	IP	TCP
	Mouse	M	ICR/Ha	Oral	-
	Mouse	F	ICR	IP	Saline
	Rat	M	Wistar	IP	-

Dose (mg/kg)	Pre-implantation losses[c]	Post-implantation losses[c]	Dominant lethality[c]	Comments	References
5	+	+	+	Postmeiotic	4
7	+	+	+		
11	+	+	+		
14	+	+	+		
22	+	+	+		
1	+	+	+	Postmeiotic	53
0.2 (x5)	+	+	+	Pre-implantation losses	
0.1 (x10)	+	+	+	in premeiotic stages	
0.16	-	-	-	Postmeiotic	41
0.32	-	-	-		
1.00	+	+	+		
1.25	+	+	+		
2.50	+	+	+		
0.001(x48)	-	-	-	Postmeiotic and pre-	4
0.01 (x48)	-	-	-	meiotic	
0.1 (x48)		+	+		
0.125(x48)	-	+	+		
0.5 (x48)	+	+	+		
1.0 (x48)	+	+	+		
0.125(x48)	-	-	-		4
0.5 (x48)	-	-	-		
2.0 (x48)	-	-	-		
1.0	-	-	-		55
2.0	+	+	+	Postmeiotic	56
2.5	+	+	+		

(continued)

Table 2. (continued)

Compound (CAS No.)	Species	Sex	Strain	Route[a]	Vehicle[b]
METEPA (57-39-6)	Mouse	M	ICR/Ha	IP	TCP
	Mouse	M	ICR/Ha	IP	TCP
	Mouse	M	CD-1	IP	TCP
ThioTEPA (52-24-4)	Mouse	M	CFLP	IP	Saline
	Mouse	M	ICR/Ha	IP	TCP
	Mouse	M	ICR	IP	HBSS
	Mouse	M	F_1D2AR	IP	Saline

Dose (mg/kg)	Pre-implantation losses[c]	Post-implantation losses[c]	Dominant lethality[c]	Comments	References
0.782	-	-	-	Postmeiotic, meiotic, and premeiotic effects at higher doses	52
1.563	-	-	-		
3.125	-	-	-		
6.25	-	-	-	Dose response from 12.5 to 100 mg/kg	
12.5	-	+	+		
25.0	+	-	-		
50.0	-	-	-		
100.0	-	-	-		
40	+	+	+	Premeiotic and post-meiotic	4
80	+	+	+		
130	+	+	+		
3.125	-	-	-	Postmeiotic	57
12.5	+	+	+	Dose response from 12.5 to 50 mg/kg	
25.0	+	+	+		
50.0	+	+	+		
5	+	+	+	Postmeiotic, meiotic, and premeiotic effect at 10 mg/kg	35
10	+	+	+		
5	+	+	+	Postmeiotic, meiotic, and premeiotic effect at 10 mg/kg	4
10	+	+	+		
2	+	+	+	Postmeiotic	23
2	+	+	+	Frequency of embryonic dependent on genotype of female	58

(continued)

Table 2. (continued)

Compound (CAS No.)	Species	Sex	Strain	Route[a]	Vehicle[b]
ThioTEPA (cont'd.)	Mouse	M	ICR	IP	HBSS
	Mouse	M	ICR	IP	HBSS
	Mouse	M	ICR	IP	HBSS
	Mouse	M	-	Oral	Olive oil
Cadmium chloride (7440-34-9)	Mouse	M	ICR/Ha	IP	-
	Mouse	M	Balb/C	IP	-
	Mouse	F	101xC H	IP	Water
	Rat	M	Sprague Dawley	Oral	-

Dose (mg/kg)	Pre-implantation losses[c]	Post-implantation losses[c]	Dominant lethality[c]	Comments	References
0.16	-	-	-	Postmeiotic	41
0.32	-	-	-		
1.0	+	+	+		
1.25	+	+	+		
2.5	+	+	+		
0.16	-	-	-	Postmeiotic	41
0.32	-	-	-		
0.64	-	+	+		
1.0	+	+	+		
1.28	+	+	+		
2.56	+	+	+		
0.2 (x5)	+	+	+	Postmeiotic	23
0.3 (x5)	+	+	+		
0.4 (x5)	+	+	+		
0.02 (x21)	o	o	o	0.1 mg/kg (x21) induced total infertility	(59
0.05 (x21)	o	o	o		
0.1 (x21)	o	o	o		
0.5 (x21)				0.05 mg/kg reduced viable embryos	
1.35	-	-	-	Possible reduction in fertility	4
2.7	-	-	-		
5.4	-	-	-		
7.0	-	-	-		
1.75	-	-	-		61
2	-	-	-		62
6.25	o	o	o	No effect on fertility	60
12.5	o	o	o		
25.0	o	o	o		

(continued)

Table 2. (continued)

Compound (CAS No.)	Species	Sex	Strain	Route[a]	Vehicle[b]
Epichlorohydrin (106-89-8)	Mouse	M	ICR/Ha	IP	-
	Mouse	M	ICR	IP	DMSO
				Oral	DMSO
				IP	DMSO
				Oral	DMSO
Ethylenimine (151-56-4)	Mouse	M	$C_{57}BL/6$	IP	-
	Mouse	M	C_3HAxC_3HF	ISc	Saline
TEM (51-18-2)	Mouse	M	-	IP	-
	Mouse	M	-	IP	Saline
	Mouse	M	ICR/Ha	IP	-
	Mouse	M	$103/C_3H$	IP	HBSS

Dose (mg/kg)	Pre-implantation losses[c]	Post-implantation losses[c]	Dominant lethality[c]	Comments	References
150	-	-	-		4
5	-	-	-		63
10	-	-	-		
20	-	-	-		
20	-	-	-		
40	-	-	-		
1 (x5)	-	-	-	Results suggest slight reduction in fertility	
4 (x5)					
4 (x5)	-	-	-		
16 (x5)	-	-	-		
5	-	+	+	Postmeiotic	64
6	-	+	+	Postmeiotic	65
0.2	+	+	+	Postmeiotic	73
0.8	+	+	+		
0.4	+	+	+	Postmeiotic	77
0.4 (x2)	+	+	+	Aspermia in premeiotic stages	
0.2	-	+	+	Postmeiotic	4
0.8	+	+	+		
0.035	-	-	-	Postmeiotic	66
0.05	-	+	+		67
0.1	+	+	+		
0.2	+	+	+		
0.3	+	+	+		
0.4	+	+	+		

(continued)

Table 2. (continued)

Compound (CAS No.)	Species	Sex	Strain	Route[a]	Vehicle[b]
TEM (cont'd.)	Mouse	M	Hybrid	IP	HBSS
	Mouse	M	Ha/ICR	IP	-
	Mouse	M	$101xC_3H$	IP	HBSS
	Mouse	M	-	IP	-
	Mouse	M	C-57	IP	Saline
	Mouse	M	$C_{57}B_1/Fe$	IP	Saline
	Mouse	M	CBA-J	IP Oral	CMC
	Mouse	M	-	Oral	Water
				Oral	Diet

Dose (mg/kg)	Pre-implantation losses[c]	Post-implantation losses[c]	Dominant lethality[c]	Comments	References
0.2	-	+	+	Postmeiotic, minor effect in premeiotic (strain: $C_3D_2F_1$/J(C_3H/HeJxDBA$_2$J)	78
0.2	-	+	+	Postmeiotic	79
0.4	-	-	-	Premeiotic (only spermatocytes tested)	67
0.8	-	+	+		
1.8	+	+	+		
1.6	+	+	+		
2.0	+	+	+		
3.0	+	+	+		
4.0	+	+	+		
0.125	+	+	+	Postmeiotic	80
0.250	+	+	+		
0.375	+	+	+		
0.500	+	+	+		
0.01 (x5)	-	+	+		68
0.05 (x5)	-	+	+	Postmeiotic	81
0.05 (x5)	+	+	+	Postmeiotic and spermatocytes	69
0.2 (x5)	+	+	+		
0.01	o	o	o	Postmeiotic (administered in drinking water/4 weeks	72
0.02	o	o	o		
0.04	o	o	o		
0.08	o	o	o	Doses with positive response not identified	
0.01	-	-	-	Administered in diet for 3 weeks	72
0.02	-	-	-		
0.04	-	-	-		
0.08	-	-	-		

(continued)

Table 2. (continued)

Compound (CAS No.)	Species	Sex	Strain	Route[a]	Vehicle[b]
TEM (cont'd.)	Mouse	M	DBA/2J	IP	-
				Oral	-
	Mouse	M	CF-1	Oral	Diet
				Oral	Diet
	Mouse	F	T-stock	IP	HBSS
			101xC H	IP	HBSS
			SECxC BL	IP	HBSS
	Mouse	F	SECxC BL	IP	HBSS
	Rat	M	JH	IP	HBSS
			SBES	IP	HBSS
	Rat	M	-	IP	-
	Rat	M	Osborne-Mendel	IP	CMC
	Rat	M	Sprague-Dawley	IP	-

Dose (mg/kg)	Pre-implantation losses[c]	Post-implantation losses[c]	Dominant lethality[c]	Comments	References
0.3	+	+	+	Oral dosing induced a lower frequency of DL than IP	70
0.3	-	+	+		
1	o	o	o	Dose given is per day; × 45 days	71
10	o	o	o		
50	o	o	o	Complete sterility	
0.1	+	+	+	Date given is per day; × 45 days	
0.2	+	+	+	Postmeiotic	
1.6	-	+	+	First period after dosing	49
1.6	-	+	+		
1.6	-	+	+	No strain differences	
1.0	o	o	o	Destruction of oocytes in early stages	10
1.6	o	o	o		
0.4	+	+	+	All postmating pre-cleavage states	26
0.6	+	+	+		
0.4	+	+	+	No strain differences	
0.6	+	+	+		
0.025	-	+	+	Postmeiotic	73
0.05	-	+	+		
0.01	-	+	+		
0.2	+	+	+		
0.4	+	+	+		
0.05 (x5)	+	+	+	Postmeiotic	69
0.5	+	+	+	Postmeiotic and spermatocytes	74

(continued)

Table 2. (continued)

Compound (CAS No.)	Species	Sex	Strain	Route[a]	Vehicle[b]
TEM (cont'd.)	Rat	M	-	IP	-
	Rat	M	-	IP	-
	Rat	M	Osborne-Mendel	IP	Water
	Hamster	M	-	IP	Saline
	G. pig	M	-	IP	Saline
Trenimon (68-76-8)	Mouse	M	C_3H	IP	Saline
	Mouse	M	C_3H	IP	Saline
	Mouse	M	CFLP	IP	Saline
	Mouse	F	C_3H	IP	Saline
	Mouse	F	C_3H	IP	Saline
	Mouse	F	C_3H CFLP NMRI	IP IP IP	Saline Saline Saline
	Mouse	F	-	IP	Single

[a]Route: IP = intraperitoneal; SC - subcutaneous; IV = intravenous;

[b]Vehicle: HBSS = Hanks balanced salt solution; PO Buff = phosphate

[c]Results: + = Data judged to represent a positive response,
- = Data judged to represent a negative response,
o = Data not specific, i.e., affected fertility or

Dose (mg/kg)	Pre-implantation losses[c]	Post-implantation losses[c]	Dominant lethality[c]	Comments	References
0.5	+	+	+	Postmeiotic	82
0.125	+	+	+	Postmeiotic	79
0.250	+	+	+		
0.375	+	+	+		
0.500	+	+	+		
0.5	-	+	+	Postmeiotic and	83
0.2 (x5)	-	+	+		
0.2	+	+	+	Postmeiotic	75
0.2	-	+	+	Postmeiotic	76
0.125	-	+	+	Postmeiotic	42
0.250	-	+	+		84
0.125	+	+	+	Postmeiotic and pre-	43
0.250	+	+	+	meiotic losses	
0.125	+	+	+	Post- and premeiotic	44
0.250	+	+	+	losses	
0.125	o	o	o	Maximum effect is when dosed at 4 weeks of age	42
0.125	+	+	+		86
0.125	+	+	+	Pre-ovulation stages	42
0.125	+	+	+	Pre-ovulation stages	87
0.125	+	+	+	Pre-ovulation stages	87
0.25	-	+	+		31

IT = intratesticular; ISC = intrascrotal.

buffer; CMC = carboxymethyl cellulose; TRC = tricaprylin.

i.e., significantly different from control values.
i.e., within control range of values.
survival.

Two papers illustrate the different sensitivities of pre- and postcopulatory stages of oogenesis. When 101 × C_3H females were dosed with 300 mg/kg EMS during the pre-ovulatory stage the evidence for dominant lethality was not convincing (22). Dosing of C_3H × 101 with 250 mg/kg after mating, i.e., during metaphase II or early pronuclear stage, a high incidence of dominant lethals was detected (26). No strain differences in response were noted when females were dosed after mating.

Only one publication (27) describes the dominant lethal effects of EMS in rats. This study demonstrated pre- and postimplantation dominant lethality in the postmeiotic cells in male mice given single IP doses of 100 mg/kg EMS. In a study of the influence of EMS in male fertility in the rat (27), it was shown that a single oral dose of 300 mg/kg and 15 daily doses of 200 mg/kg had similar results, reducing litter sizes in the period corresponding to the postmeiotic stages of spermatogenesis.

A number of facts can be deduced from the literature surveyed. Strain differences influence the incidence of dominant lethals in mice, though to a lesser degree in males than in females. There is little difference between the results of oral and IP dosing. The lowest single dose of EMS to induce dominant lethals in male mice was 150 mg/kg. The highest dose inducing no detectable dominant lethality in males was 100 mg/kg. The lowest dose to induce dominant lethals in female mice was 250 mg/kg when given during postcopulatory precleavage stages. A single IP dose of 100 mg/kg EMS produced dominant lethals in male rats. Dominant lethality in both species was confined to postmeiotic spermatogenesis. No premeiotic effects were detected.

Methyl Methanesulfonate (66-27-3)

Methyl methanesulfonate (MMS) has been used to induce dominant lethal mutations in male and female mice and in male rats.

A number of studies have demonstrated a high frequency of pre- and postimplantation dominant lethality in the spermatozoa and late spermatids of male mice after IP doses of 50 mg/kg MMS. In dose-response experiments carried out in a number of laboratories (28, 29), dominant lethality increased with dose over the range 20 to 80 mg/kg after IP dosing.

In a dose-response study in male mice given single IP doses of MMS ranging from 3.125 to 100 mg/kg body weight, dominant lethality, characterized by an increase in early fetal deaths, was detected in the first week after dosing with 12.5 mg/kg. No dominant lethality was recorded in animals given 6.25 mg/kg MMS (20). The authors concluded that a no-effect level was not clearly defined but it does appear to be in the region of 50 to 10 mg/kg.

In studies in which female mice were given 100 mg/kg MMS by IP or IM injections, dominant lethals were induced (30,31). Strain differences have been reported (9).

A single IP dose of 50 mg/kg MMS induced dominant lethals in the postmeiotic cells in male rats, and 25 daily doses of 2.5 mg/kg reduced male fertility (27,32).

Diethylnitrosamine (DEN) (55-18-5)

A search of the literature revealed only one report (38) of a dominant lethal assay using DEN. Male mice were given single IP doses of 13.5 g mg/kg of DEN which failed to induce a significant increase in dominant lethality. Examination of the data show that both pre- and postimplantation losses were increased when compared with control values but the differences were not statistically significant.

A more detailed investigation of diethylnitrosamine is necessary to allow a confident assessment of its dominant lethal potential.

Dimethylnitrosamine (DMN) (62-75-9)

Two groups have reported dominant lethal studies in mice dosed with DMN. In the first (4), no evidence of dominant lethality was recorded in male mice dosed IP with 8 or 9 mg/kg DMN. In the second study (38), subcutaneous dosing with 4.4 mg of DMN/kg induced dominant lethal mutations in postmeiotic stages characterized by reduced fetal implants and an increase in early fetal deaths. This study was carried out in male mice and the data provide evidence of weak dominant lethal induction by DMN.

The conflicting results in the two studies might be explained by a number of factors such as different strains of mice, the route of application (i.e., IP versus subcutaneous) and the different dose levels used.

The results of the second study indicate that, under the experimental conditions used, DMN induced dominant lethality in male (101 × C_3H) F_1 mice.

Cyclophosphamide (50-18-0)

Dominant lethality has been reported in male mice after IP doses of cyclophosphamide ranging from 60 to 240 mg/kg. Postmeiotic cells were the most sensitive to both pre- and postimplantation losses. At higher doses, i.e., 200 mg/kg or more, some premeiotic changes have been recorded, including fetal losses (39) and reduced fertility (40), although the induction of dominant lethality in premeiotic cells by cyclophosphamide required confirmation.

Results of unpublished studies (41) confirm that single IP doses of 60 mg/kg cyclophosphamide induced postmeiotic dominant lethality and that doses of 40 mg/kg or less gave negative results. Single oral doses ranging from 80 to 210 mg cyclophosphamide/kg also induced a high frequency of dominant lethals in males as did single IP doses of 100 mg/kg (40,41).

In a comparison between single and repeated IP doses it was shown that the dominant lethal effects of single IP doses and the same total dose fractionated into 20 applications was very similar (23). In the same paper dominant-lethality was described after male mice had been maintained on drinking water containing 0.005, 0.01, or 0.02% cyclophosphamide. Another study by the same author showed that when 0.01% cyclophosphamide was administered to males over a period of 12 weeks, dominant lethals were detected after only 1 week, reaching a maximum frequency after 2 to 4 weeks (23). No significant increase in the frequency of dominant lethals occurred in subsequent weeks.

Female mice appear to be less sensitive to the induction of dominant lethals by cyclophosphamide than males, and positive results were obtained after a single IP dose of 200 mg/kg (31).

Marginal increases in fetal losses were detected after female mice had consumed drinking water containing 0.01% cyclophosphamide for 4, 8 or 12 weeks.

The dominant lethal effects of cyclophosphamide may be summarized as follows. Single IP doses of 60 mg/kg induced postmeiotic dominant lethals in male mice. No dominant lethals were detected after single IP doses of 48 mg/kg. Dominant lethals were induced in female mice by single IP doses of 210 mg/kg, or oral doses of 200 mg/kg. No dominant lethality was detected after oral dosing of females with 150 mg/kg.

Myleran (55-98-1)

Myleran has been shown to induce dominant lethal mutations in premeiotic, meiotic, and postmeiotic stages in male mice. The lowest dose tested, 10 mg/kg by IP injection, induced dominant lethals in spermatocytes and spermatozoa, but had no effect on early spermatids. A dose of 20 mg/kg produced sterility during premeiotic stages (47).

In a second study, male mice dosed IP with 25 mg/kg Myleran showed postimplantation dominant lethality in postmeiotic and meiotic cells and pre-implantation losses after 50 mg/kg (4).

In female mice given single IP doses of 40 mg/kg Myleran (49) postimplantation dominant lethality was observed, characterized by a reduction in live embryos with a corresponding increase in the number of dead implants. The total number of implants was unchanged indicating an absence of pre-implantation losses. There was little difference in the frequency of dominant lethals at various mating intervals. In the same study only small strain-related differences in sensitivity were detected. T-stock females were slightly more resistant to dominant lethal induction by Myleran than two hybrid strains. Only one dose was used in this experiment.

The effect of a range of single IP doses from 10 to 60 mg/kg was tested in another study in female mice (10). Dominant lethality was detected at 20 mg/kg and above, and all four dose levels had a detrimental effect on fertility. The rapid cessation of reproduction in females dosed with 40 or 60 mg/kg was thought to be due to the rapid destruction of oocytes in early stages. Doses of 10 and 20 mg/kg Myleran reduced fertility throughout the reproductive life of the mice.

After dosing male rats with 4 or 6 mg/kg Myleran, pre-implantation losses occurred during the seventh and tenth weeks (27). No increase in postimplantation lethality was recorded. During the fifth and sixth weeks after dosing with 10 mg/kg, a slight reduction in the number of viable embryos was found in addition to pre-implantation losses in later weeks. The results indicate that a single IP dose of 4 mg/kg Myleran induced pre-implantation dominant lethality in premeiotic cells and that 10 mg/kg also affects meiotic cells. There was no evidence of dominant lethality in spermatids or spermatozoa.

In a repeated dosing experiment it was shown that doses of 5 mg/kg × 2 and 2 mg/kg × 5 produced a similar effect on rat spermatogenesis as one IP dose of 10 mg/kg Myleran (37).

Myleran-induced dominant lethality may be summarized as follows. A single IP dose of 10 mg/kg induced dominant lethals in spermatocytes and mature spermatozoa in male mice. A single IP dose of 20 mg/kg induced dominant lethals in female mice, with a slight reduction in fertility after dosing with 10 mg/kg. Small strain differences were detected in the response of female mice to Myleran. A single IP dose of 4 mg/kg induced postimplantation dominant lethality in premeiotic cells of male rats. A "no detectable effect" dose was not obtained for mice of either sex or for male rats.

TEPA (545-55-1)

Dominant lethal assays have been carried out on TEPA by a number of workers, and in all cases male mice were used.

After intraperitoneal dosing of male mice with TEPA, reductions in the total number of implantations and increases in early fetal deaths confirmed pre- and postimplantation dominant lethality in postmeiotic spermatogenesis. Late spermatids appear to be the most sensitive stage. Although dominant lethals were not demonstrated in premeiotic cells, evidence was presented to suggest a toxic effect to spermatogonia resulting in reduced fertility (4) or complete sterility (50) after IP doses of 5 or 10 mg/kg, respectively.

The lowest single IP doses inducing a significant degree of dominant lethality were 1.0 mg/kg (41,51) and 1.25 mg/kg (34,52). No detectable dominant lethals were recorded in male mice given 0.5 or 0.625 mg/kg.

Two repeat dosing studies have been described (4). After oral doses of 2.0 mg/kg on 48 consecutive days to male mice, no dominant lethality was detected. Repeated IP dosing of 0.1 mg/kg, again for 48 consecutive days, induced both post- and premeiotic dominant lethal mutations, as demonstrated by reduced implantations and increased fetal deaths. No detectable effects were recorded in animals given 48 repeated IP doses of 0.01 mg/kg, i.e., a total dose of 0.48 mg/kg.

When a total dose of 1.0 mg/kg was given, either as a single IP injection, or when fractionated as 5 × 0.2 mg/kg or 10 × 0.1 mg/kg, dominant lethals were detected, but fractionation of the dose reduced the frequency (53).

METEPA (57-39-6)

Three dominant lethal assays have been reported in the literature in which male mice were given single IP doses of METEPA in tricaprylin solution. In two of them (52,57) a dose-related increase in dominant lethals occurred ranging from 12.5 to 50 or 100 mg/kg. A dose of 12.5 mg/kg METEPA induced pre- and postimplantation dominant lethality in postmeiotic cells. At higher doses there was some evidence of premeiotic and meiotic effects (4,52). No detectable dominant lethals were recorded in male mice given 0.5 or 0.625 mg/kg.

ThioTEPA (52-24-4)

Assays in male mice demonstrated that IP dosing with 2, 5, or 10 mg/kg ThioTEPA induced dominant lethals in postmeiotic cells. Both pre- and postimplantation losses were detected. The higher dose of 10 mg/kg also induced dominant lethal mutations in meiotic and premeiotic stages (4,23,35).

Unpublished results (41) showed that postmeiotic dominant lethality was induced in male mice by single IP doses of 0.64 mg/kg ThioTEPA. Negative results were obtained in mice given single doses of 0.32 mg/kg. Srám (23) also demonstrated dominant lethality in male mice after 5 × 0.2 mg/kg doses of ThioTEPA.

The influence of the genotype of the female on the frequency of dominant lethals has also been investigated (58). When mated to males given IP injections of 2 mg/kg ThioTEPA, C57BL/6J Sto (B6) females produced the highest frequency of dead implants, CBA/Lac Sto (CBA) produced the lowest frequency and CBA x B6, a hybrid between the two, gave intermediate results.

In an earlier study (59), male mice were given oral doses of ThioTEPA on each of 21 successive days. Doses of 0.1 mg/kg × 21 and 0.5 mg/kg × 21 induced total infertility, and 21 daily doses of 0.02 mg/kg reduced the number of viable embryos.

Cadmium Compounds (7440-43-9)

No evidence of a dominant lethal effect has been detected in mice of either sex after IP dosing with solutions of cadmium chloride. In one experiment (4), it was suggested that the pregnancy rate was reduced in females mated to males after dosing with cadmium, but this finding requires confirmation.

The highest dose used showing no detectable dominant lethality in mice was 7 mg/kg cadmium chloride. In a study in which male rats were given single oral doses ranging from 6.25 to 25.0 mg/kg cadmium chloride, no effect on fertility was detected (60).

Epichlorohydrin (106-89-8)

No dominant lethality was detected in male mice after single IP doses of up to 150 mg/kg epichlorohydrin, after single oral doses of 20 and 40 mg/kg, or after 5 repeated doses from 1 to 16 mg/kg. In some of the repeated dosing groups there was a suggestion of an impairment of male fertility but this requires substantiation (4,63).

Ethylenimine (151-56-4)

Two assays carried out after dosing male mice with 5 or 6 mg/kg ethylenimine demonstrated postimplantation dominant lethality in postmeiotic cells. No pre-implantation losses or effects on spermatocytes or spermatogonia were recorded. The greatest frequency of dominant lethals occurred in the first week after dosing, i.e., in mature spermatozoa. A "no detectable effect" dose was not used in either study.

Triethylenemelamine (TEM) (51-18-3)

The dominant lethality of triethylenemelamine has been investigated in mice, rats, golden hamsters, and guinea pigs.

In mice, TEM has proved a potent mutagen in both sexes, and a single IP dose of 0.2 mg/kg is regularly used as a positive control in assays in male mice. Postmeiotic dominant lethality has been shown to increase with dose from 0.05 to 0.4 mg/kg after single IP injections to male mice (66,67) with early fetal deaths proving a more sensitive measure of dominant lethals than preimplantation losses. The highest frequency of dominant lethals occurred in mid-spermatids.

Premeiotic cells are more resistant to the induction of dominant lethals and more than a 10-fold increase in dose, i.e., 0.8 mg/kg, is required to demonstrate dominant lethals in spermatocytes (67). An earlier study, in which male mice were given IP doses of 0.4 mg/kg TEM, demonstrated postmeiotic dominant lethality with loss of fertility in premeiotic stages.

The effects of repeated dosing on male mice have also been studied. Five daily IP doses of 0.01 mg/kg TEM induced postimplantation dominant lethality (68), and a comparison of IP and oral dosing showed that five oral doses of 0.2 mg/kg TEM induced a similar frequency of dominant lethals to five IP doses of 0.05 mg/kg. This was confirmed in another experiment (70).

When TEM was administered to male mice in the diet at rates of 1, 10, or 50 mg/kg/day for 45 days, all the 50 mg/kg males died, 3 of the 10 mg/kg group died, and surviving males were completely sterile (71). This study was repeated using lower doses and postmeiotic dominant lethality was induced in male mice maintained for 45 days on diet containing the equivalent of 0.1 mg TEM/kg per day (71). However, when doses ranging from 0.01 and 0.08 mg/kg/day were administered to the diet for 21 days, no dominant lethality was detected (72). The same paper describes marginal dominant lethal effects in male mice when a similar range of concentration of TEM was administered in the drinking water.

Single IP doses to 1.6 mg/kg TEM to female mice induced dominant lethals in the first 5-day period after dosing. No differences in response were detected between T-stock females and two hybrid strains (49). When the total reproductive capacity of females was investigated after single IP doses of 1.0 or 1.6 mg/kg TEM destruction of oocytes was detected in early stages of follicular development (10). Exposure during postcopulation, precleavage stages proved more sensitive to dominant lethality than precopulatory oocytes, and a single IP dose of 0.4 mg/kg reduced fetal implants and increased early fetal deaths.

The dominant lethal effects of TEM have been studied in detail in male rats. A single IP dose of 0.025 mg/kg induced a low frequency of postmeiotic dominant lethality (73), and a higher dose of 0.5 mg/kg was effective in spermatocytes (74). When 0.05 mg/kg TEM was given IP on each of five successive days, postmeiotic dominant lethals in spermatocytes.

Male golden hamsters (75) and guinea pigs (76) responded to single IP doses of 0.2 mg TEM/kg with postmeiotic dominant lethality. The sensitivities of four species compared with similar, hamsters and rats showing a slightly higher frequency of dominant lethality than guinea pigs and mice.

The dominant lethal effects of triethylenemelamine in mammals may be summarized as follows. Single IP doses of 0.05 mg/kg induced postmeiotic dominant lethality in male mice. Single IP doses of 0.035 mg/kg did not result in any detectable dominant lethal mutations. Postmeiotic dominant lethality was detected in male mice after 0.1 mg/kg TEM was administered in the diet each day for 45 days. In female mice, single IP doses of 1.6 mg/kg TEM induced dominant lethal mutation in precopulatory stages of oogenesis. Only 0.4 mg/kg TEM was required to induce dominant lethality in postcopulation, precleavage stages. A single IP dose of 0.025 mg/kg induced a small frequency of postmeiotic dominant lethality in male rats. A "no detectable effect" dose was not determined in rats or female mice.

Trenimon (68-76-8)

Dominant lethal mutations have been reported in both male and female mice after intraperitoneal dosing with Trenimon. In all cases either 0.125 or 0.250 mg/kg doses were used, reflecting the clinical doses used in practice.

In male mice 0.125 mg/kg induced postmeiotic dominant lethality characterized by postimplantation fetal deaths. In two studies (43,44), marginal pre-implantation losses were detected in spermatocyte and spermatogonial stages suggesting premeiotic dominant lethality.

Postimplantation fetal deaths were also demonstrated in female mice after dosing with 0.125 mg/kg, and the most sensitive stages of oogenesis were reported as pre-ovulation metaphases I and II. Where pre-implantation losses were detected, they were thought to be caused by a cytotoxic effect on the oocytes rather than by dominant lethality.

No strain differences were suggested when CFLP and C_3H males were compared.

A "no detectable effect" dose was not used in any of the studies reported, though this was shown to be below 0.125 mg/kg.

REFERENCES

1. Röhrborn, G. In: Chemical Mutagenesis in Mammals and Man, F. Vogel and G. Röhrborn, Eds., Springer Verlag, Berlin-Heidelberg-New York, 1970, pp. 148-155.

2. Bateman, A. J., and Epstein, S. S. In: Chemical Mutagenesis, A. Hollaender, Ed., Plenum Press, New York-London, 1971, pp. 541-567.

3. Generoso, W. M. Genetics 61: 461-470 (1969).

4. Epstein, S. S., Arnold, E., Andrea, J., Bass, W., and Bishop, Y. Toxicol. Appl. Pharmacol. 23: 288-325 (1972).

5. Epstein, S. S., and Schafner, H. Nature 219: 385-387 (1968).

6. Anderson, D., Hodge, M. C. E., and Purchase, I. F. H. Mutat. Res. 40: 359-370 (1976).

7. Parkin, R., Waynforth, H. B., and Magee, P. N. Mutat. Res. 21: 155-161 (1973).

8. Ehling, U. H., Cumming, R. B., and Malling, H. V. Mutat. Res. 5: 417-428 (1968).

9. Generoso, W. M. Genetics 61: 461-470 (1969).

10. Generoso, W. M., Stout, S., K., and Huff, S. W. Mutat. Res. 13: 171-184 (1971).

11. Ehling, U. H. Mutat. Res. 11: 35-44 (1971).

12. Ehling, U. H. Mutat. Res. 13: 433-436 (1971).

13. Ehling, U. H., and Neuhauser, A. EMS Newsletter 6: 14-15 (1972).

14. Bempung, M. A., and Trower, E. C. J. Hered. 64: 324-328 (1973).

15. Kikuchi, Y., Mizutani, M., and Kaziware, K. Takeda Kenkysho Ho 30: 762-770 (1971).

16. Ehling, U. H. Mutat. Res. 26: 285-295 (1974).

17. Generoso, W. M. EMS Newsletter 6: 14 (1972).

18. Generoso, W. M., Russell, W. L., Huff, S. W., Stout, S. K., and Gosslee, D. G. Genetics 77: 741-752 (1974).

19. Ray, V. A., Holder, H. E., Salsburg, D. S., Ellis, J. H., Just, L. J., and Hyneck, M. L. Toxicol. Appl. Pharmacol. 30: 107-116 (1974).

20. Arnold, D. W., Kennedy, G. L., Jr., Keplinger, M. L., and Calandra, J. C. Toxicol. Appl. Pharmacol. 38: 79-84 (1976).

21. Soares, E. R. Mutat. Res. 37: 245-252 (1976).

22. Generoso, W. M., and Russell, W. L. Mutat. Res. 8: 589-598 (1969).

23. Šrám, R. J. Mutat. Res. 41: 25-42 (1976).

24. Anderson, D., McGregor, D. B., and Purchase, I. F. H. Mutat. Res. 40: 349-358 (1976).

25. Anderson, D., McGregor, D. B., Purchase, I. F. H., Hodge, M. C. E., and Cuthbert, J. A. Mutat. Res., 43: 231-246 (1977).

26. Suter, K. E., and Generoso, W. M. Mutat. Res. 34: 259-270 (1976).

27. Partington, M., and Jackson, H. Genet. Res. 4: 333-345 (1963).

28. Ehling, U. H., Frohberg, H., Schulze-Schencking, M., Lang, R., Lorke, D., Machemer, L., Matter, B. E., Mueller, D., Röhrborn, G., Buselmaier, W., and Roll, R. Mutat. Res. 29: 261 (1975).

29. Dean, B. J., and Johnstone, A. Mutat. Res., 42: 269-278 (1977).

30. Dean, B. J., and Blair, D. Mutat. Res. 40: 67-72 (1976).

31. Machemer, L., and Lorke, D. Mutat. Res. 29: 209-214 (1975).

32. Partington, M., and Bateman, A. J. Heredity 19: 191-200 (1964).

33. Moutschen, J. Mutat. Res. 8: 581-588 (1969).

34. Epstein, S. ., Bass, W., Arnold, E., and Bishop, Y. Food Cosmet. Toxicol. 8: 381-401 (1970).

35. Machemer, L., and Hess, R. Experientia 27: 1050-1052 (1971).

36. Beliles, R. P., Korn, N., and Benson, B. W. Res. Commun. Chem. Pathol. Pharmacol. 5: 713-724 (1973).

37. Jackson, H., Fox, B. W., and Craig, A. W. J. Reprod. Fert. 2: 447-465 (1961).

38. Propping, P., Röhrborn, G., and Buselmaier, W. Mol. Gen. Genet. 117: 197-209 (1972).

39. Brittinger, D. Humangenetik 3: 156-165 (1966).

40. Fritz, H., Mueller, D., and Hess, R. Agents Actions 3: 35-37 (1973).

41. Šrám, R. J. Unpublished studies.

42. Röhrborn, G., and Vogel, F. Dtsch. Med. Wochenschr. 92: 2315-2321 (1967).

43. Röhrborn, G. In: Chemical Mutagenesis in Mammals and Man, F. Vogel and G. Röhrborn, Eds., Springer Verlag, Berlin-Heidelberg-New York, 1970, pp. 294-316.

44. Machemer, L., and Stenger, E. G. Arzneim. Forsch. 21: 1037-1039 (1971).

45. Leonard, A., Vandesteene, R., and Marsboom, R. Mutat. Res. 26: 427-430 (1974).

46. Dean, B. J., Doak, S. M. A., and Somerville, H. Food Cosmet. Toxicol. 13: 317-323 (1975).

47. Šrám, R. J., and Zhurkov, V. S. Paper presented at the 6th Annual Meeting of the European Environmental Mutagen Society, Gernrode, G.D.R., 1976.

48. Ehling, U. H., and Malling, H. V. Genetics 60: 174-175 (1968).

49. Generoso, W. M., Huff, S. W., and Stout, S. K. Mutat. Res. 11: 411-420 (1971).

50. Joshi, S. R., Page, E. C., Arnold, E., Bishop, Y., and Epstein, S. S. Genetics 65: 483-494 (1970).

51. Šrám, R. J., Benes, V., and Zudova, Z. Folia Biol. (Prague) 16: 407-416 (1970).

52. Epstein, S. ., Arnold, E., Steinberg, K., Makintosh, D., Shafner, H., and Bishop, Y. Toxicol. Appl. Pharmacol. 17: 23-40 (1970).

53. Šrám, R. J., and Zudova, A. Folio Biol. (Prague) 19: 58-67 (1973).

54. Epstein, S. S., Bass, W., Arnold, E., Bishop, Y., Joshi, S., and Adler, I. D. Toxicol. Appl. Pharmacol. 19: 134-146 (1971).

55. Barteva, R., and Šrám, R. J. Unpublished data.

56. Benes, V., Šrám, R. J., and Tuscang, R. J. Hyg. Epidemiol. Microbiol. Immunol. 19: 163-172 (1975).

57. Wallenstein, S., Brobst, J., and Bagdon, R. E. Mutat. Res. 26: 458-459 (1974).

58. Malashenko, A. M., and Surkova, N. I. Sov. Genet. 11: 210-214 (1975); Genetika 11(2): 105-111 (1975).

59. Hershberger, L. G., Hansen, D. M., and Hansen, L. M. Proc. Soc. Exptl. Biol. Med. 131: 667-669 (1969).

60. Dixon, R. L., Lee, I. P., and Skerins, R. J. Environ. Health Perspect. 13: 59-67 (1976).

61. Gillavod, N., and Léonard, A. Toxicology 5: 43-47 (1975).

62. Suter, K. E. Mutat. Res. 30: 365-374 (1975).

63. Šrám, R. J., Cerna, M., and Kučerova, M. Biol. Zbl. 95: 451-462 (1976).

64. Malashenko, A. M. Sov. Genet. 4: 538-543 (1968); Genetika 4(4): 158-164 (1968).

65. Malashenko, A. M., and Egorov, I. K. Sov. Genet. 4: 14-18 (1968): Genetika 4(1): 21-27 (1968).

66. Matter, B. E., and Generoso, W. M. Mutat. Res. 17: 87-92 (1973).

67. Matter, B. E., and Generoso, W. M. Genetics 77: 753-763 (1974).

68. Petersen, K. W., Brandchaft, D., Turner, M., and Frigge, F. H. J. Anat. Rec. 166: 362 (1970).

69. Collins, T. F. X. Food Cosmet. Toxicol. 10: 353-361 (1972).

70. Soares, E. R., and Sheridan, W. Mutat. Res. 31: 342-343 (1975).

71. Hasting, S. E., Huffmann, K. W., and Gallo, M. A. Mutat. Res. 40: 371-378 (1976).

72. Jorgenson, T. A., Newell, G. W., Spangord, R. J., Kay, D. L., and Gribling, P . L. Mutat. Res. 26: 459 (1974).

73. Bateman, A. J. Genet. Res. 1: 381-392 (1960).

74. Green, S., Carr, J. V., Sauro, F. M., and Legator, M. S. J. Pharmacol. Exptl. Therap. 187: 437-443 (1973).

75. Lyon, M. F., and Smith, B. D. Mutat. Res. 11: 45-58 (1971).

76. Cox, B. D., and Lyon, M. F. Mutat. Res. 30: 293-298 (1975).

77. Cattanach, B. M., and Edwards, R. G. Proc. Roy. Soc. (Edinburgh) B67: 54-64 (1958).

78. Petersen, K. W., and Legator, M. S. Mutat. Res. 17: 87-92 (1973).

79. Thayer, P. S., and Kensler, C. J. Toxicol. Appl. Pharmacol. 25: 157-168 (1973).

80. Arnold, D. W., Kennedy, G. L., Keplinger, M. L. Mutat. Res. 26: 459-460 (1974).

81. Petersen, K. W., Legator, M. S., and Figge, F. H. J. Mutat. Res. 14: 126-129 (1972).

82. Palmer, K. A., Green, S., and Legator, M. S. Food Cosmet. Toxicol. 11: 53-62 (1973).

83. Green, S., Sauro, F. M., and Frideman, L. Food Cosmet. Toxicol. 13: 507-510 (1975).

84. Röhrborn, G. Humangenetik 1: 576-578 (1965).

85. Röhrborn, G., and Berrang, H. Mutat. Res. 4: 231-233 (1967).

86. Röhrborn, G. Humangenetik 2: 81-82 (1966).

87. Machemer, L., and Hess, R. Experientia 29: 190-192 (1973).

CHAPTER 19

MUTAGENICITY OF SELECTED CHEMICALS IN SISTER CHROMATID EXCHANGE ASSAYS

S. Wolff and P. Perry

Laboratory of Radiobiology and Department of Anatomy

University of California - San Francisco, San Francisco

California 94143

and A. T. Natarajan

Laboratory of Chemical Mutagenesis

University of Leiden, Leiden, The Netherlands

INTRODUCTION

New methods for staining chemically dissimilar sister chromatids have allowed the nonautoradiographic observation of sister chromatid exchanges (SCEs) with great clarity and precision. The methods, which are quite simple, have led to experiments showing that in mammalian cells alkylating agents and other mutagenic carcinogens can induce lesions in DNA that lead to the formation of sister chromatid exchanges. These exchanges are induced at far lower concentrations of the chemicals than are needed to induce ordinary chromosome aberrations. The sensitivity of the system is such that it has been suggested many times that it be used to replace the usual cytogenetic test for these chemicals.

In vitro experiments on mammalian cells in culture as well as in vivo experiments on whole animals to which the drugs have been administered have been carried out. Both systems seem capable of detecting the effects of chemicals that do not require metabolic activation, as well as those that do.

In order to make cells containing chromosomes that have chemically dissimilar sister chromatids, the cells are allowed to replicate twice in the presence of the thymidine analog, 5-bromodeoxyuridine (BrdUrd). Because DNA replicates semiconservatively, the chromosomes will now contain one chromatid that is bifilarly substituted with BrdUrd, and a sister chromatid that is unifilarly substituted. These chemically different sister chromatids will stain differently with a fluorochrome or with a fluorochrome plus Giemsa (FPG technique), resulting in preparations in which it is very easy to see exchanges between the differentially stained sister chromatids. Alternatively, the cells can be allowed to replicate for only one cycle in the presence of BrdUrd before being allowed to replicate once again in its absence. Under these conditions, one chromatid of a chromosome will be unifilarly substituted, whereas its sister will be unsubstituted. Again, the two are chemically dissimilar and will stain differently from one another. The experiments can be carried out on cultured fibroblasts or even on cultured peripheral lymphocytes from animals including man.

In vivo experiments can be carried out by injecting the chemicals into the animal to be studied. The animal's peripheral lymphocytes can then be cultured in the presence of BrdUrd, as in a typical in vitro experiment. Alternatively, the animal can be infused with BrdUrd for one round of replication, which is then followed by enough time to allow a round of replication in the absence of the analog. Metaphases that have been accumulated in bone marrow or other tissues will now have cells containing chromosomes that will stain differentially.

The total time for these experiments will vary according to the cell cycle time of the cells being studied. For instance, with Chinese hamster ovary (CHO) cells in tissue culture, two rounds of replication occur in 24 hr. The cells can be treated with the chemical either for two cell cycles, or, if the chemical needs metabolic activation, can be treated for a short period of time in the presence of S-9 Mix prior to BrdUrd addition (at the beginning of the experiments) and then cultured for 24 hr in the presence of BrdUrd. If human lymphocytes are being studied, they can be cultured for 54-56 hr, at the end of which time half of the metaphase cells will have replicated only once in the presence of BrdUrd, and half will have replicated twice.

The chromosomes are very easy to score for sister chromatid exchanges. Thus, the test is easy, quick, accurate, and, incidentally, inexpensive.

Table 1 includes the results obtained in published studies of the chemicals included in the list used for the Comparative Mutagenesis Workshop.

Table 1. Sister chromatid exchanges (SCE) induced by test mutagens

Mutagen	Cell	In vivo	In vitro	S-9	Concentration	Exposure time	SCE/cell increment over control	References
Aflatoxin B1	CHO		+	-	10^{-6}-10^{-4} M	30 min-24 hr	1.5-3×	1
	CHO							
	CHO		+	+	10^{-6} M	30 min	3×	
Cyclophosphamide	CHO		+	-	10^{-5}-10^{-3} M	24 hr	1.6×	2
	CHO		+	-	10^{-5}-10^{-3} M	24 hr	1.5×	3
	CHO		+	+	10^{-3} M	20 min	4.5×	3
	Rabbit lymphocytes	+			20-35 mg/kg	1 day	3.7×	4
	Mouse							
	Bone marrow	+			12.5-25 mg/kg		3.8-6×	5
	Spermatogonia	+		-	5-20 mg/kg		5×	6
	Mouse 3T3		+	-	20-500 μg/kg		1.5×	6
	Human lymphocytes		+	-	20-500 μg/kg		-	7

(continued)

Table 1. (continued)

Mutagen	Cell	In vivo	In vitro	S-9	Concentration	Exposure time	SCE/cell increment over control	References
DEN	CHO		+	+	4–100 mM	1 hr	2	7
			+	–	200 mM	1 hr	–	7
	Mouse bone marrow	+			25–100 mg/kg	28 hr	–	8
DMN	CHO		+	+	2.7–40 mM	1 hr	–	7
			+	–	135 mM	1 hr	–	7
	Mouse bone	+			0.1–1 mg/kg	28 hr	1.9–2.9	8
EMS	CHO		+	–	3×10^{-4}–3×10^{-2} M	24 hr	6	2
	CHO		+	–	10^{-5}–10^{-3} M	24 hr	3	3
	Rabbit lymphocytes	+		–	50–200 mg/kg	1–15 day	3	4

	Human lymphocytes						
	Normal	+	-	0.25 mg/ml	72 hr	4×	9
	Fanconi's anemia	+	-	0.25 mg/ml	72 hr	3×	9
	Human fibroblasts						
	Normal (GM637)	+	-	10^{-6}-10^{-5} M	48 hr	1.1×	10
	Xeroderma (XP12RO)	+	-	10^{-6}-10^{-5} M		1.5-2.7×	10
MMS	CHO	+	-	3×10^{-6}-3×10^{-5} M	24 hr	6×	2
	Rabbit lymphocytes	+	-	25 mg/kg	1-15 day	1.6×	3
	Human fibroblasts						
	Normal (GM637)	+	-	10^{-7}-5×10^{-6} M	48 hr	1.4×	10
	Xeroderma (XP12RO)	+	-	10^{-7}-10^{-6} M	48 hr	1.5×-2.1×	10

(continued)

Table 1. (continued)

Mutagen	Cell	In vivo	In vitro	S-9	Concentration	Exposure time	SCE/cell increment over control	References
MNNG	CHO		+	-	10^{-7}–10^{-6} M	24 hr	2–9×	2
	Mouse bone marrow				0.15–1.5 mg/kg	28 hr	2×	8
	Human fibroblasts							
	Normal (GM637)		+	-	10^{-9}–10^{-6} M	48 hr	2×	10
	Xeroderma (XP12RO)		+	-	10^{-9}–10^{-6} M	48 hr	>10×	10
ThioTEPA	Vicia root tips	+		-	200–500 μM	2 hr	3×	11
Trenimon	Mouse							
	Bone marrow	+			2.5–12.5 μg/kg	28 hr	4.6×	8
	Bone marrow	+			2.5–12.5 μg/kg	?	1.5×	5
	Human lymphocytes		+	-	10^{-8} M–	24 hr	3×	12
	Human lymphocytes		+	-	5×10^{-5} M 1×10^{-1} μg/ml	24 hr	up to 11×	13

	Human fibroblasts	+	-	10^{-6}–5×10^{-4} μg/ml	24 hr	up to 9×	13
Mitomycin C	CHO	+	-	10^{-9}–10^{-7} M	24 hr	20×	2
	Chinese Hamster 0-6	+	-	2.5×10^{-9}–10^{-6} M	8 hr	12×	14
	Human fibroblasts						
	Normal (GM637)	+		10^{-10}–10^{-8} M	48 hr	2×	10
	Xeroderma (XP12 (XP12RO)	+		10^{-10}–10^{-8} M	48 hr	5×	10
	Human						
	Normal, 4 lines	+	-	0.01 μg/ml	2 day	4×	9
	Fanconi's anemia, 2 lines	+		0.01 μg/ml	2 day	3-4×	9
	Human lymphocytes						
	Normal	+	-	0.01-0.03 μg/ml	24 hrs	2.5-3.5×	9
	Fanconi's anemia			0.01-0.03 μg/ml	24 hrs	1.3-1.5×	8
	Normal	+		0.003-0.1 μg/ml	72 hrs	up to 12×	15

(continued)

Table 1. (continued)

Mutagen	Cell	In vivo	In vitro	S-9	Concentration	Exposure time	SCE/cell increment over control	References
Mitomycin C (continued)	Mouse fibroblasts, $HGPRT^-$		+	-	0.01 μg/ml	24-48 hr	4-23×	16
	Mouse							
	Spermatogonia	+		-	0.025-0.5 mg/kg	14 hr	4×	17
	Bone marrow	+		-	0.3 mg/kg	14 hr	2.5-3.8×	6
	Bone marrow	+			1-6 mg/kg, infused		10× (less in older animals)	18
	Muntjac fibroblasts		+	-	0.01-1.0 μg/ml	18-20 hr	up to 4×	19
x-Rays	CHO		+		75-100 rad(G_1)		1×	10
			+		800 rad(G_1)		2×	2
					400 rad(G_2)		2×	2
					400 rad(G_2)			
	Human lymphocytes		+	-	50 rad(G_2)		1×	20
					100 rad(G_2)		1×	20
					150 rad(G_1)		1.8×	20

REFERENCES

1. Takehisa, S., and Wolff, S. Mutat. Res. 45: 263-270 (1977).

2. Perry, P., and Evans, H. J. Nature 258: 121-125 (1975).

3. Stetka, D. G., and Wolff, S. Mutat. Res. 41: 343-350 (1976).

4. Stetka, D. G., and Wolff, S. Mutat. Res. 41: 333-342 (1976).

5. Vogel, W., and Bauknecht, T. Nature 260: 448-449 (1976).

6. Allen, J. W., and Latt, S. A. Chromosoma 58: 325-340 (1976).

7. Natarajan, A. T., Tates, A. D., van Buul, P. P. W., Meijers, M., and de Vogel, N. Mutat. Res. 37: 83-90 (1976).

8. Bauknecht, T., Vogel, W., Bayer, U., and Wild, D. Humangenetik 35: 229-307 (1977).

9. Latt, S. A., Stetten, G., Juergens, L. A., Buchanan, G. R., and Gerald, P. S. Proc. Natl. Acad. Sci. (U.S.) 72: 4066-4070 (1975).

10. Wolff, S., Rodin, B., and Cleaver, J. E. Nature 265: 347-349 (1977).

11. Kihlman, B. A. Chromosoma 51: 11-18 (1975).

12. Beek, B., and Obe, G. Humangenetik 29: 127-134 (1975).

13. Hayashi, K., and Schmid, W. Humangenetik 29: 201-206 (1975).

14. Kato, H., and Shimada, H. Mutat. Res. 28: 459-464 (1975).

15. Latt, S. A. Proc. Natl. Acad. Sci. (U.S.) 71: 3162-3166 (1974).

16. Lin, M. S., and Alfi, O. S. Chromosoma 57: 219-225 (1976).

17. Allen, J. W., and Latt, S. A. Nature 260: 449-451 (1976).

18. Kram, D., Schneider, E. L., Senula, G. C., and Nakanishi, Y. Mutat. Res. 60: 339-347 (1979).

19. Huttner, K. M., and Ruddle, F. H. Chromosoma 56: 1-13 (1976).

20. Solomon, E., and Bobrow, M. Mutat. Res. 30: 273-278 (1975).

CHAPTER 20

MUTAGENICITY OF SELECTED CHEMICALS IN IN VIVO CYTOGENETIC ASSAYS

R. J. Preston

Biology Division, Oak Ridge National Laboratory
Oak Ridge, Tennessee 37830

I.-D. Adler

Gesellschaft fur Strahlen-und Umweltforschung
Munich, West Germany

A. Léonard

Laboratory of Genetics, Department of Radiobiology
C.E.N./S.C.K. B-2400 Mol, Belgium

and

M. F. Lyon

Medical Research Council, Radiobiology Unit, Harwell
Didcot, Oxon OX 11 ORD, England

INTRODUCTION

There are a variety of test systems available for the analysis of chromosome aberrations following in vivo treatment. There are advantages and disadvantages to each, which means that the information obtained from any one test must be critically examined. Because of the variety of test systems employed, it is also impossible to provide a concise summary of the effect of each of the chemicals under study.

Bone Marrow

The basic method is to treat the animals (usually Chinese hamster or mouse) with the chemical, either with a single dose or with two doses 24 hrs apart, and then, at different times after treatment, to sacrifice the animal, remove the femurs, and make preparations of the bone marrow cells. Since the bone marrow cells are a cycling population, in order to study the induced yield of chromosome aberrations, it is necessary to analyze cells which are in their first mitosis after treatment. This is usually achieved by sampling cells 6-12 hr after treatment. Multiple sampling should be used to obtain cells which were in different stages of the cycle at the time of treatment. Samples taken at longer times after treatment (> 24 hr) will enable the persistence of aberrations to be studied. Many variations, particularly as regards treatment regime, are described. The interpretation of data obtained from experiments where multiple treatments were given is complicated. This is due in part to the fact that the populations of cells exposed to each dose will be different, since cells in sensitive stages of the cell cycle will be killed. This more resistant population can either still be in a resistant stage of the cell cycle when the next dose is given, or it can progress into a sensitive stage. If the cells are sensitive to aberration induction as well as cell killing, which seems probable, then the results obtained will be hard to interpret. These multiple treatment regimes are not representative of a chronic treatment.

The aberrations observed following treatment with all the chemicals studied here are almost solely of the chromatid type, when analysis is made at the first mitosis after treatment. There are many different categories of aberrations listed in different papers, which makes analysis even more difficult. It is quite common to record gaps in the total aberration frequency and also to conclude that there is a significant difference between treated and control when only the difference in gap frequency is significant. It is quite legitimate to record gaps, but since their significance in terms of genetic hazard is not understood, it is better not to include them in total aberration frequencies. The types of chromatid aberrations are probably best recorded as: chromatid deletions, isochromatid deletions, chromatid intrachanges (both intra arm and inter arm) and chromatid interchanges. If chromosome-type aberrations are observed, they are best recorded as chromosome terminal and interstitial deletions, and exchanges (rings and dicentrics).

Spermatogonia

Many of the comments concerning the analysis of chromosome aberrations in the bone marrow apply to the studies on differentiating spermatogonia. The basic method is to treat the animal, and at different times after treatment to kill the animal, remove the testis and make preparations. Differentiating spermatogonia are a cycling population, and so, in order to study the induced frequency of aberration cells must be sampled at their first division after treatment, i.e., about 6-12 hours. Multiple samplings enable cells which were in different stages of the cell cycle at the time of treatment to be analyzed. It can also be advantageous to pulse with ^{3}H-thymidine to distinguish by autoradiography which cells were undergoing DNA synthesis at the time of treatment. Treatments with multiple doses present the same problems as described for bone marrow cells.

If mice are treated with a chemical and the spermatocytes analyzed some 6 weeks or more later, the cells analyzed will have been spermatogonial stem cells at the time of treatment. It is apparent that, with the possible exception of TEM, none of the chemicals under study here induce aberrations in the spermatogonial stem cells which are recovered in the spermatocytes, even though many induce aberrations in the bone marrow and differentiating spermatogonia. This is not too difficult to reconcile. There are many cell divisions between the stem cell and the spermatocyte, so that cells containing reciprocal translocations together with other aberration types will almost certainly die at the first or a subsequent division after treatment. Also the translocations produced are of the chromatid type, and because of the segregation of such an exchange, there is a likelihood that it will produce unbalanced, probably lethal, products at mitosis. The probability of recovering a reciprocal translocation in a spermatocyte following spermatogonia stem cell treatment is low. Therefore, this test highlights the fact that despite the fact that a chemical can produce considerable chromosomal damage in differentiating spermatogonia, it does not mean that a high frequency of translocations will be recovered. It is also not feasible to use this test as one to indicate the potency of a chemical.

It is, of course, also possible to treat early meiotic stages, i.e., prereplication, and analyze aberrations at diakinesis-metaphase I of the primary spermatocyte. The first division after treatment is analyzed in this case.

Oocytes and First Cleavage Divisions

In order to study the induction of chromosome aberrations in female animals, usually mice, the animal is treated, and oocytes analyzed either at metaphse I or metaphase II. Superovulation has been used to increase the frequency of analysable cells.

It is also possible to study effects on male postmeiotic cells and differentiating ova by analyzing the first or subsequent cleavage division(s). These studies are all relatively recent, and so very few results are available for the chemicals considered here.

Leukocytes

The majority of data presented for human in vivo studies have been obtained from leukocyte cultures, since this is by far the most practical cell system to use. Samples are taken at different times after treatment, in order to study the induced yield of aberrations and the persistence of these aberrations. The leukocytes are nondividing cells, and so it is not possible to study the effect of the different chemicals on different stages of the cell cycle. The cells are stimulated to enter division by adding phytohemagglutinin to the culture. The cultures should be fixed at a time when those metaphases analyzed will be in their first culture division. There is still discussion as to the most suitable fixation time, but it appears that at 44 hr the great majority of cells will be in their first culture division. The time used in the experiments reported here are usually considerably longer, and might account, in part, for the observation of chromosome-type damage following treatment with certain chemicals which could be derived from chromatid-type aberrations as a result of a second division cycle.

The wide range of techniques employed for in vivo cytogenetic studies makes it impossible to give a general description of the methods. What is described here is the principle involved in the tests and the pitfalls which should be considered when interpreting the results. The papers given as references can be used to find details of any particular test as used by a particular investigator.

RESULTS

A summary table is provided for each chemical. The diversity of tests, together with the extent of the data, and sometimes the discrepancy between results, makes it unrealistic to provide a brief overall summary.

Aflatoxin B1

Work on the mutagenic effect of aflatoxins has recently been reviewed by Ong (1) and more briefly by Clark (2). From this it would appear that, although there are extensive data on effects on mammalian cells in vitro and a variety of submammalian systems, there is relatively little information on mammalian cells in vivo.

In germ cells of the mouse, Epstein and Shafner (3) found rather weak induction of dominant lethals by a dose to males of 68 mg/kg of a mixture of aflatoxin B1 and G1. The dose used was surprisingly high, since the LD_{50} is given as 9 mg/kg (4). Léonard et al. (5) gave a dose of 5 mg/kg aflatoxin B1 dissolved in a mixture of DMSO and saline intraperitoneally (IP) to male mice. The treated males were then mated to one female each for 5 weeks and then killed for chromosomal studies of the testes. These authors found no statistically significant induction of dominant lethal mutations, as judged by the mean litter size at birth. However, there was a nonsignificant reduction in litter size in weeks 2 and 3 (treated spermatids). The proportion of females (mated to treated males) which became pregnant decreased progressively over weeks 1 to 5 from 64% (16/25) to 36% (9/25) ($t = 19.8$; $p < 0.0001$, authorsç statistics). The mean testis weight was, however, the same in the treated and control group (286 ± 9 mg). Owing to the differences in doses used and method of measuring dominant lethals there is thus no real conflict in results between Epstein and Shafner (3) and Léonard et al. (5). It is possible that with sufficiently extensive data a small yield of dominant lethals would have been detected at the dose used by Leonard et al., but on the other hand it should be noted that Epstein and Shafner used a very high dose and their data were limited.

Léonard et al. further tested for the induction of chromosome aberrations by scoring spermatocytes for the presence of translocations induced in treated spermatogonia (Table 1). In 5000 cells of 25 treated males only 1 aberrant cell was found, indicating no difference from the controls (control: 2000 cells from 10 males, 1 aberrant cell; DMSO controls: 2000 cells from 10 males, 0 aberrant cells). Similarly no translocations were found among 185 male offspring of the treated males.

By contrast, a dose such as that given by Léonard et al. is highly carcinogenic to the liver. In view of the extreme sensitivity of the liver to aflatoxin, Srivastava et al. (6) studied the induction of chromosome aberrations in the liver of Chinese hamsters by 1 mg/kg aflatoxin B1 IP in DMSO. Partial hepatectomy was performed 24 hr before injection of aflatoxin, and colchicine (4 mg/kg, IP) was injected 24 hr after aflatoxin, followed by a second dose of colchicine (1 mg/kg) 4-5 hr before killing at 70-72 hr after the hepatectomy. The controls were hepatectomised and injected with DMSO only. The paper is disappointing, in that the authors apparently do not give the control data. Of 121 metaphases from seven animals (presumably all treated ones) 82 showed aberrations, including 40 with chromatid gaps, 26 with

Table 1. Aflatoxin B1

Species	Cell type	Lowest dose to produce significant yield of aberrations	References
Chinese hamster	Liver (after partial hepatectomy)	1 mg/kg	6
Mouse	Spermatogonia (observed in spermatocytes)	1 translocation in 2000 cells after 5 mg/kg	5

"beaded appearance" and 16 with "asynchronous centromere division." One chromatid exchange was also seen. The chromatid gaps were said to be preferentially located in chromosomes X, Y, and 1, in heterochromatic regions. The proportion of breaks in heterochromatic areas was said to be higher in treated than in control animals, although as previously mentioned no control data were given. Asynchronous centromere division was said to occur 8-10 times more frequently in treated than in control cells, though again control data were not provided.

In another study on liver, in this case in rats, Rogers and Newberne (7) found mitotic inhibition after a dose of 3 mg/kg.

In view of the facts that aflatoxins are known to contaminate human food, and that aflatoxin B1 is clearly mutagenic in some test systems (1), it would appear that further tests are needed on mammalian cells in vivo. However, in view of the negative findings of Leonard et al., there is no reason to expect a strong effect in germ cells.

ICR-170

There appears to be no published information on the in vivo cytogenetic effects of ICR-170.

Ethylenimine

There appears to be only one published report (8) of the in vivo cytogenetic effects of ethylenimine. This single item also happens to be very vague. The conclusions are that following exposure to < 0.5 ppm ethylenimine, for a mean time of about 8 years, there was no significant increase in chromosome aberration frequencies in the leukocytes of chemical workers.

Methyl Methanesulfonate (MMS)

MMS has been extensively studied as a reference mutagen, and there has been considerable work on effects of various kinds on germ cells of mammals in vivo but relatively little on somatic cells.

Somatic Cells. Frei and Venitt (9) compared the effects of MMS and N-methyl-N-nitrosourea-(MNU) in inducing chromosome damage in bone marrow of female C57BL mice (Table 2). At doses which caused roughly equal toxicity and equal alkylation of DNA (i.e., 0.8 mmole/kg) MNU caused markedly more chromosome damage than MMS. The difference was explained by the known differences in pattern of alkylation of DNA by the two agents. If the animals were given 1% caffeine in the drinking water (from 3 days before treatment until killing) the yield of chromosome damage after MMS and also after a low dose (0.08 mmole/kg) of MNU was markedly increased, but not after the high dose of MNU. Three mice and 50 metaphases per mouse were scored in each treatment group, and 20 metaphases per mouse were studied in detail for types of aberration. There appears to be a misprint in the authors' Table I, (reproduced as Tables 3 and 4) which indicates no enhancing effect of caffeine at 24 hr after treatment of MMS, although the text and Frei and Venitt's Table III indicate that damage was greater after caffeine treatment at this time. The effects of caffeine were explained by its known effects in inhibiting repair. The most common types of aberration were chromatid breaks, followed by chromatid gaps, but there was also a few chromatid exchanges (Table 2).

Male Germ Cells. Léonard and Linden (10) dosed 12-week old BALB/c mice with a single intraperitoneal (IP) injection of 50 mg/kg MMS and 100 days later made chromosome preparations of spermatocytes in order to score aberrations induced at the spermatogonial stage. Altogether 700 cells were scored from seven mice (50 cells per testis). No translocations were found, and only a single cell carried a fragment. This negative result is in contrast to the marked induction of dominant lethals and heritable translocations after treatment of male postmeiotic cells with MMS, but is not surprising in that only 9 of 37 chemicals so far subjected to this type of test have given positive results (11).

Moutschen (12) injected 30 mg/kg or 60 mg/kg MMS IP into C57BL male mice aged 3-6 months, killed them at varying times from 1 to 30 days later, and made squash preparations of the testes,

Table 2. Methyl methanesulfonate

Species	Cell type	Lowest dose to produce significant yield of aberrations	References
Mouse	Bone marrow	100 mg/kg (24 hr after treatment)	9
Mouse	Spermatogonia (observations at spermatocyte metaphase)	No translocations with 50 mg/kg	10
Mouse	Meiotic prophase (observed at anaphase I and II)	30 mg/kg (3 days after treatment) Most aberrations eliminated by anaphase II	12
Mouse	Post-meiotic male germ cells	25 mg/kg ($6\frac{1}{2}$ to $8\frac{1}{2}$ days after treatment)	13
Mouse	Oocytes (post DNA synthesis)	No aberrations following 100 mg/kg	14
Mouse	Oocytes (observed at first cleavage division)	50 mg/kg	14

which he then scored for aberrations at meiotic anaphase I or II. He found chromosome or chromatid bridges and fragments, most of which were eliminated before anaphase II. There was no difference between the effects of the two doses and maximum damage (i.e., 40% aberrations) was observed 3 days after treatment, a stage believed to correspond with treatment at the end of pachytene. In each sample 300 cells were analyzed, but no control data are given.

Brewen et al. (13) developed a method by which they could study cytogenetic changes induced in postmeiotic male cells. The males were mated to normal superovulated females at varying numbers of days after treatment, and chromosome preparations were then made from the ferilized eggs at the first cleavage division.

The doses of MMS were 25, 50, or 100 mg/kg, injected intravenously, and the times of mating ranged from 1 to 23 days after

Table 3. Frei and Venitt's data on chromosome aberrations in bone marrow of C57BL female mice after MMS treatment

Agent	Dose (mmole/kg)	Caffeine	Time after treatment (hr)	Mitoses with breaks (%)	
				Mean	SD
None	-	-	-	2.0	1.6
		+	-	2.0	1.3
MMS	1.1	-	6	6.7	4.2
		+	6	8.0	5.3
		-	24	27.4[a]	8.5[a]
		+	24	27.4[a]	8.5[a]
		-	48	3.3	1.1
		+	48	33.0	19.2

[a]One of these values must be a misprint.

treatment, corresponding with spermatozoal to early spermatid stages. The aberrations found in male pronuclei were mainly isochromatid deletions, with some chromatid deletions and interchanges. At the higher doses, and times of peak sensitivity, some cells had "shattered chromosomes" (Table 5). On the assumptions that any visible aberrations would be lethal, and that any cell with more than five aberrations would lead to preimplantation death, the observed cytogenetic results agreed well with previous work of others on dominant lethality. The fact that chromatid type lesions were seen was in accord with current views that MMS produces lesions in DNA which lead to mutagenesis through repair. In late spermatids the competence for repair is lost and hence lesions induced then would persist until the first cleavage division, when chromatid type aberrations would be produced.

Female Germ Cells. Brewen and Payne (14) studied chromosome aberrations at metaphase I of meiosis, or in the female pronucleus at the first cleavage division, after treatment of females with 50 or 100 mg/kg MMS (single intravenous injection). The females were superovulated and mated at times of 0.5 to 14.5 days after treatment.

At the first cleavage division the aberrations were of the chromatid type, mainly isochromatid and chromatid deletions. Some cells had shattered chromosomes (Table 6). By contrast, in the methaphase I oocytes, only 1 isochromatid deletion was seen in

Table 4. Types of aberration 24 hr after treatment with 1.1 mmole/kg MMS (data of Frei and Venitt)

Caffeine	Gaps		Total gaps	Breaks		Chromatid exchanges	Total breaks and exchanges	Total aberrations
	Chromatid	Isochromatid		Chromatid	Chromosome			
–	39	2	41	110	3	7	120	161
+	33	1	34	175	2	13	190	224

300 cells. This was interpreted to mean that an intervening round in DNA synthesis is necessary for DNA lesions induced by MMS to be manifested as chromosome damage, and this also explains the presence of chromatid type aberrations at first cleavage. The change in frequency of aberrations with time after treatment agreed qualitatively with the changes in frequency of dominant lethals found by previous authors.

Trenimon

Trenimon has been widely studied as a reference mutagen, and there are a variety of in vivo cytogenetic studies with experimental animals.

Somatic Cells. Chomosome damage in bone marrow of Chinese hamsters after various doses of Trenimon was studied by Schmid et al. (15). The cells were scored as having 1-2 breaks and/or exchanges, multiple aberrations, or "pulverized" chromosomes, and 50-100 metaphases were scored per animal.

The types of aberration were said to show "no obvious qualitative difference" from those produced by cyclophosphamide. Data on the latter agent show that it produced predominantly chromatid breaks and interchanges.

In order to study the development and decay of aberrations with time after treatment, the hamsters were given two IP injections of 0.25 mg/kg Trenimon 24 hr apart and killed at varying intervals after the second dose. Maximum damage was found after 6-8 hr, the effect remained high at 10-18 hr, but had fallen by about half by 30 hr, with very little damage remaining at 2-3 days (Table 7). This dose regime was so severe that it led to death from intestinal damage after a few days. Marinone and Venturelli (16) reported somewhat similar studies on the bone marrow of the rat, in which they found that the peak incidence of aberrations was reached at 48 hr after a single dose of Trenimon, but they gave no numerical data.

To investigate the dose-response relationships Schmid et al. (15) dosed hamsters with 0.125 mg/kg Trenimon on 1-4 successive days. (After 4 days the treatment was lethal to all animals.) The percentage of abnormal cells increased approximately linearly over the four doses (Table 8). In a further experiment animals were given varying sizes of dose, the treatment involving two doses 24 hr apart. (In both these dose-response experiments the animals were killed 6 hr after the last dose.) The single dose ranged from 0.031 to 0.5 mg/kg (Table 9). The dose-response curve rose very steeply, and the proportion of cells with multiple

Table 5. Data of Brewen et al. on chromosome aberrations in male pronuclei after treatment of spermatid and spermatozoal stages with MMS

Dose (mg/kg)	Interval (days)	Cells with aberrations							Shattered cells	Total cells	Predicted %		
		0	1	2	3	4	5	>5			Dominant lethality[a]	Pre-implantation loss[b]	Deaths implants[c]
Control	-	99	1	0	0	0	0	0	0	100	1	0	1
100	2½	24	6	10	2	3	2	3	0	50	52	6	49
	5½ + 6½	6	0	1	2	1	2	4	45	61	90	80	50
	9½	0	0	2	2	8	5	6	27	50	100	66	100
	11½+12½	19	7	11	3	5	1	5	9	60	68	23	58
	14½	45	4	0	0	1	0	0	0	50	10	0	10
	15½	45	3	1	1	0	0	0	0	50	10	0	10
	17½	45	4	1	0	0	0	0	0	50	10	0	10
	18½	49	1	0	0	0	0	0	0	50	2	0	2
	22½	50	0	0	0	0	0	0	0	50	0	0	0
50	2½	25	0	0	0	0	0	0	0	25	0	0	0
	3½	54	9	0	0	0	0	0	0	63	14	0	14
	5½	27	14	5	1	1	0	0	2	50	46	4	44
	9½	23	12	12	1	1	0	0	1	50	54	2	52
	10½	29	11	8	1	1	0	0	0	50	42	0	42
	11½+12½	32	12	7	1	2	1	1	2	58	45	5	42

	13½	41	6	1	1	0	0	0	1	50	18	2	16
	15½	69	5	1	0	0	0	0	0	75	9	0	9
	16½	49	1	0	0	0	0	0	0	50	2	0	2
	23½	50	0	0	0	0	0	0	0	50	0	0	0
25	6½	44	5	1	0	0	0	0	0	50	12	0	12
	7½	46	3	0	0	1	0	0	0	50	8	0	8
	8½+9½	80	8	1	0	0	0	0	0	89	10	0	10

[a]Total percentage of cells carrying at least one chromosomal aberration.

[b]Proportion of cells with more than 5 chromosomal aberrations.

[c]Proportion of cells with 5 or less chromosomal aberrations that had an aberration, i.e., $\Sigma(1-5)/\Sigma(0-5)$.

Table 6. Data of Brewen and Payne on chromosome aberations at first cleavage after treatment of females with MMS

Dose (mg/kg)	Interval (days)	No. of cells scored	No. of cells with aberrations						Expected lethality (%)
			0	1	2	3	>3	Shattered	
Control	-	100	200	0	0	0	0	0	0
50	2.5	75	70[a]	2	2	0	0	1	8.0[a]
	6.5	50	47	3	0	0	0	0	6.0
	10.5	50	49	1	0	0	0	0	2.0
	14.5	50	50	0	0	0	0	0	0
100	0.5	50	40	4	0	0	0	6	20.0
	1.5	50	43	3	1	0	1	2	14.0
	2.5	50	40	5	1	1	0	3	20.0
	6.5	50	43[a]	5	0	0	1	1	16.0[a]
	10.5	50	47	3	0	0	0	0	6.0
	14.5	50	50	0	0	0	0	0	0

[a]Includes a triploid zygote.

Table 7. Data of Schmid et al. on development of chromosome damage with time after 2 × 0.25 mg/kg Trenimon to Chinese hamsters[a]

	Time after second treatment							
	6 hr	9 hr	10 hr	14 hr	18 hr	30 hr	2 day	3 day
Normal cells (%)	17.3	19.3	38.0	34.6	30.3	65.0	95.7	98.7
1-2 aberrations (%)	11.7	21.0	24.7	16.6	17.6	14.7	4.3	1.3
Multiple aberrations (%)	60.7	58.0	36.7	44.2	40.4	16.6	0	0
Pulverized cells (%)	4.0	1.7	0.7	4.6	11.7	3.7	0	0

[a]100 metaphases from each of three animals at each time.

Table 8. Data of Schmid et al. on chromosome damage after 1-4 daily IP doses of 0.125 mg/kg Trenimon[a]

	No. of doses			
	1	2	3	4
Normal cells (%)	90.7	78.7	64.2	58.4
1-2 aberrations (%)	7.3	15.5	16.9	17.6
Multiple aberrations (%)	2.0	5.8	18.9	24.0
Pulverized cells (%)	0.0	0.0	0.0	0.0

[a]150 metaphases from each of three animals per point.

aberrations rose steeply also. Cells with pulverized chromosomes were seen only at the two highest doses.

Male Germ Cells. Leonard and Linden (10) treated 12 week old BALB/c mice with 0.2 mg/kg Trenimon IP and 100 days later made chromosome preparations of spermatocytes to score for aberrations induced in spermatogonia (Table 10). Schleiermacher (17) in similar work gave C3H mice single doses of 0.0625 or 0.125 mg/kg and NMRI mice 0.125 mg/kg Trenimon and killed them at various intervals afterwards for examination of spermatocytes. Leonard and Linden scored 600 cells from six animals and Schleiermacher scored 186 cells from treated spermatogonia, but no rearrangements were seen in either series. Thus, Trenimon resembles the majority of 37 tested chemicals in giving a negative result in this test (11).

Table 9. Data of Schmid et al. on chromosome damage after two IP doses of Trenimon of varying concentrations[a]

	Single doses (mg/kg)				
	0.5	0.25	0.125	0.0625	0.031
Normal cells (%)	11.1	41.2	77.6	90.9	97.4
1-2 aberrations (%)	6.6	15.7	12.2	9.1	2.4
Multiple aberrations (%)	55.3	40.6	10.2	0.0	0.2
Pulverized cells (%)	27.0	2.5	0.0	0.0	0.0

[a]150 metaphases from each of three animals per point.

Schleiermacher (17), in the work already mentioned, killed male mice for chromosome preparations of spermatocytes at varying numbers of days after treatment and thereby searched for aberrations induced at various stages of meiotic prophase. No consistent induction of aberrations was found. Two animals showed some chromatid fragments and multivalents, but this could not be repeated. However, there was an increased frequency of univalents, either autosomal or X-Y, at all stages, including spermatogonial (Table 11). The frequencies were highest after treatment of leptotene and preleptotene. The authors state that the results fluctuated considerably, and that they found some control animals which gave as high a frequency as the treated animals. Successive batches of C3H controls gave 1.3, 3.7, and 24.6% univalents (Table 11).

As it is known that an increased frequency of univalents can arise as an artifact there must remain some doubt as to whether this effect was indeed a result of the Trenimon treatment.

Table 10. Trenimon

Species	Cell type	Lowest dose to produce significant yield of aberrations	References
Chinese hamster	Bone marrow	2 x 0.0625 mg/kg (Doses 24 hr apart, analysis 6 hr after 2nd dose)	15
Mouse	Spermatogonia (observed in spermatocytes 100 days post-treatment)	No translocations after 2 mg/kg	10,17
Mouse	Meiotic prophase (observed in 4-8 cell embryos)	No consistent induction of aberrations even at 2 x 0.125 mg/kg	17
Chinese hamster	Post-meiotic stages (observed in 4-8 cell embryos)	0.25 mg/kg	18
Mouse	Oocytes (observed at metaphase II)	0.25 mg/kg	19
Mouse	Oocytes (observed in early embryos 2 cell stage)	0.25 mg/kg	19
Chinese hamster	Oocytes (observed at metaphase II)	0.25 mg/kg (less aberrations at this dose than in the mouse)	2

Table 11. Univalents after treatment of male mice with Trenimon[a]

Strain	No. of mice	Dose (mg/kg)	No. of cells	Cells with univalents	Univalents (%)
C3H	5	0	315	4	1.3
	18	0.125	867	129	14.8
	9	0	1,820	68	3.7
	18	0.0625	2,885	135	4.6
NMRI	7	0	705	21	3.0
	11	0 + NaCl	981	49	5.0
	20	0.125	1,756	172	9.8
C3H	3	0	199	49	24.6
F1	2	0	96	7	7.3
101	2	0	140	8	5.7
NMRI	1	0	51	3	5.8

[a]Data of Schleiermacker (17).

Although there have been various studies of the cytogenetic effects of Trenimon on postmeiotic germ-cell stages of males, it seems that there are certain difficulties in the interpretation of them.

Datta et al. (21) treated five male (C3H x 101) F1 hybrid mice aged 8 weeks with 0.125 mg/kg IP Trenimon, and five similar control males with NaCl solution. At weekly intervals the males were mated with untreated females which were dissected on days 13-16 of pregnancy. Chromosome preparations were then made from the fetal livers. From the first three weeks of mating 35 control embryos were obtained, all normal. The Trenimon series gave 33 embryos, of which two sired in week 1 carried reciprocal translocations, and one in week 2 carried a trisomy. Thus, this experiment confirmed that Trenimon could cause chromosome aberrations in postmeiotic male germ cells, but clearly many aberrations resulting in genome unbalance would have been eliminated before the fetuses reached the mid-fetal stage examined, and hence one can draw no conclusions concerning the frequency of aberrations actually induced.

Röhrborn (22) and Schmid and Binkert (18) studied younger embryos, at early cleavage stages. Röhrborn (22) treated C3H male mice with 0.125 mg/kg Trenimon and examined embryos sired

Table 12. Data of Röhrborn (22) on cytogenetic findings in early cleavage stages of C3H mice after a dose of 0.125 mg/kg Trenimon to spermatozoa and late spermatids

	No. of females	No. of metaphases	Chromosome no.							Aneuploid metaphases (%)	Aneuploid embryos (%)[a]
			>38	38	39	40	41	42	>42		
Trenimon	14	24	2	1	3	14	-	3	1	41.7	55.0
Control	21	37	-	2	1	30	3	1	-	18.9	15.4[b]

[a]Two-cell.

[b]$p < 0.01$.

Table 13. Aberrations in metaphase II oocytes of mice after

Strain	Dose (mg/kg)	Time relative to dose (hr)	No. of cells	Total aberrations (%)	Structural aberrations (%)	No. of Single	No. of Multiple
F1	0.25	-8	31	90.3	48.4	10	5
	0.25	0	48	93.8	60.4	20	3
	0.25	+3	21	76.2	38.1	8	-
	1.0	+3	24	75.0	62.5	7	4
	0	-	81	29.6	12.3	9	-
C3H	0	-	99	25.3	18.2	16	1

[a]Data of Röhrborn and Hansmann (23).

[b]k-j = satellite association or centric fusion.

[c]k-c = association of a chromatid (c) with a kinetochore (k).

by treated spermatozoa and late spermatids (Table 12). The exact stage examined is not clear, since the text mentions "fist cleavage division" but the table heading reads "early cleavage stages" and the table gives the observed number of metaphases as higher than the number of embryos, indicating that in fact at least some of the embryos had progressed to second or third cleavage. An increased frequency of aneuploidy was reported, with no mention of the presence or absence of structural aberrations. Schmid and Binkert (18) treated male Chinese hamsters with 0.25 mg/kg Trenimon and mated them to normal females, with most matings taking place at around 18 days after treatment. The embryos were examined at the 4-8 cell stage; nine of a total of 33 carried chromosome structural abnormalities. These aberrations were all of the chromosome type, and no numerical abnormalities nor chromatid aberrations were found. Of 126 control embryos only two showed structural anomalies, and six had various numerical aberrations. Since these embryos were at the 4-8 cell stage there may already have been some elimination of the grosser abnormalities. Moreover, there is no inconsistency between the finding of chromosome-type aberrations here, and of chromatid aberrations in work of others with other agents at first cleavage.

Female Germ Cells. Röhrborn and Hansmann (23) and Röhrborn et al. (19) studied chromosome aberrations at metaphase II and in 2-4 cell embryos respectively, after treatment of 10-12 week

Trenimon treatment (23)[a]

cells with Chromatid breaks	Gaps	k-k[b]	k-c[c]	No. of cells	Aneuploid (%)	No. of chromatids <40 (%)	>40 (%)
-	-	2	-	24	87.5	75.0	12.5
7	-	3	2	42	76.2	50.0	26.2
-	-	-	-	21	47.6	38.1	9.5
5	-	3	1	20	55.0	35.0	20.0
2	-	3	2	81	23.5	17.3	6.2
2	-	10	3	95	7.4	6.3	1.1

old female mice with Trenimon. Röhrborn and Hansmann gave C3H or (101 × C3H) F1 females a dose of 0.25 or 1.0 mg/kg and gave controls NaCl. The animals were superovulated and treated at various times before or after the injection of human chorionic gonadotrophin (HCG) used to induce ovulation. The animals were killed for examination 16 hr after HCG, when the oocytes were at metaphase II.

Both structural and numerical aberrations were found, at higher than control frequencies (Table 13). There were no significant differences between treatments at 8 hr before, 0 hr, and 3 hr after injection of HCG, and also no difference between the effects of 0.25 and 1.0 mg/kg, when both were given 3 hr after HCG. The authors suggest that the "saturation point of genetic hazard" had been reached at 0.25 mg/kg.

Röhrborn et al. (19) used the same strains of mice as Röhrborn and Hansmann. Again the animals were superovulated, and in this work the Trenimon injection (0.25 mg/kg) was in all cases given 3 hr after HCG. The animals were mated to untreated males and killed 44 hr after detection of vaginal plugs. At this stage the C3H embryos were: two-cell, 87.2%; three-cell, 8.5%; four-cell, 4.3%. The F1 embryos were: two-cell, 94.4% and three-cell, 7.6% (authors' figures). Both structural and numerical

Table 14. Data of Röhrborn on chromosome aberrations in early embryos after treatment of maternal metaphase I with 0.25 mg/kg Trenimon

Strain	Dose (mg/kg)	No. of cells	Aberrations (%)		No. of embryos	Aberrations (%)		No. of cells	Aneu-ploid (%)	No. of embryos	Aneu-ploid (%)
			Total	Structural		Total	Structural				
C3H	0.25	26	88.5	38.5	21	85.7	33.3	22	77.3	18	77.8
	0	35	20.0	0.0	26	23.1	0.0	35	20.0	26	23.1
F1	0.25	48	60.4	21.4	42	64.3	23.8	46	63.0	40	67.5
	0	29	24.1	6.9	24	29.2	8.3	29	17.2	24	20.8

[a]Excluding gaps.

abnormalities were increased in the treated series (Table 14), although the increase in structural aberrations was not significant in the hybrid embryos. Fragments were the most common type of structural change, with chromatid exchanges, breaks and gaps also frequent (Table 15). Among the aneuploid embryos hypodiploid counts were far more frequent than hyperdiploid (Table 16). The hyperdiploid cells could not be due to artifacts, and the difference from controls was significant. The percentage of cells with structural aberrations was below, and the percentage with any aberrations was above theproportion of dead implantations in a previous dominant lethal test. The authors suggest that some of the abnormal embryos were lost before implantation.

Table 15. Data of Röhrborn et al. (19) on types of chromosome aberration in early embryos

	Experi- mental	Control
Chromatid type interchanges	7	1
Chromosome exchanges	4	1
Fragments	19	-
Chromatid breaks	1	-
Unspecific anomalies	1	-
Chromatid gaps	9	1
Isochromatid gaps	3	-

In a continuation of this work Hansmann and Röhrborn (24) gave similar F1 hybrid females the same dose (0.25 mg/kg Trenimon IP) 3 hr after HCG; and in this case killed them 68 hr after the vaginal plug, when the embryos were at the morula stage. The control embryos had a mean of 10.9 blastomeres, but the experimental series had a mean of only 8.6. The difference between the two groups was statistically significant. Micronuclei were noted in 15% of control and 31% of the experimental embryos, and again the difference was statistically significant. Both structural and numerical aberrations were more frequent in the experimental series, and mosaics were very rare (Table 17). The types of structural aberration included long marker chromosomes,

Table 16. Types of aneuploidy in early embryos

Strain	Dose (mg/kg)	No. of chromosomes (<40)	<38	38	39	40	41	42	>42	(>40)
C3H	0.25	1	3	5	4	5	3	0	0	1
	0	0	2	1	1	28	3	0	0	0
F1	0.25	1	6	8	7	17	6	1	0	0
	0	0	2	2	0	24	1	0	0	0

dicentrics, Robertsonian translocations, and chromatid deletions, breaks and exchanges. Some embryos could not be scored for numerical aberrations because of overlapping chromosomes. Hypodiploid embryos were more common than hyperdiploid, which were found only in the experimental series. The authors compared their results with their earlier data on metaphase II oocytes and two-cell embryos and concluded that little or no elimination of abnormal embryos had occurred before the morula stage. Thus, the elimination must occur after implantation, and this was consistent with the results in their previous dominant lethal studies. The types of structural aberration, however, did change with time, as chromatid-type interchanges were found in metaphase II and dicentric and deleted chromosomes later. They suggest that chromatid-type aberrations can develop "due to a secondary effect long after treatment."

Similar work, but with Chinese hamsters, was carried out by Hansmann et al. (20). The animals were given a single IP injection of 0.25 or 1.0 mg/kg Trenimon, 8 hr before, at 0 hr, or 3 hr after an injection of HCG used for superovulation and were killed 19 hr after HCG for analysis of metaphse II oocytes. Both numerical and structural aberrations were more frequent in the treated series than in the controls (Table 18). The stage 8 hr before HCG was the most sensitive, being significantly more sensitive at both dose levels than the the stage 3 hr after HCG. At both -8 and +3 hr, the frequency of structural aberrations increased with dose, but the increase was significant only at -8 hr. The chromatid breaks were not randomly distributed, since chromosomes 1 and 2 were involved in 75% of all cases, chromosomes 3, 4, and X in only 16.7%, and chromosomes 5, 6, and 7 in only 8.3% of cases. Chromosomes 8, 9, and 10 were not involved in structural aberrations, but were often involved in hypoploidies. The frequency of aneuploid cells was much lower than in previous work with mice, in both control and treated series. The treated

Table 17. Chromosome aberrations in preimplantation stages of mice after 0.25 mg/kg Trenimon to maternal metaphase I (24)

Dose (mg/kg)	No. of embryos	Structural Aberrations		No. of embryos	Aneuploid (%)			
		No.	%		Total	Hypo ploid	Hyper-ploid	Hyper-ploid × 2
0.25	88	51	58.0	56	48.2	39.3	8.9	17.9
0	42	8	19.0	28	25.0	25.0	0.0	0.0

Table 18. Chromosome aberrations in metaphase II oocytes

Dose (mg/kg)	Stage (hr)	No. of oocytes	Structural Aberrations No.	%
0	-	229	20	8.7
0.25	-8	56	28	50.0
1.00	-8	19	16	84.2
0.25	0	47	17	36.2
0.25	+3	49	10	20.4
1.00	+3	29	10	37.0

[a]Data of Hansmann et al. (20).

[b]B'=chromatid break and fragment; F'=fragment; D'=chromatid deletion without fragment; E'=chromatid exchange.

cells seemed to have definitely higher frequencies, and hyperdiploid cells were only seen in the treated series. Structural aberrations also were less frequent than in previous work with mice.

Triethylenemelamine (TEM)

In Vivo Cytogenetic Effects on Somatic Cells. Michel and Legator (25) studied in rat bone marrow cells the cytogenetic alterations induced by TEM [2,4,6-tris(1-aziridinyl)-s-triazine or tetramine]. Except for a deviation at the highest dose level tested, the extent of cytogenetic damage was directly and linearly related to TEM dose. Between the control and intermediate (0.2 mg/kg) dose levels cytogenetic damage showed a 4-fold increase (Table 19).

In Vivo Cytogenetic Effects on Germ Cells. The chromosome-breaking ability of TEM in the male germ cells has been amply demonstrated in both premeiotic and postmeiotic stages.

In premeiotic male germ cells, TEM produces chromosome fragments and bridges in anaphases I and II of male mice given 6 mg/kg and killed 20 hr or 2, 3, 5, 7, 10, or 20 days after injection. The same observations were observed in animals killed 20 hr after

of Chinese hamsters after treatment with Trenimon[a]

No. of anomalies/oocyte[b]				No. of	<11	>11	2n	Aneu-ploid
B'	F'	D'	E'	oocytes	%	%	%	(%)
0.031	0.022	0.022	0.013	220	5.0	0.0	0.0	5.0
0.196	0.125	0.089	0.161	37	5.4	5.4	0.0	10.8
0.158	0.632	0.316	0.211	13	0.0	7.7	7.7	15.8
0.030	0.128	0.021	0.191	37	16.2	0.0	0.0	16.2
0.082	0.122	0.061	0.020	46	8.7	0.0	0.0	8.7
0.111	0.074	0.111	0.074	24	0.0	0.0	8.3	8.3

2, 5, 6, 12 or 24 mg/kg. The yield of fragments was maximum after 1 day, and the frequency of anaphase bridges after 2 days and increased linearly with the dose. This ability of TEM to produce chromosome breaks in the germ cells is probably responsible for the spermatogonial killing resulting, after high doses, in a temporary sterile period as observed by Jackson and Bock (26), Cattanach and Edwards (27) and Cattanach (28). It could explain also the cell killing observed in mouse oocytes by Generoso et al. (29,30) and Cattanach (31). The spermatocyte test on the treated males has been used to detect the induction of reciprocal translocations in spermatogonia by Cattanach and Williams (32) and Adler (33). In the experiments of Cattanach and Williams (32), two dose levels (0.5 mg/kg and 2.0 mg/kg) of TEM were employed and only a low frequency (<1%) of reciprocal translocations was observed (Table 19). Adler (33) treated adult mice and 10- and 11-day-old embryos *in utero* with six different doses of TEM ranging from 0.05 to 0.5 mg/kg. The incidence of translocation multivalents was 0.16% in the prenatally treated experimental groups and 0.33% in treated adult males. According to Maliszewski and Ficsor (34), 0.5 μM/g TEM could increase the frequency of multivalents in Swiss albino mice.

Cox and Lyon (35) failed, however, to detect any translocation in the spermatogonia of guinea pigs and golden hamsters given 0.2 mg/kg TEM IP. Since the doses used in the experiments of Cox and Lyon (35) were much lower it is impossible to say whether

Table 19. Triethylenemelamine (TEM)

Species	Cell type	Lowest dose to produce significant yield of aberrations	References
Rat	Bone marrow	0.2 mg/kg (4-fold increase) linear dose response curve	25
Mouse	Spermatogonia	0.1 mg/kg (18 hr after treatment) There was a very low frequency of translocations (<1%), observed in spermatocytes following treatment of spermatogonial stem cells with 0.5 and 2.0 mg/kg. This is due to a loss of translocations as a result of cell killing, and segregation of the exchange not to a lack of induction of aber rations by TEM: No translocations observed in spermatocytes of guinea pig or golden hamster after treatment of spermatogonia with 0.2 mg/kg.	32
Mouse	Post-meiotic germ cells (Fl translocation test)	0.64 mg/kg (0.2 mg/kg also effective in small study)	37

guinea pig and hamster spermatogonia are resistant to TEM or whether there would be some effect after higher doses. On the other hand, Cacheiro and Swartout (36) were unable to detect reciprocal translocations in sterile sons derived from TEM treated spermatogonia or oocytes.

The Fl translocation test on the males derived from treated animals or on the sons from the Fl females has been widely used to study the ability of TEM to induce chromosome rearrangements in postmeiotic male germ cells. In a small-scale experiment Cattanach (37) tested 21 animals sired, the first week after treatment, by male mice given 0.8 mg/kg or 0.64 mg/kg TEM. Six or possibly seven animals of the 21 tested, that is, 28.5 or 33.3%, were carrying a translocation. This is about the same frequency of translocations as is produced by a dose of 600 R x-rays. Those results were confirmed later (30). In a further experiment, Cattanach (38) described an F1 female sired by a male given 0.2 mg TEM/kg body weight and showing a translocation between an autosome and the X chromosome. In a small sample of F1 female mice sired the first three weeks after treatment three females, 13%, (as shown by analyzing the meiotic chromosome of the sons) were found to carry reciprocal translocations. Three males from a total of 100 F1 sons sired by male mice receiving TEM (dose not indicated) in their drinking water were found to carry reciprocal translocations by Jorgenson et al. (39). The high frequency of completely sterile males showing sterility associated with chromosome rearrangements in the form of a chain in the F1 progeny of TEM- (and TEPA-)treated animals contrasts with the results obtained with ionizing radiation. This observation could suggest that TEM, as well as TEPA, induces breaks preferably close to the end of the chromosome (40).

Some other parameters such as paternal sex-chromosome losses or chromosomal inversions have been used to test the ability of TEM to induce chromosome aberrations in mouse post-meiotic germ cells. Analyzing the F1 females sired the first three weeks after treatment by males given 0.2 mg/kg TEM; Cattanach (41) got 11 matroclinous XO animals from a total of 838 (1.4%) compared to an incidence of 0.4% (5/1087) in the controls. Two other females from the treated group had 40 chromosomes but were probably carrying an X-chromosome deletion. Roderick and Hawes (42) administered, through oral and intraperitoneal routes, an average dose of 0.19 mg/kg of TEM and examined a total of 114 F1 males sired during the first two weeks after treatment. Three animals (2.6%) exhibited chromosomal inversions, as shown by a high incidence of bridges in the first meiotic anaphase. The estimate of the rate of induction of inversions was therefore 3/119/114 = 1.39 $\times 10^{-1}$ per mg/kg TEM.

In the experiments of Matter and Jaeger (43), male mice received a single IP dose of 0.4 or 0.8 mg/kg of TEM and were mated over 3 weeks with control females. Each female was checked daily for the presence of a vaginal plug and, when positive, was replaced by another female. Examination performed about 52 hr after detection of vaginal plugs showed that most of the embryos were already in the second cleavage division stage. Whereas none of the 52 control embryos with metaphases showed chromosomal aberrations, 52.3% (0.4 mg/kg) and 83.3% (0.8 mg/kg) of the embryos in the TEM groups exhibited chromosomal damage. These values are in quite good agreement with the dominant lethality results.

Conclusions. Extensive studies have been performed on the cytogenetic effects of TEM on somatic and male germ cells of mammals. All the results obtained demonstrated that TEM is, with TEPA, one of the most powerful chemical mutagens tested up to now on mammals.

In bone marrow cells the extent of cytogenetic damage is directly and linearly related to TEM dose. It could be interesting, however, to get more information on the type of chromosome aberrations obtained in somatic cells.

The yield of chromosome breaks induced in spermatocytes and spermatogonia increases linearly with the TEM dose and the induction of chromosome breaks in spermatogonia resulting in cell killing is probably responsible of the temporary sterile period induced by high doses of TEM in male mice.

TEM appears able to induce a low frequency of reciprocal translocation in mouse spermatogonia (<1%). Such ability has not been demonstrated in guinea pig and golden hamster, but the doses used for these two species were very low.

Different indices of cytogenetic changes have been used to test the ability of TEM to produce chromosome aberrations in postmeiotic male germ cells. They include the production of F1 translocated male and female animals or of chromosomal inversions as well as the paternal sex-chromosome losses. Positive results have been obtained in all the experiments. It has been shown that the stages of spermatogenesis most sensitive to damage by ionizing radiation also appear to be the ones most sensitive to damage by TEM. However, the variation of sensitivity to TEM between pre- and postmeiotic male germ cells is greater than to radiation mutagenesis. Thus a high frequency of translocations is obtained following treatment of postmeiotic germ cells with a low TEM dose (0.2 mg/kg) whereas this compound is relatively ineffective upon spermatogonial cells (up to 2 mg/kg).

TEPA [Tris(1-aziridinyl)phosphine Oxide], MeTEPA [Tris(2-methyl-1-aziridinyl)phosphine Oxide], and Thio-TEPA [Tris(1-aziridinyl)phosphine Sulfide]

In Vivo Cytogenetic Effects on Somatic Cells. A very extensive study has been performed by Adler et al. (44) on the in vivo cytogenetic effects of TEPA (triethylene phosphoramide; aphoxide; APO) on bone marrow cells of male rats (Table 20). Experiments were designed to investigate cytogenetic effects at varying times after administration of single doses, the dose-response relationships at a fixed time following administration of the drug as well as the possible cumulative effects. Time-response studies,

Table 20. TEPA

Species	Cell type	Lowest dose to produce significant yield of aberrations	References
Rat	Bone marrow	2.5 mg/kg (24 and 48 hr after treatment)	44
Mouse	Spermatogonia (translocations scored in spermatocytes)	4 mg/kg (probably needs repeating)	55
Mouse	Postspermatogonial cells (F1 translocation test)	2.5 mg/kg	54

the male rats were injected intraperitoneally with single doses of 10 mg/kg TEPA: Bone marrow samples were prepared at subsequent intervals of 6, 12, 24, 48, 72, and 96 hr. In dose-response studies, bone marrow samples were prepared 24 hr following single intraperitoneal injections of TEPA at logarithmically increasing doses of 0.65, 1.25, 2.50, 5.0, and 10.0 mg/kg. Cumulative effects were studied by injecting a dose of 5 mg/kg TEPA on four consecutive daily occasions. Marrow samples were prepared 6 and 24 hr following the termination of treatment.

The results were that most of the induced aberrations were of the chromatid type. Chromatid aberrations were maximal at 24 hr, when most affected cells exhibited more than 20 aberrations. At 96 hr these effects were still in slight excess of control levels. Mitotic activity was markedly reduced up to 72 hr, and approached control values by 96 hr following treatment. The variability between animals was maximal in the most highly affected groups. A dose-related increased incidence of cells with chromatid aberrations and a concomitant dose-related decrease in the mitotic index was found in the dose response study. The incidence of cells with chromatid aberrations were 0.5, 3.0, 11.5, 39.0, and 87.5 after 0.65, 1,25, 2.5, 5.0, and 10.0 mg/kg TEPA, respectively. Repeated TEPA treatment resulted in a relatively low number of cells with aberrations (6.0%) as compared with single injection. This is probably due to cell death as shown in the drastically reduced mitotic index.

Chromosome aberrations have also been reported in bone marrow cells of mice given MeTEPA (Mapo; methaphoxide) (45).

All the results reported with ThioTEPA (thiophosphoramide) are in very good agreement with the observations of Adler et al. (44). In the experiments of Arseneva and Golovnika (46) bone marrow of male mice given 4 mg/kg ThioTEPA were examined at different intervals of time between 4 and 24 hr after treatment. (Table 21). The incidence of cells with aberrations was 8.8% at the 4 hr interval, 37.2% at the 18 hr interval and 32.8% after 24 hr. Malashenko and Surkova (47-50) and Surkova and Malashenko (51) found also most of the aberrations to be of the chromatid type the maximum effect being observed 24 hr after injection of the mutagens. The incidences of abnormal cells were 1.4%, 42%, and 93% after 0.16, 5, and 10 mg/kg ThioTEPA, respectively. They reported also some difference between the strains. The strains 101/H, C57Bl/6 and A/Sn were the most sensitive, whereas the hybrids F1 (CBA x C57Bl/6) and F1 (C3H x 101) were more resistant.

Table 21. ThioTEPA

Species	Cell type	Lowest dose to produce significant yield of aberrations	References
Mouse	Bone marrow	1.25 mg/kg (18 hr after treatment)	46,51
Mouse	Spermatogonia (translocations scored in spermatocytes)	5 mg/kg	48,49
Mouse	Postspermatogonial cells (F1 translocation test)	5 mg/kg	48,49

In Vivo Cytogenetic Effects on Germ Cells. Reciprocal translocations are the only type of chromosome damage induced by TEPA and derivatives which has been studied in the germ cells.

According to Zudova and Šrâm (52), the injection of 4 mg/kg TEPA to male mice can induce as much as 17% of reciprocal translocations in spermatogonia. In other series of experiments the same authors (53) failed to find significant differences between the yields of reciprocal translocations produced in spermatogonia by 1 mg/kg TEPA given in a single dose or in fractionated injections. Most (78%) of the translocation configurations were in the form of a chain. The same type of translocation associated with sterility was also observed by Epstein et al. (54) in the F1 generation after treatment of postmeiotic stages. According to Sotomayor and Cumming (40) this observation could suggest that TEPA induces breaks preferably close to the end of the chromosomes. Epstein et al. (54) found also that TEPA induces more heritable translocations in epididymal sperm and later spermatids. The incidence of F1 translocated males was found by Epstein et al. (54) to be 23% (7/30) after treatment of postmeiotic male germ cells with 2.5 mg/kg and 28% (7/25) after 5 mg/kg given to spermatids (55,56). The incidence of 50% of translocated F1 males reported by Šrâm et al. (55,56) after treatment of spermatogonia with 5 mg/kg appears questionable. Furthermore Cacheiro and Swartout (36) were unable to detect reciprocal translocation configurations in sterile males derived from TEPA-treated spermatogonia or oocytes.

After 5 mg/kg ThioTEPA, Malashenko and Surkova (48) report an incidence of 10.6% of translocated male offspring in the F1 generation from treated postmeiotic stages and the production of 3% of dividing spermatocytes I carrying translocation configurations after treatment of spermatogonia.

The results of Šrâm and Zudova (56) on compounds related to TEPA are summarized in Table 22.

Conclusions. TEPA, MeTEPA, ThioTEPA and related compounds can induce chromosome aberrations in somatic cells and male germ cells of mammals, even after very low doses (0.16 mg/kg).

In somatic cells, the induced aberrations are the chromatid type, the yield of aberrations is maximal 24 hr after treatment, and there is an evident relation between the dose and the incidence of abnormal cells. Repeated treatments result in a reduction of cells with aberrations probably due to cell death.

TEPA, ThioTEPA, and related compounds can induce chromosome rearrangements such as reciprocal translocations in premeiotic male germ stages and in postmeiotic ones. The chromosome rearrangements observed in spermatocytes I after treatment of spermatogonia or in the F1 translocated sons from treated postmeiotic cells are mostly in the form of chain. The unexpected and surprising high incidence of translocations observed in spermatocytes, or recovered in F1, after treatment of spermatogonia with TEPA should be confirmed.

Table 22. Translocation frequency in spermatocytes I after treatment of spermatogonia

Compound	Dose (mM/kg)	Frequency of translocation configurations (%)
Methylaminobis (1-aziridinyl)phosphine oxide	3.0–6.0	0.5–0.9
Ethylaminobis(1-aziridinyl)phosphine oxide	3.0	0.9
Propylaminobis(1-aziridinyl)phosphine oxide	3.0	0.3
Isopropylbis(1-aziridinyl)phosphine oxide	3.0	0.8
Diethylbis(1-aziridinyl)phosphine oxide	3.0	0.4

Those compounds appear much more effective than TEM in inducing chromosome aberrations in male germ cells. For the reason given above it is, however, impossible to given a quantitative comparison.

Mytomycin C

Mitomycin C clearly induces chromatid aberrations of all types in bone marrow and spermatogonia of the mouse. There is an indication that it is less effective in the Chinese hamster, and in fact proved to be negative in spermatogonia, from limited data. In one study (61) it was shown that Mitomycin C induced aberrations in early spermatocytes, which were observed in diakinesis metaphase I as chromatid deletions and chromatid exchanges (multivalents). This is one of a very small number of compounds so far shown to be effective at this stage. No translocations were observed when the spermatocytes derived from treated spermatogonial stem cells were analyzed, but this was the case with most chemicals, and the explanation has been given already. The data are summarized in Tables 23-25.

Table 23. Mitomycin C

Species	Cell type	Lowest dose to produce significant yield of aberrations	References
Chinese hamster	Bone marrow	10 mg/kg (max. 18 hr after treatment) only dose considered in detail	57
Mouse	Bone marrow	0.35 mg/kg (12 hr after treatment)	58
Mouse	Spermatogonia	0.035 mg/kg (48 hr after treatment)	59
Mouse	Spermatogonia (observed in spermatocytes)	No translocations after 5 mg/kg	60
Mouse	Spermatocytes	2.5-3.75 mg/kg (11 days after treatment)	61

Table 24. Cytogenetic effects after treatment with Mitomycin C

Tissue	Species	Effect	References
Bone marrow	Mouse	+	58,59
	Mouse	+	62
	Mouse	+(SCE)[a]	63
	Rat	+	64
	Chinese hamster	+	57
Spermatogonia	Mouse	+	·65,66
	Mouse	+	58,59
	Mouse	+(SCE)[a]	67
	Rat	NT	
	Chinese hamster	-	68
Spermatogonia developed into spermatocytes	Mouse	-	65,66
	Mouse	-	60
	Mouse	-	69
	Rat	+?	70
	Chinese hamster	-	64
Spermatocytes	Mouse	+	61
	Rat	NT	
	Chinese hamster	NT	

[a]SCE is sister chromatid exchange.

Epichlorhydrin

Epichlorhydrin has been studied in mouse bone marrow (71) and human leukocytes (72). Results are summarized in Table 26.

Cyclophosphamide

Cyclophosphamide has been widely used in mammalian in vivo cytogenetic tests, often as a positive control clastogen (Table 27). It is of particular interest since it has been used for cytostatic therapy, so that cytogenetic data are also available for man. It is also clear that cyclophosphamide requires metabolic activation in order to be clastogenic.

Following IP or oral treatments, chromosome aberrations have been analyzed principally in bone marrow and spermatogonial cells

Table 25. In vivo cytogenetic effects after treatment with Mitomycin C

Species	Tissue	Experimental conditions	Effects: Qualitative	Effects: Quantitative	References
Chinese hamster	Bone marrow	5,10,15 mg/kg IP preparation 6,12, 18, and 24 hr after treatment	Chromatid aberrations, mainly in S-phase	With 10 mg/kg, 6 hr, 28% aberrations; 18 hr, 65%; 24 hr, low	57
Mouse	Bone marrow	5 mg/kg preparation	Chromatid aberrations, symmetric exchange	With 101 H strain, 54%; C57B1, 42%; C3H, 40%; C3Hx101, 27% aberrations	63
Mouse	Spermatogonia	3.5, 1.75, 0.35, and 0.035 mg/kg IP preparation, 12, 24, and 48 hr after treatment	Mitotic inhibition, chromatid gaps, breaks, terminal deletions, exchanges between homologous chromosomes	Maximal effects: At 3.5 mg/kg, 12 hr 135%; 1.75 mg/kg, 24 hr, 73%; 0.35 mg/kg, 48 hr, 142%; 0.035 mg/kg, 48 hr, 48%	58,59
	Bone marrow	Same	Same	Maximal effects: at 3.5 mg/kg, 12 hr, 150%; 1.75 mg/kg 24 hr, 55%; 0.35 mg/kg, 12 hr, 17%	

(continued)

Table 25. (continued)

Species	Tissue	Experimental conditions	Effects: Qualitative	Effects: Quantitative	References
Mouse	Spermatogonia	0.5, 0.3, 0.1, 0.05, and 0.025 mg/kg IP, 31-33 hr after treatment	Sister chromatid exchanges (SCE)	Increase from 0.05 (3.6 SCE/cell) to 0.5 (7.2 SCE/cell) mg/kg in spermatogonia, control 1.8 SCE/cell	67
	Bone marrow	0.5, 0.3, 0.1, 0.05, and 0.025 mg/kg IP, 11 and 19 hr after treatment		Increase between 11 hr (8.4 SCE/cell) in bone marrow, control 3.4 SCE/cell	62
Mouse	Spermatogonia	2.5, 3.75, and and 5.0 mg/kg IP preparation 12, 18, 24 hr after treatment	Mitotic inhibition, tion, symmetrical exchanges in heterochromatic centromer region	Maximum effect: 2.5 mg/kg, 18 hr 24%; 3.75 mg/kg, 24 hr, 37%; 5.0 mg/kg, 24 hr, 50%	65,66
	Spermatogonia developed into spermatocytes	2.5 and 5.0 mg/kg IP preparation 50-60 days after treatment	No translocation multivalents in 1000 cells from 10 animals per dose		

Mouse	Spermatogonia developed into spermatocytes	5 mg/kg IP preparation, 50 days after treatment	No translocation multivalents in 1400 cells from 7 animals		60
Mouse	Spermatogonia developed into spermatocytes	0.05 and 0.1 mg/kg chronically over 8 weeks, preparation 3 weeks after end of treatment	No translocation multivalents		69
Rat	Spermatogonia developed into spermatocytes	0.67 and 6.67 mg/kg preparation, 60 days after treatment	Analysis of diplotene, diakinesis metaphase I, anaphase I and telophase I; no differentiation between chromatid and chromosome aberrations	60% multivalents (21% XY-autosome) in diakinesis. Diplotene is the most sensitive phase!?	70
Mouse	Spermatocytes	2.5, 3.75, and 5.0 mg/kg prepation, 10, 11 and 12 days after	Chromatid aberrations induced in S-phase of primary spermatocytes	Maximal effects, 11 days: 2.5 mg/kg, 4%; 3.75 mg/kg 13%; 5.0 mg/kg, 20%	61
		Labeling expt		95-100% coincidence between label and aberrations	

Table 26. Epichlorhydrin

Species	Cell type	Dose (mg/kg)	Route of treatment	Observations	References
Mouse	Bone marrow	0 (control)	IP	24 hr after exposure 0.006 breaks/cell	71
		1		0.033 breaks/cell	
		3		0.084 breaks/cell	
		5		0.112 breaks/cell	
		10		0.360 breaks/cell	
		20		0.488 breaks/cell	
		50		0.360 breaks/cell	
		0 (control)	Oral	0.006 breaks/cell	
		5		0.064 breaks/cell	
		20		0.372 breaks/cell	
		40		0.288 breaks/cell	
		100		0.372 breaks/cell	
		10 (×5) (24 hr interval)	IP	6 hr after last exposure 0.350 breaks/cell	
		20 (×5)	Oral	0.450 breaks/cell	
		5 doses/7 days			
		5 (×5)	IP	0.692 breaks/cell	
		10 (×5)	IP	0.600 breaks/cell	
		20 (×5)	IP	2.176 breaks/cell	
		20 (×5)	Oral	0.460 breaks/cell	

Human	Leukocytes (52-58 hr culture)	1 year exposure dose unknown	-	Control, before exposure 3,573 cells 31 chromatid deletions 7 chromosome deletions 1 chromatid exchange 4 chromosome exchange Exposed, one year 3, 336 cells 34 chromatid deletions 21 chromosome deletions 5 chromatid exchanges 7 chromosome exchanges	72

Table 27. Summary of cyclophosphamide data

Species	Cell type	Lowest dose to produce significant yield of aberrations	Highest dose with no significant effect	References
Mouse	Spermatogonia	40 mg/kg marginal 97 mg/kg positive	13.5 mg/kg (350 mg/kg to spermatogonia gives no aberrations in spermatocytes)	74,75
Chinese hamster	Spermatogonia	96 mg/kg (×2) (24 hr apart)	-	80
Mouse	Bone marrow	5 mg/kg (24 hr after treatment) (for deletions)	20 mg/kg (for exchanges) (96 hr after treatment 200 mg/kg gave no aberrations)	78
Chinese hamster	Bone marrow	20 mg/kg (24 hr after treatment)	10 mg/kg (72 hr after treatment with 10 doses 64 mg/kg, 24 hr apart, there were no aberrations)	78
Human	Bone marrow	-	40 mg/kg	78
Human	Leukocytes	3-5 mg/day for 4-6 weeks (far different from any other study)	500 mg to 44 g shown to have no effect	86-95

of mice and Chinese hamsters (Tables 28 and 29). Treatment of humans was usually by IV injection, followed by analysis, at different times after treatment, of the bone marrow or peripheral leukocytes. The information on humans is sparse and often complicated by the fact that treatment with other cytostatic drugs was given along with or before or after cyclophosphamide (Table 30). However, it does seem reasonable to conclude that man is less sensitive in terms of total chromosome aberration frequency than the laboratory animals tested.

It is also important to note that when spermatogonia are treated, and analysis made of primary spermatocytes, no translocations are observed, although it is clear that they are induced. It is easy to explain this observation in terms of cell killing as a result of cells having to pass through several mitotic divisions before reaching the spermatocyte stage, as well as by segration of the exchange itself. There is also a strong indication that cyclophosphamide-induced aberrations are rapidly lost by cell killing at the first mitosis, so that 48-72 hr after treatment the yield is greatly reduced. Cyclophosphamide also appears to produce its effect on cells that are undergoing DNA synthesis, or in cells which pass through S before observation.

The data cannot be reported in their entirety because they are numerous, and also because of the complexity of listing all chromsome aberration types.

Myleran

In Vivo Cytogenetic Effects on Somatic Cells. The cytogenetic effects of Myleran [1,4-di(methanesulfonoxy)butane; Busulfan] have been examined in bone marrow cells of male mice by Léonard and Léonard (97). To study the dose-response relationship the male mice were injected with 5, 10, 20, or 40 mg/kg of Myleran. Bone marrow samples were prepared 24 hr later. The time response was investigated by examining bone marrow samples 1, 2, 4, or 10 days after IP injection of 40 mg of Myleran. Most of the structural aberrations were of the chromatid type, and the dose-response relationship was found to be linear. The chromatid and chromosome aberrations were maximal at 2 days and decreased sharply after longer intervals of time.

The sharp decrease in the yield of structural aberrations occurring 4 days after treatment may explain the negative findings reported in humans given Myleran for medical purposes. Thus Karjincanic et al. (98) could not detect a significant increase in the yield of chromosome aberrations in eight patients treated for chronic myeloid leukemia with 2 mg Myleran two times daily. Stavem et al. (99) also failed to observe chromosome aberrations

Table 28. Effects of cyclophosphamide in the mouse

Cell type	Strain	Dose (mg/kg)	Interval between doses	Route of treatment	Time of fixation after treating
Spermatogonia/ spermatocytes	C3H×101	350	-	IP	F1
	C3H	200	-	IP	4-22 days
	NMRI	200	-	IP	9-96 hr 6 or 12 hr 24 hr 48 hr 72 hr 96 hr Control
		13.5	-	IP	24 hr
		40	-	IP	24 hr
		120	-	IP	24 hr
	CFLP	97	-	Oral	1-21 days
	BALB/c	200	-	IP	100 days
Bone marrow	C3H×101	200	-	IP	4-96 hr
	OFI	200	-	IP	6, 24, 45 hr

Observations	References
No translocations in F1 (25% of F1 heterozygous for 1 or more translocations from post-spermatogonial treatment)	40,73
Increase in univalents for all stages of meiotic prophase, highest in pre-leptotene No significant frequency of aberrations	17
In spermatogonial metaphases: Very little 12 aberrations (7 interchanges)/150 cells 9 aberrations (4 interchanges)/150 cells 5 aberrations (1 ingerchange)/150 cells 2 aberrations /150 cells 2 aberrations /300 cells 0 aberrations /100 cells 4 aberrations (2 interchanges)/100 cells 16 aberrations (11 interchanges)/150 cells	74
Increase in univalents days 12, 13 (no metaphases on day 14) No aberrations days 1-9; days 12, 13, very vew metaphases, few cells scored; day 12, 18% aberrant metaphase I (0 translocations); day 13, 20% aberrant metaphase I (3 translocations/80 cells)	75
No aberrations in spermatocyte diakinesis-metaphase I	10
Maximum aberration yield of 24 hr; decrease at 48 hr; no aberrations at 96 hr At 24 hr 250 cells analyzed: 127 chromatid gaps, 54 chromatid deletions, 13 isolocus deletions, 9 interchanges, 108 cells with multiple aberrations	76
Percentages of abnormal metaphases in treated cells: 6 hr, 16%; 24 hr, 76.8%; 48 hr, 12.8%; control, 0%	77

(continued)

Table 28. (continued)

Cell type	Strain	Dose (mg/kg)	Interval between doses	Route of treatment	Time of fixation after treating
	ICR		-	IP	24 hr
		Control			
		5			
		10			
		20			
		40			
		80			
Oocytes		5	-	IP	10-12 wk
		70			
		100			
		30-145	-	IP	12 hr

in the lymphocytes from a 59-year-old woman treated with Myleran (4 mg daily for 1 month, then 2 mg Myleran daily for 5 days each week during one year) and examined 2 months after the treatment was terminated. Some chromosome aberrations were observed, however, in the bone marrow cells. Since she had previously received a total of 21 mCi ^{32}P for polycythemia vera and a prednisone treatment for a severe autoimmune hemolytic anemia, it is possible that the chromosome aberrations (deletions, translocations) were not the result of the Myleran treatment. Koller (100) used rats of Wistar origin implanted subcutaneously with Walker carcinoma and mice injected in the peritoneal cavity with Ehrlich or Krebs ascites tumors. The highest dose of Myleran administered to rats was 2 mg/kg body weight per week and to mice 1 mg/kg body weight

Observations	References
Of 250 cells analyzed: 0.004 deletions/cell, 0 exchanges/cell 0.116 deletions/cell, 0 exchanges/cell 0.316 deletions/cell, 0 exchanges/cell 0.400 deletions/cell, 0 exchanges/cell 0.896 deletions/cell, 0.056 exchanges/cell 3.7 deletions/cell, 0.040 exchanges/cell The 40 and 80 gm/gk data include cells with more than 10 aberrations/cell Data for Wistar rats showed significant yields of deletions at 20 mg/kg	78
At 11th day of pregnancy; 20 cells analyzed on metaphase II, 20% abnormal, not signiicantly higher than matched control 70 and 100 mg/kg at 17th day of pregnancy 20% abnormal at 70 mg/kg; 33% abnormal at 100 mg/kg; neither significantly different from control	79
Preovulatory oocytes treated: 28.1% abnormal M II (32 cells) 40.0% abnormal M II (25 cells) 40.6% abnormal M II (32 cells) 13.5% abnormal M II (37 cells) 74.0% abnormal M II (50 cells) Aneuploidy increased after 145 mg/kg	

per week. The number of injections could reach 20. Injection of Myleran result in a variable phase of mitotic suppression in the tumors, followed by the appearance of dividing cells, many of which showing fragmented chromosomes, anaphase bridges and, later, micronuclei.

In Vivo Cytogenetic Effects on Germ Cells. Myleran produces chromosome fragments and bridges in anaphase I and II of male mice given 16 mg/kg and killed 20 hr or 2, 3, 4, 5, 7, 10, 20, or 30 days after treatment. The same aberrations were observed in animals killed 20 hr after 8, 16, or 24 mg/kg. The yield of aberrations in anaphase I was maximal 2 days after treatment and increased linearly with the dose. It is possible that those chro-

Table 29. Effects of cyclophosphamide in the Chinese hamster

Cell type	Dose (mg/kg)	Interval between doses	Route of treatment	Time of fixation after treating
Spermatogonia	8–192 (×2)	24 hr	Oral	24 and 48 hr after 2nd dose
	100(×5)	24 hr	Oral	24 hr after last dose
Bone marrow	Control 32(×2) 64(×2) 128(×2)	24 hr	Oral	6 hr after 2nd dose
	64(×2)	24 hr	Oral	6 hr after 2nd dose
	64(×2)	24 hr	Oral	6 hr after 2nd dose 8 hr 10 hr 18 hr
	1–5 days 64/dose	24 hr	Oral	6 hr after last dose
	64/dose	5x in 1 week	Oral	6 hr after last dose
	64/dose	10x in 2 weeks	Oral	6 hr after last dose
	128 256 512	– – –	Oral Oral Oral	6 hr

Observations	References
48 hr samples "much better" than 24 hr 2 × 8 mg/kg significant effect for exchanges 2 × 96 mg/kg lowest significant dose for total aberrations	80
Treated (704 cells analyzed): total aberrations (not gaps) 14 (1.99%), translocations 5 (0.71%)	81
Control (809 cells): total aberrations (not gaps) 2 (0.25%), translocations 0	
For 400 cells, 15 gaps 400 cells, 37 gaps (9.25%), 8 deletions (2.0%), 5 interchanges (1.25%) 400 cells, 43 gaps (10.75%), 22 deletions (5.5%), 4 interchanges (1.0%) 400 cells, 29 gaps (7.25%), 18 deletions (4.5%), 31 interchanges (7.7%)	82
Control: 3% gaps; treated 19% chromatid-type, 6% chromosome-type aberrations	83
100 cells: 54.3% abnormal metaphases	15
62.0% abnormal metaphases 51.0% abnormal metaphases 42.7% abnormal metaphases	
6 hr, 5.0% abnormal metaphases (200 cells) 2 days, 41.4% abnormal metaphases (350 cells) 3 days, 16.2% abnormal metaphases (300 cells) 4 days, 15.8% abnormal metaphases (200 cells) 5 days, 25.8% abnormal metaphases (300 cells) 59.4% abnormal metaphases (500 cells)	15
96% abnormal metaphases	
Inconsistent data; some animals had 50% aberrant cells, some almost no chromosome damage Very few mitoses	

(continued)

Table 29. (continued)

Cell type	Dose (mg/kg)	Interval between doses	Route of treatment	Time of fixation after treating
Bone marrow (cont'd)	8/dose 2 days to 7 wk, 5 days/ wk	24 hr	Oral	6 hr after last dose
	40(×2)	24 hr	IP	6 hr after 2nd dose 24 hr after 2nd dose
	10-160 Control 10 20 40 80 160	-	IP	24 hr
	65(2)	24 hr	Oral	6 hr after 2nd dose
	1-32 (×10) over 14 days	24 hr	Oral	5-8 hr after last dose
	64(10)	24 hr	Oral	6 hr-13 days after last dose

Observations	References
Aberration yield fairly constant throughout experiment. Never more than 4% abnormal metaphases.	15
42.2% abnormal metaphases	15
22.2% abnormal metaphases	
Enhanced by 200 mg/kg caffeine given with cyclophosphamide, no effect of 35 mg/kg caffeine: 6 hr (+ 200 mg/kg caffeine), 60.2% abnormal cells 24 hr (+ 200 mg/kg caffeine), 27.0% abnormal cells	
250 cells analyzed	
0.036 deletions/cell, 0 exchanges/cell	96
0.036 deletions/cell, 0 exchanges/cell	
0.172 deletions/cell, 0.020 exchanges/cell	
0.312 deletions/cell, 0.020 exchanges/cell	
0.532 deletions/cell, 0.028 exchanges/cell	
0.172 deletions/cell, 0.016 exchanges/cell	
10-20% abnormal metaphases; 33 chromatid deletions/100 cells, 6 chromatid exchanges/100 cells, some pulverized metaphases (7%)	84
1-16 mg/kg/day no significant difference from control;	85
32 mg/kg/day, 0.1 exchanges/cell, 0.13 deletions/cell; 6 hr, 57.5% metaphases with aberrations; 38.5% pulverized metaphases	
Aberration yield falls rapidly to that of the control by 3 days after the last dose	

Table 30. Effects of cyclophosphamide (CPP) in man

Cell type	Patient diagnosis	Dose	Interval between doses	Route of treatment	Time of fixation after treating
Leukocytes	Bronchiolar carcinoma	2.6 g over 10 days	24 hr	-	2 wk + 72-96 hr culture
	Epidermoid carcinoma of lung	1200 mg over 4 days	24 hr	-	Immed. + 72-96 hr culture
	Various malignant tumors	N-Methylurea (400 to 9600 mg) + CCP (1000-7000 mg)	96 hr NMU, 7 days CCP	IV	72 hr after CCP + 54 or 72 hr culture
	Glomerulonephritis and nephrosis	3-5 mg/kg for 6-8 months	-	-	74 hr culture
	Lung cancer	Not given	5 days every 6 weeks	IV	72 hr culture
		Not given	Not given	Not given	72 hr culture

Observations	References
6 normal karyotypes, 2 with loss of C-chromosome, 1 with loss of N.1 extra C	86
3 normal karyotypes, 1 with loss of C-chromosome, 1 with loss of F-chromosome, 2 acentric fragments	
No increase of aberration frequency above controls	87
Before treatment 1.5% deletions, 0 exchanges During treatment (8-18 patients): 4-6 weeks, 8.3% chromatid deletions, 10.3% chromosome deletions, 0.5% dicentrics; 3-4 months, 6.6% chromatid deletions, 8.2% chromosome deletions, 0.4% dicentrics; 6 months, 6.3% chromatid deletions, 8.9% chromosome deletions, 0.2% dicentrics After treatment: 4-6 weeks, 4.4% chromatid deletions, 8.2% chromosome deletions, 0.6% dicentrics; 4-15 months, 0.2% chromatid deletions, 1.0% chromosome deletion	88
Pretreatment control (13 patients): 2.5% cells with "breaks," 0.2% cells with "major aberrations" Post-treatment control (14 patients): 5.6% cells with breaks, 1.5% cells with major aberrations Cyclophosphamide-treated (5 patients): 5.0% cells with breaks, 1.1% cells with major aberrations	89
Control (2 persons): 1.8% breaks/cell Treated (2 patients): 2.2% breaks/cell	90

(continued)

Table 30. (continued)

Cell type	Patient diagnosis	Dose	Interval between doses	Route of treat-ment	Time of fixation after treating
	Gynaeco-logical tumors	1.5-1.7 g	-	-	48 or 68 hr culture
	Ovarian tumors	15-30 mg/kg per dose Total 1.6-14.1 g	-	IV	48 or 68 hr culture
		1.8-18.8 g			
	Idiopathic lipoid nephrosis	150-175 mg/sgM/day		-	-
	Various malig-nancies	500 mg-44g	Varied	-	72 hr culture

Observations	References
Control (14 tumor patients): 2.2% deletions, 0 exchanges	91
Treated (23 tumor patients): 11.6% deletions, 1.5% exchanges (chromatid + chromosome)	
Control (27 patients): 2.2% deletions, 0.05% exchanges	92
Treated (30 patients): 12.2% deletions, 1.6% exchanges (chromatid + chromosome)	
Deletions 6.7-23.3% Exchanges 1.5-5.6%	
Pretreatment control (5 patients): 4.3% "abnormalities" 35-79 days treatment (5 patients): 11.2% "abnormalities" (excluding 1 patient with 106%) 50-100 days (3 patients): 5% "abnormalities"	93
These abnormalities were largely gaps; 15% of them were deletions and exchanges.	
Many treatments were with multiple drugs, including other alkylating agents.	94
Control (10 subjects): 0-2% chromatid breaks, 0-1% chromosome-type aberrations Treated: 0-4% chromatid breaks, 0-1% chromosome-type aberrations Six patients showed 5-15% cells with chromatid breaks and exchanges, no increase in chromosome-type aberrations.	

(continued)

Table 30. (continued)

Cell type	Patient diagnosis	Dose	Interval between doses	Route of treatment	Time of fixation after treating
	Ovarian carcinoma	15 mg/kg Total dose 1.2-18.2 g over 6-8 weeks	-	IV	68 hr culture, few hours-90 days after treatment
Bone marrow	Various malignancies	40 mg/kg	-	IV	24 hr after treatment

mosome aberrations induced in spermatogonia are responsible for the gradual fall in numbers of spermatogonia which can result in a temporary sterile period reported in animals given Myleran (101). No chromosome rearrangements such as reciprocal translocations were observed in dividing spermatocytes I after treatment of spermatogonia with a single dose of 16 mg/kg (10).

There is also possibly a relation between the induction of chromosome aberrations and the effect of Myleran on mammalian oocytes. Thus Generoso et al. (30) have studied histologically the effects of 40 mg/kg Myleran on mouse ovaries fixed 3, 7, 14, and 120 days after treatment. The effect on oocyte number in ovaries fixed 3 days after treatment was only slight and was considerably less than after treatment with TEM. Progressive reduction in the number of oocytes was observed at 7 and 14 days after treatment. At 14 days, the number of oocytes was down to about the level found after 1.0 to 1.6 mg/kg TEM. Most of the surviving oocytes were in growing follicles but oocytes in the earliest stages were still observed occasionally. The ovary was empty of oocytes by 120 days after treatment. These results confirm previous observations of Bollog (102), Aleksandrov and Anisimov (103), and Burkl and Kellner (104).

Conclusions. Most of the structural aberrations induced by Myleran in mouse bone marrow cells are of the chromatid type. The dose-response relationship is linear. The yield of aberra-

Observations	References
Control (37 tumor patients): 0.019% chromatid deletions Treated (20 tumor patients): Large number of gaps, 0.16 deletions/cell, 0.02 exchanges/cell	95
Rapid decrease with time after cessation of therapy	
Controls (5 persons, 250 cells): 0.4 breaks/cell, 0 exchanges/cell Treated (5 patients, 250 cells): 0.07 breaks/cell, 0.008 exchanges/cell	78

tions is maximum at 2 days and decreases sharply after longer intervals of time.

With respect to the germ cells, Myleran has been shown to induce chromosome breaks in spermatocytes and spermatogonia. Such an effect could probably explain the temporary sterile period induced in male mice by Myleran as well as the effect of Myleran on mammalian oocytes. Myleran fails, however, to induce chromosome rearrangements in spermatogonia.

It should be interesting to perform systematic studies on the effects of Myleran on the postmeiotic male germ cells by analyzing the F1 generation.

MNNG, DEN, and DMN

Results with MNNG, DEN, and DMN in the mouse, the rat, and the Chinese hamster are summarized in Tables 31-33.

Vinyl Chloride

The mutagenicity of vinyl chloride and its derivatives has recently been reviewed by Bartsch and Montesano (112) and by Fishbein (12).

The carcinogenicity of vinyl chloride in man has been demonstrated very clearly. Moreover, it is a compound which is becoming increasingly used in manufacturing processes, and in articles sold to the public for daily use. Thus exposed individuals include not only workers actually employed in factories using the compound, but also the populations living near the relevant factories, and to a lesser extent the general population (113).

Various studies have shown vinyl chloride to be mutagenic in submammalian systems, particularly after activation by mammalian cells in vitro. However, there have been relatively few tests with mammalian systems in vivo, and these have mainly involved work with human leukocytes.

There have now been several studies of the frequency of chromosome aberrations in leukocytes of workers in factories using vinyl chloride in the manufacture of PVC (Table 34). Samples of blood from exposed workers were cultured for times ranging from 48 to 72 hr in different experiments and the chromosomes were examined for gaps, breaks and rearrangements of various kinds. Little information was available concerning the level of exposure of the workers, but in some cases it was known to have been over 500 ppm at times (114-116). In some studies the duration of exposure was stated, and ranged from 9 to 34 years. Unexposed controls were also studied in each case. Kučerova's controls appeared to be the same ones as used in a study in a different factory of epichlorhydrin. In five of the six studies the blood samples from the exposed subjects showed a higher level of chromosome damage than in the controls. Furthermore, in all these five studies the predominant aberrations found were breaks and gaps. In three studies of Ducatman et al. (116), Hansteen et al. (117), and Purchase et al. (118), other types of aberration were increased also, but there was no consistency as to the types of aberration that were increased. Thus, there is good evidence from all these studies that vinyl chloride causes chromosome breakage in man, but less clear evidence as to whether it causes chromosome exchanges of types which, if they occurred in the germ cells, would give rise to genetic hazards.

In the sixth study, however, Fleig and Thiess (114) found no difference from controls in the level of chromosome aberrations in the exposed workers. Fleig and Thiess gave more details of the actual level of exposure to which the workers had been subjected than did any of the other authors. Three workers were believed to have experienced >3000 ppm at times for periods 3-5 years. However, in recent years continuous monitoring of VC levels had ensured that levels in room air had been below 500 ppm from 1968, and below 100 ppm from 1974. Eight of the 10 exposed workers had been employed before 1968, but one of these eight had

Table 31. Effects of MNNG, DEN, and DMN

Species	Mutagen	Cell type	Lowest dose to produce significant yield of aberrations	References
Mouse	MNNG	Spermatogonia (observed in spermatocytes)	No translocations after 48 mg/kg[a]	105
Rat	DEN	Liver	50 ppm in drinking water for 3, 4, or 5 weeks (maximal after 4 weeks) Only dose used	106
Chinese hamster	DMN	Liver	5 mg/g IP once a week for 6 weeks. Observed 40 or 150 days past treatment. Only one dose used.	107

[a]This does not imply that aberrations were not induced, but that those aberrations induced were lost by cell killing or segregation of the translocation at the cell divisions intervening between treatment and observation (i.e., spermatogonial stem cells were treated and observations were made in primary spermatocytes).

experienced <1 ppm. Thus, for the remaining seven workers there is no reason to suppose that the failure to detect increased chromosome aberrations was due to little exposure to VC. It is possible that the discrepancy between Fleig and Thiess' findings and those of the other authors lies in the method of scoring aberrations. Fleig and Thiess state that for each subject 50 metaphases were photographed and analyzed according to the Denver nomenclature. They make no mention of breaks, gaps, chromatid aberrations, etc. This may indicate that they were searching for gross chromosomal changes such as translocations, rings and dicentrics. If so, this could explain their negative findings since, as previously mentioned, the other five studies showed no consistency concerning aberrations of this kind, but agreed in showing increases in breaks and gaps. There is some doubt concerning Fleig and Theiss' numerical data, since they quote aberration percentages of 2.5, 3.3 and 5 for individual workers, and on the basis of 50 metaphases per

Table 32. Cytogenetic effects after treatment with DEN and DMN

Mutagen	Tissue	Mouse	Rat	Chinese hamster	Guinea pig	References
DEN	Bone marrow	-	NT	- (SCE)	NT	108,109
	Liver	NT	+	NT	NT	106,110
	Induced hepatomas	NT	+	NT	+	106,111
DMN	Liver	NT	NT	+	NT	107
	Bone marrow	NT	NT	NT	NT	

[a]SCE is sister chromatid exchange.

individual such percentages could not arise from whole numbers of cells.

There appear to have been no reported studies of chromosome breakage in experimental animals in vivo, using known doses, and little work on germ cells. Anderson et al. (119) performed an extensive dominant lethal test on mice. Groups of male mice were exposed for 6 hr/day for 5 days to levels of 3000, 10,000, or 30,000 ppm vinyl chloride and compared with negative controls (treated with air only) and positive controls treated with cyclophosphamide or EMS. All groups of animals were mated for 8 suc-

Table 33. In vivo cytogenetic effects after treatment with DEN and DMN

Mutagen	Cell type	Experimental conditions	Effects: Qualitative	Effects: Quantitative	References
DEN	Rat, liver	50 ppm in drinking water for 3, 4, or 5 weeks; preparation 3, 4, 8, and 20 weeks	Numerical changes (ploidies) and abnormal karyotypes	5% ploidies directly after treatment, increasing to 22% after 4 weeks, decreasing to 9% after 20 weeks. 5% cells with abnormal chromosomes	106
	Rat, induced hepatoma	50 ppm in drinking water, 3, 4, or 5 weeks; preparations at 3, 4, 8, 20 weeks	Numerical and morpholotical chromosome changes in primary tumors	All tumors showed chromosomal instability	
	Guinea pig, induced hepatoma	0.042 g/1. in drinking water for 6 days/week for 7-11 months	Chromosomal instabilities in tumor transplants	Of 6 transplanted tumor lines, 4 showed 10-30% aberrations, 4-40% polyploid cells and all had marker chromosomes	111
DMN	Chinese hamster, liver	5 mg/g IP once a week for 6 weeks; preparation after 40 or 150 days	Chromosome-type aberrations	0.52 aberrations/cell after 40 days, 0.29 aberrations/cell after 150 days; 87% cells with no damage, few cells with 10 aberrations	107

Table 34. Frequencies of chromosome aberrations in cultured blood leukocytes of workers exposed to vinyl chloride

Exposure details	No. of individuals		No. of cells	Aberrations No.	Aberrations %	Types of aberrations	References
Mean: 15 yr	E	11	550	116	21.1	Mainly breaks and gaps	116
>500 ppm at times	C	10	500	76	15.2	Unstable aberrations significantly more frequent in exposed	
4-34 yr	E	10	500	ns	0-5	50 cells per subject karyotyped	114
>3000 ppm to <1 ppm	C	4	200	ns	0-5	Types of aberration not stated	
9-29 yrs	E	7	1450	318	9.5	Mainly breaks and gaps	121
20-30 ppm recently	C	3	566	11	1.9	Other types of aberrations not significantly increased	
none	E	39	3900	ns			
None	E	39	3900	ns		Mainly breaks and gaps	117
	C	16	1600	ns		Stable rearrangements more common than expected	
9: 16 yr 6: >500 ppm	E	9	1769	51	2.9	Mainly breaks and gaps	115
3: 500 ppm	C	ns	3573	41	1.2	Few exchanges	

None	E	56	5600	ns	8.13	Mainly breaks and	118
	C	24	2400	ns	4.18	Stable and unstable aberrations increased also	
1-17 yr, up to 500 ppm/yr	E	18	3600		2.6 (laboratory) 1.9 (polymerization)	Severe chromosome aberrations in personnel from the polymerization probably result from repeated exposure to diagnostic x-rays	122
	C	10	2000		2.6		

[a]E = exposed; C = controls; ns = not stated.

cessive weeks after treatment. No induction of dominant lethals was detected in any of the vinyl chloride treated groups, although there were clear effects in both the positive control groups. This result contrasts with the findings of Infante et al. (120), who studied fetal mortality among the wives of workers exposed to vinyl chloride. They found a clear increase in fetal mortality among the wives of exposed workers, both when compared with other pregnancies of the same woman occurring before the husband's exposure, and when compared with suitably matched controls. The most probable explanation of such an increase in fetal wastage would seem to be genetic damage to the husband's germ cells, in other words, dominant lethal mutations. At present it is not clear why the human studies should give a positive result, while the mouse results were equally clearly negative. Further work, possibly with more prolonged exposure of animals, would appear to be needed.

Cadmium

It is difficult to assess the in vivo cytogenetic effects of cadmium, particularly because of a concurrent exposure to other heavy metals. In general terms it would appear that there is no significant increase in aberration frequency due to cadmium exposure. Exposed groups do show an increase in aberration frequency compared to controls, but only when lead and/or zinc levels are also increased. The study of Bui et al. (123) when only cadmium was involved showed no significant increase. The data of Shiraishi and Yosida (124-126) show a very high frequency of chromosomally abnormal cells (9-50%) in eight of 12 Itai-Itai patients studied (four showed small increase). However, Bui et al. showed no difference in four out of four Itai-Itai patients studied. They suggest some factor(s) other than cadmium could be involved. Shiraishi and Yosida (126) also indicate this might be the case. Data for cadmium are summarized in Table 35.

Ethyl Methanesulfonate (EMS)

Data for EMS are summarized in Tables 36 and 37.

Natulan

Some data are available for Natulan; they are summarized in Tables 38 and 39.

Table 35. Effects of cadmium

Species	Cell type	Dose	Route of exposure	Observations	References
Mouse (BALB/c)	Spermatogonia (observed in spermatocytes)	0.5–3.0 mg/kg ($CdCl_2$)	IP	No translocations observed after 0.5, 1.75, or 3.0 mg/kg cadmium chloride (total of 3000 cells, 30 mice)	127
	Postmeiotic (F1 translocation test)	1.75 mg/kg ($CdCl_2$)		120 animals, 3000 cells No translocations	
Mouse	Oocytes (superovulated)	$CdCl_2$	Subcutaneus	Control 199 cells 1 n-1 (hypoploid)	128
		3 mg/kg		155 cells 3 n-1 1 n+1	
		6 mg/kg		126 cells 2 n-1 1 n+1 2 2n Sample 12 hr after injection, analyze metaphase II (However, treatment is post DNA synthesis)	

(continued)

Table 35. (continued)

Species	Cell type	Dose	Route of exposure	Observations	References
Human	Leukocytes (48 hr culture)	Blood level 0.395 ± 0.268 μg/100 ml		Occupational exposure Lead blood level also high Control (15 individuals) 1650 cells 0.0050 deletions/cell 0.0007 dicentrics/cell Exposed (24 individuals) 4800 cells 0.0094 deletions/cell 0.0026 dicentrics/cell 0.0013 chromatid exchanges/cell Significant difference for overall aberrations/cell	129
Human	Leukocytes (48 hr culture)	Ca content in blood 0.6-17.9 μg/100 ml	Occupational inhalation	High levels of lead also Controls (12 persons) 2400 cells 0.54% deletions 0.08% chromatid exchanges Rolling mill (high zinc, low cadmium and lead) (12 persons) 4600 cells 0.61% deletions 0.24% chromatid exchanges 0.65% chromosome exchanges	130

				No specific relationship of aberration yield and blood cadmium level Significant differences for exchanges between both exposed groups and controls, but not between the two exposed groups	
Human	Leukocytes (48 hr cultures)	Low and high cadmium exposures	Inhalation	High zinc, low lead, no cadmium 1600 cells (5 persons) 0.024 deletions/cell 0.001 chromatid exchanges/cell 0.007 chromosome exchanges/cell	131
				High levels zinc, lead, and cadmium 1500 cells (5 persons) 0.025 deletions/cell 0.0007 chromatid exchanges/cell 0.003 chromosome exchanges/cell	
				High level lead and cadmium, no zinc 1300 cells (4 persons) 0.016 deletions/cell 0.0008 chromatid exchanges/cell 0.004 chromosome exchanges/cell	

(continued)

Table 35. (continued)

Species	Cell type	Dose	Route of exposure	Observations	References
				Control 900 cells (5 persons) 0.005 deletions/cell 0.0 exchanges Significant difference between total workers and controls. Possibly lead that is cause of aberration differences.	
Cow	Leukocytes (48 hr cultures)	-	Oral	Acute intoxication with a mixture of metals lead, chromium, cadmium (50x control), cobalt, zinc, copper, and arsenic did not produce significant increase in chromosome aberrations	132
Human	Leukocytes (48 or 72 hr culture)	Various	-	Controls (Japanese) 4.4-6.1 ng/g Cd in blood 380 cells (4 persons) 6.1% deletions 0 exchanges	123

Itai-itai patients 15.5-28.8 ng/g Cd in blood
396 cells (4 persons)
5.5% deletions
1.3% chromosome exchanges (4 of 5 in one individual)

Control (Swedish) 1.4-3.2 ng/g Cd in blood
300 cells (3 persons)
5% deletions
1% chromosome deletions

Exposed worker (Swedish)
24.7-61.0 ng/g Cd in blood
500 cells (5 persons)
1.4% deletions
0.6% chromosome exchanges
No significant difference between controls and Cd exposed persons

(continued)

Table 35. (continued)

Species	Cell type	Dose	Route of exposure	Observations	References
Human	Leukocytes (50 and 72 hr culture)	-	-	Control (9 persons) 5080 cells 0.014 deletions/cell 0 exchanges Itai-itai patients (12 persons) 4571 cells 0.36 chromatid deletions/cell 0.03 chromatid exchanges/cell 0.01 dicentrics + rings/cell 0.01 symmetrical translocations/cell Later study, 50 hr fixation Control (9 persons) 2200 cells 0.024 deletions/cell 0 exchanges Itai-itai patients, 7 persons, 1600 cells 0.26 chromatid deletions/cell 0.02 chromatid exchanges/cell 0.005 dicentrics/cell 0.02 translocations/cell	126

Table 36. EMS

Species	Cell type	Lowest dose to produce significant yield of aberrations	References
Mouse	Bone marrow	115 mg/kg (6 hr after treatment)	133

Table 37. In vivo cytogenetic effects after treatment with ethyl methanesulfonate (EMS)

Species and cell type	Experimental conditions	Effects: Qualitative	Effects: Quantitative	References
Mouse, CD-1; bone marrow	360 mg/kg IP preparation of bone marrow; 6, 12 and 24 hr after treatment	Chromatid aberrations in S and G_2	Highest incidence of breaks after 6 hr (16.8%)	133
	50, 70, 90, 100, 110, 115, 120, 150, 200, 250, 360, and 600 mg/kg IP preparation of bone marrow; 6 hr after treatment		Lowest effective dose was 115 mg/kg (6.4% breaks). Dose-dependent increase of breaks up to 360 mg/kg	
Mouse, (C3H (C3H×101) F1; spermatogonia developed into spermatocytes	240 mg/kg IP preparation of testes 12 weeks after	Translocation multivalents at diakinesis	No effect	32
Mouse, BALB/c; spermatogonia developed into spermatocytes	336 mg/kg IP preparation of testes 120 days after treatment	Translocation multivalents at diakinesis	No effect	105
Mouse, Swiss albino; spermatocytes	5 and 10 mole/g preparation of testes 48 hr after treatment	Univalents at diakinesis	No effect	24

Table 38. Natulan

Species	Cell type	Lowest dose to produce significant yield of aberrations	References
Mouse	Bone marrow	100 mg/kg, 24 hr after treatment	62
Mouse	Spermatogonia	200 mg/kg, 24 hr after treatment	134
Mouse	Spermatogonia (observed in spermatocytes)	50 mg/kg × 4, 2 hr intervals	134
Human	Leukocytes	4.8 g total dose (not definitive)	135

Table 39. In vivo cytogenetic effects after treatment with procarbacine (Natulan)

Species and cell type	Experimental conditions	Effects		References
		Qualitative	Quantitative	
Mouse, CBA; spermatogonia and bone marrow	100, 200, 400, 600 and 800 mg/kg IP preparation of testes and bone marrow; 24 hr after treatment	Chromatid aberrations	Dose-dependent increase in bone marrow from 100 to 600 mg/kg Dose-dependent increase in spermatogonia from 600 to 800 mg/kg	136
Mouse, (101 C3H) F1; spermatogonia	200, 400, 600, and 800 mg/kg IP preparation of testes; 6, 12, and 24 hr after treatment	Chromatid aberrations	Increase 24 hr after 200 and 400 mg/kg, after 12 hr with 600 and 800 mg/kg	134
	50, 100 and 200 mg/kg, ×4 spaced at 2 hr intervals, preparation of testes 12 and 18 hr after treatment		Increase 18 hr after 50 mg/kg × 4, after 12 hr with 100 and 200 mg/kg × 4	

Mouse, (101 C3H) F1; spermatogonia developed into spermatocytes	Treatment as for spermatogonia analysis, preparation of testes 100-140 day after treatment	Translocation multivalents at diakinesis	No increase after treatment Increase after fractionated treatment, not dose-dependent	134
Human (leukocytes)	Total doses of 4.0 to 6.5 g; samples taken 1 day to 12 months after treatment; 21 patients	Chromatid aberrations	50 cells scored/patient before and after treatment 0-14% deletions (gaps excluded control 0% deletions)	135
Human (leukocytes)	Patient treated for Peyroni's disease (6 mo). No dose given	Chromatid aberrations	6-8% cells with aberrations. Unclear how many, if any, due to procarbazine	49

REFERENCES

1. Ong, T. Mutat. Res. 32: 35-53 (1975).

2. Clark, A. M. Mutat. Res. 32: 77-82 (1976).

3. Epstein, S. S., and Schafner, H. Nature 219: 385-387 (1968).

4. Newberne, P. M., and Butler, W. H. Cancer Res. 29: 236-250 (1969).

5. Léonard, A., Deknudt, G. H., and Linden, G. Mutat. Res. 28: 137-139 (1975).

6. Strivastava, P. R., Srivastava, A. K., and Lucas, F. V. Genetica 45: 371-376 (1975).

7. Rogers, A. E., and Newberne, P. M. Cancer Res. 27: 855-864 (1967).

8. Gaeth, J., and Thiess, A. M. Z. Arbeitsmed, Arbeitschutz. 22: 357-362 (1972).

9. Frei, J. V., and Venitt, S. Mutat. Res. 30: 89-96 (1975).

10. Léonard, A., and Linden, G. Mutat. Res. 16: 297-300 (1972).

11. Léonard, A. Rad. Environ. Biophys. 13: 1-8 (1976).

12. Moutschen, J. Mutat. Res. 8: 581-588 (1969).

13. Brewen, J. G., Payne, H. S., Jones, K. P., and Preston, R. J. Mutat. Res. 33: 239-250 (1975).

14. Brewen, J. G., and Payne, H.S. Mutat. Res. 37: 77-82 (1976).

15. Schmid, W., Arakaki, D. T., Breslan, N. A., and Culbertson, J. C. Humangenetik 11: 103-118 (1971).

16. Marinone, G., and Venturelli, R. Friuli. Med. 25: 85-119 (1970).

17. Schleiermacher, E. Chem. Mutagenesis Mamm. Man 22: 317-341 (1970).

18. Schmid, W., and Binkert, F. Mutat. Res. 21: 233-234 (1973).

19. Röhrborn, G., Kuehn, O., Hansmann, I., and Thon, K. Humangenetik 11: 316-322 (1971).

20. Hansmann, I., Netter, J., Neher, R. J., and Rohrborn, G. Mutat. Res. 25: 347-359 (1974).

21. Datta, P. K., Frigger, H., and Scheiermacher, E. In: Chemical Mutagenesis in Mammals and Man, F. Vogel and G. Rohrborn, Eds., Springer, Verlag, Berlin, 1970, pp. 194-213, 1970.

22. Rohrborn, G. In: Chemical Mutagenesis in Mammals and Man, F. Vogel and G. Rohrborn, Eds., Springer, Verlag, Berlin, 1970, pp. 294-316.

23. Rohrborn, G., and Hansmann, I. Humangenetik 13: 184-198 (1971).

24. Hansmann, I., and Rohrborn, G. Humangenetik 18: 101-109 (1973).

25. Michel, T. M., and Legator, M. S. Mutat. Res. 26: 460 (1974).

26. Jackson, H., and Bock, M. Nature 175: 1037-1038 (1955).

27. Cattanach, B. M., and Edwards, R. G. Proc. Roy. Soc. (Edinburg), B67, 54-64 (1958).

28. Cattanach, B. M. Mutat. Res. 3: 346-353 (1966).

29. Generoso, W. M., Huff, S. W., and Stout, S. K. Mutat. Res. 11: 316-322 (1971).

30. Generoso, W. M., Stout, S. K., and Huff, S. W. Mutat. Res. 13: 171-184 (1971).

31. Cattanach, B. M. Intern. J. Rad. Biol. 3: 288-292 (1959).

32. Cattanach, B. M., and Williams, C. E. Mutat. Res. 13: 371-375 (1971).

33. Adler, I. D. Mutat. Res. 29: 204 (1975).

34. Maliszewski, W. M. B., and Ficsor, G. Mutat. Res. 26: 446 (1974).

35. Cox, B. D., and Lyon, M. F. Mutat. Res. 30: 293-298 (1975).

36. Cacheiro, N. L. A., and Swartout, M. S. Genetics 80: 518 (1975).

37. Cattanach, B. M. Nature 180: 1364-1365 (1957).

38. Cattanach, B. M. Z. Vererbungsl. 92: 165-182 (1961).

39. Jorgenson, T. A., Newell, G. W., Scharpf, L. G., Gribling, P., O'Brien, M., and Chu, D. Mutat. Res. 31: 337 (1975).

40. Sotomayor, R. E., and Cumming, R. B. Mutat. Res. 27: 375-388 (1975).

41. Cattanach, B. M. Mutat. Res. 4: 73-82 (1967).

42. Roderick, T. H., and Hanes, N. L. Genetics 76: 109-117 (1974).

43. Matter, G. E., and Jaeger, I. Mutat. Res. 33: 251-260 (1975).

44. Adler, I. D., Ramarao, G., and Epstein, S. S. Mutat. Res. 13: 263-273 (1971).

45. Manna, G. R., and Das, K. P. Proc. Indian Sci. Congr. 60: 676 (1973).

46. Arseneva, M. A., and Golovkina, A. V. Vliyanic. Ioniz. Izluch. Nasledstvennost. Akad. Nauk. SSR, 1966: 122-133.

47. Malashenko, A. M., and Surkova, N. I. 5th ICLA Symposium, Gustav Fisher Verlag, Stuttgart, 1973.

48. Malashenko, A. M., and Surkova, N. I. Genetika 10: 51-58 (1974).

49. Malashenko, A. M., and Surkova, N. I. Genetika 10: 92-98 (1974).

50. Malashenko, A. M., and Surkova, N. I. Genetika 11: 66-72 (1975).

51. Surkova, N. I., and Malashenko, A. M. Genetika 10: 81-87 (1974).

52. Zudova, Z., and Šràm, R. J. EMS Newsletter 4: 41-42 (1970).

53. Zudova, Z., and Šràm, R. J. Mutat. Res. 21: 53 (1973).

54. Epstein, S. S., Bass, W., Arnold, E., Bishop, Y., Joshi, S., and Adler, J. D. Toxicol. Appl. Pharmacol. 19: 134-146 (1971).

55. Šràm, R. J., Zudova, Z., Benes, V., and Symon, K. EMS Newsletter 3: 13-14 (1970).

56. Šràm, R. J., and Zudova, Z. Folia Biol. 20: 1-13 (1974).

57. Lavappa, K. S., and Yerganian, G. Genetics 71: 32-33 (1972).

58. Schleiermacher, E. Mutat. Res. 21: 47-48 (1973).

59. Schleiermacher, M. A., and Schleiermacher, E. Mutat. Res. 19: 99-108 (1973).

60. Gilliavod, N., and Leonard, A. Mutat. Res. 13: 274-275 (1971).

61. Adler, I. D. Mutat. Res. 35: 247-256 (1976).

62. Allen, J. W., and Latt, J. A. Chromosoma 58: 325-340 (1976).

63. Surkova, N. I., and Malashenko, A. M. Genetika 11: 38-45 (1975).

64. Adler, I. D. Unpublished data.

65. Adler, I. D. Mutat. Res. 21: 20-21 (1973).

66. Adler, I. D. Mutat. Res. 23: 369-379 (1974).

67. Allen, J. W., and Latt, S. A. Nature 260: 449-451 (1976).

68. Brewen, J. G. Unpublished data.

69. Savkovic, N., Pecevski, J., and Maric, N. Mutat. Res. 38: 107 (1976).

70. Bempong, M. A., and Trower, E. C. J. Hered. 64: 324-328 (1973).

71. Šràm, R. J., Cerna, M., and Kucerova, M. Biol. Zbl. 95: 451-462 (1976).

72. Kucerova, M. Mutat. Res. 41: 123-130 (1976).

73. Cumming, R. B., and Walton, M. F. Genetics 68: 514 (1971).

74. Rathenberg, R. Humangenetik 29: 135-140 (1975).

75. Rathenberg, R., and Muller, D. Agents and Actions 2: 180-185 (1972).

76. Datta, P. K., and Schleiermacher, E. Mutat. Res. 8: 623-625 (1969).

77. Mazue, A., Vindel, J. A., Landsmann, F., and Brunand, M. Azneim.-Forsch. 24: 1422-1425 (1974).

78. Goetz, P., Šràm, R. J., and Dohnalova, J. Mutat. Res. 31: 247-254 (1974).

79. Hansmann, I. Mutat. Res. 22: 175-191 (1974).

80. Machemer, L. N. S. Arch. Pharmak. (Suppl.) 282: R61 (1974).

81. Machemer, Z., and Lorke, D. Mutat. Res. 40: 243-250 (1976).

82. Muller, D., Langauer, M., Rathenberg, R., Strasser, F. F., and Hess, R. Verh. Deut. Ges. Pathol. 56: 381-384 (1972).

83. Muller, D., and Strasser, F. F. EMS Newsletter 6: 22 (1972).

84. Muller, D., and Strasser, F. F. Mutat. Res. 13: 377-382 (1971).

85. Schmid, W., and Staiger, G. R. Mutat. Res. 7: 99-108 (1969).

86. Bridge, M. F., and Melamed, M. R. Cancer Res. 32: 2212-2220 (1972).

87. Selezneva, T. G., and Korman, N. P. Sov. Genetics 9: 1575-1579 (1973).

88. Dobos, M., Schuler, D., and Fekete, G. Humangenetik 22: 221-227 (1974).

89. Kaung, D. T., and Swartzendruber, A. A. Dis. Chest. 55: 98-100 (1969).

90. Freeman, A. I., Sinks, L. F., and Cohen, M. M. J. Radiat. 77: 996-1003 (1970).

91. Schmid, E., and Bauchinger, M. Deut. Med. Wochenshr. 93: 1149-1151 (1968).

92. Bauchinger, M., and Schmid, E. Z. Krebsforsch 72: 77-87 (1969).

93. Summit, R. L., Tipton, R. E., and Cox, C. B. Clin. Res. 21: 113 (1973).

94. Schinzel, A., and Schmid, W. Mutat. Res. 40: 139-166 (1976).

95. Schmid, E., and Bauchinger, M. Mutat. Res. 21: 271-274 (1973).

96. Röhrborn, G., and Buckel, A. Hum. Genet. 33: 113-119 (1976).

97. Léonard, A., and Léonard, E. D. Mutat. Res. 56: 329-333 (1977).

98. Krajincanic, B., Lazarov, A., Zunic, Z., and Rdojicic, B. Strahlentherapie 151: 459-462 (1976).

99. Stavem, P., van der Hagen, C. B., Vogt, E., and Sandnes, K. Clin. Genet. 7: 227-231 (1975).

100. Koller, P. C. Mutat. Res. 8: 199-206 (1969).

101. Partington, M., Fox, B. W., and Jackson, H. Exptl. Cell Res. 33: 78-88 (1964).

102. Bollag, W. Experientia 9: 268 (1953).

103. Aleksandrov, V. A., and Anisimov, V. N. Bull. Exp. Med. (USSR) 73: 334-336 (1973).

104. Burkl, W., and Kellner, G. Z. Zellforsch. Mikroskop. Anat. 43: 97-103 (1959).

105. Léonard, A., Deknudt, G., and Linden, G. Mutat. Res. 13: 89-92 (1971).

106. Takayama, S., Masahito, H., and Yamada, K. Gann Monog. Cancer Res. 17: 343-354 (1975).

107. Brooks, A. L., and Cregger, V. Mutat. Res. 21: 214 (1973).

108. Tates, A. D., and Natarajan, A. T. Unpublished data.

109. Bayer. Unpublished data.

110. Grover, S., and Fischer, P. Eur. J. Cancer 7: 77-82 (1971).

111. Zbar, B., Wepsic, H. T., Rapp, H. J., Whangpeng, J., and Borsos, T. J. Natl. Cancer Inst. 43: 821-831 (1969).

112. Bartsch, H., and Montesano, R. Mutat. Res. 32: 93-114 (1975).

113. Fishbein, L. Mutat. Res. 32: 267-308 (1976).

114. Fleig, J., and Thiess, A. M. Arbeitsmed. Sozialmed. Präventimed. 9: 280-283 (1974).

115. Kucerova, M. Mutat. Res. 41: 123-130 (1976).

116. Ducatman, A., Hirschhorn, K., and Selikoff, I. J. Mutat. Res. 31: 163-168 (1975).

117. Hansteen, I. L., Hillestad, L., and Thiis-Evensen, E. Mutat. Res. 38: 112 (1976).

118. Purchase, I. F. H., Richardson, C. R., and Anderson, D. Lancet (ii): 410-411 (1975).

119. Anderson, D., Hodge, M. C. E., and Purchase, I. F. M. Mutat. Res. 40: 359-370 (1976).

120. Infante, P. F., Wagoner, J. R., and Waxweiler, R. J. Mutat. Res. 41: 131-142 (1976).

121. Funes-Cravioto, F., Lambert, B., Lindsten, J., Ehrenberg, L., Natarajan, A. T., and Osterman-Golkars, S. Lancet (i): 459 (1975).

122. Léonard, A., Decat, G., Léonard, E. D., Leteure, M. J., Decuyper, L. J., and Nicaise, C. C. J. Toxicol. Environ. Health, in press.

123. Bui, T.-H., Lindsten, J., and Nordberg, G. F. Environ. Res. 9: 187-195 (1975).

124. Shiraishi, Y. Humangenetik 27: 31-44 (1975).

125. Shiraishi, Y., and Yosida, T. M. Ann. Rept. Natl. Inst. Genetics 22: 44-45 (1971).

126. Shiraishi, Y., and Yosida, T. M. Proc. Japan Acad. Sci. 48: 248-251 (1972).

127. Gilliavod, N., and Leonard, A. Toxicol. 5: 43-47 (1975).

128. Shimada, T., Watanabe, T., and Endo, A. Mutat. Res. 40: 389-396 (1976).

129. Bauchinger, M., Schmid, E., Einbrodt, H. J., and Dresp, J. Mutat. Res. 40: 57-62 (1976).

130. Deknudt, G., and Leonard, A. Environ. Physiol. Biochem. 5: 319-367 (1975).

131. Deknudt, G., Leonard, A., and Ivanov, B. Environ. Physiol. Biochem. 3: 132-138 (1973).

132. Léonard, A., Deknudt, G., and Debackere, M. Toxicol. 2: 269-273 (1974).

133. Ray, V. A., Holden, H. E., Salsburg, D. E., Ellis, J. H., Just, G. J., and Hyneck, M. L. Toxicol. Appl. Pharmacol. 30: 107-116 (1974).

134. Adler, I. D. Annual Progress Report, EC-Contract No. 066 74 1 ENV D, 1976.

135. Vormittag, W. Wien. Klin. Wochenscht., 86: 69-75 (1974).

136. Tates, A. D., and Natarajan, A. T. EC-Contract No. 030 74 1 ENV N, (1976).

CHAPTER 21

MUTAGENICITY OF SELECTED CHEMICALS IN UNSCHEDULED DNA SYNTHESIS ASSAYS

Gary A. Sega

Biology Division, Oak Ridge National Laboratory
Oak Ridge, Tennessee 37830 (U.S.A.)

Ann D. Mitchell

SRI International, 333 Ravenswood Avenue
Menlo Park, California 94025 (U.S.A.)

INTRODUCTION

The basis for this assay rests on the fact that normal, semiconservative DNA synthesis takes place only during the S phase of the cell cycle. If a cell is treated in such a way that the DNA is damaged it is oftentimes possible to detect an unscheduled DNA synthesis (UDS) occurring during portions of the cell cycle other than S. This UDS presumably represents an attempt by the cell to repair the damaged DNA. The types of DNA damage which may be repaired include alkylated bases and phosphate groups, single-strand breaks, and pyrimidine dimers.

UDS can be studied by using either cells treated in culture or cells treated within the living animal. In the former case, cells are generally pretreated in some way to reduce the background level of scheduled DNA synthesis occurring in the culture. The use of hydroxyurea or removal of a specific growth factor from the culture medium are both commonly used for this purpose. Exposure of the cells to the suspected mutagen takes place for some fixed period of time after which they are washed and grown in a medium containing a radioactive DNA precursor, usually tritium labeled thymidine ([^{3}H]dT). Unscheduled incorporation of [^{3}H]dT into the DNA of the treated cells is then measured using either autoradiography or liquid scintillation counting (LSC). Concurrent controls also are run to determine the amount of DNA labeling resulting from residual levels of scheduled DNA synthesis.

When live animals are used the chemical agents being tested are generally injected or sometimes fed. For somatic cells, tissue samples are recovered from treated animals and incubated in media containing [^{3}H]dT. Autoradiographs showing intermediate labeling of cells which is not found in controls give an indication of UDS.

When the germ cells of male mice are studied, the [^{3}H]dT is injected directly into the testes and the chemical being tested is usually given IP (1,2). Because of the long time period (28-30 days) from the time of the last scheduled DNA synthesis in primary spermatocytes until late spermatid formation, there are many germ cell stages that can be analyzed for UDS in this way (3). Autoradiographs of germ cells recovered from the testes shortly after treatment can be used to study UDS or one can make use of the timing of spermatogenesis and spermiogenesis in the mouse. This requires waiting for a particular germ cell stage to mature into sperm in the vasa deferentia or epididymides, then recovering millions of purified sperm heads which are assayed for the unscheduled presence of [^{3}H]dT by using LSC (1). By using the whole mouse, both direct-acting chemicals and those requiring metabolic activation can be studied for their ability to induce UDS in the germ cells (4). A very important advantage of studying UDS in vivo in mouse germ cells is that one is looking at the actual cells which are of genetic consequence to future generations.

RESULTS

Results of UDS studies are summarized in Table 1. No reports of studies on unscheduled DNA synthesis (UDS) were found for the following chemicals: cadmium compounds, epichlorohydrin, Myleran, TEPA, MeTEPA, Thio-TEPA, Trenimon, and vinyl chloride.

Results for individual chemicals are given in Tables 2-8.

MNNG is able to induce UDS in somatic cells grown in culture. Microsomal activation is not required. The lowest effective exposure of MNNG was ca. 0.01 mM (Table 2).

DMN is able to induce UDS in several somatic cell types but only under conditions where metabolism of DMN to a reactive compound is possible (Table 3). No UDS has been observed in the germ cells of male mice.

TEM induces UDS in rate lymphocytes (14), human fibroblasts (13), and mouse spermatids (11). The lowest effective exposure

(Table 4) was 0.05 mg/kg (0.0002 mM).

MMS is able to induce UDS in a variety of mammalian cells in culture as well as in meiotic and several postmeiotic germ cell stages of male mice (Table 5). The lowest effective exposure appears to be somewhere betweeen 0.01 and 0.1 mM. In the mouse, an exposure of 5 mg/kg (ca. 0.05 mM) induces the uptake of approximately six times as much [^{3}H]dT into early spermatids as is found in the controls.

EMS is able to induce UDS in a variety of mammalian cells in culture as well as in meiotic and several postmeiotic germ cell stages of male mice (Table 6). The lowest effective exposure appears to be about 0.1mM. In the mouse, an exposure of 10 mg/kg (ca. 0.08mM) induces the uptake of approximately three times as much [^{3}H]dT into early spermatids as is found in the controls.

Limited UDS studies on aflatoxin B1 have been reported (Table 7). The work has been restricted to cells in culture. The lowest effective exposure is ca. 0.01 mM for ½ hr in the presence of a microsomal activating system.

Only one study has been made of the UDS induced by ICR-170 (20), and that was done in vitro with opossum lymphocytes (Table 8). The lowest effective exposure was 0.001 mM.

UDS induced by DEN has only been studied in early spermatid stages of the mouse (11). A 125 mg/kg (ca. 1 mM) exposure gave negative results (Table 8). Apparently little or no activated chemical reaches the testes.

UDS studies with Mitomycin C are limited to one in vitro test with P-388 mouse cells (12). The lowest exposure inducing UDS was 0.015 mM (Table 8).

Exposures of Natulan from 0.0001 to 100 mM do not induce measurable UDS in WI-38 cells (Table 8), either with or without microsomal activation.

UDS induced by ethylenimine has only been studied in the germ cells of male mice (11). An exposure of 5 mg/kg (ca. 0.1 mM) gave a UDS response in early spermatids that was about 65 times the control value (Table 8). An exposure as low as 0.2 mg/kg (ca. 0.005 mM) is estimated to give a readily detected UDS response.

UDS induced by cyclophosphamide has been studied only in the germ cells of male mice (4). An exposure of 200 mg/kg (ca. 0.8 mM) gave a UDS response in germ cell stages from early meiosis through midspermatids (Table 8). The UDS was three to six times the control values.

Table 1. Summary table

Chemical	Results of UDS studies	Effective exposure (mM)	References
MNNG	Positive in several fibroblast cultures	∿0.01	5–7
DMN	Positive in fibroblast culture, epithelial and endothelial lung cells, liver parenchyma and some kidney cells; negative in lymphocytes and germ cells of male mice.	∿0.5	6,8–11
TEM	Positive in lymphocyte culture and in germ cells of male mice	∿0.0002 (lymphocytes) ∿0.01 (germ cells)	14 11
MMS	Positive in several cell culture systems. Positive in germ cells of male mice.	∿0.05	1,6,8,13,15–20
EMS	Positive in Chinese hamster cells, lymphocytes and fibroblasts. Positive in the germ cells of male mice.	∿0.1	1,6,8 13,15,20
Aflatoxin B1	Positive in fibroblasts and primary rat liver cells	∿0.01	6,21
ICR-170	Positive in lymphocyte culture	∿0.001	20

DEN	Negative in germ cells of male mice	--	11
Mitomycin C	Positive in cultures of P-388 mouse cells	∿0.015	12
Natulan	Negative in fibroblast culture	--	13
Ethylenimine	Positive in germ cells of male mice	∿0.1 (Estimated that ∿0.005 would be effective)	11
Cyclophosphamide	Positive in germ cells of male mice	∿0.8	4

Table 2. UDS results for MNNG (CAS 70-25-7)

Biological material	Method of exposure	Exposure or range of exposures	Method for detecting UDS	Lowest effective exposure	Comments	References
Normal and XP fibroblasts in Eagle's media	Chemical added to culture for 90 min, then cultured 90 min more in presence of $[^3H]dT$	0.0005-0.05 mM	Autoradiography	∿0.01 mM	Both normal and XP fibroblasts showed about the same levels of UDS	5
Human fibroblasts in arginine deficient media supplemented with bovine serum	Chemical added to culture for 3 hr, then cultured 1.5 hr in media containing $[^3H]dT$	0.01-0.1 mM	Autoradiography	∿0.01 mM		6

Epidermal and fibroblastic mouse cells in Medium 199	Chemical added to culture for 1 hr, then cultured for 3 hr in media containing either $[^3H]dT$ or $[^3H]dG$	1-50 μg/ml (∿0.007-∿0.34 mM)	Density labeled DNA with BUdR, used CsCl gradients and measured ^3H incorporation into light DNA using LSC	1 μg/ml using $[^3H]dG$ 25 μg/ml using $[^3H]dT$	Using $[^3H]dG$ to measure UDS is claimed to increase the sensitivity of this assay system	7
Contact-inhibited human fibroblasts, WI-38 cells, in BME	Chemical and $[^3H]dT$ added to culture for 3 hr	0.0001-1 mM	Extraction of DNA followed by measurement of DNA content and LSC	∿0.01 mM		13

Table 3. UDS results for DMN (CAS 62-75-9)

Biological material	Method of exposure	Exposure or range of exposures	Method for detecting UDS	Lowest effective exposure	Comments	References
Human lymphocytes in Eagle's modified media	Chemical added to culture for 1 hr; cells cultured 12 hr more in media with [^{3}H]dT	0.0001-10.0 mM	TCA precipitable activity in cells measured by LSC	All doses negative	All exposures produced negative results, presumably because no activating system was present	8
Human fibroblasts in Eagle's modified media	Chemical added to culture for 1 hr; cells then cultured in media with [^{3}H]dT, time not given	0.5-100 mM	Autoradiography	0.5 mM	All exposures with microsomal activating system present were +, without the microsomes all results were negative	9

C3/H mice	Chemical injected subcutaneously using DMSO/ethanol as carrier; 2 hr after injection tissues removed and cultured in media with [^{3}H]dT for 3 hr	150 mg/kg (∿1.8 mM)	Autoradiography of tissue sections	150 mg/kg (∿1.8 mM)	UDS detected in epithelial and endothelial lung cells, liver parenchyma and some kidney cells	10
Human fibroblasts in arginine deficient media supplemented with bovine serum	Chemical added to culture for 1 hr; cells then cultured for 1.5 hr in media containing [^{3}H]dT	∿1.0-10 mM	Autoradiography	∿1.0 mM	UDS only detected when microsomal activating system was present	6

(continued)

Table 3. (continued)

Biological material	Method of exposure	Exposure or range of exposures	Method for detecting UDS	Lowest effective exposure	Comments	References
(C3Hf × 101)F1 hybrid males	IP injection of chemical; testicular injection of [^{3}H]dT	8 mg/kg (∿0.1 mM)	Purified sperm heads derived from early spermatid stages and assayed by LSC	Results were negative	No UDS detected in early spermatids at this exposure; apparently little or no activated chemical reaches the testes	11

Table 4. UDS results for TEM (CAS 51-18-3)

Biological material	Method of exposure	Exposure or range of exposures	Method for detecting UDS	Lowest effective exposure	Comments	References
Lymphocytes from CD strain, albino rats	IP injection of chemical and [^{3}H]dT; lymphocytes recovered 6 hr after injection	0.05-5.0 mg/kg (∿0.0002-0.02 mM)	Autoradiography	0.05 mg/kg (∿0.0002 mM)	UDS detected at all exposures; linear increase in UDS except at highest exposure	14
Contact-inhibited human fibroblasts, WI-38 cells, in BME	Chemical and [^{3}H]dT added to culture for 3 hr	0.0001-1 mM	Extraction of DNA followed by measurement of DNA content and assayed by LSC	∿0.1 mM		13
(C3Hf × 101)F1 hybrid males	IP injection of chemical; testicular injection of [^{3}H]dT	2 mg/kg (∿0.01 mM)	Purified sperm heads derived from early spermatid stages and assayed by LSC	2 mg/kg only exposure tested	UDS detected in early spermatid stages was ∿3× the control values	11

Table 5. UDS results for MMS (CAS 66-27-3)

Biological material	Method of exposure	Exposure or range of exposures	Method for detecting UDS	Lowest effective exposure	Comments	References
Chinese hamster cells in Eagle's media	Chemical added to culture for 3 hr; cells cultured 4 hr more in media with [^{3}H]dt	0.7, 1.5 mM	Autoradiography	0.7 mM	0.7 and 1.5 mM both gave + results	15
Normal and XP fibroblasts in Eagle's media	Chemical added to culture for 45 min; cells cultured 4 hr more in media with [^{3}H]dT	0.01-1.0 mM	Isolated DNA on CsCl gradients, then used LSC	0.01 mM	0.01-1.0 mM gave + response with normal fibroblasts, XP fibroblasts gave + response at the one exposure tested, 0.5 mM	16
P388 cells in Fisher's media	Chemical added to culture for 1 hr; cells cultured 30 min more in media with [^{3}H]dT	60, 120 μg/ml	Autoradiography	60 μg/ml	60 and 120 μg/ml both gave + result; UDS nearly complete 11 hr after end of treatment	17

Human lymphocytes in Eagle's modified media	Chemical added to culture for 1 hr; cells cultured 12 hr more in media with [^{3}H]dT	0.01-10 mM	TCA precipitable activity in cells measured by LSC	0.1 mM	0.1-10 mM gave + result; maximum UDS observed at 0.1 mM	8
Human leukocytes in saline	Chemical added to culture for 1 hr; cells incubated with [^{3}H]dT containing media for various times up to 8 hr post-treatment	∿0.1-10 mM	TCA precipitable activity in cells measured by LSC	∿1 mM	1-10 mM gave + result; maximum UDS observed at ∿5 mM which was double the control value	18
HeLa cells in Eagle's media	Chemical added to culture for 3 hr; [^{3}H]dT present the last 2 hr of incubation	0.01, 0.1, 1.0 mM	Extraction of DNA followed by LSC	1.0 mM	Only the highest exposure used showed + result	19
Opossum lymphocytes in Eagle's modified media	Chemical added to culture for 1 hr; cells cultured 4 hr more in media with [^{3}H]dT	0.1-10 mM	Extraction of DNA followed by LSC	0.1 mM	0.1-1 mM gave + response, - response at 10 mM	20

(continued)

Table 5 (continued).

Biological material	Method of exposure	Exposure or range of exposures	Method for detecting UDS	Lowest effective exposure	Comments	References
Human fibroblasts in arginine deficient media supplemented with bovine serum	Chemical added to culture for 1.5 hr; cells cultured 1.5 hr more in media with $[^3H]dT$	0.1–10 mM	Autoradiography	0.1 mM	0.1–10 mM gave + response	6
(C3Hf × 101)F1 hybrid males	IP injection of chemical; testicular injection of $[^3H]dT$	5–100 mg/kg	Purified sperm heads derived from early spermatid stages and assayed by LSC	5 mg/kg (∿0.05 mM)	5–100 mg/kg gave + response; the lowest exposure tested still gave ∿6× the control values	1
Contact-inhibited human fibroblasts, WI-38 cells, in BME	Chemical and $[^3H]dT$ added to culture for 3 hr	0.0001–1 mM	Extraction of DNA followed by measurement of DNA content and assayed by LSC	∿0.1 mM		13

Table 6. UDS results for EMS (CAS 62-50-0)

Biological material	Method of exposure	Exposure or range of exposures	Method for detecting UDS	Lowest effective exposure	Comments	References
Chinese hamster cells in Eagle's modified media	Chemical added to culture for 3 hr; cells cultured 4 hr in media with [^{3}H]dT	0.7, 1.5 mM	Autoradiography	0.7 mM	0.7 and 1.5 mM both gave + results but there was only a slight increase in UDS at the higher exposure	15
Human lymphocytes in Eagle's modified media	Chemical added to culture for 1 hr; cells cultured 12 hr more in media with [^{3}H]dT	0.1-10.0 mM	TCA precipitable activity in cells measured by LSC	0.1 mM	All exposures from 0.1-10 mM gave a + result, with the maximum UDS occurring at 1.0 mM	8
Opossum lymphocytes in Eagle's modified media	Chemical added to culture for 1 hr; cells cultured 4 hr more in media with [^{3}H]dT	0.001-10 mM	Extraction of DNA followed by LSC	10 mM	Only the highest exposure gave + result	20

(continued)

Table 6. (continued)

Biological material	Method of exposure	Exposure or range of exposures	Method for detecting UDS	Lowest effective exposure	Comments	References
Human fibroblasts in arginine deficient media supplemented with bovine serum	Chemical added to culture for 3 hr; cells cultured 1.5 hr more in media with $[^3H]dT$	∿5-10 mM	Autoradiography	∿5 mM	5 → 10 mM gave + response	6
(C3Hf 101)F1 hybrid males	IP injection of chemical; testicular injection of $[^3H]dT$	10-300 mg/kg	Purified sperm heads from early spermatid stages and assayed by LSC	10 mg/kg (∿0.08 mM)	10 → 300 mg/kg gave + response; the lowest exposure tested still gave ∿3× the control value	1
Contact-inhibited human fibroblasts, WI-38 cells, in BME	Chemical and $[^3H]dT$ added to culture for 3 hr	0.001-10 mM	Extraction of DNA followed by measurement of DNA content and assayed by LSC	∿1.0 mM		13

Table 7. UDS results for aflatxoin B1 (CAS 1162-65-8)

Biological material	Method of exposure	Exposure or range of exposures	Method for detecting UDS	Lowest effective exposure	Comments	References
Human fibroblasts in arginine deficient media supplemented with bovine serum	Chemical added to culture for 0.5 or 2 hr; cells cultured an additional 1.5 hr in media with [^{3}H]dT	∿0.01 → 1 mM	Autoradiography	0.01 mM	With no activating microsomes present a 2-hr exposure to the chemical was needed to obtain a + result With activation, UDS was detected with a 0.5 hr exposure	6
Primary culture of rat liver cells in Williams' media	Chemical added to culture for 40 min; cells cultured an additional 3 hr in media with [^{3}H]dT	∿0.001-1.0 mM	Autoradiography	0.001-0.01 mM	Lowest effective exposure was between 0.001 and 0.01 mM; no exogenous activating system was used	21

(continued)

Table 7. (continued)

Biological material	Method of exposure	Exposure or range of exposures	Method for detecting UDS	Lowest effective exposure	Comments	References
Contact-inhibited human fibroblasts, WI-38 cells, in BME	For testing without microsomal activation, the chemical and $[^3H]dT$ were added to the culture for 3 hr. For testing with microsomal activation the chemical and microsomes were added to culture for 1 hr, removed, and $[^3H]dT$ added for 3 hr	0.00001-1 mM	Extraction of DNA followed by measurement of DNA content and LSC	∿0.1 mM	∿0.1 mM was the lowest effective concentration both with and without microsomal activation	13

Table 8. UDS studies on ICR-170, DEN, Mitomycin C, Natulan, Ethylenimine and Cyclophosphamide

Compound	Biological material	Method of exposure
ICR-170 (CAS 146-59-8)	Opossum lymphocytes in Eagle's modified media	Chemical added to cul- for 1 hr; cells cultured 4 hr more in media with [^{3}H]dT
DEN (CAS 55-18-5)	(C3Hf × 101)F1 hybrid males	IP injection of chemi- cal; testicular injec- tion or [^{3}H]dT
Mitomycin C (CAS 50-07-7)	P-388 mouse cells in modified Eagle's media	Chemical added to culture for 5 hr; [^{3}H]dT present during the last 2 hr of incubation
Natulan (CAS 366-70-1)	Contact-inhibited human fibroblasts, WI-38 cells, in BME	For testing without microsomal activation, the chemical and [^{3}H]dT were added to the culture for 3 hr For testing with micro- somal activation, the chemical and microsomes were added to culture for 1 hr, removed, and [^{3}H]dT added for 3 hr

(continued)

Table 8. (continued)

Exposure or range of exposures	Method for detecting UDS	Lowest effective exposure	Comments	References
0.001-0.1 mM	Extraction of DNA followed by by LSC	∿0.001 mM	Amount of UDS decreased with increasing exposure UDS was 3× higher at 0.001 mM than at 0.1 mM	20
125 mg/kg (∿1 mM)	Purified sperm heads derived from early spermatid stages and assayed by LSC	Negative result	No UDS detected in early spermatids at this exposure; apparently little or no activated chemical reaches the testes	11
0.5-50 μg/ml (0.0015-0.15 mM)	DNA density labeled with BUdR to distinguish semiconservative synthesis from UDS; DNA isolated on CsCl gradients and assayed by LSC	5 μg/ml (∿0.015 mM)	UDS was not linear with exposure: 5 μg/ml gave 3× the control value, 50 μg/ml gave only 4× the control value	12
0.0001-100 mM	Extraction of DNA followed by measurement of DNA content and liquid scintillation counting of each sample	Negative results at all exposures	UDS was not detected at any exposure with or without microsomal activation	13

(continued)

Table 8. (continued)

Compound	Biological material	Method of exposure
Ethylenimine (CAS 151-56-4)	(C3Hf × 101)F_1 hybrid males	IP injection of chemical; testicular injection of [^{3}H]dT
Cyclophosphamide (CAS 50-18-0)	(C3Hf × 101)F_1 hybrid males	IP injection of chemical; testicular injection of [^{3}H]dT

(continued)

Table 8. (continued)

Exposure or range of exposures	Method of detecting UDS	Lowest effective exposure	Comments	References
5 mg/kg (∿0.1 mM)	Purified sperm heads derived from early spermatid stages and assayed by LSC	Only 5 mg/kg tested; this exposure gave a strong + result	A 5 mg/kg exposure gave a level of UDS ∿65× the control value. By linear extrapolation an exposure of 0.2 mg/kg should give a readily measured UDS response	11
200 mg/kg (∿0.8 mM)	Purified sperm heads derived from meiotic and postmeiotic stages; assayed by LSC	200 mg/kg (∿0.8 mM)	Germ cell stages from early meiosis through mid-spermatids underwent UDS	4

REFERENCES

1. Sega, G. A., Owens, J. G., and Cumming, R. B. Mutat. Res. 36: 193-212 (1976).

2. Sega, G. A. Genetics 92: s49-s58 (1979).

3. Sega, G. A. Proc. Natl. Acad. Sci. (U.S.) 71: 4955-4959 (1974).

4. Sotomayor, R. E., Sega, G. A., and Cumming, R. B. Mutat. Res. 62: 293-309 (1979).

5. Stich, H. F., and San, R. H. C. Mutat. Res. 13: 279-282 (1971).

6. San, R. H. C., and Stich, H. F. Int. J. Cancer 16: 284-291 (1975).

7. Hennings, H., and Michael, D. Cancer Res. 36: 2321-2325 (1976).

8. Lieberman, M. W., Baney, R. N., Lee, R. E., Sell, S., and Farber, E. Cancer Res. 31: 1297-1306 (1971).

9. Laishes, B. A., and Stich, H. F. Biochem. Biophys. Res. Commun. 52: 827-833 (1973).

10. Stich, H. F., and Kieser, D. Proc. Soc. Exptl. Biol. Med. 145: 1339-1342 (1974).

11. Sega, G. A., unpublished data.

12. Ørstavik, J. Acta Pathol. Microbiol. Scand. B81: 711-718 (1973).

13. Mitchell, A. D., unpublished data.

14. Michel, T. M., and Legator, M. S. Mutat. Res. 23: 41-45 (1974).

15. Hahn, G. M., Yang, S. J., and Parker, V. Nature 220: 1142-1144 (1968).

16. Cleaver, J. E. Mutat. Res. 12: 453-462 (1971).

17. Fox, M., and Ayad, S. R. Chem.-Biol. Interact. 3: 193-211 (1971).

18. Clarkson, J. M., and Evans, H. J. Mutat. Res. 14: 413-430 (1972).

19. Brandt, W. N., Flamm, W. G., and Bernheim, N. J. Chem.-Biol. Interact. 5: 327-339 (1972).

20. Meneghini, R. Chem.-Biol. Interact. 8: 113-126 (1974).

21. Williams, G. M. Cancer Letters 1: 231-236 (1976).

CHAPTER 22

MUTAGENICITY OF SELECTED CHEMICALS IN THE MAMMALIAN MICRONUCLEUS TEST

B. E. Matter

Biological and Medical Research Division
Sandoz Ltd., Basle, Switzerland

and

D. Wild

Zentrallabor für Mutagenitätsprüfung der Deutschen
Forschungsgemeinschaft, Freiburg, F.R.G.

INTRODUCTION

The production of micronuclei as a result of chromosome aberration has been known for many years. Only recently, however, has there been interest in micronuclei for the purpose of rapid screening of mutagens in the environment, as a result of pioneer work by Schmid et al. (1-4). In recent years, a substantial number of different classes of chemical compounds has been tested with the micronucleus test in various laboratories.

Experiments have revealed that the micronucleus test is a simple and practical in vivo cytogenetic procedure for detecting both chromosome breaking agents (clastogens) and agents causing chromosome loss due to partial impairment of the mitotic apparatus (3-5).

Although micronuclei are known to occur in different cells and tissues of various organisms, the term "micronucleus-test" (mt) refers to a procedure in which the bone marrow of in vivo-treated mammals is analyzed for the presence of micronuclei in the anucleated young erythrocytes. For a detailed description of the nature of micronuclei, methodological aspects, advantages, limitations and other features of the method, the reader is referred to the review articles by Schmid (3-5).

The aim of this paper is to review and tabulate the results obtained so far with the mt. If not otherwise stated, the tables contain only results obtained in anucleated young erythrocytes. It should be mentioned, however, that several authors have reported findings in other bone marrow cells in addition to those obtained in erythrocytes, for example by means of the "nucleus anomaly test" or other modifications of the mt (1,6-8). These additional findings are not included in the tables, as the great majority of induced micronuclei present themselves in the anucleated young erythrocyte (3,5,6) and the nearly unlimited supply of scoreable young erythrocytes as well as the ease of scoring them for the presence of micronuclei are important features rendering the mt practicable for screening purposes (5).

The mean spontaneous incidence of micronucleated erythrocytes (MNE) are very low, and their range in individual animals quite narrow. For mice and Chinese hamsters, for example, the mean incidences of MNE among the total erythrocytes scored amount to 0-0.7% for individual animals (9). Similar results have been obtained in other laboratories as well as with other species (10-12).

It is well known that--as in other cytogenetic methods--various factors can influence the outcome of an experiment. These relate either to specificities of the compound and test material in question (e.g., relative toxicity and cytotoxicity, mechanisms of action), or to the test protocol (e.g., treatment and preparation of bone marrow, mode of scoring erythrocytes). Such factors have been described in detail elsewhere (4-6,8,13).

These factors must be considered when comparing results from different laboratories. A survey of the literature has revealed that such factors influence results mainly in quantitative terms (e.g., shapes of dose-response curves), whereas the reproducibility of findings in qualitative terms is very satisfactory. For the evaluation of the mt data, therefore, the "yes-or-no-approach" has been used as follows.

Any compound revealing a statistically significant increase in MNE over the spontaneous level of concurrent controls in one or more dose groups was considered by us to be a mutagen in this test system, and marked with a plus sign (+). Such signs were also used in the absence of a statistical evaluation, as long as the treated group(s) showed an increase over the spontaneous level by a factor of 3 or more. Compounds showing no effects were marked with a minus sign (-). Results judged to be questionable were labeled plus-or-minus (±), which merely indicates that more testing needs to be done for a proper evaluation of the compound in question. Some publications do not contain raw data, in which case the symbols refer to the judgement of the respective authors as indicated in the tables.

RESULTS

Table 1 summarizes the overall effects obtained with 20 reference mutagens. Table 2 provides more detailed information and references concerning individual studies of these compounds.

In addition to mt experiments, several authors have performed comparative studies involving various mutagens and different cytogenetic procedures in order to establish correlations between different types of cytogenetic damage in different tissues and cells. These findings are summarized in Table 3.

Table 4 contains a list of results with other substances so far tested with the mt in order to clarify which classes of compounds are detectable by this method.

DISCUSSION

For a discussion of the mt results summarized in Tables 1 and 2, the 20 agents are subdivided into groups according to their structure and action mechanisms.

The alkylating agent ethylenimine itself has not been investigated, but its polyfunctional alkylating derivatives MeTEPA, TEM, Thio-TEPA, and Trenimon were active in all studies, with different routes of application, on six different mammalian species, and at doses as low as 0.03 mg/kg for Trenimon (6). Similarly, several groups reported activity of the alkylating methanesulfonate esters EMS and MMS in mice, with lowest reported effective doses of 31 mg/kg for MMS (27), and 400 mg/kg for EMS (6). Myleran, a bifunctional alkylating methanesulfonate, was studied independently by two groups. It was active following two (intraperitoneal?) applications of a total dose of 100 mg/kg (26), whereas no effect was observed following five daily administration with unspecified doses (15,16). In view of the alkylating and antileukemic activity of myleran and of its mutagenic potential in other tests, including cytogenetic tests, the positive result is considered representative. The negative result could be due to low total doses or to the dose fractionation and 5-day treatment. This could obscure an effect, since micronucleated

Table 1. Summary of results obtained with 20 reference mutagens[a]

Compound	No. of studies performed	Studies reported +	Studies reported ±	Studies reported −	Lowest reported effective dose mg/kg	Lowest reported effective dose mmole/kg	Dose range reported in which micronuclei have been obtained (mmole/kg)
Aflatoxin B1	3		1	2			
Cadmium salt	1			1			
Cyclophosphamide	15	15			5	0.04	0.04–4.0
DEN	1			1			
DMN	4	2	1	1	37	0.5	0.5–1.0
EMS	6	6			400	3.2	3.2–7.2
Epichloroyhydrin	0						
Ethylenimine	0						
ICR-170	0						
MeTEPA	2	2			0.1	0.0005	0.0005–0.04
Mitomycin C	4	4			3.5	0.01	0.01–0.08
MMS	5	5			31	0.3	0.3–2.0
MNNG	2			2			
Myleran	2	1		1	b	b	b
Procarbazine	2	2			13	0.05	0.05–0.5
TEPA	0						
Thio-TEPA	2	2			2.0	0.01	0.01–1.0
TEM	8	8			0.1	0.0005	0.0005–0.08
Trenimon	13	13			0.03	0.0001	0.0001–0.01
Vinyl chloride	0						
X-Rays	2	2			8 rad		8–230 rad

[a]The dose levels refer to the total doses applied as a single or multiple treatment.
[b]Only one dose tested.

Table 2. Detailed information and bibliography on the 20 reference mutagens

Compound	Solvent[a]	Application No.	Application route[b]	Species	Result	References
Aflatoxin B1	DMSO	2	IP	Mouse	±	14
	DMSO	2	IP	Syrian hamster	−	14
	DMSO	5	IP	Mouse	−[d]	15,16
Cadmium sal salt	H_2O	5	IP	Mouse	−	15,16
Cyclophosphamide	CMC	2	PO	Chinese hamster	+	15,16
	?	1	IP	Chinese hamster	+	8
	H_2O	2	IP	Mouse	+	6,18
	H_2O	5	IP	Mouse	+[d]	15,16, 19
	Saline	2	IP	Mouse	+	12,20
	Saline	1	IP	Mouse	+	21
	Saline	1	IV	Mouse	+	21
	Saline	1	PO	Mouse	+	21
	Saline	1	SC	Mouse	+	21
	?	1	IP	Mouse	+	17
	?	1	IP	Rat	+	17
	?	1	IV	Man	+	17
Diethylnitrosamine (DEN)	Saline	2	IP	Mouse	−	12,22
Dimethylnitrosamine (DMN)	Saline	2	IP	Mouse	+	12,22
	H_2O	2	IP	Mouse	+	14
	H_2O	2	IP	Syrian hamster	±	14
	DMSO	5	IP	Mouse	−	15,16
Ethyl methanesulfonate (EMS)	H_2O	5	IP	Mouse	+	19
	Saline	2	IP	Mouse	+	6,18
	Saline	2	IV	Mouse	+	6,18
	Saline	2	PO	Mouse	+	6,18
	DMSO	2	IP	Mouse	+	23
MeTEPA	H_2O	2	IP	Mouse	+	24
	H_2O	5	IP	Mouse	+[d]	15,16

(continued)

Table 2. (continued)

Compound	Solvent[a]	Application No.	Application Route[b]	Species		References
Mitomycin C	?	1	IP	Rat	+	25
	Saline	2	IP	Mouse	+	6,18
	H_2O	5	IP	Mouse	+[d]	15,16
	H_2O	2	IP?	Mouse	+	26
Methyl methane-sulfonate (MMS)	Saline	2	IP	Mouse	+	6,18
	Saline	1	IP	Mouse	+	27,28
	H_2O	2	IP	Mouse	+	29
	H_2O	5	IP	Mouse	+[d]	15,16
MNNG	Saline	2	IP	Mouse	-	6,18
	H_2O	5	IP	Mouse	-	15,16
Myleran	DMSO	2	IP?	Mouse	+	26
	DMSO	5	IP	Mouse	-[d]	15,16
Procarbazine	Saline	2	IP	Mouse	+	12
	DMSO	5	IP	Mouse	+[d]	15,16
Thio-TEPA	Saline	2	IP	Mouse	+	11
	H_2O	5	IP	Mouse	+[d]	15,16
TEM	Saline	2	IP	Mouse	+	6,14, 18
	Saline	2	IV	Mouse	+	6
	Saline	2	PO	Mouse	+	6
	Saline	1	IP	Mouse	+	30
	Saline	2	IP	Syrian hamster	+	14
	Saline	2	IP	Rat	+	31-33
Trenimon	Saline	2-12	IP	Chinese hamster	+	1
	Saline	2	IP	Chinese hamster	+	2,9
	Saline	2	IP	Mouse	+	6,18 2,9 34
	Saline	1	IP	Mouse	+	13
	Saline	2	IP	Rat	+	2,35
	Saline	2	IP	Syrian hamster	+	2
	Saline	2	IP	Guinea pig	+	2
	Saline	2	IV	Rhesus monkey	+	2

(continued)

Table 2. (continued)

Compound	Solvent[a]	Application No.	Route[b]	Species[c]	Result	References
X-Rays	-	1	Whole body	Mouse	+	19,27
γ-Rays	-	1	Whole body	Mouse	+	36

[a]Solvents: DMSO, dimethyl sulfoxide; CMC, sodium carboxylmethylcellulose.

[b]Administration routes: IP, intraperitoneal; IV, intravenous; PO, oral; SC, subcutaneous.

[c]Animals: Chinese hamster (Cricetulus griseus); Syrian hamster (Mesocricetus auratus).

[d]No raw data given.

erythrocytes resulting from such treatments do not accumulate during that time period, but are continuously eliminated (1).

Mitomycin C contains elements of ethylenimine and of urethane. It does not react with DNA, but is reduced in vivo into an alkylating, DNA-crosslinking agent. All studies performed showed an effect on the mt, the lowest reported effective dose being 3.5 mg/kg in mice (6). Mitomycin C was also active in rats at a similar dose (3.0 mg/kg) (25).

MNNG is an alkylating agent and potent mutagen in microbial test systems in vitro. It did not, however, show an effect in the mt, although high doses up to 200 mg/kg were assayed (6,15, 16, 41). This can probably be explained by the rapid decomposition of MNNG in vivo, resulting in insufficiently high concentrations to cause a cytogenetic effect in bone marrow. Both the result and interpretation are consistent with the reported SCE induction by MNNG in mammalian cells in vitro (50) and with the absence of this effect in vivo in bone marrow cells (22).

A total of 15 studies on the induction of micronuclei by cyclophosphamide were reported, all showing clear-cut effects. The lowest reported effective dose in mice was found to be 5 mg/kg (17,21). Effects of a single dose of 40 mg/kg were compared in

Table 3. Summary of comparative studies: results obtained in different cytogenetic test systems

Compound	Species	Cytogenetic techniques[a]						References
		mt	bm	cytes	gonia	fb	SCE	
Cyclophosphamide	Chinese hamster	+	+					8,7
	Mouse	+	+					17
	Rat	+	+					17
	Man	+	+					17
	Mouse	+					+	22
DEN	Mouse	-					-	22
DMN	Mouse	+					+	22
Mitomycin C	Rat	+	+					25
MNNG	Mouse	-					-	22
Thio-TEPA	Mouse	+				+		11
Trenimon	Chinese hamster	+	+					1,9
	Mouse	+					+	22
Bleomycin	Mouse	-				+		11
CNU-ethanol, BCNU	Mouse	+	+	+	+			37,38

Cytosine arabinoside	Mouse	+	+		11
5-Fluorouracil	Mouse	+	+		11
6-Mercaptopurine	Mouse	+	+		11
Methotrexate	Mouse	+	+		11
PDMT	Mouse	+		+	22
Vincristin	Mouse	+	-		11

[a] mt, micronucleus test; bm, bone marrow metaphase method; cytes, spermatocytes; gonia, spermatogonia; fb, Chinese hamster fibroblasts in vitro; SCE, sister chromatid exchange on in vivo bone marrow cells.

mice, rats, Chinese hamsters, and in five human cancer patients. Micronuclei were induced by this treatment in all rodents studied as well as in humans, the degree of damage decreasing in the given order of the species. A parallel pattern of effects was evident when bone marrow cells of these species were analyzed for the presence of chromosome breaks (17) (see also Table 3). These results are a clear and valuable support for a quantitative correlation between the results obtained in the mt in several species and those obtained in a traditional cytogenetic test system.

DMN was studied in mice by three groups independently. Bauknecht et al. (22) and Wild (12) reported significant and dose-related effects and a lowest effective dose of 2 × 18.5 mg/kg. Effects of DMN at doses between 5 and 140 mg/kg were also studied by Friedman and Staub (14). The frequency of MNE did not, however, exceed a value of 0.44%. Though this is a four-fold increase over the controls it is still within the control range observed in several other laboratories. The same authors also reported an unusually low incidence of micronuclei following TEM treatment. The low effects of these two mutagens observed in that laboratory may, therefore, have a common origin. Heddle and Bruce (15, 16), on the other hand, were unable to detect an effect of DMN at a dose of 30 mg/kg. This dose was presumably injected in five daily fractions of 6 mg/kg/day, and in the light of the dose-effect curves reported by Bauknecht et al. (22) and Wild (12), it is conceivable that the fractionation of relatively low dose levels and the nonaccumulation of effects during the 5-day treatment was responsible for the negative result.

DEN, in contrast to DMN, was not significantly active in the mt, even at high dose levels (i.e., 200 mg/kg) (12,22). Similar results have been obtained with the in vivo SCE technique on bone marrow cells. Whereas DMN was active at relatively low dose levels (0.2 mg/kg), DEN was inactive even at doses which, on a molar basis, were 700 times higher than the active DMN doses (22). It should be mentioned, however, that DEN was found to exert SCE-inducing potential in vitro in the presence of metabolizing liver microsomes (51).

These results can be explained as follows. We cannot exclude the possibility that different reaction mechanisms of the alkylation by the active DEN versus DMN metabolites may result in different alkylation patterns. This could contribute to a differential mutagenic activity of these agents in vivo, but it appears unlikely that one can explain the observed qualitative difference on the basis of different reaction mechanisms alone. It seems more likely that the results obtained with DEN are due to the short lifespan of the ultimately mutagenic metabolite, leading to an insufficient concentration at the target to produce cytogenetic effects in the bone marrow.

Table 4. Results obtained with other compounds

Compound	Species	Result	References
Alkylating agents			
Direct alkylating agents			
1,3-Bis(2-chloroethyl)-3-nitrosourea (BCNU)	Mouse	+	38
1(2-Chloroethyl)-1-nitroso-ureidoethanol (CNU-ethanol)	Mouse	+	37
Mechlorethamine (nitrogen mustard)	Mouse	+[a]	15,16
Trimethylphosphate	Mouse	+	30
	Mouse	+[a]	15,16
Indirect alkylating agents			
4-Bromo - PDMT	Mouse	+	12
4-Chloro - PDMT	Mouse	+	12
Cyclophosphamide/Cytembena (combination)	Mouse	+	39
	Rat	+	39
	Chinese hamster	+	39
	Man	+	39
Dibutylnitrosamine	Mouse	-[a]	15,16
3,3-Diethyl-1-phenyl-triazene	Mouse	+	12
3,3-Dimethyl-1-phenyl-triazene (PDMT)	Mouse	+	12
3,3-Dimethyl-1-(pyridyl-3)-triazene (PyDMT)	Mouse	+	12
Ifosfamide	Mouse	+	12
Sufosfamide	Mouse	+	12

(continued)

Table 4. (continued)

Compound	Species	Result	References
2,4,6-trichloro PDMT	Mouse	+	12
Trofosfamide	Mouse	+	12
Antibiotics			
Actinomycin D (Dactinomycin)	Mouse	-[a]	15,16
Adriamycin (Doxorubicin)	Mouse	+	11
Bleomycin	Mouse	-	11
Chloramphenicol (Chloromycetin)	Mouse	-[a]	15,16
Cycloheximide	Mouse	+[a]	15,16
Griseofulvine	Mouse	-[a]	15,16
Aromatic amines, aminophenols, and phenols			
Acetylaminofluorene	Mouse	±	14
	Syrian hamster	±	14
2-Aminofluorene	Mouse	-[a]	15,16
4-Amino-2-hydroxytoluene	Rat	-	10
2-Amino-4-nitrophenol	Rat	-	10
m-Aminophenol	Rat	-	10
p-Aminophenol	Rat	-	10
Benzidine	Rat	+	40
Butylated hydroxytoluene	Mouse	-[a]	15,16
4-Chlororesorcinol	Rat	-	10

(continued)

Table 4. (continued)

Compound	Species	Result	References
2-Methyl-p-phenylenediamine (2,5-diaminotoluene)	Rat	-	10
4-Methoxy-m-phenylenediamine	Rat	-	10
	Mouse	-[a]	15,16
1-Naphthol	Rat	-	10
1-Naphthylamine	Mouse	-[a]	15,16
2-Naphthylamine	Mouse	-[a]	15,16
4-Nitro-o-phenylenediamine	Rat	-	10
2-Nitro-p-phenylenediamine	Rat	-	10
p-Phenylenediamine	Rat	-	10
Resorcinol	Rat	-	10
Aromatic hydrocarbons			
Benzo-[a]pyrene	Mouse	-	15,16
9,10-Dimethyl-1,2-benzanthracene	Mouse	+	15,16
3-Methylcholanthrene	Mouse	±	14
	Syrian hamster	±	14
	Mouse	-	15,16
Drugs			
Acetylsalicylic acid	Mouse	-[a]	15,16, 41
Acriflavine	Mouse	+[a]	15,16
Aminopterin	Mouse	+[a]	15,16

(continued)

Table 4. (continued)

Compound	Species	Result	References
Ascorbic acid	Mouse	-	15,16
Chloralhydrate	Mouse	-	15,16
Clemastine (Tavegyl)	Mouse	-[a]	41
Codeine phosphate	Mouse	-[a]	15,16
Cytembena (cis-β-4-methoxyl-benzoyl-β-bromoacrylic acid, sodium salt)	Mouse	+	39
	Rat	+	39
	Chinese hamster	+	39
	Man	+	39
Epinephrine	Mouse	-[a]	15,16
Ergotamine tartrate	Mouse	-	9
	Chinese hamster	-	9
Dihydroergotoxine mesylate (Hydergine)	Mouse	-	9
	Chinese hamster	-	9
Methysergide hydrogen maleate (Deseryl)	Mouse	-	9
Hycanthone	Mouse	+	30
	Mouse	+[a]	15,16
Methotrexate	Mouse	+[b]	11
Phenobarbital	Mouse	-	15,16
Praziquantel	Mouse	-	42
Quinacrine	Mouse	-	
Thioridazine (Melleril)	Mouse	-[a]	41

(continued)

Table 4. (continued)

Compound	Species	Result	References
Fluorescent whitening agents			
Disodium 4,4'-bis[(4-anilino-6-morpholino-1,3,5-triazin-2-yl)amino] stilbene-2,2'-disulfonate	Chinese hamster	-	7
Disodium 4,4'-bis{[4-anilino-6-(N-methyl-N-2-hydroxyethyl) amino-1,3,5-triazin-2-yl] amino}stilbene-2,2'-disulfonate disulfonate	Chinese hamster	-	7
Disodium 4,4'-bis(2-sulfostyryl)biphenyl	Chinese hamster	-	7
Sodium-2-(4-styryl-3-sulfophenyl)-2H-naphtho[1,2-d] triazole	Chinese hamster	-	7
Inorganic compounds			
Lead acetate	Mouse	-	15,16, 26
Methyl mercury acetate	Mouse	-[a]	15,16
Potassium chromate	Mouse	+	12
Sodium chloride	Mouse	-[a]	15,16
Sodium nitrite	Mouse (PO)[c]	-	34
Miscellaneous compounds			
Aroclor 1254	Mouse	-[a]	15,16
Calcium cyclamate	Mouse	-[a]	15,16
1,1-Dimethylhydrazine	Mouse	-[a]	35

(continued)

Table 4. (continued)

Compound	Species	Result	References
Dimethyl sulfoxide	Mouse	-	23
Ethanol	Mouse	-	23
Ethylenethiourea	Mouse	-	34,43
Glucose	Mouse	-[a]	15,16
Hydroxyurea	Mouse	-[a]	15,16
N-Hydroxy urethane	Mouse	+	12
Monosodium glutamate	Mouse	-[a]	15,16
Urethane	Mouse	+	12,15,16
Nitrogen compounds + nitrite, nitrosation products			
Benzthiazuron + nitrite	Mouse (PO)[c]	-	44
Carbaryl + nitrite	Mouse (PO)[c]	-	44
Carbofuran + nitrite	Mouse (PO)[c]	-	44
Dimethoate + nitrite	Mouse (PO)[c]	-	44
Ethiofencarb + nitrite	Mouse (PO)[c]	-	44
Ethylenethiourea + nitrite	Mouse (PO)[c]	+	41
Formetanate + nitrite	Mouse (PO)[c]	-	44
Linuron + nitrite	Mouse (PO)[c]	-	44
Maneb + nitrite	Mouse (PO)[c]	-	44
Methabenzthiazuron + nitrite	Mouse (PO)[c]	-	44

(continued)

Table 4. (continued)

Compound	Species	Result	References
Methylurea + nitrite	Mouse (PO)[c]	+	45
Carbaryl, N-Nitroso	Mouse (PO)[c]	-	44
N-Nitrosoethylenethiourea	Mouse (PO)[c]	+	34
N-Nitroso methomyl	Mouse (PO)[c]	-	44
Propham + nitrite	Mouse (PO)[c]	-	44
Propoxuron + nitrite	Mouse (PO)[c]	-	44
Thiram (TMTD) + nitrite	Mouse (PO)[c]	-	44
Nitroheterocyclic compounds			
Formic acid 2-[4-(5-nitro-2-furyl)-2-thiazolyl]hydrazide (FNT)	Mouse	-[a]	15,16
2-(2-Furyl)-3-(5-nitro-2-furyl)-acrylamide (AF-2)	Mouse Rat	-[a] +	15,16 31
Niridazole	Mouse	-[a]	30
Nitrofurantoin	Rat	-	31
Nitrofurazone	Mouse Rat	-[a] -	15,16 31
4-Nitroquinoline 1-oxide	Mouse	+ -[a]	12 15,16
N-[5-(5]Nitro-2-furyl)-2-thiazolyl] formamide (FANFT)	Mouse	-[a]	15,16
Ronidazole	Mouse	±	29

(continued)

Table 4. (continued)

Compound	Species	Result	References
Pesticides			
Pesticidal and other benzimidazoles			
2-Aminobenzimidazole	Mouse (PO)[c]	-	46
Benomyl	Mouse (PO)[c]	+	46
Benzimidazole	Mouse	-	46
Benzimidazolecarbamonitrile	Mouse (PO)[c]	-	46
2-Benzimidazolylurea	Mouse (PO)[c]	+	46
Carbendazim (MBC)	Mouse (IP)[c] Mouse (PO)[c] Chines hamster (PO)[c]	- + +	46 46 47
Other pesticides			
Captan	Mouse	-	47
DDT	Mouse	-[a]	15,16
2,4-Dichlorophenoxyacetic acid (2,4-D)	Mouse	-	28
Pirimiphos-methyl	Mouse	-	48
2,4,5-Trichlorophenoxy-acetic acid (2,4,5-T)	Mouse	-	28
1,3,5-Tris(hydroxyethyl) s-hexahyrdotriazine (Grotan BK)	Rat	-	40

(continued)

Table 4. (continued)

Compound	Species	Result	References
Plant constituents			
Chlorogenic acid	Rat	-	35
Colcemide	Mouse	+	11
Colchicine	Mouse	+	6,18
	Mouse	-[a]	15,16
	Chinese hamster	+[a]	47
Potato extract, irradiated	Rat	-	35
Vinblastine	Mouse	+[a]	15,16
Vincristine	Mouse	+	11
Purine and pyrimidine derivatives			
Azathioprine	Mouse	+[a]	15,16
	Chinese hamster	+[a]	47
	Man	+[b]	49
5-Bromo-2'-deoxyuridine	Mouse	+[a]	15,16
5-Iodo-2'-deoxyuridine	Mouse	-[a]	15,16
Caffeine	Mouse	-	6,15, 16,18
	Chinese hamster	-	47
Cytosine arabinoside	Mouse	+	11
5-Fluorouracil	Mouse	+[b]	11
6-Mercaptopurine	Mouse	+	11

[a]No information on methodology and no data given.

[b]Modified micronucleus test protocol.

[c]Types of administration are only given in those cases where this information appears to be essential: PO = oral application; IP = intraperitoneal injection.

In one of the reported studies on aflatoxin B1, no effect was observed in mice, but again, the doses were not indicated (15,16). The other study on mice came from the group which also reported on DMN (14). It showed unclear results at IP doses of 15 and 20 mg/kg. The maximum frequency was 0.22% MNE, corresponding to a 2.7-fold increase over the concurrent controls. This result is even lower than that found with DMN and cannot with certainty be interpreted as positive. No effect was seen by the same authors in Syrian hamsters, which had been treated with relatively low doses (0.2-3 mg/kg). In summary, the available data on aflatoxin B1 are an inadequate basis for an evaluation of its potential mutagenic effect in the mt.

Cadmium chloride, at a total dose of 15 mg/kg (in five daily fractions), did not induce micronuclei in mouse bone marrow (15,16).

Procarbazine exerted a strong cytogenetic effect in mice, the lowest effective dose (2 × 6.5 mg/kg IP) being in the range of human therapeutic doses (12). Positive effects were also observed by Heddle and Bruce (15,16).

X-rays, the reference mutagen, is a potent micronuclei inducer (27). The lowest reported effective dose was 8 rad, indicating that the mt is one of the most sensitive test systems available for mutagenicity screening.

In summary, x-rays and 15 of the 20 chemicals listed have so far been studied in the mt. Besides x-rays, 11 of these chemicals were active, namely, cyclophosphamide, DMN, EMS, MeTEPA, Mitomycin C, MMS, Myleran, procarbazine, thio-TEPA, TEM, and Trenimon. No or questionable effects were reported with aflatoxin B1, cadmium chloride, DEN, and MNNG.

Several authors have performed comparative studies involving various mutagens and cytogenetic techniques. The findings are summarized in Table 3, which clearly shows that clastogens are detectable both by the mt and various metaphase methods, including SCE. Furthermore, it appears that the effects produced by a chemical can be picked up in various mammalian species, including man, at comparable dose levels, rendering the mt potentially valuable for hazard evaluations.

As shown in Table 3, certain compounds are not detectable by all cytogenetic techniques. On the one hand, spindle poisons such as Vincristin induce micronuclei but can obviously not be detected by metaphase methods (5). On the other hand, Bleomycin, a strong clastogenic agent in vitro and in vivo, is unable to induce micronuclei due to its peculiar inability to damage bone marrow cells cytogenetically (5).

In Table 4 we have listed published results obtained with various other chemical compounds. It is evident from Table 4 that the mt can detect cytogenetic effects of various classes of chemicals.

Note added in proof: This paper reviews the literature up to 1978. An updated, though somewhat less critical review, has recently been made by D. Jensen and C. Ramel (Mutat. Res. 75: 191-202, 1980).

REFERENCES

1. Boller, K., and Schmid, W. Humangenetik 11: 35-54 (1970).
2. Matter, B., and Schmid, W. Mutat. Res. 12: 417-425 (1971).
3. Schmid, W. Agents and Actions 3: 77-85 (1973).
4. Schmid, W. Mutat. Res. 31: 9-15 (1975).
5. Schmid, W. In: Chemical Mutagens, Vol. 4, A. Hollaender, Ed., Plenum Press, New York, 1976, pp. 31-53.
6. Matter, B. E., and Grauwiler, J. Mutat. Res. 23: 239-249 (1974).
7. Müller, D., Fritz, H., Langauer, M., and Strasser, F. F. In: Environmental Quality and Safety, Suppl. Vol. IV, Fluorescent Whitening Agents, F. Coulston and F. Korte, Eds., Thieme Publishers, Stuttgart, 1975, pp. 247-263.
8. Müller, D., Langauer, M., Rathenberg, R., Strasser, F. F., and Hess, R. Verh. Deut. Ges. Pathol. 56: 381-384 (1972).
9. Matter, B. E. J. Int. Med. Res. 4: 382-392 (1976).
10. Hossack, D. J. N., and Richardson, J. C. Experientia 33: 377-378 (1978).
11. Maier, P., and Schmid, W. Mutat. Res. 40: 325-338 (1976).
12. Wild, D. Mutat. Res. 56: 311-317 (1978).
13. von Ledebur, M., and Schmid, W. Mutat. Res. 19: 109-117 (1973).
14. Friedman, M. A., and Staub, J. Mutat. Res. 43: 255-262 (1977).

15. Heddle, J. A., and Bruce, W. R. In: Origins of Human Cancer, Vol. 4, H. H. Hiatt, J. D. Watson, and J. A. Winsten, Eds., Cold Spring Harbor Conferences, 1977.

16. Heddle, J. A., and Bruce, W. R. In: Progress in Genetic Toxicology, D. Scott, B. A. Bridges, and F. H. Sobels, Eds., Elsevier/North Holland Biomedical Press, Amsterdam, 1977, pp. 265-274.

17. Goetz, P., Sram, R. J., and Dohnalova, J. Mutat. Res. 31: 247-254 (1975).

18. Matter, B. E., Jaeger, I., and Grauwiler, J. In: Experimental Model System in Toxicology and Their Significance in Man. (Proc. Soc. Study Drug Toxicol. Vol. 15) Excerpta Med. Int. Congr. Ser. No. 311, Elsevier, Amsterdam, 1974, pp. 275-279.

19. Heddle, J. A. Mutat. Res. 18: 187-190 (1973).

20. Maier, P., and Schmid, W. Mutat. Res. 30: 299-302 (1975).

21. Frank, D. W., Trzos, R. J., and Good, P. I. Mutat. Res. 56: 311-317 (1978).

22. Bauknecht, T., Vogel, W., Bayer, U., and Wild, D. Hum. Genet. 35: 299-307 (1977).

23. Chaubey, R. C., Kavi, B. R., Chauhan, P. S., and Sundaram, K. Mutat. Res. 43: 441-444 (1977).

24. Richardson, J. C. Mutat. Res. 26: 391-394 (1975).

25. Miller, R. C. Environ. Health Perspect. 6: 167-170 (1973).

26. Jacquet, P., Léonard, A., and Gerber, G. B. J. Toxicol. Environ. Health 2: 619-624 (1977).

27. Jensen, D., and Ramel, C. Mutat. Res. 41: 311-320 (1976).

28. Jensen, D., and Renberg, L. Chem.-Biol. Interact. 14: 291-299 (1976).

29. Hite, M., Skeggs, H., Noveroske, J., and Peck, H. Mutat. Res. 40: 289-304 (1976).

30. Weber, E., Bidwell, K., and Legator, M. S. Mutat. Res. 28: 101-106 (1975).

31. Goodman, D. R., Hakkinen, P. J., Nemenzo, J. H., and Vore, M. Mutat. Res. 48: 295-306 (1977).

32. Jellema, M. M., and Schardein, J. L. Toxicol. Appl. Pharmacol. 31: 107-113 (1975).

33. Jellema, M. M., and Schardein, J. L. Toxicol. Appl. Pharmacol. 27: 422-430 (1974).

34. Seiler, J. P. Experientia 31: 214-215 (1975).

35. Hossain, M. M., Huismans, J. W., and Diehl, J. F. Toxicology 6: 243-251 (1976).

36. Heddle, J. A., and Carrano, A. V. Mutat. Res. 44: 63-69 (1977).

37. Tates, A. D., and Natarajan, A. T. Mutat. Res. 37: 267-278 (1976).

38. Tates, A. D., Natarajan, A. T., de Vogel, N., and Meijers, M. Mutat. Res. 44: 87-95 (1977).

39. Goetz, P., Srȃm, R. J., Kodytkova, I., Dohnalova, J., Dostalova, O., and Bartova, J. Mutat. Res. 41: 143-152 (1976).

40. Urwin, C., Richardson, J. C., and Palmer, A. K. Mutat. Res. 40: 43-46 (1976).

41. Matter, B. E., and Grauwiler, J. Mutat. Res. 29: 198-199 (1975).

42. Machemer, L., and Lorke, D. Arch. Toxicol. 39: 187-197 (1978).

43. Schüpbach, M., and Hummler, H. Mutat. Res. 56: 111-120 (1978).

44. Seiler, J. P. Mutat. Res. 48: 225-236 (1977).

45. Aeschbacher, H. U., Gottwick, D., and Würzner, H. P. Mutat. Res. 43: 71-80 (1977).

46. Seiler, J. P. Mutat. Res. 40: 339-348 (1976).

47. Tsuchimoto, T., and Matter, B. E. Mutat. Res. 46: 240 (1977).

48. Seiler, J. P. In: The Prediction of Chronic Toxicity from Short Term Studies (Proc. Soc. Study Drug Tox., Vol. 17), Excerpta Medica Int. Congr. Series No. 376, Elsevier, Amsterdam, 1975, pp. 398-404.

49. Jensen, D., and Renberg, L. Chem.-Biol. Interact. 14: 291-299 (1976).

50. Perry, P., and Evans, H. J. Nature 258: 121-125 (1975).

51. Natarajan, A. T., Tates, A. D., van Buul, P. P. W., Meijers, M., and De Vogel, N. Mutat. Res. 37: 83-90 (1976).

CHAPTER 23

MUTAGENICITY OF SELECTED CHEMICALS IN MAMMALS: THE HERITABLE TRANSLOCATION TEST

W. M. Generoso

Biology Division, Oak Ridge National Laboratory
Oak Ridge, Tennessee 37830 (U.S.A.)

B. Cattanach

MRC Radiology Unit, Harwell, Didcot, Berkshire
United Kingdom

and

A. M. Malashenko

Laboratory of Experimental Biological Models
Academy of Medical Sciences, U.S.S.R., 143412
Moscow, U.S.S.R

Male translocation heterozygotes may be either partially sterile (semisterile) or completely sterile, depending upon the position of the breaks or whether or not the sex chromosomes are involved. Since translocation heterozygotes do not show phenotypic characteristics that distiguish them from normal mice, translocation heterozygotes cannot be detected by casual observation of the progeny. On the other hand, translocation heterozygotes have two characteristic features that allow unequivocal detection of these animals. First, translocation heterozygotes produce two types of gametes: balanced and unbalanced. Complete sterility and presence of unbalanced gametes are easily revealed by fertility testing. Second, because of pairing of homologous chromosome regions and formation of chiasmata in these regions, multiple association of chromosomes can be observed among diakenesis-metaphase I meiocytes (cytological screening).

Partial sterility is attributable to the production of gametes that are unbalanced, a condition known as duplication deficiency. The unbalanced sperm are capable of fertilization, but they lead to early embryonic lethality observed primarily as resorption moles. The degree of partial sterility depends on the proportions by which balanced and unbalanced constituents are represented in the ejaculate, which in turn is assumed to be a function of the relative incidence of adjacent I and adjacent II segregations.

Sterile male translocation heterozygotes are characterized by small testes (approximately one-third normal size on the average). In most of these, sterility is attributable to failure of spermatogenesis, while in some spermatogenesis is not impaired. In any case, sterility appears to arise when one or both breakpoints are close to either end of a chromosome or when the exchange involves the Y chromosome. In addition, sterility may ensue when more than one exchange is present. There is also a slim possibility that such sterility could have arisen from the male being XXY.

All published papers and abstracts on the subject were included with the exception of preliminary reports of studies in which full papers are also available. Also included are reports made available to the reviewers that are either in press or in preparation for publication. The data are summarized in Table 1 and given in detail in Table 2.

The most extensive data on spontaneous frequency of heritable translocations are given in the paper by Generoso et al. (21). The spontaneous frequency varies from 9.1×10^{-4} (3 steriles and 1 partially sterile male in 4392 progeny tested) to 2.3×10^{-4} (1 partially sterile male in 4392 tested), depending on whether the sterile males are assumed to be translocation heterozygotes. The steriles, although typical for sterile translocations--i.e., small testes size and spermatogenesis blocked prior to diakinesis-metaphase I stage--were not analyzed cytologically and may be instead carriers of other types of genetic change such as XXY. It should be noted, however, that even if cytological analysis had been performed, failure to detect exchanges in metaphases of somatic cells does not necessarily preclude translocation heterozygosity. The partially sterile male was confirmed cytologically as a translocation heterozygote.

In the great majority of cases, complete sterility among male progeny recovered from treated meiotic and postmeiotic male germ cells is now generally known to be attributable to translocation heterozygosity. Thus, the frequencies entered in the report included both partial and completely sterile progeny. It should be noted that even if completely sterile progeny are excluded, the decisions made on each compound still hold.

With the exception of cadmium, aflatoxin B1 and mitomycin C, mutagenicity of the rest of the compounds that were studied using the heritable translocation test was clearly established.

Clearcut induction of heritable translocations was found only in male meiotic and postmeiotic stages. Although positive induction in treated spermatogonial stem cells was reported for TEPA (4) and Thio-TEPA (7), more extensive data with TEM (6), TEPA (6), and cyclophosphamide (6,13) indicate the ineffectiveness of alkylating chemicals in inducing heritable translocations in this germ cell stage.

Table 1 follows on pages 684 and 685. Table 2 appears on pages 686 through 705. References are to be found on pages 710 and 711.

Table 1. Summary on the effectiveness of selected chemicals in inducing heritable translocations[a]

Compound	Number of studies	Lowest effective dose studied (mg/kg)	Dose range studied (mg/kg)	Comments
Aflatoxin B	1			Present data for this compound are inadequate to permit decisions on its effectiveness in inducing heritable translocations
Mitomycin C	2			Present data for this compound are inadequate to permit decision on its effectiveness in inducing heritable translocations
Cyclophos-phamide	2	350	350-400	This compound is clearly effective in inducing heritable translocations; it is likely also to be effective at some lower doses
TEPA	3	5	2.5-30	This compound is clearly effective in inducing heritable translocations; it is likely also to be effective at some lower doses
Thio-TEPA	2	1.25	1.25-5	This compound is clearly effective in inducing heritable translocations; it is likely also to be effective at some lower doses

MMS	3	40	40-50	This compound is clearly effective in inducing heritable translocations; it is likely also to be effective at some lower doses
EMS	4	50	50-250	This compound is clearly effective in inducing heritable translocations both in mice and Armenian hamsters. Effects of doses lower than 50 mg/kg will be difficult to detect. A dose-effect curve may be constructed from data in Table 2.
TEM	7	0.0125	0.0125-4	This compound is clearly effective in inducing heritable translocations. Effects of doses lower than 0.0125 mg/kg will be difficult to detect. A dose-effect curve may be constructed from data in Table 2.
Cadmium	1			Present data for this compound are inadequate to permit decisions on its effectiveness in inducing heritable translocations

[a] No information is available on the following chemicals: epichlorohydrin, DEN, DMN, ICR-170, ethylenimine, MeTEPA, MNNG, Myleran, Natulan, Trenimon, and vinyl chloride.

Table 2. Heritable translocation studies on selected animals

Compound	Mammalian species	Doses	Route of administration	Number of progeny tested	Number of translocation heterozygotes
Aflatoxin B1	Male mice BALB/c	5 mg/kg single dose	IP injection	185	0
Mitomycin C	Male mice C3H	0.05 mg/kg	Not given	Not given	Not given

Germ cell stage sampled	Result reported by authors	Comment	References
Not known. Posttreatment period when progeny were conceived not indicated	Negative	Screening of male progeny was done cytologically. The only dose used is probably too low, relative to the maximum tolerated dose. The method of treatment (single injection) used and the low number of progeny tested could not have possibly allowed sampling of the most sensitive stage, if there was one. Thus, the study was not adequate.	1
All stages	Not given	The information came from a short abstract. Until a full report becomes available we cannot judge on the adequacy of the test.	2

(continued)

Table 2. (continued)

Compound	Mammalian species	Doses	Route of adminis-tration	Number of progeny tested	Number of translo-cation hetero-zygotes
Mitomycin C (cont'd.)	Male rats	6.67 mg/kg single injec-tion	IP injec-tion	Not given	Not given
TEPA	Male mice A/L	Two doses given at interval of 12 hr for total of 2.5 mg/kg	IP	25	7 (28%)
				36	18 (50%)

Germ cell stage sampled	Result reported by authors	Comment	References
Spermatids	Positive	This report described the occurrence of a range of chromosome abnormalities in individual diakinesis-metaphase I cells from testes of F_1 male rats. Since the frequencies of aberrations was relatively low, it seems clear that no single F_1 animal carried a translocation. However, although the number of cells scored was given, there was no indication of the number of F_1 animals studied. The significance of this apparently negative result is therefore in doubt.	3
Spermatids	Positive	This report has to be treated very lightly, particularly that for spermatogonia. It is likely that translocations preexisted in the strain of mice used.	4

(continued)

Table 2. (continued)

Compound	Mammalian species	Doses	Route of adminis- tration	Number of progeny tested	Number of translo- cation hetero- zygotes
TEPA (cont'd.)	Male mice ICR/Ha	5 mg/kg single injection	IP	20	1 (5%)
				38	7 (18.4%)
				39	0
				20	0
				40	0
	Male mice (101 × C3H) F_1	20–30 mg/kg single injection	IP	1031	0
Thio-TEPA	Male mice C57BL/6Y line B6	5 mg/kg single injection	IP	85	9 (10.6%)
				160	5 (3.1%)

Germ cell stage sampled	Result reported by authors	Comment	References
Spermatozoa	Positive	This report provides the more convincing evidence that TEPA induced heritable translocations in spermatids.	5
Spermatids			
Spermatocytes	Negative		
Differentiating spermatogonia	Negative		
Spermatogonial stem cells	Negative		
Spermatogonial stem cells	Negative	This negative result with spermatogonial stem cells is contrary to the positive inductions reported by Sram et al. (4) for TEPA and by Malashenko and Surkova (7) for Thio-TEPA.	6
Spermatozoa and spermatids	Positive	Data clearly indicate positive induction of heritable translocations in male postmeiotic stages.	7
Spermatogonial stem cells	Positive	In view of the negative results in spermatogonia with TEPA, cyclophosphamide and TEM (from references 6 and 13), more test with Thio-TEPA is needed to see if indeed Thio-TEPA is effective in inducing heritable translocations at this stage.	7

(continued)

Table 2. (continued)

Compound	Mammalian species	Doses	Route of adminis-tration	Number of progeny tested	Number of translo-cation hetero-zygotes
Thio-TEPA (cont'd)	Male mice C57BL/6	1.25 mg/kg single injection	IP	128	43 (33.5%)
EMS	Male Armenian hamster (*Cricetulus migratorius*)	100 mg/kg single injection	IP	15	2 (13.3%)
	Male mice (C3H × 101)F_1	250 mg/kg single injection	IP	Not given	44%
		250 mg/kg single injection given to mice that were pre-fed with butylated hydroxy-toluene (BHT)	IP	Not given	19%

Germ cell stage sampled	Result reported by authors	Comment	References
Spermatids	Positive	Data clearly indicate positive induction of heritable translocations in spermatids.	8
Spermatozoa and spermatids	Positive	Although the number tested is very small, EMS seemed to have induced heritable translocations in male postmeiotic stages, indicating a similarity between Armenian hamster and mice.	9
Spermatozoa	Positive	Published only in abstract form. No other details are available.	10
Spermatozoa	Positive		

(continued)

Table 2. (continued)

Compound	Mammalian species	Doses	Route of adminis-tration	Number of progeny tested	Number of translo-cation hetero-zygotes
EMS (cont'd.)	Male mice JU/Fa	240 mg/kg single injection	IP	197	37 (19%)
	Male mice (101 × C3H)F_1	50 mg/kg single injection	IP	853	6 (0.7%)
		100 mg/kg single injection	IP	1013	18 (1.78%)
		150 mg/kg single injection	IP	621	45 (7.25%)
		200 mg/kg single injection	IP	246	79(32.11%)

Germ cell stage sampled	Result reported by authors	Comment	References
Spermatozoa and spermatids	Positive	This is the first report which showed that heritable translocations can be induced by EMS in mice and that stage sensitivity differences are the same for dominant lethals and translocations.	11
Spermatozoa	Positive	The stage sampled is the most sensitive for dominant lethal induction. The fequency induced by 50 mg/kg dose is significantly higher than control (P = 0.022).	12
Spermatozoa	Positive	Comparative dominant lethal study in the same report showed that clear-cut dominant-lethal	12
Spermatozoa	Positive	effects were found at 150 mg/kg and higher doses but not at lower doses. Since the experi-	
Spermatozoa	Positive	mental conditions for all doses were the same, a dose-effect curve may be constructed.	12

(continued)

Table 2. (continued)

Compound	Mammalian species	Doses	Route of administration	Number of progeny tested	Number of translocation heterozygotes
Cyclophosphamide	Male mice (C3H × 101)F_1	350 mg/kg single injection	IP	35	5 (14.3%)
				84	33 (39.3%)
				187	2 (1.1%)
				202	0
	Male mice (101 × C3H)F_1	350-400 mg/kg single injection	IP	1148	1 (0.09%)
MMS	Male mice Alderly Park strain I	50 mg/kg single injection	IP	82	5 (6%)
	Male mice NMRI (Kissleg-SPF)	40 mg/kg single injection	IP	250	27 (11%)

Germ cell stage sampled	Result reported by authors	Comment	References
Spermatozoa	Positive	This report gave the first indication that heritable translocations induced by a chemical mutagen can be recovered from treated spermatocytes, in addition to the postmeiotic stages.	13
Spermatids	Positive		
Spermatocytes	Positive		
Spermatogonial stem cells	Negative		
Spermatogonial stem cells	Negative	These data show the inefficiency of cyclophosphamide in inducing heritable translocations in spermatogonial stem cells.	6
Spermatids	Positive	This report gave the first evidence for MMS induction of heritable translocations in male mice. The screening for male translocation heterozygotes was done only on F_1 males from litters of reduced size, which may have an effect on the true rate of induction.	14
Spermatozoa and spermatids	Positive	This report provides the more substantial evidence of translocation induction by MMS.	15

(continued)

Table 2. (continued)

Compound	Mammalian species	Doses	Route of adminis-tration	Number of progeny tested	Number of translo-cation hetero-zygotes
MMS (cont'd.)	Male mice	50 mg/kg single injection	IP	Not given	Not given
TEM	Male mice	0.64 mg/kg or 0.8 mg/kg given as two injec-tions on two con-secutive days	IP	28	10 (35.7%)
	Male mice; stock derived from 4 separate strains (JH, CRL, MS, RCL)	0.2 mg/kg single injection	IP	111 74	8 (7.2%) 22 (28.4%)

Germ cell stage sampled	Result reported by authors	Comment	References
Not given	Positive	This piece of information was quoted by Leonard (1973). No other details are given.	16
Spermatozoa and spermatids	Positive	This is the first report to show that TEM induced heritable translocations in mice. The frequency was from the combined data for the two doses.	17
Spermatozoa	Positive	This report provided more extensive evidence for induction of heritable translocations with TEM as well as the first to show correlation in the inducibility of dominant-lethal mutations and heritable translocations in male postmeiotic stages, i.e., the frequency of translocations increased with that of dominant lethals.	18
Spermatids	Positive		

(continued)

Table 2. (continued)

Compound	Mammalian species	Doses	Route of adminis-tration	Number of progeny tested	Number of translo-cation hetero-zygotes
TEM (cont'd.)	Male mice CD (Charles River)	0.3 mg/kg single injection	IP	66·	20 (30%)
	Male mice (101 × C3H)F_1	0.0125 mg/kg single injection	IP	927	8 (0.9%)
		0.025 mg/kg single injection	IP	732	10 (1.4%)
		0.05 mg/kg single injection	IP	597	35 (5.9%)
		0.1 mg/kg	IP	466	70 (15.0%)
		0.2 mg/kg	IP	204	59 (28.9%)

Germ cell stage sampled	Result reported by authors	Comment	References
Spermatids	Positive	This report also showed that TEM is very effective in inducing heritable translocations in mouse spermatids. It also indicates a lack of correlation between sperm head abnormalities and presence of reciprocal translocations.	19
Spermatids	Positive	The stage sampled is the most sensitive for dominant-lethal and heritable translocation induction.	6
Spermatids	Positive	The frequency induced by 0.0125 mg/kg is significantly higher than spontaneous ($p < 0.01$).	
Spermatids	Positive	Comparative dominant-lethal study (cited in the same report) showed that clear-cut dominant-lethal effects were detected at 0.05 mg/kg and higher doses but not at lower doses. Since the experimental conditions for all doses were the same, a dose-effect curve may be constructed.	
Spermatids	Positive		
Spermatids	Positive		

(continued)

Table 2. (continued)

Compound	Mammalian species	Doses	Route of adminis- tration	Number of progeny tested	Number of translo- cation hetero- zygotes
TEM (cont'd.)	Male mice COX (SW)	0.0125 mg/kg/ day for 4 weeks	Drinking water	180	3 (1.7%)
		0.025 mg/kg/ day for 4 weeks	Drinking water	130	8 (6.2%)
		0.05 mg/kg/ day for 4 weeks	Drinking water	21	2 (10%)
		0.05 mg/kg/ day for 4 weeks	Drinking water	146	3 (2.1%)
	Male mice (101 × C3H) F_1	0.025 mg/kg/ day on each of 44 week days during a 51-day period	IP	162	48 (29.6%)

Germ cell stage sampled	Result reported by authors	Comment	References
Pachytene and late spermatocytes and all postmeiotic stages	Positive	Males in the first three treatment groups were mated for one week beginning immediately after the end of treatment while males in the last treatment group were mated during the third week after the end of treatment. Comparative dominant-lethal study showed lack of dominant-lethal effects for the first and last treatments. Thus, this study also showed higher sensitivity of the heritable translocation test than the dominant-lethal test.	20
Pachytene and late spermatocytes and all postmeiotic stages	Positive		
Pachytene and late spermatocytes and all postmeiotic stages	Positive		
Pachytene and late spermatocytes and early spermatids	Positive		
All stages in male gametogenesis	Positive	A single dose of 0.025 mg/kg induced 1.4% heritable translocations in the most sensitive stage. Therefore, these results indicate that TEM-induced complete symmetrical exchanges accumulate in the germ-cell pool.	21

(continued)

Table 2. (continued)

Germ cell stage sampled	Result reported by authors	Comment	References
Pachytene spermatocytes	Positive	These data, together with those of Sotomayor and Cumming (13) showed that heritable translocations were recovered from treatment of spermatocytes with alkylating clastogens.	22
Spermatogonial stem cells	Negative	This frequency is not significantly higher than the spontaneous frequency.	6
Spermatozoa and spermatids	Negative	Screening of male progeny was done cytologically. Because of the low number of progeny tested and of the fact that the progeny were conceived over a three-week period, this test was clearly inadequate.	23

Compound	Mammalian species	Doses	Route of administration	Number of progeny tested	Number of translocation heterozygotes
TEM (cont'd.)	Male mice (101 × C3H)F_1	2 mg/kg single injection	IP	247	13 (5%)
	Male mice (101 × C3H)F_1	3-4 mg/kg single injection	IP	1633	2 (0.12%)
Cadmium (as $CdCl_2$)	Male mice BALB/c	1.75 mg/kg single dose	IP	120	0

REFERENCES

1. Lēonard, A., Deknudt, G. and Linden, G. Mutat. Res. 28: 137-139 (1975).

2. Savkovic, N., Pecevski, J., and Maric, N. Mutat. Res. 38: 107 (1976).

3. Bempong, M. A., and Trower, E. C. J. Hered. 66: 285-289 (1975).

4. Sràm, R. J., Zudova, Z., and Benes, V. Fol. Biol. 16: 367-368 (1970).

5. Epstein, S. S., Bass, W., Arnold, E., Bishop, Y., Joshi, S., and Adler, I. D. Toxicol. Appl. Pharmacol. 19: 134-146 (1971).

6. Generoso, W. M., Cain, K. T., Huff, S. W., and Gosslee, D. G. In: Advances in Modern Toxicology, Vol. 1, W. Gary Flamm and M. A. Mehlman, Eds., Hemisphere, Washington, D. C., in press.

7. Malashenko, A. M., and Surkova, N. I. Soviet Genet. 10: 51-58 (1974).

8. Malashenko, A. M., unpublished data.

9. Lavappa, K. S. Lab. Animal Sci. 24: 62-65 (1974).

10. Cumming, R. B. Genetics 64: s14 (1970).

11. Cattanach, B. M., Pollard, C. E., and Isaacson, J. H. Mutat. Res. 6: 297-307 (1968).

12. Generoso, W. M., Russell, W. L., Huff, S. W., Stout, S. K., and Gosslee, D. G. Genetics 77: 741-752 (1974).

13. Sotomayor, R. E., and Cumming, R. B. Mutat. Res. 27: 375-388 (1975).

14. Jackson, H., Partington, M., and Walpole, A. L. Brit. J. Pharmacol. 23: 521-528 (1964).

15. Lang, R., and Adler, I. D. Mutat. Res. 48: 75-88 (1977).

16. Lēonard, A., and Schroeder, X. Quoted by Lēonard, A. In: Chemical Mutagens--Principals and Methods for Their Detection, A. Hollaender, Ed., Vol. 3, Plenum Press, New York-London, 1973, pp. 21-56.

17. Cattanach, B. M. Nature 180: 1364-1365 (1957).

18. Cattanach, B. M. Z. Vererbungslehre 90: 1-6 (1959).

19. Staub, J. E., and Matter, B. E. Arch. Genet., in press.

20. Sheu, C. W., Moreland, F. M., Oswald, E. J., Green, S., and Flamm, W. G., in preparation.

21. Generoso, W. M., Cain, K. T., Huff, S. W., and Gosslee, D. G. In : Chemical Mutagens--Principles and Methods for Their Detection, Vol. 5, A. Hollaender and F. J. de Serres, Eds., Plenum Press, in press.

22. Generoso, W. M., Krishna, M., Sotomayor, R. E., and Cacheiro, N. L. A. Genetics 85: 65-72 (1977).

23. Gilliavod, N., and Léonard, A. Toxicology 5: 43-47 (1975).

CHAPTER 24

MUTAGENICITY OF SELECTED CHEMICALS IN THE MAMMALIAN SPOT TEST

R. Fahrig

Schöneckstrasse, 7800 Greigurg, Federal Republic of Germany

G. W. P. Dawson

Department of Genetics, Trinity College, Dublin, Eire

L. B. Russell

Biology Division, Oak Ridge National Laboratory Oak Ridge, Tennessee (U.S.A.)

The spot test is an in vivo method for the detection of genetic alterations, including point mutations, in somatic cells of mice. Mouse embryos which are heterozygous for different recessive coat-color genes are treated in utero at one or more stages between days 7 and 11 postconception by injection of a mutagen into the peritoneal cavity of the mother animal or by other appropriate routes of administration. If this treatment leads in a pigment precursor cell to an alteration of the wild-type allele of one of the genes under study or to its loss, a color spot in the adult coat may be seen. As each mouse observed represents many treated embryonic cells in which a mutation could have occurred, and as the whole spectrum of chromosome damage and DNA-alterations should be covered, this method is sensitive to mutagens of all types of action.

Suitable mouse embryos can be derived from the cross of any a/a strain x T (1,2) or from the cross T x HT (3,4). Embryos of the genotype a/a; b/+; $c^{ch}p$/++; d se/++; s/+ are produced by mating young, previously untreated females of the inbred C57BL strain (a/a; otherwise wild type) to fertile males of the rotation bred T-stock (a/a = nonagouti; b/b = brown; $c^{ch}p/c^{ch}p$ = chinchilla and pink-eyed dilution; d se/d se = Maltese dilution and short ear; s/s = piebald). Detectable mutations, designated below by an asterisk (*), or loss of the wild type allele of the coat color genes (designated by Df) will be expected to give spots of the following phenotypes:

b/Df(B) or b/b*	brown
c^{ch}/Df(C) or c^{ch}/c*	light brown
d/Df(D) or d/d*	gray to dark gray
p/D (P) or p/p*	gray
c^{ch}p/Df(C P)	near-white

Mouse embryos of the genotype bp a pa/+ a +; b/+; c^{ch}p/++; d se/++; s/+; ln fz/++; pe/+ are produced by mating young previously untreated females of the T-stock to fertile males of the HT-stock (bp pa/bp pa = brachypodism and pallid; ln fz/ln fz = leaden and fuzzy; pe/pe = pearl). From this cross the above spots will be observable together with:

pa/Df(Pa) or pa/pa*	light gray
ln/Df(Ln) or ln/ln*	gray to dark gray

The phenotype of pe/pe is probably too dark to be detected against the background coat color.

The mechanism by which a heterozygous recessive gene comes to expression can be either gene mutation (including the whole spectrum of possible DNA alterations), loss of the wild-type allele through deficiency or monosomy, a recombinational process such as mitotic crossing over, or mitotic gene conversion.

Of the numerical and structural chromosome aberrations that can lead to loss of the wild-type allele, only those that are able to pass the filter of several mitoses would cause a spot with expression of the recessive. However, it is known that deficiencies of considerable length can survive somatically. For a given spot, it cannot, in general, be determined which of the several causes enumerated is responsible for expression of the recessive gene. The possibility of dominant mutations affecting the melanocyte, and of differentiational anomalies in pigment, must also be taken into account, but such events are apparently rare, as determined by the treatment of homozygous wild-type embryos.

As far as white (containing no pigment) and white-gray (containing only traces of pigment) near-midventral spots are concerned, it is likely that these are not the result of a genetic alteration involving the specific locus. They can be induced in homozygous wild-type as well as heterozygous animals (1,2,5). From what is known about the origin and migration of melanocytes in mouse development, it is likely that such spots result from an insufficiency of pigmented cells due to killing of their embryonic precursors. White roughly midventral spots are therefore indicative of cytotoxicity rather than expression of the recessive.

White spots that are not confined to the ventral region could be the consequence of somatic haploidy for $\underline{c^{ch}p}$, resulting from deficiency or monosomy; or, the consequence of the somatic genotype $\underline{c^{ch}p}/\underline{c^{ch}p}$ (phenotype light gray), resulting from mitotic recombination. Scoring in the spot test must therefore be by position as well as color spots.

Results reported in the spot test are summarized in Table 1. Results for individual chemicals are given in Tables 2-11.

Tables 1 through 11 follow. References are to be found on page 727.

Table 1. Summary of results in the spot test

Relative rank	Substance (CAS no.)	Report of laboratories Postive	Negative	Lowest effective concentration (LEC) mg/kg	mmole/kg	Dose range in which positive effects have been obtained (mg/kg)	Highest dose with negative effect (mg/kg)
1	TEM 51-18-3	1	-	0.5	0.0025	0.5-0.8	Not determined
2	Mitomycin C 50-07-7	3	-	2	0.006	2-4 × 2.5	Not determined
3	Endoxan 50-18-0	1	-	5	0.02	Only 1 dose tested	Not determined
4	Natulan 366-70-1	2	-	50	0.22	50-3 × 100	Not determined
5	EMS 62-50-0	3	-	50	0.4	50-4 × 150	Not determined
6	MNNG± 70-25-57	1	-	35	0.25	Only 1 dose tested	Not determined
6	MMS 66-27-3	3	-	50	0.45	50-4 × 150	Not determined
6	DMN 62-75-9	2	-	7.5	0.1	7.5-4 5	Not determined
6	DEN 55-18-5	2	1	5	0.05	5-3 50	20 (2 laboratories)
-	X-rays	2	-	100 R		100-150 R	Not determined

Table 2. Triethylenemelamine (TEM)[a]

Dose (mg/kg)	Day(s) of fetal development on which the dose was given	Offspring examined	Animals with spots of genetic relevance[b]		Animals with spots of questionable genetic relevance		References
			No.	%	No.	%	
0.8	10¼	41	1	2.4	9	22.0	6
0.5	10¼	120	4	3.3	21	17.5	6
0	--	141	0	0	2	1.4	6

[a]LEC, 0.5 mg/kg; spot test with (C57BL/E × T)F_1; IP injection in buffered saline. Result: positive.

[b]It should be noted that in early studies with chemicals, white midventral spots were sometimes mistakenly classified as being of mutational origin.

Table 3. Mitomycin C[a]

Dose (mg/kg)	Day(s) of fetal development on which the dose was given	Offspring examined	Animals with spots of genetic relevance		Animals with spots of questionable genetic relevance		References
			No.	%	No.	%	
2	10¼	66	2	3.0	10	15.2	6
0	--	141	0	0	2	1.4	
2	10	23	1	4.3	1	4.3	7
2	10	263	4	1.5	4	1.5	
0	-	891	1	0.1	5	0.5	
2.5	7	83	0	0	2	2.1	8,9
2.5	8	176	11	6.2	4	2.3	
2.5	9	110	7	6.3	4	3.6	
2.5	10	148	5	3.2	4	2.7	

1.875	7+8+9+10	119	7	3.2	4	2.7
2.5	7+8+9+10	49	9	18.3	8	16.3
2.5	8+9+10	104	19	18.3	8	7.7
3.75	7+8+9+10	Lethal to embryos				
0	7+8+9+10	112	2	1.8	3	2.7

[a]LEC, 2 mg/kg; test systems, spot test with (C57BL/E × T)F_1 (7); and with (T × HT)F_1 (8,9); IP injection in buffered saline. Result: positive.

Table 4. Endoxan (cyclophosphamide)[a]

Dose (mg/kg)	Day(s) of fetal development on which the dose was given	Offspring examined	Animals with spots of genetic relevance		Animals with spots of questionable genetic relevance		References
			No.	%	No.	%	
5	10	78	4	5.1	1	1.3	7
5	10	102	2	2	0	0	7
0	-	891	1	0.1	5	0.5	7

[a] LEC, 5 mg/kg; spot test with (C57BL/6JHan × T)F_1; IP injection in buffered saline. Result: positive.

Table 5. Natulan (procarbazine)[a]

Dose (mg/kg)	Day(s) of fetal development on which the dose was given	Offspring examined	Animals with spots of genetic relevance		Animals with spots of questionable genetic relevance		References
			No.	%	No.	%	
50	7	112	3	2.7	1	0.9	10
50	8	111	4	3.6	1	0.9	
50	9	95	10	10.5	1	1.0	
50	10	88	6	6.8	2	2.3	
100	7	97	6	6.2	1	1.0	10
100	8	72	7	9.7	2	2.8	
100	9	101	11	10.9	35	34.6	
100	10	99	6	6.0	3	3.0	

(continued)

Table 5. (continued)

Dose (mg/kg)	Day(s) of fetal development on which the dose was given	Offspring examined	Animals with spots of genetic relevance		Animals with spots of questionable genetic relevance		References
			No.	%	No.	%	
25	7+8+9	85	7	8.2	0	0	10
50	7+8+9	43	3	7.0	2	4.7	
100	7+8+9	25	7	28	0	0	
0	7+8+9+10	112	2	1.8	3		4
50	9	121	5	4.1	6	5	
75	9	103	12	11.7	17	16.5	
100	9	78	8	10.3	19	24.4	
0	–	245	3	1.2	0		

[a]LEC, 50 mg/kg; spot test with (T × HT)F_1; IP injection in buffered saline. Result: positive.

Table 6. Ethyl methanesulfonate (EMS)[a]

Dose (mg/kg)	Day(s) of fetal development on which the dose was given	Offspring examined	Animals with spots of genetic relevance		Animals with spots of questionable genetic relevance		Comments	References
			No.	%	No.	%		
50	10	75	4	5.3	1	1.3	Experiments have also given positive results at a treatment on the 8th and 11th day of fetal development	1,7, 11
100	10	228	11	4.8	6	2.6		
0	-	891	1	0.1	5	0.5		
50	10¼	82	1	1.2	3	3.7	At 50 mg/kg nature or or location of spot prevented unequivocal diagnosis as spot of genetic relevance	6
100	10¼	163	1	0.6	4	2.5		
0	-	141	0		2	1.4		
150	7+8+9+10	152	15	9.9	0	0		9

[a]LEC, 50 mg/kg; spot test with (C57BL/6JHan × T)F_1 (2,7,11), with (C57BL/E × T)F_1 (6), or with (T × HT)F_1 (9); IP injection in buffered saline. Result: positive.

Table 7. N-Methyl-N'-nitro-N-nitrosoguanidine (MNNG)[a]

Dose (mg/kg)	Day(s) of fetal development on which the dose was given	Offspring examined	Animals with spots of genetic relevance		Animals with spots of questionable genetic relevance		References
			No.	%	No.	%	
35	10	67	1	1.5	0	0	7
35	10	108	2	1.9	0	0	7
35	10	90	2	2.2	0	0	7
0	-	891	1	0.1	5	0.5	7

[a]LEC, 35 mg/kg; spot test with (C57BL/6JHan × T)F_1; IP injection in buffered saline. Result: positive.

Table 8. Methyl methanesulfonate (MMS)[a]

Dose (mg/kg)	Day(s) of fetal development on which the dose was given	Offspring examined	Animals with spots of genetic relevance		Animals with spots of questionable genetic relevance		References
			No.	%	No.	%	
100	7	150	4	2.7	0	0	9
100	8	150	0	0	0	0	
100	9	144	6	4.2	3	2.1	
100	10	152	7	4.6	1	0.7	
150	7	106	4	3.8	0	0	
150	8	94	6	6.4	1	1.1	
150	9	106	4	3.8	6	5.7	
150	10	105	16	15.2	2	1.9	
50		72	12	16.7	1	1.2	

(continued)

Table 8. (continued)

Dose (mg/kg)	Day(s) of fetal development on which the dose was given	Offspring examined	Animals with spots of genetic relevance No.	%	Animals with spots of questionable genetic relevance No.	%	References
50	7+8+9+10	72	12	16.7	1	1.2	
100	7+8+9+10	68	8	11.8	2	2.9	
150	7+8+9+10	75	9	12.0	10	13.3	
0	7+8+9+10	112	2	1.8	3	2.7	
50	10¼	60	0	0	0	0	6
0	-	141	0	0	2	1.4	
50	10¼	115	1	0.9	2	1.7	12
100	10¼	61	3	4.9	8	13.1	
0	-	562	0	0	21	3.8	
50	11	149	6	4.0	5	3.4	2
0	-	891	1	0.1	5	0.5	

[a] LEC, 50 mg/kg; spot test with $(T \times HT)F_1$ (9), with $(C57BL/E \times T)F_1$ (6,12), and with $(C57BL/6JHan \times T)F_1$ (2); IP injection in buffered saline. Result: positive.

Table 9. Dimethylnitrosamine (DMN)[a]

Dose (mg/kg)	Day(s) of fetal development on which the dose was given	Offspring examined	Animals with spots of genetic relevance		Animals with spots of questionable genetic relevance		References
			No.	%	No.	%	
7.5	10	95	1	1	2	2	7
7.5	10	97	2	2	1	1	
0	-	891	1	0.1	5	0.5	
5.0	7+8+9+10	83	2	2.4	0	0	10
10.0	7+8+9+10	Lethal to embryos					
0	-	149	0	0	0	0	

[a]LEC, 7.5 mg/kg; spot test with (C57BL/6JHan × T)F_1 (7) and with (T × HT)F_1 (10); IP injection in buffered saline. Result: positive.

Table 10. Diethylnitrosamine (DEN)[a]

Dose (mg/kg)	Day(s) of fetal development on which the dose was given	Offspring examined	Animals with spots of genetic relevance		Animals with spots of questionable genetic relevance		References
			No.	%	No.	%	
5	8	97	2	2.1	0	0	10
5	10	56	1	1.8	0	0	
10	7	67	2	3.0	0	0	
10	8	80	3	3.8	0	0	
10	9	60	2	3.3	0	0	
10	10	108	2	1.9	0	0	
10	7+8+9+10	113	3	2.7	0	0	
15	7+8+9+10	190	3	1.6	1	0.5	
50	7+8+9+10	Lethal to embryos					

50	7+8+9	85	2	2.4	4	4.7	9
50	8+9+10	113	5	4.3	2	1.8	
0	7+8+9+10	112	2	1.8	3	2.7	
20	10¼	70	0	0	2	2.9	6.12
30	10¼	133	2	1.5	3	2.3	
50	10¼	129	2	1.6	5	3.9	
10	10	146	0	0	0	0	7
20	10	101	0	0	1	1.0	
0	-	891	1	0.1	5	0.5	

[a]LEC, 5 mg/kg; spot test with (T × HT)F_1 (9,10), with (C57BL/E × T)F_1 (6,12), and with (C57BL/6JHan × T)F_1; IP injection in buffered saline. Result: positive.

Table 11. X-Rays[a]

Dose	Offspring examined	Animals with spots of genetic relevance		Animals with spots of questionable genetic relevance		References
		No.	%	No.	%	
100 R	235	26	11.1	0	0	1
0	202	1	0.5	0	0	1
100 R	127	3	2.4	8	6.3	2
0	152	0	0	0	0	2

[a]LEC, 100 R; spot test with (C57BL × NB)F_1 (1) and with (C57BL/6JHan × T)F_1 (2); treatment time, 11th day of fetal development. Result: positive.

REFERENCES

1. Russell, L. B., and Major, M. H. Genetics 42: 161-175 (1957).

2. Fahrig, R. Molec. Gen. Genet. 138: 309-314 (1975).

3. Davidson, G. E., and Dawson, G. W. P. Mutat. Res. 38: 151-154 (1976).

4. Neuhäuser, A. GSF-Bericht B 798, München, 1977, pp. 42-43.

5. Fahrig, R., and Seiler, J. P. In preparation.

6. Russell, L. B. Arch. Toxicol. 38: 75-85 (1977).

7. Fahrig, R. Arch. Toxicol. 38: 87-98 (1977).

8. Carter, J., and Dawson, G. W. P. The establishment of a convenient test for somatic mutation in mammals. Progress Report, Jan. 1977 - June 1977, Contract 169 - 77 - 1 ENV - EIR of the EEC Environmental Research Programme.

9. Dawson, G. W. P. Personal communication.

10. Dawson, G. W. P., Davidson, G. E., and Carter, J. The establishment of a convenient test for somatic mutation in mammals. Progress Report April - December 1976, Contract 023 - 74 - ENV - EIR (G. W. P., Dawson) of the EEC Environmental Research Programme.

11. Fahrig, R. In: Chemical Mutagens, Principles and Methods for their Detection, Vol. 5, A. Hollaender and F. J. de Serres, Eds., Plenum Press, New York, 1978, pp. 151-176.

12. Russell, L. B. Personal communication.

CHAPTER 25

MUTAGENICITY OF SELECTED CHEMICALS IN INDUCTION OF SPECIFIC LOCUS MUTATIONS IN MICE

U. H. Ehling

Abteilung für Genetik der Gesellschaft für Strahlen- und Umweltforschung (GSF), D-8042 Neuherberg, Germany

The specific locus method was developed by W. L. Russell (1) and consists essentially of mating treated and untreated wild-type mice to a test stock. The multiple recessive tester stock that has been exclusively used for testing chemicals considered by the workshop is the Oak Ridge test stock. The Oak Ridge test stock is homozygous for the following markers: a/a, b/b, c^{ch}/c^{ch}, d se/d se, p/p, s/s. Other test stocks were developed in Harwell (2) and Moscow (3,4). A detailed description of the mutant genes was published by Green (5).

By use of this test it is possible to detect recessive mutations in the first generation. The recessive mutations will include small inter- or intragenic deletions affecting a specific locus as well as gene mutations at the base-pair level, which lead to a change from the dominant wild-type to a recessive allele at the locus concerned (6-10).

Generally, the induction of specific locus mutations was tested in $(101\times C3H)F_1$ male mice, 10-14 weeks old. Immediately after a single intraperitoneal injection of the test compound, each male was caged separately with an untreated Oak Ridge test stock female of the same age. The offspring were counted, sexed and carefully examined externally at birth. The litters were examined again when cages were changed, the final examination being at weaning age. The reliability of the classification by phenotype was generally confirmed by an allelism test. A detailed description of the method was published elsewhere (11-13).

RESULTS AND DISCUSSION

Induction of specific locus mutations in mice is summarized in Table 1.

Table 1. Summary of induction of specific locus mutations in mice

Compound	Dose range tested (mg/kg)	Lowest tested dose positive (mg/kg)	Spermatogenic response[a] In post-spermatogonia	In spermatogonia
Cyclophosphamide	350	350	+	nt
DEN	ns	-	ti	ti
EMS	100-400	200	+	-
ICR-170	4		ti	?
Mitomycin C	4-7	5.25	-	+
MMS	40-150	40	+	-
MNNG	50		ti	-
Myleran	20		ti	-
Natulan	200-800	400	+	+
TEM	0.5-4.0	0.5	+	+
X-rays	300-1000	300	+	+

[a]ns = not specified, nt = not tested; ti = test incomplete.

To date, the following compounds have not been tested: aflatoxin B1, cadmium, dimethylnitrosamine (DMN), epichlorohydrin, ethylenimine, MeTEPA, TEPA, thio-TEPA, Trenimon, and vinyl chloride.

Results for cyclophosphamide and DEN are given in Table 2. Cyclophosphamide induces specific locus mutations in postspermatogonial germ cell stages of mice ($p = 0.001$). No data are available from spermatogonia (Table 2).

The data for DEN are not sufficient for a definitive conclusion.

EMS induces specific locus mutations only in postspermatogonial stages of mice (Table 3, experiment 2, $p = 0.01$; experiment 3, $p = 0.01$). In addition, the induction of a dominant belly-spot mutant 2 days after treatment of male mice with 240 mg/kg of EMS and a mutant, which exhibited a quivering, shaking behavior, conceived 3 days after treatment in the same experiment was reported (19).

The difference between the mutation rate with ICR-170 in the control group and in the offspring derived from treated spermatogonia is not significant (Table 4, $p = 0.14$). The p value for the overall mutation rate of the laboratory is 0.09. IP injection of 4 mg/kg of ICR-170 does not induce dominant lethal mutations in spermatozoa and spermatids in $(101\times C3H)F_1$-male mice (16).

MMS significantly increases the frequency of specific locus mutations only in postspermatogonial germ cell stages of mice (Table 5). The lowest dose, which increases the mutation rate in spermatozoa and spermatids significantly, is 40 mg/kg of MMS. The p value for the mating interval 5-8 days after treatment is 0.0025 and for 9-12 days 0.0001. The difference in the number of offspring in the mating intervals 1-20 days post-treatment is due to the induction of dominant lethal mutations (16).

It is unlikely that MNNG is an effective mutagen for the induction of specific locus mutations in the mouse according to Table 6; the sample size is small, however. An IP injection of 50 mg/kg of MNNG does not induce dominant lethal mutations in spermatozoa and spermatids in $(101\times C3H)F_1$-male mice (16).

A dose of 5.25 mg/kg of Mitomycin C (Table 7) induces specific locus mutations in spermatogonia of mice ($p = 0.04$).

In postspermatogonial germ cell stages the induction of dominant lethal mutations by Myleran (Table 8) interferes with the possible induction of specific locus mutations (24). To avoid an erroneous statement it is necessary to repeat the experiment and use a 4-day mating schedule. It is unlikely that a dose of 20 mg/kg of Myleran induces specific locus mutations in spermatogonia of mice.

A dose of 600 mg/kg of Natulan significantly increases the mutation frequency in postspermatogonial germ cells ($p = 0.014$) and in spermatogonia ($p=0.003$) of mice (Table 9). The point estimate of the induced mutation rate in spermatogonia of 800 mg/kg of procarbazine is only one fourth of that expected from a linear extrapolation.

TEM induces specific locus mutations in spermatocytes and spermatogonia of mice (Table 10). It is interesting to note that the mutation frequency after 2 × 2.0 mg/kg of TEM in older males is higher than in younger males.

SUMMARY

Six compounds (cyclophosphamide, EMS, Mitomycin C, MMS, Natulan, and TEM) induced specific locus mutations in mice. The data base of three compounds is insufficient to make a definite statement (DEN, ICR-170, and Myleran). MNNG is probably ineffective in the specific locus test. No data are available for ten compounds: aflatoxin B1, cadmium, DMN, epichlorohydrin, ethylenimine, MeTEPA, TEPA, thio-TEPA, Trenimon, vinyl chloride.

EMS and MMS induced specific locus mutations exclusively in spermatozoa and spermatids. Cyclophosphamide also induced specific locus mutations in postspermatogonial germ cell stages, but the induction of mutations in spermatogonia was not tested. Natulan and TEM induced specific locus mutations in postspermatogonial germ cell stages and spermatogonia; Mitomycin C induced specific locus mutations exclusively in spermatogonia.

Because of the limited data base, every attempt to rank the mutagens for their efficiency in inducing specific locus mutations must be provisional. The following rankings are based on the relationship of the doses for the induction of mutations and lethality (LD_{50}). For postspermatogonial germ cell stages:

MMS > TEM > EMS > Natulan > Cyclophosphamide

For spermatogonia:

Natulan > Mitomycin C > TEM

Table 2. Cyclophosphamide and DEN

Mutagen	Dose (mg/kg)	Germ cell stage treated	No. of F1 offspring	No. of mutations at 7 loci	Mutation frequency per locus per gamete $\times 10^5$	References
Cyclophosphamide[a]	350	Postspermatogonia	3,642	3	11.8	14
DEN[b]	138	Postspermatogonia	1,702	0		15

[a] Source unknown, solvent, isotonic saline solution; route, intraperitoneal injection; volume, not given.

[b] Source, Eastman Kodak, Rochester, N.Y.; solvent, Hanks' balanced salt solution; route, intraperitoneal injection; volume, not given.

[c] Weighted average dose.

Table 3. Ethyl methanesulfonate (EMS)[a]

Expt. no.	Dose (mg/kg)	Germ cell stage treated	No. of F1 offspring	No. of mutations at 7 loci	Mutation frequency per locus per gamete × 10^5	References
1	0	--	15,625	0		21
	250	Postspermatogonia	724	0		
	400		511	0		
	100	Spermatogonia	988	0		
	200		1,964	0		
	250		5,939	0		
	400		5,896	0		
2	200	Spermatozoa Spermatids	7,784	4	7.3	17
3	Not specified	Postspermatogonia	1,922	2	14.9	18

[a]Source, Eastman Organic Chemicals, Rochester, N.Y.; solvent, Hanks' balanced salt solution; route, intraperitoneal injection; volume, 1 ml.

Table 4. ICR-170[a]

Dose (mg/kg)	Germ cell stage treated	No. of F1 offspring	No. of mutations at 7 loci	Mutation frequency per locus per gamete × 10^5	References
0	-	15,625	0		11
4	Postsperma-togonia	1,092	0		11
4	Spermatogonia	9,275	2	3.1	20

[a]2-Methoxy-6-[3-(ethyl-2-chloroethyl)aminopropyl-amino]acridine dihydrochloride; source, Dr. H. J. Creech, Institute for Cancer Research, Philadelphia, Pa.; Solvent, Hanks' balance salt solution; route, intraperitoneal injection; volume, 1 ml.

Table 5. Methyl methanesulfonate (MMS)[a]

Expt. no.	Dose (mg/kg)	Germ cell stage treated	Mating intervals, days after treatment	No. of F1 offspring	No. of mutations at 7 loci	Mutation frequency per locus per gamete × 10^5	References
1	0	-		15,625	0		21
	50	Postspermato-		637	0		
	100	gonia		1,149	1	12.4	
	150			558	1	25.6	
	50	Spermatogonia		4,529	1	3.2	
	100			9,004	1	1.6	
	150			4,996	0		
2	0	-		1,629	0		22
	50	Spermatozoa		2,478	1	5.8	
	75	and		429	1	33.3	
	100	spermatids		206	0		
3	0		-	74,037	4	0.8	22
	40		1-4	1,769	0		
			5-8	977	2	29.2	
			9-12	1,065	3	40.2	
			13-16	1,692	1	8.4	
			17-20	1,844	0		

[a]Sources: Eastman Organic Chemicals, Rochester, N.Y., and Schuchardt, München, German; solvent, Hanks' balanced salt solution or distilled water, route intraperitoneal injection; volume, 1 ml.

Table 6. N-Methyl-N-nitroso-N'-nitroguanidine (MNNG)[a]

Dose (mg/kg)	Germ cell stage treated	No. of F1 offspring	No. of mutations at 7 loci	References
0	-	15,625	0	
50	Postspermatogonia	894	0	11
50	Spermatogonia	7,408	0	20

[a]Source: Koch-Light Laboratories, Colnbrook, Great Britain; solvent, Hanks' balanced salt solution; route, intraperitoneal injection; volume, 1 ml.

Table 7. Mitomycin C[a]

Dose (mg/kg)	Germ cell stage treated	No. of F1 offspring	No. of mutations at 7 loci	Mutation frequency per locus per gamete × 10^5	References
0	-	14,708	0		
4	Postspermato-	277	0		
5.25	gonia	10,263	1	1.4	23
6		2,165	0		
7		1,580	0		
4	Spermatogonia	2,759	0		23
5.25		20,711	6	4.1	
6		2,618	1	5.5	
7		1,114	1	12.8	
2	Oocytes	2,847	1	5.0	23
4	(up to 7 weeks)	1,515	0		
2	Oocytes (more	4,956	0		
4	than 7 weeks)	1,204	0		

[a]Source, Kyowa Hakko Kogyo, Tokyo, Japan; solvent, distilled water; route, intraperitoneal injection; volume 1 ml.

Table 8. Myleran[a]

Dose (mg/kg)	Germ cell stage treated	No. of F1 offspring	No. of mutations at 7 loci	References
0	-	15,625	0	
20	Postspermatogonia	911	0	20
20	Spermatogonia	9,884	0	20

[a]Source, Burroughs Wellcome Co., Research Triangle Park, N.C.; solvent, DMSO mixed with saline (final concentration of DMSO in the solution injected was 6%); route, intraperitoneal injection; volume, 1 ml.

Table 9. Procarbazine (Natulan)[a]

Dose (mg/kg)	Germ cell stage treated	No. of F1 offspring	No. of mutations at 7 loci	Mutation frequency per locus per gamete × 10^5	References
0	-	37,150	2	0.8	
200	Postspermato-	5,849	0		25,26
400	gonia	3,394	1	4.2	
600		1,930	2	14.8	
800		1,771	0		
200	Spermatogonia	1,979	0		
400		12,466	3	3.4	25,26
600		45,413	16	5.0	
800		39,184	6	2.2	

[a]Source, Hoffmann-La Roche, Grenzach-Wyhlen, Germany; solvent, distilled water; route, intraperitoneal injection; volume, 1 ml.

Table 10. Triethylenemelamine (TEM)[a]

Dose (mg/kg)	Interval between dose fractions (hr)	Age of males at time of treatment (months)	Germ cell stage treated	No. of F1 offspring	No. of mutations at 7 loci	Mutation frequency per locus per gamete × 10^5	References
0.5	-	3-4	Spermatocytes	667	1	21.4	27
1.0	-	12		264	0		
0.5	-	12	Spermatogonia	5,427	0		27
1.0	-	12		6,496	1	2.2	
2x2.0	24	3-4		5,784	1	2.5	
2x2.0	24	12		5,360	4	10.7	

[a]Source, no information; solvent, no information; route, intraperitoneal injection; volume, doses adjusted according to the weights of animals.

Table 11. Radiation

Radiation	Approximate dose rate (R/min)	Dose (R)	Germ cell stage treated (mating intervals, days post exposure)	No. of F1 offspring	No. of mutations at 7 loci	Mutation frequency per locus per gamete × 10^5	References
-	-	0		531,500	28	0.8	
γ (Cs^{137})	0.001	86	Spermatogonia	59,810	6	1.4	28
γ (Cs^{137})	0.001	300		49,569	15	4.3	
γ (Cs^{137})	0.001	600		31,652	13	5.9	
γ (Cs^{137})	0.009	300		58,457	10	2.4	
γ (Cs^{137})	0.009	516		26,325	5	2.7	
γ (Cs^{137})	0.009	861		24,281	12	7.1	
γ (Cs^{137})	0.8	600		28,059	10	5.1	
× (250kvp)	9	600		40,326	23	8.1	
× (250kvp)	90	300		65,548	40	8.7	
× (250kvp)	90	600		119,326	111	13.3	
γ (Cs^{137})	0.009	258	Oocytes	27,174	1	0.5	
γ (Cs^{137})	0.009	400		37,049	2	0.8	
γ (Cs^{137})	0.8	400		20,827	7	4.8	
× (250kvp)	90	400		11,124	15	19.3	
γ (Cs^{137})	60	600	(1–4)	363	1	39.4	22
			(5–8)	403	0		
			(9–12)	347	1	41.2	
			(13–16)	98	0		
			(17–20)	45	0		
γ (Cs^{137})	60	1000	(1–7)	920	2	31.1	
			(8–14)	1,356	8	84.3	

REFERENCES

1. Russell, W. L. Cold Spring Harb. Symp. Quant. Biol. 16: 327-336 (1951).

2. Lyon, M. F., and Morris, T. Genet. Res. 7: 12-17 (1966).

3. Malashenko, A. M. Genetika 11 (No. 1): 146-147 (1975).

4. Malashenko, A. M. Genetika 12 (No. 3): 163-165 (1976).

5. Green, M. C. In: Biology of the Laboratory Mouse, E. L. Green, Ed., McGraw-Hill, New York, 1966, pp. 87-150.

6. Ehling, U. H. Arch. Toxicol. 32: 19-25 (1974).

7. Schlager, G., and Dickie, M. M. Mutat. Res. 11: 89-96 (1971).

8. Russell, L. B. Mutat. Res. 11: 107-123 (1971).

9. Russell, L. B., and DeHamer, D. L. Genetics 74: s236 (1973).

10. Russell, L. B., and Cacheiro, N. L. A. Genetics 86: s53-s54 (1977).

11. Ehling, U. H. In: Chemical Mutagenesis in Mammals and Man, F. Vogel and G. Röhrborn, Eds., Springer-Verlag, Berlin, 1970, pp. 156-161.

12. Cattanach, B. M. In: Chemical Mutagens, A. Hollaender, Ed., Plenum Press, New York-London, 1971, Vol. 2, pp. 535-539.

13. Searle, A. G. Mutat. Res. 31: 277-290 (1975).

14. Cumming, R. B., and Walton, M. F. Genetics 68: s14 (1971).

15. Russell, L. B. Arch. Toxicol. 38: 75-85 (1977).

16. Ehling, U. H., Cumming, R. B., and Malling, H. V. Mutat. Res. 5: 417-428 (1968).

17. Russell, W. L. John Hopkins Med. J. (Suppl.) 1: 239-247 (1972).

18. Cumming, R. B. Personal communication.

19. Cattanach, B. M., Pollard, C. E., and Isaacson, J. H. Mutat. Res. 6: 297-307 (1968).

20. Ehling, U. H. Schr. Reihe Ver. Wasser-Boden-Lufthyg. 40: 21-37 (1973).

21. Ehling, U. H., and Russell, W. L. Genetics 61: s14-s15 (1969).

22. Ehling, U. H. GSF-Bericht B798: 13-16 (1977).

23. Ehling, U. H. Mutat. Res. 26: 285-295 (1974).

24. Ehling, U. H., and Malling, H. V. Genetics 60: 174-175 (1968).

25. Ehling, U. H. Mutat. Res. 41: 113-122 (1976).

26. Ehling, U. H., and Neuhäuser, A. Mutat. Res. 46: 218 (1977).

27. Cattanach, B. M. Mutat. Res. 3: 346-353 (1966).

28. Russell, W. L. Nucleonics 23: 53-56, 62 (1965).

CHAPTER 26

COMPARATIVE MUTAGENICITY OF METHYL METHANESULFONATE AND ETHYL METHANESULFONATE

Sohei Kondo*

Faculty of Medicine, Osaka University, Osaka, Japan

INTRODUCTION

Methyl methanesulfonate (MMS) and ethyl methanesulfonate (EMS) are highly mutagenic in lower organisms but only moderately carcinogenic in mice and rats. No case reports or epidemiological studies on the hazards of MMS and EMS to humans are known to the Working Group. This review summarizes and evaluates the data on genetic effects of these compounds from the aspect of their relevance to the assessment of possible genetic effects of these or related compounds in humans. First, a general summary is given of available data on MMS and EMS and some typical examples are given of dose-related effects of EMS and MMS in mice, Drosophila, and some other organisms. Then, molecular mechanisms of mutagenesis are discussed. Third, an attempt is made to assess the human hazard of these compounds on the basis of the experimental data and finally the areas requiring further investigation are considered.

The chemical structures are:

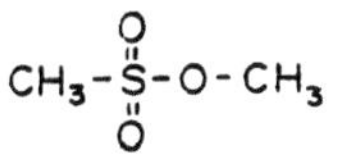

MMS

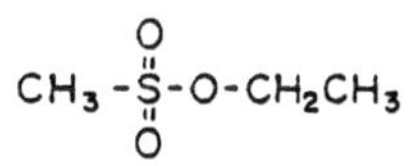

EMS

*With the collaboration of D. Anderson, H. Brockman, W. P. Lee, T. Matsushima, G. Mohn, C. Nauman, J. M. Parry, R. Pertel, R. San, B. Vig and S. Wolff

GENERAL SUMMARY OF GENETIC AND RELATED EFFECTS OF MMS AND EMS

Since the data on MMS and EMS are too numerous to be reviewed in a short paragraph, some of the more essential findings are listed in Tables 1-4. These tables show that MMS and EMS produce almost all kinds of genetic effects in all the organisms tested. Both MMS and EMS induce genetic effects without metabolic activation. It is puzzling that activation fractions prepared from the livers of mice and rats enhance some genetic effects of EMS or MMS in some microorganisms but inhibit other genetic effects.

DOSE-RELATED GENETIC EFFECTS OF MMS AND EMS IN MICE

Specific Locus Mutations Induced by MMS and EMS

Insufficient data are available for plotting reliable dose-response curves. The linear response curves A, B, and C in Fig. 1 are hypothetical ones constructed by using observed point for each curve; curves A and B are for MMS mutagenesis in postspermatogonia and spermatogonia, respectively, and curve C is for EMS mutagenesis in postspermatogonia. The dose response for MMS mutagenesis, however, is not linear (see the upper part of curve A') when plotted by using three observed points. The hypothetical quadratic curve D for EMS-mutagenesis in spermatogonia is based on data for no mutants per 14,787 offspring at weighted dose of 293 mg/kg and on the reasoning given later. These curves serve as a basis for discussion in the section on the implication of the results in considering human genetic hazards.

Comparison of Dose-Response Curves for Dominant Lethals with MMS and EMS

The data given previously on dominant lethals are quoted and replotted in the following way. Instead of the frequency of dominant lethals, F, where F = 1 - (embryos/test female) ÷ (embryos/control female), the surviving fraction S (the frequency of living embryos S = 1 - F) is plotted in logarithmic units against the dose of MMS or EMS (18). The example in Fig. 2 shows that the curve with EMS for dominant lethals is sigmoidal, whereas that with MMS is linear. Both EMS and MMS are potent inducers of dominant lethals in postmeiotic male stages but not in spermatogonial stages. Females are more resistant than males to induction of dominant lethals by EMS and MMS.

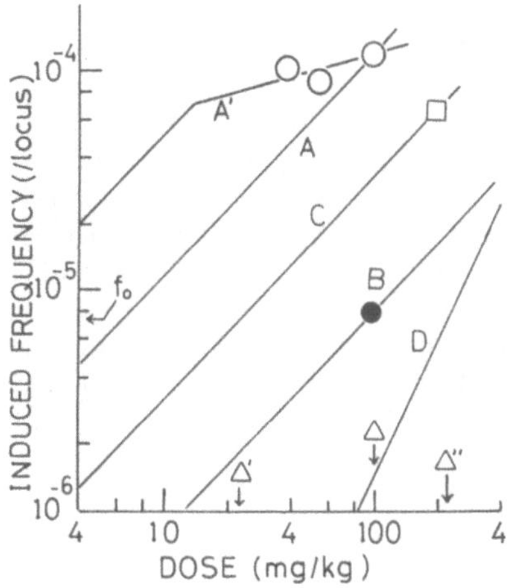

Figure 1. Hypothetical curves for the dose-response of the frequency of induction of specific locus mutations in germ cells of mice: (A, B) mutations induced in postspermatogonia and spermatogonia, respectively, by intraperitoneal injection of MMS; (C, D) mutations induced by EMS in postspermatogonia and spermatogonia, respectively. Circular and square symbols show values cl calculated from experimental data (13-16), disregarding statistical errors; Symbol f_o stands for the spontaneous mutations rate (17); (Δ, Δ', Δ") doubling doses of MMS for spermatogonia, of EMS for postspermatogonia, and of EMS for spermatogonia, respectively.

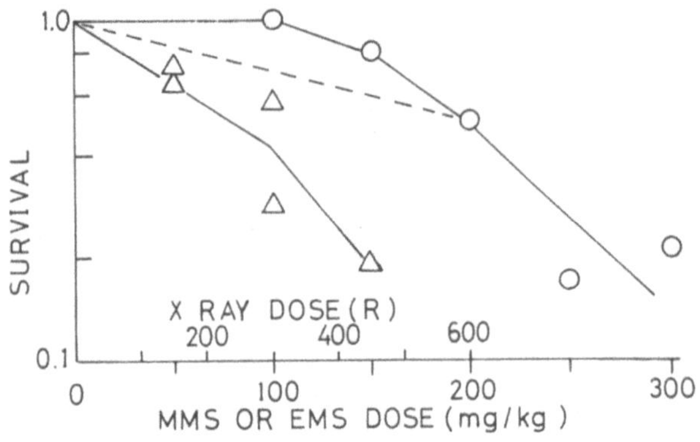

Figure 2. Effect of dose on proportion of mice without dominant lethals. Data after intraperitoneal injection of MMS (— —) and EMS (— —) and after irradiation with x-rays (--------) are quoted from Ehling et al. (19), Generoso et al. (18), and Ehling (20), respectively.

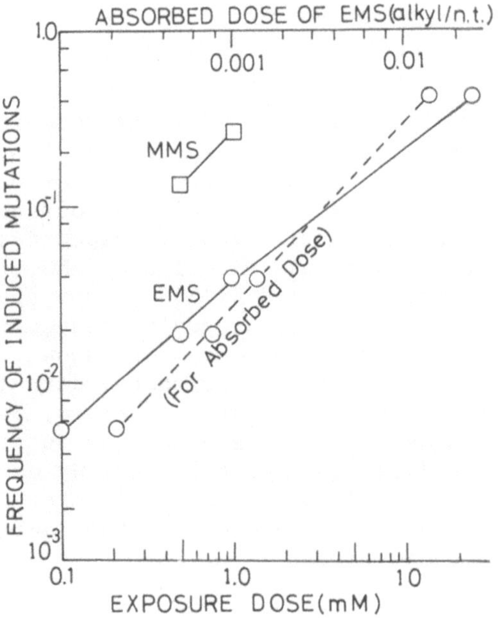

Figure 3. Effect of dose on frequency of sex-linked recessive lethals in Drosophila. Male adult flies were fed on solution containing EMS (— — ; -- --) or MMS (— —). Data for EMS and MMS are from Aaron and Lee (24) and Vogel and Leigh (25), respectively.

Table 1. Summary of MMS Test Systems[a]

Test system[b]	Genetic end points[c]	Metabolic and cell-stage factors	Exposure route[d]	Dose response[e]	LPTD[f]	Mutation rate
Salmonella	BPS		P	Linear	130 μg/plate	
	BPS		Fluctuation		0.1 μg/ml	
	BPS		HMA	NR	150 mg/kg	
E. coli	BPS		L	Concave	110 μg/kg	
	BPS		Fluctuation		0.1 μg/ml	
	BPS		HMA	Linear	130 mg/kg	
	RT		L		1100 μg/ml	
B. subtilis	RT	(S9-Inhibit)	L	NR	440 μg/ml	
Neurospora	SLM; BPS; FS;DEL		L	Concave	220 μg/ml (4×)[g]	5.5×10^{-7}/μmole/ml/hr
	HMA		IP		250 mg/kg (58×)[g]	
	HMA (in vitro)		L		330 mg/kg (58×)[g]	

(continued)

Table 1. (continued)

Test system	Genetic end points	Metabolic and cell-stage factors	Exposure route	Dose response	LPTD	Mutation rate
Schizosaccharomyces (n)	PM	(S9-Enhance)	L	NR	550 μg/ml	4.6×10^{-6}/locus/μg/ml/hr
	HMA					3.6×10^{-6}/locus/μg/ml/hr
Saccharomyces (2n)	PM		L	NR	5500 μg/ml	
	Gene conversion		L	Concave	110 μg/ml	
	Crossover		L	Convex	5500 μg/ml	
	CL		L	Concave	5500 μg/ml	7×10^{-7}/chrom/μg/ml/hr
Soybean	CM;MR		L	Linear	60 μg/ml (4 hr)	0.18/leaf/mM (4 hr)
	PM		L	Linear	60 μg/ml (4 hr)	0.37/leaf/mM (4 hr)
Barley	CA	Seeds	L	Linear	110 μg/ml(5 hr)	0.1/mM (5 hr)
	PM + CM	Seeds	L	Concave	250 μg/ml(4 hr)	0.002/mM (4 hr)

<u>Bombyx mori</u>	SLM		Injection		NR	5 μg/capita	
<u>Drosophila</u>	RL	Sperm	Feed	2d	Linear	0.05 mM $(6\times)^g$	0.2/X-chrom/mM
	CL		Feed	2d	Concave	0.5 mM $(5\times)^g$	0.01/X-chrom/mM
	TR		Feed	1d	Concave	1.0 mM	0.01/mM
Cultured mammalian cells: cytogenetic observations							
Hamster:CHO	CA	Unsynchro-nized -	L		NT	66 μg/mg	
Human	CA	G_0 lympho-cytes	L		NT	1 μg/ml	
Hamster:CHO	SCE	Unsynchro-nized	L		Convex	1.1 μg/ml	
Human	SCE	Fibroblasts	L		Convex	0.55 μg/ml (48 hr)	10/cell/μg/ml
XP	SCE	Fibroblasts	L		Convex	0.01 μg/ml (48 hr)	800/cell/μg/ml
Normal vs. XP	UDS	Fibroblasts	L		Linear	11 μg/ml	No difference

(continued)

Table 1. (continued)

Test system[b]	Genetic end points[c]	Metabolic and cell-stage factors	Exposure route[d]	Dose response[e]	LPTD[f]	Mutation rate
In vivo mammalian						
Mouse	SLM; germinal	Spermatozoa	IP	NT	40 mg/kg	2×10^{-6}/locus/mg/kg
Mouse	SLM; germinal	Spermatogonia	IP	NT	100 mg/kg	8×10^{-8}/locus/mg/kg
Mouse	SLM; somatic	Fetus Day 11	IP	NT	50 mg/kg	1×10^{-6}/locus/mg/kg
Mouse	CA	Bone marrow	?	NT	120 mg/kg	4×10^{-4}/locus/mg/kg
Mouse	MNE	Bone marrow	IP		30 mg/kg	
Mouse	DL	Postmeiosis	IP	Linear	10 mg/kg (100 mg/kg for ♀)	6×10^{-3}/mg/kg
Mouse	Abnormal	Postmeiosis	IP	Concave	250 mg/kg	

Mouse	CA	Postmeiosis	IV	Concave	25 mg/kg	0.01/cell/mg/kg
Mouse	CA	Oocytes	IV	Concave	50 mg/kg	0.0016/cell/ mg/kg
Mouse	UDS	Spermatids	IP		5 mg/kg (6×)[g]	
Rabbit	SCE	G_0 Lymphocytes		NT	25 mg/kg	0.17/cell/mg/kg

[a]MMS (methyl methanesulfonate), CH_3-O-SO_2-CH_3; MW 110; LD_{50} 200 mg/kg for mice.

[b]Test system; XP = xeroderma pigmentosum; FA = Fanconi's anemia.

[c]Genetic end points: BPS = base pair substitutions; RT = repair test with repair-defective strains; SLM = specific locus mutations; FS = frame-shifts; DEL = deletions; HMA = host-mediated assay; PM = point mutation; CL = chromosome loss; CM = chromosome mutations; MR = mitotic recombinations; CA = chromatid/chromosome aberrations; RL = recessive lethals; TR = translocation; SCE = sister chromatid exchanges; UDS = unscheduled DNA synthesis; MNE = micronucleated erythrocytes; DL = dominant lethals; XRL = X-chromosome linked recessive lethal; w = locus in Drosophila; y = locus in Drosophila.

[d]Exposure route: P = plate test; HMA = host-mediated assay; L = liquid test; IP = intraperitoneal; F = feed; IV = intravenous.

[e]Dose response: NR = not recorded; NT = not tested.

[f]LPTD = lowest positive tested dose.

[g]The indicated values times the control values were observed after the indicated "lowest positive doses" were given.

Table 2. Summary of EMS test systems[a]

Test system[b]	Genetic end points[c]	Metabolic and cell-stage factors	Exposure route[d]	Dose response[e]	LPTD[f]	Mutation rate
Salmonella	BPS		P	Concave	2000 μg/plate	
	HMA		HMA	NR	35 mg/kg	
E. coli	BPS		P		700 μg/plate	
	BPS	(S9 Enhance)	L	Concave	700 μg/ml	
	RT		L		2500 μg/ml	
B. subtilis	RT	(S9 Enhance)	L	NR	800 μg/ml	
Neurospora	SLM:BPS:FS		L	Concave	10000 μg/ml (7×)[g]	3.1×10^{-8}/μmole/ml/(0.5 hr)
	HMA		IP		300 mg/kg (25×)[g]	
	HMA *in vitro*		L		3100 mg/kg (13×)[g]	

Schizosaccharomyces (n)	PM	(S9 Inhibit)	L		62500 μg/ml	1.5×10^{-6}/locus/μg/ml/hr
	HMA				62500 μg/ml	1.9×10^{-6}/locus/μg/ml/hr
Saccharomyces (2n)	PM		Fluctuation		0.2 μg/ml	
	Gene conversion		Fluctuation		0.1 μg/ml	
	Gene conversion	(S9 Enhance)	P	Concave	3125 μg/ml	
	HMA					NE
	Crossover	(S9 Inhibit)	L	Convex	6075 μg/ml	
	CL		L	Concave	25000 μg/ml	2×10^{-7}/chrom/μg/ml/hr
Tradescantia (clone 4430)	PM+CM		Gas	Concave	40 ppm × 6 hr	5.6×10^{-4}/hair/ppm (6 hr)

(continued)

Table 2. (continued)

Test system[b]	Genetic end points[c]	Metabolic and cell-stage factors	Exposure route[d]	Dose response[e]	LPTD[f]	Mutation rate
Soybean	CM, MR		L	Linear	1000 ppm 24 hr	0.5/leaf/mM (24 hr)
	PM		L	Linear	1000 ppm 24 hr	1.23/leaf/mM (24 hr)
Barley	CA		L	Concave	6 mg/kg (5 hr)	2.6×10^{-3}/mM (5 hr)
	PM+CM		L	Concave	5 mg/ml (5 hr)	7×10^{-4}/mM (5 hr)
<u>Bombyx mori</u>	SLM	Postmeiosis	Injection	NR	5 μg/capita (5×)[g]	
<u>Drosophila</u>	XRL	Sperm	F × 2 days	Linear	0.1 mM (24 hr)	5×10^{-2}/X chrom/mM
	y	Sperm	F × 2 days			9×10^{-5}/locus/mM

Drosophila (cont'd.)	w	Sperm	F × 2 days			2×10^{-4}/locus/mM
Cultured mammalian cells: cytogenetic observations						
Hamster:CHO	SCE	Unsynchro-nized	L	Convex	12.5 μg/ml	0.4/cell/μg/ml
Human	SCE	Fibroblasts	L	Convex	1.25 μg/ml	2/cell/μg/ml
XP	SCE	Fibroblasts	L	Convex	0.125 μg/ml	70/cell/μg/ml
Human	SCE	G_0-G_1 Lymphocytes	L	NT	250 μg/ml	0.1/cell/μg/ml
FA	SCE	G_0-G_1 Lymphocytes	L	NT	250 μg/ml	0.08/cell/μg/ml
Normal vs. XP	UDS	Fibroblasts	L	Linear	37.5 μg/ml	No difference
In vivo mammalian						
Mouse	SLM; germinal	Spermatogonia	IP	NT	>300 mg/kg	No mutations

(continued)

Table 2. (continued)

Test system[b]	Genetic end points[c]	Metabolic and cell-stage factors	Exposure route[d]	Dose response[e]	LPTD[f]	Mutation rate
Mouse	SLM; germinal	Postmeiosis	IP	NT	200 mg/kg	4×10^{-7}/locus/mg/kg
Mouse	SLM; germinal	Fetus, day 11	IP	NT	100 mg/kg	1×10^{-7}/locus/mg/kg
Mouse	SLM/ somatic	Fetus, day 11	IP	NT	50 mg/kg	1×10^{-6}/locus/mg/kg
Mouse	CA	Bone marrow	IP	Concave	115 mg/kg	4×10^{-4}/cell/mg/kg
Mouse	MNE	Bone marrow	IP		400 mg/kg	
Mouse	DL	Postmeiosis	IP	Concave	150 mg/kg (250 mg/kg for ♀)	1×10^{-3}/mg/kg
Mouse	TR	Postmeiosis	IP	Concave	50 mg/kg	2×10^{-4}/mg/kg

Mouse	Abnormal sperm	Postmeiosis	IP	Concave	500 mg/kg	
Mouse	UDS	Spermatid	IP		10 mg/kg (3×)[g]	
Rabbit	SCE	Lymphocyte	IP	Linear	50 mg/kg	0.03/cell/mg/kg

[a]EMS (ethyl methanesulfonate), C_2H_5-O-SO_2-CH_3; MW 124; LD_{50} 450 mg/kg for mice.

[b]Test system; XP = xeroderma pigmentosum; FA = Fanconi's anemia; NH = normal human.

[c]Genetic end points: BPS = base pair substitutions; RT = repair test with repair-defective strains; SLM = specific locus mutations; FS = frame-shifts; DEL = deletions; HMA = host-mediated assay; PM = point mutation; CL = chromosome loss; CM = chromosome mutations; MR = mitotic recombinations; CA = chromatid/chromosome aberrations; RL = recessive lethals; TR = translocation; SCE = sister chromatid exchanges; UDS = unscheduled DNA synthesis; MNE = micronucleated erythrocytes; DL = dominant lethals; XRL = X-chromosome linked recessive lethal; w = locus in Drosophila; y = locus in Drosophila.

[d]Exposure route: P = plate test; HMA = host-mediated assay; L = liquid test; IP = intraperitoneal; F = feed; IV = intravenous.

[e]Dose response: NR = not recorded; NT = not tested; NE = no effect at highest dose.

[f]LPTD = lowest positive tested dose.

[g]The indicated values times the control values were observed after the indicated "lowest positive doses" were given.

Table 3. Mutations in cultured mammalian cells -- MMS

Cell line	Locus[a]	Selective system[b]	Expression time, hr	Dose response	LPTD[c]	Mutation rate	References
L5178($TK^+/^-$)	TK	BUdR	144	Nonlinear	7.5 μg/ml	20×10^{-6} μg/ml (4 hr)	1,2
L5178($TK^+/^-$)	HGPRT	Thioguanine	144	Linear	7.5 μg/ml	6×10^{-6} μg/ml (4 hr)	1,2
L5178Y($TK^+/^+$)	HGPRT	Thioguanine	192	Linear	13.2 μg/ml	0.22×10^{-6} μg/ml (2 hr)	3
L5178Y($TK^+/^+$)		Ouabain	48	No mutation			3
L5178Y($TK^+/^+$)		Excess thymidine	96	Nonlinear	13.2 μg/ml	0.6×10^{-6} μg/ml (2 hr)	3
L5178Y($TI^+/^+$)	HGPRT	Thioguanine	168	Linear	55 μg/ml	0.25×10^{-6} μg/ml (2 hr)	4
P388($TK^+/^-$)	TK	IUdR	48	Nonlinear	80 μg/ml	4×10^{-6} μg/ml (0.5 hr)	5

(continued)

P388($TK^+/^-$)		Excess thymidine	48	Nonlinear	80 μg/ml	0.4×10^{-6} μg/ml (0.5 hr)	5
V79		Ouabain	55	Concave	132 μg/ml	$.09 \times 10^{-6}$ μg/ml (0.5 hr)	6
V79	HGPRT	Azaguanine	100	Concave	55 μg/ml	$.72 \times 10^{-6}$ g/ml μg/ml/hr	7
V79	HGPRT	Thioguanine	192	Linear	165 μg/ml	0.43×10^{-6} μg/ml (2 hr)	8

[a] Locus: TK = thymidine kinase; HGPRT = hypoxanthine guanine phosphoribosyl transferase.

[b] Selective system: BUdR = 5-bromo-2'-deoxyuridine; IUdR = 5-iodo-2'-deoxyuridine.

[c] LPTD = lowest positive tested dose.

Table 4. Mutations in cultured mammalian cells -- EMS

Cell line	Locus[a]	Selective system[b]	Expression time, hr	Dose response	LPTD[c]	Mutation rate	References
L5178Y($TK^+/^-$)	TK	BUdR	48	Nonlinear	31 μg/ml	1×10^{-6} μg/ml (2 hr)	1,2
L5178Y($TK^+/^-$)	HGPRT	Thioguanine	144	Linear	100 μg/ml	0.81×10^{-6} μg/ml (4 hr)	1,2
L5178Y($TK^+/^+$)	HGPRT	Thioguanine	192	Linear	500 μg/ml	0.08×10^{-6} μg/ml (2 hr)	9
L5178Y($TK^+/^+$)		Ouabain	48	Linear	500 μg/ml	0.028×10^{-6} μg/ml (2 hr)	9
L5178Y($TK^+/^+$)		Excess thymadine	96	Concave	500 μg/ml	0.028×10^{-6} μg/ml (2 hr)	9
L5178Y($TK^+/^+$)		Ara C	96	Convex	500 μg/ml	0.0002×10^{-6} μg/ml (2 hr)	10
L5178Y($TK^+/^+$)	HGPRT	Thioguanine	168	Not tested	720 μg/ml	0.4×10^{-6} μg/ml (2 hr)	11

L5178Y($TK^{+}/^{-}$)	TK	BUdR	96	Linear	313 μg/ml	0.41×10^{-6} μg/ml (2 hr)	4
P388($TK^{+}/^{-}$)		Excess thymidine	48	Linear	250 μg/ml	0.6×10^{-6} μg/ml/hr	5
CHO	HGPRT	Thioguanine	168	Linear	12.5 μg/ml	4.5×10^{-6} μg/ml (16 hr)	12
V79		Ouabain	50	Linear	500 μg/ml	0.4×10^{-6} μg/ml (2 hr)	6
V79	HGPRT	Thioguanine	192	Linear	1250 μg/ml	0.18×10^{-6} μg/ml (2 hr)	8

[a]Locus: TK = thymidine kinase; HGPRT = hypoxanthine guanine phosphoribosyl transferase.

[b]Selective system: BUdR = 5-bromo-2'-deoxyuridine; IUdR = 5-iodo-2'-deoxyuridine; Ara C = cytosine arabinoside.

[c]LPTD = lowest positive tested dose.

Chromosome Aberrations after Treatment with MMS and EMS

The major cause of dominant lethals is believed to be chromosomal damage. An obvious question is whether the presumed major cause is identical with any of the chromosome aberrations detected cytologically. Brewen et al. (21) have studied chromosome aberrations at the first cleavage division after the treated male mice were mated with superovulated control females. There is a good correlation between the frequency of cells with visible aberrations and the frequency of dominant lethals (21, 22).

Chromosome aberrations are a very useful biological parameter. Thus, the fact that chromosome aberrations in cells of mouse bone marrow showed no increase in a low dose range but a sharp increase with above about 110 mg/kg of EMS (23) may indicate that EMS has a low ability to penetrate bone marrow cells in the dense population. It is important to determine whether MMS shows a similar dose-response curve.

The incidence F of translocations in F_1 mice after EMS treatment (18) follows a very clear exponential curve, $F \propto \exp\{aD\}$, where D is the dose of EMS and a is the proportionality constant. It would be interesting to know if translocations induced by MMS show a similar response curve.

GENETIC EFFECTS OF MMS AND EMS IN DROSOPHILA MELANOGASTER

Dose-Response Curves for Mutations by EMS and MMS

When Drosophila are fed on sucrose water containing either EMS or MMS or injected with 0.2 μl of EMS or MMS solution, the resultant mutations of their spermatozoa are approximately linearly related with the given concentration of EMS or MMS over a wide concentration range (cf. Fig. 3). The mutation rate of spermatogonia by EMS or MMS is severalfold lower than that of spermatozoa. Accurate dose-response curves are needed for the mutations of spermatogonia and oocytes after treatment with EMS and MMS.

Dominant Lethals and Other Chromosomal Damage with EMS and MMS in Drosophila

In regard to "lowest effective dose" of EMS or MMS, spermatozoa are one order of magnitude more resistant to the induction of dominant lethals, translocations, and chromosome loss, than to the induction of recessive lethals. The frequency of dominant lethals increases linearly with the dose of MMS above an apparent threshold concentration of about 0.1 mM, in a way similar to the frequency of recessive lethals, but chromosome losses rarely occur after MMS treatment.

Both EMS and MMS act primarily at postmeiotic stages and induce point mutations, deletions, dominant lethals, translocations, and chromosome loss, but they have not been reported to produce nondisjunctions. On treatment of spermatogonia with EMS or MMS, mutations are induced, though at a lower rate, but dominant lethals are not induced. These are only some essential points from the Group I report (26).

DOSE-RESPONSE CURVES FOR MUTATIONS INDUCED BY MMS AND EMS IN HIGHER PLANTS AND MICROORGANISMS

Mutations by MMS and EMS in Barley, Tradescantia, and Soybeans

EMS is one of the most potent known mutagens of barley. When seeds are soaked in an aqueous solution of EMS, the frequency of chlorophyll mutations induced in barley follows the relation, $F \propto D^n$ (n = 2.7-1.4), where F is the frequency of M_2 seedling mutants and D the dose of EMS. In contrast, on irradiation, the yield of chlorophyll mutants increases linearly with increase in the dose of x-rays (27). It is interesting to note that the curves for the survival of fertility of barley plants (plotted in logarithmic units against the dose of x-rays or EMS) are linear for x-ray irradiation but have a shoulder for EMS; they resemble the curves for dominant lethals induced by EMS and x-rays in mice (Fig. 2).

Somatic mutations in the stamen hairs of three clones of Tradescantia also follow dose-frequency relations of nonlinear ($F \propto D^{2.4}$) and nearly linear ($F \propto D^{1.3}$) types after treatment with EMS and x-rays, respectively, except for the radiosensitive clone 0106, which shows an unusual relationship of $F \propto D^{0.7}$ on treatment with EMS (19). This relation raises the possibility that clone 0106 may have a constitutive error-prone mode of replication (see section on Molecular Mechanism).

After soaking soybeans in aqueous solutions of MMS and EMS, light-green spots on yellow leaves of homozygotes ($y_{11}y_{11}$) can be scored as indices of somatic mutations. Similarly, dark-green and yellow-green spots can be scored on light green leaves of heterozygotes ($Y_{11}y_{11}$), and twin adjacent spots of dark-green and yellow-green are frequently seen. Some typical examples taken from the report of Group I (29) are replotted in Fig. 4. The relation of the number of spots per leaf (N) and the dose (D), i.e., $N \propto D^n$, is different for homozygotes and for heterozygotes. Moreover, the data vary from experiment to experiment (possibly due to seasonal variation or a difference in the age of the beans).

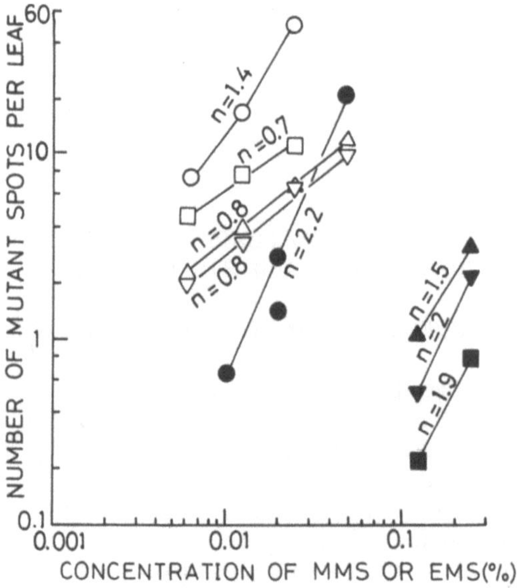

Figure 4. Effect of dose on somatic mutant spots on leaves of soybean. Seeds were soaked in an aqueous solution of MMS (○ , □ , △ , ▽) or EMS (● , ■ , ▲ , ▼) for 24 hr. The mutations scored were: (1) light-green spots (○ , ●) on yellow leaves of homozygous $y_{11}y_{11}$ plants and (2) dark-green spots (△ , ▲), yellow spots (▽ , ▼) and adjacent twin (dark-green and yellow) spots (□ , ■) on light-green leaves of heterozygous $Y_{11}y_{11}$ plants. The n values denote the indices in the equation, $N \propto D^n$ where N and D are the number of spots per leaf and the mutagen concentration, respectively. Data are from Arenaz (30), Vig and Paddock (31), and Vig et al. (32), quoted in Vig and Sung (29).

Nevertheless, the typical data in Fig. 4 indicate that MMS gives an index close to linearity (n = 0.7-1.3) and EMS a nearly quadratic value (n = 1.5-2.2) for the relationship of $N \propto D^n$. These results indicate fundamental differences between the mutagenic actions of MMS and EMS.

It would be interesting to know whether the barley and Tradescantia used to obtain EMS data also show any approximately linear response for mutations with MMS.

Dose-Response Curves for Mutations by MMS and EMS in Fungi, Bacteria and Bacteriophages

The idea that the relationship of mutation frequency (F) to the dose (D) is linear ($F \propto D$) for MMS and nonlinear ($F \propto D^n$, where $n \cong 1.5$-2.5) for EMS is supported by the figures given in the Group I Report (33) for reversions of His^- strain of *Salmonella typhimurium*, TA100, and by the majority of the available data on *E. coli* and fungi cited in the review articles of Group I reviewers. Bacteriophages are exceptional among microorganisms in that the frequency of mutations induced by EMS is linearly related to the dose of EMS (34).

MECHANISMS FOR THE BIOLOGICAL ACTIONS OF MMS AND EMS

The main effect of both MMS and EMS on DNA in vitro, and on DNA in vivo without activation by cellular components, is alkylation. However, the main product, 7-alkylguanine, is not necessarily the major cause of mutagenesis, and the actual mutagenic effects of EMS and MMS must be examined by biological means. Information on the mechanisms of mutations in eukaryotes is still very limited, and thus speculations on the mechanisms of induction of mutations in mammals are based on known mechanisms of mutagenesis in bacteria and other microorganisms.

Molecular Mechanisms of Mutagenesis by MMS and EMS in Microorganisms

One important difference between mutagenesis induced by MMS and by EMS is the fact that *E. coli* $recA^-$ strain does not mutate with MMS (or many other mutagens) but does mutate with EMS, whereas the $recA^-$ strain shows highly elevated sensitivity to killing by both MMS and EMS (Table 5). Based upon the model of UV mutagenesis proposed by Witkin (35), we can speculate that mutagenesis induced by MMS occurs via errors in the filling of gaps produced in daughter strands due to the MMS-induced lesions in the template strands or via some other misrepair of alterations in DNA produced directly or indirectly by MMS, and that this assumed error-prone repair requires the recA+ gene product. It was proposed previously (36) that mutagenesis by EMS may occur by bypass of recA+-dependent error-prone repair, probably by misreplication. If EMS mutagenesis occurs by misreplication, then the mutation frequency should increase linearly with increase in the amount of mispairing lesions induced by EMS. This is actually the case with T4 phage treated in vitro (34). However, the relationship of mutation frequency to the dose for E. coli is nearly quadratic, that is, $F \propto D^{2.5}$, where F is the frequency of base-pair substitutions and D is the dose of EMS (37). Similarly, for UV, the relation of the frequency to the dose is linear for phage T4 (38) and approximately quadratic for E. coli (35). Therefore, by analogy to inducible error-prone repair, which does not occur in undamaged cells but which is induced in UV-damaged cells (35), it is tempting to assume that EMS induces,

Table 5. Comparison of mutagenicities of MMS and EMS in uvrA⁻, polA⁻ and recA⁻ strains of E. coli[a]

DNA repair	Relative sensitivity[b]					
	Mutation by[c]			Inactivation by		
	X-rays	MMS	EMS	X-rays	MMS	EMS
Wild type	+	+	+	+	+	+
uvrA⁻	+	+	++	+	+	+
polA⁻	+	+	+(++[d])	+++	+++	+++
recA⁻	-	-	+	+++	+++	+++

[a]Data of Kondo et al. (36) and Ishii and Kondo (37).

[b]Sensitivity is classified as undetectable (-), normal (+), moderately high (++), and very high (+++).

[c]Mutations studied were reversions from Arg⁻ to Arg⁺ phenotype caused mainly by forward mutations at t-RNA structure genes due to base pair substitutions and ColB^R (from "sensitive" to "resistant" to colicin B) mutations, except for MMS which was tested only for Arg⁺ mutations.

[d]Valid only for ColB^R mutations.

via a recA-gene-product-independent process, an intracellular activity for enhancing the mutability of EMS-damaged DNA. N-Methyl-N'-nitro-N-nitrosoguanidine (MNNG) can induce mutation-enhancing activity in the $recA^-$ strain of E. coli for subsequently infecting normal or EMS-treated λ phage (39,40).

Further, a $uvrA^-$ strain shows a normal and a moderately high sensitivity to mutations by MMS and EMS, respectively (Table 5). This indicates that premutational damage by EMS, but not that by MMS, is partly excised from the DNA by excision repair. From similarity in the differential sensitivity pattern of $uvrA^-$, $polA^-$ and $recA^-$ strains to killing by x-rays and MMS (Table 5), and from the approximately linear relation of mutation frequency to dose after treatment with x-rays and MMS, we may conclude, as pointed out before, that MMS is x-ray mimetic. MMS also mimics x-rays in its action on barley (41). However, the patterns of EMS- or MMS-induced mutations are not necessarily the same in different species. In phage T4, EMS mainly induces transitions especially favoring the pathway G:C → A:T, whereas MMS, which acts via mis-repair, induces virtually all types of mutation (42). In E. coli, EMS induces not only A:T → G:C transitions but also transversions of A:T → T:A and G:C → C:G (43). In Neurospora, EMS mainly induces transitions, favoring the A:T → G:C pathway, but also produces transversions, frameshift mutations and deletions (44). As is the case for yeast (45) EMS mutagenesis in Neurospora seems to occur mainly by misrepair (46).

Changes in Mutagenicities of EMS and MMS Relative to Ionizing Radiation with Increasing Genome Complexity

In Table 6, the mutagenicities of MMS and EMS are compared with those of x- or γ-rays for mice, cultured Chinese hamster cells, barley, Drosophila and E. coli, on an equivalent lethality basis. The comparison is made with induced frequencies at 30% dominant lethals for mice and Drosophila, at about 30% reduction in seedlings for barley seeds and at 70% survival for cultured cells and E. coli. At postspermatogonial stages in mice, doses of 40 mg/kg of MMS and 200 mg/kg of EMS, which give 30% dominant lethals, cause mutation frequencies of 1×10^{-4} and 7×10^{-5} per specific locus, respectively. These yields are 60% (however, see the values in parentheses in Table 5 and Discussion) and 40%, respectively, of the mutation with a γ-ray dose producing the same level of dominant lethals. With spermatogonia, however, mutagenicity of either MMS or EMS seems very low compared to that of γ-rays. The mutagenicity of MMS is one fifth that of x-rays for chlorophyll mutations in barley, twice that of x-rays for sex-linked recessive lethals in Drosophila and 10 times the latter for base-pair substitutions in E. coli. The mutagenicity of EMS is four times that of x-rays

Table 6. Comparison of mutagenicities of MMS, EMS, and ionizing radiation on mouse, hamster, barley, Drosophila and *E. coli*

Organism	Mutations	Cell stage	Induced mutation frequency at 70% survival dose[a]: Radiation (x- or -rays)	MMS	EMS	Relative mutagenicity vs. x-rays: MMS	EMS	References
Mouse in vivo	Average at 7 loci (a, b, c^{ch}, d, se, p, s)	Spermatogonia	2.5×10^{-5}[b]	(0.3×10^{-5})[c]	$(<10^{-6})$[c]	(0.1)[c]	(<0.1)[c]	15,17
		Postspermatogonia	17×10^{-5}	10×10^{-5}	7×10^{-5}	0.6	0.4	
				$[34\times10^{-5}]$[d]		[2][d]		
Chinese hamster in vitro	HGPRT (AG-resistance)	Cultured cells	4×10^{-5}	4×10^{-5}	5×10^{-4}	1	10	12
Barley	Chlorophyll (/M_1 spike)	Seeds	8%	1.8%	34%	0.2	4	18

Drosophila	Sex-linked recessive lethals	Spermatozoa	5%	10%	>40%	2	>8	24,25, 47
E. coli	$Arg^- \rightarrow Arg^+$: base pair substitutions		0.01×10^{-5}	0.1×10^{-5}	1×10^{-5}	10	100	36,34

[a]The following values are used. For 30% dominant lethals in mice: 300 R of γ-rays, 40 mg/kg of MMS, and 200 mg/kg of EMS. For 70% survival of Chinese hamster cells: 300 R of γ-rays, 140 μg/ml of MMS for 30 min, and 1200 μg/ml of EMS for 2 hr. For 23∿25% lethality of M_1 barley seedlings: 30 kR of x-rays, 0.25 mg/ml of MMS for 8 hr and 2.8 mg/ml of EMS for 20 hr. For 30% dominant lethals in Drosophila: 2 kR of x-rays, no less than 25 mM of EMS orally for 1 day, and 0.36 mM MMS orally for 2 days. For 70% survival of *E. coli*: 2 kR of x-rays, 1 mg/ml of MMS for 1 hr and 30 mg/ml of EMS for 1 hr. These values are quoted or estimated from the references listed.

[b]At low dose rates.

[c]Very crude estimates from mutant yields (16) of 2/18529 (102 mg/kg of MMS) and 0/14787 (293 mg/kg of EMS) which are not significantly different from the control rate of 7.5×10^{-6} (Table 8).

[d]Mutations in the F_1 offspring from only the matings 5 to 12 days after treatment (see Table 8).

for barley, more than eight times the latter for Drosophila, and 100 times the latter for E. coli.

Thus, in comparison with x- or γ-rays, EMS is a very potent mutagen in E. coli, and a potent mutagen in barley and Drosophila but only a weak mutagen in mouse germ cells; MMS is a potent mutagen in E. coli and a moderate mutagen in Drosophila but only a weak mutagen in barley and mouse germ cells. On the other hand, HGPRT mutations in a Chinese hamster cell line (Table 6) and in a mouse cell line are one order of magnitude more inducible by EMS than by MMS at the same survival level, as is the case for base-pair substitutions in E. coli. These complexity patterns of mutagenesis may reflect variation in the modes of mutagenesis from base-pair substitutions by mispairing or misrepair to deletions by misrepair, since a major pathway of x-ray mutagenesis is deletion mutations by misrepair even in bacteriophage T4 (48).

Chemical Nature of Alterations Produced by MMS and EMS in DNA

Alterations in DNA caused by treatment in vitro or in vivo with MMS and EMS have been extensively studied by using ^{3}H- or ^{14}C-labeled chemicals to label the DNA (49). Using a chromatographic method, these alterations of DNA can now be identified fairly precisely, as shown in Table 7. It is clear that 7-alkylguanine is the major product (60-80% of the total) common to EMS and MMS, whereas among the minor products, O^6-alkylguanine is produced to a considerable extent by EMS but scarcely by MMS. There is some evidence that the product of premutagenic or precancerous damage may be O^6-alkylguanine, which is susceptible to error-free repair. On the other hand, the alkylation of a purine such as 7-alkylguanine and later depurination is postulated to be a major cause of inactivation and gross changes of DNA chromosome aberrations after treatment with EMS and MMS. Some types of endonucleases with specificity for alkylated purines and depurinated sites have been reported. However, there is still a wide gap between our knowledge on the chemical nature of the alterations and the biological effects produced by treatments with alkylating agents, as emphasized by Singer (49). We hope that some coordinated research projects will be carried to fill this gap.

Dosimetry

Lee and co-workers (24) first succeeded in measuring the absorbed dose of EMS to sperm DNA in units of alkylation/nucleotide by treating male Drosophila with ^{3}H-EMS. Results obtained in this way show that the exposure dose of EMS (i.e., the concentration of EMS in the feeding liquid) is not necessarily linearly related

Table 7. Distribution of alkyl products from in vitro and in vivo alkylation of DNA

DNA source	Reagent	Reaction type	Sites and relative amounts of alkylation[a]								References
			A			G			C		
			1	3	7	3	O^6	7	3	Phosphate	
In vitro											
Phage T4	MMS	S_N2	3	17				80	ND		51
Phage T4	EMS	$S_N1 + S_N2$		8.7	1.4		2.2	56.4	ND	(20)	52
Salmon sperm	MMS	S_N2	1.1	9.8	0.3	0.7	0.3	82	1	(3)	53
Salmon sperm	EMS	$S_N1 + S_N2$	0.8	5.6	1.6	1.9	2.0	69.6		(20)	54
Salmon sperm	MNNG	S_N1	1	11		2	7	72	2		55
Salmon sperm	MNU	S_N1	2.7	8.2	1.2	0.6	6.7	65.6		(18)	54
In vivo											
HeLa	MMS	S_N2	4.6	8.7		0.7	0.3	83.7	1.0	1.0	55
HeLa	EMS	$S_N1 + S_N2$	0.1	2.2	0.7	0.4	0.3	81	0.1	(15)	49

(continued)

Table 7. (continued)

DNA source	Reagent	Reaction type	Sites and relative amounts of alkylation[a]								References
			A			G			C		
			1	3	7	3	O^6	7	3	Phosphate	
HeLa	MNNG	S_N1	0.9	11.2		1.9	6.8	62.8	1.9	(15)	55
L-cells	MNU	S_N1		5.4			3.5	64			56
Rat liver	MMS	S_N2	0.6	7.2		0.9	ND	91			57

[a]These values were obtained from different publications and therefore do not all add up to 100%. Values in parentheses give approximate relative amounts of reaction products. The term ND indicates that the derivative was not detected in experiments where the author was aware of its existence and was able to detect it in other experiments.

to the frequency of EMS-induced sex-linked recessive lethals over a wide range of EMS concentrations (Fig. 3).

Upon injection of ^{3}H-MMS into mice, Segerbäck et al. (58) found that the amounts of alkylation in intracellular DNA (in units of alkylation/nucleotide) in the testis and liver are proportional to the extent of alkylated amino acids in haemoglobin. This provides a simple method for monitoring the amounts absorbed in germ cells of vertebrates exposed to MMS or similar chemicals.

The extent of alkylation of DNA by treatment with EMS in vivo for short exposure periods increases linearly with time. It is unknown whether this indicates a linear rate of reaction of EMS with intracellular DNA for premutagenic alkylation such as formation of O^6-ethylguanine, and it is also unknown whether the extent of alkylation per nucleotide at 50% lethality, i.e., LD_{50} in absorbed dose units, is the same with EMS as with MMS. Dosimetric work is needed on MMS and EMS in various organisms, especially mice and Drosophila, under the conditions used in most previous genetic experiments. This would make it possible to make more complete use of available data in assessing human hazards.

EXTRAPOLATION TO HUMAN GENETIC HAZARDS

Except in very rare cases no data were available on the genetic effects of any mutagens in humans until it became possible to culture human cells. Moreover, human cells in culture represent, at best, only part of the normal characteristics of the cells in the whole body as an integrated unit. Thus we must use the data obtained in experimental organisms to assess human hazards.

Mutations and Chromosome Aberrations Induced by MMS and EMS in Cultured Cells

From data obtained under the in vitro culture conditions usually used, it seems that the responses of human cells to various mutagens are not substantially different from those of most other mammalian cells. Of course, this does not guarantee that data obtained with experimental animals such as mice can be applied directly to humans but it shows that work should be concentrated on a single species such as mice.

Assessment of the Genetic Effects of MMS and EMS in Humans from Data in Experimental Animals

Indirect Method Using the "Doubling Dose." If we assume a linear relationship between mutation frequency and mutagen dose, then the frequency of induced mutations at a given locus, f_i, at an exposure dose D can be given by

$$f_i = f_o (D/\Delta) \quad (1)$$

where f_o is the spontaneous mutation frequency and Δ the doubling dose (Fig. 1).

An estimate for the mutation frequency at seven specific loci after treatment of spermatogonia of mice with a weighted dose of 102 mg/kg of MMS by intraperitoneal injection was 1.5×10^{-5} (per locus/gamete), whereas the spontaneous frequency was 0.75×10^{-5} (per locus/gamete) (Table 8). Thus, if we assume 100 mg/kg of MMS as the doubling dose Δ, then, from Eq. (1) we have for mouse spermatogonia (see curve B in Fig. 1):

$$f_i = f_o D/100 \quad (1a)$$

If we assume that Eq. (1a) is applicable to man, Eq. (1a) means that uptake of 1 mg of MMS per kg body weight increases the mutation rate by 1% of the spontaneous rate.

In the above calculation, we have neglected the contribution of mutations in postspermatogonial cells mutagenized by MMS. If we assume that the postspermatogonial stages last for time τ after an acute exposure to the chemical mutagen and that the effective, subsequent mating period is T, then Eq. (1) can be rewritten as in Eq. (2)

$$f_i = f_o(D/\Delta)[1 + (\tau/T)(r - 1)] \quad (2)$$

where r is the ratio of the induced mutation rate per unit dose for postspermatogonia to that for spermatogonia. Equation (2) takes the form (3) for mice:

$$f_i = (f_o D/100)[1 + 13(\tau/T)] \quad (3)$$

where we have used Eq. (1a) and inserted a crude value of r of 14 obtained from the Group I report (see Discussion).

Modifying Eq. (2), we obtain Eq. (3) for man:

$$f_i = f_o \delta (D/\Delta)[1 + (4 - 1)(\tau/T)] \quad (3)$$

where δ is the ratio of the absorbed gonad dose (alkylation/nucleotide) per unit exposure (mg/kg) to MMS in man to that in mouse and where we make the following assumptions: (1) there is a linear dose-frequency response in man; (2) the doubling dose is the same for man and mouse; (3) the mutability of spermatids relative to that of spermatogonia is the same for man as for mouse.

Table 8. Specific locus mutations induced by MMS in postspermatogonia of mice

Dose (mg/kg)	Matings after treatment (days)	No. of F_1 offspring	No. of mutations at 7 loci	References
40	1-4	1769	0	15,16
	5-8	977[a]	2	
			3	
	9-12	1065[a]	3	
	13-16	1692	1	
	17-20	1844	0	
0	--	531,500	28	17
0	--	74,037	4	16

[a]The reduction in the number of offspring after these matings is due to induction of dominant lethals.

Assumptions (1) to (3) are also made in considering ionizing radiation, and the δ value can be accurately obtained by work on radiation dosimetry. On exposure to MMS or any other chemical, however, δ is very difficult to estimate but its measurement is one of the most important, future research projects.

In the case of EMS, we may assume from the arguments made in the preceding sections that the dose-frequency response is linear for spermatozoa and spermatids but nonlinear for spermatogonia:

$$f_i = f_o \, (D/\Delta') \qquad \text{(postspermatogonia)} \qquad (4)$$

$$f_i = f_o \, (D/\Delta'')^n \quad n > 1 \qquad \text{(spermatogonia)} \qquad (5)$$

where Δ' and Δ'' are doubling doses for postmeiotic male germ cells and spermatogonia, respectively. Equaions (4) and (5) are illustrated by curves C and D in Fig. 1. From Eqs. (4) and (5) we can readily obtain the equation for the overall average frequency of mutations induced in males during the effective mating period of T after an acute dose D as follows:

$$f_i = f_o \, (D/\Delta'')^n \, [1 - (\tau/T)] + f_o \, (\tau D/T\Delta') \qquad (6)$$

where τ is the period of postmeiosis in males as defined in Eq. (2).

Equation (6) means that, over a low dose range, the second term due to the contribution from postmeiotic cells is no longer negligible, in spite of the small value of time τ compared with the large T value. This is illustrated in Fig. 1 by curve C (for mutations in postspermatogonia) to be compared with curve D (for mutations in spermatogonia). Curve D has been constructed making the following hypothetical assumptions: $n = 2$, $\Delta''/\Delta' \sim 10$, and $f_i = 6 \times 10^{-6}$ at a D value of 200 mg/kg for spermatogonia [actually no mutations in 14,787 offspring at weighted dose of 290 mg/kg (16)]. By comparison of curves C and D we find that the ratio of mutations inducible in spermatogonia to those in spermatozoa and spermatids changes from 1/10 at a dose of 200 mg/kg to 1/100 at a dose of 30 mg/kg. This may be an exaggerated hypothetical case but it helps to demonstrate the importance of chemical mutagenesis of spermatozoa and spermatids which is neglected in the assessment of radiation hazards.

If Eq. (6) describes actual EMS mutagenesis in mice, it may be used for assessment of human hazards after introducing the dosimetry factor δ, as done for Eq. (3). Probably actual mutagenesis is more complicated, but, however complicated the mutation frequency versus dose relationship is, it may be approximated, at least in a limited dose range, by the following:

$$f_i = f_o \left(\frac{D}{\Delta''}\right)^{n''} \left(1 - \frac{\tau}{T}\right) + f_o \frac{\tau}{T}\left(\frac{D}{\Delta'}\right)^{n'} \qquad (7)$$

In fact, preliminary data on MMS mutagenesis in postspermatogonia of mice (16) fit the second term of Eq. (7) with an n' value of about 0.3 (see curve A' in Fig. 1). In this case, even if the first term of Eq. (7) takes a linear form, i.e., $n'' = 1$, the contribution of the second term (due to mutagenesis in postspermatogonia) relative to the first term increases markedly with decrease in the dose (compare curve A' with curve B in Fig. 1).

Direct Method. In contrast to the 7-locus mutations in germ cells of mice (Table 6), slightly deleterious mutations are more frequently induced by EMS than by X rays for Drosophila at the same recessive-lethal level (59). EMS is 10 times more mutagenic than γ-rays or MMS for HGPRT mutations in Chinese hamster cell line V79 at 70% survival level (Table 6) and also 10 times more mutagenic than MMS for HGPRT mutations in mouse cell line L5178Y (Tables 3 and 4). On the other hand, MMS is more potent than EMS for thymidine kinase (TK) mutagenesis which is supposed to represent large chromosome deletions (60). Similarly to base-pair substitutions in *E. coli* (Table 6), ouabain-resistance mutations in line V79 at 70% survival level are 10 times more mutable by EMS than by MMS but hardly mutable by γ-rays (12), indicating that ouabain resistance mutations may be of base pair substitution type. Thus, mutagenicities of EMS and MMS relative to ionizing radiation vary in a complicated way with different types of mutation. Whether the doubling dose method is valid for chemical mutations remains open. As done for assessment of radiation genetic effects (61), direct measurements of recessive-lethal and skeletal-dominant mutations by EMS and MMS need to be studied.

DISCUSSION

In mice, treatment with an MMS dose of 100 mg/kg induced a mutation frequency of 1.1×10^{-4} in postmeiotic male germ cells and of 8×10^{-6} in spermatogonia (see Fig. 1). Thus, the relative mutation sensitivity r, i.e., ratio of induced mutation rate per unit dose in postmeiotic cells to that in spermatogonia, is 14, provided that the frequency versus dose relationship is linear. This means that the period of postmeiosis τ of about 3 weeks is no longer negligible for considerations of genetic effects compared with the remaining mating period T of about 50 weeks on the average. In mathematical terms, the contribution to the overall mutation yield from postmeiotic cells relative to that from spermatogonia is given by $(r-1)(\tau/T)$, as shown in Eq. (2), and in the present case this ratio is 0.8 (=13×3/50). For man, however, the τ/T value is much smaller than that for mouse. Further, since the above estimate for mutation frequency in spermatogonia is based on an observed value of only two mutants per 18,529 offspring, which is not significantly above the control rate of 7.5×10^{-6} (Table 8), we may only qualitatively conclude that spermatogonia are very resistant, probably one order of magnitude more than post-spermatogonia, to mutations by MMS.

Recently, Ehling (15) reported the important preliminary result that when mice were exposed to MMS, specific locus mutations in spermatozoa were most frequent in the short mating period of 5 to 12 days after treatment (Table 8) as is the case with dominant lethals (22). If this is the case, then it implies that with a higher dose of MMS a larger fraction of mutagenized spermatozoa is selectively killed, leading to underestimation of the mutation yield at high doses. Thus, on increase in the dose of MMS above a certain dose D_C, the relationship between the induced frequency f_i and the dose D is no longer linear but is given by Eq. (8):

$$f_i \propto D^n \qquad n < 1 \text{ for } D > D_c$$
$$n = 1 \text{ for } D \leqq D_c \qquad (8)$$

Equation (8) is illustrated in Fig. 1 by curve A', which is based on observed points at doses of 40, 57, and 100 mg/kg (16) and on an assumed D_c value of 15 mg/kg. In spermatogonia, however, similar selective killing does not seem to occur after treatment with MMS or EMS (62). Theoretically, it seems possible to produce high yields of mutations in spermatogonia by MMS and EMS with fractionated doses over a long period (63).

Recently, L. B. Russell (64) reported that the frequencies of presumed recessive mutations, detected by spots of altered color in the fur of heterozygous mice, were roughly parallel to those of specific locus mutations induced in spermatogonia by the same seven chemicals tested. This finding indicates that mutagenesis in spermatogonia may be simulated by somatic mutations in mice. In order to determine the relationship of mutation frequency to the dose, we need to know the number of pigment precursor cells surviving after treatment with mutagen. The mutagen is usually administered to fetuses on or around day 10.5. At this time the pigment precursor cells seem sensitive to the lethal action of the mutagen, and the presumed high sensitivity would greatly reduce deletion-type mutations, resulting in production of point mutations as a major part of the induced somatic mutations.

In contrast to the very limited stages of spermatozoa that are sensitive to EMS and MMS, all cell stages from spermaotozoa to spermatogonia are sensitive to x-rays with production of dominant lethals (65). This clearly indicates that a difference in the physicochemical action of the mutagen, which would largely be represented by a difference in the chemical nature of DNA damage, is responsible for the difference in germ cell-stage dependence of dominant lethals by MMS (or EMS) and x-rays. It is tempting to speculate that the major cause of the dominant lethals induced by MMS and EMS is chromosome aberrations (see the previous discussion in this review) and that aberrations are induced largely via DNA strand breaks resulting from depurination of 7-alkylguanine. This model is not fully confirmed, but it is compatible with the following findings, since alkylation-induced depurination of guanine is a slow process (50). First, late spermatozoa ejaculated 1 or 2 days after treatment are resistant to dominant-lethal mutations by MMS and EMS (19). Secondly, after treatment of Drosophila with MMS and EMS, dominant lethals and translocations are much more frequent in stored sperms than in unstored sperms (26). The model described above implies that spermatocytes and spermatogonia are resistant to dominant lethals by MMS and EMS because alkylation-induced depurination rarely occurs due to repair, active DNA replication, or other factors. Further, the coincidence in the high sensitivity stage of germ cells between specific-locus and dominant-lethal mutations after treatment with MMS (Table 8) may indicate that specific locus mutations by MMS (but not by mitomycin C and Natulan) are also mainly induced by chromosome aberrations.

Resistance of spermatogonia by specific locus mutations is seen with both chemicals such as EMS and MMS, and with x- or γ-rays. This appears to reflect, at least partly, a selection process operating on diploid (rather than haploid) genomes carrying mutations due to chromosome aberrations between meiotic metaphase and fertilization. In fact, L. B. Russell (66) reported that small (14%) and large (42%) fractions of the mutations induced by ionizing radiation in the se-d regions in spermatogonia and postspermatogonia, respectively, are aberrations as detected by complementation tests.

As discussed briefly in this review, the major problems in assessing the genetic effects of chemicals on humans on the basis of laboratory data are the same as those met in assessing the hazard of ionizing radiation. For the assessment, it is necessary to have: (1) reliable dose-response curves for genetic effects under various conditions, (2) data on chemical dosimetry, and (3) an understanding of the mechanism of chemical mutagenesis. As mentioned in this review we are just beginning to obtain some reliable data relevant to these problems. As pointed out in the timely review by W. L. Russell (67), mutagenesis in mammalian germ cells as a complex process which is affected by many factors even after exposure to radiation, and it is an even more complex process after treatment with chemicals. At present, there is hardly any rapid assay that can be used in place of direct measurement of transmitted mutations induced in mammalian germ cells. As pointed out by Auerbach (68) and W. L. Russell (67), it is better not to use a simplistic concept such as REC (rem-equivalent-chemical) until more information is available on mutagenesis of germ cells of mammals and other organisms. The complexity of comparative mutagenesis has been discussed in part with respect to Table 6 and in the preceding section.

Since sex-linked recessive lethals induced by MMS or EMS in Drosophila are almost free from deletions, as detected by complementation tests (26), it is tempting to assume that the potent mutagenicity of γ-rays in mice (compared with the weak mutagenicity of MMS or EMS shown in Table 6) is due to the ability of γ-rays to induce deletion-type specific locus mutations, whereas MMS and EMS are potent mutagens in Drosophila but poor mutagens in mice (Table 6) because they are poor inducers of deletions. However, if we recalculate the MMS-induced mutation frequency using only the fraction of F_1 offspring from the mating interval of 5 to 12 days after treatment (Table 8), then the frequency at 40 mg/kg of MMS is four times more, being 34×10^{-5} and MMS is twice as mutagenic as x-rays for mice as is the case for Drosophila (see Table 6). A post-treatment time of 5 to 12 days would be long enough to produce strand breaks in DNA of spermatozoa of mice via hydrolytic depurination of MMS-induced 7-methylguanine with half life of 150 hr (69). Whether EMS also induces high yields of specific locus mutations only in the early to middle stages of spermatozoa is an interesting problem since EMS is far less potent than MMS in inducing dominant lethals for spermatozoa of Drosophila (26).

The resistance of spermatozoa in the late stages to induction of dominant lethals by MMS or EMS may be, at least partly, due to reduced penetration of these chemicals simply because the late spermatozoa are in a deferent canal outside the testis. This possibility can be readily tested by dosimetric work on differential alkylation of DNA of spermatozoa in different locations and stages. Whether the unscheduled DNA synthesis occurring in spermatids after treatment with EMS or MMS (70) results exclusively from repair and, if so, whether it is efficient enough to account for the resistance of spermatids to induction of dominant lethals by the chemicals are problems for future study; no decisive conclusions can be reached without more data. At present, the method adopted in this review of comparing data on chemical mutagenesis seems to be one of the most useful ways of classifying the complicated data available in this field.

Many important subjects have been neglected in this review simply because no relevant information on them was available to the Working Group. The effects of chemical mutagens in female germ cells and chemical-induced deleterious effects on development which may be caused by somatic mutations are as important as the effects on male germ cells. Moreover, the possibility that some types of anomaly or related disease may be caused by somatic mutation is more likely with chemical mutagens than with ionizing radiation; this is because, since most human cancers seem to be related, directly or indirectly, to environmental carcinogens and since most carcinogens are mutagens, we must realize that we are living in a sea of potentially mutagenic chemicals. Furthermore, a knowledge of somatic mutations in mice is indispensable for determining whether cancer is caused by somatic mutation. Therefore, it seems of profound significance that, following the pioneering work of Russell and Major (71), the detection of somatic mutations in mice by spots of altered color in the fur has now been established (72).

As partly discussed in the previous section, EMS and probably other chemicals known as base-pair-substitution inducers might be potent for induction in mice of mutations of some types other than the visible recessive at the currently used seven loci, such as minute genomal damage. Slightly deleterous mutations, if they occur more frequently than severe mutations as shown for Drosophila (59), may impose more burdens than the latter to the human society.

SUMMARY

MMS and EMS are mutagenic in bacteriophages, bacteria, fungi, insects, higher plants, mice and mammalian cultured cells and they induce almost all kinds of genetic effects, except presumably nondisjunction. In phage T4 and E. coli, MMS induces mutations via misprepair whereas EMS can induce mutations via mispairing. In eukaryotes, EMS mutagenesis seems to occur mainly by misrepair. EMS produces mispairing DNA alteration of O^6-alkylguanine more efficiently than MMS. Because of the repairable nature of the premutational damage or for other unknown reasons, EMS mutagenesis shows an approximately two-hit response in the low dose range in many cases, whereas MMS mutagenesis is mostly of nearly single hit type. A notable exception is that EMS-treated sperm of Drosophila gives rise to an exactly single hit response for induction of recessive lethal mutations if the doses are measured as the degree of alkylation of DNA of sperm. The importance of chemical dosimetry is emphasized. When compared at the same survival level with x-rays, EMS is very potent mutagen in E. coli, considerably potent mutagen in Drosophila and barley, but weak mutagen in mice, whereas MMS is also weak mutagen in mice. After treatment of mice with MMS and EMS, spermatozoa are sensitive to specific-locus and dominant-lethal mutations, whereas spermatogonia are one order of magnitude resistant to specific-locus mutations and produce no dominant lethals. Based on the doubling dose concept, mathematical formulae are proposed to estimate genetic hazards to humans from the data on mice and it is suggested that postmeiotic male germ cells could be as important as spermatogonia in assessing genetic hazards from environmental chemicals. MMS and EMS are mainly used only in the laboratory and hence they probably constitute no appreciable genetic threat to humans but they can serve as standard chemicals when assessing hazards for humans of other environmental alkylating agents by testing their effects on experimental animals.

ACKNOWLEDGEMENTS

I am grateful to Dr. Mituo Ikenaga for preparing Table 7, to Mrs. Elizabeth Ichihara for her critical reading of the manuscript and helpful suggestions, and to Dr. Y. Ishii and Miss Y. Matsumoto for their competent help during the preparation of the manuscript. Preparation of this paper was partly supported by Grant-in-Aid from the Japanese Ministry of Education, Science and Culture.

REFERENCE

1. Clive, D. Unpublished data.

2. Clive, D. Environ. Health Perspect. No. 6: 119-125 (1973).

3. Cole, J., and Arlett, C. F. Mutat. Res., in press.

4. Knaap, A. G. A. C. EEC Progr. Rept., September 1977.

5. Anderson, D., and Fox, M. Mutat. Res. 25: 107-122 (1974).

6. Arlett, C. F., Turbill, D., Harcourt, S. A., Lehmann, A. R., and Colella, C. M. Mutat. Res. 33: 261-278 (1975).

7. Myhr, B. A., and DiPaolo, J. A. Genetics 80: 157-169 (1975).

8. Van Zeeland, A. A., and Simons, J. W. I. M. Mutat. Res. 35: 129-138 (1976).

9. Cole, J., and Arlett, C. F. Mutat. Res. 34: 507-526 (1976).

10. Rogers, A., and Arlett, C. F. EEC Progr. Rept., September 1977.

11. Knaap, A. G. A. C., and Simons, J. W. I. M. Mutat. Res. 30: 97-110 (1975).

12. Hsie, A. W., Brimer, P. A., Mitchell, T. J., and Gosslee, P. G. Somatic Cell Genet. 1: 247-261 (1975).

13. Ehling, U. H., and Russell, W. L. Genetics 61: 514-515 (1969).

14. Russell, W. L. Arch. Toxicol. 38: 141-147 (1977).

15. Ehling, U. H. GSF-Ber. B798, 13-16 (1977).

16. Ehling, U. H. Chapter 25. This volume.

17. Russell, W. L. Nucleonics 23: 53-56, 62 (1965).

18. Generoso, W. M., Russell, W. L., Huff, S. W., Stout, S. K., and Gosslee, D. G. Genetics 77: 741-752 (1974).

19. Ehling, U. H., Cumming, R. B., and Malling, H. V. Mutat. Res. 5: 417-428 (1968).

20. Ehling, U. H. Mutat. Res. 11: 35-44 (1971).

21. Brewen, J. G., Payne, H. S., Jones, K. P., and Preston, R. J. Mutat. Res. 33: 239-250 (1975).

22. Ehling, U. H. Arch. Toxicol 38: 1-11 (1977).

23. Ray, V. A., Holden, H. E., Salsburg, D. S., Ellis, J. H. Jr., Just, L. J., ana Hyneck, M. L. Toxicol. Appl. Pharmacol. 30: 107-116 (1974).

24. Aaron, C. S., and Lee, W. R. Mutat. Res. 48: 27-44 (1978).

25. Vogel, E., and Leigh, B. Mutat. Res. 29: 383-396 (1975).

26. Vogel, E., Lee, W. R., Schalet, A., and Würgler, F. Chapter 9. This volume.

27. Konzak, C. F., Nilan, R. A., Wagner, J., and Foster, R. J. In: The Use of Induced Mutations in Plant Breeding (Proc. FAO-IAEA Sympl, Rome 1964), Pergamon Press, Oxford, 1965, pp. 49-70.

28. Nauman, C. H., Sparrow, A. H., and Schairer, L. A. Mutat. Res. 38: 53-70 (1976).

29. Vig, B. K., and Sung, A. R. Chapter 10. This volume.

30. Arenaz, P. M. S. Thesis, Univ. Nevada, 1976, 102 pp.

31. Vig, B. K., and Paddock, E. F. Theor. Appl. Genet. 40: 316-321 (1968).

32. Vig, B. K., Nilan, R. A., and Arenaz, P. Env. Exptl. Bot. 76: 223-239 (1976).

33. Haroun, L., and Ames, B. N.

34. Krieg, D. R. Genetics 48: 561-580 (1963).

35. Witkin, E. M. Bacteriol. Rev. 40: 869-907 (1976).

36. Kondo, S., Ichikawa, H., Iwo, K., and Kato, T. Genetics 66: 187-217 (1970).

37. Ishii, Y., and Kondo, S. Mutat. Res. 27: 149-161 (1975).

38. Drake, J. W. The Molecular Basis of Mutation, Holden-Day, San Francisco, 1970.

39. Kondo, S., and Ichikawa, H. Molec. Gen. Genet. 126: 319-324 (1973).

40. Yamamoto, K., Kondo, S., and Sugimura, T. J. Mol. Biol. 118: 412-429 (1978).

41. Nilan, R. A., and Veleminsky, J. Chapter 11. This volume.

42. Green, R. R., and Drake, J. W. Genetics 78: 81-89 (1974).

43. Yanofsky, C., Ito, J., and Horn, V. Cold Spring Harbor Symp. Quant. Biol. 31: 151-162 (1966).

44. Malling, H. V., and de Serres, F. J. Mutat. Res. 6: 181-193 (1968).

45. Prakash, L. Genetics 78: 1101-1118 (1975).

46. de Serres, F. Personal communication.

47. Lea, D. E. Action of Radiations on Living Cells. Cambridge Univ. Press, 2nd ed., 1955.

48. Conkling, M. A., Grunau, J. A., and Drake, J. W. Genetics 82: 565-575 (1976).

49. Singer, B. Progr. Nucl. Res. Mol. Biol. 16: 219-332 (1976).

50. Lawley, P. D., and Brookes, P. Biochem. J. 89: 127-138 (1963).

51. Lawley, P. D., and Martin, C. N. Biochem. J. 145: 85 (1975).

52. Lawley, P. D., and Shah, S. A. Chem.-Biol. Interact. 5: 286 (1972).

53. Lawley, P. D., Orr, D. J., and Jarman, M. Biochem. J. 145: 73 (1975).

54. Lawley, P. D., and Thatcher, C. J. Biochem. J. 116: 693 (1970).

55. Strauss, B., Scudiero, D., and Henderson, E. In: Molecular Mechanisms for Repair of DNA, Plenum Press, New York, 1975, Part A, p. 13.

56. Walker, I. G., and Ewart, D. F. Mutat. Res. 19: 331 (1973).

57. Craddock, V. M. Biochim. Biophys. Acta 312: 202 (1973).

58. Segerbäck, D., Calleman, C. J., Ehrenberg, L., Löfroth, G., and Osterman-Golkar, S. Mutat. Res. 49: 71-82.

59. Mukai, T. Genetics 65: 335-348 (1970).

60. Clive, D. EMS Abstracts, 1977.

61. U.N. Scientific Committee on the Effects of Atomic Radiation. Sources and Effects of Ionizing Radiation (report to the General Assembly, with annexes), U.N. Publication, New York, 1977.

62. Ehling, U. H. Personal communication.

63. Preston, R. J., and Brewen, J. G. Mutat. Res. 36: 333-344 (1976).

64. Russell, L. B. Arch. Toxicol. 38: 75-85 (1977).

65. Bateman, A. J. Heredity 12: 213-232 (1958).

66. Russell, L. B. Mutat. Res. 11: 107-123 (1971).

67. Russell, W. L. Arch. Toxicol. 38: 141-147 (1977).

68. Auerbach, C. Mutat. Res. 33: 3-10 (1975).

69. Lawley, P. D. In: Molecular Mechanisms for Repair of DNA. P. C. Hanawalt and R. B. Setlow, Eds., Plenum Press, New York, 1975, Part A, pp. 25-28.

70. Sega, G. A., Owens, J. G., and Cumming, R. B. Mutat. Res. 36: 193-212 (1976).

71. Russell, L. B., and Major, M. H. Genetics 42: 161-175 (1957).

72. Fahrig, R., Dawson, W. P., and Russell, L. B. Chapter 24. This volume.

CHAPTER 27

COMPARATIVE MUTAGENICITY OF DIMETHYLNITROSAMINE AND DIETHYLNITROSAMINE

K. Sankaranarayanan

Department of Radiation Genetics and Chemical Mutagenesis

Sylvius Laboratories, State University of Leiden

Wassenaarseweg 72, Leiden, and The J. A. Cohen Institute

Interuniversity Institute for Radiopathology and Radiation

Protection, The Netherlands

INTRODUCTION

The dialkyl nitrosamines, of which dimethylnitrosamine and diethylnitrosamine are good examples, belong to the larger class of N-nitroso compounds, many of which are potent carcinogens in a large variety of animal species. In 1954 Barnes and Magee (1) demonstrated that dimethylnitrosamine was acutely hepatotoxic to laboratory animals. Subsequently, in 1956 they showed (2) that dimethylnitrosamine fed to rats produced a very high incidence of malignant liver tumor. These studies provided the impetus for extensive research on a whole range of nitroso compounds including their carcinogenity, pharmacology, chemistry, biochemistry, and mutagenic potential. The prodigious amount of literature that has accumulated in these different areas has been summarized in a number of excellent reviews (3-15).

The mammalian and submammalian species susceptible to carcinogenesis by DMN or DEN and their organ specificities are listed in Tables 1 and 2. The importance of DMN, DEN and other nitrosamines as possible human carcinogens was first suggested in 1963 by Druckrey et al. (16), who postulated that secondary aliphatic amines could react in the body with sodium nitrite to form

Table 1. Mammalian and submammalian species susceptible to carcinogenic action by DMN and DEN[a]

	Species
Mammals	Monkey
	Rat
	Mouse
	Guinea pig
	Syrian hamster
	Chinese hamster
	European hamster
	Rabbit
	Dog
	Pig
	Mink
	Mastomys
	Gerbil
Birds	Fowl
	Grass parakeet
Fish	Rainbow trout
	Aquarium fish (Brachydanio rerio, Lebistus reticulatus)

[a]Data of Montesano and Bartsch (12).

nitrosamines by nitrosation. This suggestion has been amply corroborated (9,11) and extended to tertiary amines (7,17,18). Nitrosatable substances in the environment include secondary and tertiary amines, quartenary ammonium compounds, ureas, carbamates and guanidines (10). The evidence that N-nitroso compounds are formed in the body from amine precursors is based on their chemical detection in gastric juices in vitro and in the mammalian (including human) stomach in vivo and the observation of acute toxic and carcinogenic effects as well as damage to cellular macromolecules after simultaneous administration of nitrite and various amines and amides (3,19).

The above considerations strongly suggest that nitrosamines have a high potential for being human carcinogens. This has been emphasized by, among others, Lijinsky and Epstein (20) who pointed out that nitrates occur widely in nature, can be enzymatically reduced to nitrites that are found in green vegetables, and are extensively used as preservatives for fish and meat. Efforts devoted to the analysis of foods and other materials for nitrosamines have revealed that they are in fact present although in very small quantities. Table 3 summarizes some of the available information.

Table 2. Carcinogenicity of DMN and DEN: species and their organ specificities

Chemical	Species	Target organs
DMN	Rat	Liver, kidney, nasal cavities
	Mouse	Lung, liver and kidney
	Syrian hamster	Liver, nasal cavities
	European hamster	Liver, kidney
	Rabbit, Mastomys, guinea pig, trout, newt, aquarium fish, and mink	Liver
DEN	Rat	Liver, esophagus, nasal cavities, kidney
	Mouse	Liver, lung, forestomach, esophagus, nasal cavities
	Syrian hamster	Trachea, larynx, nasal cavities, lung, liver
	Chinese hamster	Esophagus, forestomach, liver
	European hamster	Nasal cavities, trachea, bronchi, larynx
	Guinea pig, rabbit, dog, pig, trout, *Brachidonio rerio*, grass parakeet, monkey	Liver

Table 3. Occurrence of N-nitrosamines in various food and other products[a]

Source	N-Nitrosamines[b]	Concentration (ppb)
Meat products		
Salami, dry sausages	DMN	10-80
Fried bacon	DMN, DEN, NPYR DPIP	1-40
Luncheon meat, salami, and pork chops (Danish)	DMN, DEN	1-4
Uncooked and fried bacon	DMN, NPYR	2-30
Frankfurters	DMN	11-84
Fish		
Chinese marine salted fish	DMN	50-100
Raw and smoked nitrite-nitrate treated salmon, shad, sable	DMN	0-26
Fresh, salted, fried cod; fresh, fried hake	DMN	1-9
Other foods		
Fruit (Solanum incanum)	DMN	ND
Cheese	DMN	1-4
Soybean oil	DMN, DEN	ND

(continued)

Table 3. (continued)

Source	N-Nitrosamines[b]	Concentration (ppb)
Miscellaneous		
Tobacco smoke condensate	DMN, DEN, MEN, MBN, DPN, NPYR, NPIP, DBN, NNN	0-18 ng/ cigarette
Tobacco	NNN	0.3-88 ppm
Stratosphere (Surveyor III)	DMN	ND

[a]Data of Montesano and Bartsch (12).

[b]NPYR = N-nitrosopyrolidine; NPIP = N-nitrosopiperidine; MEN = N-nitrosoethylmethylamine; MBN = N-nitrosobutylmethylamine; DPN = N-nitroso-dipropylamine; DBN = N-nitrosodibutylamine; NNN = N^1-nitrosonornicotine.

CHEMICAL AND PHYSICAL PROPERTIES AND USES

Dimethylnitrosamine (CAS 62759; synonyms: dimethylamine, N-nitroso; N,N-dimethylnitrosamine; abbreviations: DMN, DMNA) has a molecular weight of 74.1 and the chemical structure of:

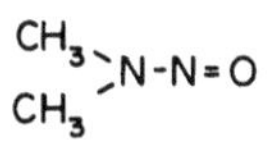

It is a yellow liquid of low viscosity and with a boiling point of 149-150°C (755 mm). It is miscible with water in all proportions and is also soluble in all common organic solvents and in lipids. It is very volatile. It is stable at room temperature for more than 14 days in aqueous solution at neutral and alkaline pH in the absence of light. It is photosensitive, undergoing denitrosation on exposure to ultraviolet light (21).

DMN has been used as an industrial solvent and also in the synthesis of the rocket fuel 1,1-dimethylhydrazine. There are patents for its use as a solvent in the fiber and plastic industry, as an antioxidant, as a softener for copolymers, as an additive for lubricants, and in condensers to increase the dieletric constant. DMN has also been used as a nematocide.

Diethylnitrosamine (CAS 55185; synonyms: diethylamine, N-nitroso; N,N-diethylnitrosamine; abbreviations: DEN, DENA) has a molecular weight of 102.1 and a chemical structure of:

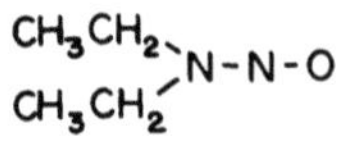

It is a yellow volatile liquid with a boiling point of 177°C (760 mm). It is soluble in water (to the extent of about 10%) and in organic solvents and lipids. It is stable at room temperature for more than 2 weeks in aqueous solution at neutral and alkaline pH in the absence of light; as in the case of DMN, DEN is less stable at strongly acid pH at room temperature. It is also light-sensitive, especially to ultraviolet light (21).

There are patents for the use of DEN as a solvent in fiber industry, as a softener for copolymers, as an additive for lubricants, in condensers to increase th dielectric constant and for the synthesis of 1,1-diethylhydrazine.

METABOLISM AND MECHANISMS OF BIOLOGICAL ACTION

The metabolism and possible mechanisms of biological action of N-nitroso compounds have been the subject of several extensive studies (4,9,19,22). Both DMN and DEN are metabolized in vivo and in vitro, and require metabolic activation to exert their carcinogenic and mutagenic activities. When DMN is administered to rats systemically at a "dose" of 30 mg/kg body weight, it is distributed uniformly among the various organs and is metabolized completely in about 5 hr (23), whereas DEN, however, is metabolized in rats at a rate of about 4% of the administered "dose" (200 mg/kg)/hr (24). In both cases metabolism occurs predominantly in the liver but also in other organs, e.g., kidney and lung (22,25, 26).

Metabolism of DMN in vitro was demonstrated with rat liver slices and rabbit liver homogenates by Magee and Vandekar (27), the activity being localized in the microsomal fraction with a requirement for reduced pyridine nucleotides and oxygen. A comparative in vitro study showed the DMN is metabolized in human liver slices at about the same rate as in rat liver slices (28). The formation of formaldehyde from DMN incubated with rat liver microsomes was reported by Brouwers and Emmelot (29), and there has been extensive subsequent work on the microsomal metabolism of this and other nitrosamines. In mammals, the available evidence indicates that in the metabolic activation of DMN and some other nitrosamines studied, the microsomal enzyme system involves the cytochrome P-450 dependent mixed function oxidases (MFO). In insect species, particularly Musca domestica, where extensive work on biochemical characterization of microsomal enzymes had been carried out, "the nature and diversity of the reactions catalysed indicate that insect microsomes possess a similar degree of metabolic versality and substrate non-specificity as those from the mammalian liver" (30). In this context, it is of interest to note that in Drosophila, Baars et al. (31) have recently obtained biochemical evidence for the presence of P-450 and in Saccharomyces cerevisiae; Callen and Philpot (32) have demonstrated that microsomal preparations have cytochrome P-450-dependent MFO activity.

There is strong evidence that DMN, DEN and other nitrosamines and nitrosamides behave as aklylating agents in biological systems. Magee and Farber (33) reported in 1962 that ^{14}C-labeled DMN interacts with rat liver nucleic acids, especially DNA in vivo, to form 7-methylguanilic acid residues that could be released as 7-methylguanine by hydrolysis. These findings were considered to support the hypothesis that an active methylating agent is formed metabolically from DMN. The alkylation of guanine was very extensive--as much as 2% of the guanine present in the nucleic acid. The idea grew that nitrosamines acted through this mechanism.

The mechanism of metabolic activation and alkylation by DMN and DEN was established in elegant experiments by Lijinsky et al. (34) and by Ross et al. (35). They injected completely deuterated DMN or DEN into rats, isolated the liver DNA and RNA, hydrolyzed them, and isolated the 7-methylguanine (7-Me-G). The 7-Me-G was analyzed by mass spectrometry, and three (in the case of DMN) or five (in the case of DEN) deuterium atoms were present. Earlier, it had been thought that diazoalkanes were the alkylating intermediates; however, if this were the case, the N-7-alkylguanine would contain only two (in the case of DMN) or four (in the case of DEN) deuterium atoms and not three or five as were actually found.

The metabolic pathway of DMN and DEN (and other dialkylnitrosamines) that seems consistent with current evidence (8,12,36) can be briefly summarized as follows (Fig. 1). Oxidative dealkylation of the dialkylnitrosamine by microsomal mixed function oxidases leads to the transient formation of monoalkylnitrosamine. The first step in the process is α-C-hydrosylation in the case of DMN; with longer chain dialkylnitrosamines, β-oxidation (37) or ω or ω_1-oxidation (38,39) may occur in addition to α-C-hydroxylation. The monoalkylderivative undergoes molecular rearrangement after the so-called concerted reaction (40,41) to the corresponding alkyldiazohydroxide, leading to the alkyldiazonium cation and subsequent release of molecular nitrogen and formation of the alkyl carbonium ion. The latter reacts with nucleic acids, proteins and other cellular components. The aldehyde generated after monodealkylation is further oxidized to yield CO_2.

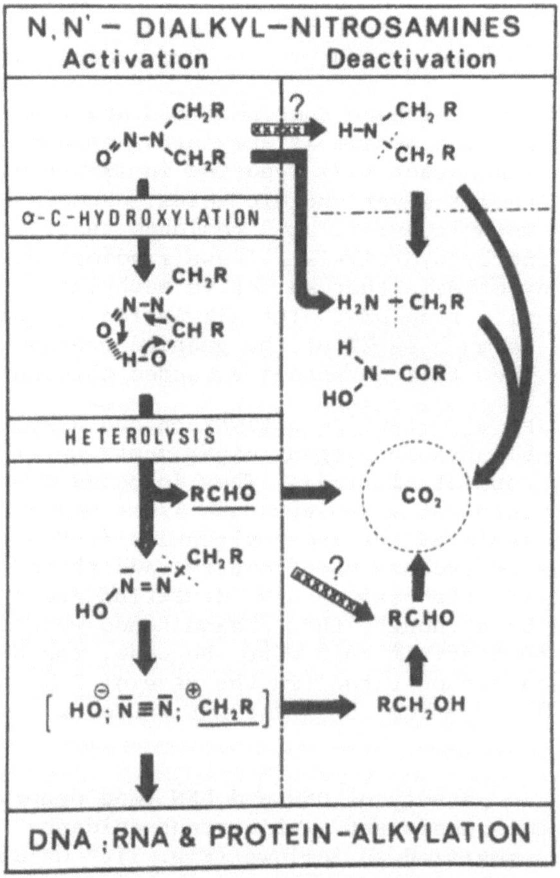

Fig. 1. Metabolic activation (left) and detoxification (right) of N,N-dialkylnitrosamines. From Montesano and Bartsch (12).

The main site in nucleic acids of base alkylation by alkylnitroso compounds as well as with other alkylating agents, is the N-7 position of guanine; however, a number of other sites are also attacked as shown in Fig. 2 (14). In addition to base alkylation, phosphate alkylation may also take place and there is evidence that this may be the major site of alkylation in the case of ethylnitrosourea (ENU). Unlike the nitroso compounds already described, agents such as MMS and others acting by a bimolecular substitution reaction (S_N2) are more selective for N-alkylation and produce little reaction at the O position (42). Although 7-alkylguanine is the major alkylated base in DNA following treatment with DMN or DEN, a number of investigators (8,12,14, 15) have presented evidence that this alkylation (N-7) may not be relevant in the processes of mutagenesis and carcinogenesis. For instance, the extent of alkylation at the N-7 position of guanine in different organs of the treated rats do not correlate well with the sites at which tumors actually appear (25,26). Data that have become available in recent years (43-47) have shown that alkylation at the 0-6 position of guanine in DNA may persist for very variable periods of time in different organs in the rat and these findings correlate well with the organ distribution of the induced tumors. It seems probable that 7-methylguanine is removed from DNA in the animal by chemical depurination while the removal of 0-6-methylguanine is probably catalyzed by enzyme systems that have markedly differing distributions in the body.

Singer (14) has stressed the point that "all of the 0-alkylpyrimidines and the 0-6-alkylguanine can be considered as promutagens; that is alkylation at each of the oxygens changes the tautomeric state so that normal base-pairing is unlikely to occur and instead, there exists the possibility that other base pairs may occasionally occur. In the case of 0-6-alkylguanine, this is indeed true (48), and there is strong presumption that this alkylation leads to mutation in vivo. The base-pairing of the 0-alkylpyrimidines (0-2 and 0-4 of alkyl U and T and 0-2-alkyl C) has not yet been examined, but it is probably safe to assume that they are similar to 0-6 alkyl G in nonspecific pairing. Since pyrimidine-oxygen alkylation accounts for a considerable proportion of alkylation by the ethylating carcinogens, it is quite possible that mutations can result from any of these modifications and not only from formation of 0-6 alkylguanine."

It is worth pointing out that the extent of alkylation at the various sites in DNA is dependent on the alkylating species: the methylcarbonium cations derived from N-nitroso compounds such as DMN or MNU are highly reactive and less selective and thus produce a wide spectrum of products, including 0-alkyl derivatives, by a unimolecular nucleophilic substitution (S_N1 type).

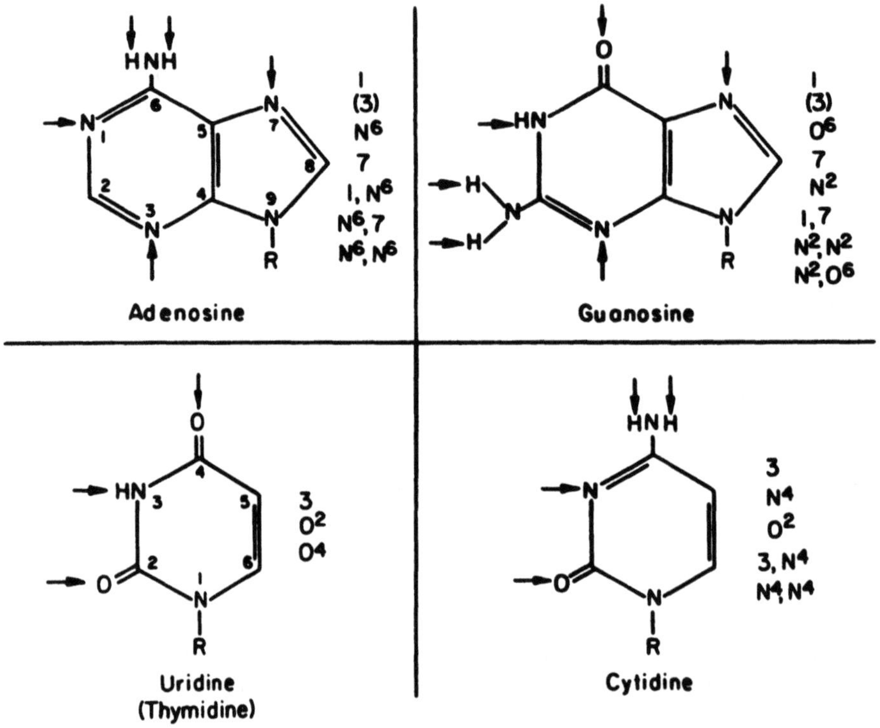

Fig. 2. Sites of alkylation of bases in nucleic acids. The arrows point to positions found to be available for nucleophilic displacement and addition reactions. The list of numbers on the side of each structural formula refers to the alkyl nucleosides which have been isolated from a nucleoside or polynucleotide reacted with a variety of alkylating agents. The numbers in parentheses indicate the two sites of reaction of a polynucleotide where only the alkyl base has been isolated (14).

Magee (8) has emphasized the fact that the details of the postulated short-lived intermediates have not been established and the metabolic pathway to the alkylating intermediate is essentially speculative. He cites three recent experimental observations, however, relating to DMN metabolism which strengthen the premise for thinking that the postulated metabolic pathway is essentially correct. First, the formation of methanol from DMN incubated with rat liver microsomal preparations in vitro has been demonstrated (49) indicating reaction between the methylating agent and water. Second, methylacetoxymethylnitrosamine has been prepared (50,51) and shown to be a more potent carcinogen than DMN (52,53). (The rationale for the preparation of the acetoxy derivatives was to obtain evidence indicative of the probably biological behavior of hydroxymethylnitrosamine which is thought to be the first metabolic product of DMN, but which has not yet been synthesized.) Third, Magee and colleagues (54) have obtained unequivocal evidence for the formation of molecular nitrogen as a metabolic product of DMN.

Finally, while DNA alkylation is central to the action of nitrosamines, it is worth pointing out that the repair of DNA lesions and DNA replication are clearly important factors in the fixation and expression of the damage induced. The interplay between these different factors is difficult to assess quantitatively at present.

GENETIC EFFECTS

The genetic effects of DMN and DEN have been studied in a number of test systems and with a variety of endpoints and test methodologies. Work carried out prior to 1970 showed that in Drosophila, both these compounds were mutagenic and in Arabidopsis, DMN was mutagenic; however, no such activity was obtained in assays that used bacteria, yeasts and fungi. In retrospect, these negative findings are attributable to the fact that both these compounds require metabolic activation and the microbial assay systems as standardly used do not provide the metabolism necessary to activate these compounds to mutagenic metabolites. In 1966, Malling (55) demonstrated that DMN and DEN were mutagenic to Neurospora crassa when the conidia were incubated in the presence of Udenfriend's hydroxylation system, which is thought to dealkylate N,N-dialkylnitrosamines to monoalkylnitrosamines in a way similar to that which occurs in mammals (Fig. 1). In 1969, Gabridge and Legator (56) using thehost-mediated assay were the first to show that DMN administered to mice induced mutations in Salmonella typhimurium which had been injected beforehand and were then reisolated from the periotneal cavity. Since then, a number of reports on the mutagenicity of DMN and DEN have appeared. Montesano and Bartsch

(12) have reviewed the data accumulated up to 1975. For the sake of comprehensiveness, the present report will include these data in addition to those that have become available since then.

An overview of the effects of DMN and DEN in the different test systems is given in Table 4, while the main results are summarized in Tables 5-17. Since these tables are self-explanatory, only the main conclusions are summarized below.

Both DMN and DEN require metabolic activation for their mutagenicity.

DMN is mutagenic in all prokaryotic and eukaryotic microbial test systems investigated whether it be the induction of forward and/or reverse mutations in *Escherichia coli*, *Bacillus subtilis*, *Salmonella typhimurium*, *Serratia marcescens*, *Saccharomyces cerevisiae*, *Neurospora crassa* or preferential growth inhibition in DNA repair deficient strains of bacteria (*E. coli*, *B. subtilis*) of the induction of somatic crossing over or mitotic gene conversion in *S. cerevisiae*. In almost all cases, mutagenic activity was recorded in the presence of mammalian microsomes. In Neurospora and yeast, the activation could be carried out either by mammalian liver chromosomes or by Udenfriend's hydroxylation system. Recently, it has been demonstrated (32) that in *S. cerevisiae* (log phase cells) DMN induces gene conversion and somatic crossing over without the addition of any activation system.

In liquid tests on the effects of DMN on survival of *E. coli B/r* and its repair-deficient derivatives *without* mammalian activation, DMN produces a concentration-dependent decrease of colony forming ability, with the strain B/s-1 being more sensitive than B/r (63).

In host-mediated assays involving bacteria, fungi and mammalian cells (mmamlian hosts), DMN records as positive.

In Drosophila, which has the enzyme capabilities to activate DMN to the mutagenic metabolites, it is a potent mutagen in all male germ-cell stages tested (the potency varying with the cell stage) for the induction of sex-linked recessive lethals, the spermatids being the most sensitive stage. There is a concentration-dependent increase in mutation frequency, but the increase is less than linear.

DMN has been found to induce somatic crossing over in *Glycine max* (soybean) and forward mutations in *Arabidopsis thalania* (a cruciferous plant) without any addition of microsomes.

In mice, the evidence for the induction of dominant lethals by DMN is weak and is based on one study; however, the compound

Table 4. Summary of effects observed with DMN and DEN in the different test systems

Test system	Assay	Results with[a] DMN	DEN	Remarks
Bacteria				
E. coli	Forward and reverse mutations	+	+	Mutagenic only in the presence of mammalian microsomes
	Differential cell-killing in DNA-repair-deficient strains	+	+	Same as above.
	Preferential growth inhibition in DNA-repair-deficient strains	+	+	Same as above.
B. subtilis	Forward mutations	+	NT	Same as above.
	Preferential growth inhibition in DNA-repair-deficient strains	+	+	Same as above; DEN > DMN
S. typhimurium	Reverse mutations	+	+	Same as above.
Fungi				
S. cerevisiae	Forward mutations and cytoplasmic petites	+	+	Mutagenic in presence of Udenfriend's hydroxylation mixture; DEN > DMN

(continued)

Table 4. (continued)

Test system	Assay	Results with[a] DMN	DEN	Remarks
Fungi (cont'd.)				
S. cerevisiae	Mitotic crossing-over	+	+	Mutagenic in presence of Udenfriend's hydroxylation mixture or mammalian microsomes DEN > DMN
	Mitotic gene conversion	+	+	Mutagenic in presence of mouse liver microsomes
N. crassa	Forward mutations	+	NT	Same as above
	Reverse mutations	+	+	Mutagenic in presence of Udenfriend's hydroxylation mixture
Bacteria	Host-mediated assays			
E. coli	In mouse, forward and reverse mutations; in baboon, reverse mutations	+	-	
S. typhimurium	In mouse or rats, reverse mutations	+	(+) or -	
S. marcescens	In mouse, reverse mutations	+	-	

(continued)

Table 4. (continued)

Test system	Assay	Results with[a] DMN	DEN	Remarks
Fungi				
S. cerevisiae	In mouse or rat, mitotic crossing-over and/or gene conversion	+	(+)	
S. pombe	In mouse and rat, forward mutations	+	NT	
N. crassa	In mouse, forward mutations	+	-	
Mouse lymphoma cells	Host-mediated assay			
L5178Y; asn⁻	In mouse; forward mutations	+	NT	
D. melanogaster	Recessive lethals	+	+	Mutagenic in all germ cells; adult feeding more effective than injection (DMN); DMN > DEN; chromosome breakage only at very high concentrations
Mammals				
Mouse	Forward mutations at specific loci; male germ cells	NT	-	
	Spot tests (somatic mutations)	+	+	
	Dominant lethals	(+)	-	
Rat lymphocytes in vivo	Chromosome aberrations	+	NT	

(continued)

Table 4. (continued)

Test system	Assay	Results with[a] DMN	DEN	Remarks
Mammals (cont'd.)				
CHO cells in vitro	Chromosome aberrations	+	+	Effective in presence of rat liver microsomes; DMN > DEN
Human lymphocytes	Chromosome aberrations	+	NT	Effective in presence of mouse liver microsomes
Rat lymphocytes in vivo	Sister chromatid exchanges	+	-	
Mouse bone marrow cells in vivo	Sister chromatid exchanges	+	-	
Chinese hamster bone marrow cells in vivo	Sister chromatid exchanges	NT	-	
CHO cells in vitro	Sister chromatid exchanges	+	+	Effective in presence of rat liver microsomes; DMN > DEN
L5178Y cells in vitro	Forward mutations to TK^-	+	+	Mutagenic in presence of mouse/rat liver microsomes; DMN > DEN
	Forward mutations to $HG\text{-}PRT^-$	+	+	Mutagenic in presence of rat liver microsomes

(continued)

Table 4. (continued)

Test system	Assay	Results with[a] DMN	DEN	Remarks
Mammals (cont'd.)				
V79 Chinese hamster cells in vitro	Forward mutations to HG-PRT$^-$	+	+	Mutagenic only after activation
Mouse bone marrow cells	Micronucleus test	+	-	
Human fibroblasts	Unscheduled DNA synthesis	+	NT	Effective only in the presence of microsomes
C3H mice	Unscheduled DNA synthesis in lung, liver and kidney cells; subcutaneous injection	+	NT	
C3H × 101 F_1 male mice	Unscheduled DNA synthesis in testis (spermatids) IP injection	-	-	
Plant systems				
V. faba lateral roots	Chromosome aberrations	-	-	
Barley seeds	Chlorophyll deficient mutations and chromosome aberrations	-	NT	
G. max (soybean) seeds	Somatic crossing-over	+	NT	
A. thaliana	Forward mutations	+	-	

[a]Symbols: + = positive; - = negative; (+) = weakly positive; NT = not tested.

Table 5. Effects of DMN and DEN in bacterial test systems (excluding host-mediated assays)

System	Endpoint and methodology	DMN	DEN	References
E. coli	Mutation assays			
	FM and/or RM at the *gal*, *arg* and *MTR* loci; LT with mouse liver S-9; 180 min treatment	Concentration-dependent increase in mutation frequency; no mutagenic activity without activation; mutations presumed to be BP substitution	Concentration-dependent increase in mutation frequency; no mutational activity without activation; mutations presumed to be BP substitution	57,58
		LEC: 1×10^{-4} mole/1. = 7.4 μg/ml	LEC: 1×10^{-4} mole/1. = 10.2 μg/ml	
	RM to *str* prototrophy; LT with rat liver S-9; 30 min treatment	Concentration-dependent increase in mutation frequency no activity without activation; reduced effect with 105,000 g supernatant	Concentration-dependent increase in mutation frequency no activity without activation; reduced effect with 105,000 g supernatant	59,60
		LEC: 6×10^{-4} mole/1. = 44.4 μg/ml	LEC: 3×10^{-4} mole/1. = 30.6 μg/ml	

RM to trp^+ in strains with different DNA repair capacities; LT with rat liver S-9; 30 min treatment	Concentration-dependent increase in mutation frequency; similar mutability in both strains; no effect without activation LEC: 1.35×10^{-1} mole/l. = 10 mg/ml	---	61
Differential cell killing in DNA-repair-deficient strains			
LT with rat liver S-9	$polA^-$ more sensitive than wild type 0.1 μg/ml	$polA^-$ more sensitive than wild type 0.1 μg/ml	62
LT without activation	Concentration-dependent decrease of CFA; strain B_{s-1} more sensitive than B/r	---	63

(continued)

Table 5. (continued)

System	Endpoint and methodology	DMN	DEN	References
	Preferential growth-inhibition in DNA repair-deficient strains			
	Disc assay with rat or hamster liver S-9	recB recC strain +ve with 0.05 μg/disc; $polA^-$ +ve with 50 μl/disc in one study and -ve in another	recB recC strain +ve with 1 μg/disc; both +ve and -ve results with $polA^-$ (5-50 μl/disc)	64-67
B. subtilis	Mutation assay			
	Rm from ilv^- to ilv^+; LT with mouse liver microsomes; 15 min and 30 min treatment	Significant induction at 200 mM; 2-3-fold increase in mutation frequency with microsomes from "induced" (PCB) animals	---	68

	Preferential growth inhibition in DNA-repair deficient strains			
	A variation of the disc assay method with and without S-9 activation	Sensitivity of Rec$^-$ > Rec$^+$; LEC: 12500 μg/ml and 25000 μg/ml for the with activation, one-half of these values	Sensitivity of Rec$^-$ > Rec$^+$; LEC: 25000 μg/ml and 50000 μl/ml, respectively (without activation)	69,70
S. typhimurium	Mutation assays			
	RM from his$^-$; S-9 activation (origin different mammals or mammalian organs); in some cases, animals, pretreated with enzyme inducers	LT: 5-500 mM, +ve in hisG46 (7 studies), TA1530 (5 studies), TA92 (1 study). Mutation induction increased in presence of S-9 from "induced" animals.	LT: 15-200 mM, +ve in hisG46 (4 studies), -ve or weak (2 studies); also +ve in TA1530 (2.5-250 mM; 2 studies) and in TA100 (7-86 mM; 1 study); increased activity in presence of S-9 from "induced" animals.	71-78 (DEN)

(continued)

Table 5. (continued)

System	Endpoint and methodology	DMN	DEN	References
	Strain TA1535 and TA100 detect missense mutations; TA1538, TA98 frameshifts; TA92, his G46 and TA1530 detect mutagens that need an intact <u>uvrB</u> system to be detected.	In TA100, +ve results, with DMN twice as effective as DEN; mutagenic potency (RM/nmole) 0.02	In TA100, DEN is one-half as effective as DMN; mutagenic potency: 0.01	72,73, 75–77, 79–83 (DEN)
	Lt and plate incorporation assays	Plate tests: negative in TA1530 (1 study), TA100 (2 studies); +ve results in hisG46 with 100 μmoles/ml in top agar overlay, or 1000–10000 μg/plate	Plate tests: -ve in TA1530 (10–50 μmol/ml agar overlay; 1 study); +ve in hisG46 and TA1530 (940–9400 μg/plate); in both cases, S-9 from "induced" animals	74,77, 83,84 (DEN)
		LEC: ±1500 μg/plate	LEC: ±1000 μg/plate	

is positive in micronucleus tests and in spot-tests in mice (the latter, for the induction of somatic mutations). Furthermore, it is potent in inducing chromatid aberrations and sister chromatid exchanges both in vivo and in vitro when assayed on lymphocytes (the in vitro studies in presence of liver microsomes) and in one study, in rat bone marrow cells in vivo. In Chinese hamster ovary cells (CHO) in vitro, the concentration of DMN required for the production of significant increases in SCE frequencies is far lower than for the induction of chromatid aberrations.

DMN induces mutations in somatic cells in vitro where there is a linear increase in mutation frequency with concentration followed by a plateau. The absolute frequencies, however, vary, depending on the test system and the microsomal activation system used. Specific locus tests or translocation tests with germ cells have not yet been carried out.

In experiments involving unscheduled DNA synthesis (UDS) with human fibroblasts in vitro, DMN yielded positive results in presence of microsomes; likewise positive results were obtained in in vivo studies with mice where UDS was measured in epithelial cells, endothelial lung cells, liver parenchyma and some kidney cells. However, negative results were recorded for human lymphocytes in vitro (no activation system present) and for spermatids in mice (in vivo) when DMN was administered intraperitoneally and ^{3}H-T, intratesticularly. One possible reason for a lack of effect in the testes (spermatids) may be that the active metabolite(s) did not reach the target cells and germ cell dosimetry with labeled DMN can establish whether this is in fact true:

DEN registers as positive in a number of assays, but is also negative in a number of others. Thus DEN has yielded negative results (or at best indications for weak positive effects) in host-mediated assays with bacteria and fungi, recessive specific locus mutations in mouse male germ cells, dominant lethal tests in mice, micronucleus tests, UDS in mouse spermatids and for the induction of sister chromatid exchanges in some mammalian in vivo systems.

In Neurospora crassa (ad-3A and ad-3B loci), 99% of the DEN induced mutations studied are point mutations (and not deletions); complementation tests have demonstrated (148) that these are base-pair substitutions.

When both DMN and DEN show mutagenic activity in a given system, DMN is generally the more potent chemical. For instance, in Drosophila, both compounds induce recessive lethals and very low frequencies of heritable chromosome breakage effects (and the latter only at relatively high concentrations). However,

Table 6. Effects of DMN and DEN in fungal test systems (excluding host-mediated assays)

System	Endpoint and methodology	Main observations with		References
		DMN	DEN	
Yeast (*S. cerevisiae*)	Mutation assays			
	RM at the *ade-6* locus; 1 M, pH 7.5, 30 min treatment; no activation	No effect	---	85
	FM to *can* resistance; LT (100 μmole/ml) in presence of Udenfriend's hydroxylation system and O_2; 6 hr treatment	Approximate doubling of mutagenic frequency. No induction in presence of N_2	5-6-fold increase; no induction in presence of N_2	86
	Cytoplasmic petites; other details as above, but 1 to 5 hr treatment	Treatment-time-dependent increase	Treatment-time-dependent increase	86
	FM to *ca*^{n} resistance and RM from *met*$^{-}$; plate test, no activation	No effect	No effect	87

Mitotic crossing-over			
LT with strain D3; 0.1 M in Udenfriend's hydroxylation mixture and O_2, 6 hr treatment	Positive evidence for induction; LEC: 0.1 M	Positive evidence for induction; LEC: 0.1 M	88
LT with strains D3 and D5; 0.77-770 mM with post-mitochondrial supernatant from mouse liver, 3 hr treatment, oxygenated suspension	Less than linear increase with increasing concentration	---	89
LT with strain D3; 100 μmole/ml in Udenfriend's hydroxylation mixture in presence of O_2, 6 hr treatment	Positive evidence for induction; treatment-time-dependent increase	Positive evidence for induction; treatment-time-dependent increase; DEN > DMN (6-fold difference with 6 hr)	87
D5 strain; 0.2-0.47 M at 37°C in buffer of log phase cells under shaking and flushing with O_2	Positive evidence for induction	---	32

(continued)

Table 6. (continued)

System	Endpoint and methodology	Main observations with DMN	Main observations with DEN	References
	Mitotic gene conversion			
	trp-5 locus in strain D4; plate test without activation mixture	No effect	No effect	87
	LT with strain D4, trp-5, and ade-2 loci; 0.77-770 mM with post-mitochondrial supernatant from mouse liver; 3 hr treatment, oxygenated suspension	Positive; concentration-dependent (but less than linear) increase; overall curve "humped"; LEC > 0.77 mM	---	89
	trp-5 and ad-2 loci, strain D4; 0.29 and 0.47 M, 2-6 hr at 37°C in exponential growth (O_2 flushed)	Positive response increasing with time at both loci	---	32

N. crassa	Mutation assays				
		RM of *ad-3B*; conidial treatment with 0.1 mM/ml solution dissolved in Udenfriend's hydroxylation mixture in presence of O_2, up to 4 hr	Positive evidence for induction; no effect in presence of N_2	Positive evidence for induction; no effect in presence of N_2; DEN > DMN	90
		FM at *ad-3A* and *ad-3B*; conidial treatment (0.09-0.36 M), 2 hr in presence of mouse liver microsomes either from phenobarbitol-treated or untreated mice	Significant increase in mutation frequency. Microsomes from "induced" mice slightly more effective	---	91

Table 7. Mutation studies with bacteria in host-mediated assays:

Host and strain	Indicator	Administration route indicator	Indicator recovered from
Mouse; random-bred, Swiss albino	E. coli, K-12	IV (tail vein)	Liver and spleen
Mouse, NMRI	E. coli, K-12	IV (tail vein)	Liver and spleen
Mouse, NMRI	E. coli, K-12	IV (tail vein)	Liver and spleen
Mouse, COBS/CD	E. coli, K-12	IP	Peritoneum
Baboon (Papio-cyanocephalus)	E. coli, K-12	IP	Peritoneum
Mouse; random-bred, Swiss albino	S. typhimurium G46, c340, c207	IP	Peritoneum
Mouse, NMRI	S. typhimurium G46, c340, c207	IP	Peritoneum
Mouse, NMRI and C3H × 101 F_1 hybrids	S. typhimurium G46, c340, c207	IP	Peritoneum
Mouse; NMRI ♂♂	S. typhimurium G46, c340, c207	IP	Peritoneum
Mouse; NMRI ♂♂	S. typhimurium TA1950	IP	Peritoneum
Mouse; C3H × 101 F_1 ♂♂	S. typhimurium G46	IP	Peritoneum

DMN

Administration route compound	Incubation time in host (min)	Endpoint studied[a]	LED or LTD (mg/kg)[b]	References
SC	Up to 210	FM, $MTR^s \rightarrow MTR^r$	LED: 30 LTD: 30	92
SC	15	FM, $MTR^s \rightarrow MTR^r$	LTD: 0.1 ml of 1% solution	57
SC	15	RM, $arg^- \rightarrow arg^+$	LTD: 0.1 ml of 1% solution	57
Oral	240	RM, $lac^- \rightarrow arg^+$	LTD: 100	93
Oral	240	RM, $lac^- \rightarrow arg^+$	LTD: 100	93
IM	150	RM: BS, FS	LTD: 3×0.1 ml of 1% solution for G46	56
Oral	180	RM: BS	LTD: 100	94
Oral and SC	180	RM: BS	LTD: 50	95
Oral, SC, IM	180	RM: BS	LTD: 182	96
Oral	300	RM: BS	LED: 5.5	96
IM	180	RM: BS	LTD: 25	72

(continued)

Table 7. (continued)

Host and strain	Indicator	Administration route indicator	Indicator recovered from
Mouse; ICR ♂♂	S. typhimurium G46	IP	Peritoneum
Mouse; ICR ♂♂	S. typhimurium G46	IP	Peritoneum
Mouse	S. typhimurium G46	IP	Peritoneum
Mouse	S. typhimurium G46	IP	Peritoneum
Mouse	S. typhimurium G46	IP	Peritoneum
Mouse	S. typhimurium G46	IP	Peritoneum
Rat, Sprague-Dawley	S. typhimurium G46	IP	Peritoneum
Rat	S. typhimurium G46	IP	Peritoneum
Mouse	S. marcescens	IP	Peritoneum

[a]FM = forward mutations; RM = reverse mutations; BS = base-pair substitutions; FS = frameshift.

[b]LED = lowest effective dose; LTD = lowest tested dose with positive results.

Administration route compound	Incubation time in host (min)	Endpoint studied[a]	LED or LTD (mg/kg)[b]	References
IM	180	RM: BS	LTD: 16.5	97
IM	180	RM: BS	LTD: 165	98
IM	5	RM: BS	LTD: 20	99
IM		RM: BS	LTD: 25	81
SC	180	RM: BS	LED: ±20	100
Oral		RM: BS	LED: ±100	100
IM	180	RM: BS	LTD: 25	72
IM		RM: BS	LTD: 100	81
SC	180	RM: BS	LTD: 3 × 100	100

Table 8. Mutation studies with fungi and mammalian cells in

Host and strain	Indicator	Administration route indicator	Indicator recovered from
Mouse; Swiss albino[b]	S. cerevisiae D3	IP	Peritoneum
Mouse; C3H[b]	S. cerevisiae D4	IP	Peritoneum
Mouse; C3H[b]	S. cerevisiae D4	IP	Peritoneum
Rat; BD ♂♂[b]	S. cerevisiae D4	IP	Peritoneum
Rat; BD ♂♂[b]	S. cerevisiae D4	Testes	Testes
Rat; Wistar ♂[c]	S. cerevisiae D4	IV (tail vein)	Liver
Rat; Wistar ♂[c]	S. cerevisiae D4	IV (tail vein)	Lung
Rat; Wistar ♂[c]	S. cerevisiae D4	Testes	Testes
Mouse and rat	S. pombe, P1	IP	Peritoneum
Mouse	N. crassa	IV (tail vein)	Liver
Mouse; BD2F$_1$	L5178Y (mouse lymphoma cells)	IP	Peritoneum
Mouse; BALB/c (athymic and their normal litter-mates)	CHO cells	IP	Peritoneum

[a]FM = forward mutations; RM = reverse mutations.

[b]Effectivity in this study: rat peritoneum > mouse peritoneum > rat testes

[c]Effectivity in this study: liver > lung > testes.

host-mediated assays: DMN

Administration route compound	Incubation time in host (min)	Endpoint studied[a]	LTD (mg/kg)	References
SC	180	Mitotic crossing-over	1300	87
SC	240	Mitotic gene conversion	740	101
SC	240	Mitotic gene conversion	740	102
SC	240	Mitotic gene conversion	740	102
SC	240	Mitotic gene conversion	740	102
SC	240	Mitotic gene conversion	740	103
SC	240	Mitotic gene conversion	740	103
SC	240	Mitotic gene conversion	740	103
Oral	180	FM at 5 ade loci	12.5	104
IM		FM at ad-3A and ad-3B loci	75	81
SC	360	RM; $asn^- \to asn^+$	2.5	105
SC	120	FM: to $HG\text{-}PRT^-$	33	106

Table 9. Mutation studies with bacteria and fungi in host-mediated

Host and strain	Indicator	Administration route indicator	Indicator recovered from
Mouse; ICR ♂♂	S. typhimurium G46	IP	Peritoneum
Mouse; C57Bl × C3H, F_1 ♂♂	S. typhimurium G46	IP	Peritoneum
Mouse; NMRI ♂♂	S. typhimurium G46	IP	Peritoneum
Mouse; NMRI ♂♂	S. typhimurium G46	IP	Peritoneum
Mouse	S. typhimurium	IP	Peritoneum
Mouse; NMRI	S. typhimurium TA1950	IP	Peritoneum
Rat; Sprague-Dawley	S. typhimurium G46	IP	Peritoneum
Rat; Sprague-Dawley	S. typhimurium G46	IP	Peritoneum
Mouse; NMRI ♂♂	E. coli 343	IP	Peritoneum
Mouse; NMRI ♀♀	E. coli 343/113	IV (tail vein)	Liver
Mouse; NMRI ♀♀	E. coli 343/113	IV (tail vein)	Liver
Mouse; NMRI ♂♂	S. marcescens; a13 his^-	IP	Peritoneum

assays: DEN

Administration route compound	Incubation time in host (min)	Genetic endpoint[a]	LED (mg/kg)	References
IM	180	RM: BS	>500	98
IM	180	RM: BS	±400	72
SC	Up to 120	RM: BS	>500	100
SC, IM, oral		RM: BS	>250	107
IM		RM: BS	±400	81
SC,IM		RM: BS	∿250	107
IM		RM: BS	>400	72
IM		RM: BS	>400	81
SC	Up to 120	FM at two loci	>500	100
SC		RM	± 66.6	58
SC		FM	> 66.6	58
SC	Up to 120	RM	>500	100

(continued)

Table 9. (continued)

Host and strain	Indicator	Administration route indicator	Indicator recovered from
Mouse; NMRI ♂♂	S. marcescens; a21 leu$^-$	IP	Peritoneum
Mouse; Swiss albino	S. cerevisiae D3	IP	Peritoneum
Mouse	N. crassa	IP	Peritoneum
Mouse	N. crassa	IV (tail vein)	Liver

[a]FM = forward mutations; BS = base-pair substitution; RM = RM = reverse mutations.

the ratio of recessive lethals to translocations, is of the order of 300:1 in the case of DEN whereas it is of the order of 15:1 in the case of DMN. Similar comparisons in mammals are hampered by the lack of adequate mutation data.

PUBLIC HEALTH IMPLICATIONS

It is now well established that small amounts of several N-nitroso compounds (including DMN and DEN) are present in the human environment and that these may be formed in the body from nitrosatable substances and nitrites. Nitrite in low concentrations is a normal constituent of food as well as of saliva which contains 1-10 mg/l.; nitrite in the environment is mostly formed by bacteria via a reduction of nitrate which is abundantly present in water and plants. The levels of nitrosamines found in different types of food and food products are variable and in food products processed with nitrite, relatively high concentrations of nitrosamines may be found (3,12,149). In contaminated meat products such as ham, bacon, salami, and frankfurters, the amounts of DMN or nitrospyrrolidine were found to vary between 1 and 10 ppb. It would thus appear that the predominant mode of exposure of humans to nitrosamines is most likely to result from food contaminated from such compounds formed from often naturally occurring precursors in the stomach (149). As Montesano and Bartsch (12) stressed, "the N-nitroso compounds comprise one of the most potent

Administration route compound	Incubation time in host (min)	Genetic endpoint[a]	LED (mg/kg)	References
SC	Up to 120	RM	>500	100
SC		Mitotic cross-crossing-over	±1900	87
IM		FM at ad-3A and ad-3B	>200	81
IP		FM at ad-3A and ad-3B	±150	81

groups of carcinogens, most of them inducing cancer after a single dose and in a large variety of animal species" and that "man is probably not resistant to the mutagenic and carcinogenic effects of N-nitroso compounds." Epidemiological studies (3,12) in which the possible role of N-nitrosamine intake in the etiology of a number of human cancers has been examined, have provided no unequivocal answers. Since estimating the extent of exposures of humans to different kinds of nitrosamines is difficult, on the basis of experimental evidence, it would seem prudent to abstain from and/or limit consumption of food and foodstuffs known to contain these substances to amounts practically achievable.

Table 10. Mutagenicity of DMN and DEN in *Drosophila melanogaster*[a]

Compound and concentration	Mode of administration	Germ cell stage(s)	Endpoint	Effect observed	Other comments	References
DMN; 1.0-10 mM in 0.4% saline	Adult injection	Spermatozoa to spermatogonia	Sex-linked recessive lethals and visibles	Mutagenic in all stages; concentration-dependent, but less than linear increase in mutation frequency; yield low in spermatozoa and high in meiotic and premeiotic stages	Brood-pattern too rough to permit definition of stages of peak sensitivity	108, 109
DMN; 5.0 mM in 0.4% saline	Adult injection	Spermatozoa to spermatogonia	Sex-linked recessive lethals	Mutagenic in all stages	Brood-pattern too rough to permit definition of stages of peak sensitivity	110

DMN; 6.8 × 10^{-2} and 6.8 × 10^{-1} mM	3-day adult feeding	Spermatozoa to spermatogonia	Sex-linked recessive lethals	Mutagenic in all stages; concentration-dependent, but less than linear increase in mutation frequency; spermatocytes probably most sensitive	Adult feeding more effective than injection	111
DMN; 1.0 mM	2-day adult feeding	Spermatozoa to spermatocytes	Sex-linked recessive lethals	8.6% recessive lethals (mean value for all stages); spermatids most sensitive for mutation induction	Adult feeding more effective than injection	112
DMN; 10 mM	27-hr adult feeding	Spermatozoa	Sex-linked recessive	18.9% recessive lethals	Adult feeding more effective than injection	112
			Loss of ring-X chromosome	1.7% (vs. 1.0% in controls)		112

(continued)

Table 10. (continued)

Compound and concen-tration	Mode of adminis-tration	Germ cell stage(s)	Endpoint	Effect observed	Other comments	Refer-ences
DEN; 7.0–30 mM in 0.4% saline	Adult injection	Spermatozoa to spermatogonia	Sex-linked recessive lethals	Mutagenic in all stages; stage of peak sensitivity varied in the different experiments	Brood-pattern not precise; DEN < DMN	113, 114
DEN; 1.0 mM	3-day adult feeding	Spermatozoa to spermatogonia	Sex-linked recessive lethals	Mutagenic in all stages; stage of peak sensitivity varied in the different experiments	Brood-pattern not precise	115, 116
DEN; 1.0×10^{-2}–5 mM	1-day adult feeding	Spermatozoa in some expts and all stages in others	Sex-linked recessive lethals	All concentrations except the lowest mutagenic; LEC: $\sim 1 \times 10^{-1}$ mM. Concentration-dependent increase in mutation frequency; spermatids most sensitive for mutation induction; spermatocytes killed at moderate and high concentrations	LC_{50} for killing of treated males: 5 mM DEN for 3 days $LC_{90} \cong 10$ mM	117

DEN; 5×10^{-2}–5×10^{-1} mM (4 levels)	2-day adult feeding	Spermatozoa in some experiments and all stages in others	Sex-linked recessive lethals	2×10^{-1} and 5×10^{-1} mM clearly mutagenic; spermatids most sensitive	118
DEN; 5 and 10 mM	3-day adult feeding	Spermatozoa to spermatocytes at 5 mM and spermatozoa at 10 mM	Sex-linked recessive lethals	Highly mutagenic; spermatids most sensitive	117
DEN; 1 mM	Larval feeding for 6 days	Spermatocytes and spermatogonia	Sex-linked recessive lethals	(8.7±4.0)% recessive lethals	118
DEN; 5×10^{-3}–5×10^{-1} mM (5 levels)	Larval feeding for 6 days + adult feeding for 2-3 days	Predominantly premeiotic and meiotic stages and some exposure to other stages (in adults)	Sex-linked recessive lethals	All concentrations except 5×10^{-3} and 1×10^{-2} mM clearly mutagenic	118

[a] General remarks: Both compounds are clearly mutagenic in Drosophila although DMN is more effective. The LEC for DMN is between 1×10^{-2} and 5×10^{-3} mM (3-day adult feeding); for DEN (2-day adult feeding, it is 2×10^{-1} mM; for the latter compound, for larval feeding, the LEC is considerably lower, i.e., 5×10^{-3} mM or thereabouts. At biologically usable and meaningful concentrations, no chromosome breakage effects. With DEN at LC_{60} and LC_{90} for killing of treated males, translocations can in fact be induced but at rather low frequencies. Both compounds induce point mutations.

Table 11. Induction of specific locus mutations (male germ cells), somatic mutations (spot tests; mutations in melanocyte precursors), and dominant lethals in mice by DMN and DEN

Strain and sex	Chemical	Dose (mg/kg)	Mode of treatment	Main observations[a]	References
			Specific Locus Mutations		
(101 × C3H)F_1 ♂♂	DEN	100 and 150	Single IP injection	No mutations in 6529 offspring from treated post-spermatogonial stages; 2 mutations in 46,9191 progeny from treated spermatogonia. The latter frequency is slightly below but not significantly different from controls.	119
			Somatic Mutations (Spot Test)		
Embryos of the cross T × HT	DMN	In buffered saline; 5	Single IP injection into mother on days 7, 8, 9, 10 of fetal development	2% of color spots of genetic relevance (control, 0%)	120
Embryos					

Embryos of the cross C57Bl × T	DMN	In buffered saline, 7.5	Same as above, IP injection on day 10 of fetal development	1-2% color spots of genetic relevance (control, 0%)	121
Embryos of the cross T × HT	DEN	In buffered saline; 5, 10, 15, and 50	Same as above; IP injection days 7-10 or 7, 8, 9, and 10 of fetal development	All treatments +ve; 3.8% color spots of genetic relevance after 10 mg/kg treatment on day 8; control frequency 0%	120
Embryos of the cross C57Bl × T	DEN	In buffered saline; 20, 30	Same as above; injection on day 11 of fetal development	1.5 to 1.6% color spots of genetic relevance at 30 and 50 mg/kg doses respectively; no induction at 20 mg/kg; control frequency 0%	122
Embryos of the cross C57Bl × T	DEN	In buffered saline; 10 and 20	Same as above; injection on day 10 of fetal development	No evidence for induction of color spots of genetic relevance	121

(continued)

Table 11. (continued)

Strain and sex	Chemical	Dose (mg/kg)	Mode of treatment	Main observations[a]	References
			Dominant Lethals		
ICR/HA ♂♂	DMN	In water; 8 or 9	Single IP injection	No evidence for dominant lethal induction; one ♂ dosed with 9 mg/kg died	123
(C3H × 101)F_1 ♂♂	DMN	In saline; 4.4	Single SC injection	Weak effect (reduced fetal implants and an increase in early fetal deaths)	100
(C3H × 101)F_1 ♂♂	DEN	In saline; 13.5	Single IP injection	Nonsignificant increase in both pre- and postimplantation losses	100

[a] +ve = postive; -ve = negative.

Table 12. Induction of mutations by DMN and DEN in mammalisn somatic cells in vitro[a]

Type	Selective agent concn (μg/ml)	Metabolic activation	Expression time (hr)	Mutagen concn	Exposure time (min)	Main observations	References
			DMN; L5178Y; FM to TK^-				
BUdR	50	Liver microsomes ($CaCl_2$-aggregated) from noninduced mice; 1.6 mg/ml NADPH in activation mixture	48-60	7.4-74 μg/ml (= 0.1 to 1 mM) 5 levels	15	Approx. linear increase in mutation frequency with conc· MP ≅ 85; LD_{50} ± 18 μg/ml; at LD_{90}, the number of mutants in is $1600/10^6$ survivors; LEC ± 5 μg/ml	124
BUdR	50	Rat liver S-9 from noninduced animals	72	1-10 μg/ml (0.014-0.14 mM) 3 levels	240	Linear increase in mutation frequency with concentration; MP ≅ 33; LD_{50} ± 3 μg/ml; at LD_{90}, the number of induced mutants is about $1300/10^6$ survivors; LEC ≅ 0.5 μg/ml	125

(continued)

Table 12. (continued)

Type	Selective agent concn (μg/ml)	Metabolic activation	Expression time (hr)	Mutagen concn	Exposure time (min)	Main observations	References
BUdR	50	Rat liver S-9 from Aroclor-induced animals	72	9.2-74 μg/ml (0.13-1 mM); 4 levels	240	Linear increase in mutation frequency with concentration; MP ≅ 3; LD_{90}, the number of induced mutants is $600/10^6$ survivors; LEC ± 4 μg/ml	
			DMN; L5178Y; FM to $HG\text{-}PRT^-$				
6-TG	5	Rat liver S-9 from Aroclor-induced animals	168	37 g/ml to 1.85 mg/ml (0.5-25 mM); 4 levels	60	Concentration-dependent linear increase in mutation frequency followed by a plateau; mutation induction/M ranged from 14×10^{-3} (0.5 mM) to 2×10^{-3} (25 mM); MP (from linear part of curve) = 14; LD_{50} ≅ 15 mM	126

DMN; V79 Chinese Hamster Cells; FM to HG-PRT$^-$							
6-TG	5	Mouse liver microsomes, 10,000g, 100,000g, and $CaCl_2$-aggregated	90, 114, 118	14.8 and 37 mg/ml (200 and 500 mM)	60	Concentration-dependent linear increase in mutation frequency; ∿5.5 × 10^{-3} (10,000g), 3.8 × 10^{-3} (100,000g) mutations/M ∿ and 1.6 × 10^{-3} ($CaCl_2$-aggregated); approx. MP's for the above three conditions: 0.074, 0.05, and 0.02	127
DEN; L5178Y; FM to HG-PRT$^-$							
6-TG	5	Rat liver microsomes from Aroclor-induced animals	168	100 μg/ml to 5 mg/ml (1-50 mM); 4 levels	60	Concentration-dependent increase in mutation frequency followed by a plateau; mutation induction/M range	126

(continued)

Table 12. (continued)

Type	Selective agent concn (μg/ml)	Metabolic activation	Expression time (hr)	Mutagen concn	Exposure time (min)	Main observations	References
						from 28 × 10^{-3} (1 mM) to 2 × 10^{-3} (50 mM); MP (from linear part of the curve) = 28; nontoxic	
6-TG	5	Rat liver S-9 from induced animals	144	50-2500 μg/ml (0.5 to 25 mM); 6 levels	60	Same as above; MP decreased from 0.49 (50 μg/ml) to 0.23 (100 μg/ml) and 50 0.04 (2500 μg/ml). LEC (interpolated value): 25 μg/ml	125
DEN; V79 Chinese Hamster Cells; FM to HG-PRT$^-$							
6-TG	5	Rat liver S-9 from Aroclor-induced animals	192	250 μg/ml to 5 mg/ml (1-50 mM); 4 levels	60	Same as above; mutation induction/M is 6 × 10^{-3} (2.5, 5, and 10 mM) and 2 × 10^{-3} (50 mM); MP (from linear part of the curve) = 6; nontoxic	126

			DEN; L5178Y; FM to TK^-				
BUdR	50	Rat liver S-9 from Aroclor-induced animals	96-168	100 μg/ml to 5 mg/ml (1-50 mM)	60	Same as above; mutation induction/M decreased from about 56×10^{-3} (1 mM) to 4×10^{-3} (50 mM); MP = 7; not toxic; same cell population used for $HG\text{-}PRT^-$ mutations	126
BUdR	50	Rat liver S-9 from induced animals	72	50-2500 μg/ml (0.5-25 mM); 6 levels	240	Same as above; the MP decreased from 0.43 (50 μg/ml) and 0.36 (100 μg/ml) to 0.04 (2500 μg/ml); saturation effect above 500 μg/ml; LEC 50 μg/ml	

[a]General remarks: In comparing Mutagenic Potencies (MP: no. of induced mutants/10^6 survivors/μg/ml/hr of treatment) it should be remembered that (i) the dependence of mutation frequency on duration of treatment has not been established, but assumed; and (ii) differences in MP between different test systems are, to a considerable extent, related to differences in metabolic activation under the different conditions employed. The lowest effective concentration (LEC) given are valid only for the particular conditions obtained in a given study and consequently, vertical comparisons in the table (penultimate column) are of very limited value.

Table 13. Induction of chromosome aberrations in mammalian somatic cells (in vivo and in vitro) by DMN and DEN

Compound	Cell type	Concentration and treatment conditions	Recovery time (hr)	Main observations	References
DMN	Rat lymphocytes (Wistar rat)	30 mg/kg, IP, blood drawn after 6 hr and used for culturing	48	3-fold increase in chromatid aberrations relative to controls	128
DMN	Rat lymphocytes (Wistar rats)	0.09 mM in vitro treatment		No effect	
DMN	Human lymphocytes	50 mM; in vitro treatment in presence of mouse liver microsomes from induced animals; 45 min treatment	24	2-fold increase in chromatid breaks and high frequency of dicentrics	129

DMN	CHO cells	8-135 mM in vitro treatment for 60 min in presence of rat liver S-9 from induced animals	16	Concentration-dependent increase in chromatid exchanges and breaks; nonlinear response; no effect in the absence of S-9	
DMN	CHO cells	0.135-13.5 mM in vitro treatment, 60 min in presence of rat liver S-9 from induced animals	16	Significant increase in aberrations at 1.35 mM; caffein posttreatment decreases the yield of chromatid breaks but increases that of exchanges	66
DMN	Chinese hamster liver cells	In vivo treatment, 5 mg/g, IP, once a week for 6 weeks	Preparations after 40 or 150 days	Chromosome type aberrations; 0.52 aberr/cell (40 days) and 0.29 aberr/cell (150 days); 87% of cells, no aberrations and a few cells with 10 aberrations	130
DEN	CHO cells	4-200 mM in vitro treatment, 60 min, in presence of rat liver S-9 from induced animals	16	Concentration-dependent increase in the frequency of chromatid aberrations and breaks; significant at 4 mM; DEN > DMN	131

(continued)

Table 13. (continued)

Compound	Cell type	Concentration and treatment conditions	Recovery time (hr)	Main observations	References
DEN	Rat liver cells	In vivo treatment; 50 ppm in drinking water for 3, 4, 5 weeks	Preparations after 3, 4, or 8 weeks	Numerical changes and abnormal karyotypes; ploidy changes, 5% of cells directly after treatment, increasing to 22% after 4 weeks, falling off to 9% after 20 weeks	132
DEN	Guinea pig; induced hepatomas	In vivo treatment; 0.042 g/l. in drinking water, 6 days/week for 7-11 weeks	Preparations after 3, 4, or 8 weeks	Chromosomal instabilities in tumor transplants; 4/6 transplanted tumor lines showed 10-30% aberrations, 4-40%, polyploid cells and all had marker chromosomes	133

Table 14. Induction of sister chromatid exchanges in mammalian cells in vivo and in vitro by DMN and DEN

Compound	Cell type	S-9	Concn range	Exposure time (hr)	Main results	Other remarks	References
DMN	Rat lymphocytes, in vivo	-	60-240 mg/kg; 3 levels	6	2-3-fold increase over controls except at 60 mg/kg	SCE's scored in 56 hr cultures	134
DMN	Rat lymphocytes, in vivo	-	300 mg/kg	6	No increase	SCE's scored in 56 hr cultures	134
DMN	Mouse bone marrow cells, in vivo	-	200 μg/kg and 2 mg/kg	28	1.9 to 2.9-fold increase over controls; LEC: 200 μg/kg	Chemical given in 2 equal half doses; 1st injection was 2 hr after BrdUrd injection; 2nd, 6 hr after BrdUrd	135
DEN	Mouse bone marrow, in vivo	-	50 and 200 mg/kg	28	No increase	Chemical given in 2 equal half doses; 1st injection was 2 hr after BrdUrd injection; 2nd, 6 hr after BrdUrd	

(continued)

Table 14. (continued)

Compound	Cell type	S-9	Concn range	Exposure time (hr)	Main results	Other remarks	References
DEN	Chinese hamster bone marrow cells, in vivo	-	25-200 mg/kg (4 levels)	24	No increase	Chemical given in a single IP dose along with the 1st injection of BrdUrd	136
DMN	CHO cells, in vitro	+	2.7-40 mM (3 levels)	1	8-fold increase at the lowest concentration (>85 SCE's/cell); accurate counting was difficult	DMN is much more effective	131
DMN	CHO cells, in vitro	-	135 mM	1	No increase		131
DEN	CHO cells, in vitro	+	4-100 mM; 3 levels	1	About a 2-fold increase with 4 mM and a 3-fold increase at higher concentrations		131
DEN	CHO cells, in vitro	-	200 mM	1	No increase		131
DMN	CHO cells, in vitro	+	0.135-1.35 mM; 3 levels	1	2-10-fold increase; significant already at the lowest concentration		66

Table 15. Effects of DMN and DEN in micronucleus tests

Chemical	Organism	Treatment conditions and test protocol	Main observations	References
DMN	Mouse	18.5-74 mg/kg saline solution, IP in two doses with a 24 hr interval; bone-marrow preparations after 30 hr using Schmid's protocol and screened for micronucleated polychromatic erythrocytes among polychromatic erythrocytes	1-1.5% micronucleated erythrocytes at dose levels of 37-74 mg/kg; significantly higher than in controls	137
DEN	Mouse	Treatment conditions and test protocol same as above but dose range, 25-200 mg/kg	No effect	137

Table 16. Induction of unscheduled DNA synthesis by DMN and DEN in mammalian systems

Chemical	Test system	Exposure and methodology	Main observations	References
DMN	Human lymphocytes in Eagle's modified media	1 hr; 0.001-10 mM; cells cultured in 12 more hr in media with ^{3}H dT	No unscheduled DNA synthesis	138
DMN	Human fibroblasts in Eagle's modified media	1 hr 0.5 to 100 mM; cultured subsequently (time not given) in presence of ^{3}H dT; with and without mammalian microsomes; autoradiography	All exposures with microsomal activation positive; LEC: 0.5 mM	139
DMN	Human fibroblasts in arginine-deficient media supplemented with bovine serum	1 hr; ∿1 to ∿10 mM; cells cultured then for 1.5 hr in media with ^{3}H dT; with and without microsomal activation; autoradiography	All exposures with microsomal activation positive; LEC: ∿1 mM	

DMN	C3H mice	150 mg/kg (∿1.8 mM), SC injection in DMSO/ethanol, 2 hr after injection tissues removed and cultured in media with ^{3}H dT for 3 hr; autoradiography of tissue sections	Positive results in epithelial cells and endothelial lung cells, liver parenchyma and some kidney cells; LEC: 150 mg/kg	140
DMN	(C3H × 101)F_1 ♂♂	8 mg/kg (∿0.1 mM) IP, testicular injection of ^{3}H dT; purified sperm heads derived from early spermatids assayed for unscheduled DNA synthesis by liquid scintillation counting	No unscheduled DNA synthesis in early spermatids; apparently no activated chemical reaches testes	141
DEN	(C3H × 101)F_1 ♂♂	Same as above except the dose was 125 mg/kg (∿1 mM)	No unscheduled DNA synthesis in early spermatids; apparently no activated chemical reaches testes	141

Table 17. Effects of DMN and DEN in plant systems

Chemical	System	Treatment conditions	Scored for	Main observations	References
DMN	V. faba lateral roots	45 min with 2×10^{-3} M and 120 min with 10^{-3} M, pH 7.2, 13-20°C followed by a recovery at 19°C for 26 hr without the chemical	Chromosome aberrations	No effect	142, 143
DEN	V. faba lateral roots	Same as above	Chromosome aberrations	No effect	142, 143
DMN	G. max; L65-1237 variety; seed treatment	Up to 24 hr with 12.5-50 ppm (in aqueous solution; 3 levels) at room temp; seeds washed and sown	Single and twin spots in leaves which are light green; single spots indicative of chromosome breakage, aberrations; twin spots: somatic crossing-over	5-10-fold increase in spot frequency absolute frequency of twin spots lower than that of singles	144

DMN	*G. max*; L65-1237 variety; seed treatment	Same as above; concn range: 60-240 ppm and up to 48 hr treatment	Single and twin spots in light green leaves; single spots indicative of chromosome aberrations; twin spots; somatic crossing-over	±20-fold increase at the lowest concentration and no further increase; LEC may be around 60 ppm with 48 hr treatment	144
DMN	Barley; Cultivar Ekonom variety; dry seed treatment	24 hr treatment with 50-1000 mM aqueous solution at 24-25°C; seeds washed and sown to raise M_1; seeds from M_1 sown to raise M_2 seedlings	Chlorophyll-deficient mutations	No significant effect; 1000 mM lethal	145

(continued)

Table 17. (continued)

Chemical	System	Treatment conditions	Scored for	Main observations	References
DMN	*A. thaliana*; dry seed treatment	3 hr (50-1600 mM) or 24 hr (50-1000 mM) treatment; seeds washed and sown; M_2 raised from M_1 seeds	Recessive lethals that act in the embryo and those producing chlorophyll-deficient mutants	Concentration-dependent increase in mutation frequency (up to 600 mM, 3 hr treatment; 400 mM, 24 hr) followed by a falling-off	146
DEN	*A. thaliana*; dry seed treatment	Same as above but 24 hr treatment with up to 225 mM	Recessive lethals and those producing chlorophyll-deficient mutants	No significant induction	147

ACKNOWLEDGMENTS

I wish to thank the authors of all the Group I papers for providing the information base for this review. In addition, I am grateful to the members of my Working Group (Drs. R. Fahrig, D. Frezza, G. Hoffmann, W. L. Russell, G. Sega, J.W.I.M. Simons, E. Vogel, D. Wild, F. Würgler) and others (Drs. C. S. Aaron and G. Mohn) for their constructive comments on the first draft of this paper, to Professor F. H. Sobels for his enthusiastic encouragement and support and to Drs. Montesano, B. Singer and Elsevier Publishing Company for permission to reproduce the figures, to Dr. A. W. Hsie for permission to quote his paper in press and to all those colleagues who made available unpublished results. The writing of this paper was supported in part by the U.S. Public Health Service (DHEW) Grant ESO 1027-03 to the University of Leiden.

REFERENCES

1. Barnes, J. M., and Magee, P. N. Brit. J. Ind. Med. 11: 167-174 (1954).

2. Magee, P. N., and Barnes, J. M. Brit. J. Cancer 10: 114-122 (1956).

3. Bogovski, P., Preussmann, R., and Walker, E. A. N-Nitroso compounds: Analysis and Formation. International Agency for Research on Cancer, IARC Scientific Publ. 9, Lyon, 1974.

4. Druckrey, H., Preussmann, R., Ivanovic, S., et al. Z. Krebsforsch. 69: 103-201 (1967).

5. Heidelberger, C. Ann. Rev. Biochem. 44: 79-121 (1975).

6. Lijinsky, W. In: Chemical Mutagens: Principles and Methods for Their Detection, Vol. 4, A. Hollaender, Ed. Plenum Press, New York, 1976, pp. 193-215.

7. Lijinsky, W., and Singer, G. M. In: N-Nitroso Compounds in the Environment, P. Bogovski and E. A. Walker, Eds., IARC Scientific Publ. No. 9, Lyon, 1974, pp. 111-000.

8. Magee, P. N. In: In Vitro Metabolic Activation in Mutagenesis Testing, F. J. de Serres et al., Eds. North-Holland Publishers, Amsterdam, 1976, pp. 213-216.

9. Magee, P. N., and Barnes, J. M. Adv. Cancer Res. 10: 163-246 (1967).

10. Mirvish, S. S. In: Topics in Chemical Carcinogenesis, W. Nakahara et al., Eds., Univ. Tokyo Press, Tokyo, 1972, pp. 279-295.

11. Mirvish, S. S. J. Toxicol. Environ. Health 2: 1267-1277 (1977).

12. Montesano, R., and Bartsch, H. Mutat. Res. 32: 179-228 (1976).

13. Neale, S. Mutat. Res. 32: 229-266 (1976).

14. Singer, B. Trends Biochem. Sci. 2: 180-183 (1977).

15. Singer, B. J. Toxicol. Environ. Health 2: 1279-1295 (1977).

16. Druckrey, H., Steinhoff, D. et al. Arzneimittelforsch. 13: 320-323 (1963).

17. Lijinksy, W., Conrad, E., and Van de Bogart, R. Nature 239: 165-176 (1972).

18. Lijinsky, W., Keefer, L., Conrad, E., and Van de Bogart, R. J. Natl. Cancer Inst. 49: 1239-1249 (1972).

19. Magee, P. N., Montesano, R., and Preussmann, R. In: Chemical Carcinogens, C. Searle, Ed. American Chemical Society Monograph 173, Washington, D.C., 1976.

20. Lijinsky, W., and Epstein, S. Nature 225: 21-23 (1970).

21. IARC Monographs on the Evaluation of Carcinogenic Risk of Chemicals to Man, Vol. 1, IARC, Lyon, 1972.

22. Montesano, R., and Magee, P. N. In: Chemical Carcinogenesis Assays, R. Montesano and L. Tomatis, Eds., IARC Scientific Publ. 10, Lyon, 1974, pp. 39-56.

23. Knecht, M. Z. Krebsforsch. 63: 293-296 (1967).

24. Heath, D. F. Biochem. J. 85: 72-91 (1962).

25. Swann, P. F., and Magee, P. N. Biochem. J. 110: 39-47 (1968).

26. Swann, P. F., and Magee, P. N. Biochem. J. 125: 841-847 (1971).

27. Magee, P. N., and Vandekar, M. Biochem. J. 70: 600-605 (1958).

28. Montesano, R., and Magee, P. N. Nature 228: 173 (1970).

29. Brouwers, J. A. J., and Emmelot, P. Exptl. Cell Res. 19: 467 (1960).

30. Wilkinson, C. F., and Brattsen, L. B. Drug Metab. Rev. 1(2): 153-228 (1972).

31. Baars, A. J., Zijlstra, J. A., Vogel, E., and Breimer, D. Mutat. Res. 44: 257-268 (1977).

32. Callen, D. F., and Philpot, R. M. Mutat. Res. 45: 309-324 (1977).

33. Magee, P. N., and Farber, E. Biochem. J. 83: 114-124 (1962).

34. Lijinsky, W., Loo, J., and Ross, A. E. Nature 218: 1174-1175 (1968).

35. Ross, A. E., Keefer, L. K., and Lijinsky, W. J. Natl. Cancer Inst. 47: 789-795 (1971).

36. Druckrey, H., Schildbach, A., Schmähl, D. et al. Arzneimittelforsch. 13: 841 (1963).

37. Krüger, F. W. Z. Krebsforsch. 79: 90-97 (1973).

38. Blattman, L., and Pressmann, R. Z. Krebsforsch. 79: 3-5 (1973).

39. Okada, M., and Suzuki, E. Gann 63: 391-392 (1972).

40. Druckrey, H. Chemische Struktur und Reaktionsmechanismen Kreserzeugender Substanzen. Umsch. Wiss. Tech., 1972, p. 94.

41. Lawley, P. D., and Thatcher, C. J. Biochem. J. 116: 693-707 (1970).

42. Lawley, P. D. Mutat. Res. 23: 283-295 (1974).

43. Goth, R., and Rajewsky, M. F. Z. Krebsforsch. 82: 37-64 (1974).

44. Nicoll, J. W., Swann, P. F., and Pegg, A. E. Nature 254: 261-262 (1975).

45. Goth, R., and Rajewsky, M. F. Proc. Natl. Acad. Sci. (U.S.A.) 71: 639-643 (1974).

46. Kleihues, P., and Margison, G. P. J. Nat. Cancer Inst. 53: 1839-1841 (1974).

47. Margison, G. P., and Kleihues, P. Biochem. J. 148: 521-525 (1975).

48. Gerchman, L. L., and Ludlum, D. B. Biochem. Biophys. Acta 308: 310-316 (1973).

49. Lake, B. G., Minski, M. J. et al. Life Sci. 17: 1599-1606 (1976).

50. Roller, P. P., Shimp, D. R., and Keefer, L. K. Tetrahedron Letters 1975: 2065.

51. Wiessler, m. Angew. Chem. 86: 817 (1974).

52. Rice, J. M., Joshi, S. R. et al. Proc. Am. Assoc. Cancer Res. 16: 32 (1975).

53. Wiessler, M., and Schmaehl, D. Z. Krebsforsch. Klin. Onkol. 85: 47 (1976).

54. Magee, P. N. et al. Unpublished work.

55. Malling, H. V. Mutat. Res. 3: 537-540 (1966).

56. Gabridge, M. G., and Legator, M. S. Proc. Soc. Exptl. Biol. Med. 130: 831-834 (1969).

57. Mohn, G., Ellenberger, J., and McGregor, D. Mutat. Res. 25: 187-196 (1974).

58. Mohn, G., Ellenberger, J., et al. Mutat. Res. 29: 221-233 (1975).

59. Nakajima, T., and Wahara, S. Mutat. Res. 18: 121-127 (1973).

60. Nakajima, T., Akira, T., and Tojyo, K. I. Mutat. Res. 26: 361-366 (1974).

61. Green, M. H. L. et al. Chem. Biol. Interact. 11: 63-65 (1975).

62. Rosenkranz, H. S. In preparation.

63. Read, J. Mutat. Res. 33: 107-112 (1975).

64. Longnecker, D. S., Curphey, T. J., Susan, J. T., et al. Cancer Res. 34: 1658-1663 (1974).

65. Fluck, E. R., Poirier, L. A., and Ruelius, H. W. Chem. Biol. Interact. 15: 219-231 (1976).

66. Natarajan, A. T. Mutat. Res. 46: 54 (1977).

67. Slater, E. E., Anderson, M. D., and Rosenkranz, H. S. Cancer Res. 31: 970-973 (1971).

68. Popper, H., Czygan, P. et al. Proc. Soc. Exptl. Biol. Med. 142: 727-729 (1973).

69. Kada, T., Tutikawa, K., and Sadaie, Y. Mutat. Res. 16: 165-174 (1982).

70. Sadaie, Y., and Kada, T. J. Bacteriol. 125: 489-500 (1976).

71. Malling, H. V. Mutat. Res. 13: 425-429 (1971).

72. Malling, H. V. Mutat. Res. 26: 465-472 (1974).

73. Bartsch, H., Malaville, C., and Montesano, R. Cancer Res. 35: 644-651 (1975).

74. Gletten, F., Weekes, U., and Brusick, D. Mutat. Res. 28: 113-122 (1975).

75. Weekes, U., and Brusick, D. Mutat. Res. 31: 175-183 (1975).

76. Frantz, C. N., and Malling, H. V. Cancer Res. 35: 2307-2314 (1975).

77. Ames, B. Unpublished results.

78. Guttenplan, J. B., Huetterer, F., and Garro, A. J. Mutat. Res. 35: 415-422 (1976).

79. Bartsch, H., Malaville, C., and Montesano, R. Chem. Biol. Interact. 10: 377-382 (1975).

80. Frantz, C. N., and Malling, N. V. Mutat. Res. 31: 365-380 (1975).

81. Malling, H. V., and Frantz, C. N. Environ. Health Perspect. 6: 71-82 (1973).

82. Sugimura, T., Yahagi, T., et al. In: Screening tests in Chemical Carcinogenesis, R. Montesano, H. Bartsch, and L. Tomatis, Eds., IARC, Lyon, 1976.

83. Yahagi, T., M., Nagao, et al. Mutat. Res., in press.

84. Bartsch, H., Camus, A. et al. Mutat. Res. 37: 149-162 (1976).

85. Marquardt, H., Zimmermann, F. H., and Schwaier, R. Z. Vererbungsl. 95: 82-96 (1964).

86. Meyer, V. W. Mol. Gen. Genet. 112: 289-294 (1971).

87. Brusick, D., and Mayer, V. W. Environ. Health Perspect. 6: 83-96 (1973).

88. Mayer, V. W. Genetics 74: 433-442 (1973).

89. Brusick, D., and Andrews, H. Mutat. Res. 26: 491-500 (1974).

90. Malling, H. V. Mutat. Res. 3: 537-540 (1966).

91. Ong, T. M., and Malling, H. V. Mutat. Res. 31: 195-196 (1975).

92. Mohn, G., and Ellenberger, J. Mutat. Res. 19: 257-260 (1973).

93. Mazue, G., Vindel, J. A. et al. Arzenium Fosch. 24: 1422-1425 (1974).

94. Muenzer, R., and Renner, H. W. Int. J. Rad. Biol. 27: 371-375 (1975).

95. Propping, P., and Buselmaier, W. Arch. Toxikol. 28: 129-134 (1971).

96. Braun, R., and Schöneich, J. Unpublished results.

97. Couch, D. B., and Friedman, M. A. Mutat. Res. 26: 371-376 (1974).

98. Couch, D. B., and Friedman, M. A. Mutat. Res. 38: 89-96 (1976).

99. Zeiger, E. Dissert. Abstr. Int. B 34: 2108 (1973).

100. Propping, P., Röhrborn, G., and Buselmaier, W. Mol. Gen. Genet. 117: 197-209 (1972).

101. Fahrig, R. Mutat. Res. 21: 30 (1973).

102. Fahrig, R. Mutat. Res. 26: 29-36 (1974).

103. Fahrig, R. Mutat. Res. 31: 381-394 (1975).

104. Fumero, S., Mondino, A. et al. Paper presented at 2nd Int. Conf. Environ. Mutagens, Edinburgh, July 11-15, 1977.

105. Capizzi, R. L., Smith, W. J. et al. Mutat. Res. 21: 6 (1973).

106. Hsie, A. W., Machanoff, R. et al. Mutat. Res., in press.

107. Braun, R., and Schoeneich, J. Biol. Zentralbl. 94: 661-669 (1975).

108. Fahmy, O. J., and Fahmy, M. J. Chem. Biol. Interact. 11: 395-412 (1975).

109. Fahmy, O. J., and Fahmy, M. J. Cancer Res. 35: 3780-3785 (1975).

110. Ondrej, M. Folia Biol. (Prague) 15: 17-25 (1969).

111. Pasternak, L. Arzneimittelforsch. 14: 802-804 (1964).

112. Vogel, E. Unpublished results.

113. Fahmy, O. G., and Fahmy, M. J. Mutat. Res. 6: 139-154 (1968).

114. Fahmy, O. J., Fahmy, M. J., et al. Mutat. Res. 3: 201-217 (1966).

115. Pasternak, L. Acta Biol. Med. Ger. 10: 436-438 (1963).

116. Pasternak, L. Arzneimittelforsch. 14: 802-804 (1964).

117. Vogel, E., and Leigh, B. Mutat. Res. 29: 383-396 (1975).

118. Vogel, E. In: The Origins of Human Cancer, Cold Spring Harbor Symp. Quant. Biol., to be published.

119. Russell, W. L. Unpublished data.

120. Dawson, G. W. P. Unpublished data.

121. Fahrig, R. Arch. Toxicol. 38: 87-98 (1977).

122. Russell, L. B. Arch. Toxicol. 38: 75-85 (1977).

123. Epstein, S., and Arnold, E., et al. Toxicol. Appl. Pharmacol. 23: 288-325 (1972).

124. Frantz, C. N., and Malling, H. V. J. Toxicol. Environ. Health 2: 179-187 (1976).

125. Clive, D. Unpublished results.

126. Knaap, A. G. A. C. Unpublished results.

127. Abbondandolo, A., Bonatti, S., et al. Mutat. Res., in press.

128. Lilly, L. J., Bahner, B., and Magee, P. N. Nature 258: 611-612 (1975).

129. Bimboes, D., and Greim, H. Mutat. Res. 35: 155-160 (1976).

130. Brooks, A. L., and Grogger, V. Mutat. Res. 21: 214 (1973).

131. Natarajan, A. T., Tates, A. D., et al. Mutat. Res. 37: 83-90 (1976).

132. Takayama, S., Hitachi, M., and Yamada, K. Gann Monographs on Cancer Res. 17: 343-354 (1975).

133. Zbar, B., Wepsic, H. T. et al. J. Natl. Cancer Inst. 43: 821-832 (1969).

134. Natarajan, A. T., van Buul, P. P., and Raposa, T. Paper presented at the Symposium on Actions of Physical and Chemical Mutagens on the Somatic Chromosomes of Man, Edinburgh, July 7 and 8, 1977.

135. Bauknecht, T., Vogel, W. et al. Hum. Genetics 35: 299-307 (1977).

136. Bayer, U. Mutat. Res., in press.

137. Wild, D. Mutat. Res. 38: 105 (1976).

138. Lieberman, M. W., Baney, R. N. et al. Cancer Res. 31: 1297-1306 (1971).

139. Laishes, B. A., and Stich, H. F. Biochem. Biophys. Res. Commun. 52: 827-833 (1973).

140. Stich, H. F., and Kieser, D. Proc. Soc. Exptl. Biol. Med. 145: 1339-1342 (1974).

141. Sega, G. A. Unpublished results.

142. Kihlman, B. A. Rad. Botany 1: 35-41 (1961).

143. Kihlman, B. A. Rad. Botany 1: 43-50 (1961).

144. Arenaz, P. M. S. Thesis, Univ. Nevada, 1976, 102 pp.

145. Gichner, T. Unpublished results.

146. Veleminsky, J., and Gichner, T. Mutat. Res. 12: 65-70 (1971).

147. Veleminsky, J., and Gichner, T. Mutat. Res. 5: 429-431 (1968).

148. Malling, H. V., and de Serres, F. J. Cancer Res. 32: 1273-1277 (1972).

149. Neubert, D. In: Progress in Genetic Toxicology, D. Scott et al., Eds., Elsevier/North-Holland, Amsterdam, 1977, pp. 95-115.

150. San, R. H. C., and Stich, H. F. Int. J. Cancer 16: 284-291 (1975).

151. Ichinotsubo, D., Mower, H. F., et al. Mutat. Res. 46: 53-62 (1977).

CHAPTER 28

COMPARATIVE MUTAGENICITY OF AFLATOXIN B1 AND VINYL CHLORIDE

B. J. Kilbey

Department of Genetics, University of Edinburgh, Scotland

AFLATOXIN B1

Introduction

Aflatoxin B1 is one of a group of toxins produced by the mold *Apsergillus flavus* and implicated as the common factor in outbreaks of Turkey X disease and similar types of disease in ducklings, pigs and cattle (1). It is the most abundant member of the group; it is also the most toxic as well as being a potent hepatocarcinogen and mutagen. In early studies it was often used not as a pure compound but as part of a mixture.

Chemically aflatoxin B1 is a highly substituted coumarin with a molecular weight of 312. In common with other coumarins, it absorbs in the ultraviolet and fluoresces at wavelengths in excess of 420 nm. It decomposes if heated above 268°C (2). Its chemical structure is

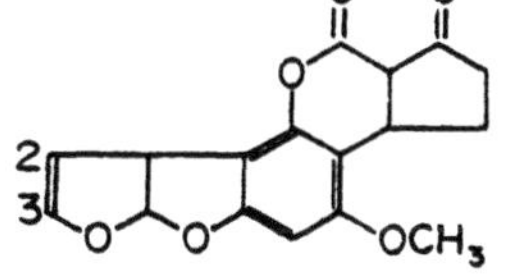

Genetic and Related Effects

Table 1 lists the organisms which have been used to test aflatoxin for its genetic effects and provides a summary of the conditions and results of each test.

Table 1. Response of tested organisms to Aflatoxin B1

Test system	Metabolic factor	Exposure (solvent, route, time, concn.)	Mutation response	Genetic endpoint	LERC[a]	References
E. coli K-12 (lambda)	+	5-70 μg/ml, 16-200 μM	+	Lambda induction	16 μM	3
B. subtilis trans. DNA	-	Concn 2:1 10 = 1 liquid; DMSO ethanol 20-220 min	+	Forward mutation		4
E. coli	+	2×10^{-2} μg per plate	+	Rec^+/rec^-	2×10^2 μg/plate	5
B. subtilis	+	6 μg/ml, 0.019 mM	+	Inhibition repair	19 μM	6
	-	12.5 μg/ml; 0.04 mM	-	Rec^+/rec^-	40 μM	6
B. subtilis	-	50 μg/ml, 0.16 mM	-	Forward mutation		4

E. coli	+	25-100 μg/ml, 80-300 μM	+	Point mutation reversion	80 μM	7
Salmonella	+	0.01-1 μg/plate	+	Reversion frameshift		8
Aspergillus			-			9
Neurospora	Growing cells	40 μg/ml, 120 μM; 2-12 hr	+	Forward specific locus		10,11
Neurospora	Conidia S9	0.67 mM; 670 μM; 2 hr; DMSO	+	Forward specific locus		4
S. cerevisiae	+	3 hr; 100 μg/ml; 300-600 μM	-	Gene conversion		7
V79 cells	-	75 μg/ml, 0.24 mM; 90 min	+	Azaguanine resistance	240 μM	12

(continued)

Table 1. (continued)

Test system	Metabolic factor	Exposure (solvent, route, time, concn.)	Mutation response	Genetic endpoint	LERC[a]	References
Human fibroblasts	S9	10^{-5} to 0.5×10^{-3} μM; 30 min	+	UDS	10^{-5} 10 μM	13
Human leukocytes	-	0.003-0.16 μM; 8 hr	+	Chromosome breaks	0.015 μM; 8 hr	14
Human leukocytes	-	0.16 μM; 24 hr	+	Chromosome breaks	16 μM, 24 hr	15
Rat liver cells in vitro in vitro	-	1% = 30 mM; DMSO; DMSO; 40 min	+ +	UDS (^{3}HTdR)	AFB1 + Af B2	16
CHO cells	+	1 μM; 30 min 100 μM, 24 hr	+ +	SCE	1 μM 10 μM	17

Drosophila recessive lethals	-	5-10^{-5} M; 50 µM; feeding DMSO 5% aflatoxin	+	Recessive lethals in chromosome II		18
Allium	-	1-250 µg/ml DMFA; 4, 24, 48 hr 0.003-0.64 µM	+	Chromosome breaks	<200 µg/ml 48 hr	19
Chinese hamster regenerating liver	-	1 mg/kg, IP; DMSO	+	Chromosome aberration	No control	20
Mouse heritable translocations	-	5 mg/kg, IP; DMSO	-	Heritable translo-cation	11	21
Mouse dominant lethal		68 mg/kg, IP; tricaprilin		Dominant lethal		22
Mouse	-	100 mg/kg, IP	-	Abnormal sperm		23

[a]Lowest effective concentration.

Subcellular Systems. Aflatoxin inhibits DNA-dependent RNA synthesis in mammalian cells in vivo, it also inhibits DNA synthesis. It binds to adenine and guanine in single- or double-stranded DNA (24-26). Evidence of a reaction with bacterial DNA in vitro has been reported by Maher and Summers (4), who showed that it produces mutational changes in the transforming DNA of Bacillus subtilis. Of these mutations, 36% they estimated, might be deletions in view of their stability.

Aflatoxin B1 can also produce damage to DNA which is subject to repair in a variety of cell types. Kada (6) has demonstrated this using the Rec^{+}/-assay in Bacillus subtilis. Ichinotsubo et al. (5) have also shown it using E. coli rec B, rec C strains and making a comparison of their sensitivity to the compound with that of the equivalent wild-type strain. Recently Goze et al. (3) reported that metabolites of B1 induce E. coli K12 (λ). The production of reparable damage in the DNA of mammalian cells also occurs: Unscheduled DNA synthesis (UDS) is induced by aflatoxin B1 in human fibrobalsts (13) and in primary cell cultures from rat liver treated with aflatoxin B1 approaching 1.0 mM (16). Sister chromatid exchange (SCE) has also been reported to occur in cultured Chinese hamster cells at levels around 10^{-6} M; 1.5-3 times the control frequencies were obtained by Wolff and Takehisa (17).

Gene Mutations. Aflatoxin B1 causes gene mutations although most of the evidence of its action comes from studies with microbial systems. In an abstract published in 1973, however, Velasquez (12) reported that doses of aflatoxin B1 giving 70-90% killing in cultured Chinese hamster cells (i.e., approaching 75 μg/ml) increased the frequency of mutations to 8 azaguanine resistance by a factor of about 20. A sample of these were shown to be cytologically normal. Lamb and Lilly (18) reported that treatment of Drosophila melanogaster with aflatoxin B1 brought about an increase in the frequency of second chromosome-linked recessive lethals. The largest increase was in the third brood probably indicating that the most sensitive germ cell stages are the spermatocytes or late spermatids. In fungi the most detailed information has been provided by Ong's study of the induction of adenine-3 mutations in Neurospora. In growing cultures 40 μg/ml of Aflatoxin B1 enhanced the frequency of forward mutations to between 120 and 150 times the control level (10,11). Unpublished data also show that growing yeast responds positively to aflatoxin, although the details are not yet available. Studies with Aspergillus conidia have proved negative so far (9). Reversion tests with bacteria also show that aflatoxin B1 or its metabolites are mutagenic. Ames et al. (8,27) report positive responses with strains of the Salmonella group of histidine auxotrophs. Strains TA100, TA1538, TA98 and TA1537 respond to aflatoxin in the presence of S9 mix. This suggests that frame shift mutagenesis only takes place. Garner (28), on the other hand, has reported that TA1530 responds to metabolites of Aflatoxin B1. This may mean that both frame shift and base pair substitution mutations are induced by the agent and would agree with Ong's conclusions based on an analysis of aflatoxin-induced ad-3 mutants in Neurospora (10). Recent results of Callen et al. (7) show that aflatoxin B1 induces reversion of an arg^- auxotroph of E. coli, and gene conversion in Saccharomyces cerevisiae in the presence of a metabolic activation system. Finally, studies with Chlamydamonas have shown that aflatoxin B1 can induce streptomycin-resistance mutations in both the nuclear and extracellular DNA components (29).

Chromosomal Damage. A number of studies show that as well as causing gene mutations aflatoxin B1 can break chromosomes both of plants and animals. Lilly (30) showed that a mixture consisting mainly of B1 and G1 could induce chromosomal aberrations in the root tips of Vicia faba. At 6.7×10^{-4} M for 3 hr at 21°C, abnormal anaphases were found for periods of up to 48 hr after termination of treatment. Maximal production occurred at 3 and 30 hr. "Abnormality" in these experiments referred to mitoses with fragments and bridges. The latter were particularly noticeable at 3 and 6 hr after treatment. A similar study by Reis (19) using Allium cepa made use of treatment times of up to 48 hours followed by immediate fixation. Concentrations of 3.2×10^{-6} M to 6.4×10^{-4} M aflatoxin B1 were used. The principal damage scored was anaphase bridge induction, and at the highest dose for 48 hr 50% of the cells were aberrant. Several studies with animal cells support the view that aflatoxin is a potent chromosome breaking agent. Three involve the use of human leukocytes. Withers et al. (31) reported that a solution containing 4.7×10^{-4} M aflatoxin mixture (37.7% B1) produced chromatid breaks as the most frequent aberration. There were no chromosome breaks. Dolimpio et al., using concentrations of a mixture of aflatoxins ranging from 1-50 μg/ml with a treatment time of 48 hr, found chromatid breaks, gaps, fragments, deletions and translocations (14). The chance that any particular chromosome would be broken bore a direct relationship to its size. Promchainant et al. (15) similarly showed that 50 μg/ml of a mixture of aflatoxins produced chromatid aberrations although chromosome aberrations were also found. This study is interesting, since the authors report that γ-rays and aflatoxin treatment do not interact in any way with each other (15). In a cell line derived from rat kangaroo kidney, aflatoxin has also been shown to be effective in producing chromosome-type aberrations. An 8 hr treatment time at 12.5 or 25.0 μg/ml aflatoxin resulted in a maximum of 42% of the cells bearing chromosomal aberrations, and the most sensitive regions of the chromosomes were reported as being heterochromatic (32). Hela cells (33) have been used to demonstrate that aflatoxin can cause DNA strand breakage at 32 μg/ml for 24 hr. Dependence of mutagenesis on active metabolism may be indicated since 100 μg/ml for 1 hr were ineffective.

These results leave no doubt that aflatoxin B1 can break chromosomes in plants and in vitro in animals but to date there is only little evidence of genetic effects in vivo in mammals. Leonard et al. (21) failed to demonstrate the induction of heritable translocations in the offspring of mice treated IP with 5 mg/kg dissolved in DMSO and saline. However, Epstein and Shafner (22) were able to show a rather weak induction of dominant lethals by using 68 mg/kg of a mixture of aflatoxins B1 and G1. The LD_{50} has been given as 9 mg/kg by Newberne and Butler (34) so the negative findings of Leonard may have arisen because he used too low a dose. Wyrobek and Bruce have reported that doses of aflatoxin of up to 100 mg/kg produce no additional sperm abnormalities in the mouse (23). Srivastava et al. (20) have studied the in vivo production of chromosomal aberrations in the liver cells of Chinese hamsters. In view of the organ specificity of carcinogenesis with aflatoxin B1 this approach is potentially most interesting. A dose of 1 mg/kg of aflatoxin B1 was administered intraperitoneally in DMSO. Partial hepatectomy was performed 24 hr prior to treatment and two post-treatment IP injections of 4 mg/kg colchicine were administered at 24 and 65 hr. Killing took place at 72 hr. Of 121 metaphases from seven animals, 82 showed aberrations. Unfortunately, the control data were not included in the communication and therefore the significance of this finding cannot be finally decided. One unpublished negative result has been provided from preliminary studies with the micronucleus test. The experimental details have not been made available. These results show good consistency and agree in demonstrating that aflatoxin B1 is a potent inducer of gene and chromosomal damage in a wide range of organisms.

Metabolic Activation

The carcinogenic action of aflatoxin B1 is highly organ-specific, the main site of action being the liver. Bacterial mutation systems appear without exception to require the presence of in vitro metabolizing systems before they respond positively to aflatoxin B1. Resting cells of yeast and conidia of Neurospora and Aspergillus fail to respond positively to the compound. In Neurospora and yeast the substance causes mutations in actively growing cells. Recently tests with microsomal systems have shown that they can stimulate resting cells of both yeast and Neurospora to respond to aflatoxin B1 (35,36). Some cell lines, for example, fibroblasts, became noticably more susceptible to aflatoxin B1 when activation systems are supplied. This was less appparent when liver cells were used in a parallel determination. Finally, aflatoxin has a variable effect on Drosophila from cells at different stages in their development. The most sensitive stages are those in which the endoplasmic reticulum is best developed.

In this respect, aflatoxin resembles other procarcinogens and promutagens.

These observations clearly point to the need for metabolism of aflatoxin B1 into an active form.

The active metabolite has not been positively identified but indirect evidence implicates the 2,3-oxide.

Swenson et al. (37,38) were able to show that in vivo in rats and in vitro using microsomal systems from rats, hamsters and human livers aflatoxin B1 binds covalently to rRNA and hepatic DNA. Mild hydrolysis of the adducts yielded a derivative which is indistinguishable from 2,3-dihydro-2,3-dihydroxy aflatoxin B1.

Swenson et al. (39) have recently shown that aflatxoin B1-2,3-dichloride, a molecule electronically similar to the 2,3-oxide can act as a carcinogenic and mutagenic agent.

Callen and Ong (35) have been able to potentiate the mutagenic action of aflatoxin B1 in both Neurospora and yeast by using aflatoxin in the presence of microsomes, cofactors and the inhibitor 1,1,1-trichloropropane 2,3-oxide (TCPO). The authors assume that aflatoxin B1 is normally converted into the active 2,3-oxide, which is itself metabolized by the enzyme epoxide hydrase to the inactive dihydrodiol. TCPO inhibits epoxide hydrase allowing the levels of the active metabolite to be increased.

Quantitative Considerations for Microbial Systems

Dose-Effect Curves. The only data which are suitable for the construction of dose-effect curves come from studies with microbial systems or single-cell cultures of mammalian material. In the studies with whole mammals so far only single dose experiments have been performed.

Dose-effect data have been published for the induction of mutations in transforming DNA of *B. subtilis* (14), the induction of *E. coli* K12 (λ) (3), the induction of resistance in Chlamydamonas (29), the induction of histidine revertants in TA100 (8), the induction of gene conversion in Saccharomyces and arginine reversion in *E. coli*. There has in addition been one study with human leukocytes which furnishes useful dose response data (14). The first of these (4) tells us little about the true response of transforming DNA since data are presented as a function of the number of lethal hits. However, the data from *E. coli*, Chlamydamonas and Salmonella cover dose ranges from 0-300 μg/ml. In most cases mutations arise in a near linear fashion with no detectable threshold. Each of them extrapolates to zero. Arginine reversions in _E. coli_ may also conform to this pattern but the curve for gene conversion is upward bending (35). Metabolic activation was provided for the bacteria and for Saccharomyces, but for these and for Chlamydamonas the simplest conclusion is that there is no restriction or saturation of the metabolizing systems as the dose increases. This is not so certain in the case of the data of Dolimpio et al. (14) on human leukocytes. It is plain from these that a nonlinear response is obtained which might well reflect the exhaustion or the saturation of the metabolizing machinery of these cells. Here too, however, no evidence of a threshold _per se_ can be found.

Hazards to Man--Problems with Quantitation. Although no direct evidence shows that aflatoxin B1 can induce either mutation in the human germ line or cancer, there are several good reasons for taking this possibility seriously. First, aflatoxins are ingested. They have been found in a wide variety of foodstuffs, sometimes at high levels, particularly in those geographical areas with climatic conditions which favor the growth of Aspergillus (40). Secondly, the frequency with which certain types of malignancy occur is geographically correlated with high levels of aflatoxin contamination of foodstuff (41). Although there are no data on the fate of aflatoxin B1 in the human body, human microsomes are known to be able to convert it into active metabolites in vitro. There is no doubt that aflatoxin B1 is a potent mutagen in a wide range of test systems. It is also able to break chromosomes including those of man himself in in vitro cell cultures. A number of studies have shown it to react with DNA and RNA.

These are good reasons for anticipating a threat from aflatoxin B1 both to man's health in the long- and short-term. However, to switch from a qualitative statement of this type to a quantitation of risk is at present impossible for three main reasons: (1) information concerning levels of exposure in populations at risk is lacking; (2) data on the relationship between applied dose and mutagenic effect in any whole-mammal system for germ-cell damage of any sort are needed; (3) we are ignorant of the kinetics of conversion of aflatoxin B1 to its active metabolite(s) in the body and their concentration in the target cells. This information will be crucial for a correct interpretation of dose-effect data once they have been obtained. If the conversion rate stays unchanged at different exposure levels, or when other experimental conditions are varied, the interpretation of dose-effect curves will be facilitated. If, however, it falls or rises with increasing dose the kinetics of the dose-effect relationship may be complex. Dolimpio et al., using human leukocytes and scoring chromosome breakage induced by aflatoxin, have produced data which show that such considerations are important. An effect of exposure level on the efficiency of break formation per dose increment is very apparent. At an exposure level of 1 μg/ml, for example, 1 μg/ml for 48 hr produces an average of 0.8 breaks/cell. At an exposure level of 10 μg/ml the same increment, 1 μg/ml for 48 hr produced only 0.14 breaks/cell. This may have several explanations which apply singly or in combination, e.g., elimination of damaged cells with increasing dose, induction or inactivation of repair systems with increasing dose, or overloading the metabolic machinery of the cell, but it shows that for Aflatoxin B1, as is being found for more and more other metabolizable mutagens, the efficiency of the mutagenic exposure can be very different at different rates of application. An empirical description of this relationship at the very least and, if possible, a deeper understanding of its causes, must be achieved before the data can be put to good use.

VINYL CHLORIDE MONOMER (VCM)

Introduction

Vinyl chloride ($CHCL = CH_2$) has been produced commercially for about 40 years. Its main use is in the production of homopolymers and copolymers, particularly PVC. The estimated production levels for 1975 were approximately 10^7 tons. Rather more than 70,000 workers are probably employed in VCM industries while those involved with PVC production probably number millions (42).

Clear information exists to implicate VCM as a carcinogen in man (42); the present paper reviews the current status of the evidence that VCM is mutagenic also.

Methods Used in Mutagenicity Studies

Table 2 gives a list of those test systems which have been used so far to test VCM or its metabolites for mutagenicity.

Results Reported

Bacterial Studies. Rannug et al. (43) reported that 20% (v/v) VCM in air in the presence of 9000g (S9 mix) supernatant of liver homogenate produced three times the spontaneous level of histidine revertants in a Salmonella mutant which reverted with base-pair substitutions. Bartsch and his co-workers (44) similarly reported that strain TA1530 exposed to 0.2, 2.0 and 20% VCM in air for 48 hr at 37°C gave 6-, 12- and 28-fold increases over the spontaneous his^+ revertant frequency. In these experiments

Table 2. Response of tested organisms to vinyl chloride monomer

No.	Test system	Metabolic factors[a]	Exposure (route; solvent; time; dose)	Mutagenic response	Genetic endpoints[b]	LREC[c]	References
1	Salmonella	S-9; R	Gas; air; 1.5 hr 20%	6-20 10^{-8}/surv.	BPS	20%; 30 min	43
	Salmonella	S-9; M	Gas; air; 6 hr; 20%	200 rev/;late	BPS	0.2%; 48 hr	44
	Salmonella	S-9; R	Gas; air; 6 hr; 20%	150 rev/plate	BPS		44
	Salmonella	S-9; H	Gas; air; 6 hr; 20%	100-330 rev/plate	BPS		44
	Salmonella	S-9; R	Gas; air; 3-9 hr;	128 rev/plate	BPS	20%; 3 hr	45
2	*E. coli*	S-9; R	Liq; buffer 2 hr; 10 mM	Positive	SLM; FM		46
3	*E. coli*	S-9; R	Gas; air; 0.1%	Reduced growth in pol A⁻	DD		47

4	Neurospora	Microsomes	Liq.; (?); 3-5 hr; NR	Negative	SLM		48
5	S. cerevisiae	-	Liq.; DMSO; 4-48 hr; 40 mM	Negative	FM		49
	S. pombe	Microsomes	Liq.; buffer; 1 hr; 16-48 mN	$25\times10^{-6}\times locus^{-1}\times mM^{-1}\times hr^{-1}$	SLM	16 mM; 1 hr	50,51
	S. cereivisiae (D5)	-	Liq.; buffer; 4-48 hr; 40 mM	Negative	MR		49
	S. cerevisiae (D4)		Liq.; buffer; 3 hr; 48 mM	$3\times10^{-8}\times locus^{-1}\times mM^{-1}\times hr^{-1}$	GC		50,51
	S. pombe	HMA	Gavage; oil; 3-12 hr; 700 mg/kg	$1.4\times10^{-9}\times locus^{-1}\times mg/kg\times hr^{-1}$	SLM	700 mg/kg; 6 hr	50
6	Mammalian cells in culture	S-15	Liq.; 5 hr; 1-6 mM	$1\times10^{-5}\times locus^{-1}\times mM^{-1}\times hr^{-1}$	SLM; AZA^R	4 mM; 6 hr	52
	V-79	S-15	Liq.; 5 hr; 1-6 mM	$0.7\times10^{-6}\times locus^{-1}\times mM^{-1}\times hr^{-1}$	SLM OUA^R	1 mM; 5 hr	52

(continued)

Table 2. (continued)

No.	Test system	Metabolic factors[a]	Exposure (route; solvent; time; dose)	Mutagenic response	Genetic endpoints[b]	LREC[c]	References
7	Drosophila		Gas; 2-17 days; 30-50,000 ppm	$2\text{-}60 \times 10^{-6} \chi ppm^{-1}$	RL	30 ppm; 2 days	53
	Drosophila		Gas; 3 hr; 1%	$1 \times 10^{-6} \times ppm^{-1}$	RL		54
	Drosophila	+Pheno-barbital	Gas; 3 hr; 1%	$1 \times 10^{-5} \times ppm^{-1}$	RL		54
8	Tradescantia		Gas; air; 6 hr; 10-150 ppm	$0.05 \times 10^{-2} \times ppm^{-1}$	Som. Mut.	10 ppm; 6 h4	55
9	Mouse		Gas; 5 days; 3,000-30,000 ppm	Negative	DL		56,57
10	Human		Occupational	Positive	CA		57-63
	Human		Occupational	Negative	CA		64
	Human		Occupational	Positive (?)	DL		65

[a] S-9R = S-9 mix with rat liver microsomes; S-9M = S-9 mix with mouse liver microsomes; S-9H = S-9 mix with human liver microsomes; HMA = host-mediated assay; NR = not reported.

[b] BPS = base-pair substitution; SLM = specific locus mutations; FM = forward mutations; MR = mitotic mutations; MR = mitotic recombination; GC = gene conversion; RL = recessive lethal; Som. Mut. = somatic mutation; CA = chromosome aberration; DL = dominant lethality; DD = DNA damage.

[c] LREC = lowest recorded effective concentration.

a metabolic activating system was present. It was also shown, however, that a substantial part of the effect of VCM is apparently direct. In the absence of a metabolizing system exposure of TA1530 to VCM at 20% for periods of up to 48 hr at 37°C produced up to 20 times the control frequencies of histidine revertants. In the presence of a metabolizing system, however, these figures were increased 2-4 times. A similar direct action of VCM on TA100 and TA1535 was shown by McCann et al. (45). Andrews et al. (66) also showed that VCM is mutagenic in TA1535, and the addition of S9 mix to the system apparently made no significant difference to the results. However, in this case no independent evidence was provided to show that the S9 preparation used retained its activity. Attempts to use solutions of VCM with Salmonella have failed even though concentrations as high as 0.083 M were used (42). This possibly results from the rapid loss of VCM vapour from the solution.

In a study with Escherichia coli a minimal dose of 1.1×10^{-2} mole/l. for 120 min was reported to be effective as a mutagen in the presence of S9 preparation. No effect was obtained without it (46). The mutations scored in this study were gal^+, arg^+ and MTR. Data from Rosenkranz' laboratory (47) also show that the pol A^- assay gives a negative result with VCM unless microsomes are present. In this case VCM was applied as a 0.1 v/v gas mixture.

Fungal Studies. Several studies have been reported using fungi as the test objects.

Drozdowicz and Huang (48) failed to demonstrate mutagenicity of VCM in Neurospora. This may in part have been a consequence of their use of ethanolic solutions from which vaporization of VCM is rapid. However, even when aqueous solutions were used in the presence of an atmosphere which was initially composed of 25% to 50% VCM in O_2 a negative result was obtained. Unfortunately, these data are inconclusive; since although an ancillary metabolizing system was provided, no independent evidence of its activity was given.

Loprieno et al. (50,51) used an aqueous solution containing 48 μg/ml of VCM (produced by bubbling a 50% v/v mixture of the gas in air continuously through the cell suspension) to show that it is able to produce forward mutations in Schizosaccharomyces pombe and gene conversion in Saccharomyces cerevisiae. In both cases the presence of a microsomal activating system was essential for an effect to be obtained. These workers also showed that S. pombe responds to VCM in the host-mediated assay (50). The microorganisms were placed in the peritoneal cavity and the mice given 700 mg/kg VCM orally in olive oil.

Mammalian Cells. Recently VCM has also been shown able to produce mutations to 8-azaguanine and ouabain resistance in Chinese hamster V79 cells. For azaguanine resistance doses of as low as 4 mM/5 hr and for ouabain of 1 mM/5 hr were detectable. Activation was provided by an S9 preparation (52).

Drosophila. Several studies with Drosophila have been conducted, and in one case (53) a wide range of doses were used. Inhalation doses of 850, 10,000 and 20,00 and 30,000 ppm for 2 days resulted in significant induction of sex-linked recessive lethal mutations compared with the controls. Positive results could also be obtained if the dose was lowered to 30 ppm and the exposure time increased to 17 days. It was also found that spermatids were the most responsive stage of spermatogenesis although all stages were found to respond to some extent. Above 10,000 ppm the level of VCM had no additional effect on the production of recessive lethals. This is interpreted by the authors to mean that the metabolizing activity of the germ cells is saturated at these doses. In spite of the ease with which recessive lethal mutations are produced in Drosophila there is no evidence that chromosome loss, dominant lethality or transmissible chromosome damage is induced. Since there is independent evidence that damage of this type requires higher doses of a number of mutagens to induce it than recessive lethals, the authors suggest that the concentration of active metabolites in the germ cells is never high enough for their induction.

Plants. Very little information is available concerning the effects of VCM in plants. At present data have been obtained only with Tradescantia. A small but significant induction of stamen hair variants has been found when gas mixtures containing from 10 to 72 ppm VCM were used for a 6-hr period. Above 72 ppm no further increase in the frequency variants was found.

Whole Mammals. With whole mammals the results have been more difficult to interpret. At present there is no direct evidence that VCM or its metabolites are active on the germ cells of an experimental mammal. However, only one test has been performed in which evidence of this might have been found. Male mice were allowed to inhale gaseous mixtures containing 3,000, 10,000 and 30,000 ppm VCM for 6 hr each day for 5 days before mating them repeatedly to virgin females (56,57). No increase in the frequency of dominant lethality was detectable although matings were continued for long enough to include all germ cell stages. In contrast to this are the data of Infante et al. on the rates of fetal wastage in wives of VCM factory workers (65) Fetal wastage was enhanced significantly in these women as compared to the rates of fetal wastage in the same women before their husbands entered the VCM industry and compared with the rates in a matched control sample. These results have been challenged on the basis of data from parallel studies (67) and the method of analysis (61). It may also be noted that the significant difference observed by Infante et al. appears to be caused not so much by a high abortion rate in the wives of VCM workers, as by a low rate in the matched controls. As the data stand, they appear to constitute a demonstration of a dominant lethal effect in man but their extension is highly desirable. Unfortunately, it is impossible to provide any reliable estimates of the doses of VCM received by these workers.

There is good evidence that VCM or a metabolic product causes chromosomal damage in man. Peripheral leukocytes from several sets of VCM workers have been studied on several occasions (57-64,67,68). Most of the results agree in finding a higher incidence of breaks and gaps in the exposed populations compared with the controls.

In follow-up studies, aberration rates in peripheral lymphocytes of workers appear to fall to normal levels following changes in VCM production routines leading to drastically reduced exposures (69). The time between the observations was two to three years.

Conclusions Concerning Mutagenicity of VCM. The experiments with microbial or single-celled mammalian systems show a high level of consistency in their response to VCM. Although Salmonella is probably able to respond in the absence of an additional metabolizing system, most workers agree that activating systems do enhance the mutagenic effectiveness of the compound. The yeasts and possibly *E. coli* seem to lack any metabolizing capacity and must rely upon the metabolizing system of the microsomes. Neurospora may also be in this category although for the reason already given the only results are inconclusive. Collectively these data apply to point or gene mutation induction. There are no test-system data which at present show VCM to be a clastogen, although it is clearly able to act in this way on the basis of the results with VCM workers.

Metabolism of VCM. There is ample evidence for concluding that VCM is converted into one or more mutagenic metabolites: As already noted, Rannug et al. (43) showed that a microsomal activation NADP dependent system is necessary for the production of histidine reversions in Salmonella. Also, although Bartsch et al. showed that VCM apparently acts directly in Salmonella (42), its mutagenic effect could be increased by the use of activating systems derived from the rat, from mice and from men (75). Furthermore, Bartsch et al. were able to show that activating systems prepared from phenobarbitone-treated rats are more effective in enhancing the response to Salmonella to VCM than extracts from untreated rats. A decrease in the effectiveness of a microsomal preparation can be brought about by agents with a depressing activity of microsomal mixed function oxidase, e.g. aminoacetonitrile or pregnenolone-16-α-carbonitrile (70).

E. coli requires activation for its response to VCM and in contrast to Salmonella this requirement appears to be absolute. Yeasts also are completely unaffected by VCM unless activating systems are included and Neurospora may also fall into this category.

Recent evidence from Drosophila also supports the earlier suggestion from germ-cell stage sensitivity data that metabolic activation occurs in this organism: Magnusson and Ramel (54) have shown that the Karsnäsbo strain of Drosophila may be made markedly more sensitive to the mutagenic effects of VCM by pretreatment with low concentrations of phenobarbitol.

In contrast to these reports, Garro et al. (71) favor an explanation of VCM mutagenesis in Salmonella which does not rely on the metabolic activity of the endoplasmic reticulum: although in their hands the effects of VCM on TA1530 are enhanced in the presence of rat liver extracts, the enhancement was the same whether livers were taken from animals which had been pretreated with Arochlor or VCM itself or not pretreated at all. In contrast, DMN mutagenesis was greatly enhanced with Arachlor-stimulated extracts and actually depressed with VCM-treated extracts compared to the unstimulated samples. These workers also present evidence to show that NADPH is not necessary for the effect of the rat liver extracts on VCM mutagenesis and that even heat-inactivated extracts are partially enhancing. They favor a free-radical mechanism for the action of VCM. The significance of these last observations needs further evaluation.

Two main routes have been proposed for the metabolism of VCM. As a result of his studies with rats, Hefner (72) has suggested that at doses below 100 ppm, VCM is converted first to 2-chloroethanol and then by the action of alcohol dehydrogenase to 2-chloroacetaldehyde. This is converted into monochloroacetic acid. The basis for these suggestions is that at doses below 100 ppm coadministration of ethanol inhibits the metabolism of VCM in the rat but that inhibition is far less pronounced at VCM doses exceeding 1000 ppm. Furthermore, SKF-525a an inhibitor of microsomal mixed function oxidase depresses the metabolism of VCM somewhat at high dose levels (1038 ppm) but has no effect at low levels of exposure (65 ppm). At high doses, therefore, the alcohol dehydrogenase-dependent pathway may be saturated leaving the microsomal activation solely responsible for metabolism. Attempts to show stimulation of VCM mutagenesis by alcohol dehydrogenase have, however, so far failed (44).

The second suggested route of metabolism involves only microsomal systems and is thought to proceed as in Eq. (1).

$$ClCH{=}CH_2 \xrightarrow[\text{NADPH}]{\text{microsomes}+O_2} ClCH\overset{O}{-}CH_2 \longrightarrow ClCH_2\text{-}CHO \longrightarrow ClCH_2\text{-}COOH \tag{1}$$

There is evidence both that some of these intermediates are mutagenic and that they can be formed by liver metabolizing systems. The information on the mutability of the metabolites is given in Table 3.

Independent evidence that chloroethylene oxide and chloroacetaldehyde is formed from VCM in the liver has also been obtained. When VCM is passed through a reaction mixture containing microsomes from phenobarbitone pretreated mice and an NADPH generating system, a volatile metabolite is produced which can be trapped by its reaction with 4-(p-nitrobenzyl)pyridine in ethylene glycol. The absorption spectrum of the adduct is identical with that obtained when chloroethylene oxide reacts with 4-(p-nitrobenzyl)pyridine. Chloroacetaldehyde gives a different spectrum. If the same metabolizing system is used with 3,4-dichlorobenzenethiol as the trapping agent, 3,4-dichlorophenylthioacetaldehyde is produced, indicating that chloroacetaldehyde is produced in the mixture (74). This is most simply accounted for by assuming that chloroethylene oxide is first formed from VCM and that it undergoes internal rearrangement to form chloroacetaldehyde.

VCM Compared with Other Mutagens. VCM or its metabolites are clearly able to induce point mutations of the base pair substitution type in Salmonella. Although there is no record in the literature of an analysis, it also seems probable that the forward mutations it produces in *S. pombe* also include this class of alteration. In common with most other mutagens VCM induces gene conversion in yeast and chromosomal damage in human leukocytes.

At the present time very few data exist which permit a quantitative comparison to be made between VCM and other mutagens. From their careful study of the mutagenic effects of VCM in *S. pombe*, Loprieno et al. (51) estimate that 2 mM VCM is equivalent to 9900 rads x-rays, 28.0 mM EMS, and 0.25 mM MMS, insofar as these are the doses required to double the spontaneous mutation frequency. However, as these authors point out the calculations do not take into account the extent to which VCM is converted into a much more mutagenic metabolite. If the efficiency of conversion is affected by the exposure to VCM the equivalence figures will be changed also. The authors were careful to use only the initial

Table 3. Comparative mutagenic effect of vinyl chloride and its metabolites

Compound	Salmonella (reverse mutation)[a] (rev/moles/ml)	*S. pombe* (forward mutation)[b] $\times 10^{4}$ (locus-mM-hr)$^{-1}$	*S. cerevisiae* (gene conversion)[b] $\times 10^{-4}$ (locus-nM-hr)$^{-1}$	Chinese hamster (V79 forward mutation)	
				Aza^{r}	Oua^{r}
Vinyl chloride	5.4	0.25	0.03	20/10 [c]	20/10 [c]
Chloroethylene oxide	125	338.0	80.0	81/10 +[d]	81/10 +[d]
Chloroacetal-dehyde	50	0.05	0.2	131/10 [e]	7/10 [e]
Chloroethanol	3	0	0	0[f]	0[f]
Chloro-aceticacid	0	NT	NT	0[f]	0[f]

[a]Data of McCann et al. (45).

[b]Data of Loprieno et al. (50,51).

[c]At 0 mM; data of Kuroki (52).

[d]At 13 μM; data of Huberman (73).

[e]At 6.4 μM; data of Huberman (73).

[f]At 2500 μM; data of Huberman (73).

linear portions of the dose-effect curves in making their estimations, but this may also lead to difficulties because S. pombe is relatively resistant to ionizing radiation at low doses.

Mutagenic Hazard from VCM. Although it has been estimated that the daily human oral intake of VCM is considerably less than 100 μg (42, 51), no estimates are at present available for total uptake and this is likely to vary considerably depending on the location of the population being studied. Even if 100 μg is taken as the average figure we cannot at the present time estimate the number of mutations this is likely to produce. To do this we should either require accurate dose-effect data in mammalian germ cell material or an estimate of the radiation equivalence of VCM in the target tissue which took into account the rates of conversion of VCM into metabolites as well as the persistence and distribution of the active metabolites in the body. This information is not available. Hefner has shown that rats exposed to a relatively low dose of VCM metabolize the substance fairly rapidly (72). Of a dose of ^{14}C-VCM calculated to be on average 0.49 mg/kg body weight, 70.5% of the radioactivity was lost in the first 15 hr, 84.8% after 70 hr. The loss was mainly in the urine presumably as polar metabolites but significant amounts were also lost in the faeces and as expired CO_2. After 75 hr only 0.2% of the radioactivity remained in the kidneys, 1.6% in the liver, 3.0% in the skin and 7.6% in the carcass. Although these figures may give some impression of the speed of the VCM metabolism, they do not help to decide what proportion of the administered dose reached the gonads either as unchanged VCM or as active metabolites. The latter proportion may be expected to be small in view of their high reactivities. Finally, one other factor must be remembered when these data are collected. There is evidence which may be interpreted as indicating that the conversion of VCM into active metabolites becomes less efficient as the loading of the metabolic systems is increased. Both the data from Drosophila and from Tradescantia have been interpreted in these terms. If this proves correct, dose rate is another variable which needs to be kept in mind when the mutagenic hazard from VCM is evaluated.

REFERENCES

1. Ong, T.-M. Mutat. Res. 32: 35-53 (1975).

2. Wogan, G. N. Fed. Proc. 27: 932-938 (1968).

3. Goze, A., Sarasin, A., Moule, Y., and Devoret, R. Mutat. Res. 28: 1-7 (1975).

4. Maher, V. M., and Summers, W. C. Nature 225: 68-70 (1970).

5. Ichinotsubo, D., Mower, J. S., and Mandel, M. Mutat. Res. 46: 53-62 (1977).

6. Kada, T. Unpublished data.

7. Callen, D. F., Mohn, G. R., and Ong, T.-M. Mutat. Res. 45: 7-11 (1977).

8. McCann, J., Springarn, E., Kobori, J., and Ames, N. Proc. Natl. Acad. Sci. (U.S.A.) 72: 979-983 (1975).

9. Lilly, L. J. Aspergillus Newsletter 8: 14-15 (1967).

10. Ong, T.-M. Molec. Gen. Genet. 111: 159-170 (1971).

11. Ong, T.-M., and de Serres, F. J. Cancer Res. 32: 1890-1893 (1972).

12. Velazquez, A., de Nava, C., Coutino, R., and Pulido, I. Mutat. Res. 21: 241-242 (1973).

13. San, R. H. C., and Stich, H. F. Int. J. Cancer 16: 284-291 (1975).

14. Dolimpio, D. A., Jacobson, C., and Legator, M. Proc. Soc. Exptl. Biol. Med. 127: 559-562 (1968).

15. Promchainant, C., Baimai, V., and Nondasuta, A. Mutat. Res. 16: 373-380 (1972).

16. Williams, M. Cancer Lett. 1: 231-326 (1976).

17. Wolff, S., and Takehisa, S. Proceedings 2nd International Conference on Environmental Mutagens, Scott, Sobels, and Bridges, Eds., Elsevier, Amsterdam, 1977, pp. 193-200.

18. Lamb, J., and Lilly, J. Mutat. Res. 11: 430-433 (1971).

19. Reiss, J. Experientia 27: 971-972 (1971).

20. Srivastava, P. K., Srivastave, A. K., and Lucas, F. V. Genetica (Hague) 45: 371-376 (1974/5).

21. Leonard, A., Deknudt, G. H., and Linden, G. Mutat. Res. 28: 137-139 (1975).

22. Epstein, S., and Shafner, H. Nature 219: 385-387 (1968).

23. Wyrobek, A. et al., personal communication.

24. Clifford, J. I., and Rees, K. R. Nature 209: 312-313 (1966).

25. Clifford, J. I., and Rees, K. R. Biochem. J. 103: 467-471 (1967).

26. Sporn, B., Dingman, C., and Phelps, H. Science 151: 1539-1541 (1966).

27. Ames, B. N., Durston, W. E., Yamasaki, E., and Lee, F. D. Proc. Natl. Acad. Sci. (U.S.A.) 70: 2281-2285 (1973).

28. Garner, R. C., and Wright, C. M. Brit. J. Cancer 28: 554-551 (1973).

29. Schimmer,)., and Werner, R. Mutat. Res. 26: 423-425 (1974).

30. Lilly, L. J. Nature 207: 443-435 (1965).

31. Withers, R. F. J. In: Proceedings Symposium on Mutational Processes, Prague, 1965, Z. Landa, Ed., Czech. Academy of Sciences, 1966, pp. 359-364.

32. Green, S., Legator, M., and Jacobson, C. Mamm. Chromosomes Newsletter 9: 36 (1977).

33. Umeda, M., Yamamoto, T., and Saito, M. Japan. J. Exptl. Med. 42: 527-535 (1972).

34. Newberne, P. M., and Butler, W. H. Cancer Res. 29: 236-250 (1969).

35. Callen, D. F., and Ong, T.-M. Mutat. Res. 49: 371-376 (1978).

36. Matzinger, P. K., and Ong, T.-M. Mutat. Res. 37: 27-32 (1976).

37. Swenson, D. H., Miller, J. A., and Miller, E. L. Biochem. Biophys. Res. Commun. 53: 1260-1267 (1973).

38. Swenson, D. H., Miller, E. C., and Miller, J. A. Biochem. Biophys. Res. Commun. 60: 1036-1043 (1974).

39. Swenson, H., Miller, J. A., and Miller, E. C. Cancer Res. 35: 3811-3823 (1975).

40. Shank, R. C., and Wogan, G. N. Food Cosmet. Toxicol. 10: 61-69 (1972).

41. Wogan, G. N. Fed. Proc. 27: 932-938 (1968).

42. Bartsch, H., and Montesano, R. Mutat. Res. 32: 93-114 (1975).

43. Rannug, U., Johansson, A., Ramel, C., and Wachtmeister, C. A. Ambio 3(5): 194-197 (1974).

44. Bartsch, H., Malavielle, C., and Montesano, R. Int. J. Cancer 15: 429-437 (1975).

45. McCann, J., Simmon, V., Streitweiser, and Ames, B. Proc. Natl. Acad. Sci. (U.S.A.) 72: 3190-3193 (1975).

46. Greim, H., Bonse, G., Radwan, Z., Reichert, D., and Henschler, D. Biochem. Pharmacol. 24: 2013-2017 (1975).

47. Rosenkranz, H. S., in press.

48. Drozdowicz, B. Z., and Huang, P. C. Mutat. Res. 48: 43-50 (1977).

49. Shahin, M. M. Mutat. Res. 40: 269-272 (1976).

50. Loprieno, N., Barale, R., Baroncelli, S., Bauer, C., Bronzetti, G., Cammellini, A., Cercignani, G., Corsi, C., Gervaki, G., Leporini, C., Nieri, R., Rossi, A. M., Stretti, G., and Turchi, G. Mutat. Res. 40: 85-96 (1976).

51. Loprieno, N., Barale, R., Baroncelli, S., Bronzetti, G., Cammellini, A., Corsi, G., Leporini, C., Nieri, R., and Rossi, A. M. IARC Monograph on the Evaluation of Carcinogenic Risk of Chemicals to Man, IARC, Lyon.

52. Kuroki, D., personal communication.

53. Vogel, E., and Berburgt, F. G. Mutat. Res. 48: 327-336 (1977).

54. Magusson, J., and Ramel, C. Mutat. Res. 57: 307-312 (1978).

55. Nauman, C., and Grant, W. Personal communication.

56. Anderson, D., Hodge, M. C. E., and Purchase, I. F. H. Mutat. Res. 40: 359-370 (1976).

57. Purchase, I. F., Richardson, C. R., and Anderson, D. Lancet ii: 410-411 (1975).

58. Ducatman, A., Hirschhorn, K., and Selikoff, I. J. Mutat. Res. 31: 163-168 (1975).

59. Funes-Cravioto, F., Lamberg, B., Lindsten, J., Ehrenberg, L., Natarajan, A. T., and Osterman-Golkar, S. Lancet i: 459 (1975).

60. Hansteen, I. L., Hillestad, L., and Thiis-Evensen, E. Mutat. Res. 38: 112 (1976).

61. Heath, C. W., Dumont, C. R., Gamble, J., and Maxweiler, R. J. Environ. Res. 14: 68-72 (1977).

62. Kucerova, M., Polivkova, Z., and Batora, J. Mutat. Res. 67: 97-100 (1979).

63. Leonard, A., Decat, G., Leonard, E. D., Lefevre, M. J., Decuyper, L. J., and Nicaise, C. L. J. Toxicol. Environ. Health, 2: 1135-1141 (1977).

64. Fleig, I., and Thiess, A. M. Arbeitsmed. Sozialmed. Praeventivemed. 9(12): 280-283 (1974).

65. Infante, P. F., Wagoner, J. K., McMichael, A. J., Waxweiler, R. J., and Falk, H. Lancet i: 734-735 (1976).

66. Andrews, A. W., Sawistowski, E. S., and Valentine, C. R. Mutat. Res. 40: 273-276 (1976).

67. Edmonds, L. D., Falk, H., and Nissim, J. E. Lancet ii: 1098 (1975).

68. Paddle, G. M. Lancet i: 1079 (1976).

69. Hansteen, I. L., Hillestad, L., Thiis-Evensen, E., and Storetvedt-Hellaas, S. Mutat. Res. 57: 271-278 (1978).

70. Malavielle, G., Barbin, A., Bresil, H., Montesano, R., and Bartsch, H. Abstr., Meeting of the European Society of Toxicology, Montpellier, France, 1975.

71. Garro, A. J., Guttenplan, J. B., and Milvy, P. Mutat. Res. 38: 81-88 (1976).

72. Hefner, R. E., Jr., Watanabe, P. G., Gehring, P. J. Ann. N. Y. Acad. Sci. 246: 135-148 (1975).

73. Huberman, E., Bartsch, H., and Sachs, L. Int. J. Cancer 15: 539-544 (1975).

74. Göthe, R., Callerman, C. J., Ehrenberg, L., and Wachtmeister, C. A. Ambio 3: 234-236 (1974).

75. Malavielle, C., Barsch, H., Barbin, A., Camus, A. M., Montesano, R., Croisy, A., and Jacquignon, P. Biochem. Biophys. Res. Comm. 63: 363-370 (1975).

CHAPTER 29

COMPARATIVE MUTAGENICITY OF N-METHYL-N'-NITRO-N-NITROSOGUANIDINE (MNNG) AND ICR-170

John W. Drake*

National Institute of Environmental Health Sciences
Research Triangle Park, North Carolina (U.S.A.)

Seymour Abrahamson*

Department of Zoology, University of Wisconsin
Madison, Wisconsin 53715 (U.S.A.

N-METHYL-N'-NITRO-N-NITROSOGUANIDINE

N-Methyl-N'-nitro-n-nitrosoguanidine (CAS No. 70-25-7) is a widely used laboratory mutagen and carcinogen known by a number of abbreviations, now most commonly MNNG. It has been applied to a wide array of organisms from prokaryotes through mammals and has been clearly demonstrated to produce a spectrum of genetic alterations in nearly all test systems, with some putative exceptions to be discussed below.

MNNG is a solid of molecular weight 147.1 (1 μM = 0.1471 μg/ml; 1 μg/ml = 6.798 μM), mp = 188°C, bp = 123.5°C; and having the structural formula

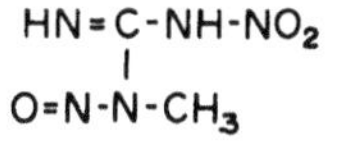

*With the collaboration of Errol Zeiger, William Sheridan, A. Schalet, Paul E. Perry, Ethel Moustacchi, Ralph C. Harvey, and E. H. Y. Chu.

It is a light-sensitive compound, changing from yellow to orange-green upon exposure. It is soluble in polar solvents, but only poorly soluble in water (to less than 0.5%). Its half-life in water at room temperature is about 200 hr, but this falls to only about 90 min in culture fluid at 37°C.

Unlike most simple monofunctional alkylating agents, MNNG methylates DNA efficiently in vivo but only very weakly in vitro. It readily decomposes in the presence of hydroxyl ions to form diazomethane, itself a potent mutagen, but this process is unlikely to be the mutagenically significant mechanism, because MNNG works best in vivo at a slightly acid pH. Since even single-stranded DNA is not readily mutagenized by MNNG in vitro, whereas the mutagenically significant O^6 position of guanine and O^4 of thymine are efficiently methylated in vivo, since MNNG often induces mutations in tightly linked clusters in the neighborhood of the replication fork, and since the in vivo correlation between MNNG alkylating and mutagenizing efficiencies is not very strong, one should suspect either that it acts via a very special pathway of metabolic activation (very different from that carried out by the standard S9 preparation, for instance, which inactivates MNNG), and/or upon a special configuration of the target molecule (not just single-stranded DNA, and perhaps not DNA alone).

Bacterial Systems

MNNG is a potent mutagen when tested with the missense (base-pair substitution) Salmonella tester strains TA1535 and TA100, but is only weakly active with the frameshift (base-pair addition or deletion) tester strain TA1537. Over a narrow dose range (0.5 to 2 μg/plate) the mutational response increases faster than linearly. Mutagenic activity is markedly decreased in the presence of liver microsome preparations, an effect dependent upon the presence of an NADPH-generating system and therefore unlikely to reflect simply methyl scavenging by miscellaneous S9 proteins.

MNNG also promotes predominantly base pair substitutions in *E. coli*, with frameshift mutagenesis by both addition and deletion mechanisms again occurring at much lower frequencies. Mammalian microsomal fractions antagonize mutagenic activity here too. In *E. coli* the lowest effective reported concentration for mutation induction was 0.16 μg/ml in the absence of a microsomal fraction.

Differential sensitivities of DNA repair-defective bacteria are readily observed in *E. coli recA*, *recB*, *recC*, *recF*, *recL*, *polA1* and *sbc* strains and in *Bacillus subtilis rec45*; but mutagenesis is reduced at most marginally in *E. coli recA*. These results reveal that the in vivo effects of MNNG include lethal

DNA damage, and also that the recA-dependent mode of error-prone repair does not contribute substantially to MNNG mutagenesis in E. coli (although error-prone repair is probably the main determinant of MNNG mutagenesis in fungi).

Fungal Systems

Studies of specific-locus MNNG mutability have been conducted at a number of loci in Neurospora, most extensively with the purple-adenine loci ad3A and ad3B. The vast majority of induced mutations behaved like point mutations of the base pair substitution type, although a few putative base pair deletion mutations were also obtained. The lowest effective dose was the equivalent of 3.7 μg/ml for 15 min. MNNG mutagenesis is reduced in N. crassa uvs-3 and uvs-5, indicating a requirement for error-prone repair (in contrast to E. coli).

MNNG induces both forward and reverse mutations in Saccharomyces cerevisiae and Schizosaccharomyces pombe, with a linear dose response relationship recorded for S. cerevisiae. The lowest effective dose tested was 10 μg/ml (for the induction of reverse mutations in S. cerevisiae). In addition to point mutations, MNNG induces both mitotic crossing over and gene conversion in S. cerevisiae; here the lowest effective dose tested was 0.1 mM (14.7 μg/ml) for both endpoints. Since crossing over was increased about 150-fold and gene conversion about 25-fold over the spontaneous backgrounds at this dose, the lowest effective concentration is clearly much lower than the lowest tested concentration. Since MNNG mutagenesis is drastically reduced in S. cerevisiae rad-6 and rad-9, an error-prone repair system is required for mutagenesis here too.

Plant Systems

In the soybean Glycine max, the induction of single and double spots in heterozygous leaves indicates the production of somatic recombination, point mutations or deletions. In contrast to a number of other mutagens, very low concentrations of MNNG were employed (6 to 50 ppm; 1 ppm ≅ 1 μg/ml); and while the yield of single and double spots was reported to be significantly increased at the lowest tested concentration in four replicates, there was no further increase with increasing dose. Since the Group I report suggests that most of the phenotypic alterations scored result from genetic recombination, it is not clear that critical tests for mutagenesis have been conducted. Moreover, the report suggests pertinent reasons for concluding that the relative inactivity of MNNG in this system resulted from inactivation of the mutagen by heat and light. An independent set of measurements with this system is therefore warranted before conclusions can be drawn concerning mutational spectrum or potency.

A variety of mutational responses have been reported in different cultivars of the barley Hordeum vulgare, wherein either dry or pre-soaked and germinated seeds can be treated. Gross nuclear chromosomal changes have been demonstrated, and also chlorophyll mutations (which are likely to be a mixture of deletions and intragenic changes). The lowest effective tested concentration was 0.07 mM (about 10 μg/ml). While a dose dependency was observed, the shape of the response has not been adequately established.

With both Allium cepa and Vicia faba, a variety of genetic end points were observed to be enhanced by MNNG, including chromatid breaks and interchanges, chromosome fragmentation and high frequencies of aberrant metaphases.

Insect Systems

Recessive lethals are widely considered to be the most sensitive indicator of genetic alteration in Drosophila melanogaster. Exposure to supposedly high but in practice indeterminant doses of MNNG induced recessive lethals in each male germ cell stage studied, namely spermatozoa, spermatids, spermatocytes and spermatogonia. Moreover, subsequent detailed analysis of 34 selected recessive lethals demonstrated that none were associated with multilocus deletions. Translocations, full and partial chromosome losses and nondisjunction were also induced, although usually at lower frequencies than were recessive lethals. Translocations were not detected in gonial cells, but this is not surprising since the sample size was small, and even with x-rays such rearrangements are rarely detected in gonial cells since (in contrast to inversions) they appear to be selectively eliminated. Finally, it should be noted that a high frequency of MNNG-induced mutations were expressed as mosaics.

Specific locus mutations are readily scored in the silkworm Bombyx mori. MNNG induced such mutations in both male and female germ cells after injection of 5 μg or more.

Cultured Mammalian Cell Systems

Three specific-locus test systems have been exposed to MNNG, namely $TK^+ \rightarrow TK^-$, $HGPRT^+ \rightarrow HGPRT^-$ and $ASN^- \rightarrow ASN^+$ (forward, forward, and reverse mutations, respectively). All systems responded, and linear dose dependencies were obtained when care was taken to allow mutation expression. As with the microbial systems, incorporating a liver microsomal metabolic activating system dramatically reduced the mutagenic potency of MNNG by orders of magnitude. The ASN reversion system was less sensitive to MNNG than were the two forward mutation systems; while the reverse pattern might have been expected as in many bacterial systems, it is possible that the particular revertible base pair(s) of the ASN system were by happenchance only mildly responsive to MNNG, and/or that numerous deletions or chromosomal mutations were being detected in the forward but not the reverse mutation systems.

A variety of mammalian cell lines, including human cells, have been employed for cytogenetic analysis of MNNG mutagenicity. Both chromatid breaks and complex exchanges have been detected in all of these systems, with differences as expected from differences in time of treatment throughout the cell cycle and other experimental variables. Clear dose/effect relationships were established over a 100-fold dose range. In addition, both sister chromatid exchanges and unscheduled DNA synthesis were induced in a variety of cell types in vitro. It is particularly interesting that the SCE assay was about two orders of magnitude more sensitive an indicator of MNNG activity than were any of the other cell culture methods: the lowest effective concentration was 0.00015 μg/ml for SCEs, versus 0.015 μg/ml for the other endpoints.

Intact Mammalian Systems

In contrast to all the previous systems, a very uncertain picture has evolved thus far with respect to MNNG mutagenicity in the mouse.

The classical germ-line specific-locus test was negative at a single dose of 50 mg/kg (ca. 50 μg/ml): no mutations were observed in a sample of 894 postspermatogonial plus 7408 spermatogonial cells tested. However, this is an extremely small sample size; for instance, a mutagenically very significant dose of 86 R of x-rays delivered chronically produces about one mutation per 10,000 spermatogonia and two per 10,000 postspermatogonial cells, values not distinguishable from the MNNG results.

Translocation figures were not observed in spermatocytes treated as spermatogonia with 48 mg/kg. Heritable translocation studies have not yet been reported. The dominant lethal test was negative at 50 mg/kg, but was positive both for sperm at 100 mg/kg and for oocytes at 70 mg/kg. The micronucleus test using bone marrow cells tested in vivo was negative over the dose range from 50 to 200 mg/kg. Although positive in vitro, the sister chromatid exchange test was negative in mouse bone marrow in vivo. The spot test, performed on ten-day-old fetuses, showed a significant increase following injection of 35 mg/kg of MNNG (1.8% of animals with spots versus 0.1% in the controls).

It is noteworthy that host-mediated assays in mice employing bacteria, yeast, or mammalian cells were all positive in the dose ranges in which the germ cell and micronucleus tests were performed; only Neurospora failed to respond, and *N. crassa* is a notoriously unreliable organism. It is thus difficult to attribute the lack of response in some of the in vivo mouse tests to a rapid inactivation of MNNG. Instead, the picture that emerges is that some of the intact mammlian systems are simply not sufficiently sensitive as routinely employed, either because of inadequate sample sizes or because of limited exploration of dose ranges. Further testing of MNNG is therefore highly desirable with increased attention to these factors, and would be mandatory before any firm conclusions could be drawn about the mutagenic potential of MNNG in the mammal.

Summary

MNNG was chosen for this comparative exercise in part because, being the first compound to be designated a "supermutagen," it has been tested in a large number of systems and is representative of a class of chemicals which are potentially dangerous because small amounts accidently released into the environment might produce extensive genotoxic effects without other major toxic indicators. While extensively employed as a laboratory mutagen, this application is probably often unwise, both because of the very dangerous nature of the compound and because of its propensity to induce clusters of tightly linked mutations.

Potencies and lowest effective tested concentrations provided in the Group I reports varied hugely (by roughtly 100,000-fold) from system to system, and especially from mutational endpoint to endpoint. It seems unlikely to us that these differences could be substantially dissipated by further and more refined measurements, so that the prospects for quantitative cross-phylum and cross-kingdom extrapolations appear dim. Risk estimation will therefore most probably have to be based upon insect or in vitro and/or in vivo mammalian data.

Based on the very limited data available from in vivo mammalian systems, one can readily calculate from the spot test an equivalence between MNNG and x-rays, and can in turn estimate that 1 mg/kg of MNNG would produce the equivalent of 0.01 of the spontaneous mammalian mutation rate. Tiptoeing further, this implies that an exposure of this magnitude to the entire U.S. population would introduce about 20 new cases of serious dominant disease per million liveborn in the first generation, plus a much greater number of recessive mutations whose deleterious effects would only become manifest many generations in the future.

ICR-170

2-Methoxy-6-chloro-9-[3-(ethyl-2-chloroethyl)-aminopropylamino]acridine (CAS No. 146-59-8) is commonly called ICR-170, although it was also called ICR-100 by some Drosophila workers in the 1960s. The commonly available form is the dihydrochloride, whose molecular weight is 479.3 (1 μM = 0.4793 μg/ml; 1 μg/ml = 2.086 μM). The mutagenesis literature contains numerous incorrect molecular weights for ICR-170, and is usually ambiguous as to whether molarities were based on the free base (MW = 406.35) or the dihydrochloride. ICR-170 is photosensitive; red or yellow light should be employed to avoid both photoinactivation and photodynamic genotoxic effects. Its structural formula is:

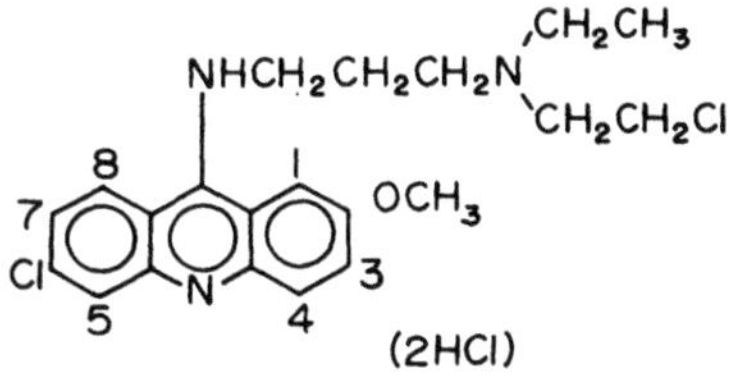

ICR-170 is a strong frameshift mutagen in a variety of systems, and is also carcinogenic. It is probably a DNA intercalator and a weak monofunctional alkylating agent whose mutagenic activity requires both chemical components simultaneously.

Bacterial Systems

The only Salmonella reversion test applied to ICR-170 was the frameshift tester strain TA1537. A linear dose response was observed over a very narrow dose range (2 to 6 μg/ml) in the absence of a metabolic activating system; in fact, the effect of such a system upon ICR-170 remains unknown. The mutagenic potency in TA1537 was 475 revertants/μg.

Numerous studies in *E. coli* have established that ICR-170 induces predominantly frameshift mutations; a small fraction of apparent base pair substitution mutations could be readily attributed to additions or deletions of three pairs at a time. The *E. coli recA*-dependent error prone repair system is not required for ICR-170 mutagenicity, but since the *B. subtilis rec* test was positive (both with and without metabolic activation), lethal as well as mutagenic DNA damage is induced.

Fungal Systems

ICR-170 induced both forward and reverse mutations in both Saccharomyces cerevisiae and Schizosaccharomyces pombe, sometimes with linear dose-response curves. Mitotic crossing over and gene conversion were also induced, but the shape of the dose-response curves were uncertain because of the limited dose ranges employed.

The purple adenine ad-3 system in Neurospora crassa has been used to detect both forward and reverse mutations induced by ICR-170. The forward mutation frequency rose faster than linearly over a fivefold dose range; at 1 µg/ml it was about 500-fold over the spontaneous background. Multilocus deletions were not induced, and most of the mutations could be explained as ordinary frame-shift mutations. The lowest effective dose tested was 0.03 µg/ml in an ad-3 reversion test. A number of other loci also responded to ICR-170 mutagenesis.

Neither the acridine nor the alkylating side-chain moieties of ICR-170 were at all strongly mutagenic of N. crassa when tested separately. The high mutagenicity of the compound might therefore result either from stabilization of the stacking of the acridine on DNA bases due to alkylation of these same bases; or to a stimulation of enzymatic conversion of the stacked interaction into a mutation heteroduplex heterozygote by means of the side-chain (which might, for instance, either inhibit gap closure during DNA replication or stimulate gap formation via excision repair).

Plant Systems

ICR-170 has been poorly characterized in the presently available plant systems. In Vicia faba, however, exposure of root tips produced an array of chromatid deletions and exchanges: about 20% of the cells had chromatid breaks after 4 hr of exposure to 100 µg/ml, and the lowest effective concentration tested was 10 µg/ml for 4 hr.

Insect Systems

ICR-170 induced recessive lethals in Drosophila melanogaster in all male germ cell stages following abdominal injection of adult males; while the injected dose was given as about 285 µg/g, leakage of the injected dose is a common problem with this procedure. In contrast to most mutagens, the highest induced recessive complete-lethal frequency was observed in the spermatogonial stages; it should also be noted that a substantial increase in mosaic lethal mutation frequencies was obtained in all post-gonial stages. A significant increase of whole and partial chromosome losses was induced in mature sperm. A presumed increase in reciprocal translocations was also induced in mature sperm; while no concurrent controls were conducted, the three events scored in 1253 tests were considerably above the historical spontaneous frequency. It is clear, however, that ICR-170 induces recessive lethal mutations more efficiently than chromosomal mutations in Drosophila.

The injection of doses of 40 μg or greater into the silkworm Bombyx mori resulted in the recovery of specific-locus mutations from treated oocytes.

Cultured Mammalian Cell Systems

ICR-170 induced forward mutations at the HGPRT locus with a lowest effective concentration of about 0.02 μg/ml and a mutagenic potency of about 3 mutants per 10^4 survivors/μg-hr/ml treatment. No metabolic activation component was present.

In vitro tests for chromosomal aberrations have not been reported with ICR-170, but unscheduled DNA synthesis has been induced in opposum lymphocytes. Over an exposure range of 1 to 100 μM, however, the response was inversely related to dose: three times more UDS occurred at the lowest dose than at the highest. Both the induction and the inhibition of excision repair is therefore implied by these results.

Intact Mammalian Systems

ICR-170 produced irritating results in male germ cells in the classical mouse specific locus system. At a single dose of 4 mg/kg, no mutations were recovered from 1092 postspermatogonial cells and two were recovered from 9275 treated spermatogonia. The Group I report interpreted these results as a doubtful positive, but we risk taking issue with this conclusion, not only because the sample size was small but also and more importantly because only 10 mutants were recovered after a dose of 300 R chronically administered to 58,457 tested spermatogonia. The latter result was clearly recognized as a positive response compared to the control frequency of approximately 1/20,000. If ICR-170 were to become an environmental pollutant, it would be reasonable to indict it as a mammalian mutagen on the basis of the specific locus test if and until the time that greatly expanded testing revealed otherwise.

Both male and female mice were employed in dominant lethal tests with ICR-170, again at 4 mg/kg. No induced dominant lethals could be detected, perhaps suggesting that ICR-170 is at most weakly active in producing chromosomal mutations. A Neurospora host-mediated assay was negative in both mice and rates at 8 mg/kg; this result might suggest mutagen inactivation in vivo, a transport deficiency, or the same sort of insensitivity seen in similar circumstances with MNNG. Unfortunately, neither the spot test nor any of numerous possible cytogenetic tests have been employed in intact mammals with ICR-170.

Summary

ICR-170 was chosen for this comparative exercise at least in part because of its strength and presumed frameshift specificity. Although extensively (if rather qualitatively) tested in microbial systems, it has been much less thoroughly characterized in Drosophila and the mouse than have many of the other compounds surveyed. Nevertheless, it was virtually always active in sensitive test systems.

The available values for lowest concentrations and potencies vary widely from system to system. Since it is unlikely that these differences could be significantly reduced by continued testing, the prospects for quantitative extrapolation among other than higher animals appear poor.

ICR-170 is primarily a laboratory mutagen, but can be instructively contemplated as an hypothetical environmental pollutant. It is clearly strongly mutagenic in a variety of systems ranging from prokaryotic microbes to plants and insects, but gives rather mixed results in the dearth of mammalian tests conducted to date. We nevertheless conclude that it would present a serious mutagenic hazard to humans, particularly since it strongly induces mutational damage in spermatogonial stem cells in Drosophila, a result supported by the very limited mouse specific locus data. If one were to attempt a risk estimation for this compound, one might conclude that an average exposure of 1 mg/kg to the population as a whole would fall within the range of doses capable of doubling the spontaneous incidence of human genetic disease resulting from simple gene mutation. This would mean introducing some 2000 cases of serious dominant disease per million liveborn in the first generations, to say nothing of the much greater number of recessive mutations whose irreversible deleterious effects would only begin to surface many generations in the future.

CHAPTER 30

COMPARATIVE MUTAGENICITY OF MYLERAN AND CYCLOPHOSPHAMIDE*

John A Heddle†

MRC Cell Mutation Unit, University of Sussex, Falmer
Brighton BN1 9QG, England

K. C. Bora and Douglas R. Stoltz

Health and Welfare, Canada, Tunney's Pasture
Ottawa, Ontario, Canada

MYLERAN

Background Information

Chemical Data. Myleran, whose proper chemical name is 1,4-butane dioldimethanesulfonate has the chemical structure (I):

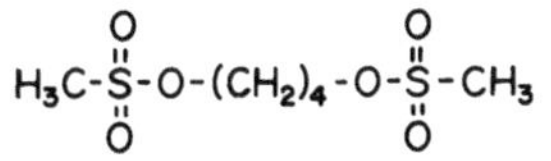

Its chemical formula is $C_6H_{14}O_6S_2$, molecular weight 246.3. It is a a white crystalline powder. It is almost insoluble in water, but 2.4 g will dissolve in 100 ml acetone 25°C, or 0.1 g in 100 ml of ethanol. It is an active alkylating agent which hydrolyzes in water. Annual production is probably less than 500 kg (1).

Uses. This compound is used as a chemotherapeutic agent for

*With the collaboration of J. Bishop, B. J. Dean, A. Mitchell, A. T. Natarajan, J. Preston, J. Schoneich, and D. Wild.

†On sabbatical leave from York University, Toronto, Canada.

the treatment of some forms of chronic myeloid leukemia, polycythemia vera, and myelosclerosis. Typical dosage level is 4-8 mg daily taken orally for several weeks. Myleran may be used for maintenance therapy at 0.5 to 2 mg/day.

Physiological Activity. In the rat and mouse 50-60% of a single dose of Myleran-^{35}S (10 mg/kg body weight) injected intraperitoneally in arachis oil was excreted within 24 to 48 hr, mainly as methanesulfonic acid; a small amount of unchanged Myleran and two unidentified components were present.

Carcinogenicity. The IARC report indicates that administration of 1,4-butanediol dimethanesulfonate (Myleran) to the mouse by intraperitoneal injection and to the rat by oral administration did not significantly increase the incidence of tumors. Intravenous administration in the mouse significantly increased the incidence of thymic lymphomas and ovarian tumors. Myleran, in conjunction with x-rays further augmented the incidence of thymic lymphomas. The increased incidence of thymic lymphomas and ovarian tumours are difficult to assess with respect to the carcinogenicity of Myleran in the mouse. "Although there is evidence that histological and cytological changes are associated with Myleran therapy, there is no firm evidence of an increased cancer risk among those treated" (1).

Summary of Mutagenicity

Mutagenicity in Microorganisms. In *E. coli*, Myleran is detected as a base-pair substitution mutagen without mammalian activation (mutation to streptomycin independence in B/sd-4) (Table 1). In strain K-12, Myleran induces prophage λ, also without mammalian activation. Similarly in the differential killing assay *polA*$^-$ bacteria are killed more efficiently than *polA*$^+$ without added S-9 (5).

In *S. typhimurium*, Myleran is detected without added S-9 on TA1535 and TA100 indicating base-pair substitution mutations are induced (6). Serum from treated rats has been reported to be effective also.

Mutagenicity in Drosophila and Plants. In Drosophila X-linked recessive lethal mutations are induced in sperm and spermatids by Myleran whether fed or injected. No other stages or genetic end points have been examined. In barley Myleran induces very low frequencies of chlorophyll-deficient mutations, but relatively high frequencies of chromosomal aberrations. In root tips of *V. faba*, Myleran produces chromatid aberrations, probably from S-phase cells. Both deletions and exchanges are induced although some studies have found relatively few exchanges. Break-points are

Table 1. Activity of Myleran detected in individual assay systems

Test system[a]	Result[b]	Comments	References
E. coli, bm	+ 100 μg/plate	Base-pair substitutions	2,3
E. coli λ induction	+ 500 μg/ml, 120 min		4
E. coli, *polA*$^+$/*polA*$^-$ growth			
Disc	+ 10 μg/disc		5
Suspension	+ 2 μg/ml	Not the lowest positive	5
S. typhimurium, fm TA1535, TA100	+ 50 μg/plate	Base-pair substitutions	6
D. melanogaster			
fm, postmeiotic	+ 0.4% at 23 μg/ml, 3 days	Feeding, slrl/ chromosome	7
fm at 14 loci	+ 2.3×10^{-3} mg/l. at doses that produce 2% slrl	Visibles	8
H. vulgare, aberr.	+ 400 mg/l.	Dry seed	9
H. vulgare, chlorophyll mutations	+ 400 mg/l.	Dry seed	10
A. cepa, aberr.	2 μg/ml, 24 hr	Chromatid aberrations	11
V. faba, aberr.	+ 25 mg/l.	Chromatid aberrations	12

(continued)

Table 1. (continued)

Test system[a]	Result[b]	Comments	References
Mouse lymphoma, TK locus locus	+ 1-2 × 10^{-3} mg/1. at 30 μg/ml, 4 hr		13
Human leukocytes	+ 5 μg/ml, 24 hr 24 hr + 2.4 μg/ml		
Mouse bone marrow, aberr.	+ 0.04/cell at 5 mg/kg	IP, 24 hr sample	14
Mouse bone marrow, micronuclei	+ 50 mg/kg	IP, 30 hr sample	15
Human lymphocytes treated in vivo	- 2 mg 2 daily (or 6 mos) + 0.002 mg/kg, chronic	(0.07 mg/kg/day)	16
Mouse males			
Spermatocytes aberr.	+ 8 mg/kg	Max. 2 days after	17
Spermatogonia aberr. at AI and AII	+ 8 mg/kg	Treatment, linear dose response curve	17
Stem cells, translocations at metaphase I	- 16 mg/kg IP	Control 20%	18
Mouse-dominant lethals			
Sperm	+ 10 mg/kg IP		19
Spermatids	- 10 mg/kg IP		19
Spermatocytes	+ 10 mg/kg IP		
Rat-dominant lethal, male premeiotic	+ 4 mg/kg IP	Sample at 10 weeks may include stem cells	20

(continued)

Table 1. (continued)

Test system[a]	Result[b]	Comments	References
Mouse, females, oocyte numbers	+	No evidence of mutation	21
dominant lethals	+ 20 mg/kg IP	40 mg/kg causes sterility	26
Mouse, specific locus			
Sperm	- 0/911 at 20 mg/kg IP	Too few tested to be considered negative	22
Spermatogonia	0/3844 × 7 at 20 mg/kg IP		22
Mouse, sperm abnormally	+ 2 × at 5 × 14.4 mg/kg daily		23

[a]No in vitro activation required. Abbreviations: fm = assay for forward mutations; bm = back mutations; slrl = sex-linked reces-recessive lethal mutations.

[b]Result column shows the lowest positive result obtained <u>or</u> the highest dose tested and not found positive.

usually in or near heterochromatin. Cells may have chromatid aberrations even at the second mitosis indicating either the chemical or a metabolite is effective long after treatment (>12 hr) or else lesions that can give rise to aberrations are still present at these late times.

<u>Mammalian Cells in Vitro</u>. Specific locus mutations are induced by Myleran in mouse lymphoma cells (TK locus) with or without metabolic activation. The mutagenic potency is 8×10^{-6} mutations/locus/survivor/μg/ml/hr. This is much lower than many other mutagens in this system. The mutation rate at 10% survival, however, is one of the highest ($\sim 1200 \times 10^{-6}$ mutations/locus). In cultured human leukocytes exposed to doses between 20 and 400 μM chromatid deletions are induced in the S phase of the cell cycle. Exchange aberrations are quite rare but are produced.

Mammalian Cells in Vivo. Myleran is an effective clastogen in somatic cells in vivo. It produces chromatid aberrations when injected intraperitoneally (IP) into mice at doses of 40 mg/kg or less. The dose response curve appears to be linear but the frequency of aberrations declines rapidly after two days. Micronuclei are induced in the bone marrow. Lymphocytes from treated patients have not shown aberrations when cultured in vitro in some cases. This may be due to the rapid loss from the bone marrow or the insensitivity of G_1 cells to damage, but is, nevertheless, surprising. Some positive results showing an increase in aberrations that was independent of dose have been reported.

In germ cells Myleran is also a clastogen in at least some stages. Chromosomal bridges and fragments are induced in meiotic cells of male mice at doses as low as 8 mg/kg. On the other hand, no reciprocal translocations were observed at metaphase I after treatment of spermatogonia with 16 mg/kg. The clastogenic activity of myleran in male mice is also indicated by the dominant lethal test which detected 10 mg/kg IP at the sperm and spermatocyte stages. In male rates 4 mg/kg was positive for premeiotic cells in this test.

Although there is no direct evidence of mutagenicity in oogenesis by Myleran in female mice Myleran decreased the number of oocytes present. While Group I reviewers believe that this may be attributable to chromosome aberrations induced in the oocytes, the observation makes it likely that Myleran does reach the germ cells and so is likely to mutate them.

In the specific locus test dominant lethality has reduced the number of offspring from treated males so much that the number examined (911) is too small to be meaningful according to the Group I reviewers. No mutants have been detected in these mice whose fathers received 20 mg/kg IP. Among the offspring arising from treated spermatogonia at this dose, no mutants have been detected in 9884 offspring. Sperm abnormalities are induced.

Summary

Myleran is a bifunctional alkylating agent that has been detected as a mutagen in all nonmammalian systems tested and in mammals, although not in all systems. It may be a carcinogen. It is capable of producing base-pair substitution mutations in bacteria, sex-linked recessive lethal mutations in Drosophila, and chlorophyll-deficient mutations in barley. Nevertheless, no specific locus mutations have been detected in a large sample of mice arising from treated spermatogonia. The specific locus test in postmeiotic stages is inadequate because of dominant lethality.

Chromosomal aberrations have been detected in plant root tips, mammalian bone marrow, and in male meiotic cells at low doses. It is not clear, however, that transmissible damage is induced. There is an effect on oocyte numbers, so the chemical probably reaches the oocytes where it might be able to cause transmissable genetic damage, although no positive evidence of this exists. Attempts to detect chromosomal damage in treated people have given mixed results.

Evaluation

When one is evaluating the risk of genetic damage, there are three components to be considered: the risk of disease such as cancer from somatic mutation, which is not considered here; the risk of an affected offspring; and the population risk. Since Myleran is administered primarily to leukemic patients, the population directly at risk is small although a significant proportion may be in their reproductive years of life. Direct evidence that Myleran causes transmissable genetic damage in mammals is lacking although dominant lethals are produced. Because of this and the known clastogenic ability of this compound in both somatic and germ cells there is a risk that chromosomally abnormal offspring could be produced. For treated males this risk is probably low if no offspring are fathered until postspermatogonially treated cells have been shed. The uncertainty of the specific locus mutation data for postspermatogonial stages also suggests this as a strategy, although given the sterility induced at these stages it may be unnecessary. The situation for exposed women is less certain, although Myleran probably reaches the oocytes. Given the clastogenic ability of this compound, treated women may wish amniocentesis to detect cytogenetically abnormal fetuses.

The population risk is difficult to evaluate. Clearly the appearance of no mutants among offspring of treated mice in the specific locus mutation test suggests a low risk. Nevertheless, the fact that a direct-acting mutagen reaches the testis and probably the ovaries of mice indicates that some mutations are to be expected. It is appropriate therefore to ask how great the mutation rate could have been and still have given this result and what the impact of such a mutation rate might be. Since upper 95% confidence limit on 0 in a Poisson is 3, the postspermatogonial rate of $0/9884 \times 7$ loci at a unit dose of 20 mg/kg has only a 5% chance of being 3/69,199, i.e., 4.3×10^{-5} mutations/locus/gamete. The spontaneous rate is 0.75×10^{-5} mutations/locus/gamete, which means the induced rate has an upper 95% confidence limit of 3.6×10^{-5} mutations/locus/gamete/unit dose. To produce an exposure of 20 mg/kg in a 70 kg person requires 1400 mg, so that the annual production of 500 kg may be reckoned as 3.6×10^{5} dose units

Thus the mutations produced would be 1.3×10^1 locus/year. Since this exposure continues over 30 years (the average reproductive life span), the mutations produced will be 3.9×10^2/locus. The population as a whole from whom the treated persons are drawn is probably 500×10^6 which gives a spontaneous frequency of $(500 \times 10^6) \times (0.75 \times 10^{-5}) = 3.8 \times 10^3$ mutations/locus. Thus the upper limit of the induced frequency would be $[(3.9 \times 10^2)/(3.8 \times 10^3)] \times 100\% = 10\%$ of the spontaneous rate. (In this calculation the dose response curve is assumed to be linear, and human germ cells are assumed to have the sensitivity of mouse spermatogonia.) The impact may in fact be much less given the likelihood of Myleran-induced sterility and dominant lethality, the fact that not all the Myleran synthesized may be used in this way, and that the exposed population is probably older than the population as a whole and less likely, on average, to have offspring. Given our lack of knowledge concerning the linear dose response curve for specific locus mutations in males or females, the effect of treatment schedules and the spectrum of mutations produced in comparison with the spontaneous mutational spectrum, the proportion of people who have children after treatment, and many other implicit assumptions, this calculation must be considered to be the crudest of estimates.

Conclusions

Myleran is a direct-acting mutagen capable of producing both point mutations and chromosomal aberrations in test organisms. In mice the chemical reaches the testis and probably the ovaries also. The population risk cannot be estimated with any reliability but the crudest of estimates based on many unproven assumptions is that the impact will be less than 10% of the spontaneous mutation rate. The risk to individuals of chromosomally abnormal offspring is uncertain, but may be significant. Experiments with mice with doses and schedules comparable to those used in chemotherapy would resolve this point. It is evident that we are in great need of mammalian data including the nature of the dose response relations and the effect of exposure schedules for chemicals generally.

CYCLOPHOSPHAMIDE

Background Information

Chemical Data. Cyclophosphamide, whose proper chemical name is N,N-bis (2-chloroethyl) tetrahydro-2H-1,3,2-oxaphosphorin-2-amine, 2-oxide monohydrate has the chemical structure II:

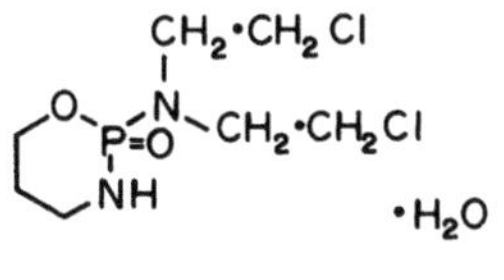

II

Its molecular weight is 279.1 ($C_7H_{15}Cl_2N_2O_2P \cdot H_2O$). It is a fine, white, odorless or almost odorless, crystalline powder. It is soluble at 20°C in 25 parts of water plus 1 part of ethanol, in benzene, chloroform, dioxane, and glycols. Aqueous solutions may be kept for a few hours at room temperature, but hydrolysis occurs at temperatures above 30°C, with removal of chlorine atoms (26).

Production. To produce cyclophosphamide, N,N-bis(2-chloroethyl)phosphamide dichloride is treated with propanolamine in the presence of trimethylamine and dioxane. Production is probably about 1000 kg annually.

Uses. Cyclophosphamide is only used as a drug. It is widely used in the treatment of various cancers. It is also used as an immunosuppressive agent in a variety of nonmalignant diseases, including rheumatoid arthritis, systemic lupus erythematosus, scleroderma, glomerulonephritis, chronic interstitial pneumonia, psoriatic arthritis, multiple sclerosis and a number of other diseases. Patients suffering from chronic hepatitis have also been treated with cyclophosphamide for long periods of time. It has been used as an immunosuppressive agent following organ transplantation.

Physiological Properties. The single intravenous LD_{50} has been reported to be 160 mg/kg body weight (bw) for rats, 400 mg/kg bw for guinea pigs, 130 mg/kg bw for rabbits and 40 mg/kg bw for dogs. The single oral LD_{50} is 180 mg/kg bw for rats. In mice, rats, dogs and man the predominant hematologic effect is leucopaenia; some depression of the bone marrow and thrombocytes also occurs. In rats the drug is rapidly absorbed from blood, the specific activity in various tissues being highest within the first 20-30 minutes following its injection. The compound is rapidly metabolized; up to 75% of injected radioactivity is excreted in the urine within 5-8 hr. The distribution of cyclophosphamide in the body and its metabolism appear to be similar in man and animals. After its injection IV, the drug is rapidly absorbed from the blood: in patients receiving 6-80 mg/kg bw/day of radiolabeled cyclophosphamide IV, the radioactivity was distributed rapidly to all tissues. Its half-life in the plasma was 6.5 hr. Sixty-eight percent of the injected label was excreted in the urine within 4 days but no radioactivity was found in the expired air or in the feces. Ten to 14% of the drug is excreted unchanged; 56% of the reactive metabolites were bound to plasma proteins. Cyclophosphamide and several of its metabolites have also been found in bile, milk, sweat, saliva and cerebrospinal fluid. Both the metabolism of intravenously administered cyclophosphamide and the rate of excretion of its metabolites show large individual variations.

Cyclophosphamide is not an alkylating agent as such but rather it requires metabolic activation in vitro and in vivo. Similar reactions may occur in bone marrow, kidney, and tumor tissues. Other drugs, such as barbiturates or corticosteroids, have been found to affect cyclophosphamide metabolism and may consequently alter its toxicity and cytostatic activity. Reactive cyclophosphamide metabolites have been shown to crosslink nucleic acids; in vitro and in vivo studies indicate that the major reaction site is the 7-position of guanine.

Teratogenicity and Effects on Fertility. Cyclophosphamide is teratogenic in several species, including mice, rats, rabbits, and chickens. It produces a variety of skeletal, soft tissue and other malformations and an increased number of resorptions; the type and frequency of malformations are strictly dose- and time-dependent. Placental transfer of ^{14}C-cyclophosphamide has been demonstrated in mice. The drug is most toxic to the human fetus during the first three months, and congenital abnormalities have been detected after IV injection of large doses to pregnant women during this period of pregnancy. Cyclophosphamide can cause sterility in either sex in man. It can damage germinal cells in prepubertal and pubertal males and accounts for premature ovarian failure in females.

Carcinogenicity. According to the IARC report (26), "Cyclophosphamide is carcinogenic in mice and rats following its intraperitoneal injection, in rats following its intravenous injection and in mice following its subcutaneous injection, in doses similar to those used in clinical practice. It produced mainly lung and lymphoreticular tumors, and also tumors of the liver and reproductive organs, sarcomas and squamous-cell carcinomas of the skin. The available case reports in which tumors were reported to have occurred in patients treated with cyclophosphamide provide insufficient evidence to determine if there is an increased risk of cancer following the therapeutic use of this drug, with the possible exception of cancer of the bladder" (26).

Summary of Mutagenicity

Mutagenicity in Microorganisms. Cyclophosphamide requires activation to mutate *E. coli* (Table 2). After activation it produces back mutations at arg^- and forward mutations to methotrexate

Table 2. Activity of cyclophosphamide detected in individual assay systems[a]

Test system	Activation[b]	Result[c]	Comments	References
E. coli, forward mutation	+	+	Base pair substitutions	27,28
E. coli, fm, ip HMA (mouse, SC)		- 870 mg/kg		29
E. coli polA^{+}/*polA*$^{-}$				
Disc	+	+ 10 µG/disc		5
Liquid	+	+ 250 µg/ml		5
B. subtilis, Rec^{+}/Rec^{-}	+	- 1000 µg/ml	Under reexamination	30
	-	+ >4 at >1000 µg/ml		30
S typhimurium				
bm	+	+ 1.4 rev/nmole	TA1535 base-pair substitution	31
IP HMA (♂ mouse, SC)		+ 870 mg/kg	G46	29
		+ 200 mg/kg	TA1950	32
Serum, mouse, IP		+ 200 mg/kg	Dose to mouse; TA1535	33
S. marcescens, IP HMA (mouse SC)		+ 870 mg/kg	Strain a21	29
		- 870 mg/kg	Strain a13	29
Yeast, chromosome loss	+	250 µg/ml	Activation required	34
S. cerevisciae				
Mitotic co		± 10,000 µg/ml	Active in D3,D5 but not D4	35
Mitotic co		+	Increasing activity as the compound aged in buffer at 30°C	35

(continued)

Table 2. (continued)

Test system	Activation[b]	Result[c]	Comments	References
Gene conversion IP HMA (mouse, PO)	+	500 mg/kg	Renal excretion blocked	36
Gene conversion testes HMA (mouse, SC)	+	380 mg/kg		37
Gene conversion in vitro				38
urine, rat, PO	+	340 mg/kg		38
Gene conversion in vitro				
urine, human, IV	+	3.1 mg/kg		39
V. faba, chromosomal chromosomal aberrations	+	34 μg/ml		40
D. melanogaster, slrl	+	Adult feeding	Postspermatogonial stages	41
Mouse lymphoma TK^-	+	$(15\text{-}27)\times10^{-6}$ m/s/μg/ml/hr	Actual data not available	13
Human lymphocytes, in vitro with rat plasma, IV	+	500-1000 mg/kg	Diluted 10-50×	42
HeLa cells, c. aberr.	+	84 μg/ml, 48 hr		
Human lymphocyte, SCE	-	500 μg/ml		43
Human lymphocyte, SCE	+	3.7 at 100 μg/kg		44
Mouse spermatogonia, SCE	+	5 × 20 mg/kg		43

(continued)

Table 2. (continued)

Test system	Activation[b]	Result[c]	Comments	References
Rabbit bone marrow, SCE	+	2.4/cell at 20 mg/kg, 24 hr	Curve probably bending up	45
Mouse bone marrow, aberr.	+	5 mg/kg IP	Curve approximately linear	46
Mouse bone marrow	+	5 mg/kg IP		
Mouse bone marrow, micronuclei	-	200 mg/kg, 96 hr	Positive at earlier times	47
Chinese hamster, bone marrow aberrations	+	20 mg/kg,	Complex dose response curve	46
Chinese hamster bone marrow	-	10×64 mg/kg daily	72 hr after last treatment, positive results at 6 hr	48
Human bone marrow	-	40 mg/kg IV 24 hr		49
Human leukocytes, in vivo, aberr.	+	3-5 mg/day, 4-6 weeks	~0.065 mg/kg/day = ~2 mg/kg total 74 hr culture	50
Human leukocytes	-	500 mg-44g total dose	72 hr culture	51
Human leukocytes, SCE	+	3× at 100 mg/day × 13	No increase prior to 13 days; decreases to 0 over 2 weeks	

(continued)

Table 2. (continued)

Test system	Activation[b]	Result[c]	Comments	References
Mouse spot test	+	30× at 5 mg/kg		52
Mouse oocytes, aneuploidy at Meta II	+	74% at 145 mg/kg IP	Spontaneous 14%; 12 hr post-treatment	53
Mouse spermatogonia/spermatocytes aberrations	+	20% abnormal cells at 97 mg/kg, oral	Metaphase I, 13 days post-treatment; 3 translocations in 80 cells	54
Chinese hamster spermatogonia, aberrations	+	2 × 96 mg/kg, oral	24 hr between treatments	55
Mouse spermato-ytes cytes, treated as stem cells	-	0/400 at 200 mg/kg IP		18
UDS mouse sperm, treated at meiosis or later stages	+	3-6× at 200 mg/kg IP		56
Mouse translocation test				
Sperm	+	5/35 at 350 mg/kg IP		57
Spermatids	+	34/84 at 350 mg/kg IP		57
Spermatocytes	+	2/287 at 350 mg/kg IP		57
Spermatogonial stem cells	-	1/1350 at 350 mg/kg IP		57

(continued)

Table 2. (continued)

Test system	Activation[b]	Result[c]	Comments	References
Dominant lethals, mouse	+	40 mg/kg, IP		58
Postmeiotic males	+	80 mg/kg, oral		59
Various stages, males	+	0.005% in drinking water for 4 weeks		60
Various stages, males	+	0.01% in drinking water for 20 days		61
Premeiotic males	+	210 mg/kg, IP		62
Premeiotic males	-	210 mg/kg, IP		63
Premeiotic females	+	210 mg/kg, IP		63
Premeiotic females	+	200 mg/kg, oral		64
Mouse, sperm abnormality	+	2× at 5 × 130 mg/kg daily IP		23
Mouse, specific locus spermatogonia	+	3/(3642×7) at 350 mg/kg		22

[a]Abbreviations: fm = assay for forward mutations; bm = assay for back mutations; co = assay for crossing over; c = chromosome; aberr. = chromosomal aberrations; uds = unscheduled DNA synthesis; hma = host-mediated assay; slrl = sex-linked recessive lethals.

[b]Shows the use of in vitro activation by +.

[c]Result column shows the lowest positive result obtained or the highest dose tested and not found positive. In some cases many other studies have been carried out at other doses in the same test system.

(MTR) resistance (in strains K-12/343/113) which indicates the production of base-pair substitution mutations. All known metabolites are positive without further activation for mutations to arg$^+$, gal$^+$ and MTR in 343/113. In some cases this mutagenicity has not been detected at comparable doses. The differential killing assay (polA$_1^-$ versus polA$_1^+$ in E. coli) is positive in both disc (250 μg) and liquid (10 μg) assays.

In S. typhimurium cyclophosphamide requires activation. It is positive on TA1535 (1.4 revertants/nmole) (31), but it required ten times the usual amount of S-9. Others have found it positive with the usual amounts of S9 (65), while still others have reported it to be inactive (27). A positive result in the absence of activation has been reported (65).

In S. cerevisciae, cyclophosphamide does not induce mitotic gene conversion without activation, but positive results on mitotic crossing over can be obtained by leaving the compound in buffer at 30°C. This result may indicate mutagenic activity arising by hydrolysis. In the host-mediated assay for mitotic crossing over both a period in the peritoneal cavity of mice treated with 500 mg/kg (but only if renal excretion is blocked) and the urine of treated mice used in vitro give positive results. Urine from treated patients is also positive.

Mutagenicity in Plants and Drosophila. Chromosomal aberrations are produced in both A. cepa roots and V. faba roots. Again, however, indications of an active breakdown (nonmetabolic) product were found, viz., the aberration rate increased with 8 min hydrolysis at 90°C but decreased with longer hydrolysis.

In Drosophila, sex-linked recessive lethal mutations were detected in treated sperm, spermatids and spermatocytes but not spermatogonia. There seemed to be a maximum at 3-5%. Neither dominant lethals nor X, Y chromosomal loss were detected in treated sperm.

Mammalian Cells in Vitro. Specific locus mutations were detected in mouse lymphoma cells treated with cyclophosphamide. Activation is required (no significant increase in mutations at 1000 μg/ml without activation), but then a positive result is obtained [(15-27) × 10^{-6} mutations/locus/survivor/μg/ml/hr]. Expressed as a rate per μg/ml/hr at 10% survival, cyclophosphamide is the most active of the compounds under consideration that has been tested to date.

Chromosomal aberrations have been detected in cultured human cells. To be effective in cultured human leukocytes, cyclophosphamide must be activated. Plasma from treated animals is effective as is cyclophosphamide incubated with rat liver slices. An effect without activation (3-4 × control) has been observed in HeLa cells and Burkitt's lymphoma. The activated form is also effective on HeLa cells. Although chromatid deletions are the most common aberration, chromatid exchanges and chromosome aberrations have been observed. The kinetics for total aberration yield are approximately linear although a "multihit" component has been reported.

Most breakage occurs in the S phase of the cell cycle. Mitotic delay occurs, so the quantitative interpretation of aberration data is difficult.

Mammalian Cells in Vivo. Bone marrow cells of mouse, rat and Chinese hamster show chromatid aberrations, chromosome aberrations, and numerical changes in animals treated with cyclophosphamide. Significant damage has been observed at 5 mg/kg in the mouse, which is in the clinical dose range. Micronuclei have been detected in the mouse at 5 mg/kg, in man at 40 mg/kg. In treated humans, aberrations similar to those induced in experimental animals have been found in the base marrow. Lymphocytes cultured from treated patients also exhibit chromosomal damage. Variations in sensitivity have been observed from one mouse strain to another and also among species. The order of decreasing sensitivity seems to be mouse, rat, Chinese hamster, and man. The mammalian spot test gave a positive result at 5 mg/kg (6/180 = 3.3×10^{-2}, treated vs. 1/891 = 0.1×10^{-2}, control).

Mouse oocytes treated premeiotically with doses up to 100 mg/kg and examined at metaphase II did not show a significant increase in abnormalities, but this may be due primarily to the small number of cells analyzed and the high spontaneous rate. Preovulatory oocytes showed a dose related increase at 30, 50, and 145 mg/kg. A large number of positive reports of chromosome abnormalities at doses of 100 mg/kg or greater exist for males.

A variety of tests which involve male germ cells have been positive. Increases of 3 to 6 times control values of unscheduled DNA synthesis have been detected in male mice at 200 mg/kg (IP) from early meiosis to midspermatids. Translocations have been recovered from male mice treated with 350 mg/kg IP at several germ cell stages: sperm (5/35), spermatids (33/84) and spermatocytes (2/287) but not spermatogonial stem cells (0/202 at 350 mg/kg, 1/1148 at 350-400 mg/kg). Cyclophosphamide is positive in the sperm abnormality assay. It produces dominant lethal effects in both males and females. In males although 13.3 mg/kg was not detectable, 40 mg/kg was positive. In females 150 mg/kg (oral) was not detected by 210 mg/kg was positive. Fractionation of the dose did not change the frequency significantly. In the specific locus mutation test, 350 mg/kg was positive in postspermatogonial stages (3/3642 offspring = 118×10^{-6} mutations per locus, versus 1.1×10^{-6} mutations per locus in the control.

Summary

Cyclophosphamide is an indirect mutagen capable of causing base pair substitutions and chromosomal aberrations. The positive results obtained without activation in *S. cerevisciae* (mitotic crossing over) may indicate either the presence of an active impurity, a residual activating ability of the yeast or the existence of an active product formed spontaneously under conditions that promote hydrolysis (e.g., high temperatures). The latter possibility has been advanced to explain data obtained on chromosomal aberrations in plant root tips. Cyclophosphamide is carcinogenic in rodents and possibly in man. It is detected as positive in almost all tests if in vitro activation is used. Since chromosomal, dominant lethal and specific locus mutations are induced in mammals in vivo, cyclophosphamide or an active metabolite must reach the gonads of both male and female mice. There can be little doubt that germ cell mutational effects are also produced in man, as active metabolites exist in the urine and chromosomal aberrations are detectable both in bone marrow and in peripheral leukocytes cultured from treated patients. The only use of the chemical is as a drug for various human diseases. The potential harm that this drug may cause to the patients involved and any future offspring should be borne in mind while alternate therapies are being considered.

Evaluation

In evaluating the risk of genetic damage there are three separate risks to be considered. The first is the consequences of somatic mutation which may lead to cancer or other disease in the treated individual. This is not considered here. The second risk is that of a transmitted mutation affecting the immediate rearrangement. The third risk is that of the population as a whole from recessive mutations which may often not cause human ill health until many generations after their induction.

The data on translocations produced by treatment of male mice at various stages of spermatogenesis and then detected in the F_1 are shown in Table 3. Although this single acute exposure is much greater than the daily human exposure, it is of interest to calculate the dose accumulated by people in each of the germ cell stages. Assuming a 10 mg/kg bw daily maintenance dose, and also assuming that the rate of translocation induction is the same per unit dose (i.e., a linear response curve and no fraction effect), one can compute an estimate of the translocation frequency that would be

Table 3. Translocation frequencies observed in male progeny of males treated with an acute dose of cyclophosphamide and the frequency expected from a daily maintenance dose of 10 mg/kg bw[a]

Germ cell stage	Stage duration (days)	Rate at 350 mg/kg bw (%)	Dose to stage (mg/kg)	Translocations expected (%)
Sperm	30	14	300	12
Spermatids	18	39	180	20
Spermatocytes	14	1	140	0.4
			Sum accumulated:	32

[a]Data of Sotomayer and Cumming (57).

produced in a cell during differentiation into a sperm. The results of such a caclulation are also shown in Table 3 where it can be seen that a very substantial frequency (32%) results. At the very least, these data indicate that an experiment with mice at doses and schedules comparable to those used in chemotherapy may well give a positive result. Furthermore, the possibility of inherited chromosomal damage in man must be considered if people of reproductive age are being treated. The murine data also suggest that males treated with cyclophosphamide might avoid much or all of the risk of an induced translocation (and perhaps other aberrations) being transmitted if they avoid having children until all of the sperm present have been generated from cells treated no later than as spermatogonial stem cells. If a pregnancy occurs in a woman whose husband has been exposed recently enough that the sperm was treated during spermatogenesis, then amniocentesis is a possibility to determine the chromosomal constitution of the fetus. The extent of exposure and the rate of chromosomal damage found in the mouse could be factors in this decision, bearing in mind that there is some evidence that man may be less sensitive than the mouse. Obviously there is need for translocation data from female mice that have been treated with cyclophosphamide so that similar risk estimates can be made.

In the case of cyclophosphamide which has no other uses than the medical ones (except for experiments and some veterinary usage) and for which there is mouse specific locus data, a novel calculation of the mutation load can be made.

The specific locus mutation rate observed in the mouse after a dose of 350 mg/kg bw is 1.2×10^{-4} m/l./gamete. If we let 1 mg/kg bw be one unit of dose (μ) this rate can be expressed as 3.4×10^{-7} m/l./gamete/μ. Since the average adult person is about 70 kg

$$1\ \mu = 70\ \text{mg/person; i.e., } 70\ \text{mg} = 1\ \text{person}\ \mu$$

and the annual production of about 1000 kg (10^9 mg) is equivalent to

$$\frac{10^9\ \text{mg}}{70\ \text{mg/person}\ \mu} = 1.4 \times 10^7\ \text{person}\ \mu$$

If there is a linear dose response curve, the distribution of these units among people will not influence the number of mutations produced which, using the mouse rate, will be

$$(3.4{\times}10^{-7}\ \text{m/l./gamete/}\mu){\times}(1.4{\times}10^7\ \text{person}\ \mu) = 5\ \text{(m/l./gamete)} \times \text{person}$$

Allowing a reproductive rate of Z gametes/person there will be a 6 Z m/l. produced.

If we assume that these mutations will arise in a population of about 500 millions (i.e., approximately that of Western Europe and North America), the spontaneous number is

$$500{\times}10^6\ \text{persons}{\times}\text{Z gametes}{\times}(7.5{\times}10^{-6}\ \text{m/l./gamete}) = 3.8{\times}10^3\ \text{Z m/l.}$$

The annual production of cyclophosphamide is thus equivalent to

$$\frac{5\ \text{Z m/1} \times 100}{3.8 \times 10^3\ \text{Z m/1}} = 0.13\%$$

of the spontaneous rate. If we consider that this rate of exposure continues for 30 years (an estimate of the average reproductive life span), the effect would be 30 × 0.13 = 4% of the spontaneous rate. Clearly this assumption is an upper limit.

This estimate may be an overestimate because (1) not all of the cyclophosphamide is used for human exposure, (2) many of the exposed persons may be beyond the reproductive age or may not have children because of their disease or other reasons, or (3) the population at risk may be somewhat larger particularly now that the manufacturers are recommending its use only in life-threatening situations and are warning physicians that a genetic risk may exist. Furthermore, the estimate depends not only on the usual assumptions of the similarity of mouse and man, and of the similarity of the induced and spontaneous mutation spectra, but also upon the assumptions of a linear dose response curve and no fractionation effect whereas no data exist on either point. In absolute terms the impact of an altered mutation rate upon human health cannot be estimated yet. It is interesting nevertheless to make even a very crude estimate of the relative impact of a mutagenic exposure. Clearly some of the uncertainties could be answered by research into the actual extent of the use of the drug and the reproductive success of the treated population. Perhaps the most significant point, however, is that for any quantitative evaluation of risk at the moment, mammalian data are essential.

Conclusions

Cyclophosphamide is a mutagen after activation that is capable of producing both point mutations and chromosomal aberrations in test organisms. In mice, the active form reaches the gonads in both males and females. Similar effects are expected in exposed people because active metabolites are excreted in the urine of exposed people and because chromosomal aberrations can be found in their bone marrow or in their lymphocytes (after culturing). The population hazard from induced mutations cannot be estimated with any reliability from the mouse data since the doses used in people differ enormously as do the treatment schedules. On the tenuous assumption that the dose response is linear, with no fractionation effect, the impact of the known annual production of cyclophosphamide is likely to be less than 4% of the spontaneous rate. Similar assumptions about the production of translocations in males suggests that there might be a significant risk of a chromosomally abnormal child if conception occurs within two months of the end of some commonly used treatment schedules. Experiments on mice at doses and schedules closer to the human situation seem feasible for both male and female mice with regard to translocation production. It seems unlikely that the specific locus mutation test would give positive results under these conditions.

REFERENCES

1. International Agency for Research on Cancer Monographs on the Evaluation of Carcinogenic Risk of Chemicals to Man, Lyon.
2. Sybalski, W. Ann. N. Y. Acad. Sci. 76: 475-489 (1958).
3. Iyer, V. N., and Sybalski, W. Appl. Microbiol. 6: 23-29 (1958).
4. Specht, I. Arch. Mikrobiol. 51: 9-171 (1965).
5. Rosenkranz, H. S., in preparation.

6. McCann, J., Choi, E., Yamasaki, E., and Ames. B. N. Proc. Natl. Acad. Sci. (U.S.) 72: 5135-5139 (1975).

7. Röhrborn, G. Zeitschr. Vererb. 90: 116-131 (1959).

8. Browning, L. S., and Altenburg, F. Genetics 46: 1317-1321 (1961).

9. Moutschen, J., and Moutschen-Dahmen, M. Hereditas 44: 415-446 (1958).

10. Heslot, H., Ferrary, R., Levy, R., and Monard, C. C. R. Acad. Sci. (Paris) 248: 729-732 (1959).

11. Deysson, G., and Truhaut, R. C. R. Acad. Sci. (Paris) D268: 83-85 (1969).

12. Michaelis, A., and Rieger, R. Zuchter 30: 150-163 (1960).

13. Clive, D Group I Report.

14. Léonard, A., and Léonard, E. D. Mutat. Res., in press.

15. Jacquet, P., Léonard, A., and Gerber, G. B. J. Toxicol. Environ. Health 2: 619-624 (1977).

16. Krajincanic, B., Lazarov, A., Zunic, Z., and Rdojicic, B. Strahlentherap. 151: 459-462 (1976).

17. Partington, M., Fox., B. W., and Jackson, H. Exptl. Cell Res. 33: 78-88 (1964).

18. Léonard, A., and Linden, G. Mutat. Res. 16: 297-300 (1972).

19. Ehling, U. H., and Malling, H. V. Genetics 60: 174-175 (1968).

20. Partington, M., and Jackson, H. Genet. Res. 4: 333-345 (1963).

21. Generoso, W. M., Stout, S. K., and Huff, S. W. Mutat. Res. 13: 171-184 (1971).

22. Ehling, U. Group I Report.

23. Bruce, W. R., and Heddle, J. A. Can. J. Genet. Cytol. 21: 319-334 (1979).

24. International Agency for Research on Cancer. Monographs on the Evaluation of Carcinogenic Risk of Chemicals to Man. Vol. 9, Lyon, 1975.

25. Jackson, H., Fox, B. W., and Craig, A. W. J. Reprod. Fert. 2: 447-465 (1961).

26. International Agency for Research on Cancer. Monographs on the Evaluation of Carcinogenic Risk of Chemicals to Man, Volume 9, Lyon, 1975, pp. 135-156.

27. Ellenberger, J., and Mohn, G. Arch. Toxikol. 33: 225-240 (1975).

28. Ellenberger, J., and Mohn, G. Mutat. Res. 39: 120-121 (1976).

29. Propping, P., Röhrborn, G., and Buselmaier, W. Moo. Gen. Genet. 117: 197-209 (1972).

30. Kada, T., Group I report.

31. McCann, J., Choi, E., Yamasaki, E., and Ames, B. N. Proc. Natl. Acad. Sci. (U.S.) 72: 5135-5139 (1975).

32. Braun, R., and Schoneich, J. Mutat. Res. 31: 191-194 (1975).

33. Braun, R., and Che, unpublished data.

34. Parry, J. M., personal communication.

35. Mayer, V. W., Hybner, C. J., and Brusikc, D. J. Mutat. Res. 37: 201-212 (1976).

36. Siebert, D. Mutat. Res. 28: 57-61 (1975).

37. Fahrig, R. Mutat. Res. 31: 313 (1975).

38. Sieberg, D. Z. Krebforsch 81: 261-267 (1974).

39. Siebert, D., and Simon, D., quoted in group I report (host mediated assay).

40. Rieger, R., and Michaelis, A. Biol. Zentralbl. 80: 301-317 (1961).

41. Vogel, E. Mutat. Res. 33: 221-228 (1975).

42. Hampel, K. E., Kober, B., Rosch, D., Gerhartz, H., and Meinig, K. Blood 27(6): 816-823 (1966).

43. Allen, J. W., and Latt, S. A. Nature 260: 449 (1976).

44. Bauknecht, T., Vogel, W., Bayer, V., and Wild, E. Human Genet. 35: 299 (1977).

45. Stetka, D. G., and Wolff, S. Mutat. Res. 41: 333-342 (1976).

46. Goetz, P., Šràm, R. J., and Dohnalova, J. Mutat. Res. 31: 247-254 (1975).

47. Datta, P. K., and Schleiremacher, E. Mutat. Res. 8: 623-628 (1969).

48. Schmid, W., and Staiger, G. R. Mutat. Res. 7: 99-108 (1969).

49. Kaung, D. T., and Swartzendruber, A. Dis. Chest 55: 98-100 (1969).

50. Dobos, M., Schuler, D., and Fekete, G. Human Genet. 22: 221-227 (1974).

51. Schinzel, A., and Schmid, W. Mutat. Res. 40: 139-166 (1976).

52. Fahrig, R. Arch. Toxicol. 38: 87-98 (1977).

53. Hansmann, I. Mutat. Res. 22: 175-191 (1974).

54. Rathenberg, R., and Miller, D. Agents Actions 2: 180-185 (1972).

55. Machemer, L. N. S. Arch. Pharmak. 282(S): R61 (1974).

56. Sotomayer, R. E., Sega, G. A., and Cumming, R. B. Mutat. Res., in press.

57. Sotomayer, R. E., and Cumming, R. B. Mutat. Res. 27: 375-388 (1975).

58. Ehling, U., personal communication.

59. Röhrborn, G., and Vogel, F. Dtsch Med. Wochenschr. 92: 2315-2321 (1967).

60. Šràm, R. J. Mutat. Res. 41: 25-42 (1976).

61. Šràm, R. J., and Zhurkov, V. S., quoted in Group I report.

62. Brittinger, D. Humangenetik 3: 156-165 (1966).

63. Röhrborn, G. Chem. Mutagenesis Mem. Man. 294-316 (1970).

64. Machemer, L., and Lorke, D. Mutat. Res. 29: 209-214 (1975).

65. Hanna, P. J., and Dyer, K. F. Mutat. Res. 28: 405-420 (1975).

CHAPTER 31

COMPARATIVE MUTAGENICITY OF TEPA, THIOTEPA, AND METEPA

N. P. Bochkov

Institute of Medical Genetics, Academy of Medical Sciences, Moscow, 115478, U.S.S.R.

INTRODUCTION

All the three chemicals are typical of alkylating compounds, and their molecules contain three ethylenimine functional groups. None of these are naturally occurring compounds; their chemical formulas are as follows:

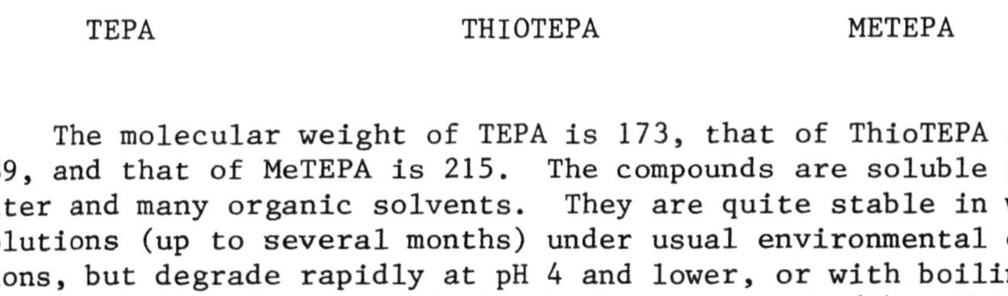

TEPA THIOTEPA METEPA

The molecular weight of TEPA is 173, that of ThioTEPA is 189, and that of MeTEPA is 215. The compounds are soluble in water and many organic solvents. They are quite stable in water solutions (up to several months) under usual environmental conditions, but degrade rapidly at pH 4 and lower, or with boiling. They are believed to degrade in the human stomach (1). ThioTEPA does not degrade in 24 hr at 37°C either in a culture medium or in a lymphocyte culture (2), and it is therefore very convenient to work with under experimental conditions.

In vivo, ThioTEPA is metabolized to TEPA because the sulfur atom is replaced by oxygen. Ultimately, ThioTEPA degrades to non-organic phosphate, which is excreted.

Originally, TEPA was produced in the U.S.A. and used in the sixties for treatment of cotton fabrics, but production was discontinued for economic reasons. The total amount of TEPA produced in the U.S.A. was 400 tons. In addition, a Japanese company, in the sixties and early seventies, produced several hundred kilograms of TEPA a month, also for the treatment of textile goods. TEPA was produced in West Germany as well.

TEPA, ThioTEPA, and MeTEPA were used as effective hemosterilants for insects, but they were not in a wide use because of their toxic effects on the environment.

TEPA was tested as a drug for cancer, but a less toxic analog was found; ThioTEPA is used for this purpose in small amounts in the U.S.A., Belgium, Sweden, and U.S.S.R. (several kilograms per year). ThioTEPA is used to treat breast adenocarcinoma, ovarian adenocarcinoma, malignant lymphomas, and cancer of the bronchi.

TEPA and ThioTEPA induce benign tumors in mice and rats. The literature includes several reports of leukemia in patients treated by ThioTEPA for malignant tumors. These data are not, however, sufficient to evaluate the carcinogenic risk of the chemical.

MUTAGENIC EFFECT OF TEPA, THIOTEPA, AND METEPA

A summary of data on mutagenic properties of TEPA, ThioTEPA, and MeTEPA are given in Tables 1-3. Some of the most significant results and conclusions are given below in detail.

Microorganisms

The mutagenic effect of all the three compounds was studied on Salmonella (3,4). All three compounds showed mutagenic properties without metabolic activation. Investigation was carried out on the test of revertant mutations in strains TA1535 and TA100. Quantitative data on dose-effect curves were not obtained for TEPA and MeTEPA. For ThioTEPA a linear dependence of dose was found (4). The mutagenic activity of ThioTEPA was 0.86 g/revertant colony.

TEPA, ThioTEPA, and MeTEPA were active in *E. coli* without metabolic activation. Very marked quantitative differences are observed with this test system, depending on the strain of bacteria, on the genetic damage recorded, and on the testing method; therefore, it is not reasonable to describe the individual experiments. All the necessary data are to be found in literature (3,5-7).

Plants

All the three mutagens were tested on cells of lateral root tips of *Vicia faba*. They all induce chromatid aberrations of a delayed type (8,9). The roots were treated for 2 hr with 7.5×10^{-4} M TEPA. Under these conditions 31.8% aberrant metaphases were found. Cells of Chinese hamster were found to be 23 times as sensitive as cells of *Vicia faba* (10,11).

ThioTEPA was used in various concentrations, from 5×10^{-4} M to 2.5×10^{-3} M for 2 hr. The number of aberrant metaphases ranged from 17.5% to 80.5%. The dose dependence was of the exponential type. The mutagenic effect of ThioTEPA was strongly modified by caffeine. In Chinese hamster cells doses less by a factor of 20-25 induced the same number of aberrations. Information on a mutagenic activity of ThioTEPA for roots of *Vicia faba* is given in literature (12-19). In addition to chromosomal aberrations, ThioTEPA induced sister chromatid exchanges. This test is more sensitive (20-23). The frequency of sister chromatid exchanges did not change with caffeine treatment.

MeTEPA has been studied less than its analogs on *Vicia faba* (9). Its mutagenic effect, however, is beyond doubt. On treatment for 0.5 hr with a MeTEPA concentration of 0.233×10^{-2} M, a maximum of aberrant metaphases (63%) was observed in 28 hr. Of other plant substrates for testing ThioTEPA, only *Allium proliferum* was used. This mutagen induces chromatid aberrations of a delayed type. At a dose of 5×10^{-4} M on treatment for 2 hr, 11% of aberrant metaphases were observed. This effect is very similar to that on *Vicia faba* (24).

Drosophila

All three compounds induce mutations in germ cells of Drosophila. The effects of TEPA were studied in the laboratory of R. Šram (25,26). According to these studies, recessive lethals and translocations were induced by TEPA at all stages from spermatogonia to spermatozoa. Dominant lethals were evaluated only at the exposure of mature spermatozoa. The compound in solution was injected intraperitoneally at a dose of $1.0\text{-}3.0 \times 10^{-1}$ mmole/l. or 10 mg/kg. Under these conditions 6-8% of new recessive lethals were induced.

Table 1. TEPA

Test system	Genetic end points	Metabolic and cell stage factors	Exposure route	Dose response
Bacterial Salmonella	Reversions	-S9, growing	Spot test	
E. coli	Reversions	-S9, growing	Spot test; liquid	
Saccharomyces	Reversions			
Schizosaccharomyces pombe	Reversions at locus and suppressors	Stationary phase		1.0
Vicia faba	Chromatid aberrations	Root tips	Grown in solution	1.0
Drosophila	Dominant lethals	Sperm and spermatogonia	Injected	2.0
	Recessive			1.0
	Translocations			2.0

LCT[a]	LEC[b]	Frequency	Comments
		+	TA1535
2.3×10^{-4} M/1		+	B/sd-4; K-39; Lambda induction WP-2
		No effect	Tested on undefined ade-1 allele
	2.3×10^{-3} M (0.1 mg/kg)	1×10^{-5}/mg/kg	Tested d19-revertants probably mainly suppressors
130 mg/kg		10×10^{-2}/mg/kg	Cell stage specific
5 mg/kg	0.5 mg/kg	0.05×10^{-2}/mg/kg	LEC extrapolated; no effect on spermatogonia strong storage effect on translocation
		+	
		+	

(continued)

Table 1. (continued)

Test system	Genetic end points	Metabolic and cell stage factors	Exposure route	Dose response
Mammals				
Chromosome aberrations				
In vivo	Chromatid aberrations	Bone marrow	Intraper-itoneal	1.5
In vitro	Chromatid aberrations	Chinese hamster fibroblasts	Solution	One dose
	Human lymphocytes	Human lym-phocytes	Solution	1.3
Translo-cation	Heritable transloca-tion (F_1)	Late spermatids; spermato-gonia	Intraper-itoneal	One dose
Dominant lethals		6 male mice; 1 female mouse; 1 rat	Intraper-itoneal; oral	

[a]Lowest concentration tested.

[b]Lowest effective concentration.

LCT[a]	LEC[b]	Frequency	Comments
0.6 mg/kg	2.25 mg/kg	+	
3.3×10^{-5} M (6.6 mg/l.)	6.6 mg/l	+	Cell stage specific; long-lived lesion
4.4×10^{-5} m (8.8 mg/l)	8.8 mg/l	+	
5 mg/kg		Spermatids 28%/2.5 mg/kg	
2.5 mg/kg		Spermatogonia-50%/2.5 mg/kg	
0.16-0.32 mg/kg	1.0 mg/kg	+	Postmeiotic stages are sensitive; oral treatment gave no effect; fraction gives a larger effect

Table 2. ThioTEPA

Test system	Genetic end points	Metabolic and cell stage factors	Exposure route	Dose response
Bacterial				
Salmonella	Reversions	-S9, growing	Plate test	1.0
E. coli	Reversions	-S9, growing	Spot test; liquid	
Plant				
Nigella	Chromatid aberration (delayed)	Root tips	Grown in solution	One dose
Allium	Chromatid aberration (delayed)	Root tips	Grown in solution	One dose
Vicia faba	Chromatid aberration (delayed)	Root tips	Grown in solution	Aberrant metaphases 1.0; concave for aberrations
	Sister chromatid exchanges	Root tips	Grown in solution	One dose
Drosophila	Dominant lethals Recessive lethals Translocations	Sperm, spermatocytes	Fed	

LCT[a]	LEC[b]	Frequency	Comments
100 mg/plate	100 mg/plate		+TA1535; +TA100
2.6×10^{-3} M/l		+	B/Sd-4; K-39; Lambda induction WP-2
18.9 mg/l	18.9 mg/l	18 breaks per hundred cells	
95 mg/l	95 mg/l	11% aberrant metaphases	
		17% aberrant metaphases (lowest conc.); 85% aberrant metaphases (highest conc.)	
38 mg/l	38 mg/l	+	
		+	Mutations highest in mature sperm
		+	
		+	

(continued)

Table 2. (continued)

Test system	Genetic end points	Metabolic and cell stage factors	Exposure route	Dose response
Mammals				
Chromosome aberrations				
In vivo	Chromatid aberrations	Bone marrow (mouse)	Intraperitoneal	1.0
	Chromatid aberrations	Leukocytes (man)	Intramuscularum	?
In vitro	Chromatid and chromosome aberrations	Leukocytes (man)	Solution	Concave
	Chromatid aberrations	Fibroblasts, all stages (Chinese hamster)	Solution	One dose
Translocation (mouse)	Heritable translocation	Late spermatids; spermatogonia	Intraperitoneal	One dose ea. expt.

LCT[a]	LEC[b]	Frequency	Comments
0.16 mg/kg	1.25 mg/kg	8×10^{-2}/mg/kg	Mouse, (×6 times for sensitive strains)
0.14 mg/kg	0.6 mg/kg	Double spontaneous level-30/10 mg/kg	
10 mg/kg			
3.8 mg/l	3.8 mg/l	0.14 breaks/cell/3.8 mg/l	
1.25 mg/kg	1.25 mg/kg	33% translocations (semisterile and sterile)	
5 mg/kg	5 mg/kg	10.6% translocations (spermatids) 3.1% translocations (spermatogonia)	

(continued)

Table 2. (continued)

Test system	Genetic end points	Metabolic and cell stage factors	Exposure route	Dose response
Dominant lethals		Premeiotic stages	Intraperitoneal; oral	
Host-mediated assay	Reversions in Salmonella		Intraperitoneal	
Sperm abnormality				

[a]Lowest concentration tested. [b]Lowest effective concentration.

ThioTEPA was studied only with regard to the induction of recessive lethals in mature spermatozoa and spermatids (27-29). The mutagenic effect of ThioTEPA is similar to that of TEPA. After the treatment of males for 3 days at a dose of 1.2×10^{-2} mmole/l., about 7% recessive lethals appeared; at a dose of 1.2×10^{-1} mmole/l. for 6 hr about 11% recessive lethals were noted.

MeTEPA was injected into mature males at a dose of about 285 μg/kg (25). Under these conditions, 4-5% recessive lethals were induced in postmeiotic cells. This study was carried out on a small number of flies and was the only study of this chemical.

Laboratory Mammals

None of these compounds was tested on any mammal for the possibility of inducing gene mutations. Thus, the question remains open. The main investigations were carried out on tests of chromosomal aberrations, dominant lethals, and heritable translocations. Separate investigations were carried out with the evaluation of the frequency of micronuclei and morphologic anomalies of sperm.

LCT[a]	LEC[b]	Frequency	Comments
0.16 mg/kg	0.64 mg/kg	+	10 mg/kg, a pre-meiotic effect is observed
25 mg/kg		+	Braun (unpublished)
		Doubling dose/ 2 mg/kg	

Somatic Cells

The cytogenetic effect of TEPA in bone marrow cells was thoroughly investigated on random-bred CD rats (30). With the use of one dose of 10 mg/kg (one intraperitoneal injection of the mutagen solution) the yield of aberrations was studied as a function of the time of harvesting cells. On harvesting cells at 6, 12, 24, 48, 72, and 96 hr, the maximum effect was found at 24 hr; 87.5% of aberrant cells being observed. Thereafter the number of aberrant cells decreases to 12%, though some excess of the experimental group over t was observed at 96 hr. At the level of maximum effect (24 hr) the dose dependence was studied after TEPA injections at the doses of 0.65, 1.25, 2.5, 5.0, and 10 mg/kg; chromatid aberrations were noted in 0.5, 3.0, 11.5, 39.0, and 87.5% of cells, respectively. The main types of damage were chromatid aberrations; the least effective dose was 2.5 mg/kg.

The frequency of chromosomal aberrations in bone marrow cells of mice and hamsters after intraperitoneal injection of TEPA was studied by Šram (32). The dose used for mice was 1 to 5 mg/kg; that for hamsters, 5 mg/kg. The maximum mutagenic effect for all doses was observed 24 hr after the injection. The frequency of aberrations in mice was twice as high as in rats and hamsters.

First data on the induction of chromosomal aberrations by ThioTEPA in bone marrow cells of non-inbred mice were obtained

Table 3. MeTEPA

Test system	Genetic end points	Metabolic and cell stage factors	Exposure route	Dose response
Bacterial				
Salmonella	Reversions	Growing, -S9	Spot test	1.0
E. coli	Reversions	Growing, -S9	Spot test	1.0
Vicia faba	Chromatid (delayed)	Root tips	Grown in solution	One dose
Drosophila	Recessive lethals	Sperm	Injected	
Mammals				
Micro-nucleus			Intraperi-toneal	
Dominant lethals	Dominant lethals	Pre- and postmeiotic male stages	Intraperi-toneal	
Sperm abnormality test				

[a]Lowest concentration tested. [b]Lowest effective concentration.

LCT[a]	LEC[b]	Frequency	Comments
5×10^{-7}M/plate		+	TA 1535
5×10^{-7}M/plate		+	WP-2, exr^{-}
500 mg/1	500 mg/1	+ (63% aberrant anaphases) (6.3×10^{-4} / mg/kg)	
285 mg/kg	2.6 10 mg/kg	$+1.6 \times 10^{-4}$ mg/kg	One experiment
Split dose of 0.1 mg/kg each at 24 hr		+	
0.782 mg/kg	12.5 mg/kg		Most effective in postmeiotic stages
		Doubling dose 12 mg/kg	

by anaphase analysis (33,34). The authors inferred that the effect of a dose of 8 mg/kg was comparable with the effect of exposure to 100 R of γ-rays. All treated doses (1, 4, and 8 mg) were genetically effective.

Metaphase analysis of bone marrow cells of C57BL/6 mice (35) showed that the maximum effect was observed 24 hr after intraperitoneal injection of the mutagen. The effect of seven doses, from 0.16 to 10 mg/kg was tested. The minimal dose showing statistically significant increase of the number of cells with structural aberrations was 0.32 mg/kg. ThioTEPA induced chromatid aberrations almost exclusively. In the tested range of doses there was observed a linear dependence of the yield of the damaged cells upon the dose. The dose of 10 mg/kg had the maximum effect, inducing 89.6% aberrant cells.

Differences in sensitivity to cytogenetic effect of ThioTEPA (dose, 5 mg/kg) were found between mice of different genotype (36). Mice of lines 101//HY, C57BL/6JY, AISnY, I/StY, BA1B/cY, DBA/2Y, AKR/JY, C3H/SnY, CBA/CaY, F_1(CBA × B6), and F_1(C3H × 101), in order of decreasing sensitivity, were studied. The maximum effect -- 45.9% of cells with structural aberrations -- was found in 101/H mice; the minimum effect -- 8.4% of aberrant cells -- was seen in F_1(C3H × 101) hybrids. Males were somewhat more sensitive than females; in 101/H, I/st, and BALB/c mice, differences between sexes were significant at $p < 0.01$.

Bone marrow cells of mice ICR exposed to ThioTEPA at doses of 0.2-200 mg/kg showed from 2.0 to 5.5% of cells to have micronuclei (37). The lowest effective dose was 2.0 mg/kg. Thus, the test gives a distorted picture of the response to the mutagen, only a small part of induced chromosomal anomalies being detected.

MeTEPA induced chromosomal aberrations in bone marrow of mice (38). A quantitative evaluation, however, was not carried out. At the same time there was a quantitative evaluation of the frequency of micronuclei in the bone marrow cells of mice CFLP (39). The doses used were from 0.1 to 0.8 mg/kg. The maximum effect (20.65% of cells with micronuclei) was found at a dose of 0.4 mg/kg. The minimal effective dose was 0.1 mg/kg.

Germ Cells

The ability of TEPA to induce dominant lethal mutations was evidenced by the results of several experiments (40-46). A more complete investigation (41) was carried out with the use of eight doses from 0.156 mg/kg to 20 mg/kg. The effect was studied for 8 weeks after a single intraperitoneal injection of the mutagen, which makes it possible to evaluate the effect on all the stages of spermatogenesis. The minimal dose which showed an effect by the second week (late spermatozoa) was 0.625 mg/kg. With the information from this experiment it is difficult to evaluate the effect quantitatively, however. Quantitative evaluation is possible on the data (46) obtained with the use of mice ICR. TEPA induced both pre- and postimplantation lethals in postmeiotic cells. Late spermatids turned out to be most sensitive, 49% of dominant lethals appearing with a dose of 1 mg/kg. Lethals were not found in spermatogonia, and this is conditioned by their sensitivity to cytotoxic activity of TEPA (42, 47). The effect on A/L mice of the injection of doses of 0.5, 1, and 2.5 mg/kg divided into two fractions at 12-hr intervals was also studied (43,44). Dividing the dose makes the yield of dominant lethals lower (48), a preliminary treatment with chloramphenicole increases it, and injection of puromicine lowers the frequency of lethals in postmeiotic cells (45). Dominant lethals are supposed to present quantitative and structural chromosomal aberrations. This assumption is confirmed by a direct investigation of early embryogenesis (47,49). Thus, the minimal effective dose was 1 mg/kg in intraperitoneal injection.

Investigation of meiosis at metaphase stage I in spermatocytes of mice after exposure of spermatogonia showed a very low yield (0.3%) of translocations (48). The effect of TEPA in germ male cells was also determined by the frequency of translocation heterozygotes among the progeny of the first generation (43,44,50). The most precise data were obtained on ICR males with the use of 5 mg/kg dose of TEPA (50). Among the progeny conceived within the first three weeks the frequency of carriers of translocations was 5, 28, and 10%, respectively. The induction of translocations in premeiotic and meiotic cells was not recorded. With an increase of the dose to 20-30 mg/kg heterozygotes for translocations were not observed in a study of a large number of progeny (31).

ThioTEPA as well as TEPA induced a high percentage of dominant lethals and translocations in postmeiotic male cells. Induction of dominant lethals was reported in ICR mice (51,52). This was studied in more detail in C57BL/6 males (53,54) mated with hybrid females. A single intraperitoneal injection at doses of 0.16, 0.32, 1.25, 2.5, and 5 mg/kg was tested on mice C57BL/6. The minimal effective dose for exposure of postmeiotic cells was 0.32 mg/kg. In the most sensitive late spermatids, ThioTEPA in a dose of 2.5 mg/kg induced 82.7% of dominant lethals. A dose of 5 mg/kg induced dominant lethals in about 50% of spermatozoa and early spermatids. The amount of damage in late spermatids (the second week) was so great, that all the embryos died before implantation (55). Similar frequencies of dominant lethals were observed in C57BL/6 mice (53,54) and ICR mice (52). If we compare maximum effects (for ThioTEPA, in late spermatids; for x-rays, in early ones), then the effect of 2.5 mg/kg of ThioTEPA (54) is equivalent to 400 R of x-radiation (56).

All the mutagens considered, TEPA, ThioTEPA, and MeTEPA, induced the maximum quantity of dominant lethals in late spermatids.

Data were obtained which might be interpreted as evidence of the repair by mice oocytes of chromosomal damages induced by ThioTEPA (dose of 2 mg/kg) in late spermatids of F_1AKD2 males (57). When males were mated with females of lines C57Bl/6 J Sto F_1 (CBA × C57Bl/6) and CBA/Ca Sto, significant differences were found between frequencies of dominant lethals in females of different genotypes. The highest frequency was found in C57Bl/6 females and the lowest, in CBA females; hybrids occupied an intermediate position.

ThioTEPA induced a high frequency of translocations in postmeiotic male germ cells. The study of F_1 progeny born after the exposure of C57Bl/6 males to a ThioTEPA dose of 5 mg/kg showed that 10.6% of the progeny were carriers of translocations. This dose turned out to be excessively high: in late spermatids, so much damage appeared that all the progeny conceived within the second week died before implantation. After a decrease of the dose to 1.25 mg/kg, that is, to the level at which the frequency of dominant lethals in late spermatids was about 40%, of the offspring conceived within the second week, 33.5% were heterozygotes for translocations (58).

After the exposure of spermatogonia, spermatocytes in treated males showed very few translocations, the maximum being 0.245% in the F_1 generation, 3.1% of heterozygotes for translocations were found after a dose of 5 mg/kg (53).

The data on mutagenic effect of ThioTEPA obtained on mice of the same genotypes (C57Bl/6) in the laboratory permit comparison of the sensitivity of somatic cells of bone marrow and later spermatids: at a dose of 1.25 mg/kg ten times as many dominant lethals and translocations appear in late spermatids as in bone marrow cells with structural aberrations. The spontaneous level, however, was only one tenth as great as bone marrow cells.

The effect of MeTEPA in germ cells was studied by the method of dominant lethals (40,51,59). A number of doses of MeTEPA in solution in tricapriline, from 0.782 to 50 mg/kg, were tested on ICR mice. The minimal effective dose was reported to be 1.56 mg/kg. A quantitative evaluation of the effect in this experiment is complicated.

For CD-1 mice the minimal effective dose was reported to be 12.5 mg/kg. The effect of the mutagen was significant within the first three weeks (dominant lethals appeared in postmeiotic cells). Because of the lack of information and the discrepancy of results of different experiments, it is not possible to use these data on MeTEPA for quantitative comparisons.

Man

The main information on the mutagenic activity of TEPA and ThioTEPA for man was obtained from experiments with a culture of lymphocytes in vitro. Only one paper (60) reports on the frequency of chromosomal aberrations in lymphocytes cultures from patients treated for tumors with ThioTEPA. The cumulative dose of (IM) of the drug was 40-100 mg. The frequency of chromosomal aberrations doubled, independent of the dose and duration of treatment.

TEPA induced a marked mutagenic effect on human lymphocytes (61-63). It is difficult, however, to judge the dose and exposure dependences by these papers, or to assess peculiarities of effects of TEPA. The dependence of the cytogenetic effect of TEPA upon the concentration was studied by Selezneva and Chebotarev (64). The incidence of aberrant metaphases was shown to increase from 6.0 to 61.0% with increasing concentration of mutagen from 0.125 to 16.0 μg/ml, and the total number of breaks increased from 7.97 to 116.3 per 100 cells. The lowest effective concentration was 0.120 μg/ml.

The mutagenic effect of ThioTEPA in the culture of human cells depends markedly upon the stage in the cell cycle (65-67). The most significant effect in the culture of leukocytes was shown on exposure to the mutagen 24-34 hr before harvesting, independent of the duration of cell cultivation. Hence, the parameters of the dose curve would depend upon the stage of the cell cycle at which the treatment with the mutagen was carried out. The dependence of the effect upon the concentration of ThioTEPA was described by a similar analytical expression of the exponential type (68,69) for all stages. Numerical coefficients in the equations, however, depend upon the stage of the cell cycle.

The effect of ThioTEPA is determined not only by the concentration, but also by the duration of the contact of cells with it. The dependence was studied at the G_0 stage (70,71). The general character of the dependence, as in the case of the concentration, is of an exponential type. Thus, to characterize the dose dependence of the effects of ThioTEPA on human chromosomes it is necessary to take into account both the concentration of the mutagen and the duration of the exposure. In other words, the dose of ThioTEPA involves an interaction of concentration and exposure.

The frequency of chromosomal aberrations on exposures to ThioTEPA is strongly dependent on temperature at the time of treatment (72). Thus, 50% of aberrant metaphases at 37°C appear at the G_0 stage as on treatment at a dose of 32 μg/ml for 1 hr. To reach the same level of aberrant metaphases at 5°C, a concentration of 840 μg/ml would have to be used.

Experiments carried out under the same conditions with the blood of different individuals showed that the cytogenetic effect of ThioTEPA might vary two- to fourfold (73). There is a possibility that in individuals of different ages the cytogenetic effect of ThioTEPA might be different (74). In the same study the frequency of chromosomal aberrations was shown to be independent of sex.

CONCLUSIONS

In all the test systems studied (different substrates, different kinds of mutations) a mutagenic effect of all three compounds was found (Table 4). A mutagenic effect was absent in only one experiment: that with TEPA on Saccharomyces.

As numerous experiments show, all three compounds, TEPA, ThioTEPA, and MeTEPA, have a direct mutagenic effect. They do not need metabolic activation and are, without doubt, strong mutagens. With increasing dose of the mutagens, the number of mutant cells may reach 80% and over. At the same time, the number of mutations per cell increases. The data on the spectrum of induced mutations are not yet complete. According to the studied tests, mutagens induce gene mutations (in microorganisms and Drosophila), chromosomal aberrations (plants, mammals), sister chromatid exchanges (plants, cells of Chinese hamster and human cells), and dominant lethals (Drosophila, mammals). However, there is no information at present on the number of genetic end points (interstitial deletions, nondisjunctions, mitotic recombinations, gene conversions, and others). Thus, conclusions about the full spectrum of genetic changes from all the three mutagens cannot yet be drawn, the results follow the same trend on other genetic end points for substrates.

The comparative mutagenic effect has not yet been studied in detail. A comparison is possible according to frequencies of induced chromosomal aberrations and dominant lethals.

Expressed fluctuations of frequencies of aberrations and dominant lethals between animals of different lines of the same species, between different animals, and between separate individuals were established. Interspecies (interline) differences might be higher than intraspecies differences.

Differences were detected in sensitivity to cytogenetic effect of ThioTEPA (dose, 5 mg/kg) between mice of different lines. The

Table 4. Results on studies of mutagenic effect of TEPA, ThioTEPA, and MeTEPA on different substrates

Substrate	Effect	TEPA	Thio-TEPA	MeTEPA
Bacteria	Gene mutations	+	+	+
Host-mediated assay		+		
Fungi	Gene mutations	±		+
Plants	Chromosomal aberrations	+	+	+
	Sister chromatid exchanges		+	
Drosophila	Dominant lethals	+	+	
	Recessive lethals	+	+	+
	Translocations	+	+	
Mammals	Chromosomal aberrations			
	In vivo	+	+	+
	In vitro	+	+	
	Sister chromatid exchanges		+	
	Translocation test	+	+	
	Micronuclei		+	+
	Dominant lethals	+	+	+
	Anomalies of spermatozoa		+	+

maximum effect of 45.9% of cells with structural aberrations was found in 101/H mice, the minimal effect -- 8.4% of aberrant cells -- was found in F_1(C3H × 101) hybrids.

The data on mutagenic effect of ThioTEPA obtained on the same line of mice (C57Bl/6) in one laboratory, permit a comparison of the sensitivity of bone marrow cells and late spermatids. But this comparison is conventional because the spontaneous levels are very different in these types of cells.

The frequency of dominant lethals in mice is twice as high as in rats (a dose of 1 mg/kg for mice being equivalent in effect to 2 mg/kg in rats). These exposures took place at the stage of late spermatids.

The frequency of chromosomal aberrations in the bone marrow of mice is twice as high as in rats and hamsters on the intraperitoneal injection of TEPA at the doses of 1 to 5 mg/kg in hamsters. The animals were killed after different periods of time (12-168 hr) after the exposure to the mutagen. The maximum mutagenic effect for all the doses was observed 24 hr after the injection. The experiments on mice and rats were carried out in the laboratory of Dr. R. Sram and those with rats were carried out in the laboratory of Dr. I. Adler.

Individual fluctuations of the frequency of chromosomal aberrations on exposure to ThioTEPA were studied on human cells. The frequency of chromosomal aberrations between individuals was established to differ by a factor of two- to threefold.

The established culture of human tumor cells was two to three times as high as the initial lymphocyte culture to cytogenetic activity of ThioTEPA.

The importance of different tests on mammals or cells in the culture is variable. The rule, evidently might be used, that the estimation is made on the most sensitive tests. For instance, the most sensitive test for TEPA and ThioTEPA is the culture of human lymphocytes (according to the majority of works).

There are no data on quantitative evaluations of the effects of TEPA, ThioTEPA, and MeTEPA in humans in vivo except for a study (60) in which the frequency of chromosomal aberrations in five patients with malignant tumors receiving no other treatment was studied. The total doses of the drug were in the range 40-100 mg, in 10 mg increments. There was an increase in the frequency of chromosomal aberrations, but no higher than twice as much.

Because of the heterogeneity of the human population, it is better to use random-bred animals in testing mutagens on mammals. Lines showing defects in metabolism or repair systems should not be used in testing on cells in a culture. Negative data on mutagenic activity of a chemical on any substrate for which the chemical shows positive activity on another substrate should be thoroughly checked.

REFERENCES

1. Mellet, L. B., and Woods. L. A. Cancer Res. 20: 524-532 (1960).

2. Kirichenko, O. P., Chebotarev, A. N., and Yakovenko, K. N. Byull. Eksptl. Biol. Med. 81: 552-553 (1976).

3. Hanna, P. J., and Dyer, K. F. Mutat. Res. 28: 405-420 (1975).

4. McCann, J., Choi, E., Yamasaki, E., and Ames, B. N. Proc. Natl. Acad. Sci (U.S.A.) 72: 5135-1539 (1975).

5. Specht, J. Arch. Microbiol. 51: 9-17 (1965).

6. Price, K. E., Buck, R. E., and Lein, J. Antimicrob. Agents Chemotherap. 1965: 505-517.

7. Letterberg, G. EMS Newsletter 3: 14-15 (1970).

8. Kihlman, B. In: Chemical Mutagens-Principles and Methods for Their Detection, A. Hollaender, Ed., Plenum Press, New York, 1971, Vol. 2, pp. 489-515.

9. Ninan, T., and Wilson, G. B. Genetika (Hague), 40: 103-119. (1969).

10. Sturelid, S., and Kihlman, B. EMS Newsletter 3: 15-17 (1970).

11. Sturelid, S. Hereditas 68: 255-276 (1971).

12. Sidorov, B. N., Sokolov, N. N., and Andreev, V. S. Sov. Genet. 2: 81-87 (1966); translated from Genetika 2: 124-133 (1966).

13. Asp, B. Mutat. Res. 21: 22 (1973).

14. Kihlman, B. A., Hartley-Asp, B., Nilssen, K., and Sturelid, S. Mutat. Res. 21: 191-192 (1973).

15. Kihlman, B. A., Sturelid, S., Hartley-Asp, B., and Nilsson, K. Mutat. Res. 17: 271-275 (1973).

16. Kihlman, B. A., Sturelid, S., Hartley-Asp, B., and Nilsson, K. Mutat. Res. 26(2): 105-122 (1974).

17. Deysson, G. C. R. Soc. Biol. 168: 687-693 (1974).

18. Sturelid, S., and Kihlman, B. A. Hereditas 79: 29-42 (1975).

19. Sturelid, S., and Kihlman, B. A. Hereditas 80: 233-246 (1975).

20. Kihlman, B. A. Chromosoma 51(1): 11-18 (1975).

21. Kihlman, B. A., Sturelid, S., Palitti, F., and Becchetti, A. Mutat. Res. 46: 130-131 (1977).

22. Sturelid, S., and Kihlman, B. A. Mutat. Res. 53(2): 270 (1978).

23. Kihlman, B. A., and Sturelid, S. Hereditas 88(1): 35 (1978).

24. Hartley-Asp, B. Hereditas 83: 223-236 (1976).

25. Benes, V., and Šram, R. Ind. Med. 38: 50-52 (1969).

26. Kočisova, J., and Šram, R. J. Folia Biol. (Prague) 20: 325-332 (1974).

27. Fahmy, O. G., and Fahmy, M. J. Cancer Res. 30: 195 (1970).

28. Lüers, H., and Röhrborn, G. Proceedings XIth Intern. Congr. Genetics 1: 64-65.

29. Lüers, H., and Röhrborn, G. Mutat. Res. 2: 29-44 (1965).

30. Adler, I. D., Ramarao, G., and Epstein, S. S. Mutat. Res. 13(3): 263-270 (1971).

31. Generoso, W. M., Cain, K. T., Huff, S. W., and Gosslee, D. T. In: Advances in Modern Toxicology, 1 (1977).

32. Šram, R. Unpublished data.

33. Arseniyeva, M. A., And Golovkina, A. V. In: Vliyanie Ioniz. Izluch. Nasledstvennost, Akad. Nauk SSSR, 1966, pp. 1-122 (Russian).

34. Arsenieva, M. A., Bakulina, E. D., Golovkina, A. V., and Lander, E. S. Genetika 5: 111-121 (1967).

35. Surkova, N. I., and Malashenko, A. M. Genetika 10(2): 81-89 (1974).

36. Surkova, N. J., and Malashenko, A. M. Genetika 11(11): 66-72 (1975).

37. Maier, P., and Schmid, W. Mutat. Res. 40: 325-338 (1976).

38. Manna, G. K., Das, K. P., and Pradeep, K. Proc. Indian Sci. Congr. 60: 676 (1973).

39. Richardson, J. C. Mutat. Res. 26(5): 391-394 (1974).

40. Epstein, S. S., and Shafner, H. Nature 219: 385 (1968).

41. Epstein, S. S. Toxicol. Appl. Pharmacol. 74: 653 (1969).

42. Epstein, S. S., Arnold, E., Andrea, J., Bass, W., and Bishop, Y. Toxicol. Appl. Pharmacol. 23: 288 (1972).

43. Šram, R. J., Zudova, Z., and Benes, V. Folia Biol. 16: 367 (1970).

44. Šram, R. J., Benes, V., and Zudova, Z. Folia Biol. 16: 407 (1970).

45. Šram, R. J. Folia Biol. 18: 367 (1972).

46. Šram, R. J. Folia Biol. 18: 139-148 (1972).

47. Joshi, S. R., Page, E. C., Arnold, E., Bishop, Y., and Epstein, S. S. Genetics 65: 483 (197).

48. Šram, R. J., and Zudova, Z. Folia Biol. 19: 58 (1973).

49. Epstein, S. S., Joshi, S. R., Arnold, E., Page, E. C., and Bishop, Y. Nature 225: 1260 (1970).

50. Epstein, S. S., Bass, W., Arnold, E., Bishop, Y., Joshi, S., and Adler, I. D. Toxicol. Appl. Pharmacol. 19: 134 (1971).

51. Epstein, S. S., Arnold, E., Steinberg, K., Mackintosh, D., Shafner, H., and Bishop, Y. Toxicol. Appl. Pharmacol. 17: 23 (1970).

52. Šram, R. J. Mutat. Res. 41: 25 (1976).

53. Malashenko, A. M., and Surkova, N. I. Genetika 10(1): 71-79 (1974).

54. Malashenko, A. M., and Surkova, N. I. Genetika 10(8): 92 (1974).

55. Semenov, Kh. Kh., and Malashenko, A. M. Byull. Eksptl. Biol. Med. 80: No. 10, 107-110 (1975).

56. Ehling, V. H. Mutat. Res. 11: 35 (1971).

57. Malashenko, A. M., and Surkova, N. I. Genetika 10(2): 105-111 (1975).

58. Malashenko, A. M., Semenov, Kh. Kh., Selezneva, G. P., and Surkova, N. I. Genetika (U.S.S.R.), 14(1): 52-55 (1978).

59. Wallenstein, S., Brobst, J., and Bagdon, R. E. Mutat. Res. 26: 458 (1974).

60. Selezneva, T. G., and Korman, N. P. Genetika (U.S.S.R.) 9: No. 12, 112-118 (1973).

61. Chang, T.-H., and Klassen, W. Chromosoma 24: 314-323 (1968).

62. Hampel, K. E., and Stopic, D. Haematol. 46: 136-141 (1971).

63. Kucerova, M., and Polivkova, Z. Mutat. Res. 34: 279-290 (1976).

64. Selezneva, T. G., and Chebotarev, A. N. Byull. Eksptl. Biol. Med. 82: 1265-1267 (1976).

65. Dubinina, L. G. Sov. Genet. 10: 1428-1436 (1974); translated from Genetika 10(11): 129-137 (1974).

66. Zhurkov, V. S., and Yakovenko, K. N. Mutat. Res. 41: 1107-1112 (1976).

67. Bochkov, N. P., and Yakovenko, K. N. In: Mutagen-Induced Chromosome Damage in Man. Evans and Lloyd, Eds, Edinburgh University Press, 1978, pp. 290-295.

68. Bochkov, N. P., Yakovenko, K. N., Chebotarev, A. N., Kravioto, F. F., Zhurkov, V. S. Sov. Genet. 8: 1595-1601 (1972); translated from Genetika 8(12): 160-168 (1972).

69. Chebotarev, A. N., and Yakovenko, K. N. Sov. Genet. 10: 1048-1054 (1974); translated from Genetika, 10(8): 150-157 (1974).

70. Kirichenko, O. P. Sov. Genet. 10: 1172-1175 (1974); translated from Genetika 10(9): 139-143 (1974).

71. Kirichenko, O. P., and Chebotarev, A. N. Genetika 12(6): 142-149 (1976).

72. Chebotarev, A. N. Sov. Genet. 10: 1178-1182 (1974); translated from Genetika 10(9): 147-153 (1974).

73. Yakovenko, K. N., Tarusina, T. O., and Kuleshov, N. P. Genetika 12(12): 139-143 (1976).

74. Bochkov, N. P., and Kuleshov, N. P. Mutat. Res. 14: 345-353 (1972).

CHAPTER 32

COMPARATIVE MUTAGENICITY OF TRIETHYLENEMELAMINE, TRENIMON, AND ETHYLENIMINE

Claes Ramel

Wallenberglaboratoriet, Stockholms Universitet
Stockholm 50, Sweden

TRIETHYLENEMELAMINE

Introduction

Triethylenemelamine (TEM), also referred to as 2,4,6-tris(1-aziridinyl)-*s*-triazine, 2,4,6-tris(ethylenimino)-*s*-triazine, 2,4,6-triethylenimino-1,3,5-triazine, Tretamine, triethanomelamine has the structure I (MW 204.2).

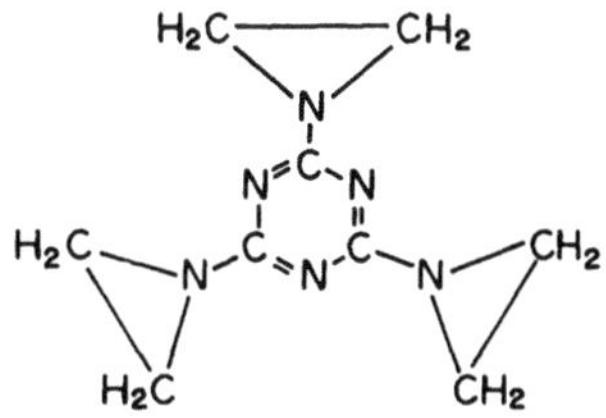

I

It is obtained as minute crystals, of neutral reaction, which melt at 139°C with decomposition. Its solubility in water is 40% (w/v) at 26°C, in methanol 12.5%, in ethanol 7.7%, and in acetone 10.6%.

TEM is relatively stable in neutral solution, but requires precautions to avoid acid conditions, because of the fast hydrogen ion-catalyzed reaction. An accidental lowering of pH may lead to rapid destruction. According to Ehrenberg and Wachtmeister (1) the time for destruction of 10% TEM in water is 4 hr at pH 7 and 13 sec at pH 3. The corresponding destruction time in 0.1 M phosphate buffer at pH 7 is 1.7 hr for 10% TEM. TEm is stable for about 3 months in an enclosed ampoule as the water solution at 4°C (2). Crystalline TEM polymerizes to an inactive material at room temperature.

TEM is used in manufacturing of resinous products and in textile industry for improving acidic dyes and in printing pastes. In plastics it is used for end-to-end linking and crosslinking of polymers. The use of TEM in laminar and other adhesives is indicated, but the extent of this use is not clear (3). Further, it has been employed as an anticancer drug and as an antiprotist against Bacillus megaterium and protozoa (Plasmodium) causing malaria. The high mutagenicity of TEM has been used to advantage for chemosterilization of insects, birds, and rodents.

TEM is a trifunctional alkylating agent which does not need activation. Presumably, the reaction within the cell is appropriate for conversion of TEM to the quaternary ethylenimonium form (4).

Results with Different Test Systems

TEM has been extensively used as a model substance or a positive control in practically all standard mutagenicity test systems and has proved active throughout. Extrapolation from one test system to another or to man therefore constitutes unusually little problem with TEM because of its common mutagenic activity. From a practical viewpoint of risk evaluation, however, TEM hardly represents an issue of primary interest, as its polyfunctional property must be considered rather special and not often encountered in the human environment. Nevertheless, the wide basic knowledge may give illustrative examples of the procedure of extrapolation and risk evaluation.

The test results have been summarized with reference to species, sex, route of administration, and doses in Table 1.

Results with Mammals

Dominant Lethal Tests. TEM has induced dominant lethals in male meiotic and postmeiotic germ cells of several species as well as maturing oocytes of female mice. Over 30 such tests are recorded in the literature, including experiments on mice, rats, golden hamsters and guinea pigs. The form of the dose-effect curve is discussed below. Conclusions to be drawn from these tests are given in the Group 1 report.

Heritable Translocations. TEM induces heritable translocations in meiotic and postmeiotic stages, and seven reports are available. The lowest effective dose reported for the most sensitive stage (spermatids) is 0.0125 mg/kg IP (12). There is a correlation between the induction of heritable translocations and dominant lethals (42), but simultaneous investigations of dominant lethals and translocations (12,43). Results by Generoso et al. (12) are in accordance with a linear dose-effect curve at higher dosages, but a concave curve at lower dosages.

In Vivo Cytogenetics. Three experiments (one on somatic cells and two on germ cells) with positive results are available, and summary and conclusions are found in the Group 1 report.

In Vitro Cytogenetics. Three studies have been performed on human leukocytes. Chromatid aberrations are induced which, according to Hampel et al. (16) follow a multi-hit kinetics in the dose-response curve.

Micronucleus Test. Eight tests have been performed on mice, rats, or Syrian hamsters. An increase in the incidence of micronuclei has been found in all tests, in mice with three routes of application: intraperitoneal, intravenous, and oral. The rank of decreasing mutagenic (as well as toxic) potential found -- intraperitoneal > intravenous > oral -- indicates that pharmacokinetic factors play an important role in mammalian experiments in vivo (19). Similar findings were reported for the dominant-lethal test (44).

Complex shapes of dose-response curves were often obtained which may partly be due to different testing protocols used in different laboratories. Of particular interest is the observation by Weber et al. (45). These authors found a linear dose-related increase if micronuclei and a drastic decrease at low and high dose levels, respectively. An explanation of such a phenomenon is given in a recent review by Schmid (46).

Specific Locus in Vivo. Seven loci tests after intraperitoneal injection has been performed by Cattanach (22) for spermatocytes and spermatogonia. Two doses of 2.0 mg/kg each given to spermatogonia resulted in 5 mutations in 11,144 offspring. An effect at lower doses is indicated by the fact that treatment with 1.0 mg/kg to spermatogonia gave one mutation on 6496 and 0.5 mg/kg to spermatocytes one mutation in 667. A dose of 0.2 mg/kg given to postmeiotic germ cells gave 3/1701 specific locus mutations (23).

Table 1. Summary of test results for TEM

Test system	Species	Sex	Route of adminis-tration	Concn. range (mg/kg)	LEC (mg/kg)	Rate (mg/kg)	Refer-ences
Mammals							
Dominant lethals (Postimplantation loss)							
Postmeiotic stages	Mouse	M	IP	0.035–4.0	0.05	176–805	5,6
	Rat	M	IP	0.025–0.4	<0.025		7,8
	Golden hamster	M	IP	0.2	<0.2		9
	Guinea pig	M	IP	0.2	<0.2		10
Oocytes	Mouse	F	IP	0.4–1.6	<0.4		11
Heritable translo-cations							
Postmeiotic stages	Mouse	M	IP	0.0125–4	<0.0125	$72–145\times10^{-2}$	12
In vivo cytogenetics							
Bone marrow	Rat			0.2–	<0.2		13
Germ cells							
Premeiotic	Mouse	M	IP	0.5–2	0.1		14
Postmeiotic	Mouse	M	IP	0.2–0.64	<0.2		15

In vitro cytogenetics	Human leukocytes			0.02-2	<0.02	16-18
Micronucleus test	Mouse		IP	0.1-10	<0.1	19
			IV	0.25-2.0	0.5	19
			Oral	0.5-16.0	1.0	19
	Rat		IP	0.1-2.5	<0.1	20
	Syrian hamster		IP	0.016-0.25	0.25	21
Specific locus in vivo						
Premeiotic and meiotic stages	Mouse	M	IP	0.5-4.0	4.0	22
Postmeiotic stages	Mouse	M	IP	0.2	0.2	23
Specific locus cell cultures	L51784 ($TK^+/-$)				<0.2	24
Spot test	Mouse		IP	0.5-0.8	<0.5	25

(continued)

Table 1. (continued)

Test system	Species	Sex	Route of adminis- tration	Concn. range (mg/kg)	LEC (mg/kg)	Rate (mg/kg)	Refer- ences
Drosophila							
Sex linked lethals	*D. melano- gaster*	M	Inj.	2.6–29.5	<2.6	0.5	26–28
Translocations	(prestorage)	M	Inj.	4.3–17.4	<4.3	0.06	27
Somatic recombination		F	Larval feeding	3000/0.5 hr	<3000		29
Loss of X or Y		M	Inj.	17.4	<17.4		30
Loss of $Y^L + Y^S$		M	Inj.	17.4	<17.4		30
Dumpy mutant		M	Inj.	17.4	<17.4		30
Chromosome breaks							
Salivary glands							
Unstored			Inj.	64	<64		31,32
Stored			Inj.	64	<64		31,32
Plants (Roots)	Vicia		In water	0.2–4000/ 0.5–1 hr	2		33,34
	Allium		In	2–200/0.5 hr	20		35
	Barley				0.1		36
Yeast							
Mitotic recombination	Saccaromyces cerevisiae		In medium	0.3–58	<0.3		37

Bacteria						
TA1575 his$^-$	Salmonella	In medium	2.5-12.5	2.5	2.5	38
Forward and back mutation	_E. coli_	In medium		0.2		39
Prophage induction	_E. coli_	In medium		500		40
Repair test	_T. subtilis_	In medium		250		41
Unscheduled DNA -synthesis, lymphocytes	Rat		0.05-5.0	0.05		13

Cell Cultures. Tests on mouse lymphoma cells (L5187Y) have an increased mutation rate at the TK-locus with 0.2 μg/ml (24).

Mouse Spot Tests. One experiment has been reported (25) showing a significant increase at 0.5 mg/kg both for spots of genetic significance and for spots of questionable genetic significance.

Results with Other Systems

Drosophila. TEM has been shown to induce recessive and dominant lethals, translocations, entire and partial loss of chromosomes, specific locus mutations and unequal and sister-strand recombination. The experiments have been summarized and thoroughly discussed in the group 1 paper.

Plants. In *Vicia faba* root tip cells, TEM caused chromatid aberrations of the delayed type. Most frequent aberration types were isochromatid breaks and chromatid translocations. A nearly linear increase of aberrations were observed after 0.5 hr treatment with 10^{5} - 5×10^{-4} M (2 - 100 mg/l.-hr) in the water. Concentrations higher than 1 g/l.-hr were lethal. Similar observations were made on *Allium cepa*.

Yeast. Gene conversion was studied in stationary phase cells of strain D4 of *Saccharomyces cerevisiae* for the dose range 5×10^{-6} - 1×10^{-3} M at an exposure time of 210 min. The lowest concentration gave a frequency slightly above the spontaneous.

Bacteria. In *Salmonella* a linear dose-effect relation was obtained with the strains TA1535 and TA100, responding to base substitutions.

In *E. coli*, mutagenic effects of TEM are indicated in all test systems used: mutations to streptomycin independence in B/Sd-4, back mutations in auxotrophic strains of K-12, prophage induction in K-12 (λ) and forward and back mutations to *gal*$^+$, *arg*$^+$ and MTR in K-12/343.

In *B. subtilis*, TEM caused a higher growth inhibition with *Rec*$^-$ than *Rec*$^+$, indicating an effect on DNA.

Unscheduled DNA Synthesis. TEM induces unscheduled DNA synthesis in rat lymphocytes and mouse spermatids. The lowest effective exposure was 0.05 mg/kg.

Summary of Genetic Alterations by TEM

TEM produces a wide spectrum of effects. The data indicate that it is active in all test systems and on all test organisms used.

Point Mutations. Evidence for the induction of point mutations by TEM has been obtained from bacteria, insects, and mammals. Both forward and back mutations are induced in bacteria, as well as prophage induction. As expected from an alkylating agent it causes base substitutions in Salmonella. Prophage induction with TEM has also been observed.

In Drosophila, several investigations have demonstrated the induction of recessive lethals. Particularly because of the polyfunctional nature of TEM, recessive lethals can be expected to be due also to structural changes, e.g. small deletions. The storage effect with recessive lethals observed by Snyder (47) and Watson (28) is in accordance with this expectation. A characteristic effect of TEM in Drosophila is the high frequency of mosaic mutations expressing themselves after one generation.

In mice the induction of point mutations by TEM has been shown by specific locus test on male germ cells and also by spot test on somatic cells after exposure in utero.

Dose-effect curves can be constructed for Drosophila (26) and Salmonella (38). In both cases the data are in accordance with a linear response. It may be pointed out, however, that the lowest dose in the study by Fahmy and Fahmy (26) gave 10-20 times the spontaneous rate of recessive lethals. The effect of TEM at lower dosages, where the repair system may operate more efficiently, is therefore not known. The study in Salmonella was performed on the repair-deficient strains TA1535 and TA100 and therefore may not be representative for repair proficient strains.

The efficiency of TEM as a mutagenic agent can be illustrated by comparison with two other well known mutagens, the monofunctional ethylene oxide (EO) and the bifunctional diepoxybutane (DEB) (Table 2).

Chromosome Alterations. As expected from the polyfunctional nature and the crosslinking properties of TEM, a particularly pronounced chromosome-breaking effect and a low LEC have been observed in the dominant lethal and translocation tests as well as in vivo and in vitro cytological tests.

The dose-related effects are in most cases not sufficiently detailed to characterize the shape of the curves. However, in the dominant lethal test the dose-effect curve based on the data by Matter and Generoso (5) allowed them to demonstrate a deviation from linearity at low dosages. They found that the dose effect curve for TEM deviated much less from linearity than that for EMS and resembled more the dose effect curve for x-irradiation.

Table 2.[a]

System	Action	Relative mutation frequency		
		EO	DEB	TEM
Salmonella G46	Reverse mutations	1	1	17
Solmonella 1513	Reverse mutations	1	1.3	60
E. coli Sd-4	Forward mutations	1	16	5000
Drosophila	Forward mutations	1	43	2500
Barley	Forward mutations	1	30	3000

[a]Data of Hussain and Ehrenberg (48).

Chromosome Alterations. As expected from the polyfunctional nature and the crosslinking properties of TEM, a particularly pronounced chromosome-breaking effect and a low LEC have been observed in the dominant lethal and translocation tests as well as in vivo and in vitro cytological tests.

The dose-related effects are in most cases not sufficiently detailed to characterize the shape of the curves. However, in the dominant lethal test the dose-effect curve based on the data by Matter and Generoso (5) allowed them to demonstrate a deviation from linearity at low dosages. They found that the dose effect curve for TEM deviated much less from linearity than that for EMS and resembled more the dose effect curve for x-irradiation.

A similar deviation from linearity is indicated in the translocation study by Generoso et al. (12).

A more complicated dose effect curve is found with micronuclei, presented by Weber et al. (45) as mentioned above.

The distribution of chromosome breaks is not random. For example an excess has been reported to occur in distal regions in human leukocytes (16), and observations on mice point in the same direction (49,50). Vicia faba aberrations are clustering in chromosomes segments containing heterochromatin (51,52).

In Drosophila there is a delayed opening of TEM-induced chromosome breaks and a manyfold increase of induced translocations in sperms by the storage in female sperm receptacle (27).

Recombination. Recombination induction has only been studied with respect to gene conversion in yeast by Fahrig (37). A log/log plot of conversions against eight concentrations gave a linear dose-effect relation.

Nondisjunction. No experiment or observation on nondisjunction has been reported.

Stage Specificity. TEM has been shown to give rise to chromosome breakage in premeiotic, meiotic, and postmeiotic stages during spermatogenesis. Postmeiotic stages, particularly spermatids, are the most sensitive ones. In dominant lethal tests on female mice postcopulatory precleavage stages were more sensitive to the induction of dominant lethality than precopulatory oocytes.

Risk Assessment

Radiation Equivalent Doses. In order to make use of the information and the experience concerning risk assessment from ionizing radiation, it has been suggested by several research workers and also by the Committee 17 report (53) that the dose-effect relations for chemical mutagens are expressed as radiation equivalent doses. For the present evaluation of TEM, as well as Trenimon and ethylenimine, it was, however, decided by the work group to omit such calculations of radiation equivalent doses, because of the uncertainty attached to them, due to factors like differences in stage sensitivity and dose effect curves for chemicals as compared to x-irradiation. A serious problem concerning radiation equivalence calculations is the lack of knowledge of the dose for chemicals. Without information on the integrated dose reaching the target cells, it is particularly difficult to compare widely different test systems, for which different routes of administration of the chemical have been employed.

Also it should be emphasized that there is no recommended set of standard irradiation data available for the various test systems, which could serve as a basis for the calculations.

Bases for Genetic Risk Assessments

Translocation Data. With doses of 0.0125-0.2 mg/kg to spermatids, Generoso's results give rates of induction of heritable translocations which vary between 56×10^{-2} and 150×10^{-2} per dose unit (1 mg/kg). The mean is 108×10^{-2}.

With doses of 3-4 mg/kg to spermatogonia, Generoso found 2 transmitted translocations in 1633 offspring, which is 12.25×10^{-4}. Subtracting the control frequency ($1/4392 = 2.28 \times 10^{-4}$) and dividing by the average dose (3.5 mg/kg) gives an induced rate of 2.85×10^{-4} per dose unit.

Dividing the control frequency by each rate of induction gives a rate doubling concentration (assuming linearity) of 2×10^{-4} mg/kg (= 0.2 μg/kg) for translocation induction in spermatids and 0.8 mg/kg (= 800 μg/kg) for translocation induction in spermatogonia.

Specific Locus Mutation Data. Cattanach's data (22) for spermatogonia are consistent with linearity (assuming that doses of 2 × 2.0 mg/kg are equivalent to 4.0 mg/kg) and with a rate of induction of 1.1×10^{-5} mutations per locus per dose unit. The Oak Ridge control frequency of 0.75×10^{-5} was used.

Cattanach (23) obtained 3/1701 specific locus mutations after a dose of 0.2 mg/kg to postmeiotic male germ cells, a frequency of 25.2×10^{-5} per locus. Assuming linearity, the rate of induction is 122×10^{-5} per locus per dose unit.

On the basis of a spontaneous frequency of 0.75×10^{-5}, the rate doubling concentrations are 0.68 mg/kg for spermatogonia and 0.006 mg/kg for postmeiotic male germ cells.

Dominant mutations in the mouse behave in essentially the same way as specific locus mutations.

Genetic Load in Man. An ICRP Committee I Task Group (98) concluded that (i) the frequency of harmful simple dominants and sex-linked conditions in man is about 1%. It was thought that these were maintained by recurrent mutations and that they would persist for an average of five generations. (ii) The frequency of mutationally maintained dominants of incomplete penetrance was taken to be about 1.6%, with persistence over 10 generations on average. (iii) Recessive mutations should be disregarded in the present context of risk estimation because deleterious heterozygous effects would be included mainly in category (ii) above and because the frequency of homozygotes which actually manifested themselves in future generations would probably be low, because of advances in diagnostic techniques etc.

The UNSCEAR Report (54) documented a frequency of 0.05% for newborn infants with unbalanced structural rearrangements resulting from translocations, etc., and generally leading to severe malformations. These would not be inherited, but some would arise from previously existing balanced rearrangements. It seems likely that most of those induced by postmeiotic treatment would be the secondary products of balanced translocations since other unbalanced products (due to asymmetrical interchanges, etc.) would probably not survive to birth.

Predictions of Genetic Risk. If we use the calculated rate doubling concentrations (RDC) and the estimates of human genetic load given above we can calculate the expected numbers of extra affected individuals per million offspring born after paternal exposure to 1 μg/kg, as shown in the Tables 3 and 4.

Table 3. Exposure of premieiotic male germ cells

Condition	Normal frequency per 10^6	Extra number affected	
		Equilibrium	1st generation
Dominant and sex-linked	10,000	15	3
Dominants with incomplete penetrance	16,000	24	2
Malformation due to chromosome structural imbalance	500	1	0-1
Malformations due to autosomal trisomy	1,400[a]	?	?
	Total	40+	5+

[a]From UNSCEAR (54).

Conclusions

TEM is a trifunctional alkylating agent which does not need metabolic activation. It can be classified as one of the most potent mutagens. Although it is highly active, also in inducing point mutations through intragenic alterations, TEM is particularly efficient for intergenic alterations and chromosome aberrations.

Table 4. Exposure of postmeiotic male germ cells

Condition	Normal frequency per 10^6	Extra number affected	
		Equilibrium	1st generation
Dominant and sex linked	10,000	1670	330
Dominants with incomplete penetrance	16,000	2670	267
Malformations due to chromosom structural imbalance	500	2500	?
Malformations due to autosomal trisomy	1,400	Nil[a]	Nil
	TotaTotal	6840	597+

[a]Only chromosome loss would be expected, causing early embryonic lethality (except in some XO).

The extensive experimental data gives no doubt about the fact that TEM must be considered a mutagen also in man. The data available from mice furthermore permit a rough prediction also of the quantitative genetic risk of TEM.

TRENIMON

Introduction

Trenimon; triaziquone, also referred to as 2,3,5-tris(1-aziridinyl)-p-benzoquinone, 2, 3, 5-tris(ethyleneimino)benzoquinone; tris, Bayer 3231, has the structure II (MW 231.25).

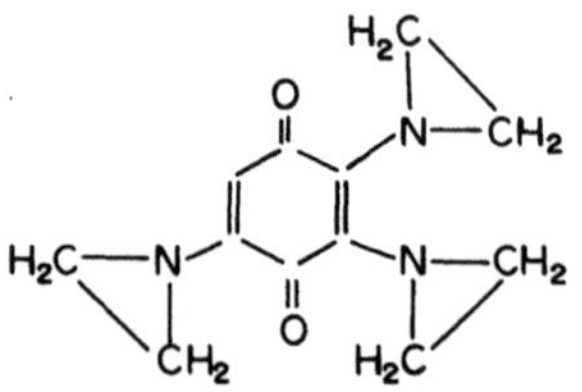

II

It is obtained from ethyl acetate as purple, needle-like crystals which melt at 163°C. It is sparingly soluble in cold water and soluble in acetone, benzene, chloroform, ethyl acetate, methanol, and warm acetic acid.

Trenimon is used as an antineoplastic agent. In West Germany a yearly number of 280 treated persons has been reported. In comparison 28 were treated with TEM (55).

A directly acting trifunctional alkylating agent with properties presumably resembling those of TEM, Trenimon decomposes in mammals within a few hours. Therefore, there is no reason to assume slow activation or inactivation (56).

Results with Different Test Systems

Trenimon has been tested in most test systems and seems to be positive in the test systems included in this report. It is chemically and functionally closely related to TEM and general comments given in the section dealing with TEM apply to Trenimon. Results are summarized in Table 5.

Mammals. Eleven tests on dominant lethals are recorded, all on mice, both males and females. Summary and conclusion from these experiments is given by group 1.

The induction in vivo of cytological aberrations with Trenimon has been recorded in Chinese hamster in bone marrow cells, in postmeiotic stages of males, and in oocytes. In mice, chromosome aberrations were furthermore observed in oocytes, while no effect was found after treatment of spermatogonia or male germ cells in meiotic prophase.

Ten in vitro tests on human leukocytes, one on human fibroblasts, and one on Chinese hamster cells in vitro are recorded, all giving positive effects with Trenimon. Several investigations involve the protection effect of radioprotectors like L-cysteine.

Two experiments with human lymphocytes and one human fibroblasts gave an increase of sister chromatid exchange.

Positive effects with Trenimon in the micronucleus test have been obtained in all 13 studies recorded, including the following species: mouse, rat, Chinese hamster, golden hamster, guinea pig, and rhesus monkey.

An induction of mutations in host-mediated assays with mice as host animals has been reported with Salmonella TA1950 (1.5 mg/kg

Table 5. Summary of test results for Trenimon

Test system	Species	Sex	Route of adminis- tration	Concn. range (mg/kg)	LEC (mg/kg)	Refer- ences
Mammals						
Dominant lethals (postimplantation loss)	Mouse	M	IP	0.125-0.25	<0.125	57
		F	IP	0.125-0.25	<0.125	58,59
In vivo cytogenetics chromosome aberration						
Bone marrow	Chinese hamster		IP	0.06-1.0	<0.03	60
Spermatogonia	Mouse	M	IP	0.06-2	-	61,62
Postmeiotic	Chinese hamster	M		0.25	<0.25	63
Oocytes	Mouse	F		0.25-1	<0.25	64
Oocytes	Chinese hamster	F		0.25-1	<0.25	65

In vitro cytogenetics Chromsome aberrations	Human leukocytes		0.3×10^{-4}–0.2 (24 hr)	6.5×10^{-4}	66
	Human fibroblasts		2.5×10^{-6}–10^{-2}	2×10^{-5}	66
	Chinese hamster		1.25×10^{-6}–4×10^{-2} (24 hr)	10^{-5}	66
Sister chromatid exchange	Human leukocytes		5×10^{-5}–1×10^{-1}		67,68
	Human fibroblasts		10^{-6} 5×10^{-4}		67,68
Micronucleus	Mouse	IP	0.03–1.0	0.06	19
	Rat	IP	0.06–0.5	0.125	69
	Golden hamster	IP	0.03–1.0	0.25	69

(continued)

Table 5. (continued)

Test system	Species	Sex	Route of adminis-tration	Concn. range (mg/kg)	LEC (mg/kg)	Refer-ences
	Chinese hamster		IP	0.03–1.0	<0.03	69
	Guinea pig			0.06–0.5	<0.06	69
	Rhesus monkey			0.125–0.25	0.125	69
Host-mediated assay	Mouse					
Salmonella TA1950			SC		0.5	70
Ascites tumor S_2			IP		0.25	71
Ascites tumor S_2			PO		12.5	71

Drosophila						
Sex-linked recessive lethals	D. melanogaster	M	Inj.	0.7	0.7	72
		M	Adult feeding, 17-24 hr	2.3-4.6	2.3	73
Translocations		M	Adult feeding, 17-24 hr	2.3-4.6	2.3	74
Dominant lethals		M	Adult feeding, 17 hr	2.3	2.3	75
Chromsome loss			Adult feeding, 17 hr	2.3	2.3	75
Plants						
Root tips	Vicia		In water, 0.5-1.0 hr	0.46-0.76	0.46	76

(continued)

Table 5. (continued)

Test system	Species	Sex	Route of administration	Concn. range (mg/kg)	LEC (mg/kg)	References
Seed						
Presoaked	Barley		In water, 0.5-1.0 hr	1.54-3.85	1.54	77
Dry	Barley			0.009-0.027	0.009	
Spots	Soybean		In water, 18-24 hr	0.25-50	0.25	78
Yeast						
Gene conversion	*Saccharomyces cerevisiae*		In medium, 24 hr	0.001-0.05	0.001	79
Bacteria						
Rev. mut.	*E. coli*		In medium		1.2	80
Prophage induction K12 (γ)	*E. coli*		In medium		0.5	40

SC), Serratia marcescens (3 × 4 mg/kg SC), E. coli (3 × 16 mg/kg SC) and ascites tumor cells (0.25 mg/kg IP and 12.5 mg/kg PO).

Drosophila. Trenimon has been shown to induce recessive and dominant lethals, complete and partial loss of choromosomes, and translocations in Drosophila males. No increase of nondisjunction and X chromosome loss was obtained after treatment of mature and immature oocytes even with high dosages of Trenimon. The experimental results on Drosophila have been summarized and discussed in the group 1 report.

Plants. In Vicia faba root tip cells, Trenimon caused chromatid aberrations of the delayed type, mostly isochromatid breaks and chromatid translocations. Preferentially isolocus breaks, mainly located in regions of primary and secondary constrictions, were recorded in barley. Trenimon is efficient in inducing point mutations in soybeans, as judged by experiments on spots on leaves after treatment of seeds.

Yeast. Mitotic gene conversion was induced according to a linear dose effect curve in the dose range 1-50 × 10^{-3} mg/1. for absolute numbers as well as conversion per 10^5 survivors.

Bacteria. In Salmonella only a spot test with his G46 is available, and this was negative. Tests with other strains are required to determine whether Trenimon has any mutagenic effects or not on Salmonella. In E. coli, a positive effect was obtained with Trenimon, both for reverse mutations in the gal locus and prophage induction in K-12 (λ).

Summary of Genetic Alterations by Trenimon

Point Mutations. In bacteria induction of point mutations by Trenimon has been demonstrated in E. coli for reverse mutations in Gal R^s. A dose-dependent increase of prophage induction also has been shown in E. coli K-12 with λ.

In host-mediated assays, Trenimon has been shown to induce base substitution in Salmonella his^- Ta 1950 after subcutaneous application. Positive results have also been obtained with Serratia marcescens his a 13 and leu a 21 as well as E. coli 343 Gal.

In Dorsophila, recessive lethals are produced by Trenimon, the most sensitive stages being sperms and spermatids. In conformity with TEM, particularly at higher doses of Trenimon, recessive lethals can be expected to emanate also from structural changes.

Chromosome Alterations. The chromosome breaking efficiency of Trenimon is very high, as expected from its polyfunctional property and in conformity with the data for the related compound TEM. The dose ranges giving positive effects are roughly of the same order of magnitude as with TEM. Induced chromatid aberrations are of the delayed type with chromatid breaks. In dominant lethal tests the peak sensitivity is in postmeiotic stages.

Recombination. Zimmermann (79,81) demonstrated the induction of mitotic gene conversion with strain D4 of yeast.

Nondisjunction. No clearcut effect on nondisjunction is reported. The occurrence of a significantly higher frequency of hyperdiploid cells in embryos after superovulation and treatment in females by Röhrborn et al. (82), however, suggests the induction of nondisjunction.

Stage Specificity. In male germ cells in mammals Trenimon induces dominant lethals primarily in postmeiotic stages. The frequency of aberrations produced in spermatogonia seems to be very low as judged by the lack of effects seen in cytological investigations of mouse spermatocytes treated as spermatogonia (61,62).

Dose Response. A dose-response curve has been described for in vivo cytogenetics on Chinese hamster bone marrow cells (60). Trenimon was given intraperitoneally in two applications separated by 24 hr, the single doses ranging between 0.031 and 0.5 mg/kg. Animals were sacrificed and chromosome preparations made 6 hr after the second drug application.

At the lowest dosages there is an indication of a relatively smaller effect (i.e., incidences of aberrant metaphases) than at higher dosages. From 0.062 mg/kg to 0.5 mg/kg every doubling of the dose practically resulted in a doubling of the effect. Using the same technique and protocol, Matter has recently confirmed these findings

It should be emphasized, however, that bone marrow is a proliferating tissue. Consequently, every deviation from a given test protocol and application schedule may lead to an alteration in the shapes of curves.

In yeast, Zimmermann (79) found a linear dose-response curve in the dose range from 1 to 50×10^{-3} μg/ml.

Conculsions

Trenimon is a trifunctional alkylating agent which does not

need metabolic activation. It is a highly potent mutagen, inducing point mutations, chromosome aberrations, and recombinations. Its properties are rather similar to the more studied TEM, with which it also is related chemically.

Trenimon can be assumed to have the same spectrum of mutagenic effects in man as in other organisms. The doses used therapeutically are similar to the ones used in experiments giving positive results.

ETHYLENIMINE

Ethylenimine, also referred to as aziridine, azacyclopropane, and dimethyleneimine, has the structure III (MW 43.07).

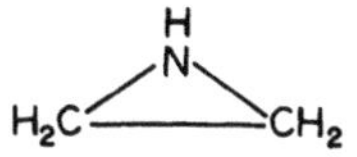

III

It is a liquid (boiling point 56.7°C/760 mm Hg) with an intense ammoniacal odor and is strongly alkaline (pK_A = 8 at 25°C). It polymerizes easily.

Ethylenimine is miscible with water and soluble in alcohol.

Unlike other alkylating agents, which usually are destroyed by alkali, ethylenimine is protected by alkali and is therefore usually stored in the presence of a some solid NaOH. The alkylating reactivity depends on the ethylenimonium ion, and at pH 7 about 91% of the total concentration is available as the imonium cation (1). The time for destruction of 10% ethylenimine at pH 7 and 27°C in water is 8 hr and in 0.1 M phosphate buffer 1 hr (1). Because of the sensitivity of ethylenimine to acid conditions, experiments without careful control of pH are unreliable.

Ethylenimine is often used as the chloride, which is unstable at room temperature. The sulfate, however, is more stable.

Ethylenimine is a monofunctional alkylating agent. While the alkylating property resides in the ethylenimonium ion, the free base is responsible for the transport through membranes (84). Ethylenimine is extremely reactive, undergoing two major types of reactions, ring-opening reactions similar to those in ethylene oxide and ring-preserving reactions in which ethylenimine acts as a secondary amine.

Table 6. Summary of test results for ethylenimine

Test system	Species	Sex	Route of adminis- tration	Concn. range (mg/kg)	LEC (mg/kg)	Refer- ences
Mammals						
Dominant lethals (postimplantation loss)	Mouse	M	IP	5-6	<5	85,86
In vitro cytogenetics	Human fibroblasts (WI38)			0.04-43.7	<0.4	87
	Human leukocytes			4.4-43.7 (1 hr)	<4.4	87
Drosophila						
Sex-linked recessive lethals	<u>D. melano-gaster</u>	M	Inj.	78-156	<78	28,88
Translocations (prestorage)		M	Inj.	39-312	<39	28,88

Dominant lethals		M	Inj.	156-312	<156	89
Plants						
Chromosomal aberrations	Vicia					
Roots			In water	8.7-437 (1 hr)		90
Seeds			In water	21.9-219 (1 hr)		90
Chromosomal aberrations, seeds	Barley		In water	83.1-269.8 (1 hr)	<83	91,92
Point mutations, seeds	Barley			83.1-460 (1 hr)	<83	91,92
Yeast						
Mitotic recombination	Saccharomyces cerevisiae			172-4300 (0.5 hr)	<172	93

(continued)

Table 6. (continued)

Test system	Species	Sex	Route of adminis-tration	Concn. range (mg/kg)	LEC (mg/kg)	Refer-ences
Mitotic gene conversion				861-4300	<861	79
Point mutation						94
Bacteria						
TA1535 his^-	Salmonella			0.2-0.8	<0.2	38
TA100 his^-	Salmonella			0.08-0.24	<0.08	95
Back mutation str^- - str^+	*E. coli*			333 (2 days)	<333	96
Unscheduled DNA synthesis, sperm	Mouse	M	Test	6	<5	97

Ethylenimine and its derivatives have a wide range of practical applications. In the textile industry it is used for crease-proofing, dyeing and printing, flameproofing, shrinkproofing, and waterproofing. It is used in adhesives, as a lubricant additive, in rocket fuels, coatings, in photographic chemicals, in curing and vulcanizing polymers. In agriculture it is applied in insecticides, chemosterilants and soil conditioner. It is also employed for chemotherapy and for antimicrobials (4).

Results with Different Test Systems

Ethylenimine has been extensively used as a mutagen both for basic and applied purposes. The wide experimental use of ethylenimine is not fully evident from the test systems reported by group 1, as the application covers many other organisms and test sytems outside that range. An indication of that is the fact EMIC indexes about 150 species which have been tested with ethylenimine while the corresponding number of species for TEM is about 30 and for Trenimon 17.

Ethylenimine is a very potent, directly acting mutagen giving rise both to point mutations and chromosomal aberrations (Table 6). It can be assumed to be mutagenic in all organisms including man. In the context of risk evaluation ethylenimine appears to be of more general interest than the other aziridines dealt with, TEM and Trenimon, because of its wide practical use.

Mammals. Two dominant lethal tests, both on male mice, are recorded, in which doses of 5 or 6 mg/kg given IP gave rise to an increase of postimplantation dominant lethals. The greatest effect occurred in mature spermatozoa. No effects on preimplantation losses and no effects on premeiotic stages were recorded.

Quantitative data are recorded for two in vitro tests on human leukocytes, one on human fibroblasts (WI 38) and one on Chinese hamster cells. Pretreatment with 0.02 μg/ml Actinomycin D increased the aberrations induced by ethylenimine with Chinese hamster cells.

Drosophila. Ethylenimine has been shown to induce sex-linked recessive lethals, translocations and dominant lethals. The results are summarized and discussed in the group 1 report.

Plants. Chromosome aberrations and gene mutations have been observed in various plants after treatment with ethylenimine. Of the test systems dealt with by group 1 records of chromosome aberrations are found for Vicia roots and seeds and barley seeds. Point mutations were observed after treatment of barley seeds.

Yeast. Mitotic recombination was induced by ethylenimine in *Saccharomyces cerevisiae* probably with a sigmoid dose response curve up to a plateau of 40 mM (93). Gene conversion was also induced. Forward mutations induced by 0.05 M ethylenimine after 10 min treatment has been recorded for the *ade* 1 and *ade* 2 loci (94).

Bacteria. Ethylenimine causes reverse mutations in his^- in Salmonella according to a linear dose response curve in the strain TA1535 and TA100, responding to base substitutions (38,95). An increase of back mutations in the streptomycin locus has also been recorded in spot tests of *E. coli*.

Unschedules DNA Synthesis. The induction of unscheduled DNA synthesis by ethylenimine has been recorded with mouse sperm after testicular injection of 5 mg/kg.

Summary of Genetic Alterations by Ethylenimine

Ethylenimine is a very potent mutagen probably effective in all organisms. It has the same wide spectrum of effects as the complex aziridines TEM and Trenimon.

Point Mutations. In bacteria back mutations are recorded in Salmonella and *E. coli*. The reversion of TA1535 and TA100 his^- in Salmonella shows as expected that the mechanism of mutation induction by ethylenimine is base substitution. A linear dose effect response is found both for TA1535 and TA100. Gene mutations in the adenine loci were observed in yeast, Saccharomyces. In higher eukaryotes, the induction of point mutations has been studied in Drosophila with sex-linked recessive lethals and in barley with chlorophyll-deficient mutants after treatment of seeds. The nature of those point mutations are, however, not defined as they contain both intragenic changes and structural changes.

Alexander and Glanges (88,98) showed a high rate of autosomal recessive lethals in Drosophila at all stages of spermatogenesis (average 19.8% for all stages). The frequency of lethals induced in chromosome 2 as compared to X (10-38 times more) was considerably higher than could be accounted for by the difference in length of the chromosomes.

Chromosome Alterations. The frequency of translocations in Drosophila as compared to sex-linked recessive lethals is about 1:2 or even higher while the corresponding ratio recorded for TEM is about 1:7 and for Trenimon 1:10 or less.

Recombination. An increase both of mitotic recombination and gene conversion has been demonstrated in yeast with ethylenimine.

Nondisjunction. No tests on nondisjunction are available.

Stage Specificity. As with the other aziridines and most other chemicals, postmeiotic stages in males both of mice and Drosophila show the highest sensitivity to ethylenimine. So far dominant lethals in mice have only been reported after treatment of spermatozoa and spermatids. The data of Alexander and Glanges (88) on autosomal recessive lethals in Drosophila, however, did not point to a pronounced stage specificity. They found a high effect in all stages of spermatogenesis.

Conclusions

Ethylenimine is a monofunctional alkylating agent which does not require metabolic activation. It has a high mutagenic potency both for point mutations and chromosome aberrations. However, in comparison to TEM and Trenimon, the data indicate a much lower effectiveness, at least with regards to chromosomal aberrations. This could be expected with a monofunctional agent, which lacks the ability to form crosslinks. As pointed out above ethylenimine chloride is easily degraded and experimental data with this compound should be interpreted with caution.

In the same way as with the other two aziridines, data on ethylenimine can be extrapolated to man, at least qualitatively.

REFERENCES

1. Ehrenberg, L., and Wachtmeister, C. A. In: Handbook of Mutagenicity Test Procedures, B. J. Kilbey, et al., Eds. Elsevier, Amsterdam: 1977, pp. 411-418.

2. Windholz, M., (ed.). The Merck Index, 9th ed., Merck, Rahway, N.J., 1976, p. 1241.

3. Dermer, O. C., and Ham, G. E. Ethylenimine and Other Aziridines: Chemistry and Applications, Academic Press, New York, 1969.

4. Fishbein, L., Flamm, W. G., and Falk, H. L. Chemical Mutagens: Environmental Effects on Biological Systems, Academic Press, New York, 1970.

5. Matter, B. E., and Generoso, W. M. Mutat. Res. 21: 41-42 (1973).

6. Matter, B. E., and Generoso, W. M. Genetics 77: 753-763 (1974).

7. Bateman, A. J. Genet. Res. 1: 381-392 (1960).

8. Arnold, D. W., Kennedy, G. L., and Keplinger, M. L. Mutat. Res. 26: 459-460 (1974).

9. Lyon, M. F., and Smith, B. D. Mutat. Res. 11: 45-58 (1971).

10. Cox, B. D., and Lyon, M. F. Mutat. Res. 30: 293-298 (1975).

11. Suter, K. E., and Generoso, W. M. Mutat. Res. 34: 259-270 (1976).

12. Generoso, W. M., Cain, K. T., Huff, S. W., and Gosslee, D. G. In: Chemical Mutagens, Vol. 5, A. Hollaender and F. J. de Serres, Eds., Plenum Press, New York, 1980, pp. 55-77.

13. Michel, T. M., and Legator, M. S. Mutat. Res. 23: 41-45 (1974).

14. Cattanach, B. M., and Williams, C. E. Mutat. Res. 13: 371-375 (1971).

15. Cattanach, B. M. Nature 180: 1364-1365 (1957).

16. Hampel, K. E., Kober, B., Roesch, D., Gerhartz, H., and Meinig, K. H. Blood 27: 816-823 (1966).

17. Chang, T. H., and Klassen, W. Chromosoma 24: 314-323 (1968).

18. J. G. Brewen, unpublished data.

19. Matter, B. E., and Grauwiler, J. Mutat. Res. 23: 239-249 (1974).

20. Jellema, M. M., and Schardein, J. L. Toxicol. Appl. Pharmacol. 27: 422-430 (1974).

21. Friedman, M. A., and Staub, J. Mutat. Res. 43: 255-262 (1977).

22. Cattanach, B. M. Mutat. Res. 3: 346-353 (1966).

23. Cattanach, B. M. Mutat. Res. 4: 73-82 (1967).

24. Clive, D., et al. Chapter 16, this volume.

25. Russell, L. B. Arch. Toxicol. 38: 75-85 (1977).

26. Fahmy, O. G., and Fahmy, M. J. Genetics 53: 566-584 (1955).

27. Ratnayahe, W. E. Mutat. Res. 5: 271-278 (1968).

28. Watson, W. A. F. Vererb. 95: 374-387 (1964).

29. Becker, J. H. Mol. Gen. Genet. 138: 11-24 (1975).

30. Snyder, L. A., and Oster, I. I. Mutat. Res. 1: 437-445 (1965).

31. Slizynska, H. Mutat. Res. 8: 165-175 (1969).

32. Slizynska, H. Genet. Res. 4: 248-257 (1963).

33. Rieger, R., and Michaelis, A. Unpublished data.

34. Ockey, C. H. J. Genetics 55: 525-549 (1957).

35. Biesele, J. J., Berger, R. E., Clarke, M., and Weiss, L. Exptl. Cell Res. (Suppl.) 2: 279-303 (1952).

36. Arnason, T. J., and Wakonig, R. J. Genet. Cytol. 1: 16-20 (1959).

37. Fahrig, R. Molec. Gen. Genet. 144: 131-140 (1976).

38. Ames, B. Unpublished data.

39. Mohn, G. R., and Ellenberger, J. In: Handbook of Mutagenicity Test Procedures. B. J. Kilbey et al., Eds., Elsevier, Amsterdam, 1977, pp. 95-118.

40. Specht, I. Arch. Mikrobiol. 51: 9-17 (1965).

41. Kada, T. Chapter 3, this volume.

42. Cattanach, B. M. Z. Vererb. 90: 1-6 (1959).

43. Sheu, C. W., Moreland, F. M., Oswald, E. J., Green, S., and Flamm, W. G. In preparation.

44. Soares, E. R., and Sheridan, W. Mutat. Res. 31: 342-343 (1975).

45. Weber, E., Bidwell, K., and Legator, M. S. Mutat. Res. 28: 101-206 (1975).

46. Schmid, W. In: Chemical Mutagens, Vol. 4, A. Hollaender, ed., Plenum Press, New York, 1976, pp. 31-53.

47. Snyder, I. A. Z. Vererb. 94: 182-189 (1963).

48. Hussain, S.,and Ehrenberg, L. Proc. Sixth Ann. Meet. Europ. Envir. Mutag. Soc. Abhd. Akad. Wissensch. DDR, Berlin, 1977, pp. 95-99.

49. Jorgenson, T. A., Newell, G. W., Scharpf, L. G., Gribling, P., O'Brien, M., and Chu, D. Mutat. Res. 31: 337 (1975).

50. Sotomayor, R. E., and Cumming, R. B. Mutat. Res. 27: 375-388 (1975).

51. Ockey, C. H. In: Chemische Mutagenese. H. Stubbes, Ed., Abhandl. DAW, Klasse Medizin: 1960, pp. 47-53.

52. Rieger, R., Michaelis, A., Schubert, I., Döbel, P., and Jank, H. W. Mutat. Res. 27: 69-79 (1975).

53. Committee 17 Report. Environmental Mutagenic Hazard. Science 187: 503-514 (1975).

54. UNSCEAR. Sources and Effects of Ionizing Radiation. United Nations, New York, 1977.

55. Vogel, P., and Jäger, P. Humangenetik 7: 287-304 (1969).

56. Röhrborn, G. In: Chemical Mutagenesis in Mammals and Man, F. Vogel and G. Röhrborn, Eds., Springer, Berlin, 1970, pp. 294-316.

57. Machemer, L., and Stenger, E. G. Arzneim. Forsch. 21: 1037-1039 (1972).

58. Röhrborn, G., and Berrang, H. Mutat. Res. 4: 231-233 (1967).

59. Machemer, L., and Hess, R. Experentia 29: 190-192 (1973).

60. Schmid, W., Arakaki, D. T., Breslau, N. A., and Culbertson, J. C. Humangenetik 11: 103-118 (1971).

61. Schleiermacher, E. In: Chemical Mutagenesis in Mammals and Man, F. Vogel and G. Rohrborn, Eds., 1970, pp. 317-341.

62. Léonard, A., and Linden, G. Mutat. Res. 16: 297-300 (1972).

63. Schmid, W., and Binkert, F. Mutat. Res. 21: 233-234 (1973).

64. Rohrborn, G., Kuehn, O., Hansmann, I., and Thon, K. Humangenetik 11: 316-322 (1971).

65. Hansmann, I., Neher, J., and Röhrborn, G. Mutat. Res. 25: 347-359 (1974).

66. Arakaki, D. T., and Schmid, W. Humangenetik 11: 119-131 (1971).

67. Beek, B., and Obe, G. Humangenetik 29: 127-134 (1975).

68. Hayashi, K., and Schmid, W. Humangeneti, 29: 201-206 (1975).

69. Matter, B. E., and Schmid, W. EMS Newsletter 6: 13 (1972).

70. Braun, R. et al. Chapter 15, this volume.

71. Braun, R. et al., ibid.

72. Lüers, H., and Röhrborn, G. Mutat. Res. 2: 29-44 (1965).

73. Mollet, P. Mutat. Res. 21: 137-148 (1973).

74. Vogel, E. et al. Chapter 9, this volume.

75. Büchi, R., and Burki, K. Arch. Genetik 48: 59-67 (1975).

76. Biesele, J. J., Berger, R. E., Clarke, M., and Weiss, L. Exptl. Cell Res. (Suppl. 2): 279-303 (1952).

77. Nicoloff, H., Rieger, R., Künzel, G., and Michaelis, A. Mutat. Res. 30: 149-152 (1975).

78. Vig, B. K., and Zimmermann, F. K. Environ. Exptl. Botany, in press.

79. Zimmermann, F. K. Mutat. Res. 11: 327-337 (1971).

80. Propping, P., Rohrborn, G., and Buselmaier, W. Mol. Gen. Genet. 117: 197-209 (1972).

81. Zimmermann, F. K. In: Chemical Mutagenesis: Principles and Methods for Their Detection, Vol. 3, A. Hollaender, Ed., Plenum Press, New York, 1973, pp. 209-239.

82. Rohrborn, G., and Hansmann, I. Humangenetik 13: 184-198 (1971).

83. Matter, B. E. J. Int. Med. Res. 4: 382-392 (1976).

84. Osterman-Golkar, S., Ehrenberg, L., and Wachtmeister, C. A. Radiation Bot. 10: 303-327 (1970).

85. Malashenko, A. M. Sov. Genet. 4: 538-543 (1968); Genetika 4(4): 158-164 (1968).

86. Malashenko, A. M., and Egorov, I. K. Sov. Genet. 4: 14-18 (1968); Genetika 4(4): 21-27 (1968).

87. Chang, T. H., and Elequin, F. T. Mutat. Res. 4: 83-89 (1967).

88. Alexander, M. L., and Glanges, E. Proc. Natl. Acad. Sci. (U.S.) 53: 282-288 (1965).

89. Sràm, R. J. Mol. Gen. Genet. 106: 286-288 (1970).

90. Sjödin, J. Hereditas 67: 155-180 (1971).

91. Wagner, J. H., Nawar, M., Konzak, C. F., and Nilan, R. A. Mutat. Res. 5: 57-64 (1968).

92. Aretisov, V. A., and Valeva, S. A. Sov. Genet. 11: 281-287 (1975); Genetika 11(3): 12-20 (1975).

93. Zimmermann, F. K., and von Laer, U. Mutat. Res. 4: 377-379 (1967).

94. Aleksandrova, N. N. Sov. Genet. 9: 87-89 (1973); Genetika 9(1): 122-125 (1973).

95. Hedenstedt, A. and Ramel, C. Unpublished data.

96. Szybalski, W. Ann. N. Y. Acad. Sci. 76: 243-244 (1970).

97. Sega, Gary A., and Mitchell, Ann D. Chapter 21, this volume.

98. Oftedal, P., and Searle, A. G. J. Med. Genet. 17: 15-20 (1980).

CHAPTER 33

COMPARATIVE MUTAGENICITY OF PROCARBAZINE (NATULAN)

Ingo Hansmann

Institut für Humangenetik der Universität Göttingen

Nikolausberger Weg 5a, D-3400 Göttingen, Germany

Procarbazine (Natulan) is an antitumor substance with an apparently different mechanism of action compared to other cytotoxic agents. It is a derivative of the basic formula: CH_3-NH-NH-CH_2R and represents a N-isopropyl-a(2-methylhydrazino)-p-toluamide monohydrochloride. It is used in clinical therapy, specifically in cases of Hodgkin's disease, but also in patients with polycythemia vera, malignant melanoma and multiple myeloma (1). Detailed reviews on procarbazine have been given by Plattner (2) Jellife and Marks (3), Stock (4), Sartorelli and Creassey (5), Carter (6), Oliverio (7), and Reed (8).

CHEMICAL PROPERTIES

Procarbazine is soluble but unstable in aqueous solutions. It is a white to pale yellow crystalline substance and has a molecular weight of 257.8.

As an intact molecule procarbazine is cytostatically not active. Autoxidation of procarbazine to azoprocarbazine and hydrogen peroxide is observed in aqueous solutions containing dissolved oxygen (9). Azoprocarbazine isomerizes to the hydrazone of monomethylhydrazone (MMH) and p-formyl N-isopropylbenzamide (10). A detailed review of the degradation of procarbazine has been given by Reed (8).

TUMOR INHIBITION

Procarbazine is used for the treatment of lymphogranulomatosis (Hodgkin's disease), alone or in combination with other agents (1) and a combined chemotherapy including vincristine, cyclophosphamide and prednisone was proposed. Recommended dosages of procarbazine, singly or divided are 100-200 mg daily in the first week for adults. Maintenance dosage is at 300 mg or less and this may be lowered after maximal response to 50-100 mg daily (8). This agent is also used for the treatment of other malignant lymphomas, chronic myelosis, and also for solid tumors like bronchogenic carcinoma (1).

PHARMACOLOGY

Procarbazine, like a variety of cytotoxic agents, causes a number of biological side effects such as leukopenia, thrombopenia, immunsuppression, carcinogenesis, teratogenesis and depression of spermatogenesis (8). The therapeutic effect is also accompanied by toxic side effects such as nausea and vomiting (11). The agent is absorbed well after parenteral or oral application and is rapidly distributed throughout the body. A remarkably short half-life time of 7 min on the average was observed in man after IV injection. The analogous half-life in dogs and cats was 12 and 14 min, respectively (12). This indicates that procarbazine is metabolized also very quickly in these animal species. The LD_{50} is given for mice, rats, and rabbits as 1320 ± 66 mg/kg body weight, 785 ± 34 mg/kg and 145 ± 11.5 mg/kg, respectively. Potentiating capabilities of procarbazine on the effects of barbiturates, antihypertensive and symptomimetic drugs have been reported (13,14).

CARCINOGENICITY

Procarbazine is a multipotential carcinogen and produces leukemia, lung adenoma and mammary adenocarcinoma in rodents. Rhabdomyosarcoma and fibrosarcoma were observed in rats (15,16). Two of 26 rhesus monkeys developed acute myologenous leukemia after chronic administration (17). Secondary acute myeloid leukemia and lymphomas in some patients treated for Hodgkin's disease were reported by some authors (18-20).

TERATOGENICITY AND ANTIFERTILITY

Chaube and Murphy (21,22) observed malformations in rats after single injections of procarbazine at doses ranging from 5 to 550 mg/kg body weight. Malformations included shortened and malpositioned tail and appendages, cleft palate and defects of digits, jaw, and face. There are also indications that procarbazine may be teratogenic in man (23,24). In vivo application caused a prominent depression of spermatogenesis followed by testicular atrophy in rats (25,26).

Beginning sterility was observed after two weeks without any signs of recovery. Similar antifertility effects on spermatogenesis in mice were reported by Lee and Dixon (27-29). In this study (27-29) most cell types involved in spermatogenesis proved to be sensitive, with the exception of mature spermatozoa and late spermatids. On comparing rat spermatogenesis by in vivo and invitro studies the conclusion can be drawn that metabolic activation is required for the antifertility effects of procarbazine (30).

Similar effects on spermatogenesis are reported in man: Azoospermia and germinal aplasia have been found in many patients treated for Hodgkin's disease (31,32). These patients, however, were treated by a combined chemotheraphy, and the antifertility effects can therefore not be ascribed solely to procarbazine.

MODE OF ACTION OF PROCARBAZINE

The very mechanisms of the various biological activities of procarbazine are not yet fully understood. An influence on the mitotic cycle by an interphase prolongation (33) supposedly due to hydrogen peroxide was observed cytologically. Hydrogen perioxide is produced catabolically from procarbazine (10). Both hydrogen peroxide and formaldehyde inhibit DNA-polymerase and DNA-dependent RNA-polymerase (10). Inhibition of DNA, RNA, and protein synthesis was observed in Ehrlich ascites cells and in male mouse meiotic cells (10,27-29,33-36). The latter observation suggests an effect of procarbazine not only on highly proliferating but also on nonproliferating tissue.

An alkylating effect of procarbazine on DNA and RNA due to the presence of metabolic intermediates was suggested by Weitzel et al. (37,38). Methylation of cytoplasmic RNA and DNA is reported as well (39,40).

MUTAGENICITY

Salmonella Mutation Assay

Procarbazine is nonmutagenic in the standard plate incorporation assay (41,42), the liquid assay and the host-mediated assay

(42), either with or without rat liver microsome activation system (S-9), induced by phenobarbital. The strains used are the standard tester strains TA1535, TA100, TA1538, TA98, TA1537 or the strains His G46, His G46/R^+, TA1530, TA1530/R^+, TA1975, TA1975/R^+, TA1535, TA1535/R^+, TA98. In one study (42), the mutagenicity of urine metabolites from procarbazine (400 mg/kg, 24 hr before urine sampling in rats) was tested on strain TA1530/R^+. No increase in revertants was observed. Also no effect was observed in the host-mediated assay up to a dose of 400 mg/kg body weight (42). The standard plate incorporation assay gave negative results up to a dose of 10,000 µg/plate (42), and the liquid assay up to a dose of 19,500 µg/ml (41).

Vicia faba

No chromosomal aberrations were induced within the range of experimental conditions (43). Lateral roots were treated for 2 hr at 20°C, pH 7.2, and a recovery time of 24 hr with doses ranging from 5.2 to 1300 µg/ml.

Drosophila

Procarbazine was administered orally to adult males. In all stages of spermatogenesis a high response for induction of sexlinked recessive lethals were observed. In spermatids a concentration-effect relationship was observed at low levels, but a plateau phase was reached for the induction of recessive lethals at higher concentrations (44). No clear-cut concentration effect was apparent after treatment of sperm.

A low response for total sex-chromosome loss (X,Y) and dominant lethals is reported for metabolically active spermatids. Procarbazine did not cause chromosome breaking events leading to partial sex-chromosome loss (Y^L,Y^S) and II-III translocations (44). The feeding dose for 2 days ranged from 2.6 to 7.800 µg/ml and the strain tested was Berlinky. The lowest effective concentration (LEC) is given as 130 µg/ml for the induction of recessive lethals. The mutational pattern described resembles that given by most precarcinogens in Drosophila, reflecting the important role of intragonadal metabolism.

Spot Test

Treatment on all days of mouse fetal development (days 7-10) was positive. The maximum response was from a dose of 100 mg/kg on day 9 of fetal development which gave over 10% of offspring showing spots compared with much less than 1% in the controls

(45). A significant increase of spots with genetic relevance was observed in offspring treated on day 9 of fetal development at a dose of 75 mg/kg body weight (46). The doses tested ranged from 50 to 100 mg/kg body weight. The yield of spots in one study (46) was lower after 100 mg/kg than could be expected from the response after 75 mg/kg.

Dominant Lethals in Mice

Dominant lethality was induced mainly in spermatids and spermatocytes (47,48). An effect after the application of 400 mg/kg is reported also for B- and intermediate spermatogonia in mice from strain CD_2F_1 (29). Mainly postimplantation loss occurred after treating spermatids and spermatocytes with 400 mg/kg body weight, but after 800 mg/kg the loss was mainly manifested before implantation (48). No dominant lethals occurred on treating spermatozoa. The lowest dose showing a significant increase of dominant lethality over the control value was 200 mg/kg (48).

Sperm Morphology Assay

An increase in mouse sperm head abnormalities was observed after procarbazine treatment. The doubling dose is given with 120 mg/kg body weight, and the highest dose tested was 1000 mg/kg body weight (49). A hump-shaped slope of abnormal sperm with increasing dose is also observed with the sperm assay. The yield after 100 mg/kg was decreased compared to the lower dose.

Micronucleus

Procarbazine is a potent inducer of micronuclei in mouse bone marrow (50). Each dose was given twice intraperitoneally separated by an interval of 24 hr. Sampling was done 6 hr after the second injection. The doses given ranged from 2×6.45 to 2×51.6 mg/kg body weight. The frequency of micronucleated polychromatic erythrocytes increased significantly even after the lowest dose (2×6.45). A plateau of response is reached with 2×12.9 mg/kg.

Sister Chromatid Exchanges (SCE)

SCE's are produced in bone marrow cells from male Chinese hamsters in the in vivo SCE method. Positive effects, which are significantly different from the control value are reported for doses between 10 and 25 mg/kg body weight. Six doses ranging from 10 to 300 mg/kg were tested, and the response of SCE was of the plateau type (51).

Chromosome Aberrations in Mammals (in Vivo)

Chromosome breakage is reported from mouse Ehrlich-ascites tumors cells after 200 and 400 mg/kg (33). Interchanges and triradials occurred predominantly, and the incidence was increased significantly even after 200 mg injected intraperitoneally. These findings have been corroborated with three other mouse carcinoma lines in vivo (52). Procarbazine was given intraperitoneally in doses of 0.004 or 0.016 mg/mouse. The bone marrow and spleen cells of the same mouse did not however show an increased frequency of chromosome anomalies. There was also no effect of procarbazine without metabolic activation in mouse Ehrlich ascites cells, mouse spleen, Hela cells and human lymphocytes (52). Adler (53) found a significant increase of chromatid aberrations in mouse spermatogonia after 800 mg/kg body weight. The dose range tested was 200-800 mg/kg.

There have been some reports on chromosome anomalies in peripheral blood from patients receiving procarbazine and various other cytostatic agents (54,55). This makes it difficult to aver that procarbazine given alone causes chromosome breakage in human lymphocytes in vivo. Lymphocytes of one patient, however, who was treated only with procarbazine for acute myelofibrosis for six months showed a significant increase of aneuploid cells and cells with structural anomalies (56).

Specific Locus Mutations in Somatic Cells of Mammals

There have been three sets of observations in mammalian cells, the first (57) involving L5178Y/$TK^{+/-}$. Three experiments were performed, one without and two with metabolic activation. In the former, a doubling of the spontaneous mutation frequency was observed up to around 100 µg/ml. Thereafter the mutation frequency decreased with increasing dose. In these experiments an extensive cell killing was observed.

In a separate set of experiments with L5178Y/$TK^{+/-}$ mouse lymphoma cells (58) in both with azaguanine resistance and Budr resistance, there was again an unusual dose response curve in the presence of metabolic activation (Fig. 1). These observations are corroborated by the results obtained with azaguanine resistance in W97 Chinese hamster cells (58). In these studies there was no evidence of cytotoxicity.

In a third series of experiments with mouse lymphoma cells, in the absence of metabolic activation procarbazine was not shown to be mutagenic in the azaguanine-resistant, ouabain-resistant, thymidine-resistant, and cytosine arabinoside-resistant system (59). This study was extended to include the fluctuation experimental design where procarbazine was shown to be mutagenic at 5 μg/ml. The lowest effective concentrations tested (LECT) for L5178Y are 560 μg/ml (range 560-2800 μg/ml) (58) and 0.5 μg/ml (57) at a range of 0.5-1600 μg/ml.

In W97 Chinese hamster cells the LECT is 560 μg/ml (58) for a dose range of 560-14,000 μg/ml.

Specific Locus Mutations in Male Germ Cells

Using the specific locus method, the ability of procarbazine to induce gene mutations in mice was examined. Different sublethal doses of the test compound were injected intraperitoneally. The effectiveness of procarbazine to induce mutations in offspring derived from different cell stages was as follows: 200 mg/kg, 0 mutations in 6,722 offspring derived from postspermatogonial stages (pg) and 4 mutations in 33,380 offspring derived from spermatogonia (g); 400 mg/kg, 1 mutation in 3,394 pg-offspring and 10 mutations in 35,047 g-offspring; 600 mg/kg, 2 mutations in 1,930 pg-offspring and 16 mutations in 45,413 g-offspring; 800 mg/kg, 0 mutations in 1,771 pg-offspring and 7 mutations in 40,013 g-offspring. The control frequency in this experiment was 3 mutations in 53,473 offspring (60). The mutation rates of 14.8×10^{-5} in the 600 mg/kg group in pg-stages and of 4.1×10^{-5} in spermatogonia after injection of 400 mg/kg of procarbazine are significantly different from the control frequency ($p \leqq 0.01$) (61). The author claims that these experiments fulfill the basic assumption for the doubling dose concept with a doubling dose for specific locus mutations in spermatogonia of mice with 114 mg/kg body weight.

CONCLUSIONS

Procarbazine is nonmutagenic in microbial systems, with and without mammalian metabolic activation (Table 1). Different strains of Salmonella were used in the standard plate incorporation assay, the liquid assay, and the host-mediated assay and various concentrations even up to 10,000 μg/plate have been tested. This agent did not cause chromosome breakage in *Vicia faba*. Procarbazine is mutagenic in Drosophila, mammalian somatic cells in vivo and in vitro with metabolic activation and in mammalian male germ cells (Table 1). A broad spectrum of genetic alterations is induced by this agent: recessive lethals, dominant lethals, 2,3

Table 1. Concentration and dose ranges of procarbazine tested in different systems for mutagenicity and lowest reported reported concentration/dose giving positive response

Test system	Concentration	Lowest concn reported positive	References
Bacteria			
Salmonella typhimurium	10-500 μg/plate	Negative	41
	10-10,000 μg/plate	Negative	42
	0.195-19,500 μg/plate	Negative	42
Host-mediated assay	80-400 mg/kg μg/ml	Negative	42
Plants			
Vicia faba	5.2-1300 μg/ml	Negative	43
Drosophila			
Males, recessive lethals	26-7800 μg/ml	130 μg/ml	44
Mammalian cells in vitro (+S9)			
Mouse lymphoma L5178Y $TK^{+/-}$	0.5-1600 μg/ml	0.5 μg/ml	57
	560-2800 μg/ml	560 μg/ml	58
	5 μg/ml	5 μg/ml	59
Chinese hamster W97	560-1400 μg/ml	560 μg/ml	58

(continued)

Table 1. (continued)

Test system	Concentration	Lowest concn reported positive	References
Mammals			
Spot test, mouse	50-100 mg/kg	75 mg/kg	45,46
Dominant lethals, mouse	22-800 mg/kg	200 mg/kg	47,48
Sperm morphology assay, mouse	0-1.000 mg/kg	120 mg/kg (doubling dose)	49
Micronucleus, mouse	2×6.45 to 2×51.6 mg/kg	2×6.45 mg/kg	50
SCE, in vivo, Chinese hamster	10-300 mg/kg	25 mg/kg	51
Chromosome aberrations			
Ehrlich ascites cells	200-400 mg/kg	200 mg/kg	33
Spermatogonia, mouse	200-800 mg/kg	800 mg/kg	53
Specific locus mutations mouse, male germ cells	200-800 mg/kg	400 mg/kg	60,61

translocations, entire X-Y loss in Drosophila and in mammals: sperm morphology alterations, genetic alterations in the spot test, dominant lethals, chromosome breakage, specific locus mutations in somatic cells as well as specific locus mutations in male germ cells.

Clarification is required on the different response and cytotoxicity in assays for specific locus mutations in mammalian cells observed in different laboratories.

Further information is required on the effect on female germ cells at different stages of development as well as on the potential induction of nondisjunction. Procarbazine is a chemical mutagen which shows a complicated dose-effect correlation in many systems tested. The dose response is either from the plateau type or the response even decreased at higher concentrations (e.g., Fig. 1). This peculiar effect may complicate not only an extrapolation from high to low doses/concentrations but have to be considered for risk estimations.

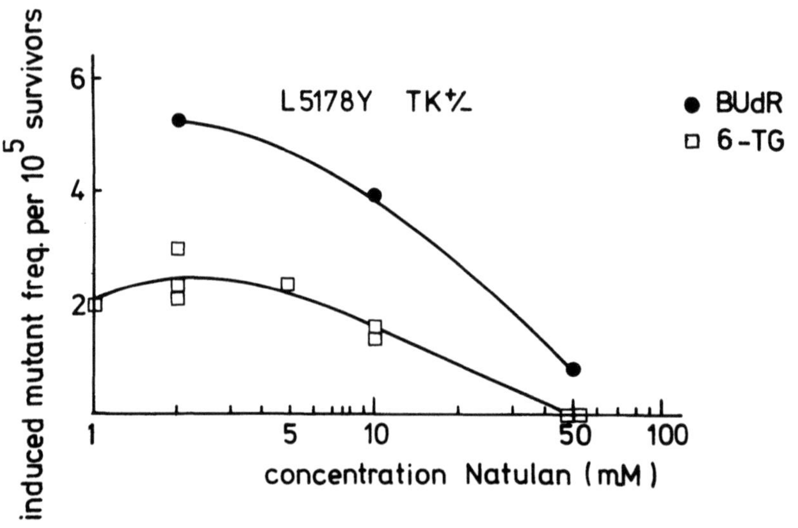

Figure 1. Mutagenicity of Natulan in L5178Y mouse lymphoma cells.

Procarbazine is mutagenic in Drosophila and in all mammalian systems tested so far and therefore potentially mutagenic in man. On the basis of these experiments genetic counselling and prenatal diagnosis for chromosome aberrations should be recommended individually.

ACKNOWLEDGEMENTS

The assistance in reviewing, discussing, and updating the manuscript of the group III members at the workshop on Comparative Chemical Mutagenesis, Raleigh, North Carolina, October 31-November 4, 1977 is acknowledged. The members were: I. D. Adler (Gesellschaft für Strahlen- und Umweltforschung, München, Germany), C. Arlett (University of Sussex, Brighton, England), G. W. P. Dawson (University of Dublin, Ireland), U. M. Ehling (Gesellschaft für Strahlen- und Umweltforschung, Munchen, Germany), W. F. Grant (McGill University, Quebec, Canada), D. McGregor (Midlothian, Scotland), P. Lohman (Medical Biological Laboratory, TNO, Rijswijk, The Netherlands), H. V. Malling (NIEHS, Research Triangle Park, North Carolina, U.S.A.), H. Ott (Directorate General for Research, Science and Education, Commission of the European Communities, Brussels, Belgium), and J. W. Simons (Universität Leiden, The Netherlands).

REFERENCES

1. Spivack, S. D. Ann. Intern. Med. 81: 795-800 (1974).

2. Plattner, P. A., Ed. Proceedings of the International Symposium on the Chemotherapy of Cancer, Lugano, April 28-May 1, 1964, Elsevier, Amsterdam-London-New York, 1964, p. 324.

3. Jellife, A. M., and Marks, J., Eds. Natulan (Ibenzmethyzin). John Wright and Sons, Ltd. Bristol, 1965.

4. Stock, J. A. Exptl. Chemotherapy 5: 333-416 (1967).

5. Sartorelli, A. C., and Creasey, W. A. Ann. Rev. Pharmacol. 9: 51-72 (1969).

6. Carter, S. K., Ed. Procarbazine (Natulan: NSC 77213): Development and Application, Cancer Therapy Evaluation Branch, National Cancer Institute, Bethesda, Md., 1970, p. 109.

7. Oliverio, V. T. In: Cancer Medicine, J. F. Holland and E. Frei, III, Eds., Lea and Feribiger, Philadelphia, 1973.

8. Reed, D. J. In: Handbook of Experimental Pharmacology, A. C. Sartorelli and D. G. Johns, Eds., Vol. 38, Part 2, Springer, Berlin-Heidelberg-New York, 1975, pp. 747-765.

9. Zeller, P., Gutmann, H., Hegedus, B., Kaiser, A., Langemann, A., and Mueller, M. Experientia 19: 129 (1963).

10. Weitzel, G., Schneider, F.. Kummer, D., and Ochs, H. Z. Krebsforsch. 70: 354-365 (1968).

11. Strickstock, K.-M., and Obrecht, P. In: Krebsforschung und Krebsbekämpfung, H. E. Bock, Ed., Vol. 6. Urban und Schwarzenberg, München-Berlin-Wien, 1967, pp. 366-375.

12. Raaflaub, J., and Schwartz, D. E. Experientia 21: 44-45 (1965).

13. de Vita, V. T., Hahn, M. A., and Oliverio, V. T. Proc. Soc. Exptl. Biol. Med. 120: 561-565 (1965).

14. Lee, I. P., and Lucier, G. W. J. Pharmacol. Exptl. Therap. 196: 586-593 (1975).

15. Kelly, M. G., O'Gara, R. W., Yancey, S. T., and Botkin, C. J. Natl. Cancer Inst. 40: 1027-1051 (1968).

16. Kelly, M. G., O'Gara, R. W., Gadekar, K., Yancey, S. T., and Oliverio, V. T. Cancer Chemotherapy Report 39: 77-80 (1964).

17. O'Gara, R. W., Adamson, R. H., Kelly, M. G., and Dalgard, D. W. J. Nat. Cancer Inst. 46: 1121-1130 (1971).

18. Andersen, R. H., and Vidaek, A. Scand. J. Haematol. 7: 201-207 (1970).

19. Michelmayr, G., Gunther, R., Lederer, B., and Huber, H. Med. Klin. (München) 68: 180-182 (1973).

20. Sneddon, I., and Wishart, J. M. Brit. Med. J. 4: 235 (1972).

21. Chaube, S., and Murphy, M. L. Proc. Amer. Assoc. Cancer Res. 5: 11 (1964).

22. Chaube, S., and Murphy, M. L. Teratology 2: 23-32 (1969).

23. Well, J. H., Marshall, J. R., and Carbone, P. P. J. Am. Med. Assoc. 205: 398-400 (1968).

24. Mennuti, M. T., Shepard, T. H., and Mellman, W. J. Obst. Gynecol. 46: 194-196 (1975).

25. Bollag, W., and Theiss, E. In: Proceedings of the International Symposium on the Chemotherapy of Cancer, P. A. Plattner, Ed., Lugano 1964, pp. 311-313.

26. Hilscher, W., and Reichelt, P. Beitr. Pathol. Anat. 137: 452-478 (1968).

27. Lee, I. P., and Dixon, R. L. Toxicol. Appl. Pharmacol. 28: 20-41 (1972).

28. Lee, I. P., and Dixon, R. L. Toxicol. Appl. Pharmacol. 23: 20-41 (1972).

29. Lee, I. P., and Dixon, R. L. J. Pharmacol. Exptl. Therap. 181: 219-226 (1972).

30. Lee, I. P., and Dixon, R. L. NIH/NIEHS Scientific Exhibition, June 2-3, 1977.

31. Van Thiel, D. H., Sherins, R. J., Meyers, G. H., Jr., and de Vita, V. T., Jr. J. Clin. Invest. 51: 1009-1019 (1972).

32. Sherins, R. J., and de Vita, V. T., Jr. Am. Intern. Med. 79: 216-220 (1973).

33. Rutishauser, A., and Bollag, W. Experientia 19: 131-132 (1963).

34. Koblet, H., and Diggelmann, H. Eur. J. Cancer 4: 45-58 (1968).

35. Gutterman, J. A., Huang, A. T., and Hochstein, P. Proc. Soc. Exptl. Biol. Med. 130: 797-802 (1969).

36. Gales, G. R., Simpson, J. G., and Smith, A. B. Cancer Res. 27: 1186-1191 (1967).

37. Weitzel, G. L., Schneider, F., Fretzdorff, A. M., Seynsche, K., and Fuiger, H. Z. Physiol. Chem. 336: 271-282 (1964).

38. Weitzel, G. L., Schneider, F., Fretzdorff, A. M., Durst, J. and Hirschmann, W. Z. Physiol. Chem. 348: 433-442 (1967).

39. Kreis, W., Piepho, S. B., and Bernhard, H. B. Experientia 22: 431-433 (1966).

40. Kreis, W., Burchenal, J. H., and Hutchison, D. J. Proc. Am. Assoc. Cancer Res. 9: 38 (1968).

41. McCann, J., Choi, E., Yamasaki, E., and Ames, B. N. Proc. Natl. Acad. Sci. (U.S.A.) 72: 5135-5139 (1975).

42. Thompson, S., and Kellicutt, L. Final report 1974-1975, Contract 023-74-ENV-EIR of the EEC Environmental Research Programme.

43. Rieger, R., Michaelis, A., Schubert, I, Kaina, B., and Heindorff, Heindorff, K. Chap. 14, this volume.

44. Blijleven, W. G. H., and Vogel, E. Mutat. Res. 45: 47-59 (1977).

45. Dawson, G. W. P., Davidson, G. E., and Carter, J. Progress Report April-December 1976, Contract 023-74-ENV-EIR, Environmental Research Program.

46. Neuhäuser, A. Annual Report, Gesellschaft für Strahlen- und Umweltforschung m.b.H., München, GSF-Bericht B 798, 1977, pp. 42-43.

47. Epstein, S. S., Arnold, E., Andrea, J., Bass, W., and Bishop, Y. Toxicol. Appl. Pharmacol. 23: 288-325 (1972).

48. Ehling, U. H. Mutat. Res. 26: 285-295 (1974).

49. Wyrobek, A. J., and Bruce, W. R. In: Chemical Mutagens: Principles and Methods of Their Detection, A. Hollaender and F. J. de Serres, Eds., Vol. V, Plenum Press, New York-London, 1978, pp.

50. Wild, D. Mutat. Res. 56: 319-327 (1978).

51. Bayer, U. Mutat. Res. 56: 305-309 (1978).

52. Therman, E. Cancer Res. 32: 1133-1136 (1972).

53. Adler, I. In: Cytogenetic Test Systems for Environmental Mutagens. T. C. Hsu, Ed., Allenheld Publishing Co., 1931, in press.

54. Vormittag, W. Wiener Klin. Wochenschr. 86: 69-75 (1974).

55. Schinzel, A., and Schmid, W. Mutat. Res. 40: 139-166 (1976).

56. Pinedo, H. M., van Hemel, J. O., Vrede, M. A., and van der Sluys Veer, J. Brit. Med. J. 3: 525 (1974).

57. Clive, D. Personal communication.

58. Knaap, A. G. A. C., and Simons, J. W. I. M. Progress Report, EEC-Contract No. 030-74-1 ENV-N, September 1977.

59. Rogers, A. M., and Arlett, C. T. Progress Report, April-December 1976, Contract No. 089-74-10 ENV, Environmental Research Programme, U.K.

60. Ehling, U. H., and Neuhäuser, A. Jahresbericht der Gesellschaft für Strahlen- und Umweltforschung m.b.H. München, GSF-Bericht B 798, 1977, pp. 17-18.

61. Ehling, U. H. In: Chemical Mutagens-Principles and Methods for Their Detection, Vol. V, A. Hollaender and F. J. de Serres, Eds., Plenum Press, New York-London, 1978, pp. 233-256.

CHAPTER 34

COMPARATIVE MUTAGENICITY OF MITOMYCIN C

Ilse-Dore Adler

Institut für Genetik, Gesellschaft für Strahlen- und Umweltforschung, D-8042 Neuherberg bei München

Federal Republic of Germany

CHEMISTRY

Mitomycin C (MC) was first isolated from Streptomyces caespitosus (1). It was demonstrated to have antibiotic and antitumor properties. Closely related antibiotics, designated mitomycin A and B and porfiromycin have also bee isolated (2). Mitomycins are colored, basic substances forming reddish-violet, plate or needle-shaped crystals. The antibiotic family was described as having unique, naturally occurring structural configurations, in that it represents the first recorded example of the formation by a microbial system of an aziridine group, the pyrrol-(1,2-α)-indole ring system, an aminobenzoquinone, and a pyrrolizine. The common mitomycin nucleus is termed Mitosane. The different compounds only vary by minor substituent groupings (3). MC has the formula $C_{15}H_{28}N_4O_5$ and a molecular weight of 334.341. Its chemical structure is:

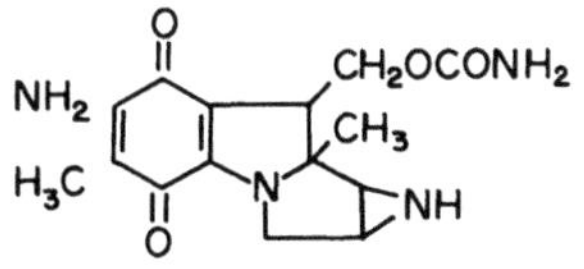

It is water-soluble and unstable under mildly acid conditions. The reduction of MC can be accomplished in vitro by several reducing agents such as dithionite. The reduction of the quinone moiety of MC is followed by spontaneous elimination of the tertiary methoxy group and formation of an aromatic indole system with two active centers, i.e., the aziridine ring and the CH_2-O-CO-NH_2 group (4). A third reactive site has been discussed for the C-7 position.

MOLECULAR ACTIVITY

The activated MC molecule selectively inhibits DNA synthesis in susceptible organisms and leads to extensive breakdown of the DNA in vitro. At high levels of MC, cellular RNA and protein synthesis are also suppressed. MC acts both as a monofunctional and as a bifunctional alkylating agent (5). The activated form of MC crosslinks complementary DNA strands as demonstrated by denaturation and renaturation of DNA or by radioactive labelling of MC (6,7). It also binds to a single DNA base by alkylation (8). Only about 1/5 - 1/10 of the total MC alkylation is attributable to crosslinkage, but the cross links are more stable than those of nitrogen mustard or other alkylating agents (5). The exact point of MC attachment to DNA is still unknown but, on the basis of model building and thermal stability of the alkylated DNA, it was postulated that MC causes links between 0-6 groups of guanines (6,7). The in vitro alkylation of DNA or RNA can only occur if DNA or RNA are present during the reduction of MC to its active form which is very unstable and rapidly loses its capacity to crosslink (9). The maximum alkylation corresponds to one antibiotic molecule per 500 nucleotide residues. The susceptibility of DNA to alkylation increases with increasing GC content (8).

TOXICITY

Median lethal doses given parenterally in rats, cats, dogs, and monkeys vary within the narrow range of 1.0-2.5 mg/kg. This compares closely with the total dose of 0.15-1.5 mg/kg likely to produce serious signs of toxicity in cancer patients. The LD_{50} for mice and rats was 8.5 and 2.5 mg/kg, respectively, after IP injection and during an observation period of 14 days (10).

MC is used for treatment of adenocarcinoma, chronic myeloic leukemia, and Hodgkin's disease as well as for prophylactic purposes following surgery or irradiation of sarcomas, epitheliomas and carcinomas of various origins. Daily doses for adults of 1-2 mg amount to a total of 40-60 mg given over a period of 1-4 months. Children are dosed with 0.03-0.09 mg/kg per day over a period of 3-4 or 6-8 weeks. MC is given by IV injection or orally. The oral doses for adults are about double the IV doses, i.e., doses of 2-6 mg/day amount to a total dose of 100-150 mg (11).

In five laboratory species (mice, rats, cats, dogs, and monkeys) lethal doses cause a protracted intoxication that is delayed in onset and characterized by anorexia, steady loss in weight, diarrhea, dehydration, and delayed death. Lesions in the hematopoetic tissues and the intestinal epithelium probably account for most fatalities in these species as well as the major signs of toxicity in patients (10).

PHARMACOKINETICS

MC requires to be activated before it can function as an alkylating agent. The activation process is a NADH-dependent enzymatic reduction of MC to its hydroquinone derivative by one of the enzymes belonging to the group of quinone reductases, also referred to as diaphorases. This facilitates a protonation of the aziridine nitrogen thereby promoting intracellular alkylation of nucleophilic centers. Attack by the activated agent on a vital cellular moiety, presumably DNA, would result in cytotoxicity; conversely, reaction with noncritical moieties or water would lead to detoxification (9). The activation process can be accomplished by any intact cell, bacterial to mammalian, as is shown by the independence of in vitro assay systems from the addition of S-9 mix. A detoxification property of the liver microsome system was demonstrated colorimetrically and biologically (12).

There is no indication of barrier effects in the testes, placenta, or brain.

MUTAGENICITY

Salmonella Mutation Assay

MC does not revert the commonly used tester strains, but does revert strains TA92 (his G46) and TA94 (his D3052) which have excision repair (Table 2). S-9 mix is not required. The concentrations ranged from 2×10^{-2} to 0.1 μg/plate. The dose response was linear with strain TA94 (13,14).

E. coli Mutation Assay

MC induces base-pair substitutions, deletions and crosslinking of DNA as well as prophage induction. S-9 mix is not required. There was a dose-dependent increase in all systems (Table 3). Crosslinking of DNA in vitro occurs only after activation of MC by cell lysates, e.g., from *Sarcina lutea*, *B. subtilis*, or *E. coli* (7).

Table 1. Effects of mitomycin C in different assay systems and organisms[a]

	Bacteria	Fungi	Plants	Drosophila	Mammalian cells	Mammals
Route of administration	Solution	Solution	Solution	Feeding and injection	Solution	IP
Activation	+	+	+	+	+	+
S-9 mix required	-	-	-	NA	-	NA
Cell killing	+	-	NA	-	+	+
Gene mutations	+	NTK	-	+	+	+
Chromosome aberrations	NA	NA	+	+	+	+
Nondisjunction and chromosome loss	NA	+	NRI	+	NA	NTK
Unscheduled DNA synthesis and DNA repair replication	NTK	NA	NTK	NA	NRI	NTK
Sister chromatid exchanges	NA	NA	+	NA	+	+
Mitotic recombination	NA	+	+	NTK	(+)	NTK
Gene conversion	NA	+	NTK	NA	NA	NA
Cell cycle specificity	NTK	NTK	+	NTK	+	+

[a]Abbreviations: NA = not applicable; NRI = no relevant information; NTK = no test known.

Table 2. Concentration ranges tested and lowest tested concentrations that gave positive responses in different assay systems applied to test mitomycin C

Assay system	Concn/dose range tested (μg/ml or mg/kg)	Lowest tested concentration positive (μg/ml)	Lowest tested dose positive (mg/kg)	References
Bacteria				
S. typhimurium, mutations (TA92, TA94) (mg/plate)	2.0×10^{-2}–1.0	2.0×10^{-2}		13,14
E. coli, mutations (arg$^+$F→ arg$^-$F)	2.5×10^{-2}–2.0	2.5×10^{-2}		15
E. coli, DNA repair				
Liquid suspension	6.0×10^{-2}–0.8	6.0×10^{-2}		16
Disc assay (mg/disc)	1.0–10	1.0		17,18
B. subtilis, rec assay	6.0×10^{-3}–0.1	6.0×10^{-3}		19,20
Fungi				
Yeast, mitotic gene conversion and crossing over	a	400		21,22
Yeast, monosomy	4–40	4		23

(continued)

Table 2. (continued)

Assay system	Concn/dose range tested (μg/ml or mg/kg)	Lowest tested concentration positive (μg/ml)	Lowest tested dose positive (mg/kg)	References
Plants				
Soybean, somatic crossing over	0.3-5.0	0.3		24
Tradescantia paludosa, chromosome aberrations	a	10		25
Allium cepa, chromsome aberrations	a	0.4		26
Vicia faba, chromosome aberrations	0.4-1200	0.4		27,28
Vicia faba, SCE	a	0.75		29
Drosophila				
Males, recessive lethals	50-200	50		30
Females, recessive lethals	100-600	100		31

(continued)

Table 2. (continued)

Assay system	Concn/dose range tested (μg/ml or mg/kg)	Lowest tested concentration positive (μg/ml)	Lowest tested dose positive (mg/kg)	References
Mammalian cells, in vitro				
Specific locus mutations				
Chinese hamster V79, thioguanine resistance	0.17-1.7	0.17		32
Chinese hamster CHO, thioguanine resistance	0.2-2.0	0.2		33
Mouse lymphoma L5178Y $TK^{+/-}$, TFT resistance	0.1-0.3	0.1		34
Chromosome aberrations				
Human leukocytes, normal	1.0×10^{-2}-1.0	0.1		35-37
Human leukocytes, Fanconi's anemia	1.0×10^{-2}-3.0×10^{-2}	1.0×10^{-2}		38,39
Human fibroblasts, normal	2.0×10^{-2}-500	b		40,41

(continued)

Table 2. (continued)

Assay system	Concn/dose range tested (µg/ml or mg/kg)	Lowest tested concentration positive (µg/ml)	Lowest tested dose positive (mg/kg)	References
Chinese hamster fibroblasts, pseudodiploid	8.4×10^{-4}-0.3	8.4×10^{-3}		42
SCE				
Human lymphocytes	3.0×10^{-3}-0.1	1.0×10^{-2}		43
Xeroderma, fibroblasts	3.0×10^{-5}-3.0×10^{-3}	3.0×10^{-4}		43
Chinese hamster hamster CHO	3.3×10^{-4}-3.3×10^{-2}	3.3×10^{-3}		44
Mammals				
Micronucleus test, mouse	0.9-2×14		3.5	45
SCE				
Mouse bone marrow	a		0.3	46
Mouse spermatogonia	0.25-0.5		0.5	46

(continued)

Table 2. (continued)

Assay system	Concn/dose range tested (μg/ml or mg/kg)	Lowest tested concentration positive (μg/ml)	Lowest tested dose positive (mg/kg)	References
Chromatid aberrations				
Chinese hamster bone marrow	5.0-15.0		10.0	47
Mouse bone marrow	0.035-3.5		0.35	48
Mouse spermatogonia	0.035-3.5		0.035	48
Mouse spermatocytes	2.5-5.0		2.5	49
Dominant lethals, mouse males	1.75-7.0		1.75	50
Sperm morphology assay, mouse	1.3-12.0		1.3	51
Host-mediated assay				
B. subtilis/ mouse, IP	a		3.0	19
Ehrlich ascites tumor/ mouse, IP	0.11 and 0.53		0.11	52
Ehrlich ascites tumor/ mouse, SC	2.13 and 10.67		2.13	52
Mammalian spot test	a		2.0	53

(continued)

Table 2. (continued)

Assay system	Concn/dose range tested (μg/ml or mg/kg)	Lowest tested concentration positive (μg/ml)	Lowest tested dose positive (mg/kg)	References
Specific locus test				
Postsperma-togonia	4.0-7.0		c	54
Spermatogonia	4.0-7.0		5.25	54

[a]One dose only.

[b]No information in group 1 report.

[c]No dose significantly positive.

E. coli DNA Repair Assay

Growth inhibition of Pol A^- bacteria is observed with 1.0 μg of MC per disc (17,18). A mammalian metabolic activation system is not required. With the liquid suspension test a positive response was obtained at 6.0×10^{-2} μg/ml (16).

Bacillus subtilis DNA Repair Assay

Growth inhibition of the 45 Rec^- bacteria was about 16× higher than for 11A Rec^+ wild type bacteria. With addition of S-9 mix the difference was only 8-fold. The lowest effective concentration recorded was 6.0×10^{-3} μg/ml (19,20).

Yeast Mutation Assay

The group 1 report does not contain pertinent information.

Table 3.

E. coli strain	Mutation type	Test method	LEC (μg/ml)	References
H/30	$arg^- \rightarrow arg^+F$	Liquid, 60 min	0.025	15,55
WP2	$trp^- \rightarrow trp^+$	Fluctuation, 3 days	0.05	56
K12/AB1157	$arg^- \rightarrow arg^+$	Liquid, 60 min	0.10	57
	$his^- \rightarrow his^+$			
C600	$lac^- \rightarrow lac^+$	Plate, 2 days	0.20	58
B/C16, pol A^-,	$\rightarrow$ Col B res.	Liquid, 60 min	0.18	16
S	$lac^+ \rightarrow lac^-$	Liquid, 60 min	0.30	59
K-12	Prophage induction	Liquid, 60 min	0.60	59-61

Neurospora Mutation Assay

The group 1 report does not contain pertinent information.

Yeast Mitotic Recombination

Both mitotic gene conversion as well as mitotic crossing over were induced by 400 μg/ml of MC. Mitotic monosomy was tested, and a significant result was obtained at 4 μg/ml (23). Mammalian metabolic activation was not required (21,22).

Barley

There is no relevant information.

Soybean

The main effect of MC is the induction of somatic crossing over. It also induced chromosome aberrations (62-64). The lowest effective concentration tested was 0.33 μg/ml of MC given for 24 hr. MC had no systemic or residual effects (24). No gene mutations are produced at the Y_{11} locus (65). MC shows additive effects with sodium azide, synergistic effects with colchicine (66) and is inhibited in its effect by deoxyribose cytidine (67).

Tradescantia paludosa

Chromatid aberrations are induced by MC. The maximum effect occurs at 24 hr. MC causes mitotic delay. The only concentration used was 10 μg/ml. Very few chromosome type aberrations were observed (25).

Allium cepa

At the only concentration used of 0.4 μg/ml for 1 hr the maximum number of chromatid aberrations was observed 26 hr later. Mainly isochromatid breaks and chromatid interchanges were observed (26).

Vicia faba

Chromatid aberrations of the delayed type are induced, the most frequent being isochromatid breaks and chromatid interchanges. Intercalary deletions, duplication-deletions and open chromatid breaks were less frequently observed. Concentrations ranged between 0.4 and 1200 μg/ml. No data on dose response are available (25-28, 68-71). An increase of SCE was observed at a concentration of 0.75 μg/ml (29).

Drosophila

MC was tested in both sexes. In males, sex-linked recessive lethals were induced in all germ-cell stages (31,72,73) and interchanges occured in spermatocytes and spermatogonia (74). Dominant lethals and 2:3 translocations were obtained for sperm, spermatids and spermatocytes. Chromosome loss and partial loss have not been tested in the male. The rate of nondisjunction remained unaffected in the male (74). There is information on a dose-dependent increase in recessive lethal mutations in males (30).

In immature oocytes and in oogonia MC caused sex-linked recessive lethals (74,75), X-chromosome loss (76), detachment of a compound X-chromosome (77), Y-fragmentation (76), and various types of interchanges between nonhomologous chromosomes (74,76,78). There were conflicting reports on the induction of nondisjunction (72,74-76). Chromosome rearrangements are produced at concentrations which give rise to 1-5% lethals. Dose-response experiments with immature oocytes indicate that the frequency of detachment bearing progeny induced with 0.6 mg/ml of MC corresponds to a dose of 250-500 R of x-rays (420 R/min) (77).

Cultured Mammalian Cells, Specific Locus Mutations

MC was mutagenic at low doses in the Chinese hamster cell lines V79 and CHO when assayed with the $HGPRT^-$ thioguanine-resistance system. The lowest effective concentration was 0.17 and 0.2 μg/ml, respectively. With V79 cells the presence of S-9 mix prevents the mutagenic action of MC (32,33). In the L5178 $Y/Tk^{+/-}$ mouse lymphoma system, selecting for trifluorothymidine resistant mutants, MC was also mutagenic at low concentrations. The lowest tested effective concentration was 0.1 μg/ml. Again, S-9 mix largely prevents the mutagenic action of MC (34).

In Vitro Chromosome Aberrations

MC-induced chromatid aberrations were observed in M1 which showed a dose-effect relationship (42) and gave rise to chromosome aberrations of the derived type in M2 (79). The MC-induced aberrations were preferably localized in the constitutive heterochromatin and in the secondary constrictions of human chromosomes No. 1, 9, 16 (37,41,80,81). In human cells chromatid translocations between chromosomes No. 1, 9, 16 were frequently found between the homologes at homologous sites, i.e., the secondary constrictions: somatic crossing over (35,36,40,82-86). Cells of Fanconi's anemia were more sensitive to the chromosome breaking activity of MC than normal human cells (39,87). The frequencies of MC induced chromosomal aberrations were enhanced by post-treatment with caffeine (88-93). Pseudodiploid Chinese hamster fibroblasts showed a dose-effect relationship for different chromatid aberrations with a lowest effective concentration of 8.4×10^{-3} μg/ml (42). The aberrations were localized in constitutive heterochromatin of the autosomes and the X-chromosome (94).

Unscheduled DNA Synthesis

There is no relevant information.

Sister Chromatid Exchanges (SCE)

MC produces SCE's in normal human leukocytes and normal human fibroblasts as well as in xeroderma and Fanconi's anemia leukocytes and fibroblasts (39,95). Xeroderma fibroblasts were the most sensitive human cells with a lowest effective concentration of 3.0×10^{-4} µg/ml (43).

MC was also effective in Chinese hamster (42,44), Indian muntjac (96), and mouse fibroblasts (97).

Addition of S-9 mix was not required. In vivo SCEs could be induced in mouse bone marrow and spermatogonial mitoses by 0.3 and 0.5 mg/kg of MC, respectively (46,98).

Micronucleus Test

In rat and mouse bone marrow erythrocytes, micronuclei were induced by 0.3 and 3.5 mg/kg of MC, respectively. The compound was given as single or double treatments separated by 24 hr. In the mouse study the frequency of micronuclei increased with dose from 1 × 3.5 to 2 × 14 mg/kg (45).

In Vivo Chromosome Aberrations

MC induced chromatid aberrations in Chinese hamster and mouse bone marrow as well as in mouse spermatogonia. The lowest positive dose tested for Chinese hamster was 10 mg/kg of MC as single IP injection (47). In mice chromatid breaks but no interchanges occurred in bone-marrow cells after 0.35 mg/kg by the same route of administration, while spermatogonia revealed 1% interchanges with only 0.035 mg/kg (48). The majority of the interchanges were derived from breaks in the centromeric heterochromatin block of the mouse chromosome (99,100). Although rather high frequencies of symmetrical interchanges were observed in spermatogonia there was no recovery of reciprocal translocations in primary spermatocytes at metaphase (100,101). Chromatid aberrations could also be induced in early prophase of meiosis (preleptotene) of spermatocytes. The lowest dose tested was 2.5 mg/kg (49). The effects of MC in all systems increased with dose.

Heritable Translocation Test

The three reports available do not permit conclusions as to what extent MC induces reciprocal translocations in mouse germ cells although there is an indication that translocation carriers can derive from MC-treated spermatocytes (102).

Dominant-Lethal Test

MC is one of the few chemicals which have been shown to induce dominant lethal mutations in premeiotic germ cells in the male mouse. In a dose-response study a single IP injection of 3.5 mg/kg of MC-induced pre- and postimplantation dominant lethals in spermatocytes. IP doses of 5.25 and 7.0 mg/kg induced dominant lethals also in spermatids, but at a lower frequency than those detected in premeiotic cells at the same doses (103). A single dose of 1.75 mg/kg induced a significant reduction in live embryos as compared to the concurrent control in the mating interval 35-46 days after treatment. The calculations were based on 20 females with implants in both groups in each mating interval of 4 days (50).

Sperm Morphology Assay

An increase in sperm head abnormalities was observed 4 weeks after treatment with 5 daily IP injections of MC. The single doses ranged from 0.2 to 12 mg/kg; the experimentally determined doubling dose was 1.3 mg/kg (51,104,105).

Host-Mediated Assay

The results with the host-mediated assay showed induction of DNA damage in *B. subtilis* (19) and *E. coli* (58), but were reported negative for revertant mutations in *S. typhimurium* G46 (106). The lowest tested positive dose was 3 mg/kg by intramuscular injection (19). Chromatid aberrations were induced in Ehrlich ascites tumor cells by 0.11 mg/kg intraperitoneally and 2.13 mg/kg subcutaneously (52).

Mammalian Spot Test

After treatment of fetal mice on days 10 or 11 of pregnancy by IP injection of females with 2 mg/kg of MC, the frequency of spots with genetic relevance varied between 1.5 and 4.3% as compared to 0.11% in the control (53,107). Injection of 2.5 mg/kg on days 8, 9, or 10 of fetal development indicated maximum response on days 8 or 9. The effect of a sequence of doses on days 8, 9, and 10 was close to being cumulative: about 18% of the offspring carried spots compared with less than 1% in the controls (108,110).

Specific Locus Test

The induction of specific locus mutations was tested in female and male mice (54). In oocytes up to 7 weeks IP injection with 2.0 mg/kg induced 5×10^{-5} mutations per locus per gamete ($p = 0.16$). It is unlikely that MC induced mutations in post-spermatogonia. Only 1 mutant was observed in 14.285 offspring derived from treated postspermatogonia. This frequency is similar to the control rate. A dose of 5.25 mg/kg MC increased significantly the mutation rate in spermatogonia (6 mutations in 20711 offspring, $p = 0.04$). The doubling dose for the induction of specific locus mutations by MC is 0.4 mg/kg (108). This value can be used for the estimation of the human risk.

ACKNOWLEDGMENTS

The author is very grateful for the assistance of the working group III members of the Workshop on Comparative Chemical Mutagenesis held in Raleigh, North Carolina, October 31-November 4, 1977, in reviewing and updating the manuscript.

The members of the working group were: C. Arlett, (University of Sussex, England), G. W. P. Dawson (University of Dublin, Ireland), U. H. Ehling (Gesellsch. f. Strahlen- u. Umweltforschung, Neuherberg, Germany), W. F. Grant (MacDonald Campus, McGill University, Quebec, Canada), I. Hansmann (Universität Göttingen, Germany), D. McGregor (Inveresk Research International, Musselburgh, Midlothian, Scotland), P. Lohman (TNO, Rijswijk, The Netherlands), H. V. Malling (NIEHS, Research Triangle Park, N.C., U.S.A.), H. Ott (Directorate General for Research, Science & Education, Commission of the European Communities, Brussels, Belgium), J. W. Simons (University Leiden, The Netherlands).

REFERENCES

1. Wakaki, S., Marumo, H., Tomoika, K., Shimiza, G., Kato, E., Kamado, H., Kudo, S., and Fujimoto, Y. Antibiot. Chemotherapy 8: 228 (1958).

2. Hata, T., Sano, Y., Sugawara, R., Matsumae, A., Kanamori, K., Shima, T., and Hoshi, T. J. Antibiotics A9: 141-146 (1956).

3. Kirsch, E. J. Mitomycins, Antibiotics II: Biosynthesis. Springer-Verlag, Berlin, Heidelberg, New York, 1967, pp. 66-76.

4. Stevens, C. L., Taylor, K. G., Munk, M. E., Marshall, W. S., Noll, K., Shah, G. D., Shah, L. G., and Uzu, K. J. Med. Chem. 9: 1-10 (1964).

5. Waring, M. J. Nature 219: 1320-1325 (1968).

6. Iyer, V. N., and Szybalski, W. Proc. Natl. Acad. Sci. (U.S.) 50: 355-362 (1963).

7. Iyer, V. N., and Szybalski, W. Science 145: 55-58 (1964).

8. Weissbach, A., and Lisino, A. Biochemistry 4: 196-200 (1965).

9. Goldberg, I. H. Am. J. Med. 39: 722-752 (1965).

10. Philips, F. S., Schwartz, H. S., and Sternberg, S. S. Cancer Res. 20: 1354-1362 (1960).

11. Molter, Dr., GmbH. (Serium Institute, P.O. Box 1210, Heidelberg, Germany). Information leaflet on mitomycin C produced by Kyowa Hakko Kogyo Co., Tokyo, Japan.

12. Schwartz, H. S., Sodergren, J. E., and Philips, F. S. Science 142: 1181-1183 (1963).

13. McCann, J., Spingarn, N. E., Kobori, J., and Ames, B. N. Proc. Natl. Acad. Sci (U.S.) 72: 979-983 (1975).

14. McCann, J., Choi, E., Yamasaki, E., and Ames, B. N. Proc. Natl. Acad. Sci. (U.S.) 72: 5135-5139 (1975).

15. Kondo, S., Ichikawa, H., Iwo, K., and Kato, T. Genetics 66: 187-217 (1970).

16. Ishii, Y., and Kondo, S. Mutat. Res. 27: 27-44 (1975).

17. D'Alisa, R. M., Carden III, G. A., Carr, H. S., and Rosenkranz, H. S. Mol. Gen. Genet. 110: 23-26 (1971).

18. Bamford, D., Sorsa, M., Gripenberg, U., Laamanen, I., and Meretoja, T. Mutat. Res. 40: 197-202 (1976).

19. Kada, T., Tutikawa, K., and Sadaie, Y. Mutat. Res. 16: 165-174 (1972).

20. Kada, T., Moriya, M., and Shirasu, Y. Mutat. Res. 26: 243-248 (1974).

21. Holliday, R. Genetics 50: 323-335 (1964).

22. Davies, P. J., and Parry, J. M. Mol. Gen. Genet. 148: 165-170 (1976).

23. Parry, J. (Department of Genetics, University College of Swansea, Swansea SA2 8PP, U.K.), personal communication.

24. Vig, B. K., and Paddock, E. F. J. Hered. 59: 225-229 (1968).

25. Utsumi, S. Japan. J. Genet. (Idengaku Zasshi) 46: 125-134 (1971).

26. Roy, S. C. Indian J. Exp. Biol. 10: 244-246 (1972).

27. Bempong, M. A. Bull. Torrey Bot. Club 99: 113-118 (1972).

28. Merz, T. Science 133: 329-330 (1971).

29. Sturelid, S., and Kihlman, B. A. Mutat. Res. 53: 270 (1978).

30. Blijleven, W. G. H. (Department of Radiation Genetics and Chemical Mutagenesis, State University of Leiden, Leiden, The Netherlands), personal communication.

31. Suzuki, D. T. Science 170: 695-706 (1970).

32. Knaap, A. G. A. C., Simons, J. W. I. M., and van Zeeland, A. A. Progress Report, September 1977, EEC-Contract No. 030-74-1 ENV-N.

33. Lohman, P. H. M. Progress Report, September 1977, EEC-Contract No. 1972-77-1 ENV-N.

34. Clive, D., (Burroughs Wellcome Co., Research Triangle Park, N.C. 27709, U.S.A.), personal communication.

35. Shaw, M. W., and Cohen, M. M. Genetics 51: 181-190 (1965).

36. Cohen, M. M. Can. J. Genet. Cytol. 11: 1-24 (1969).

37. Morad, M., Jonasson, J., and Lindsten, J. Hereditas 74: 273-282 (1973).

38. Sasaki, M. S., and Tonomura, A. Cancer Res. 33: 1829-1836 (1973).

39. Latt, S. A., Stetten, G., Juergens, L. A., Buchanan, G. R., and Gerald, P. S. Proc. Natl. Acad. Sci. (U.S.) 72: 4066-4070 (1975).

40. German, J., and La Rock, J. Tex. Rep. Biol. Med. 27: 409-418 (1969).

41. Sinkus, A. G. Tsitologiya 11: 933-940 (1969).

42. Kato, H., and Shimada, H. Mutat. Res. 28: 459-464 (1975).

43. Wolff, S. Ann. Rev. Genet. 11: 183-201 (1977).

44. Perry, P., and Evans, H. J. Nature 258: 121-125 (1975).

45. Matter, B. E., and Grauwiler, J. Mutat. Res. 23: 239-249 (1974).

46. Allen, J. W., and Latt, S. A. Nature 260: 449-451 (1976).

47. Lavappa, K. S., and Yerganian, G. Genetics 71s: 32-33 (1972).

48. Manyak, A., and Schleiermacher, E. Mutat. Res. 19: 99-108 (1973).

49. Adler, I.-D. Mutat. Res. 35: 247-256 (1976).

50. Ehling, U. H. Mutat. Res. 13: 433-436 (1971).

51. Wyrobek, A. J. Genetics 92s: 105-119 (1979).

52. Wobus, A. M., Thieme, R., and Schoeneich, J. Mutat. Res. 46: 242 (1977).

53. Russell, L. B. Arch. Toxicol. 38: 75-85 (1977).

54. Ehling, U. H. Mutat. Res. 26: 285-295 (1974).

55. Kondo, S. Genetics 73: 109-122 (1973).

56. Green, M. H. L., Muriel, W. J., and Bridges, B. A. Mutat. Res. 38: 33-42 (1976).

57. Murayama, I., and Otsuji, N. Mutat. Res. 18: 117-119 (1973).

58. Mazue, G., Vindel, J. A., Landsmann, F., and Brunaud, M. Arzneim. Forsch. 24: 1422-1425 (1974).

59. Zampieri, A., and Greenberg, J. Genetics 57: 41-51 (1967).

60. Lein, J., Heinemann, B., and Gourevitch, A. Nature 196: 783-784 (1962).

61. Endo, H., Ishizawa, M., Kamiya, T., and Sonoda, S. Nature 198: 258-260 (1963).

62. Vig, B. K. Theor. Appl. Genet. 43: 27-30 (1973).

63. Vig, B. K. Mutat. Res. 31: 49-56 (1975).

64. Vig, B. K. Genetics 73s: 583-596 (1973).

65. Vig, B. K. Genetics 75: 265-277 (1973).

66. Vig, B. K. Theor. Appl. Genet. 41: 145-149 (1971).

67. Vig, B. K. Mol. Gen. Genet. 116: 158-165 (1972).

68. Shah, V. C., Rao, S. R. V., and Arora, O. P. Nucleus (Calcutta) 25: 92-96 (1972).

69. Shah, V. C., Rao, S. R. V., and Arora, O. P. Indian J. Exp. Biol. 10: 431-435 (1972).

70. Shah, V. C., Rao, S. R. V., and Arora, O. P. Indian J. Biochem. Biophys. 9: 251-253 (1972).

71. Wu, T.-P. Taiwania 17: 248-254 (1972).

72. Suzuki, D. T. Genetics 51: 635-640 (1965).

73. Suzuki, D. T., Piternick, L. K., Hayashi, S., Tarasoff, M., Baillie, D., and Erasmus, U. Proc. Natl. Acad. Sci. (U.S.A.) 57: 907-912 (1967).

74. Schewe, M. J., Suzuki, D. T., and Erasmus, U. Mutat. Res. 12: 255-267 (1971).

75. Ostertag, W., and Haake, J. Z. Vererbungsl. 98: 299-308 (1966).

76. Walker, V. K., and Williamson, J. H. Mutat. Res. 28: 227-237 (1975).

77. Parker, D. R., and Williamson, J. H. Mutat. Res. 9: 273-286 (1970).

78. Schewe, M. J., Suzuki, D. T., and Erasmus, U. Mutat. Res. 12: 269-279 (1971).

79. Kato, H. Exptl. Cell Res. 85: 239-247 (1974).

80. Funes-Cravioto, F., Yakovienko, K. N., Kuleshov, N. P., and Zhurkov, V. S. Mutat. Res. 23: 87-105 (1974).

81. Broegger, A. Hereditas 77: 205-208 (1974).

82. Cohen, M. M., and Shaw, M. W. J. Cell Biol. 23: 386-395 (1964).

83. Nowell, P. C. Exptl. Cell Res. 33: 445-449 (1964).

84. Broegger, A., and Johansen, J. Chromosoma 38: 95-104 (1972).

85. Bourgeois, C. A. Chromosoma 48: 203-211 (1974).

86. Lakshmy, G. V., and Singh, S. J. Anat. Soc. India 23: 71-75 (1974).

87. Latt, S. A., Juergens, L. A., Dubin, H. G., Buchanan, G. R, and Gerald, P. S. Pediatr. Res. 9: 314 (1975).

88. Hartley-Asp, B., and Kihlman, B. A. Hereditas 69: 326-328 (1971).

89. Asp, B. Mutat. Res. 21: 22 (1973).

90. Kihlman, B. A., Sturelid, S., Hartley-Asp, B., and Nilsson, K. Mutat. Res. 17: 271-275 (1973).

91. Broegger, A. Genetics 74s: 31 (1973).

92. Kihlman, B. A., Sturelid, S., Hartley-Asp, B., and Nilsson, K. Mutat. Res. 26: 105-122 (1974).

93. Broegger, A. Mutat. Res. 23: 353-360 (1974).

94. Natarajan, A. T., and Schmid, W. Chromosoma 33: 48-62 (1971).

95. Latt, S. A. Proc. Natl. Acad. Sci. (U.S.) 71: 3162-3166 (1974).

96. Huttner, K. M., and Ruddle, F. H. Chromosoma 56: 1-13 (1976).

97. Lin, M. S., and Davidson, R. L. Nature 254: 354-356 (1975).

98. Allen, J. W., Latt, S. A. Mutat. Res. 38: 401 (1976).

99. Adler, I.-D. Mutat. Res. 21: 20-21 (1973).

100. Adler, I.-D. Mutat. Res. 23: 369-379 (1974).

101. Gilliavod, N., and Léonard, A. Mutat. Res. 13: 274-275 (1971).

102. Adler, I.-D., and Neuhäuser, A. Mutat. Res. 53: 143-144 (1978).

103. Ehling, U. H. Mutat. Res. 11: 35-44 (1971).

104. Wyrobek, A. J., and Bruce, W. R. Proc. Natl. Acad. Sci. (U.S.) 72: 4425-4429 (1975).

105. Wyrobek, A. J., and Bruce, W. R. In: Chemical Mutagens: Principles and Methods of their Detection, A. Hollaender and F. de Serres, Eds., Vol. V, Plenum Press, New York-London, 1978, pp. 257-285.

106. Gaeridge, M. G., and Legator, M. S. Proc. Soc. Exptl. Biol. Med. 130: 831-834 (1969).

107. Fahrig, R. Arch. Toxicol. 38: 87-98 (1977).

108. Dawson, G. W. P., Davidson, G. E., and Carter, J. Progress Report, April-December 1976, EEC-Contract No. 023-74-1 ENV-EIR.

109. Ehling, U. H. Mutat. Res. 41: 113-122 (1976).

110. Carter, I., Dawson, G. W. P.: Progress Report, January-June 1977, EEC-Contract No. 169-77-1 ENV-EIR.

CHAPTER 35

COMPARATIVE MUTAGENICITY OF EPICHLOROHYDRIN AND CADMIUM

B. A. Bridges

MRC Cell Mutation Unit, University of Sussex, England*

EPICHLOROHYDRIN

Epichlorohydrin, $\overline{O-CH_2-CH}-CH_2Cl$, is an important industrial chemical used in the production of many chemical compounds and polymers. Commercial U.S. production of epichlorohydrin was 550 million pounds in 1975, a 70% increase over the amounts produced in 1969. Its major uses are in the manufacture of glycerol and epoxy resins; other products in which it plays a role include emulsifiers, lubricants, dyestuffs, surfactants, and adhesives.

The U.S. National Institute of Occupational Safety and Health estimates a 50,000 population exposure to epichlorohydrin, probably all involved in its production and manufacture. The U.S. Occupational Safety and Health Administration has recommended that maximum exposures to epichlorohydrin over a 10-hr period should not exceed 0.5 ppm (2 mg/m^3); and for 15-min periods the concentration should not exceed 5 ppm (20 mg/m^3). The compound is a weak carcinogen to mice; subcutaneous injections produced local sarcomas. There are no case reports or epidemiological studies on cancer in man.

Epichlorohydrin (molecular weight 92.5) is soluble in ethanol, ethyl ether, and chlorinated aliphatic hydrocarbons, and slightly soluble in water (6.4% at 20°C). It has a freezing point of -58°C and a boiling point of 116°C. The saturation concentration in air at 25°C is 22,390 ppm and its vapor pressure is 20 mm Hg at 29°C.

---

*With the collaboration of B. Gledhill, T. Kada, A. Leonard, M. F. Lyon, V. Mayer, G. Newell, R. A. Nilan, G. Obe, I. F. Purchase and F. Zimmerman.

The structure of epichlorohydrin leads one to suspect that it would behave as an alkylating agent, and all the biological information is consistent with this property. Evaluation of qualitative risk is thus in some respects straightforward, although the problems of quantification are as severe as with any other agent.

Screening Tests with Bacteria

Repair-Deficient Bacteria. Epichlorohydrin is positive in both the Pol assay (1) and the Rec assay (2,3) without the need for metabolic activation, indicating that it produces DNA damage characteristic of an alkylating agent. In the Pol assay 1 µl was active with the disc protocol and 0.01 µl/ml was detectably positive in liquid. In the Rec assay the minimum detectable concentration was 0.1 µg/ml in liquid. The effect of epichlorohydrin largely disappeared in the presence of S9 mix, due presumably to competition for

Mutation in Escherichia coli. It has been shown since 1960 that epichlorohydrin induces Trp^+ mutations in *E. coli* B/r (4). The lowest concentration reported effective was 5×10^{-2} M (4600 µg/ml). It is direct-acting and less effective per lethal hit than x-rays or EMS as a mutagen in this organism.

Mutation in S. typhimurium. Epichlorohydrin is a direct-acting base pair substition mutagen in *S. typhimurium* (3,5-7), mutating strains TA1535, G46, TA100, and TA1950 to His^+. In the spot test protocol 0.5 to 50 µl was detectable (5). In liquid, 1.08 to 108 mM (100 µg/ml to 10,000 µg/ml) was detectable (5). In the vapor phase 0.025 to 10 µl per desiccator was detectable (7). The response was somewhat nonlinear with dose but without evidence of a threshold (Fig. 1). The vapor phase result is important because it illustrates well the frequently made observation that mutagenesis can be detected with very much lower amounts of chemical with vapor-phase application than with conventional liquid or gel-phase application. In this case approximately 0.003 µg/ml of air was detectable in a desiccator compared with around 100 µg/ml of agar and around 100 µg/ml of liquid. These differences probably reflect the partition of the substance between the liquid and gas phases. It has been shown, for example, that epichlorohydrin added to plates is mutagenic at very much lower concentrations if the plates are sealed in a jar to prevent evaporation (8). The gas/liquid partition will vary with each experimental system and in our opinion renders invalid any attempt at quantitative comparative mutagenicity between in vitro systems. The high efficiency of vapor phase epichlorohydrin may well also be relevant to occupational exposure which is commonly by the inhalation route.

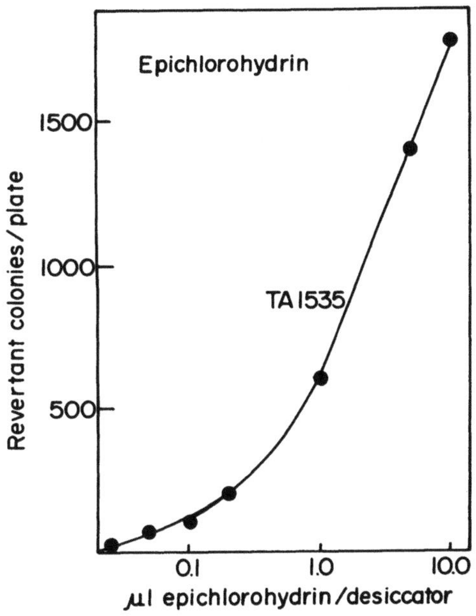

Fig. 1. Mutagenicity of epichlorohydrin, assayed in a desiccator experiment (see Table 2), on *S. typhimurium* TA1535. Spontaneous revertant colonies have been subtracted (V. Simmon, personal communication).

Mutation in K. pneumoniae. Epichlorohydrin is reported positive in inducing streptomycin resistance mutations although no details are available (9).

Host-Mediated Assay. In the mouse G46, TA100, and TA1950 were positive in the peritoneum after subcutaneous or intramuscular injection of epichlorohydrin (5); 100 mg/kg was detectable. At first sight this figure (which could be of the order of 100 μg/ml) is surprisingly low, since it is almost identical to the minimum concentration detectable in published liquid tests in vitro, and apparently shows no effect of scavenging by other organic components in vivo. We have given above, however, our reasons for believing that the actual concentration in liquid in vitro may have been very different from the applied concentration thus making any comparison very difficult, if not impossible.

Screening Tests with Eukaryotes

Mutation in Yeast. Arg^+ mutations are induced in yeast with an exponential response curve. The rate of induction is 40×10^{-7} mutations/mM-min (10).

Mutation in Neurospora crassa. Reversions of purple Ad^- strain 38701 are induced by 0.15 M (ca. 14 mg/ml) epichlorohydrin in a time-dependent manner. Epichlorohydrin was the most potent of a series of monoepoxides tested (11).

Mutation in Barley. Two eceriferum mutants have been reported to be induced by epichlorohydrin in barley (12). No experimental details, however, were given.

Chromosome Damage in Vicia faba. Many types of chromosome damage are produced in *V. faba*, particularly chromatid aberrations of the delayed type and isochromatid breaks (13).

Mutation in Drosophila. Positive results have been reported in Drosophila (14).

Gene Mutation in Culture Mammalian Cells. Knaap (15) has demonstrated a dose-dependent induction of thioguanine-resistant mutations in L5178Y mouse lymphoma cells (Table 1). Because of uncertainties about dosimetry with epichlorohydrin we are unwilling to use these data for quantitative comparison with other systems. It is clear, however, that epichlorohydrin is a fairly efficient mutagen in this system.

Table 1. Induction of $HGPRT^-$ mutations (resistant to thioguanine at 5 μg/ml) by epichlorohydrin in L5178Y mouse lymphoma cells[a]

Concentration (2 hr application)		Surviving mutation	Induced mutation frequency (7-day expression time)
mM	μg/ml		
0.5	46	0.68	2.9×10^{-5}
1.0	92	0.21	5.1×10^{-5}

[a]Data of Knaap (15).

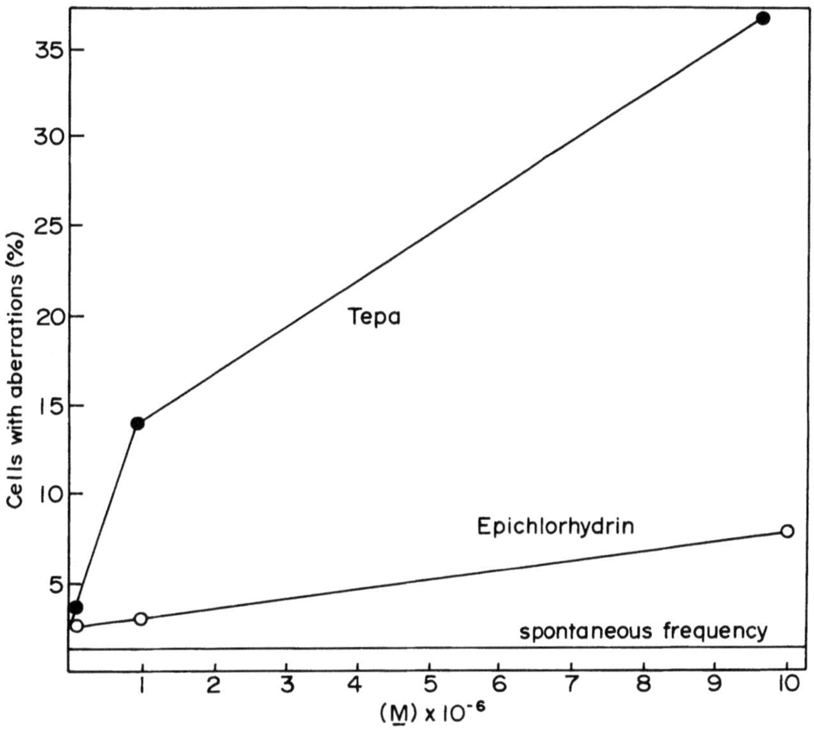

Fig. 2. Changes in chromosomes of human lymphocytes in culture when grown in the presence of epichlorohydrin or TEPA during the last 24 hr: change in percentage of aberrations.

Chromosome Damage in Cultured Human Lymphocytes. Epichlorohydrin induces breaks and aberrations in human lymphocytes of both chromatid and chromosome types although less efficiently than TEPA (16,17). Few aberrations are produced after treatment of G_0 cells which is typical of monofunctional alkylating agents. Some response curves are shown in Figures 2 and 3 with TEPA as a positive control. The lowest detectable concentration was between 10^{-6} and 10^{-7} M, and there was no evidence of a threshold. The doubling dose was probably less than 10^{-7} M with a 24 hr exposure (although this concentration would not have been maintained for 24 hr). The shape of the dose-response curve is very unusual and gives us some anxiety about the experiments. The explanations, however, may well lie in the problems involved in applying known concentrations of this mutagen.

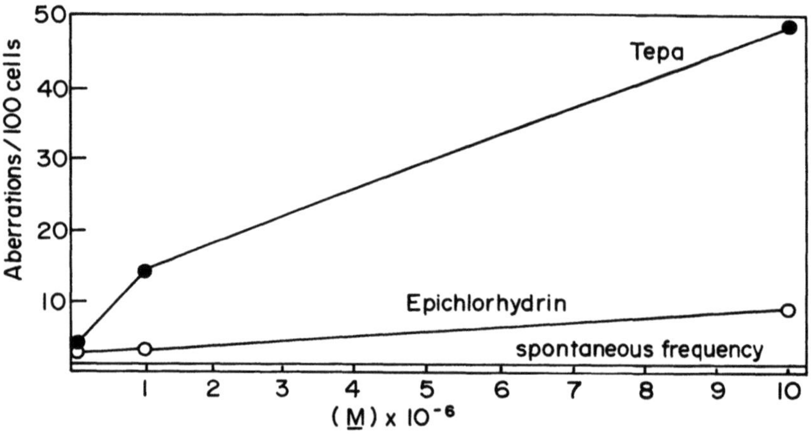

Fig. 3. Changes in chromosomes of human lymphocytes in culture when grown in the presence of epichlorohydrin or TEPA during the last 24 hr: change in aberrations per 100 cells.

Studies in the Whole Animal

In Vivo Cytogenetics. In the mouse, positive results have been obtained with metaphase analysis of bone marrow cells following 1 to 50 mg/kg IP and 5-100 mg/kg oral administration (5). The dose response curves (Figs. 4 and 5) show no evidence of a threshold. The doubling dose is between 1 and 2.5 mg/kg IP and between 1 and 7.5 mg/kg orally. With acute exposures, the dose response leveled off over about 20 mg/kg but considerably higher responses were observed if the same total dose above 20 mg/kg was given in a series of five fractions separated by 24 hr.

Positive results have also been reported in the Chinese hamster (5) although no details were given.

Transplacental Mutagenicity in Somatic Cells. Inui and Kada (18) have shown potent mutagenic activity in the Syrian hamster transplacental mutagenicity assay. For methodology Inui et al. (19). There was a dose-dependent induction of HGPRT (azaguanine-resistant) mutants among embryonic cells cultured 72 hr later from animals treated IP on the 11th day of pregnancy (Table 2).

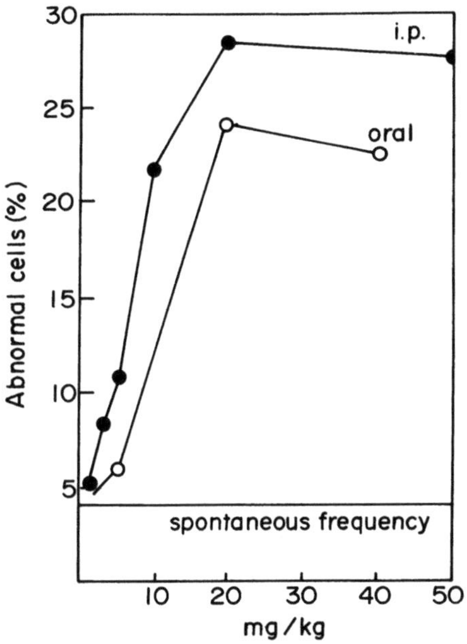

Fig. 4. Changes in chromosomes of mouse bone marrow cells following oral or IP application of epichlorohydrin: change in percentage of abnormal cells. Sampling was 24 hr after administration; 250 cells scored.

Dominant Lethal Test. Epichlorohydrin was negative in the mouse at 5, 10 and 20 mg/kg IP and at 20 and 40 mg/kg orally (5,20). Repeated applications of 5 × 10 mg/kg IP or 5 × 4 mg/kg and 5 × 16 mg/kg orally also had no dominant lethal effect. Occasional effects on male fertility were found.

Effects in Man

A prospective cytogenetic study has been reported (21) in 35 workers in an epichlorohydrin production plant. The workers were exposed to relatively high air concentrations (0.5 to 5 mg/m^3) for two years. The authors report a slight increase in the frequency of open chromatid breaks and possibly also of chromatid and chromosome exchanges. These effects are small and independent confirmation would be desirable.

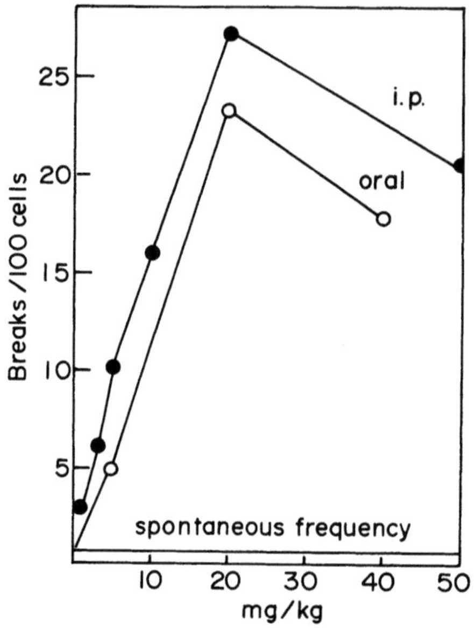

Fig. 5. Changes in chromosomes of mouse bone marrow cells following oral or IP application of epichlorohydrin: change in number of breaks per 100 cells. Sampling was 24 hr after administration; 250 cells scored.

Conclusions

From the viewpoint of qualitative comparative mutagenesis, epichlorohydrin provides a good example of a low molecular weight direct acting monofunctional alkylating agent (Table 3). It gives uniformly strong positive results in all tests for DNA repair and for gene mutation utilizing a wide range of systems from bacteria to mammalian cells treated both in vitro and in vivo. As expected of a small monofunctional agent its ability to cause chromosome damage is less pronounced. Nevertheless, positive (although relatively weak) effects have been reported for all systems examined cytogenetically, including mouse bone marrow following both IP and oral administration. The only exception is the dominant lethal test, but it should be borne in mind that this assay detects chromosome aberrations primarily in nondividing cells. Thus the possibility of effects in dividing spermatogonia cannot be excluded. The observation of site-of-injection sarcomas in mice is also a property expected of a direct acting alkylating agent.

Table 2. Induction of mutations to 8-azaguanine resistance in embryonic cells of the Syrian golden hamster after transplacental application of epichlorohydrin[a]

Dose (mg/kg)	Survival (%)	Mutant colonies: Total mutants per 10^7 cells plated	Mutant colonies: Induced mutants per 10^7 cells plated
0	100	2.2	-
25	92.0	19	19.5
50	96.0	16	15.5
100	91.1	176	189.8
200	5.9	36	606.4

[a]Data of Inui and Kada (18).

We have strong reservations about the dosimetry in most of the in vitro studies following the clear difference in mutagenic response in Salmonella after vapor-phase and liquid-phase treatment. Because of the great uncertainty about the concentrations existing around the exposed cells in most in vitro systems, we feel that attempts to produce figures for mutation rates or lowest effective concentrations could be very misleading, and a variable underestimate of the potency of the substance probably exists in most systems. The conventional explanation for such an affect is that the substance rapidly moves from aqueous solution to the vapor phase whence it may be rapidly lost. In view of the fact that the volatility of epichlorohydrin is not greatly different from that of water, however, we are not totally happy with this explanation and feel that the question merits further research.

Thoughts on the possible risk to man have centered largely on the chromosome damaging effect for which there is preliminary evidence from human subjects. We are, however, more concerned about a possible gene mutation risk, particularly in the view of the transplacental induction of forward mutations in Syrian hamster embryo cells where the doubling dose was well below 25 mg/kg.

Table 3. Epichlorohydrin: summary chart

Test	Result	Remarks
Bacteria		
Escherichia coli		
polA assay	+	Enhanced sensitivity
Bacillus subtilis		
rec assay	+	Enhanced sensitivity
Escherichia coli		
WP2 reverse mutation	+	Suspension test
WP2 uvrA reverse mutation	+	Suspension test
Salmonella typhimurium		
Reverse mutation		
TA1535, G46, TA100, TA1950	+	Plate test
TA1535 at 37°C	+	Plates exposed to vapor
Klebsiella pneumoniae		
Mutation to stretpomycin resistance	+	
Host-mediated assay, mouse intraperitoneal with Salmonella typhimurium G46, TA100, TA1950 reverse mutation of IM injection	+	
Fungi		
Schizosaccharomyces pombe, reverse mutation	+	Suspension test
Neurospora crassa, adenine reverse mutation of mutant 38701	+	Suspension test
Plants		
Barley forward mutation	+?	Only two mutants scored
Vicia faba root tips chromosome damage	+	

(continued)

Table 3. (continued)

Test	Result	Remarks
Insects		
Drosophila recessive lethals	+	
Bombyx mori, oocyte specific locus mutation	+	Injection
Cultured mammalian cells		
Human lymphocytes	+?	Weak effect, no good dose response
HGPRT mutation in mouse lymphoma cells	+	Suspension test
Mammals in vivo		
Mouse bone marrow	+	Oral ≦ IP
Dominant lethal mouse	-	Oral and IP same dose range as in bone marrow test
Transplacental somatic cell mutagenesis (HGPRT in Syrian Golden Hamster)	+	IP application
Human epidemiology		
Prospective program with workers exposed to agent lymphocytes	+ (W)	

There is good evidence that epichlorohydrin can reach the testes of rats and cause temporary and sometimes permanent sterility (22-24). In man it has been reported that mutagenic activity was detectable in the urine of a worker exposed to 25 ppm epichlorohydrin (25). There is thus a prima facie case for a heritable gene mutation risk to man and consideration should be given to a mouse sperm morphology abnormality test and perhaps a specific locus mouse study. It would also be advisable to explore possible methods of monitoring exposed populations for gene mutations rather than employing the conventional cytogenetic methods. Two possible approaches are the examination of blood cells of exposed subjects for presumed HGPRT mutations using autoradiography and the examination of sperm for morphological defects.

In view of the large annual production of epichlorohydrin (2.5×10^5 tons/annum in USA alone) and its relatively uncontrolled use, we consider that a thorough re-evaluation of the genetic risk to man should be initiated. Inhalation is likely to be a major route of human exposure and we conclude by noting without comment the following facts. The OSHA maximum permissible occupational exposure per 10 h period is 2 ng/ml in the atmosphere. The minimum air concentration of epichlorohydrin detectably mutagenic in a 7 hr exposure of Salmonella is about 3 ng/ml.

CADMIUM

Cadmium is widely distributed among the earth's minerals and is mined primarily from zinc and lead ores. Cadmium enters the environment chiefly through manufacturing operations such as smelting and refining of ores, recovery of scrap metals, combustion of coal and oil, and disposal of sewage, sludge and waste plastics. Natural sources of cadmium from weathering and erosion in the human environment are comparatively insignificant. Emission into the atmosphere from industrial sources results in its deposition on land and water in the nearby area where it is taken readily into biological systems. Cadmium is accumulated over the lifetime and is furthermore concentrated by the food chain.

Humans accumulate cadmium primarily by ingestion of food and secondarily from cigarette smoke, while air and water are relatively insignificant sources. The human intake from all sources is estimated to be 20-50 μg/day, reaching a maximum burden of approximately 30 mg at age 50-60 years. From 3 to 8% of the dietary intake of cadmium is absorbed by the body, while respiratory absorption is more efficient, from 15-30% of the intake. This latter consideration might be of some significance in human exposure in an industrial setting. Blood concentrations of cadmium for humans in nonindustrial settings reach 1 μg/100 ml, from which it is distributed to most tissues of the body. It is concentrated, however, in the liver and kidneys, which contain 50-75% of the body burden.

Acute exposure of experimental animals to cadmium results in many degenerative processes including testicular necrosis, abortion, fetal malformations, renal and liver damage, osteomalacia, anemia, and emphysema, some of which have been observed in humans as well, and has led to the prediction that low-level chronic exposure might be a significant health hazard. Subcutaneous injection of rats with several cadmium salts gave rise to local sarcomas, while certain soluble cadmium salts injected subcutaneously gave rise to testicular tumors as well. No tumors were observed in the prostate gland of these rats. Other studies of oral or inhalation administration to rats were inconclusive. Two human studies suggest that occupational exposure may increase the risk of prostate cancer; however, these studies were too small to be definitive.

It is impossible to make a satisfactory evaluation of the genetic risk to man of exposure to cadmium, partly because of inadequacies in the data available and partly because there is little indication as to how it might act at the molecular level.

Screening Tests with Bacteria

Kada (2) has found a 4-fold difference in minimal inhibitory concentration without microsomes in the Rec assay, confirming Nishioka (26). This is a particularly intriguing observation, since positive results in the Rec assay usually indicate a covalent bonding to, or chemical breakage of DNA, and neither of these properties is to be expected for cadmium. A few compounds are known which appear specifically to affect growth of Rec^- bacteria without reacting covalently with DNA. In some cases (e.g., 8-methoxypsoralen) these are known to cause frameshift mutations in bacteria. Cadmium, however, does not induce either frameshift mutations in Salmonella strains TA1538 or TA98 (2) or base-pair substitution mutations in E. coli WP2, WP2 uvrA, Salmonella TA1535 or TA100 (2,27). The effect in the Rec assay is at present inexplicable although consistent with some yeast data (see below).

Screening Tests with Eukaryotes

Yeast. A strain deficient in double-strand break repair was more sensitive to the growth inhibitory action of cadmium than a wild-type strain (28).

Drosophila. A sex-linked recessive lethal study in Drosophila (29) which failed to detect any effect of exposing males to 50 mg/l. $CdCl_2$ in the medium is regarded as preliminary by the Group I reviewers, although a similar negative result was reported by Ramel and Friberg with 62 mg/l $CdCl_2$ in the substrate (30). A further negative experiment has been reported in the Russian literature (31).

Studies in Plants. It seems clear that the genetic material of plants is susceptible to cadmium.

In barley, a weak chromosome-breaking effect (between 0.3 and 0.5% aberrations at anaphase) was observed with a number of cadmium salts (chloride, bromide, iodide, nitrate, sulfate, acetate) at 10^{-5} M for 2 hr (32). A chromosome-breaking effect has also been reported in pea seedlings (31,33). In *Crepis capillaris*, cadmium nitrate (0.1 to 0.01 M for 1 hr) increased the number of aberrations two- to three-fold (34). It tended to produce chromosome rather than chromatid aberrations, and its effect was independent of the stage of the cell cycle at the moment of treatment. Interestingly, it has been reported that cadmium nitrate reduced the yield of chromosome aberrations induced by ethylenimine in Crepis (35).

In *Vicia faba*, concentrations of cadmium nitrate between 10^{-2} and 10^{-3} M caused total breakdown of the mitotic system. At lower concentrations there were chromosomal structural changes, chromatid translocations, fragmentation at metaphase, and gaps and bridges at anaphase. There was a nearly linear concentration effect curve between 10^{-5} M and 10^{-4} M. The lowest effective concentrations were 10^{-5} M for 24 hr with a 2-hr recovery period (36-38).

Mutation in Plants. It has been reported (33) that cadmium induces mutations in Pisum, but detailed examination reveals that the cadmium treatment was combined with copper, zinc, chromium, manganese, iron, cobalt, nickel, aluminum, lithium, silver, mercury, and tellurium. The report is therefore irrelevant to our present consideration. In Chlorella, a mutagenic action has also been reported with cadmium chloride (39), but the mutants were of the "speckled" type due to unstable lethal sectoring. The genetic basis of these mutations is not fully understood but they could well arise due to aberrant segregation of genetic material.

Cytogenetic Studies with Mammalian Cells in Vitro. Chinese hamster cells show a pronounced interference with mitosis when treated with cadmium sulfate. During the recovery period (> 12 hr) aberrations mainly of the chromatid type were observed following exposure to 10^{-5} M for 1 hr (40). Chromosome aberrations have also been reported in cultured human leukocytes following treatment with cadmium sulfide in vitro at 6.2×10^{-2} μg/ml (ca. 0.5 μM) (41). CdS treatment caused an increase in chromatid and isochromatid breaks, and in translocations and dicentrics (Table 4). Moreover, there was an increase in the number of cells with a loss of 1 or 2 chromosomes with increasing time of treatment. We are, however, surprised by the very high aberration values obtained after only 4 hr treatment and have reservations about the validity of these data.

Table 4. Cadmium: summary chart[a]

Species	Assay	Effect	Salt	Concentration and remarks
Screening tests				
Bacteria				
Bacillus subtilis	Rec^-	+	$CdCl_2$	4 μg/ml LEC
E. coli	WP2 (BPS)	-	$CdCl_2$)	
	WP2 UvrA	-	$CdCl_2$)	Up to aqueous
			)	solution
Salmonella	TA1535, TA100 (BPS)	-	$CdCl_2$)	concentration
	TA1538, TA98 (FS)	-	$CdCl_2$)	
Yeast	Repair deficient (growth inhibition)	+		
Drosophila	Sex-linked lethal	-	$CdCl_2$	50 mg/l., males exposed
	Sex-linked lethal	-	$CdCl_2$	62 mg/l. in substrate
Plants				
Barley	Chromosome breaks	W	Several salts	10^{-5} M for 2 hr
Pea	Chromosome breaks	+	NR	NR
Crepis	Chromosome breaks	+	$CdNO_3$	0.1-0.01 M for 1 hr
	Chromatid breaks	+	$CdNO_3$	10^{-5} M, 10^{-4} M LEC 10^{-5} M for 24 hr with 2 hr recovery

(continued)

Table 4. (continued)

Species	Assay	Effect	Salt	Concentration and remarks
Mammalian cells				
CHO	Chromatid breaks	+	$CdSO_4$	10^{-4} M for 1 hr; recovery period 12 hr
Human leukocytes	Chromatid breaks Aneuploidy	+?) +?)	CdS	6.2×10^{-2} mg/ml
Human leukocytes	Chromosome breaks	−	$CdCl_2$	5×10^{-5} M, 5×10^{-6} M
Whole mammal				
Mouse	Dominant lethal	−	CdS	1.35-7 mg/kg, IP
	Translocations (postmeiotic) (heritable)	−?	$CdCl_2$	1.75 mg/kg, IP IP
	Translocations (premeiotic)	−	$CdCl_2$	0.5-3 mg/kg, IP
	Chromosome breaks Aneuploidy	−) +?)	$CdCl_2$	306 mg/kg subcutaneous
Rat	Dominant lethal	−	CdS	6.25, 12.5, and 25 mg/kg, oral
Cow	Chromosome breaks	−	Numerous NR; metals including Cd	ate contaminated hay
Sheep	Aneuploidy	+?	NR	NR

(continued)

Table 4. (continued)

Species	Assay	Effect	Salt	Concentration and Remarks
Human				
Leukocytes	Chromosome breaks	+?	Cd++	Blood levels 0.395±0.268 μg/100 ml; blood lead high
Leukocytes	Chromosome and chromatid breaks	+?	Cd++	Blood levels 1.6 to 17.9 μg/100 ml; blood lead high
(Itai Itai) Leukocytes	Chromosome breaks	-	Cd++	NR; no lead exposure
(Itai Itai) Leukocytes	Chromosome breaks	+?	Cd++	NR; other factors possible

[a]NR = not recorded; W = weakly induced.

Conflicting observations have been made by DeKnudt and Deminetti (42) on human lymphocytes. The duration of the cultures was 48 or 72 hr, and cadmium chloride was added at the beginning of the culture or 24 hr later. No increase in the incidence of chromosome or chromatid aberrations was observed with the dose levels used (5×10^{-5} M and 5×10^{-6} M).

An interesting report in the Russian literature claims that cadmium chloride increases the frequency of chromosome and chromatid deletions induced in rat embryo cells by Kilham virus (43).

Studies in the Whole Mammal

Dominant Lethal Test. No dominant lethal effect has been observed in three studies with mice using IP administration of cadmium chloride at doses between 1.35 and 7 mg/kg. In one study around 80 mice per group were employed with a dose of 2 mg/kg. In a smaller study some suggestion was found of reduced fertility (20,44,45).

A rat study in which a single dose of 6.25, 12.5, or 25 mg/kg cadmium chloride was administered orally was also negative, and no effect was observed on male fertility during the first 90 days after treatment (46).

Cytogenetic Studies on Germ Cells. A small study in mice failed to find heritable translocations after treatment of male mice with 1.75 mg/kg IP (44). The male progeny were screened cytologically. In view of the low number of progeny tested (120) and of the fact that the progeny were conceived over a three-week period, we regard these observations as preliminary although they probably exclude a potent mutagenic effect.

In the same study possible effects on spermatogonia were tested by observation of spermatocytes following IP administration of 0.5 to 3 mg/kg cadmium chloride. No translocations were observed in 3000 cells (30 mice).

In a study of superovulated mouse oocytes no structural anomalies were observed after subcutaneous injection of 3 or 6 mg/kg cadmium chloride (47). Metaphase II analysis was carried out 12 hr after injection. There was, however, a dose-dependent increase in numerical anomalies: the control showed 1/199; at 3 mg/kg, there were 4/155; at 6 mg/kg, 6/126 ($p = 0.095$). In addition, one blocked metaphase I figure consisting of 20 bivalents was observed from a mouse treated with 6 mg/kg. We have considerable reservations about this effect although, if real, the implications could be important.

Cytogenetic Studies (Somatic Cells) on Animals Other than Man. A relevant study (48) was carried out on nine cows accidentally intoxicated with a variety of metals including lead, chromium, cadmium, cobalt, zinc, copper and arsenic. No significant increase in chromosome aberrations was found in 48 hr cultured leukocytes although the study would have been better able to detect a possible small increase if the number of control cells (400) and animals (2) had been as great as the number exposed (1800 and 6, respectively).

There is also a report that cadmium induces chromosomal hypoploidy in sheep leukocytes (49). The data indicate a significant excess of cells with extreme hypoploidy ($n < 19$) but it is noticeable that the spread of chromosome numbers is large even in the controls. In our opinion these data cannot be considered as usable in view of the many possible sources of error.

Studies in Man. Studies in man are usually bedevilled by the fact that the exposed subjects are concurrently exposed to other metals such as lead and/or zinc.

In a study by Bauchinger et al. (50) of 24 occupationally exposed workers with blood cadmium levels of 0.395 ± 0.268 μg/100 ml, examination of leukocytes after 48 hr culture showed increases in deletions/cell dicentrics/cell and chromatid exchanges/cell. Overall the increase in aberrations per cell relative to 15 unexposed individuals was significant, but the blood lead level was also high.

In a study in which workers were exposed by occupational inhalation (51,52) blood cadmium levels of 1.6 to 17.9 μg/100 ml were recorded, but again lead levels were also high. There were significant increases in both chromatid and chromosome exchanges in 48 hr cultured leukocytes. An earlier study (53) had yielded similar results. In the study of Bui et al. (54), cadmium-exposed subjects did not have associated lead exposure, and there were no significant differences between controls and exposed subjects when 48 or 72 hr leukocyte cultures were examined. Shiraishi (55) and Shiraishi and Yosida (56) did, however, show increases in chromatid deletions and exchanges and in dicentrics and translocations. Both Bui et al. and Shiraishi therefore studied Itai-Itai patients with conflicting results. Both groups suggest that factors other than cadmium could be responsible for positive results.

We are concerned that the possibility of medical exposure to mutagens (e.g., x-rays or drugs) in these patients does not appear to have received adequate attention. Overall, our opinion is that there is no clear evidence to implicate cadmium specifically an a mutagen in any of these studies, although the possibility cannot be excluded that it may interact with one or more other environmental factors.

Conclusions

The only genetic effect that can be ascribed to cadmium with any confidence is the production of chromosome damage in plant material. There is, however, no information about the mechanism by which the effects in plants occur. Studies in microorganisms have not been helpful so far in this regard. The only positive effects have been enhanced growth inhibition in DNA repair-deficient bacteria and yeast. In the absence of confirmatory data (gene mutation studies in bacteria are uniformly negative) this cannot be ascribed to an effect on the genome of these organisms, nor is there any reason at present to believe that it is related to the chromosome damaging effect in plants. Cadmium is thus an example of a small class of substances which show as positive in tests such as the Rec assay but where the mode of action is obscure and perhaps unrelated to genetic damage.

There is no evidence for the induction of gene mutations by cadmium in any organism.

In contrast to the situation in plants, evidence for the production of chromosome damage in animals has been negative or, where positive results have been reported, they are highly questionable due either to contaminating exposure to other heavy metals or to poor experimental design and execution. Chromosome damage has been reported in some (but not all) studies on human subjects exposed to cadmium, but the contribution of other contaminating heavy metals, diet, and possibly medical treatment is unclear. There is a need for adequately controlled and well executed experimental studies into the role of these factors and the way (if any) in which they interact with cadmium to cause chromosome damage in animals.

At present there does not seem to be enough evidence to conclude that cadmium alone is likely to induce genetic effects in man, although it may prove to have a weak effect in conjunction with certain other factors as yet unknown.

REFERENCES

1. Rosenkranz, H. Personal communication.

2. Kada, T. Personal communication.

3. Elmore, J. D., Wong, J. L., Laumbach, A. D., and Streips, U. N. Biochem. Biophys. Acta 442: 405-419 (1976).

4. Strauss, B., and Okubo, S. J. Bacteriol. 79: 464-473 (1960).

5. Sram, R. J., Cerna, M., and Kucerová, M. Biol. Zentralbl. 95: 451-462 (1976).

6. Mukai, F. H., and Hawryluk, I. Mutat. Res. 21: 228 (1973).

7. Simmon, V. Personal communication.

8. Bridges, B. A. Unpublished data.

9. Voogd, C. E. Mutat. Res. 21: 52-53 (1973).

10. Baranowska, H., Prazmo, W., and Putrament, A. Acta Microbiol. Pol. A7: 25-32 (1975).

11. Koelmark, G., and Giles, N. H. Genetics 40: 890-902 (1955).

12. Lundqvist, U., von Wettstein-Knowles, P., and von Wettstein, D. Hereditas 59: 473-504 (1968).

13. Loveless, A. Nature 167: 338-342 (1951).

14. Rapoport, I. A. Dokl. Akad. Nauk SSR 60: 469-472 (1948).

15. Knaap, A. Personal communication.

16. Kucerová, M., Polivkova, Z., Sram, R., and Matousek, V. Mutat. Res. 34:271-278 (1976).

17. Kucerová, M., and Polivkova, Z. Mutat. Res. 34: 279-290 (1976).

18. Inui, N., and Kada, T. Personal communication.

19. Inui, N., Kaketoni, M., and Nishi, Y. Mutat. Res. 41: 351-360 (1976).

20. Epstein, S. S., Arnold, E., Andrea, J., Bass, W., and Bishop, Y. Toxicol. Appl. Pharmacol. 23: 288-325 (1972).

21. Kucerová, M., Zhurkov, V. S., Polivkova, Z., and Ivanova, J. E. Mutat. Res. 48: 355-360 (1977).

22. Hahn, J. d. Nature 226: 87 (1970).

23. Cooper, E. R. A., Jones, A. R., and Jackson, H. J. Reprod. Fertil. 38: 379-386 (1974).

24. Jones, A. R., Davies, P., Edwards, K., and Jackson, H. Nature 224: 83 (1969).

25. Kilham, D. J. Criteria for a Recommended Standard for Occupational Exposure to Epichlorhydrin. NIOSH, U.S. Department of Health, Education and Welfare, 1976.

26. Nishkioka, H. Mutat. Res. 26: 437-438 (1974).

27. Venitt, S., and Levy, L. S. Nature 250: 493-495 (1974).

28. Parry, J. Personal communication.

29. Sörsa, M., and Pfeifer, S. Hereditas 74: 273-277 (1973).

30. Friberg, L., Piscator, M., Nordberg, G. F., and Kjellstroem, T. In: Cadmium in the Environment, CRC Press, Cleveland, Ohio, 1974, pp. 131-135.

31. Von Rosen, G. Hereditas 40: 258-263 (1954).

32. Degraeve, N. Rev. Cytol., Biol. Veg. 31: 233-244 (1971).

33. Von Rosen, G. Hereditas 43: 644-664 (1957).

34. Ruposhev, A. R. Genetika 12(3): 37-43 (1976).

35. Ruposhev, A. R. L. Tsitol. Genet. 10: 111-113 (1976).

36. Gläss, E. Z. Botanik 43: 359 (1955).

37. Gläss, E. Z. Botanik 44: 1-58 (1956).

38. Gläss, E. Chromosoma 8: 260-284 (1956).

39. Anikeeva, I. D., Valuina, E. N., and Kogan, J. g. Genetika 11(12): 78-83 (1975).

40. Roehr, G., and Bauchinger, M. Mutat. Res. 40: 124-130 (1976).

41. Shiraishi, Y., Kurahashi, H., and Yosida, T. H. Proc. Japan. Acad. 48: 133-137 (1972).

42. Deknudt, G. H., personal communication.

43. Zasukhina, G. D., Shalunova, N. V., Shvetsova, T. P., and Lomanova, G. A. Dokl. Biol Sci. 224: 425-427 (1975); Dokl. Akad Nauk SSSR 224: 1189-1191 (1975).

44. Gilliavod, N., and Léonard, A. Toxicology 5: 43-47 (1975).

45. Suter, K. E. Mutat. Res. 30: 365-374 (1975).

46. Dixon, R. L., Lee, I. P., and Sherins, R. J. Environ. Health Perspect. 13: 59-67 (1976).

47. Shimada, T., Watanabe, T., and Endo, A. Mutat. Res. 40: 389-396 (1976).

48. Léonard, A., Deknudt, G., and Debackere, M. Toxicology 2: 269-273 (1974).

49. Doyle, J. J., Pfander, W. H., Crenshaw, D. B., and Shethen, J. M. Interface 31: 9 (1974).

50. Bauchinger, M., Schmid, E., Einbrodt, H. J., and Dresp, J. Mutat. Res. 40: 57-62 (1976).

51. Léonard, A., Deknudt, G., and Gilliavod, N. Ann. Gebloux 80(2): 87-93 (1974).

52. Deknudt, G., and Léonard, A. Environ. Physiol. Biochem. 5: 319-327 (1975).

53. Deknudt, G., Léonard, A., and Ivanov, B. Environ. Physiol. Biochem. 3: 132-133 (1973).

54. Bui, T.-H., Lindsten, J., and Nordberg, G. F. Environ. Res. 9: 187-195 (1975).

55. Shiraishi, Y. Humangenetik 27: 31-41 (1975).

56. Shiraishi, Y., and Yosida, T. H. Ann. Rept. Natl. Inst. Genetics 22: 44-45 (1971).

CHAPTER 36

COMPARATIVE CHEMICAL MUTAGENICITY: CAN WE MAKE RISK ESTIMATES?

D. Clive

Wellcome Research Laboratories, Research Triangle Park

North Carolina, 27709 (U.S.A.)

INTRODUCTION

The goal of risk estimation is to extrapolate toxicity data from model test systems to some system at risk--usually man--and in the process, to gain some quantitative insight into the magnitude of health hazards involved. For example, fairly early in the long and expensive process of drug development, some idea of the therapeutic index of a candidate drug is desired. If its toxic effects occur at doses too close to the therapeutic dose, further investment in that compound is usually contraindicated. This process is not perfect, of course; but despite their defects, such risk estimations from data in nonhuman species and model systems continue to be of service.

The difficulties associated with mutagenic risk estimation cannot be belittled. Estimating almost any type of toxicological risk involves threading an interdisciplinary labyrinth of pharmacodynamic barriers and intricate evolutionary paths separating the in-test and the at-risk species. And once these tricky passages have been negotiated with the help of whatever scientific threads there are to guide toxicologists, there still remains the Minotaur of Murphy's Law to contend with at the end of the maze. For mutagenesis, the issue is even more complex, since most of the mutagenic models in common use are in vitro systems. It is widely affirmed--with little or no supporting evidence--that mutagenic risk estimation in general and such risk estimation from in vitro test systems in particular either is not possible or should not even be attempted:

Thou mayst not wander in that labyrinth;
There Minotaurs and ugly treasons lurk. (1)

Why, then, are we making the present effort? The primary reason is that the Comparative Chemical Mutagenicity Workshop efforts made available an expertly evaluated data base on a score of reference mutagens in approximately 30 different mutagenicity test systems. There were close to 100 active participants in this world-wide effort and nearly half of these prepared the Group I literature review and evaluation documents (see next section). With approximately 90% of the work being done by others it was not difficult for Dr. de Serres to persuade us to do the last bit.

We do not presume to attempt estimating risk to man in this work. Our intent is simply to answer two questions. First, what form of informational edifice are the various in vivo mammalian mutagenicity assays shaping? Is it a Taj Mahal or is it a Tower of Babel? And second, do any of the in vitro assays show promise in predicting the in vivo rodent results, if the latter be uniform? If one or both of these questions are answered in an encouraging manner, these efforts will be a fitting finale to the nearly man-century amount of effort which went into this Comparative Chemical Mutagenicity Workshop.

And even if it should turn out that our enthusiasm has over-compensated for a weak data base, Tables VII-X should at least define the gaps in our present knowledge regarding the mutagenicity of these reference mutagens.

THE DATA BASE

Our sole sources of data for the mutagenicity studies analyzed in this Comparative Chemical Mutagenicity summary are the various draft versions of the Group I reports (except for Yeast Specific Loci) presented in these volumes. These documents include the best studies available from the published literature for each of the 20 reference mutagens in the 30 distinguishable mutagenicity systems discussed below. They also include, for the simpler, more rapid assays, supplementary, hitherto unpublished data produced specifically for this effort.

A decision was made during the week of the Workshop presentations to include oncogenicity as a 31st (probably) mutagenic end point. Although there is no Group I report for these studies, most of the reference mutagens comprising these efforts have been summarized (2); only in one instance (ICR-170) have we had to employ an oncogenicity study (3) not covered in these monographs.

As in most undertakings of the scope of this comparative chemical mutagenicity project, there are deficiencies and limitations in the data base. From the present restricted point of view these fall into the categories of statistical deficiencies in the original studies (e.g., assaying one dose of a mutagen in the mouse specific locus test until a single mutant offspring is detected); insufficient information included in the Group I reports for our purposes (none will be singled out); paucity of studies on these mutagens (one system had tested only 2 of these mutagens; two mutagens, i.e., cadmium salts and vinyl chloride, had not been sufficiently tested to warrant inclusion in these comparisons; only in Drosophila were there usable studies for all mutagens); and we had to work from preliminary draft documents for most of these systems and these were often updated during the workshop presentations.

To these extents, then, the conclusions reached in the present studies may be deficient. In addition, rare decisions on our part altered the interpretations of the Group I draft reports. In one of these, a positive result for mitomycin C in a rarely used repair-competent strain of Salmonella (TA 94) was rejected and replaced by the uniformly negative results obtained in the standard repair-deficient tester strains. In the other alteration, we called ICR-170 positive in the mouse specific locus test (2 mutants out of 9275 progeny vs. a spontaneous frequency of 0 mutants out of 15,625 progeny) despite the Group I report's verdict that it was not significant.

And finally, for the oncogenicity studies (2,3) all of the estimations of lowest effective dose and oncogenic potency are our own responsibility, but follow the same format as described for mutagenicity in a subsequent section.

DEFINING MUTAGENIC RISK

For our purposes we will assume that the maximum number of mutants inducible in a given population by a unit amount of a particular mutagen is a measure of that chemical's mutagenic risk to that population.

We will first show that the ratio of induced mutant frequency f^i to dose d, which we have elsewhere (4) referred to as mutagenic potency, is proportional to this measure of mutagenic risk, and that lowest effective dose (LED) is also a good approximation to mutagenic risk.

Consider a quantity Q of mutagen evenly and completely distributed at a mutagenically effective dose d over a population

N_Q of test (or other) organisms. For example, if we are administering Q mg of mutagen to whole animals of average weight, W (kg) at the dose of d (mg/kg), then the number of treated animals, N_Q, would be

$$N_Q = Q/dW \tag{1}$$

If we are treating cells in culture at a concentration of c organisms/ml with Q μg of mutagen at d (μg/ml), then the size of the exposed population N_Q would be

$$N_Q = Qc/d \tag{2}$$

In all such instances,

$$N_Q \propto Q/d \tag{3}$$

Among the progeny of these N_Q treated organisms, a number m_Q of detectable mutants will be induced by this quantity Q of mutagen; these m_Q mutants are assumed to be over and above the number expected to have occurred spontaneously among the progeny of N_Q organisms. The induced mutant frequency f^i is given by

$$f^i = m_Q/N_Q \tag{4}$$

If we consider a unit amount of mutagen, instead of our quantity Q we have $m = m_Q/Q$ mutant progeny induced from a treat population of $N = N_Q.Q \propto 1/d$ [Eq. (3)] organisms at the same frequency:

$$f^i = \frac{m_Q/Q}{N_Q/Q} = \frac{m_Q}{N_Q} = \frac{m}{N}$$

At some dose d, we will maximize m, the number of mutant progeny. This m_{max} is our defined measure of mutagenic risk. Now since $m = f^i N$ and $N \propto 1/d$, it follows that

$$m \propto f^i/d \tag{5}$$

In particular, m_{max} is proportional to the maximum value of mutagenic potency (4) f^i/d) observed over the entire dose-response curve. Hence mutagenic potency is proportional to mutagenic risk, as we set out to show.

To demonstrate the utility of LED as a measure of mutagenic risk, consider that, at the LED, f^i will equal the lowest significant effect (LSE). Any higher dose d can be expressed as a multiple of LED and the resulting mutant frequency as a (usually different) multiple of LSE. Thus,

$$d = \beta(\text{LED}) \tag{6}$$

$$f^{i} = \gamma(\text{LSE}) \tag{7}$$

and, from Eq. (5)

$$m \propto \frac{\gamma(\text{LSE})}{\beta(\text{LED})} \tag{8}$$

Since LSE is characteristic for each system and, in particular, is independent of mutagen, then, for any particular system, $(\gamma/\beta)(\text{LSE})$ is constant, and

$$m \propto 1/(\text{LED}) \tag{9}$$

and LED (more precisely, 1/LED) can be used to approximate mutagenic risk, as we wished to show.

In the following sections we shall attempt to develop estimates of mutagenic risk for 18 reference mutagens (vinyl chloride and cadmium salts were excluded because of the paucity of their data base) in 31 mutagen assay systems. Both maximum mutagenic potency and LED were used as estimators of mutagenic risk. However, because of deficiencies in the data base necessary to calculate mutagenic potency, and of large-scale agreement between the two sets of risk estimates where both could be calcualted, only the LED-derived values will be presented here in any detail.

ESTIMATING LED (LEDE)

For all practical purposes, LED's are not available for most of these mutagens in most of these systems. What are available are values for lowest effective dose tested (LEDT). In most instances it has been possible to determine various estimates of LED (denoted as LED-estimated, or LEDE) from the available data. All of these LEDE's have weaknesses, but for each mutagen in each system the method for determining LEDE was chosen with that system's and mutagen's characteristics in mind. In all instances the LEDE was chosen so as to produce some statistically significant increase in mutant frequency over the spontaneous frequency.

For example, in Salmonella, linear dose-responses are quite common, permitting the use of a linearly extrapolated LED (LELED) as LEDE. Since dose-response curves are presented in the Salmonella Group I report, those mutagens yielding nonlinear dose-response curves can have an LEDE taken directly from the curve itself, making conservative interpolations where doubts exist.

In some systems dose-response curves are not available, usually due to practical considerations relating to the particular system. Thus, in the heritable translocation assay, cyclophosphamide (CP) was tested only at 350 mg/kg (Table 1). This dose produced 33 translocation heterozygotes out of 84 progeny tested (39%) as compared with a spontaneous frequency ≦ 0.1%. This dose clearly overestimates the LED of cyclophosphamide, but the question is: By how much?

Table I. Calculation of various LED estimates for EMS and cyclophosphamide in the heritable translocation assay

Mutagen	LEDT (mg/kg)[a]	No. translocation heterozygotes/ no. progeny	Heterozygotes (%)	LELED (mg/kg)[b,c]	LEDG (mg/kg)[d]
EMS	50	6/853	0.7	35	42
EMS	(200)	79/246	32	3.1	25
CP	350	33/84	39	4.5	40

[a]LEDT (lowest effective dose tested) = the lowest dose reported with positive results.

[b]Spontaneous frequency = 0.1%; LSE (lowest significant effect) = 0.5%.

[c]LELED (linearly extrapolated lowest effective dose) = dose tested/ (mutant frequency/LSE).

[d]LEDG (lowest effective dose--geometric) = $\sqrt{(\text{LELED}) \times (\text{LEDT})}$.

If we assume the lowest significant effect to be 0.5% we can calculate an LELED of (350 mg/kg)/(39%/0.5%) = 4.5 mg/kg, a value only 1.3% of the LEDT. But how legitimate is linear extrapolation in a system which measures two-hit events? As we shall see below we can usually afford approximately three-fold errors in our LEDE's without significantly compromising our conclusions. But these two values differ by a factor of 80, with both values probably being substantially in error. What we need is some compromise estimate between these two extremes. The heritable translocation data on EMS are informative in this regard (see Table I).

EMS at 50 mg/kg gave 0.7% translocation heterozygotes, a value quite close to our assumed lowest significant effect (Table I). Thus 50 mg/kg is close to a real LED for EMS.

At 200 mg EMS/kg, 32% translocation heterozygotes were induced. From these results an LELED of (200 mg/kg)/(32%/0.5%) = 3.1 mg/kg can be calculated, closely paralleling our above analysis of the cyclophosphamide results. Somewhere between this LELED of 3.1 mg/kg and our high dose of 200 mg/kg lies the measured LED of approximately 50 mg/kg. Perhaps fortuitously in this instance, the geometric mean of 3.1 and 200 mg/kg is 25 mg/kg ($= \sqrt{3.1 \times 200}$), a value within a factor of 2 of our measured LED. This LED-geometric (LEDG) has been widely used in obtaining LEDE's where it is felt that the mutagen and system are insufficiently characterized to justify linear extrapolation of the data. It is defined, in general, as

$$\text{LEDG} = \sqrt{(\text{LEDT})(\text{LELED})} \quad (10)$$

and involves the intermediate calculation of LELED.

Table I illustrates the above steps for determining LEDG estimates for both the single (high) dose of cyclophosphamide and the high dose of EMS, if we had had only that high dose to proceed from.

The LEDE's, derived by one of the processes just described, are presented in Table II for those of the 18 reference mutagens which have been adequately studied in the ten in vivo mammalian systems. Subsequent sections and tables will follow these data as they undergo transformations into the graphic summary of mutagenic risk. The remaining 21 mutagen assay systems are summarized in the risk estimate Tables 8-10. ("Mutagenic risk" implies only risk to the system in question, at this point.)

As a final caution, it should be mentioned that in a few instances there was insufficient information supplied in the Group I reports to make anyy estimate of LED. For such cases LEDT's were accepted with the full realization of their shortcomings.

THE NUMBERS PROBLEM

Table II summarizes the LEDE's estimated from the Group I reports by the procedures described above for the in vivo mammalian systems, including oncogenicity. If we examine one mutagen across all test systems (as is done in the Group II reports, published elsewhere in these volumes) we are struck by the amount of quantitative variation that exists in the LEDE's. Thus, for example,

Table II. LEDE's of 18 reference mutagens in in vivo mammalian systems

Chemical	LEDE (mg/kg)[a]									
	Germinal				Somatic					
				Cytogenetics						
	Mouse specific locus	Heritable translocation	Dominant lethal	Germinal	Somatic	Unscheduled DNA synthesis	Sister chromatid exchanges	Micronucleus test	Spot test	Oncogenicity
Trenimon			0.125	0.10	0.08		0.08	0.03		1.6
TEM	0.15	0.009	0.05	≦0.2	0.14	0.05		0.1	0.28	6.0
Aflatoxin B1					0.24					15
Ethylenimine			5			1				4.6
Mitomycin C	3.2		3.5	2.5	0.17		0.2	3.5	1.5	2.6
TEPA		0.84	1.0	0.37	1.5					
Thio-TEPA		0.16	0.64	1.1	1.25			2		47
METEPA			12.5					0.1		

ICR 170	2.7		NEG(>4)							4.8
DMN			4.4		8.3	65	0.8	37	5.3	18
Myleran			4	8	5			100		48
MMS	15	8.7	20	16	46	3.2	28	31	25	40
Cyclo-phospha-mide	130	40	60	49	2.9	150	14	5	2.8	60
Epichlo-rohydrin			NEG (>150)		2.1					
MNNG	NEG(>50)		50				1.5	NEG (>200)	25	20
Natulan	230		200	≤200	≤100			13	15	250
EMS	90	42	100		94	9.1	160	400	22	100
DEN			NC (>13.5)		60	NEG (>125)	NEG (>100)	NEG (>200)	3.5	110

[a] NEG (>x): No significant effect at doses up to and including x mg/kg; NC: not conclusive, testing recommended in Group I report to permit "a confident assessment of its dominant lethal potential."

the LEDE's for TEM range between 0.009 mg/kg in the heritable translocation assay and 0.28 mg/kg in the spot test, a 30-fold difference. An even larger range in LEDE's exists for epichlorohydrin (2.1 mg/kg in somatic cytogenetics; negative at 150 mg/kg in inducing dominant lethality: a range of ≥75-fold).

When we extend these comparisons to include in vitro systems, we see even wider variations, such as the nearly 1000-fold difference in LEDE's for aflatoxin B1 in inducing specific locus mutations in cultured mammalian cells (9×10^{-7} M-hr) and in Neurospora (7×10^{-4} M-hrs) (see appropriate Group I reports for data leading to these values). And as a last example, consider Natulan in cultured mammalian cells: ineffective at instigating unscheduled DNA synthesis at 1×10^{-1} M-hr, it produces specific locus mutations at an LEDE of 4.7×10^{-6} M-hr, a ≥20,000-fold difference! Such variation is the rule, not the exception, and must be circumvented before we can meaningfully compare chemical mutagenicity.

Another difficulty to overcome is the different units in which doses, including LEDE's, are reported. These units include mg/kg, M-hr, and μg/plate. This variability is compounded by the variable exposure times characteristically employed in each assay: these can be as indeterminately variable as the mutagens' biological and/or chemical half-lives in vivo and/or in vitro, or as precisely variable as 30 min to 48 hr (assuming stability of mutagen) in in vitro studies.

ITS RESOLUTION

Figure 1 expresses, in graphic form, the ranges in LEDE's shown in Table II for the 18 reference mutagens where adequately tested in the 10 in vivo mammalian systems. Similar figures can be constructed for the other 21 mutagen assay systems if we maintain comparability of dose units within each separate figure.

Each of these 31 distinct range scales can now be divided into, say, five uniformly sized subranges, each characterized by a different risk factor, which span the entire observed LEDE range for that system. This has been done, diagrammatically, in Figure 2, for the heritable translocation test. In the course of this transition it is important to note that we have eliminated our absolute chemical LEDE units (e.g., 5 mg/kg in Fig. 2; but elsewhere 10^{-5} M-hr; 3 μg/plate) which vary from system to system. These have been replaced by relative, unit-less but quantitative comparisons of LEDE's (e.g., one-fiftieth; 500 times greater; nearly equal) within each system. The highest risk factor (Fig. 2) is 5 (equivalent to a semiquantitative +++++, and represented by the color red in Tables VII-X) for those mutagens having the lowest LEDE's and the greatest mutagenic risk. This is followed by risk factors of 4 (++++; orange), 3 (+++; yellow), 2 (++; green) and, for those reference mutagens having the highest LEDE's and lowest mutagenic risk, 1 (+; blue). A negative result in a particular system is represented by a zero risk factor or, on the color charts, black. Blank (white) spaces on all of the LEDE or risk estimation charts represent either inadequate testing of a chemical in a particular system, or no test results presented in the Group I reports.

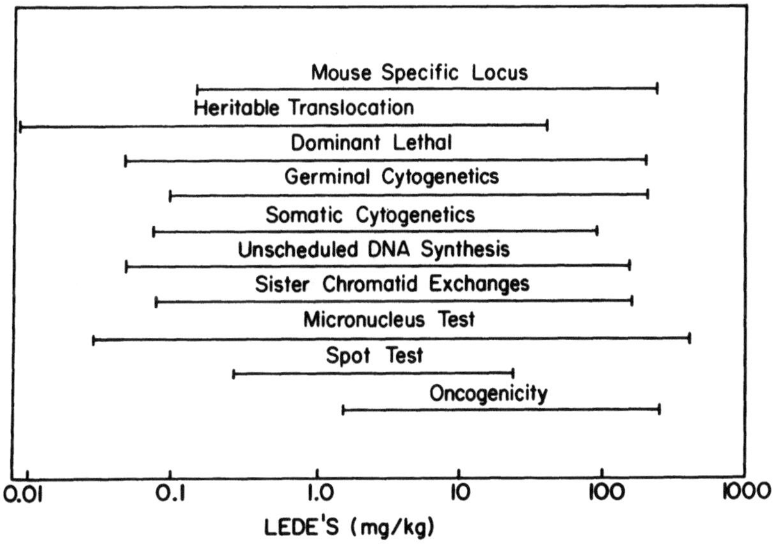

Fig. 1: Ranges in lowest effective doses (LEDE's) seen among 10 in vivo mammalian systems for 18 reference mutagens. LEDE's (mg/kg) are from acute or subacute dosing only. The data are from Table II.

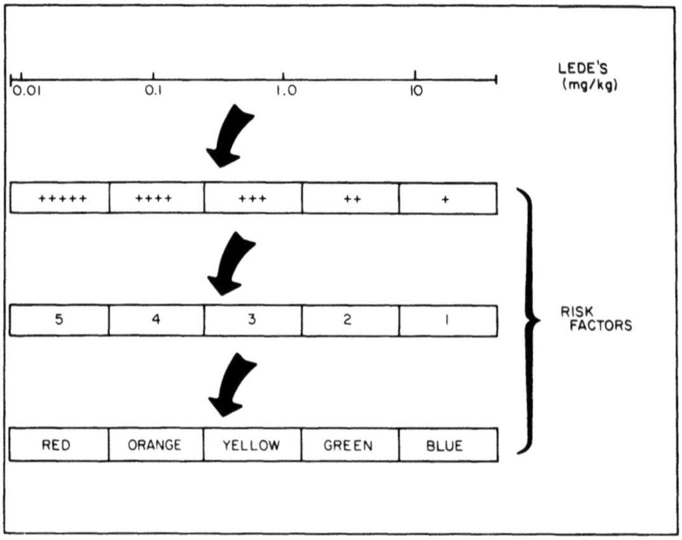

Fig. 2. Transformation from LEDE's to risk factors for the heritable translocation test.

At this point, it is appropriate to emphasize the immense variation in LEDE's typically seen within each of these 31 test systems (as opposed to the variation in the LEDE's seen for each mutagen, which we described above). In the spot test LEDE's vary from 0.28 mg/kg for TEM up to 25 mg/kg for MNNG, an 89-fold range (Table II). At the other extreme, in B. subtilis repair assays LEDE's vary from 0.00075 µg/ml for mitomycin C to 3125 µg/ml for DMN, a 4,000,000-fold range (data not shown, but present in Group I reports)! Among the 10 in vivo mammalian systems shown in Table II the range in LEDE's perceived by each system varies from the previously mentioned low of 89-fold for the spot test to 13,000-fold for the micronucleus test; these ten ranges in LEDE's average out to be 3200-fold.

Table III. Transformation from LEDE's to mutagenic risk factors for the heritable translocation assay.

Chemical	LEDE (mg/kg)	Mutagenic risk factor[a]
TEM	0.009	5
TEP	0.84	3
Thio-TEPA	0.16	4
MMS	8.7	1
Cyclophosphamide	40	1
EMS	42	1

[a] LEDE range: 0.009-42 mg/kg; risk category magnitude: $\rho = \sqrt[5]{42/0.009} = 5.4$. For a LEDE range of 0.009-0.049 mg/kg, the mutagenic risk factor is 5; for a LEDE range of 0.049-0.26 mg/kg, 4; for LEDE 0.26-1.4 mg/kg, 3; for LEDE 1.4-7.8 mg/kg, 2; and for LEDE 7.8-42 mg/kg, 1.

Because the range of LEDE's in any particular system typically spans orders-of-magnitude, the division into risk factor categories is done on a geometric--rather than on an arithmetic--basis (Fig. 2 and Table III). Each risk category magnitude ρ is a measure of the range of LEDE's contained in each risk category for a particular system, and is defined as

$$\rho = \sqrt[5]{\text{highest LEDE/lowest LEDE}} \qquad (11)$$

The LEDE boundaries of successive risk categories are (proceeding from risk factor 5 to risk factor 1): lowest LEDE; $\rho \times$ lowest LEDE; $\rho^2 \times$ lowest LEDE; $\rho^3 \times$ lowest LEDE; $\rho^4 \times$ lowest LEDE; and $\rho^5 \times$ lowest LEDE = highest LEDE. Thus, for the heritable translocation assay (Table III, footnote), $\rho = \sqrt[5]{42\ \text{mg/kg}/0.009\ \text{mg/kg}} = 5.4$. The successive risk factor boundaries are LEDE's of 0.009, 0.049 (= 5.4×0.009), 0.26 (= $5.4^2 \times 0.009$), 1.4 (= $5.4^3 \times 0.009$), 7.8 (= $5.4^4 \times 0.009$) and 42 (= $5.4^5 \times 0.009$ = highest LEDE) mg/kg. Risk factor 3, for example, represents all LEDE's between 0.26 and 1.4 mg/kg in this system. The remaining four risk factors

Table IV. Mutagenic risk factors for 18 reference mutagens in in vivo mammalian systems

	Germinal				Somatic					
				Cytogenetics						
Chemical	Mouse specific locus	Heritable translocation	Dominant lethal	Germinal	Somatic	Unscheduled DNA synthesis	Sister chromatid exchanges	Micronucleus test	Spot test	Oncogenicity
Tremimon			5	5	5		5	5		5
TEM	5	5	5	5	5	5		5	5	4
Aflatoxin B1					5					3
Ethylenimine			3			4				4
Mitomycin C	3		3	3	5		5	3	4	5
TEPA		3	4	5	3					
Thio-TEPA		4	4	4	4			3		2
METEPA			2					5		
ICR-170	4		0							4

DMN			3		2	1	4	2	2	3
Myleran			3	3	3			1		2
MMS	2	1	2	2	1	3	2	2	1	2
Cyclo-phospha-mide	1	1	1	1	3	1	2	3	3	2
Epichlo-rohydrin			0		3					
MNNG	0		1				4	0	1	3
Natulan	1		1	1	1			2	1	1
EMS	1	1	1		1	2	1	1	1	1
DEN					1	0	0	0	3	1

are also assigned, each to its appropriate range of LEDE's, as in the footnote to Table III. And finally, in the top portion of Table III each reference mutagen is assigned a risk factor on the basis of its LEDE in this system.

In this fashion Table II can be converted into Table IV, in which each reference mutagen is associated with a mutagenic risk factor in whichever systems it has been adequately tested. Three additional tables (not shown here), similar to Table IV, can be constructed for the rest of the 31 test systems, one for four in vitro cultured mammalian cell systems, one for six insect and plant systems, and finally, one for eleven microbial systems. These three tables and Table IV are presented in color form in Tables VII-X. We shall return to these tables after we describe how the reference mutagens in Table IV were ranked in the sequence shown.

COMPARATIVE CHEMICAL MUTAGENICITY

In Vivo Mammalian Systems

A second set of four tables (not shown here) showing mutagenic risk factors has been constructed from mutagenic potencies instead of from LEDE's; these raise to eight the number of tables of mutagenic risk factors. As mentioned previously, the LEDE-derived and the mutagenic potency-derived risk factors are rather similar; therefore only the LEDE set is presented here in detail. However, for some in vivo mammalian systems analyses, to be discussed in this section, we will be utilizing some statistics of both sets of Risk Factors.

Tables II, IV, and VII-X have as a standard format the ranking of 18 reference mutagens in their first column. This ranking is in descending order of mean risk factor, obtained in part from Table IV for each mutagen in a fashion discussed below in this section and presented in Table V. Simply put, because of the importance attached to this primary mutagen ranking, it has been based upon as much in vivo mammalian mutagenicity information (e.g., LEDE and mutagenic potency) as possible and represents in itself a measure of relative mutagenic risk in in vivo mammalian systems.

This being the case, we can state the following assumption upon which most of the remaining analyses rest. If test system X ranks a sufficiently large number of reference mutagens in exactly the same sequence as our primary mutagen ranking, then system X has "predicted" mutagenic risk to the whole animal as well as have the 10 in vivo mammalian systems; the extent to which the two rankings differ is a measure of the deficiencies of system X in "predicting" risk to the whole animal. These ranking differences can be quantitated by means of a Spearman rank correlation analysis (5), which will be described later in this section, after a detailed description of how our primary mutagen ranking was achieved.

Table V has been constructed in order to best determine the relative mutagenic risk of these 18 reference mutagens (our primary mutagen ranking). Our principal statistic throughout this table is a mean risk factor, as derived from all or portions of Table IV or its mutagenic potency-derived equivalent. For example, in column 3 of Table V is listed a mean risk factor for each mutagen, derived directly from each row of Table IV, together with a measure (standard deviation) of how variably each mutagen behaved throughout those in vivo mammalian systems in which it was tested. Thus, Trenimon earned risk factors of 5 in all 6 in vivo mammalian systems in which it was tested (top row of Table IV); in Table V, its mean risk factor ± 1 s.d. is, therefore, 5.00 ± 0.00. In like fashion, the rest of column 3 of Table V has been constructed from Table IV.

Column 4 of Table V was constructed in precisely the same fashion as was column 3, but from the in vivo mammalian risk factors assigned from mutagenic potency--instead of from LEDE--calculations. As mentioned earlier, columns 3 and 4 differ but slightly, and potency calculations have been for the most part ignored, except for Table V and its corollaries.

Other mean risk factors can be calculated by combining LEDE- and potency-derived risk factors. This has been done in column 6 for all 10 in vivo mammalian systems, including oncogenicity, and in column 5 for the nine strictly mutagenic systems, excluding oncogenicity. And finally, in column 7, the means of the two (LEDE and potency) oncogenicity risk factors for 15 of the 18 reference mutagens are listed. (These 15 mutagens all had acute or subacute oncogenicity studies similar to the mutagenicity dosing, performed on them. The remaining three--TEPA, METEPA, and epichlorohydrin--had only chronic dosing studies which resulted in disproportionately high LEDT's; these were excluded from further consideration.)

From Table V we wish to determine whether we can pool various sets of data in order to simplify our risk estimation task. First, columns 3 and 4 are clearly comparable. A Spearman rank correlation analysis (see explanation in next paragraph) indicates that these two columns have a rank correlation coefficient, $r = 1.00$ ($p \leqq 0.00002$). These, then, can be combined, as has been done in columns 5-7.

Table V. Summary of various mean in vivo mammalian mutagenic risk factors[a]

Rank	Mutagen	Derived from: LEDE	Derived from: Potency	Combined LEDE and potency: Without oncogenicity	Combined LEDE and potency: With oncogenicity[b]	Oncogenicity (LED and potency)
1	Trenimon	5.00 ± 0.00	5.00 ± 0.00	5.00 ± 0.00	5.00 ± 0.00	5.00 ± 0.00
2	TEM	4.89 ± 0.33	5.00 ± 0.00	5.00 ± 0.00	4.94 ± 0.24	4.50 ± 0.71
3	Aflatoxin B1	4.00 ± 1.41	4.50 ± 0.71	5.00 ± 0.00	4.25 ± 0.96	3.50 ± 0.71
4	Ethylenimine	3.67 ± 0.58	4.00 ± 1.00	3.75 ± 0.96	3.83 ± 0.75	4.00 ± 0.00
5	Mitomycin C	3.88 ± 0.99	3.75 ± 1.04	3.64 ± 0.93	3.81 ± 0.98	5.00 ± 0.00
6	TEPA	3.75 ± 0.96	3.75 ± 0.50	3.75 ± 0.71	3.75 ± 0.71	---
7	Thio-TEPA	3.50 ± 0.84	3.50 ± 0.55	3.70 ± 0.48	3.50 ± 0.67	2.50 ± 0.71
8	MeTEPA	3.50 ± 2.12	3.50 ± 2.12	3.50 ± 1.73	3.50 ± 1.73	---
9	ICR-170	2.67 ± 2.31	2.67 ± 2.52	1.75 ± 2.06	2.67 ± 2.16	4.50 ± 0.71
10	DMN	2.43 ± 0.98	2.43 ± 0.98	2.42 ± 1.00	2.43 ± 0.94	2.50 ± 0.71
11	Myleran	2.40 ± 0.89	2.20 ± 1.30	2.38 ± 1.19	2.30 ± 1.06	2.00 ± 0.00

12	MMS	1.80 ± 0.63	1.80 ± 0.79	1.83 ± 0.71	1.80 ± 0.70	1.50 ± 0.71
13	Cyclophosphamide	1.80 ± 0.92	1.70 ± 0.95	1.78 ± 0.94	1.75 ± 0.91	1.50 ± 0.71
14	Epichlorohydrin	1.50 ± 2.12	1.50 ± 2.12	1.50 ± 1.73	1.50 ± 1.73	---
15	MNNG	1.50 ± 1.64	1.33 ± 1.37	1.10 ± 1.37	1.42 ± 1.44	3.00 ± 0.00
16	Natulan	1.14 ± 0.38	1.29 ± 0.49	1.25 ± 0.45	1.22 ± 0.43	1.00 ± 0.00
17	EMS	1.11 ± 0.33	1.11 ± 0.33	1.13 ± 0.34	1.11 ± 0.32	1.00 ± 0.00
18	DEN	0.83 ± 1.17	1.00 ± 1.26	0.80 ± 1.23	0.92 ± 1.16	1.50 ± 0.71

[a] Mean ± standard deviation of indicated assessments [from Table IV and its equivalent for mutagenic potency (not shown)].

[b] Spearman's rank correlation coefficient (column 5 vs. column 7) = 0.65 ($p \leq 0.007$) (see Table VI). This justifies pooling these results to give column 6. Column 6 is used to establish mutagen sequence on all charts; overall mean risk assessments decrease from top to bottom.

Is oncogenicity correlated with mutagenicity (columns 5 and 7), at least insofar as ranking by mean risk factors is concerned? Table VI addresses this question again by means of a Spearman rank correlation analysis (5). The rankings of these 15 mutagens (column 2) and oncogens (column 3) are determined from columns 5 and 7, respectively, of Table V. Thus, from Table V, column 7, trenimon and mitomycin C both have the highest mean oncogenic risk factors of 5.00, and are each ranked 1.50 (mean of 1 and 2); next come TEM and ICR-170, tied with mean oncogenic risk factors of 4.50, and each ranked 3.5 (mean of 3 and 4); and so on. And likewise for column 5 of Table V. These rankings are listed in Table VI, columns 2 and 3. Column 4 presents the squares of the differences between these rankings of each mutagen, to serve as a measure of how differently the mutagenic and oncogenic end-points ranked them. And finally, a rank correlation coefficient of $r = 0.65$ ($p \leqq 0.007$) was calculated from these values. (Note: r varies between -1, a perfect inverse correlation, through 0, a completely random association, to +1, a perfect direct correlation.) Such a degree of correlation suggests that we would not be far amiss if we were to consider oncogenicity as simply a 10th mutagenic end-point. This we have done in column 6 of Table V. This combined ranking of all of the in vivo mammalian systems serves as our ultimate estimate of relative mutagenic risk; it also defines the primary mutagen rankings in Tables II and IV-X.

At this stage of our analysis, it is apparent that we are speaking of mutagenic risk, not in terms of each particular test system, but rather in the context of an overall mutagenic risk to mice (and occasionally other rodent species). For, of these 10 in vivo mammalian systems, each except the spot test ranks these reference mutagens in essentially our primary mutagen ranking sequence. This conclusion can be gleaned from Table XI (which we shall discuss below). Here, a Spearman rank correlation analysis has been performed on the rankings as determined by each in vivo mammalian system individually, compared with the mean ranking of all 10 systems (i.e., our mutagen sequence in these tables). Only those systems having r values which are significant at $p \leqq 0.05$ are tabulated in Table XI, and nine of the ten in vivo mammalian systems have values of r between 0.79 and 0.89 (column 4). These nine systems, therefore, are responding quite similarly to relative mutagenic risk, suggesting, in a somewhat circuitous fashion, that it is feasible to address the problem of mutagenic risk in the manner we have selected and that most of the in vivo mammalian systems are telling us the same story concerning the relative risks of these reference mutagens.

Table VI. Spearman's rank correlation for oncogenicity versus mean of nine in vivo mammalian mutagenicity systems

Chemical	Ranking according to		(Δ^2)[c]
	Mutagenicity[a]	Oncogenicity[b]	
Trenimon	2	1.5	0.25
TEM	2	3.5	2.25
Aflatoxin B1	2	6	16
Ethylenimine	4	5	1
Mitomycin C	6	1.5	20.25
TEPA	-	No acute study	-
Thio-TEPA	5	8.5	12.25
METEPA	-	No acute study	-
ICR-170	11	3.5	56.25
DMN	7	8.5	2.25
Myleran	8	10	4
MMS	9	12	9
Cyclophosphamide	10	12	4
Epichlorohydrin	-	No acute study	-
MNNG	14	7	49
Natulan	12	14.5	6.25
EMS	13	14.5	2.25
DEN	15	12	9

[a]From Table V, column 5.

[b]From Table V, column 7.

[c](Column 3-Column 2)². For N = 15 and $\sum_{i=1}^{N} \Delta_i^2 = 194.00$, Spearman's rank correlation coefficient, r = 0.65 ($p \leq 0.007$), thereby justifying the use of column 6, Table V in determining mutagen rankings.

Other Systems

Up to this point, we have described a method for quantitating mutagenic risk in each of 31 test systems in a fashion which avoided test system idiosyncracies; as a result, we have obtained a best estimate of relative in vivo mammalian mutagenic risk and shown that nine out of the ten in vivo mammalian end-points considered agree on this scale of mutagenic risk. In the present section we wish to display all of the risk estimates derived for all 31 mutagen assay systems in a series of color charts (Tables VII-X)* and to discuss the potential of in vitro systems for estimating mutagenic risk to whole mammals.

Tables VII-X are spectral representations of all of the mutagenic risk (to each particular system) data following transformation from LEDE's to numerical and then to colored risk factors as described for the in vivo mammalian systems in Figure 2 and Table III, and described above. Thirty-one test systems are represented in these four charts; the LEDE data for these were derived from 22 Group I reports plus references (2,3) for oncogenicity. (The yeast specific locus data in the Group I report were not presented in a consistent format that was compatible with our purposes, and often lacked critical information on dose, exposure time and mutation frequencies (spontaneous or induced). These two columns in Table X, therefore, are based on our own evaluation of the literature cited in this Group I report.)

Since the primary mutagen ranking is in itself a measure of relative risk to the whole animal we are interested in how well each of the individual test systems "predicts" this sequence. To this end, a Spearman's rank correlation analysis has been performed for each system, testing the mutagenic risk ranking defined by that system against our primary ranking shown on each chart. A rank correlation coefficient r and a measure of its significance p were calculated for each of the 31 systems. Those systems for which $p \leq 0.05$ are listed in Table XI in order of decreasing values of r (given in column 4). Also tabulated are the number of reference mutagens tested (N, column 3) and the number of "false" negatives (column 5) for each system.

Only 18 of the 31 test systems considered are able to rank these reference mutagens in a fashion statistically similar to our primary mutagen ranking. Soybean makes the best showing in this regard ($r = 0.90$) followed closely by nine of the ten in vivo mammalian systems (r values between 0.79 and 0.89), as mentioned above. As a matter of statistics, these values of r are not likely to be significantly different one from the other and we may conclude that (a) in vivo mammalian test systems, with the possible exception of the spot test ($r = 0.57$, $p \leq 0.053$,

*Tables VII-X follow page 1064.

N = 9), are surprisingly single-minded in their relative risk estimations, and (b) the presence of soybean at the top of our list indicates a need for caution in accepting these results without further questioning.

This further questioning can proceed along two lines. First, since these are all potent reference mutagens in most systems, even a single negative result in a given system speaks seriously against that system's general utility. Thus, DEN was negative in vivo in the micronucleus test, in inducing sister chromatid exchanges and in instigating unscheduled DNA synthesis, at doses between 100-200 mg/kg (Tables II, IV, and VII); yet it was detected as a mutagen in in vivo cytogenetics at an LEDE of 60 mg/kg, and in the spot test at 3.5 mg/kg, and it was oncogenic at 110 mg/kg. Such false negatives have been used to discriminate against the systems in Table XI which gave rise to them, and a test system ranking of systems which gave no false negative results is presented in column 6; only 11 systems survived this second cut.

The second question which can be addressed to the data summarized in Table XI is the extent to which each system has examined the 18 reference mutagens. For example, many of these mutagens do not require metabolic activation and thus should be readily detected in all systems. By either chance or design, a system could have been used to examine only those mutagens not requiring metabolic activation and thereby present a misleading picture of its overall potential.

To overcome this essentially statistical deficiency a third restriction was applied to the systems still ranked in column 6, to wit, that each system have tested at least 50% (9/18) of the reference mutagens. Column 7 ranks the adequately tested systems which did not give any false negative results. Only two of the systems from column 6 were eliminated with this restriction: soybean and the heritable translocation test. It would be interesting to see further work done with the soybean system to see if its DMN-activating prowess extends to other promutagens.

Of these nine "best risk-estimating" systems ranked in column 7, three are in vivo mammalian systems (germinal and somatic cytogenetics, and oncogenicity). A question which has often been raised and challenged in the past few years concerns the ability (or lack thereof) of in vitro and/or submammalian systems to make risk estimations. This is particularly important once we leave our present realm of potent reference mutagens and carcinogens and proceed to weaker ones where system sensitivity becomes limiting. (Perhaps such sensitivity limitations of mammalian systems can be partially overcome by chronic, reproductive life-time dosing similar to the now traditional oncogenicity assays. Even though

Table XI. Test system rankings on basis of Spearman's rank correlation coefficients[a]

No.	System	N[b]	r[c]	No. missed	Ranking[d] False negatives	N≥9	In vitro
1	Soybean	6	0.90	0	1		
2	Germinal cytogenetics	9	0.89	0	2	1	
3	In vivo SCE	8	0.89	1			
4	Heritable translocation	6	0.89	0	3		
5	Somatic cytogenetics	14	0.88	0	4	2	
6	Mouse specific locus	8	0.86	1			
7	Oncogenicity	15	0.83	0	5	3	
8	Micronucleus	13	0.82	2			
9	Dominant lethal	16	0.80	2			
10	In vivo UDS	7	0.79	1			
11	Cultured mammalian cells, SCE	9	0.76	0	6	4	1
12	*Vicia faba*	17	0.74	3			
13	Cultured mammalian cells, specific locus	12	0.73	0	7	5	2
14	Cultured mammalian cells, cytogenetics	15	0.67	0	8	6	3
15	Barley	9	0.63	0	9	7	4
16	*E. coli* pol A^+/pol A^-, disc	11	0.54	0	10	8	5

(continued)

Table XI. (continued)

17 *Salmonella typhimurium*	17	0.50	2			
18 Drosophila	18	0.46	0	11	9	6

[a]Individual system vs. mean of in vivo mammalian systems.

[b]N = number of mutagens tested.

[c]r = Spearman's rank correlation coefficient, individual system vs. mean of in vivo mammalian systems. Only those systems are tabulated for which r is significant at $p \leqq 0.05$.

[d]Successive restrictions are: (1) Did it give any false negatives? (2) Has it tested a sufficient number of these mutagens ($N \geqq 9$)? (3) Is it an in vitro test (not part of in vivo rankings)?

the latter tests usually involve long-term dosing, the oncogenicity studies cited in Table II were chosen because their dosing regimens were similar to the mutagenicity protocols. It is not oncogenicity per se which requires chronic dosing; rather it is the fact that most carcinogens appear to be weak mutagens which would not be detected with acute dosing.)

The final column of Table XI ranks only the six in vitro and/or submammalian systems which have adequately sampled our reference mutagens, given no false negatives and "predicted" mutagenic risk in a fashion statistically similar to that found by ten in vivo mammalian systems. Three of these six systems are in vitro cultured mammalian cell systems--sister chromatid exchanges, specific locus and cytogenetics--with r values between 0.67 and 0.76. A fourth system is barley (r = 0.63); a fifth is the *E. coli* $polA^+/polA^-$ disc assay (r = 0.54); and the last system is Drosophila (r = 0.46).

Salmonella and the *E. coli* polA disc assay are the only microbial systems which found a place in Table XI (r = 0.50 and 0.54, respectively). However, both Natulan and mitomycin C were negative in the standard Ames' strains. (Mitomycin C was positive only in TA94, a repair-competent strain; this result was considered irrelevant to the way this system is used in the real world and was thus rejected.)

Thus, for whatever else Table XI is worth, it is significant that 16/18 systems which can potentially estimate mutagenic risk (column 2) and 10/11 of those which did not give any false negative results (column 6) had eukaryotic genomes and appear capable of detecting chromosomal events. (The only microbial system in column 6 measures DNA damage instead of point mutations.) This relative importance of genomal architecture in estimating mutagenic risk to mammals is underscored by the fact that 3 plant systems--soybean, Vicia faba and barley--fare surprisingly well in this regard.

Perhaps with appropriate attention to the design of in vitro systems (activation systems, genetic end-points) the day might not be too far away when such rapid systems can provide us with at least a preliminary mutagenic and oncogenic risk estimate. Studies of this present type can hopefully point out the current deficiencies in these assays so that credible results can be someday achieved.

WHERE TO NEXT?

The blank spaces in Tables VII-X identify areas of ignorance concerning these 18 reference mutagens in 31 mutagenicity test systems. There are 75 of these blanks among the in vivo mammalian systems out of a total of 180 mutagen-system combinations, indicating that the in vivo mammalian systems' data base is only 58% complete for these 18 mutagens. Among the in vitro mammalian systems this figure is 62%; it is 54% for insect and plant systems and 58% for the microbial systems. The overall average saturation of the data base is only 58%.

If the approach we have outlined here is to be pursued further, and if in vitro systems are to be considered as a component of the risk estimation process, this skeletal data base must be fleshed out. In particular, the spot test was the only in vivo mammalian system which did not correlate well with the other nine such systems; this failure resulted from the combination of a disproportionately low LEDE for DEN in one laboratory which was not reflected in other experiments, and a small number (9) of tested reference mutagens. It would be surprising if further testing did not improve the ability of the spot test to rank these reference mutagens in line with the other in vivo systems.

Equal interest should be paid to the plant systems which fared so well in Table XI except for two limitations. First, all but Vicia faba had tested nine or fewer of these mutagens, a condition which contributes greatly to downgrading the statistical significance of Spearman's rank correlation coefficient. And second, promutagens are likely to present a problem in these systems--despite the success of soybean with DMN--until methods of metabolic activation can be applied. Considering the hardiness

Table VII. Risk assessment of 18 reference mutagens in in vivo mammalian systems

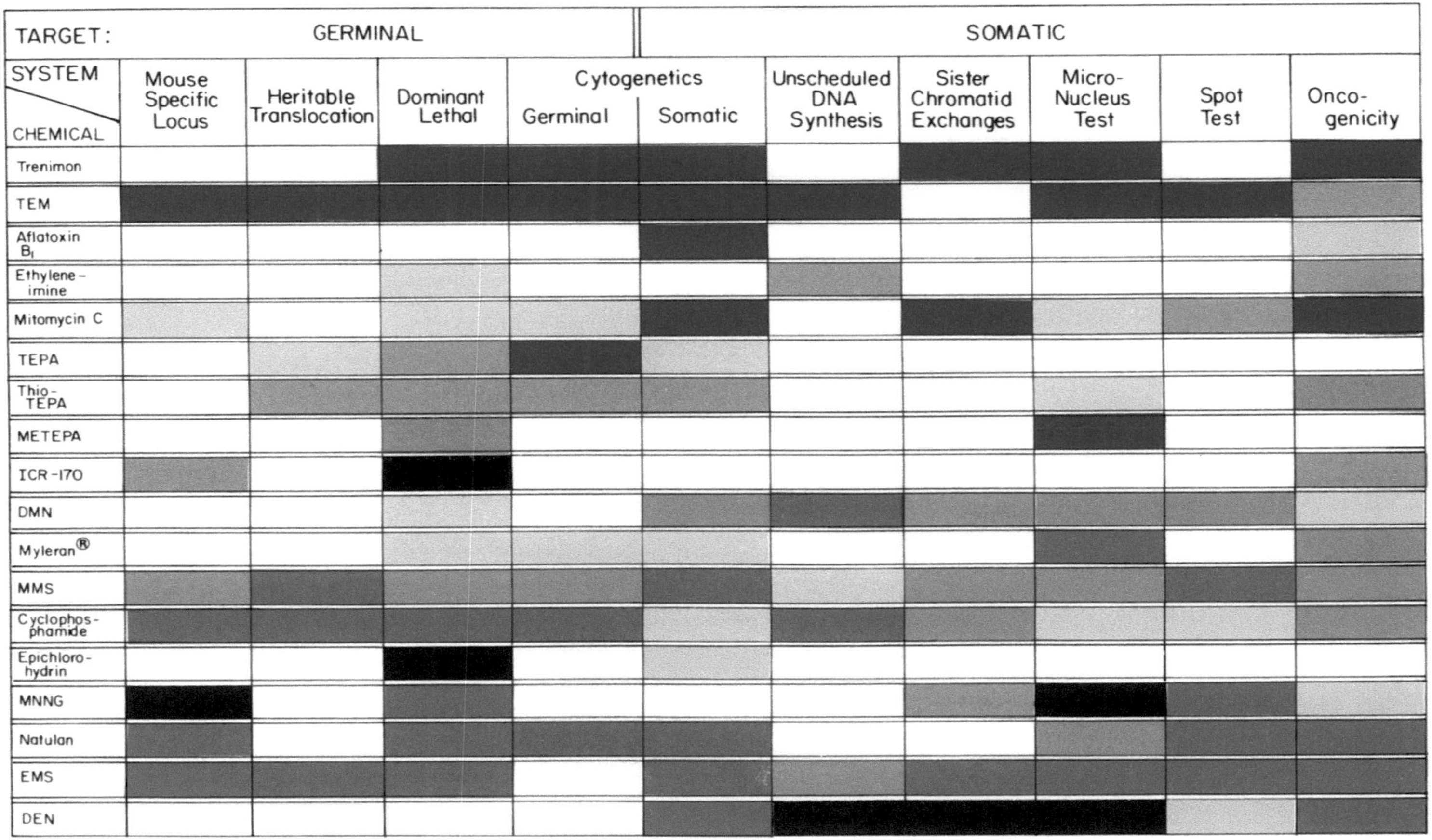
TARGET:
GERMINAL
SOMATIC
SYSTEM
CHEMICAL
Mouse Specific Locus
Heritable Translocation
Dominant Lethal
Cytogenetics
Germinal
Somatic
Unscheduled DNA Synthesis
Sister Chromatid Exchanges
Micro-Nucleus Test
Spot Test
Onco-genicity
Trenimon
TEM
Aflatoxin B1
Ethylene-imine
Mitomycin C
TEPA
Thio-TEPA
METEPA
ICR-170
DMN
Myleran®
MMS
Cyclophos-phamide
Epichloro-hydrin
MNNG
Natulan
EMS
DEN

Table VIII. Risk assessment of 18 reference mutagens in in vitro mammalian systems

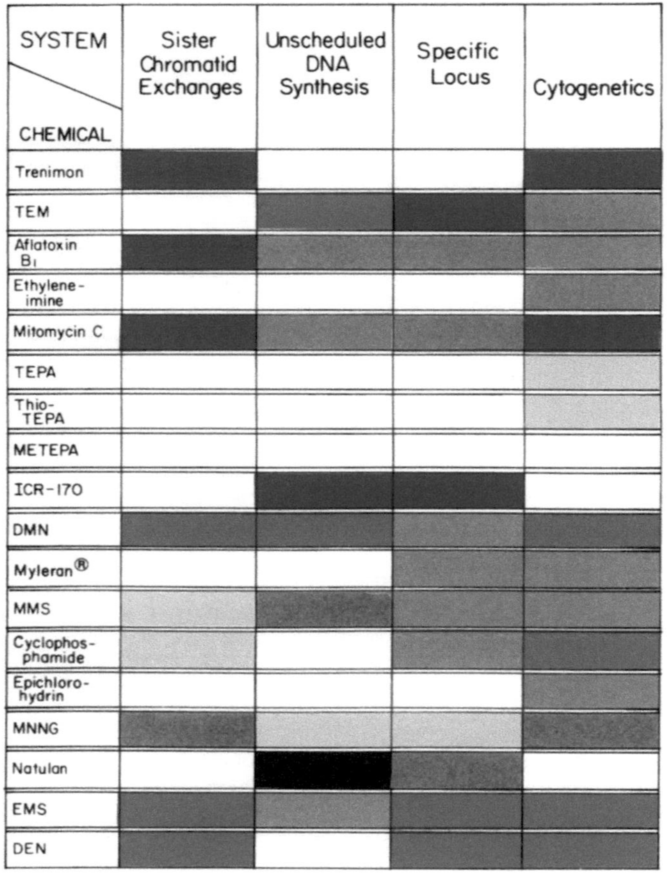
SYSTEM
CHEMICAL
Sister Chromatid Exchanges
Unscheduled DNA Synthesis
Specific Locus
Cytogenetics
Trenimon
TEM
Aflatoxin B1
Ethylene-imine
Mitomycin C
TEPA
Thio-TEPA
METEPA
ICR-170
DMN
Myleran®
MMS
Cyclophos-phamide
Epichloro-hydrin
MNNG
Natulan
EMS
DEN

Table IX. Risk assessment of 18 reference mutagens in insect and plant systems

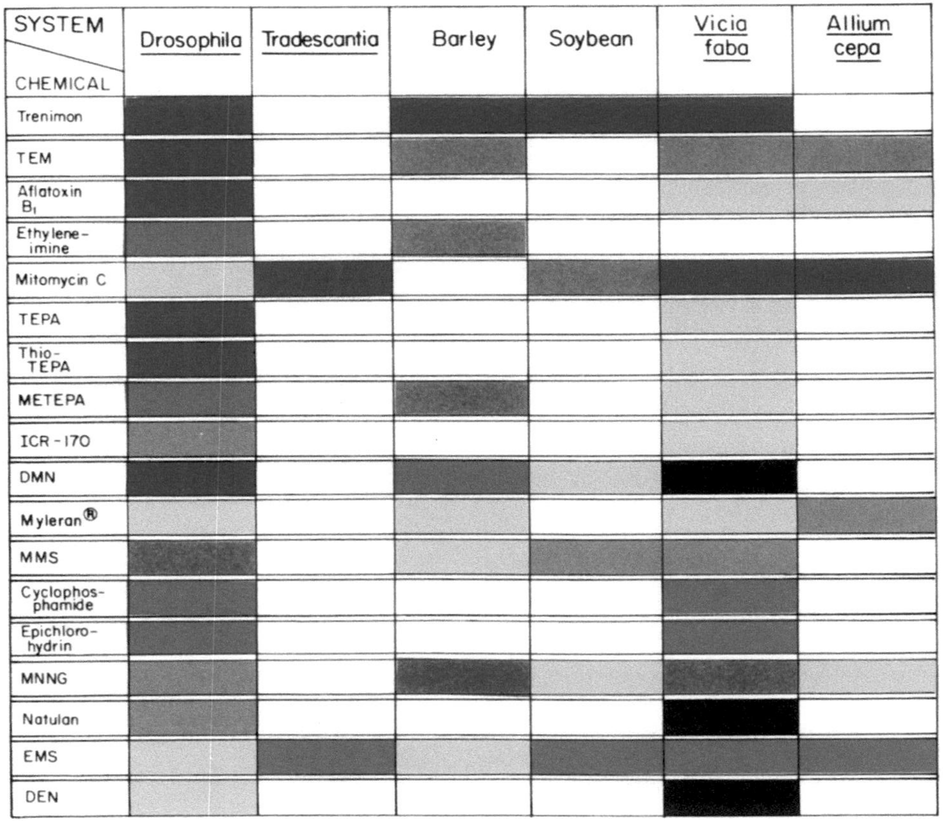
SYSTEM
CHEMICAL
Drosophila
Tradescantia
Barley
Soybean
Vicia faba
Allium cepa
Trenimon
TEM
Aflatoxin B1
Ethylene-imine
Mitomycin C
TEPA
Thio-TEPA
METEPA
ICR-170
DMN
Myleran®
MMS
Cyclophos-phamide
Epichloro-hydrin
MNNG
Natulan
EMS
DEN

Table X. Risk assessment of 18 reference mutagens in microbial systems

SYSTEM / CHEMICAL	Host-Mediated Assay	*Salmonella typhimurium*	*E. coli* Specific Locus	*E. coli* pol A^+/pol A^-		*B. subtilis* Repair	*Neurospora*	Yeast			
								Specific Locus		Gene Conversion	Crossing over
				Disc	Liquid			Forward	Reverse		
Trenimon											
TEM											
Aflatoxin B_1											
Ethylene-imine											
Mitomycin C											
TEPA											
Thio-TEPA											
METEPA											
ICR-170											
DMN											
Myleran®											
MMS											
Cyclophos-phamide											
Epichloro-hydrin											
MNNG											
Natulan											
EMS											
DEN											

of seeds, it is surprising that treatment with promutagens in the presence of S-9 has not already succeeded.

Even if this fleshing out continues to support the principal tenets implied by Table XI, we are still faced with one critical objection to applying risk estimates obtained in those systems to man. Table XI is strongly biased toward chromosomal end-points, especially the in vivo mammalian systems. It is this same biased set of systems which determined our primary ranking sequence against which each system was compared and from which comparison Table XI was constructed. So there is a logical deficiency here, especially when one considers that about half of human genetic diseases appears to be point-mutational in nature (6). It is only partly reassuring that the mouse specific locus test--which might be detecting primarily point mutations--is in agreement with the other in vivo mammalian systems in ranking these mutagens.

Most of the microbial systems appear to be genetically too foreign to mammalian genomes and, perhaps for this reason, fare distinctly poorer in estimating relative mutagenic risk than do eukaryotic systems. Of the two microbial systems with the best track record, Salmonella failed to detect both mitomycin C and natulan in the standard tester strains and was cut in columns 6 and 7 of Table XI for this reason. The deficiencies in the Salmonella system which are responsible for these false negatives are apparently not a problem in the E. coli polA system, the other microbial assay to achieve Table XI status. For mitomycin C, the reason appears to be that this mutagen's effects are a result of misrepair, and are lethal if not repaired; the Salmonella tester strains are designed repair-deficient while the polA system measures repair.

It is not clear why Natulan is missed in Salmonella and not in the polA system, but it might relate to the exclusivity of the genetic end-point (point mutations of the base substitution or frameshift varieties) in the former system and the broader end-point (repairable DNA damage) in the other. In this sense the polA system would be expected to detect those mutagens whose sole effect is at the chromosomal level in eukaryotes, without inducing any point mutations--should such mutagens exist.

Before ending this discussion we might pay a little attention to the question of the role, if any, of in vitro mutagenicity assay systems. For these 18 reference mutagens it appears possible that a small number of such systems are capable of "predicting" mutagenic risk to whole animals. However, there are some in vivo assays which are not much more difficult, time-consuming or expensive to run than, say, cultured mammalian cells or plants. Why not be satisfied with these?

The answer can be glimpsed in Table VII. DEN is detected in somatic cytogenetics and is oncogenic; it is missed in the three other in vivo mammalian systems in which it was tested--unscheduled DNA synthesis, sister chromatid exchange, and the micronucleus test. Other mutagens towards the bottom of this table are sporadically negative (e.g., epichlorohydrin in the dominant lethal assay and MNNG in the mouse specific locus and the micronucleus tests). Yet these are potent carcinogens by the sensitivity standards of current oncogenicity assays. The implications from this are that (1) we are looking at the Goliaths of the mutagen world, the so-called "super mutagens"; (2) that a large army of weaker mutagens is also out there; and (3) that some of these in vivo systems--as presently employed--lack the sensitivity to pick them up. Saccharim might be considered as an example of such a weak mutagen/carcinogen. It is in such a service that the appropriately validated in vitro test systems might find their true role in risk estimation.

ACKNOWLEDGMENTS

The authors are indebted to all of the participants in the Comparative Chemical Mutagenicity Workshop, held in the Research Triangle, N.C., October 30-November 5, 1977, particularly to those who prepared the Group I reports. We are especially grateful to Liz von Halle at EMIC for providing us with an apparently unending stream of revised reports; to Dr. Gerald Hajian for his statistical insights and encouragement; and to Linda Byrd, Mary Marr, and Robert White for their expert advice on data presentation.

REFERENCES

1. Shakespeare, W. Henry VI, Part I, Act V, Scene iii.

2. IARC Monographs on the Evaluation of Carcinogenic Risk of Chemicals to Man, Vols. 1-15, Lyon.

3. Peck, R. M., Tan, T. K., and Peck, E. B. Cancer Res. 36: 2423-2427 (1976).

4. Clive, D. In: Progress in Genetic Toxicology, D. Scott, B. A. Bridges, and F. H. Sobels, Eds., Elsevier/North Holland, Amsterdam, 1977, pp. 241-247.

5. Siegel, S. Nonparametric Statistics for the Behavioural Sciences, McGraw-Hill, New York, 1956, pp. 202-213.

6. Trimble, B. K., and Doughty, J. H. Ann. Hum. Genet. 38: 199-223 (1974).

CHAPTER 37

ESTABLISHMENT OF REQUIREMENTS FOR ESTIMATION OF RISK FOR THE HUMAN POPULATION

F. H. Sobels

Department of Radiation Genetics and Chemical Mutagenesis

University of Leiden, The Netherlands

INTRODUCTION

Originally I had been scheduled to present some sort of summary and synthesis. When it appeared to take longer than expected to receive the Group II papers, this assignment was changed into "Requirements for Risk Estimation," and to determine in what respect we need more or a different kind of data than have been exposed to us at this workshop.

What I propose to do first, is to use ionizing radiation as an example, since this is the mutagenic agent for which more data for risk estimation are available than for any chemical mutagen. This will, I believe, help us to define what kind of data and assumptions have been used to estimate risks for various end-points of genetic damage. Subsequently, we will examine whether similar data are available for any of the chemical mutagens under discussion, and in what respect we might expect essential differences from the situation with ionizing radiation.

GENETIC RADIATION HAZARDS

The germ cell stages of greatest importance for the assessment of genetic risks are the spermatogonia in males and the oocytes in females, since these cells form the permanent germ cell population throughout the reproductive life span. It is assumed that in these stages the genetic damage accumulates from conception until the end of the reproductive life span. The conditions of radiation most applicable to man concern small doses of acute radiation, given at a high dose rate, as for example from diagnostic x-rays, or protracted chronic exposures, like those from applications of atomic energy. Various genetic end-points will be considered separately. Three broad classes of genetic effects can be distinguished: (1) gene or point mutations, including both changes of one or a few nucleotide bases, and small deficiencies of up to a few bands of Drosophila salivary chromosomes missing; (2) reciprocal translocations; (3) nondisjunctional events leading to aneuploidy (gains or losses of chromosomes). In the absence of suitable data I will refrain from the discussion of nondisjunction in general. Since I have been rather closely associated with the group of geneticists that prepared the new 1977 Report of the United Nations Scientific Committee on the Effect of Atomic Radiation, I will use these data for our considerations regarding radiation. For a more detailed treatment of the various facts and assumptions, I may refer the reader to the recent UNSCEAR Report (1). For the estimation of genetic risk of radiation in man, two different methods can be employed. With the "direct" method, genetic damage is expressed in the expected number of mutations and translocations from a particular dose. With the other "doubling dose" method, the induced genetic changes are expressed as a percentage increase of known forms of genetic disease and malformations in man.

Recessive Mutations

In the absence of human data, all risk estimates regarding point mutations, including small deletions, are based on mutation frequencies from irradiated mouse spermatogonia. There are no obvious reasons to believe that the mutational response to radiation of human spermatogonia will be vastly different from that of the spermatogonia of the mouse. For estimating the total frequency of recessive mutations per genome, the data on recessive autosomal lethals, as obtained by Lüning and Eiche (2) and Lüning and Searle (3) are used. They are induced at the rate of 0.9×10^{-4} per gamete per rad at high doses of acute x-irradiation. These mutations that result in death of the embryo before birth, are estimated to represent somewhat less than one half of the total number of recessive mutations (4). Consequently, the total rate of induction of recessive mutations is at least $2 \times 0.9 \times 10^{-4} = 1.8 \times 10^{-4}$ per rad per gamete. On the basis of studies with mutations at specific loci for different dose rates by Russell (5) and for different fractionation schedules (6), it is assumed that under conditions applicable to man of low dose rates or low doses, the above figure needs to be divided by a factor of 3. The figure thus obtained is 0.6×10^{-4} per rad per gamete for spermatogonial irradiation. Since autosomal recessive lethals represent mutational events at unselected loci throughout the entire genome, they offer a more reliable basis for estimation of risks than multiplication of a specific locus rate at pre-selected loci which is then multiplied by the assumed number of loci (a method used in the UNSCEAR Report 1972). Of particular importance, obviously, is the degree of dominance of these "recessives," or in other words, the amount of damage that can be expected in the immediate descendants of the irradiated parental generation. In what follows under the heading of "Dominant Mutations," we will consider how this estimate can be obtained.

Dominant Mutations

Recent observations by Selby and Selby (7) on induced dominant mutations affecting the skeleton of the mouse make it possible in combination with Ehling's earlier data (8), to arrive at a risk estimate for both dominant and recessive mutations with partial dominance. They are induced at the rate of 4×10^{-6} per rad per gamete (when appropriately corrected for dose fractionation and dose rate effects). In order to translate the rate of induction of these mutations causing skeletal abnormalities in the mouse into an overall rate for mutations with dominant effects resulting in serious disability in man, the following information is required. First, what proportion of all dominant conditions in man affect the skeleton. Second, in what measure skeletal abnormalities, as can be recorded in the mouse, may constitute a serious handicap in humans. With regard to the first point, McKusick's data (9) show that of 583 autosomal dominants recorded in man, 328 are clinically important, and of these, 74 involve the skeleton to some extent. However, it is thought that this figure is an overestimate because of the ease of diagnosis of such abnormalities and according to Carter and McKusick (10) the true figure is of the order of 10%. With respect to what proportion of the observed skeletal abnormalities would impose a serious disability, the figure of 50% is thought to represent the best estimate. It follows that to estimate the rate of induction of the total number of mutations with deleterious dominant effects in man, the induction rate of 4×10^{-6} obtained for the mouse has to be multiplied by 10 (so as to represent all organ systems) and to be divided by 2 (to correct for dominant mutations reflecting really serious disability). This figure of 20×10^{-6} per gamete per rad, thus means that following 1 rad of irradiation of the spermatogonia, 20 per million of the direct offspring will manifest a "dominant" genetic disease (1).

Now, in context, it is of particular importance to note that this figure of 20 per million also includes those mutations showing partial dominance of the 60 per million recessive mutations discussed earlier. This is so because of 160 dominant mutations, known in the mouse, at least 45 or 27% are lethal when homozygous (i.e., received from both parents). Moreover, homozygotes for deleterious human dominants, though rarely observed, are lethal.

For irradiation received at low doses or dose rates, the mutational risk in the progeny of females is low, provided the human ovary responds to irradiation as does the mouse ovary (11).

Translocations

For the induction of translocations, material is available from some nine human volunteers whose testes had received x-ray doses of 78, 200, or 600 rad (12). Since the human figures do not differ from those in a certain primate species, the marmoset, the data are combined for purposes of risk estimates, and this provides us with a translocation rate of 6.94×10^{-4} per rad per cell, on the basis of a linear dose-response curve (13). From experimental data obtained in the mouse with dose fractionation (14) and low dose rates (15,16), it appears that for these conditions of human exposure, the above figure needs correction to 3.47×10^{-4} for low dose x-rays and to 0.69×10^{-4} for chronic gamma irradiation. These figures reflect the translocation frequencies induced in spermatogonia and scored in the metaphase of the first meiotic division. Furthermore, for estimating the frequency of translocations in the progeny, experimental evidence suggests (as expected from theory), that here the rate of recovery will be one quarter of that seen in the spermatocytes (17). In the F_1 the number of unbalanced translocation-carrying zygotes will be twice that of the balanced translocation-ones so that, altogether, 87 and 174 or 17 and 34 balanced and unbalanced translocations per million conceptions can be expected. Moreover, it has been assumed that 6% of the unbalanced zygotes will result in children with multiple congenital abnormalities. Thus there will be 2 or 10 of such children per million conceptions after one rad of paternal (spermatogonial) exposure, and about five times this number, of recognizable abortions and furthermore about twice this number will die so early as to go undetected. The risk of translocation induction after irradiation of females is thought to be low.

In summary then, the genetic damage in the first generation following irradiation of the parents with 1 rad will amount to 20-30 seriously disabled individuals per million liveborn (20 from mutations and 2-10 resulting from translocations) (see Table 1). The induction of numerical chromosome aberrations, involving gains or losses of whole chromosomes, is not included in this estimate, since no reliable data on this phenomenon are available.

Table 1. Estimated risk of induction per rad of low LET irradiation of various kinds of genetic damage in man at low doses or low dose rates[a]

End point	Expected number per million gametes resulting from irradiation of: Spermatogonia	Oocytes[b]	Expression in first generation per million births[b]
1. Autosomal mutations	60[c]	-	20[d] (items 1–3)
2. Dominant visibles	Very low	-	
3. Skeletal mutations	4	-	
4. Balanced reciprocal translocations	17-87	Low	Low
5. Unbalanced products of item 4 above	34-174	Low	2-10[e]
6. X-chromosome loss	Very low	Low	-
7. Other chromosome anomalies	-	-	-

[a]Data from UNSCEAR Report (1).

[b]Dashes indicate that inadequate or no information is available.

[c]Based on the rate of induction of mutations in mice that are lethal in the homozygous condition.

[d]Overall rate of dominant effects, based on skeletal mutations and presumably including dominant visibles and heterozygous effects of autosomal recessive mutations.

[e]Expressed as congenital malformations; in addition there would be 11-55 recognizable abortions and 22-109 early embryonic losses.

PREREQUISITES FOR RISK ESTIMATION

There are two additional assumptions that have been so often reiterated in the past, that they are not even mentioned in the latest UNSCEAR Report (1). These are: (1) linear proportionality between radiation dose and the number of induced mutations, and (2) cumulative effects; that is, it is assumed that once a mutation has been induced in the germ cell stages at risk, recovery of this change or elimination of the cell bearing the mutated gene does not occur. Linear proportionality means of course that there is no threshold, or safe dose, below which mutations are no longer induced, and that when considering large populations, it is only the total radiation dose to which the reproductive childbearing section of the population is exposed, which is of importance. In other words, the same number of mutations are expected to arise from exposure of 100 million individuals to 0.10 R reach, as from exposure of 10 million people to 1 R; in both cases the total population exposure was 10 million R.

Both these principles were already defined many years ago by Muller (18) on the basis of data collected for Drosophila. For the mouse proportionality of mutation frequency with dose is based on data of W. L. Russell and L. B. Russell (19) for acute and chronic exposures. The accumulative effects for the mouse have been demonstrated for specific locus mutations by Russell (20), for dominant lethals by Sheridan (21), and for translocations by Léonard (22), many years ago. In brief, what counts, therefore, for assessment is the total dose that has been accumulated over the period from conception until usage of the germ cell in question for generating new progeny; genetically significant doses thus take account of this total, accumulated dose, including a weighting factor for child expectancy.

DOUBLING DOSE METHOD

I will now briefly discuss the kind of estimates that can be derived by expressing risks in relation to the natural incidence of genetic disease in man. An advantage of the doubling dose method is that the risks are measured in terms of known genetic disease. The doubling dose method requires the following kinds of information: the natural incidence of the different classes of genetic disease; an estimate of to what extent these genetic abnormalities are maintained by recurrent mutation; and the doubling dose. The method involves the assumption that the irradiation induced mutation rate for each kind of genetic defect is proportional to the rate with which it originates spontaneously and that the various genes respond with the same kind of mutational changes, as that which arise spontaneously. These concepts are by no means proved. Moreover, we are faced with a deplorable lack of data on the nature of spontaneous mutations. This is a particular handicap when using the doubling-dose approach, because if one particular mutagen increases one fraction of the spontaneous damage, and another mutagen some other fraction, the ultimate risk estimates thus obtained will be totally different, and cannot be compared.

Data on the natural incidence of the various diseases are based on a recent survey by Trimble and Doughty (23) for British Columbia with modification for dominant diseases and chromosomal anomalies (1). The total figure of 10.5% comprises: 1% dominant and x-linked diseases, 0.1% recessive diseases, 0.4% chromosomal diseases and 9.0% diseases with complex etiology (see second column in Table 2). Of the latter category, 5% (9.0% × 5% = 0.45%) is thought to arise by recurrent mutation.

On the basis of mouse data, 100 rad seems the best estimate for a doubling dose covering such different genetic endpoints as point mutations and translocations (under conditions of low doses and chronic exposure) (1,13,24). Findings on the mortality of children born to survivors of the atomic bombings of Hiroshima and Nagasaki (26) suggest that the doubling dose in man is not likely to be lower than 100 rad. Using these figures it can be estimated that following irradiation at a rate of one rad per generation, 185 additional cases per million liveborn will be expected when the population has reached a new equilibrium. The first generation increment will be about one-third of this, or, about 63 cases per million (see Table 2).

It is of interest to note that the figure for the radiation-induced mutational effects in the first generation of 20 individuals per million per rad, arrived at by the direct method, does not differ much from that of 25 per million as estimated by the doubling dose method. New is the approach to consider all effects in the immediate progeny collectively by lumping those of mutations with dominant and recessive mutations, and to express them in terms of disability in man. This has been made possible through the new data of Selby and Selby on the transmissibility and phenotypic manifestation of skeletal mutations in the mouse and their consideration in consultation with clinical geneticists familiar with hereditary syndromes in man, like McKusick and Carter. This method now combines the advantages of the two (direct and doubling dose) methods in that it enables the prediction, with reasonable precision, of tangible genetic damage in man. In conclusion it can be stated that the present risk estimates are not so totally different from the previous ones, but, thanks to a lot of new data, their reliability seems to have increased considerably.

Table 2. Estimated effect of 1 rad per generation of low dose, or low dose-rate low LET irradiation on a population of one million live births (assumed doubling dose: 100 rad)[a]

Disease classification[b]	Current incidence[c]	Effect of 1 rad/generation: First generation[d]	Effect of 1 rad/generation: Equilibrium
Autosomal dominant and X-linked diseases	10,000	20	100
Recessive diseases	1,100	Relatively slight	Very slow increase
Chromosomal diseases	4,000[e]	38[f]	40
Congenital anomalies	90,100[g]	5	45
Anomalies expressed later			
Constitutional and degenerative diseases			
Total	105,200	63	185
Percent of current incidence		0.06	0.17

[a] Data from UNSCEAR Report (1).

[b] Follows that given in the BEIR Report (24).

[c] Based on Trimble and Doughty (23) with certain modifications as outlined in the UNSCEAR Report.

[d] The first generation incidence is assumed to be one-fifth of that at equilibrium for autosomal dominant and x-linked diseases; for congenital anomalies etc. it is one-tenth of that at equilibrium [see BEIR report (24)].

[e] Includes all numerical anomalies (sex-chromosomal and autosomal) and aneuploid structural abnormalities but excludes balanced translocations.

(continued)

Table 2. (continued)

[f]The first generation incidence is assumed to include all numerical anomalies and three-fifths of the unbalanced translocations (the remaining two-fifths being derived from a balanced translocation in one parent); for rationale see UNSCEAR (1).

[g]Of these diseases 5% are presumed to respond in a manner similar to simple dominants (=5% mutational component).

SUITABLE MATERIAL AND PERTINENT CONDITIONS FOR ESTIMATION OF RISKS FROM CHEMICAL MUTAGENS

In the light of this exposure on genetic radiation effects, I want to examine, along the lines of the above criteria, what data are available for chemical mutagens and what can possibly be done with them in terms of risk evaluation. First of all it may be noted that for man himself no data are available on chemically induced genetic damage. Since man is most closely related to other mammals it is obvious that high significance should be assigned to data obtained with mammalian in vivo assay systems. From a general toxicological point of view, experimentation with the intact mammal will help to solve the question of whether mutagenic concentrations indeed reach the target cells to invoke genetic damage. Moreover, information regarding response to different modes of administration, pharmacokinetics, absorption, distribution, metabolic transformation and excretion, cannot be obtained without studies on mammals. Other factors that encumber extrapolation from test systems at a lower level of biological organization, concern differences in the organization of genetic material, in the enzymatic regulation of repair processes, and in the sensitivity differences of various stages of development of the germ cells. Furthermore, it is not always realized that in comparing the results obtained with different assay systems, we are largely ignorant of what the comparison actually involves in terms of genetic endpoints, since each assay system may in fact register essentially different kinds of genetic damage. Thus, how results for a Salmonella assay, or tests with Neurospora, yeast, Drosophila, DNA repair, or sister-chromatid exchanges should be translated in terms of human suffering poses problems that are not easy to solve.

Risk estimates for genetic radiation damage are mainly based on mutation studies with the mouse. The question of whether, in general, extrapolation from mouse to man is indeed permissible for chemical agents, is difficult to answer. Some doubt is, in this respect, perhaps not entirely unjustified in the light of observations of extreme specificity of mutagenic activity as is the case with formaldehyde and urethane. Urethane, which produces tumors in one species of rodents but not in another; is a rather extreme example. In any case, it would seem that at the present stage of our methodology, mutation studies with the mouse offer the only approach from which extrapolation to man is possible. Consequently, it seems appropriate to explore to what extent the required data can in fact be obtained along these lines.

The conditions of exposure most relevant to the human population are, apart from accidents, or exceptional cases of very high professional or medical exposures, presumably, as in the case of radiation, low doses and chronic exposures. Also, with regard to the stages in which accumulation of genetic changes from conception until the end of reproductive age can be expected, spermatogonia and oocytes, as was the case with radiation, are presumably of greatest significance. An exception, in this respect, has to be made for substances such as TEM, with an extremely high sensitivity of induction in spermatids (in the order of 4000 times that of psermatogonia), so that despite the short life span of these stages, it cannot be neglected for risk estimates. The chances of contamination of the environment with such polyfunctional alkylated agents is, however, presumably very small.

With regard to the other conditions required for estimation of risks, it seems appropriate to distinguish between the induction of events relating to chromosome breakage and mutations.

CHROMOSOME ABERRATIONS AND THE POSSIBILITY OF FALSE NEGATIVES

Numerical chromosome aberrations will not be considered in the absence of suitable data, as was the case with radiation. In view of their high spontaneous incidence of 0.6% among liveborn, and 50% among abortuses, relevant studies to explore in what measure these spontaneous frequencies are increased by exposure to mutagenic agents, are urgently required.

The chromosome aberrations under discussion are a consequence of chromosome breakage, such as translocations. Before one proceeds to tests involving whole mammals, testing for the detection of chromosome-breaking activities, with short-term cultures of human lymphocytes in vitro, will often be used (27,28). Chromosome breakage events induced in the intact mammal can be detected in somatic cells by cytogenetic studies on cell populations contained in the bone marrow or peripheral blood. The micronucleus assay, as developed by Schmid and co-workers (29) and Heddle (30), provides the simplest and fastest assay. Occurrence of chromosome breakage can in this way be determined without the tedious metaphase analysis. For damage incurred in the germ cells, tests for dominant lethals (31) and heritable translocations (32) are used. Translocations can be scored in spermatocytes of the treated animals, or more laboriously, progeny are screened for semisterility. In contrast with predominantly negative findings for induction of reciprocal translocations, after treatment of the spermatogonia with various chemicals, chromatid interchanges were observed at high frequency following treatment with mitomycin (33).

In this context it may be pointed out that all routine screening systems using the intact mammal rely on the detection of chromosome-breakage effects. The definite advantage of results obtained with these tests is that the measured damage can, with appropriate qualification, be translated into tangible genetic damage in man. The shortcoming of these tests derives from the fact that chromosome breakage could well be a less sensitive indicator than gene mutations, and thus they may for certain compounds generate false negatives.

Support for this conclusion is derived from the recent observations of Vogel on the "two-level effect" in that chromosome breakage effects in Drosophila, such as dominant lethals, translocations, and chromosome loss, invariably require higher concentrations for their induction than do recessive lethals. These findings are documented in Tables 3 and 4. Mutagenic effectiveness is expressed as the LEC/LD_{50} ratio; LEC, the lowest effective concentration, stands for the concentration that results in a significant increase in the induced frequency of the endpoint under consideration over the control frequencies. The data in Table 3 show that for the indirect carcinogens Natulan, 2,4,6-tri-Cl-PDMT, a triazene, and DEN, chromosome aberrations are either not produced or only at very much higher concentrations than recessive lethals. This observation then stands in sharp contrast to the high efficiency of these compounds in inducing recessive lethal mutations. Even with powerful chemical mutagens, such as MMS or TEB, much higher concentrations are required for the production of dominant lethals or chromosome loss than for recessive lethal induction (Table 4). For mutagens of the oxazaphosphorine type, vinyl chloride or chloroprene, the concentrations of the reactive metabolites apparently are not sufficient to elicit any chromosome breakage at all. A characteristic of the dose effect curves obtained with these substances is a flattening off at higher concentrations, so that even with very high concentrations, no greater mutation frequencies are observed than about 1-3%. The interpretation is presumably an overloading or saturation of the enzyme systems required to convert these compounds into the genetically active agents. At a level of 1-3% recessive lethals, even substances such as MMS do not induce chromosome breakage phenomena, and Vogel concludes that the concentration

Table 3. LEC:LD_{50} ratio in Drosophila for various genetic endpoints[a,b]

Chemical[b]	LEC:LD_{50} Recessive lethals	Dominant lethals	Chromosome loss	Translocation
Natulan	1:100	1:2 (+?)	1:2	1:2
2,4,6-triCl-PDMT	1:1000	1:4	1:2	1:4
DEN	1:100	Inactive	Inactive	Inactive
MMS	1:100	1:10	1:5	1:10
TEB	1:5000	1:1000	1:200	1:1000

[a]Data of Vogel and Leigh (34) and Blijleven and Vogel (35).

[b]Natulan (procarbazine); N-isopropyl-ω-(2 methylhydrazino)-p-toluamide; 2,4,6-triCl-PDMT, 2,4,6-trichloro-1-phenyl-3,3-dimethyltriazene; DEN, diethylnitrosamine; MMS, methyl methanesulfonate: TEB, 2,3,5,6-tetraethylenimino-1,4-benzoquinone.

Table 4. Specific mutagenic activity of procarcinogens in Drosophila

Compound	Recessive lethals	Dominant lethals	Chromosome loss Entire (X,Y)	Chromosome loss Partial (Y)
Vinyl chloride	+	0	0	0
Cyclophosphamide	+	0	0	0
Trofosfamide	+	0	0	0
Ifosfamide	+	0	0	0
3-PyDMT[b]	+	0	0	0

[a]Data of Vogel (36) and Verburgt and Vogel (37).

[b]3,3-Dimethyl-1-(3-pyridyl)-triazene.

of the above agents at the genetic material in the target cells has remained below the level required for the induction of chromosome breakage. Altogether, such results give further experimental support to the notion that there is an essential difference between gene mutations and chromosome breakage phenomena, a concept that was proposed by the late H. J. Muller since the very early stages of mutation research. In the context of estimation of risks it is important to realize that simply changes in concentration thus may result in drastic changes in the spectrum of the various kinds of induced damage.

The finding that the effective concentration for the production of chromosome breakage differs from that required to increase the number of point mutations, raises the question of the general validity of this observation, and in particular, whether there exists a similar situation in mammals. A definite answer to the problem is not yet possible, simply because of the lack of practicable methods to score efficiently for point mutations in mammals in vivo. The significance of these observations should, we believe, not be underestimated because, as has been noted before, all routine assay systems with the intact mammal rely on the detection of chromosome breakage phenomena. In the case of negative results there is the possibility that the agent or its metabolite did not reach the required concentration to break chromosomes, but nevertheless might well have been effective in inducing gene mutations (for example, see under epichlorohydrin).

MUTATIONS IN MAMMALS

For the detection of heritable gene mutations in mammals one has had until now to rely entirely on specific locus tests. More than a quarter of a century of intensive research with this method has generated a wealth of data that contributed in a highly significant manner to the understanding of the various physical (effects of dose, dose rate, dose fractionation, LET) and biological (sex, cell stage, age, interval between conception and irradiation) parameters governing the induction of mutations by ionizing radiation. These factors have in fact been taken into account in the assessment of genetic radiation hazards in man.

In view of the labor and high costs, the application of this method has remained restricted to only a few laboratories in the world. The fact that mutations at any of these loci is a rare event, places severe limitations on the number of chemicals and concentrations that can be investigated (see, for example, Table 5). This creates problems in the evaluation of compounds with low mutagenic activity. Undoubtedly, most environmental agents seem to belong to this category, as has been amply demonstrated by employing other test systems.

Table 5. Summary of the data obtained with mammalian assay estimation of risks are included[a]

Compounds	Mutations SP.-gonia	Post SP.-gonia	Spot test	IV
Aflatoxin	nt	nt	nt	+
Cadmium	nt	nt	nt	nt
Cyclophosphamide	nt	+	+	nt
DEN	-	-	+	+
DMN	nt	nt	+	+
EMS	-(4C)	+(2C)	+	+
Epichlorohydrin	nt	nt	nt	+
Ethylenimine	nt	nt	nt	+
ICR-170	+[b]	nt	nt	+
MeTEPA	nt	nt	nt	nt
Mitomycin	+(4C)[b]	-(4C)	+	+
MMS	+[b]	+(4C)	+	+
MNNG	-	-	+	+
Myleran	-	(-)	nt	nt
Procarbazine	+(4C)[b]	+(4C)	+	+
TEM	+(3C)[b]	it	+	+
TEPA	nt	nt	nt	nt
Thio-TEPA	nt	nt	nt	nt
Trenimon	nt	nt	nt	nt
Vinyl chloride	nt	nt	nt	nt

[a]Abbreviations: nt = not tested; it = inadequate test; - = negative; IV = in vitro mammalian cells; HT = herita

[b]Most relevant data

systems for twenty compounds; data most relevant for the

	In vivo cytogenetics					
HT	Sp.-gonia	Post Sp.-gonia	Oocytes	Bone marrow	MN	Liver
it	-	?	nt	-	-	+
it	-	dl-	nt	human?	-	nt
+	±	+	+[b]	+human	+	nt
nt	nt	nt	nt	-	-	nt
nt	nt	nt	nt	+	+	nt
+	-	dl	nt	+	+	nt
nt	nt	dl-	nt	mouse+	nt	nt
nt	nt	nt	+[b]	human	nt	nt
nt	nt	-	nt	nt	nt	nt
nt	?	dl	nt	+	+	nt
it	+[b]	+	nt	+	+	nt
+	-	dl	+[b]	+	+	nt
nt	nt	HT	nt	-	-	nt
nt	-	dl	nt	nt	+	+
nt	+[b]	nt	nt	+human	+	nt
+	(+)	+[b]	+[b]	+	+	nt
+	+[b]	nt	nt	+	nt	nt
+	+[b]	HT	nt	+	+	nt
-	+[b]	+[b]	+[b]	+	+	nt
nt	-	dl	nt	+human	nt	nt

C = concentration; 4C = 4 concentrations tested; + = positive translocation; dl = dominant lethal; MN = micronuclei.

The fact that for such chemicals specific locus studies by necessity will have to remain restricted to high concentrations, poses the problem of extrapolation, that is what damage can be expected at low realistic concentrations of human exposure. At the risk of being redundant, it may, in this context, be stated once again that the basic principle underlying the evaluation of genetic radiation hazards is the linear dose-effect relationship. Proportionality forms the basis that only the total genome dose to the population counts, irrespective of how this is distributed over the various individuals. The same holds for the principle of cumulation that likewise takes account of the total dose received from conception until the end of child-bearing age.

In the absence of relevant data for all agents under study, it seems to remain a matter of conjecture whether extrapolation from high concentrations to what can be expected at low concentrations is permissable. A priori, one cannot exclude the possibility, however, that all kinds of factors, such as repair, resorbtion, interaction with other cell constituents, activation, inactivation or saturation of enzyme systems may lead to different yields of mutation per unit dose at different concentrations or different modes of exposure. More specific examples of such modifying factors will be presented below for the various substances under consideration.

For recessive lethals and dominant skeletal mutations that formed the basis for the estimation of radiation risks, no chemical data are available. For TEM, Sheridan (38) obtained a highly positive response in scoring for recessive lethals in mouse spermatids. Selby recently reported (39) how the amount of labor involved in the scoring for dominant mutations affecting the skeleton can be reduced by employing the so-called "sensitive indicator approach." Instead of scoring for all skeletal abnormalities, this method is restricted to only a few of the most conspicuous aberrations which comprise, however, one quarter of the total.

MAMMALIAN DATA FOR THE TWENTY COMPOUNDS: RELEVANCE AND POSSIBLE COMPLICATIONS IN THE ASSESSMENT OF RISKS

In an attempt to assess in what measure the material presented in this workshop could be used for the estimation of risks, observations in mammalian assay systems have been summarized in a tabular form (see Table 5). The most relevant data have been indicated as noted. In addition, observations that may present possible complications will be briefly discussed, taking the chemicals in alphabetical order.

Aflatoxin

An environmental mutagen 100 times more effective than EMS. Data for mammalian germ cells that would permit a risk estimation are not available. Dolimpio's observations (40) for chromatid breaks show that at low exposure levels, the dose increment is more effective than the same increment at high exposure levels. Similarly, chronic exposures are more effective than acute. The saturation observed in both cases presumably results from overloading of a metabolic activation system. An analogous situation has been recorded for cyclophosphamide and vinyl chloride by Vogel in Drosophila.

Cadmium

Very little convincing data for chromosomal aberrations and none for gene mutations. Occasional positive results in mammals could be due to presence of other heavy metals.

Cyclophosphamide

Chromosome aberrations are induced at clinical doses of 5 mg/kg. Of interest is the pronounced species specificity; in order of decreasing sensitivity for the induction of chromosome aberrations, the sequence is: mouse, rat, Chinese hamster, and man. Positive results have been recorded for oocytes and postmeiotic stages of male mice. Human males having received therapeutic treatment should be counselled to refrain from generating progeny over a period of time required to eliminate the most sensitive, postmeiotic stages of spermatogenesis from the testes. For women becoming pregnant following treatment, amniocentesis is recommended, as Heddle pointed out (41), to determine the chromosomal constitution of the fetus.

EMS

The main contribution is from postgonial stages. The induction of chromosome aberrations in bone marrow suggests a threshold at low concentrations, and a saturation at high concentrations.

Epichlorohydrin

Epichlorohydrin is an environmental mutagen of considerable

concern. Mouse in vivo cytogenetics show no evidence for a threshold. There are low doubling doses of 1-2.5 mg/kg IP and of 1-7.5 mg/kg orally. Following acute exposures, the response levels off at 20 mg/kg, but considerably higher responses are observed if the same total dose is given in five fractions, 24 hr apart. Epichlorohydrin acts as a gene (point) mutagen in mammalian cells both in vitro and with in vitro assay following in vivo (transplacental) exposure; there is a positive dose-response relation but no evidence for threshold. It acts as a weak clastogen in man, but, as the review by Bridges (42) indicates, more concern about its capacity of inducing gene mutations is warranted. The available data suggest that epichlorohydrin represents one of those typical monofunctional alkylating agents that is relatively more effective in inducing gene mutations than chromosome aberrations. In view of widespread industrial exposure, further studies are obviously appropriate.

ICR-170

Positive data are reported for the induction of mutations both in mouse spermatogonia (only one concentration has been tested) and in vitro. A negative dose-response curve was observed in cultured mammalian cells.

Mitomycin C

Even with a low concentration of 5.25 mg/kg, mitomycin C produced mutations in spermatogonia. The fact that this compound acts as an inhibitor of DNA synthesis explains why postmeiotic stages yielded negative results in the specific locus test. The induction of dominant lethals in these stages is thought to result from inhibition of RNA and protein synthesis at high concentra- interchanges were observed. In Drosophila oocytes, mitomycin induces interchanges between nonhomologous chromosomes. It is surprising that there was no evidence for induced nondisjunction, since this might have been expected on the basis of Parker's and Williamson's observations (44) that such interchanges may lead to abnormal segregation.

MMS

A concentration of 5 mg/kg significantly increases the frequency of dominant lethals in mice. As Kondo (45) reported, postmeiotic stages are 1000-fold more sensitive for the induction of chromosome aberrations than premeiotic ones. Chromosome aberrations in pronuclei, following treatment of spermatids and spermatozoa suggest repair, a threshold, or multi-hit kinetics.

Myleran

Myleran imposes higher risks for human males. Treated women may wish amniocentesis to identify cytogenetically abnormal fetuses.

Procarbazine (Natulan)

In general, there is an increase of mutagenic activity up to a given dose of Natulan, followed by a decline or plateau at higher doses. An inverse dose-effect relationship, that is, diminishing yield with increase in dose, for mutations at the HGPRT-locus in V79 Chinese hamster cells and both the HGPRT- and TK-loci in L5178Y mouse lymphoma cells in the complete absence of an increase of toxicity is reported by Knaap (46). Higher concentrations are thought to inhibit the enzymes responsible for the conversion of Natulan to its mutagenic metabolites.

This observation is of importance, in that it illustrates that, in the absence of dose-effect curves, linear extrapolation downward from yields obtained at high concentrations could possibly lead to a drastic underestimation of the mutagenic potency. In specific locus experiments Ehling (47) finds a doubling dose of 114 mg/kg.

TEM

Extremely low doses are effective in the mouse, heritable translocations being induced at 1/400 of the lethal dose. The dose-effect relationship for dominant lethals in the mouse is linear, unlike that for many other compounds, such as EMS (48). In postmeiotic stages TEM is 3800 times more effective in inducing translocations than in spermatogonia (49). For the evaluation of risks this factor has obviously to be taken into account for substances of this kind, despite the shorter life span of the postmeiotic stages.

TEPA

Species-specific differences are seen in yield of chromosome aberrations in bone marrow, the mouse being twice as sensitive as the rat.

Thio-TEPA

Differences in response have been reported between different species, tissues and strains of mice.

Trenimon

At low doses a relatively smaller effect was observed for in vivo cytogenetics than at high doses.

Vinyl Chloride

Chronic exposures are more effective than acute, presumably as a result of saturation of activation systems. This is analogous to the recessive lethal response in Drosophila, which levels off at increasing concentrations (37).

In considering now the data most relevant for risk estimation, that is, those indicated by circled plus signs in Table 5, it can be seen that for gene mutations induced in spermatogonia, data are available for five compounds and none for oocytes. For cytogenetic damage in spermatogonia there are data for four or five compounds. Comparatively greater weight should perhaps be given to damage in postmeiotic stages by TEM and Trenimon. For chromosomal damage in oocytes, data are available now for five compounds. In view of the differences in cytological structure between mouse and human oocytes, the question whether extrapolation from mouse oocytes to those in the human is a valid procedure cannot yet be answered.

In summary then, it appears that for this group of compounds which probably represents a selection of substances that have been most intensely studied, data that could be used for evaluation of risks are presently available for some five different agents. Moreover, as has been outlined above, the mammalian data on the various substances, reveal a number of unusual dose-effect curves, shoulders, effects of chronic or fractionated exposures, or species-specific differences in response, all factors that could present problems in the quantitative evaluation of risks.

MONOFUNCTIONAL ALKYLATING AGENTS

The assumption is often made that alkylating agents and agents that act via alkylating reaction products predominate among chemicals with mutagenic activity in the human environment. For estimating the risks of such compounds we are in a somewhat more favorable position than would be the case for other classes of mutagens. Ehrenberg and co-workers (50) have developed an elegant method for direct estimation of risk in man by determining the tissue dose from the degree of electrophilic substitution in certain macromolecules, such as haemoglobin. The tissue dose, defined as the time integral of the concentration of the proximal mutagen is related through certain reaction-kinetic parameters--which can be determined experimentally--to the mutation frequency. This method has been successfully used by Ehrenberg to assess the genetic risk arising from industrial exposure to ethylene oxide. It circumvents the difficulty of determining risk in man by extrapolation from animal tests and is, according to Ehrenberg, many orders of magnitude more sensitive than mutation tests using biological material, so that one is not faced with the problem of whether a negative test indeed excludes mutagenicity in man.

Another advantage with regard to alkylating agents is the methodology developed by Lee, Aaron, and Sega and coworkers (51-56) to measure the degree of alkylation of the DNA within the germ cells. Only values determined by means of this method give a reflection of the real dose within the target cells and not the concentration of the chemical as it has been applied. Aaron and Lee (55) recently showed that the EMS-induced frequency of recessive lethals in Drosophila sperm increases in a strictly linear fashion with the degree of alkylation of the DNA, a perfect demonstration of the absence of a threshold.

Of particular significance is the finding that for monofunctional alkylating agents a certain prediction can now be made of the kind of genetic effects they will generate, on the basis of what is known of their reaction pattern (S_N1 or S_N2) and the so-called Swain-Scott substrate constant s, which expresses the dependence of the reaction rate on the nucleophilicity n of the receptor molecule (57,58). In a somewhat simplified manner, one could state that agents of low- Swain-Scott s factor, following S_N1 reaction mechanism, as, for example, ENU (ethyl nitrosourea) alkylate relative more extensively at the O_6 site of guanine than do typical S_N2 alkylating agents, characterized by a high s factor, such as MMS, which comparatively alkylate predominantly the N_7 of guanine (59). Recent results of Vogel (60,61) with Drosophila demonstrate that S_N1 type agents like ENU or DEN with a low s factor predominantly induce mutations and very few chromosome breakage effects, and then only so at high concentrations. Agents of the S_N2 type like MMS, characterized by a high s factor, produce high breakage frequencies, particularly at moderate and high concentrations when the mutation to translocation ratio approaches unity. However, increase in the frequencies of both chromosome aberrations and SCE's follow similar kinetics, irrespective of the type of alkylating agents used, indicating that these events arise from similar primary lesions. The striking correlation between chemical reaction pattern, on the one hand, and the spectrum of the induced genetic changes, on the other, suggests that at least for monofunctional alkylating agents, prediction will be possible on the basis of reaction pattern, whether the agent under study is likely to generate chromosomal aberrations. Or whether alternatively, this will require concentrations one or two magnitudes higher than those that are tolerable by the animal in question. False negatives may thus be avoided for agents of the S_N1 type in (mammalian) assay systems that depend on the detection of chromosome breakage phenomena; tests detecting gene mutations will be more helpful to define the mutagenic potential of such agents.

METHODOLOGY AVAILABLE IN MAN

Apart from Ehrenberg's approach to relate mutation induction to alkylation proteins, described under alkylating agents, there are a few other methods to assess the presence of mutagenic agents in the human body.

Strauss and Albertini (63) recently reported an autoradiographic method to detect 6-thioguanine-resistant lymphocytes in human peripheral blood quantitatively. Elevated frequencies of the variant lymphocytes were found in cancer patients that were treated with cytotoxic agents. The method has the advantage of indeed being capable of detecting somatic cell mutations occurring in vivo.

Another approach to assess the presence of mutagenic agents in human body fluids consists of testing samples of blood or urine with sensitive microbial assay systems. In view of the fact that some metabolites are concentrated 1000-fold in the excreted urine, the detection capacity of this method is great, provided one is dealing with long-lived metabolites (63). The method has been successfully employed with yeast by Siebert and Simon (64) for patients treated with cyclophosphamide, and by Legator et al. (65) with Salmonella to test blood and urine samples of patients that received treatment with niridazole and metronidazole.

For various chemicals chromosomal aberrations in short-term lymphocyte cultures drawn from peripheral blood have been measured (66). It is impossible, however, to correlate these aberration yields in somatic cells with frequencies of translocations to be expected in the germ cells. This is even so for radiation, as exemplified by studies of van Buul (67-69). In view of negative findings in patients that have been treated with potent mutagens, Schinzel and Schmid (70) consider this assay system inadequate for monitoring human populations that have been exposed to weak mutagens.

COMPARISONS WITH THE PARALLELOGRAM

One possible approach to satisfy our ignorance on the induction of mutations in mammals consists of a series of stepwise comparisons for different endpoints of genetic damage at different concentration levels; this is referred to as the parallelogram (71,72). The underlying principle of the parallelogram is to obtain information on genetic damage that cannot be directly measured, by correlating different genetic end-points that can be determined experimentally. For compounds that have elicited a positive mutagenic response in submammalian assay systems, one could first determine the induction of SCE's and chromosome aberrations in mammalian cells in vitro. These frequencies can now profitably be compared with the induction of mutations at different loci in in vitro mammalian cell lines (thioguanine resistance, TK deficiency, or ouabain resistance). The induction of cytogenetic damage in the intact mammal would permit a comparison of the induction frequencies under in vitro and in vivo conditions, and these frequencies could now be used as a calibration to assess the frequency of mutation induction to be expected under in vivo conditions (Fig. 1).* The underlying assumption is that the induction frequencies of chromosome aberrations and point mutations are correlated under the different conditions.

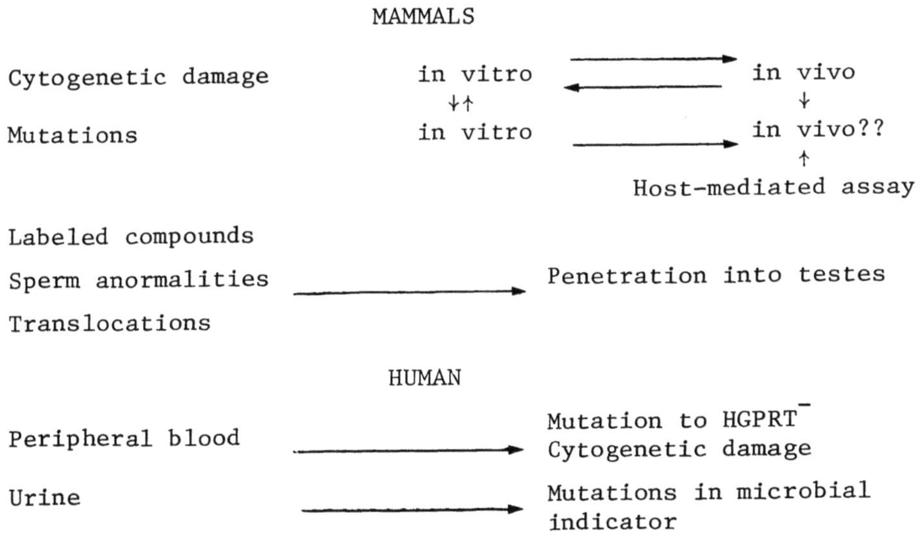

Figure 1. A "parallelogram" approach for estimating mutational damage.

To assay the extent of penetration of the mutagen into tne testes first simple tests with labeled compounds would be helpful. These could then be profitably complemented by the determination of sperm abnormalities or translocation induction in postmeiotic cells. A host-mediated assay might further augment the reliability of the data on mutation induction. From the data that have been obtained in the experimental test systems, a more precise extrapolation to man could be made by the use of some chemical that can be assayed directly in the human perhiperal blood or urine. The use of human sperm samples or of mouse follicular fluid from the oocytes for assaying mutation induction by means of microbial tests systems were recently proposed by Lyon (74).

For carcinogens, further extension to transformation in vitro, coupled with assays for malignancy in experimental mammals might help in achieving better quantitative correlation between mutation induction and actual carcinogenic risks. In this context, I want to mention recent data reported by Neal of the Courtauld Institute (75), who developed an ingenious technique that allows the determination of mutation induction in E. coli strains in a mammalian host at nitrosamine concentrations which lead to tumor formation in a later stage. Thus a direct correlation between mutation induction in the microorganism and the incidence of tumors should be possible and this then significantly increases the predictive value of microbial test systems for the induction of tumors.

The parallelogram approach refers to another perspective of what can possibly be achieved, apart from defining the sensitivity of our assay systems, by systematic studies on "comparative chemical mutagenesis." It could at least help, I believe, as a complement to specific locus studies to obtain some more information on the possible mutagenic risks involved.

CONCLUSION

In returning now once more to the comparison with the effects of radiation, there is one obvious advantage in analyzing the effects of chemicals which has not been spelled out before, and this is that ionizing radiation is characterized by what one might call a transient effect in that the ionizations or activations, and even the radiochemical reactions proceed in a matter of micro- or nanoseconds, so nothing in fact is left behind in the irradiated biological material. By contrast, chemicals can be further analyzed on the basis of their reactions with the nuclear DNA in the germ cells under consideration. For real quantiative estimates of the mutagenic potency of the various agents in the different test systems, simple comparisons of the induction frequencies for mutation and chromosome aberrations is a useful exercise, though obviously not good enough. What we need are quantitative data on the amount of reaction products that are formed in the germ cell DNA and their relation at different concentration levels to the induction frequencies of the various genetic events. The work of Aaron, Lee and Sega (51-56) has shown the way how such data can be obtained. For a proper comparison of the effectiveness of chemical mutagens, specifically for such quantitative comparisons as outlined under the paralellogram, it would be of great importance if more extensive data of this kind were collected and then correlated with the mutagenic capacity of the agents under study.

Perhaps I may end with expressing a word of caution. As geneticists we are often urged to make quantitative risk estimates. The arguments being used are that if we fail to do so, genetic hazards will no longer be taken seriously, governmental agencies will lose their concern about the hereditary effects of exposure to mutagenic chemicals, or at best that people with fewer qualifications and reservations will take over what should have been our duty.

At the present state of the art, I would personally prefer to retain my scientific integrity, state the outcome of various test systems, but try to resist the temptation to come up with a quantified estimate having such large margins of uncertainty so as to become practically meaningless (71,76,77).

This attitude should not be construed as defeatist, but reflects for the environmental chemical agents at large our knowledge "at this point in time." One should, I quote W. D. Rowe (78): "Beware of giving the 'ball park' estimate; any estimate within the realm of credibility, given by anyone considered an expert, will immediately be accepted by the receiver and promulgated as fact, or, to put it in another way: Credible estimates are propagated as facts. One does not have to be an expert or authority on the subject being discussed, but only considered one."

The risk inherent to such "risky estimates" might well be that we geneticists could lose our "credibility" in the future which in the long run might do more harm to generations yet to come.

ACKNOWLEDGMENTS

I want to thank my colleague, Dr. Ekkehart Vogel, for quoting his data and Dr. L. K. Sankaranayanan for his help with the initial radiation paper. The Drosophila work cited in this paper received support from the National Institute of Environmental Health Sciences, Contract No. ESO 1827-01 and from Contract No. 030-71-1 ENV N1 of the Environmental Research Program of the European Community to the University of Leiden. I greatly appreciated receiving so many constructive comments on the mammalian data for the twenty compounds from the various authors responsible for the Group II papers; any errors are my own responsibility. The efficient help of John Wasson and Mike Shelby both during and after the conference in Raleigh is greatly acknowledged.

I take pleasure in thanking Professor M. M. Green for providing me with the opportunity to work at the Department of Genetics, University of California, Davis, while part of the final draft of this paper was completed.

DISCUSSION

Diana Anderson: You are assuming the sensitivity in one of the approaches that chromosome damage and gene mutations are the same. What happens when you have agents which are exclusively chromosome breaking and don't cause gene mutations?

Sobels: This is a question I asked one of the first days: whether any of you knew of substances that exclusively break chromosomes. I know the case of 8-ethoxyaffeine (Kihlman and Gustavson, in plants); I don't know whether this has ever really been proved in other organisms. I believe that there is also daunomycin and bleomycin, and some person in the audience mentioned to me that benzene only breaks chromosomes; these cases, however, are rare exceptions. As far as I know practically all clastogens or mutagens produce mutations, but if you can test them in the same system as Drosophila, it invariably requires higher concentrations to produce chromosome aberrations than mutations. There are examples in Drosophila where the activation system gets saturated, like with cyclophosphamide or vinyl chloride, so that you simply don't achieve the concentrations required to break the chromosomes. Now it is conceivable that if this occurs in mammals you would require such high concentrations to break chromosomes that either the animal or the target cells are killed. Working out the S_N1 and S_N2 reaction constants might have some predictive value to define whether we are dealing with an agent inducing mainly point mutations or one which, in addition, has a high capacity for breaking chromosomes.

Bryn Bridges: Once you go outside the alkylating agents, of course, you can find examples of complete specificity. One that occurs to me is 8-methoxypsoralen in the dark, which is a gene mutator both in mammalian cells and bacteria but is totally negative on all chromosome tests including SCE's right up to completely toxic levels, so that is a gene mutagen and it doesn't react covalently with DNA in the dark.

Sobels: This is a very frightening substance because of its widespread use for psoriasis.

Bridges: We are trying to develop a risk evaluation procedure for that substance. Mary Lyon and I have been involved in this, and it's going to be very difficult.

Sobels: Well, the dermatologists are afraid that the patients they now cure of this horrifying disease, might develop cancer at a later stage, because the substance is quite mutagenic, also in our in vitro mammalian cell systems.

Bridges: The problem then is the immediately obvious one--that with light on the skin you get an additional photochemical reaction which will break chromosomes and can quite possibly cause cancer, but at least you can monitor the skin. You have mutations you can see whether they are growing skin tumors but I am worried about what happens to the germ cells inside the body where there is no light but where I am quite convinced there will be gene mutations.

Sobels: What is even more frightening is that 8-methoxypsoralen in combination with visible light produces a beautiful tan in the treated patients and that there are some cosmetic manufacturers who include this in their suntan lotions.

Mary Lyon: I thought I would like to reinforce Bryn's point that compounds might differ in their relative potencies at inducing gene mutations and chromosome structural alterations and also the numerical abnormalities. The numerical abnormalities of course are very important in man in inducing malformations. I think one should try to consider the three types of mutational change, gene, numerical, and structural, separately. To me the need to do that makes the use of the doubling dose method rather inconvenient. It means you have to have three separate doubling doses for the three types of anomalies. To me the system is rather easier to handle if you use the other approach and work out your rates for the three types of aberration; then you can simply add them together. That is one point.

Sobels: The answer to that one is that I am in full agreement with you; the direct method certainly has great advantages.

Lyon: My second point was, you were reluctant to jeopardize your scientific reputation on trying to produce precise estimates of risks but I wonder whether some of us are going to find ourselves pushed into it by the patients. It already happens with radiation that, although tremendous work exists on the risks of radiation and very detailed recommendations as to upper limits of exposure and so on, whenever a radiation worker has the misfortune to have an abnormal baby, the parents will start asking whether this is due to their work and you have got to find out what dose they have had and work out what the risk was and so on and so on. I wonder whether this is not going to happen also with chemicals and, of course, with not only occupational exposures but also medical exposures.

Sobels: I agree that it does happen and I have several times been called by doctors who had treated patients with high doses of cytostatics and then suddenly the female patient found herself pregnant. One solution is such cases is, as Heddle suggested, amniocentesis, although this will probably only identify chromosomal aberrations. Likewise, I am frequently called by doctors who have exposed patients to relatively high doses of radiation. When advising in such cases, one should bear in mind that any possible increase of genetic defects should be viewed against a background of 10.6% spontaneous incidence of diseases and malformations with some genetic etiology. The problem is, I believe, that the doctor in question can never disprove that even a small amount of radiation hasn't contributed to the birth of an abnormal child, against such high background figures. If pregnancy is detected sufficiently early, I would usually advise for an abortion.

Marvin Legator: I must say that I find this a very pessimistic kind of outlook where the chances of making any real quantitative estimate on most materials in the near future may be rather doubtful. I quite agree that it is better not to come up with any figures than to come up with meaningless figures. This obviously raises a terrible dilemma: what do you do when you are faced with a compound that indeed seems like a hazard and yet you can't put any quantitative terms on it. Now is it possible that getting a positive in any of our relevant systems--the heritable translocation, what have you--that because of the insensitivity of most of these systems, this is enough to say here we have a mutagen. Now we may have some additional information, say, that we have not swamped the metabolic system for instance, and we know something of the distribution of the compound. We are possibly now in the period in which the people in the field of carcinogenicity found themselves so many years ago, where they say we simply cannot make a risk estimate; the only thing we can say is this is a carcinogen and therefore based on the use of the compound or the intended use; we simply have to make our decision on that particular basis. In other words, if we are talking about a nonnutritive food additive, do we really have to go to a quantitative risk estimate? Is it enough to say that here, given the best we can do at the present time, at the present state-of-the-art, this compound is mutagenic if we are talking about a nonnutritive food additive? This, of course, raises some horrible thoughts in most of our minds, because this then becomes the Delaney Clause for mutagenicity, and I know that most of us are unwilling to face up to that fact. But it just very well may be that we are going to have to face up to that fact because we are now in a definite dilemma where we have TOSCA with us now; we are going to be forced to make decisions for which we will not have the proper data base, and I think it may not be too far off to say that given the insensitivity--and I think most of our systems indeed are insensitive--given the number of unknown factors for so many of our chemicals, the best we can say is that this compound is mutagenic and it reaches the germinal cells, and then make our decision on this basis. It is not a very satisfying scientific approach, but it may be the real world.

Sobels: You are not saying something entirely different from what I said when I distinguished four phases in the evaluation of environmental mutagens (74). The system consists of starting with a quick detection system for mutagenic activity using, for example, Salmonella or E. coli, and a subsequent verification stage using yeast, Neurospora, Drosophila, or mammalian systems in vitro. If all these verifying tests have produced positive results, my advice to the company or regulating agency in question would be: "You have to be extremely careful in this case." Such advice might well be enough to ban or at least restrict the distribution of that particular compound, but it is something entirely different to quantify the hazards and come up with a human risk assessment. This is the point, where I said that I feel one is forced to a certain amount of scientific dishonesty, if one really is forced to make a quantitative risk estimate which is not based on solid data of mouse spermatogonia or oocytes, provided one can, in fact, extrapolate and that there are not species-specific differences. I have tried to show two alternatives of what I hoped were constructive approaches, and the prospects for alkylating agents seem definitely better. For agents that are nonalkylating, the situation seems more difficult, but by the system of building bridges; i.e., the paralellogram kind of comparisons, one might well get closer to having at least some idea of what the risk could be. However, before all this work is done--and I think it is worthwhile doing it--I would not be a very enthusiastic supporter of coming up with a real quantitative risk in terms of so many mutations per million conceptuses.

Legator: I quite agree; the only thing I disagree with is your initial detection phase but that is another problem with that fast system.

William L. Russell: I might make the remark that might soften what I see as a potential disagreement between the committee reports and Frits Sobels' presentation. Namely, in the committee reports it seems to be a feeling that transmitted gene mutation tests in mammals were going to be necessary, and I think Dr. Sobels left the impression that it was too difficult for a while. Under certain conditions the specific locus test is nowhere as difficult as it has seemed for radiation experiments. I presented this view before, and some of you may have heard it. One example I would like to use is 5-chlorouracil, which was considered a potential hazard of the chlorinated organic in drinking water. At our lab we did a test for other reasons but ran a small specific locus test along with it in which we obtained a total of slightly over 300 offspring. On the basis of that we were able to exclude 5-chlorouracil as a mutator causing more than 2% of the spontaneous rate in humans. The reason for this is that one can give 5-chlorouracil in a much higher concentration than will ever be reached in drinking water with no toxic effects in mice, and so by simple multiplication and assuming linearity all the way down--which I think is probably justified in this case--one can make this exclusion. So, it is not always necessary in making a risk estimate to get the specific locus test on scale. Zero frequency with adequate sample size can give you an exclusion of a risk higher than a certain amount. So, under certain conditions, the specific locus test is extremely easy to run. This was slightly over 300 offspring with 11 treated males, and we reached this conclusion.

Sobels: Suppose you had a substance with a negative dose effect slope?

Russell: I am certainly willing to entertain that concept, but I would like to have a single case in mammals where this occurs at levels where there is no toxic effect on the spermatogonia. In the case of 5-chlorouracil, we also ran it at one tenth of the dose on which this calculation was made, and there was no effect at that dose either.

Sobels: It sounds very good.

F. de Serres: I would like to explore myself some of the remarks that Marvin Legator made as well as Mary Lyon. We may well find ourselves in a position with the Delaney Amendment type of approach to certain types of chemicals: those that are put in our foods or those that are used in cosmetics. There are other areas where we are subjected to a variety of chemicals to more or less improve the quality of life, and it seems in many cases we can well do without these. But there is undoubtedly an exposure to workers in industry as well as other types of industrial concerns where there are chemicals that are produced that we really can't do without. There will have to be some kind of quantitative risk estimation both for the people who make those chemicals and the people who use them. The ground rules may well change as a function of how they are used and how they are distributed. So, I don't think either one approach or the other are mutually exclusive alternatives.

Bill Lee: I would like to make a comment about the very valuable suggestion of Professor Ehrenberg in the use of hemoglobin as a means of measuring the level of alkylation. I do wish this would be referred to as a secondary receptor, because we are interested in the dose at the area of interest at the germline. The question then comes as to how close there is a correlation between the alkylation level of hemoglobin and that in the germline. This is the type of information that can be obtained with present methodologies. I do hope that one of the things that comes to the list of needs is the determining of the alkylation level to germ cells and their correlation with alkylations levels in tissues that are available for monitoring such as hemoglobin.

de Serres: If there are no further questions, I would like to thank Professor Sobels for outlining so very nicely some of the problems we are going to have in this very difficult area of estimation of risk for the human populations. Thank you again.

*Note added in proof: In the three years that have elapsed between presentation and appearance of this paper, my views concerning risk estimates have somewhat changed. It would now appear that specific DNA adducts offer a more reliable basis for estimating mutation frequencies according to the parallelogram procedure (73). Appropriate correlation with measurements on alkylated haemoglobin in human erythrocytes would, according to Ehrenberg's suggestions, then afford extrapolation to man (ICPEMC Committee-4 Report, in preparation).

REFERENCES

1. UNSCEAR Report 1977, Report of the United Nations Scientific Committee on the Effects of Atomic Radiation, Annex H, United Nations, New York, 1977.

2. Lüning, K. G., and Eiche, A. Mutat. Res. 34: 163-174 (1976).

3. Lüning, K. G., and Searle, A. G. Mutat. Res. 12: 291-304 (1971).

4. Russell, W. L., and Russell, L. B. Radiation Res. Suppl. 1: 296-305 (1959).

5. Russell, W. L. Suppl. Japan. J. Genet. 40: 128-140 (1965).

6. Lyon, M. F., and Phillips, R. J. S., and Bailly, H. J. Mutat. Res. 15: 185-190 (1972).

7. Selby, P. B., and Selby, P. R. Mutat. Res. 43: 357-375 (1977).

8. Ehling, U. H. Genetics 54: 1381-1389 (1966).

9. McKusick, V. A. Mendelian Inheritance in Man, 5th Ed., Johns Hopkins University Press, Baltimore, 1978.

10. Carter, C. O., and McKusick, V. A. Personal communication to UNSCEAR, 1977.

11. Russell, W. L. Proc. Natl. Acad. Sci. (U.S.A.) 74: 3523-3527 (1977).

12. Brewen, J. G., Preston, R. J., and Gengozian, N. Nature 253: 468-470 (1975).

13. Sankaranarayanan, K. Mutat. Res. 35: 371-414 (1976).

14. Lyon, M. F., Morris, T., Glenister, P., and O'Grady, S. E. Mutat. Res. 9: 219-223 (1970).

15. Searle, A. G., and Beechey, C. V. Mutat. Res. 15: 89-81 (1972).

16. Searle, A. G., Beechey, C. V., Green, D., and Humphreys, E. R. Mutat. Res. 41: 297-310 (1976).

17. Generoso, W. M., Cain, K. G., and Huff, S. W. Biol. Div. Ann. Rep. Oak Ridge Natl. Lab. ORNL-4993, 1974, pp. 136-138.

18. Muller, H. J. In: Radiation Biology, A. Hollaender, Ed., McGraw Hill, New York, 1954, pp. 351-626.

19. Russell, W. L. Pediatrics 41: 223-230 (1968).

20. Russell, W. L. Proc. Int. Conf. on Peaceful Uses of Atom. Energy, Vol. 11, p. 382, United Nations, New York, 1956.

21. Sheridan, W. Mutat. Res. 5: 163-172 (1968).

22. Léonard, A., and Deknudt, G. Mutat. Res. 9: 127-133 (1970).

23. Trimble, B. K., and Doughty, J. H. Ann. Hum. Genet. (London) 38: 199-223 (1974).

24. BEIR Report, Report of the Advisory Committee on the Biological Effects of Ionizing Radiation, National Academy of Sciences, National Research Council, Washington, D.C., 1972.

25. Searle, A. G. In: Seminars on Radiation Biology and Protection, Orsay, France, May 22-27, 1976, European Economic Community, 1977.

26.

26. Neel, J. V., Kato, H., and Schull, W. L. Genetics 76: 311-326 (1974).

27. Evans, H. J. In: Chemical Mutagens, A. Hollaender, Ed., Vol. 4, Plenum Press, New York, 1976, pp. 1-29.

28. Evans, H. J., and Riordan, M. L. Mutat. Res. 31: 135-148 (1975).

29. Matter, B., and Schmidt, W. Mutat. Res. 12: 417-425 (1971).

30. Heddle, J. A. Mutat. Res. 18: 187-190 (1973).

31. Bateman, A. J., and Epstein, S. S. In: Chemical Mutagens, A. Hollaender, Ed., Vol. 2, Plenum Press, New York, 1971, pp. 541-568.

32. Léonard, A. Mutat. Res. 31: 291-298 (1971).

33. Adler, I. D. Mutat. Res. 23: 369-379 (1974).

34. Vogel, E., and Leigh, B. Mutat. Res. 29: 383-396 (1977).

35. Blijleven, W. G. H., and Vogel, E. Mutat. Res. 45: 47-59 (1977).

36. Vogel, E. Mutat. Res. 33: 221-228 (1975).

37. Verburght, F. G., and Vogel, E. Mutat. Res. 48: 327-336 (1977).

38. Sheridan, W. Personal communication, 1978.

39. Selby, P. B. 11th Meeting Environmental Mutagen Society, Nashville, March 1980.

40. Dolimpio, D. A., Jacobson, C., and Legator, M. Proc. Soc. Exptl. Biol. Med. 127: 559-562 (1968).

41. Heddle, J. A., Bora, K. C., and Stoltz, D. R. (this volume).

42. Bridges, B. A. (this volume).

43. Ehling, U. (this volume).

44. Parker, D. R., and Williamson, J. H. In: The Genetics and Biology of Drosophila. M. Ashburner and E. Novitski, Eds., Vol. 1C, Academic Press, New York, 1976, pp. 1252-1264.

45. Kondon (this volume).

46. Knaap, A. G. A. C. Mutat. Res., in press.

47. Ehling, U. H. In: Chemical Mutagens, A. Hollaender and F. J. de Serres, Eds., Plenum Press, New York, Vol. 5, 1978, pp. 233-253.

48. Generoso, W. M., Russell, W. L., Huff, S. W., Stout, S. K., and Gossleer, D. G. Genetics 77: 741-752 (1974).

49. Brewen, J. G. Mutat. Res. 41: 15-24 (1976).

50. Ehrenberg, L. In: Banbury Report. 1. Assessing Chemical Mutagens: The risk to Humans, V. K. McElheny and S. Abrahamson, Eds., Cold Spring Harbor Laboratory, 1979, pp. 157-190.

51. Lee, W. R. Mutat. Res. 38: 311-316 (1976).

52. Lee, W. R. In: Chemical Mutagens, A. Hollaender and F. J. de Serres, Eds., Vol. 5, Plenum Press, New York, 1978, pp. 177-202.

53. Aaron, C. S. Mutat. Res. 38: 303-310 (1976).

54. Sega, G. A., Cumming, R. B., and Walton, M. F. Mutat. Res. 24: 317-333 (1974).

55. Aaron, C. S., and Lee, W. R. Mutat. Res. 49: 27-44 (1978).

56. Aaron, C. S., van Zeeland, A. A., Mohn, G. R., Natarajan, A. T., Knaap, A. G. A. C., Tates, A. D., and Glickman, B. W. Mutat. Res. 69: 201-216 (1980).

57. Osterman-Golkar, S., Ehrenberg, L., and Wachtmeister, C. A. Radiation Bot. 10: 303-327 (1970).

58. Veleminsky, J., Osterman-Golkar, S., and Ehrenberg, L. Mutat. Res. 10: 169-174 (1970).

59. Lawley, P. D. Mutat. Res. 23: 283-295 (1974).

60. Vogel, E., and Natarajan, A. T. Mutat. Res. 62: 51-100 (1979).

61. Vogel, E., and Natarajan, A. T. Mutat. Res. 62: 101-123 (1979).

62. Strauss, G. H., and Albertini, R. J. Mutat. Res. 61: 353-379 (1979).

63. Zimmermann, F. K. Mutat. Res. 41: 163 (1976).

64. Siebert, D., and Simon, U. Mutat. Res. 21: 257-262 (1973).

65. Legator, M. S., Truong, L., and Connor, T. H. In: Chemical Mutagens, A. Hollaender and F. J. de Serres, Eds., Vol. 5, 1978, Plenum Press, New York, pp. 1-23.

66. Kilian, D. J., and Picciano, D. In: Chemical Mutagens. A. Hollaender, Ed., Vol. 4, Plenum Press, New York, 1976, pp. 321-329.

67. van Buul, P. P. W. Mutat. Res. 20: 369-376 (1973).

68. van Buul, P. P. W. Mutat. Res. 36: 223-236 (1976).

69. van Buul, P. P. W. Mutat. Res. 45: 61-68 (1977).

70. Schinzel, A., and Schmid, W. Mutat. Res. 40: 139-166 (1976).

71. Sobels, F. H. Mutat. Res. 46: 245-260 (1977).

72. Sobels, F. H. In: Progress in Genetic Toxicology, D. Scott, B. A. Bridges, and F. H. Sobels, Eds., Elsevier/North-Holland Biomedical Press, 1977, pp. 175-182.

73. Sobels, F. H. J. Environ. Pathol. Toxicol., in press.

74. Lyon, M. F. Personal communication.

75. Neal, S. E.E.C. Contact Group on Chemical Mutagenesis, Brighton, Sept. 1977.

76. Auerbach, C. A. Mutat. Res. 33: 3-10 (1975).

77. Sobels, F. H. Mutat. Res. 38: 361-366 (1976).

78. Rowe, W. D. Research/Development, Dec. 1977.

INDEX

MIX
Papier aus verantwortungsvollen Quellen
Paper from responsible sources
FSC® C105338
FSC
www.fsc.org

If you have any concerns about our products,
you can contact us on
ProductSafety@springernature.com

In case Publisher is established outside the EU,
the EU authorized representative is:
Springer Nature Customer Service Center GmbH
Europaplatz 3, 69115 Heidelberg, Germany

Printed by Libri Plureos GmbH
in Hamburg, Germany